Food Science Text Series

Fifth Edition

For other titles published in this series, go to
www.springer.com/series/5999

Food Analysis

Fifth Edition

Edited by

S. Suzanne Nielsen
Purdue University
West Lafayette, IN, USA

 Springer

Editor
S. Suzanne Nielsen
Department of Food Science
Purdue University
West Lafayette
Indiana
USA

ISSN 1572-0330   ISSN 2214-7799   (electronic)
Food Science Text Series
ISBN 978-3-319-45774-1   ISBN 978-3-319-45776-5   (eBook)
DOI 10.1007/978-3-319-45776-5

Library of Congress Control Number: 2017942967

Printed on acid-free paper

This Springer imprint is published by Springer Nature
The registered company is Springer International Publishing AG
The registered company address is: Gewerbestrasse 11, 6330 Cham, Switzerland

Preface and Acknowledgments

The intent of this fifth edition book is the same as that described in the Preface to the first four editions – a text primarily for undergraduate students majoring in food science, currently studying the analysis of foods. However, comments from users of the first four editions have convinced me that the book is also a valuable text for persons in the food industry who either do food analysis or interact with analysts.

The big focus of this edition was to do a general update on methods and to make the content easier for readers to compare and contrast methods covered. The following summarize changes from the fourth edition: (1) general updates, including addition and deletion of methods, (2) three new chapters ("Determination of Total Phenolics and Antioxidants Capacity in Food and Ingredients," "Food Microstructure Techniques," "Food Forensic Investigation"), (3) rewrote and/or reorganized some chapters, (4) added tables to some chapters to summarize and compare methods, and (5) added some colored figures.

As stated for the first four editions, the chapters in this textbook are not intended as detailed references, but as general introductions to the topics and the techniques. Course instructors may wish to provide more details on a particular topic to students. Chapters focus on principles and applications of techniques. Procedures given are meant to help explain the principles and give some examples, but are not meant to be presented in the detail adequate to actually conduct a specific analysis. As in the first four editions, all chapters have summaries and study questions, and keywords or phrases are in bold type, to help students focus their studies. The grouping of chapters by category is similar to the fourth edition. However, due to the increased use of spectroscopy and chromatography for many basic analyses, chapters on these topics are covered early in the book. Instructors are encouraged to cover the topics from this text in whatever order is most suitable for their course. Also, instructors are invited to contact me for access to a website I maintain with additional teaching materials related to this textbook and the accompanying laboratory manual.

Starting with the third edition, the competency requirements established by the Institute of Food Technologists were considered. Those requirements relevant to food analysis are as follows: (1) understanding the principles behind analytical techniques associated with food, (2) being able to select the appropriate analytical technique when presented with a practical problem, and (3) demonstrating practical proficiency in food analysis laboratory. This textbook should enable instructors to meet the requirements and develop learning objectives relevant to the first two of these requirements. The laboratory manual, now in its third edition, should be a useful resource to help students meet the third requirement.

I am grateful to all chapter authors for agreeing to be a part of this project. Authors have drawn on their experience of teaching students and/or experience with these analyses to give chapters the appropriate content, relevance, and ease of use. I wish to thank the authors of articles and books, as well as the publishers and industrial companies, for their permission to reproduce materials used here. Special thanks is extended to the following persons: Baraem (Pam) Ismail for valuable discussions about the content of the book and reviewing several book chapters, Ben Paxson for drawing/redrawing figures, and Telaina Minnicus and Mikaela Allan for word processing assistance. I am also very grateful to Bill Aimutis, Angela Cardinali, Wayne Ellefson, Chris Fosse, and David Plank who were valuable for discussions and arranged for me to visit with numerous scientists in the analytical laboratories at the following companies/institute: Cargill, ConAgra Foods, Covance, and General Mills in the USA, and Bonassisa Lab and the Institute of Science of Food Production in Italy.

West Lafayette, IN, USA   S. Suzanne Nielsen

Abbreviations

2-D	Two-dimensional
3-D	Three-dimensional
3-MCPD	3-Monochloropropane 1,2-diol
AACC	American Association of Cereal Chemists
AACCI	AACI International
AAS	Atomic absorption spectroscopy
AAPH	2,2′-Azobis (2-amidinopropane) dihydrochloride
ABTS	2,2′-Azino-bis (3-ethylbenzenothazoline-6-sulfonic acid)
ADI	Acceptable daily intake
ADP	Adenosine-5′-diphosphate
AE-HPLC	Anion exchange high-performance liquid chromatography
AES	Atomic emission spectroscopy
AFM	Atomic force microscopy
AMS	Accelerator mass spectrometer
AMS	Agricultural Marketing Service
AOAC	Association of Official Analytical Chemists
AOCS	American Oil Chemists' Society
AOM	Active oxygen method
APCI	Atmospheric pressure chemical ionization
APHA	American Public Health Association
API	Atmospheric pressure ionization
APPI	Atmospheric pressure photoionization
AQC	6-Aminoquinolyl-N-hydroxysuccin-imidyl carbamate
ASE	Accelerated solvent extraction
ASTM	American Society for Testing Materials
ATCC	American Type Culture Collection
ATP	Adenosine-5′-triphosphate
ATR	Attenuated total reflectance
AUC	Area under the curve
a_w	Water activity
B_0	External magnetic field
BAW	Base and acid washed
BCA	Bicinchoninic acid
BCR	Community Bureau of Reference
Bé	Baumé modulus
BHA	Butylated hydroxyanisole
BHT	Butylated hydroxytoluene
BOD	Biochemical oxygen demand
BPA	Bisphenol A
BSA	Bovine serum albumin
BSDA	*Bacillus stearothermophilus* disk assay
BSE	Backscattered electrons
BSTFA	N,O-Bis(trimethylsilyl) trifluoroacetamide
CAD	Collision-activated dissociation
CAST	Calf antibiotic and sulfa test
CAT	Computerized axial tomography
CCD	Charge-coupled device
CDC	Centers for Disease Control
CFR	Code of Federal Regulations
CFSAN	Center for Food Safety and Applied Nutrition
cGMP	Current Good Manufacturing Practices
CI	Confidence interval
CI	Chemical ionization
CID	Collision-induced dissociation
CID	Commercial item description
CID	Charge injection device
CIE	Commission Internationale d'Eclairage
CLA	Conjugated linoleic acid
CLND	Chemiluminescent nitrogen detector
CLSM	Confocal laser scanning microscopy
CMC	Critical micelle concentration
COA	Certificate of analysis
COD	Chemical oxygen demand
C-PER	Protein efficiency ratio calculation method
CPG	Compliance policy guidance
CP-MAS	Cross-polarization magic angle spinning
CQC	2,6-Dichloroquinonechloroimide
CRC	Collision reaction cells
CSLM	Confocal scanning laser microscopy
CT	Computed technology
CT	Computed tomography
CV	Coefficient of variation
CVM	Center for Veterinary Medicine
DAL	Defect action level
DART	Direct analysis in real time
DDT	Dichlorodiphenyltrichloroethane
DE	Degree of esterification
dE*	Total color difference
DF	Dilution factor
DFE	Dietary folate equivalent
DHHS	Department of Health and Human Services
DIAAS	Digestible indispensable amino acid score
DIC	Differential interferential contrast

DMA	Dynamic mechanical analysis	Fc	Fragment crystallizable
DMF	Dimethylformamide	FCC	Food Chemicals Codex
DMD	D-Malate dehydrogenase	FD&C	Food, Drug, and Cosmetic
DMSO	Dimethyl sulfoxide	FDA	Food and Drug Administration
DNA	Deoxyribonucleic acid	FDAMA	Food and Drug Administration
DNFB	1-Fluoro-2,4-dinitrobenzene		Modernization Act
dNTPs	Deoxynucleoside triphosphates	FDNB	1-Fluoro-2,4-dinitrobenzene
DON	Deoxynivalenol	FFA	Free fatty acid
DRI	Dietary references intake	FID	Free induction decay
DRIFTS	Diffuse reflectance infrared Fourier	FID	Flame ionization detector
	transform spectroscopy	FIFRA	Federal Insecticide, Fungicide, and
DRV	Daily reference value		Rodenticide Act
DSC	Differential scanning calorimetry	FNB/NAS	Food and Nutrition Board of the
DSHEA	Dietary Supplement Health and		National Academy of Sciences
	Education Act	FOS	Fructooligosaccharide
DSPE	Dispersive solid-phase extraction	FPA	Focal plane array
DTGS	Deuterated triglycine sulfate	FPD	Flame photometric detector
DV	Daily value	FPIA	Fluorescence polarization immunoassay
DVB	Divinylbenzene	FPLC	Fast protein liquid chromatography
DVS	Dynamic vapor sorption	FRAP	Ferric reducing antioxidant power
dwb	Dry weight basis	FSIS	Food Safety and Inspection Service
E_a	Activation energy	FT	Fourier transform
EAAI	Essential amino acid index	FTC	Federal Trade Commission
EBT	Eriochrome black T	FT-ICR	Fourier transform ion cyclotron
ECD	Electron capture dissociation		resonance
ECD	Electron capture detector	FTIR	Fourier transform infrared
ECD	(Pulsed) electro-chemical detector	FTMS	Fourier transform mass spectrometry
EDL	Electrodeless discharge lamp	G6PDH	Glucose-6-phosphate dehydrogenase
EDS	Energy dispersive spectroscopy	GATT	General Agreement on Tariffs and
EDTA	Ethylenediaminetetraacetic acid		Trade
EEC	European Economic Community	GC	Gas chromatography
EFSA	European Food Safety Authority	GC-AED	Gas chromatography-atomic emission
EI	Electron impact ionization		detector
EIE	Easily ionized elements	GC-FTIR	Gas chromatography-Fourier transform
ELCD	Electrolytic conductivity detector		infrared
ELISA	Enzyme-linked immunosorbent assay	GC×GC	Comprehensive two-dimensional gas
EM	Electron microscopy		chromatography
EPA	Environmental Protection Agency	GC-MS	Gas chromatography-mass spectrome-
EPSPS	5-Enolpyruvylshikimate-3-phosphate		try
	synthase	GC-O	Gas chromatography-olfactory
Eq	Equivalents	GFC	Gel-filtration chromatography
ERH	Equilibrium relative humidity	GIPSA	Grain Inspection, Packers and
ES	Electrospray		Stockyards Administration
E-SEM	Environmental scanning electron	GLC	Gas-liquid chromatography
	microscopy	GMA	Grocery Manufacturers of America
ESI	Electrospray ionization	GMO	Genetically modified organism
ESI	Electrospray interface	GMP	Good manufacturing practices (also
ETD	Electron transfer dissociation		current good manufacturing practice in
ETO	Ethylene oxide		manufacturing, packing, or holding
EU	European Union		human food)
Fab	Fragment antigen-binding	GOPOD	Glucose oxidase/peroxidase
FAIMS	Field-asymmetric ion mobility	GPC	Gel-permeation chromatography
FAME	Fatty acid methyl esters	GRAS	Generally recognized as safe
FAO/WHO	Food and Agricultural Organization/	HACCP	Hazard analysis and critical control
	World Health Organization		point
FAS	Ferrous ammonium sulfate	HAT	Hydrogen atom transfer
FBs	Fumonisins	HCL	Hollow cathode lamp

HETP	Height equivalent to a theoretical plate	KFR	Karl Fischer reagent
HFS	High fructose syrup	KFReq	Karl Fischer reagent water equivalence
HIC	Hydrophobic interaction chromatography	KHP	Potassium acid phthalate
		LALLS	Low-angle laser light scattering
HILIC	Hydrophilic interaction liquid chromatography	LC	Liquid chromatography
		LC-MS	Liquid chromatography-mass spectroscopy
HIS	Hyperspectral imaging		
HK	Hexokinase	LFS	Lateral flow strip
H-MAS	High-resolution magic angle spinning	LIMS	Laboratory information management system
HMDS	Hexamethyldisilazane		
HPLC	High-performance liquid chromatography	LM	Light microscopy
		LOD	Limit of detection
HPTLC	High-performance thin-layer chromatography	LOQ	Limit of quantitation
		LTM	Low thermal mass
HQI	Hit quality index	LTP	Low-temperature plasma probe
HRGC	High-resolution gas chromatography	m/z	Mass-to-charge ratio
HRMS	High-resolution accurate mass spectrometry	MALDI	Matrix-assisted laser desorption ionization
		MALDI-TOF	Matrix-assisted laser desorption time-of-flight
HS	Headspace		
HVP	Hydrolyzed vegetable protein	MALLS	Multi-angle laser light scattering
IC	Ion chromatography	MAS	Magic angle spinning
IC_{50}	Median inhibition concentration	MASE	Microwave-assisted solvent extraction
ICP	Inductively coupled plasma	MCL	Maximum contaminant level
ICP-AES	Inductively coupled plasma-atomic emission spectroscopy	MCT	Mercury:cadmium:telluride
		MDGC	Multidimensional gas chromatography
ICP-MS	Inductively coupled plasma-mass spectrometer	MDL	Method detection limit
		MDSC™	Modulated Differential Scanning Calorimeter™
ICP-OES	Inductively coupled plasma-optical emission spectroscopy		
		mEq	Milliequivalents
ID	Inner diameter	MES-TRIS	2-(N-morpholino)ethanesulfonic acid-tris(hydroxymethyl)aminomethane
IDF	Insoluble dietary fiber		
IDK	Insect damaged kernels	MLR	Multiple linear regression
IEC	Inter-element correction	MRI	Magnetic resonance imaging
Ig	Immunoglobulin	MRL	Maximum residue level
IgE	Immunoglobulin E	MRM	Multiple-reaction monitoring
IgG	Immunoglobulin G	MRM	Multiresidue method
IMS	Ion mobility mass spectrometry	MS	Mass spectrometry (or spectrometer)
IMS	Interstate Milk Shippers	MS/MS	Tandem MS
InGaAs	Indium-gallium-arsenide	Ms^n	Multiple stages of mass spectrometry
IR	Infrared	MW	Molecular weight
IRMM	Institute for Reference Materials and Measurements	NAD	Nicotinamide-adenine dinucleotide
		NADP	Nicotinamide-adenine dinucleotide phosphate
ISA	Ionic strength adjustor		
ISE	Ion-selective electrode	NADPH	Reduced NADP
ISFET	Ion sensitive field effect transitor	NCM	N-Methyl carbamate
ISO	International Organization for Standardization	NCWM	National Conference on Weights and Measures
IT	Ion trap	NDL	Nutrient Data Laboratory
ITD	Ion trap detector	NFDM	Nonfat dry milk
IT-MS	Ion trap mass spectrometry	NIR	Near-infrared
IU	International Units	NIRS	Near-infrared spectroscopy
IUPAC	International Union of Pure and Applied Chemistry	NIST	National Institute of Standards and Technology
JECFA	Joint FAO/WHO Expert Committee on Food Additives	NLEA	Nutrition Labeling and Education Act
		NMFS	National Marine Fisheries Service
kcal	Kilocalorie	NMR	Nuclear magnetic resonance
KDa	Kilodalton		

NOAA	National Oceanic and Atmospheric Administration	qMS	Quadruple mass spectrometry
NOAEL	No observed adverse effect level	QqQ	Triple quadrupole
NPD	Nitrogen phosphorus detector or thermionic detector	Q-TOF	Quadrupole-time-of-flight
		Q-trap	Quadruple-ion trap
NSSP	National Shellfish Sanitation Program	QuEChERS	Quick, easy, cheap, effective, rugged, and safe
NVC	Nonvolatile compounds	RAC	Raw agricultural commodity
NVOC	Nonvolatile organic compounds	RAE	Retinol activity equivalents
OC	Organochlorine	RASFF	Rapid alert system for food and feed
OD	Outer diameter	RDA	Recommended daily allowance
ODS	Octadecylsilyl	RDI	Reference daily intake
OES	Optical emission spectroscopy	RE	Retinol equivalent
OMA	Official Methods of Analysis	R_f	Relative mobility
OP	Organophosphate/organophosphorus	RF	Radiofrequency
OPA	O-Phthalaldehyde	RF	Response factor
ORAC	Oxygen radical absorbance capacity	RI	Refractive index
ORAC	Optimized Rowland circle alignment	RIA	Radioimmunoassay
OSI	Oil stability index	R_m	Relative mobility
OT	Orbitrap	RMCD	Rapidly methylated β-cyclodextrin
OTA	Ochratoxin A	ROSA	Rapid one-step assay
PAD	Pulsed-amperometric detector	RPAR	Rebuttable Presumption Against Registration
PAGE	Polyacrylamide gel electrophoresis		
PAM I	*Pesticide Analytical Manual, Volume I*	RS	Resistant starch
PAM II	*Pesticide Analytical Manual, Volume II*	RVA	RapidViscoAnalyser
P_c	Critical pressure	SAFE	Solvent-assisted flavor evaporation
PCBs	Polychlorinated biphenyls	SASO	Saudi Arabian Standards Organization
PCR	Principal components regression	SBSE	Stir-bar sorptive extraction
PCR	Polymerase chain reaction	SD	Standard deviation
PDA	Photodiode array	SDF	Soluble dietary fiber
PDCAAS	Protein digestibility-corrected amino acid score	SDS	Sodium dodecyl sulfate
		SDS-PAGE	Sodium dodecyl sulfate-polyacrylamide gel electrophoresis
PDMS	Polydimethylsiloxane		
PEEK	Polyether ether ketone	SEC	Size-exclusion chromatography
PER	Protein efficiency ratio	SEM	Scanning electron microscopy
PFPD	Pulsed flame photometric detector	SERS	Surface-enhanced Raman scattering
pI	Isoelectric point	SET	Single electron transfer
PID	Photoionization detector	SFC	Solid fat content
PLE	Pressurized liquid extraction	SFC	Supercritical fluid chromatography
PLOT	Porous-layer open tabular	SFC-MS	Supercritical fluid chromatography-mass spectrometry
PLS	Partial least squares		
PME	Pectin methylesterase	SFE	Supercritical fluid extraction
PMO	Pasteurized Milk Ordinance	SFE-GC	Supercritical fluid extraction-gas chromatography
PMT	Photomultiplier tube		
ppb	Parts per billion	SFI	Solid fat index
PPD	Purchase product description	SI	International System of Units
ppm	Parts per million	SKCS	Single kernel characteristics system
ppt	Parts per trillion	SMEDP	Standard Methods for the Examination of Dairy Products
PSPD	Position-sensitive photodiode detector		
PTV	Programmed temperature vaporization	SO	Sulfite oxidase
PUFA	Polyunsaturated fatty acids	SOP	Standard operating procedures
PVDF	Polyvinylidine difluoride	SPDE	Solid-phase dynamic extraction
PVPP	Polyvinylpolypyrrolidone	SPE	Solid-phase extraction
Q	Quadrupole mass filter	SPME	Solid-phase microextraction
QA	Quality assurance	SRF	Sample response factor
QC	Quality control	SRM	Standard reference materials

SRM	Selected-reaction monitoring	TQ	Triple quadrupole
SRM	Single-residue method	TS	Total solids
SSD	Solid state detector	TSQ	Triple stage quadrupole
STOP	Swab test on premises	TSS	Total soluble solids
SVC	Semi-volatile compounds	TSUSA	Tariff Schedules of the United States of America
SVOC	Semi-volatile organic compounds		
SXI	Soft x-ray imaging	TTB	Alcohol and Tobacco Tax and Trade Bureau
TBA	Thiobarbituric acid		
TBARS	TBA reactive substances	TWI	Total weekly intake
TCA	Trichloroacetic acid	TWIM	Traveling wave
TCD	Thermal conductivity detector	UHPC	Ultra-high pressure chromatography
TCP	Tocopherols	UHPLC	Ultra-high performance liquid chromatography
TDA	Total daily intake		
TDF	Total dietary fiber	UPLC	Ultra-performance liquid chromatography
TDU	Thermal desorption unit		
T-DNA	Transfer of DNA	US	United States
TD-NMR	Time domain nuclear magnetic resonance	USA	United States of America
		USCS	United States Customs Service
TEAC	Trolox equivalent antioxidant capacity	USDA	United States Department of Agriculture
TEM	Transmission electron microscopies		
TEMED	Tetramethylethylenediamine	USDC	United States Department of Commerce
Tg	Glass transition temperature	USP	United States Pharmacopeia
TGA	Thermogravimetric analysis	UV	Ultraviolet
Ti	Tumor-inducing	UV-Vis	Ultraviolet-visible
TIC	Total ion current	Vis	Visible
TLC	Thin-layer chromatography	VC	Volatile compounds
TMA	Thermomechanical analysis	VOC	Volatile organic compounds
TMCS	Trimethylchlorosilane	WDS	Wavelength dispersive x-ray
TMS	Trimethylsilyl	wt	Weight
TOF	Time-of-flight	wwb	Wet weight basis
TOF-MS	Time-of-flight mass spectrometry	XMT	X-ray microtomography
TPA	Texture profile analysis	XRD	X-ray diffractometer
TPTZ	2,4,6-Tripiyridyl-s-triazine	ZEA	Zearalenone

Contents

Contributors

William R. Aimutis Intellectual Asset Management, Cargill, Inc., Wayzata, MN, USA

Huseyin Ayvaz Department of Food Engineering, Canakkale Onsekiz Mart University, Canakkale, Turkey

James N. BeMiller Department of Food Science, Purdue University, West Lafayette, IN, USA

Robert L. Bradley Jr. Department of Food Science, University of Wisconsin, Madison, WI, USA

Mirko Bunzel Department of Food Chemistry and Phytochemistry, Karlsruhe Institute of Technology, Karlsruhe, Germany

Sam K.C. Chang Department of Food Science, Nutrition, and Health Promotion, Mississippi State University, Starkville, MS, USA

Christopher R. Daubert Department of Food, Bioprocessing and Nutritional Sciences, North Carolina State University, Raleigh, NC, USA

Hulya Dogan Department of Grain Science and Industry, Kansas State University, Manhattan, KS, USA

Jinping Dong Global Food Research, Cargill Research and Development Center, Cargill, Inc., Plymouth, MN, USA

Ronald R. Eitenmiller Department of Food Science and Technology, The University of Georgia, Athens, GA, USA

Wayne C. Ellefson Nutritional Chemistry and Food Safety, Covance Laboratories, Madison, WI, USA

Yong D. Hang Department of Food Science and Technology, Cornell University, Geneva, NY, USA

G. Keith Harris Department of Food, Bioprocessing and Nutritional Sciences, North Carolina State University, Raleigh, NC, USA

Y-H. Peggy Hsieh Department of Nutrition, Food and Exercise Sciences, Florida State University, Tallahassee, FL, USA

Baraem P. Ismail Department of Food Science and Nutrition, University of Minnesota, St. Paul, MN, USA

Helen S. Joyner (Melito) School of Food Science, University of Idaho, Moscow, ID, USA

Jerrad F. Legako Department of Nutrition, Dietetics, and Food Sciences, Utah State University, Logan, UT, USA

Maurice R. Marshall Department of Food Science and Human Nutrition, University of Florida, Gainesville, FL, USA

Lisa J. Mauer Department of Food Science, Purdue University, West Lafayette, IN, USA

Lloyd E. Metzger Department of Dairy Science, South Dakota State University, Brookings, SD, USA

Dennis D. Miller Department of Food Science, Cornell University, Ithaca, NY, USA

Rubén O. Morawicki Department of Food Science, University of Arkansas, Fayetteville, AR, USA

Michael A. Mortenson Global Food Research, Cargill Research and Development Center, Cargill, Inc., Plymouth, MN, USA

S. Suzanne Nielsen Department of Food Science, Purdue University, West Lafayette, IN, USA

Sean F. O'Keefe Department of Food Science and Technology, Virginia Tech, Blacksburg, VA, USA

Ronald B. Pegg Department of Food Science and Technology, The University of Georgia, Athens, GA, USA

Michael H. Penner Department of Food Science and Technology, Oregon State University, Corvallis, OR, USA

Devin G. Peterson Department of Food Science and Technology, The Ohio State University, Columbus, OH, USA

Oscar A. Pike Department of Nutrition, Dietetics, and Food Science, Brigham Young University, Provo, UT, USA

Joseph R. Powers School of Food Science, Washington State University, Pullman, WA, USA

Michael C. Qian Department of Food Science and Technology, Oregon State University, Corvallis, OR, USA

Qinchun Rao Department of Nutrition, Food and Exercise Sciences, Florida State University, Tallahassee, FL, USA

Gary A. Reineccius Department of Food Science and Nutrition, University of Minnesota, St. Paul, MN, USA

Bradley L. Reuhs Department of Food Science, Purdue University, West Lafayette, IN, USA

José I. Reyes-De-Corcuera Department of Food Science and Technology, The University of Georgia, Athens, GA, USA

Luis Rodriguez-Saona Department of Food Science and Technology, The Ohio State University, Columbus, OH, USA

Michael A. Rutzke School of Integrative Plant Science, Cornell University, Ithaca, NY, USA

George D. Sadler PROVE IT LLC, Geneva, IL, USA

Var L. St. Jeor Global Food Research, Cargill Research and Development Center, Cargill, Inc., Plymouth, MN, USA

Rachel R. Schendel Department of Food Chemistry and Phytochemistry, Karlsruhe Institute of Technology, Karlsruhe, Germany

Shelly J. Schmidt Department of Food Science and Human Nutrition, University of Illinois at Urbana-Champaign, Urbana, IL, USA

Senay Simsek Department of Plant Sciences, North Dakota State University, Fargo, ND, USA

Daniel E. Smith Department of Food Science and Technology, Oregon State University, Corvallis, OR, USA

Denise M. Smith School of Food Science, Washington State University, Pullman, WA, USA

J. Scott Smith Department of Animal Sciences and Industry, Kansas State University, Manhattan, KS, USA

Bhadrirju Subramanyam Department of Grain Science and Industry, Kansas State University, Manhattan, KS, USA

Rohan A. Thakur Bruker Daltonics, Billerica, MA, USA

Leonard C. Thomas DSC Solutions, Smyrna, DE, USA

Catrin Tyl Department of Food Science and Nutrition, University of Minnesota, St. Paul, MN, USA

Robert E. Ward Department of Nutrition and Food Sciences, Utah State University, Logan, UT, USA

Randy L. Wehling Department of Food Science and Technology, University of Nebraska, Lincoln, NE, USA

Ronald E. Wrolstad Department of Food Science and Technology, Oregon State University, Corvallis, OR, USA

Vincent Yeung Department of Animal Science, California Polytechnic State University, San Luis Obispo, CA, USA

Yan Zhang Department of Cereal and Food Sciences, Department of Food Science, Nutrition, and Health Promotion, Mississippi State University, Starkville, MS, USA

General Information

Introduction to Food Analysis

S. Suzanne Nielsen

Department of Food Science,
Purdue University,
West Lafayette, IN 47907-2009, USA
e-mail: nielsens@purdue.edu

S. Nielsen (ed.), *Food Analysis*, Food Science Text Series,
DOI 10.1007/978-3-319-45776-5_1, © Springer International Publishing 2017

1.1   INTRODUCTION

Investigations in food science and technology, whether by the food industry, governmental agencies, or universities, often require determination of food composition and characteristics. Trends and demands of consumers, national and international regulations, and realities of the food industry challenge food scientists as they work to monitor food composition and to ensure the quality and safety of the food supply. As summarized by McGorrin [1] in a review of food analysis history, "the growth and infrastructure of the model food distribution system heavily relies on food analysis (beyond simple characterization) as a tool for new product development, quality control, regulatory enforcement, and problem solving." All food products require analysis of various characteristics (i.e., chemical composition, microbial content, physical properties, sensory properties) as part of a **quality management** program, from raw ingredients, through processing, to the final product. Of course, food analysis is also used extensively for research on foods and food ingredients. The nature of the sample and the specific reason for the analysis commonly dictate the choice of analytical methods. **Speed, precision, accuracy, robustness, specificity**, and **sensitivity** are often key factors in this choice. **Validation** of the method for the specific **food matrix** being analyzed is necessary to ensure usefulness of the method.

Making an appropriate choice of analytical technique for a specific application requires a good knowledge of the various techniques (Fig. 1.1). For example, your choice of method to determine the salt content of potato chips would be different if it is for nutrition labeling compared to quality control. The success of any analytical method relies on the proper selection and preparation of the food sample, carefully performing the analysis, and doing the appropriate calculations and interpretation of the data. Methods of analysis developed and endorsed by several nonprofit scientific organizations allow for standardized comparisons of results between different laboratories and for evaluation of less standard procedures. Such **official methods** are critical in the analysis of foods, to ensure that they meet the legal requirements established by governmental agencies. **Government regulations** and **international standards** most relevant to the analysis of foods are mentioned here but covered in more detail in Chap. 2, and nutrition labeling regulations in the United States are covered in Chap. 3.

1.2   REASONS FOR ANALYZING FOODS AND TYPES OF SAMPLES ANALYZED

1.2.1   Overview

Consumer trends and demands, national and international regulations, and the food industry's need to manage product quality dictate the need for analysis of food ingredients and products (Table 1.1) and explain the types of samples analyzed.

1.2.2   Consumer Trends and Demands

Consumers have many choices regarding their food supply, so they can be very selective about the products they purchase. They demand a wide variety of products that are safe, nutritious, and of high quality and offer a good value. Consumer demand has driven significant growth in products making claims, many of which bring an increased need for food

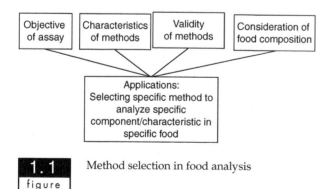

1.1 figure   Method selection in food analysis

1.1 table   Reasons for analyzing foods

1. Food safety
2. Government regulations
   (a) Nutrition labeling
   (b) Standards – mandatory and voluntary
   (c) Food inspection and grading
   (d) Authenticity
3. Quality control
4. Research and development

analysis. For example, the consumer-driven, gluten-free claim has led to increased raw material/ingredient testing and finished product testing, required to comply in the United States with the definition of this claim established by the Food and Drug Administration. Many consumers are interested in the relationship between diet and health, including functional foods that may provide health benefits beyond basic nutrition. The use of social media by consumers has changed expectations and raised questions about the food supply. Such trends and demands by consumers increase the need for food analysis and present some unique challenges regarding analytical techniques [2, 3].

1.2.3 Government Regulations and International Standards and Policies

For food companies to market safe, high-quality foods effectively in a national and global marketplace, they must pay increasing attention to government regulations and guidelines, and to the policies and standards of international organizations. Food scientists must be aware of these regulations, guidelines, and policies related to food safety and quality, and must know their implications for food analysis. Government regulations related to the composition of foods include nutrition labeling, product claims, standards, inspection and grading, and authenticity. This latter issue of authenticity is a challenge for the food industry, given the constant threat of economic adulteration of food products and ingredients. Detecting untargeted compounds in foods and determining their identity are challenges that require advanced analytical techniques that are powerful, sensitive, and fast [2]. The industry is also challenged and forced to "chase zeros" when laws are written stating zero as the level of certain compounds allowed. US government regulations relevant to food analysis are covered in Chaps. 2 and 3. Also covered in Chap. 2 are organizations active in developing international standards and safety practices relevant to food analysis.

1.2.4 Food Industry Management of Product Quality

1.2.4.1 *Raw Ingredients to Final Product*
To compete in the marketplace, food companies must produce foods that meet consumer demands, comply with government regulations, and meet quality standards of the company. Whether it is new or existing food products or ingredients, the key concern for the food industry is safety of the food, but quality management goes well beyond safety. The management of product quality by the food industry is of increasing importance, beginning with the raw ingredients and extending to the final product eaten by the consumer. Analytical methods must be applied across the entire

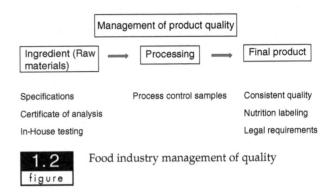

1.2 figure Food industry management of quality

food supply chain to achieve the desired final product quality (Fig. 1.2). It is obvious that the food processor must understand the entire supply chain to successfully manage product quality.

In some cases, the cost of goods is linked directly to the composition as determined by analytical tests. For example, in the dairy field, butterfat content of bulk tank raw milk determines how much money the milk producer is paid for the milk. For flour, the protein content can determine the price and food application for the flour. These examples point to the importance for accurate results from analytical testing.

Traditional quality control and quality assurance concepts are only a portion of a comprehensive quality management system. Food industry employees responsible for quality management work together in teams with other individuals in the company responsible for product development, production, engineering, maintenance, purchasing, marketing, and regulatory and consumer affairs.

Analytical information must be obtained, assessed, and integrated with other relevant information about the food system to address quality-related problems. Making appropriate decisions depends on having knowledge of the analytical methods and equipment utilized to obtain the data related to the quality characteristics. To design experiments in product and process development, and to assess results, one must know the operating principles and capabilities of the analytical methods. Upon completion of these experiments, one must critically evaluate the analytical data collected to determine whether product reformulation is needed or what parts of the process need to be modified for future tests. The situation is similar in the research laboratory, where knowledge of analytical techniques is necessary to design experiments, and the evaluation of data obtained determines the next set of experiments to be conducted.

1.2.4.2 *Types of Samples Analyzed*
Chemical and physical analysis of foods is an important part of a quality assurance program in food processing, from ingredients and raw materials, through

table 1.2 Types of samples analyzed in a quality assurance program for food products

Sample type	Critical questions
Raw materials	Do they meet your specifications? Do they meet required legal specifications?
	Are they safe and authentic?
	Will a processing parameter have to be modified because of any change in the composition of raw materials?
	Are the quality and composition the same as for previous deliveries?
	How does the material from a potential new supplier compare to that from the current supplier?
Process samples control	Did a specific processing step result in a product of acceptable composition or characteristics?
	Does a further processing step need to be modified to obtain a final product of acceptable quality?
Finished product	Does it meet the legal requirements? What is the nutritive value, so that label information can be developed? Or is the nutritive value as specified on an existing label?
	Does it meet product claim requirements (e.g., "low fat," "gluten free")?
	Will it be acceptable to the consumer?
	Will it have the appropriate shelf life?
	If unacceptable and cannot be salvaged, how do you handle it? (trash? rework? seconds?)
Competitor's sample	What are its composition and characteristics? How can we use this information to develop new products?
Complaint sample	How do the composition and characteristics of a complaint sample submitted by a customer differ from a sample with no problems?

Adapted and updated from [6, 7]

processing, to the finished products [4, 5]. Chemical and physical analysis also is important in formulating and developing new products, and evaluating new processes for making food products, and in identifying the source of problems with unacceptable products (Table 1.2). Competitor's samples are increasingly relevant, e.g., store brands versus national brands. For each type of sample listed in Table 1.2 to be analyzed, it may be necessary to determine either just one or many components. The nature of the sample and the way in which the information obtained will be used may dictate the specific method of analysis. For example, process control samples are usually analyzed by rapid methods, whereas nutritive value information for nutrition labeling generally requires the use of more time-consuming methods of analysis endorsed by scientific organizations. Critical questions, including those listed in Table 1.2, can be answered by analyzing various types of samples in a food processing system.

1.2.4.3 Increasing Dependence on Suppliers

Downsizing in response to increasing competition in the food industry often has pushed the responsibility for ingredient quality to the suppliers. Companies increasingly rely on others to supply high-quality and safe raw ingredients and packaging materials. Many companies have **select suppliers**, on whom they rely to perform the analytical tests to ensure compliance with detailed specifications for ingredients/raw materials. These specifications, and the associated tests, target various chemical, physical, and microbiological properties, as appropriate based on the nature of the ingredient. Such specifications for raw materials/ ingredients come in various forms within the food industry, with three commonly used forms described below:

1. **Technical/Product Data Sheet**: supplier uses when salesperson is selling ingredients; gives maximum, minimum, and/or range of values, as appropriate, and methods of analysis
2. **Specifications**: internal company document that defines company (processor) requirements (minimum, maximum, and/or target values) and links these to specific methods of analysis; much of the data come from the Technical/ Product Data Sheet; give the requirements for the Certificate of Analysis
3. **Certificate of Analysis** (COA): includes results of analytical tests related to predetermined specifications for specific shipment sent to customer; gives actual values and methods of analysis

As an example of specification information, Table 1.3 gives such information for the ingredient semolina, which is coarse-ground flour from durum wheat (high protein), used especially to make pasta, but also for other cereal grain-based products. For the purposes of this table, specific numbers are not given, as they would be in the actual documents for an ingredient.

Ingredient specifications and COAs are important for making specific food products. For example, with the wrong starch (i.e., with wrong specifications), a specific food product may not process correctly nor have the desired finished product quality attributes. Also, for example, if the COA indicates the granulation size of a specific lot of rolled oats is "out of spec," the finished granola bars may not have the desired properties, which can result in increased

1.3 table Example of information included in technical data sheet, specification, and certificate of analysis (COA) for semolina

Properties	Technical/product data sheet	Specification	Certificate of analysis
Chemical			
Moisture	Max value; AACCI	Max value; AOAC/AACCI	Actual value; AACCI
Protein	Min value; AACCI	Min value; AOAC/AACCI	Actual value; AACCI
Ash	Max value; AACCI	Max value; AOAC/AACCI	Actual value; AACCI
Falling number value	Target value, +/−; AACCI		
Enrichment			
Niacin		Max, min, target value; AOAC/AACCI	
Thiamine mononitrate		Max, min, target value; AOAC/AACCI	
Riboflavin		Max, min, target value; AOAC/AACCI	
Ferrous sulfate		Max, min, target value; AOAC/AACCI	
Folic acid		Max, min, target value; AOAC/AACCI	
Physical			
Bran specks	Max value; Internal	Max value; AOAC/AACCI	Actual value; Internal
Black specks	Max value; Internal	Max value; Internal	Actual value; Internal
Color L, a, b		Max L value, Min b value; AOAC/AACCI	Actual values (Hunter)
Color (linear E)			Actual value (calculated from Hunter LAB)
Extraneous matter	Complies with FDA regulations; AACCI		
Insect fragments		Max value; AOAC/AACCI	
Rodent hair		Max value; AOAC/AACCI	
Granulation	Value, +/−. % over #40, 60, 80, 100 sieve. % thru #100 sieve; Rotap	Min, max, target value. % over #40, 60, 80, 100 sieve. % thru #100 sieve; AOAC/AACCI	Actual value. % over #40, 60, 80, 100 sieve. % thru #100 sieve; Rotap
Microbiological			
Standard plate count; Total plate count	Product is considered not ready to eat and requires further processing, so no microbiological guarantees provided	Target value; FDA BAM	
Yeast		Target value; FDA BAM	
Mold		Target value; FDA BAM	
Vomitoxin	Complies with FDA advisory max level	Max value; FDA BAM	
Shelf life	Number of days at recommended storage conditions		

Actual, minimum, maximum, and/or target value; source/type of method

Source of method: *AOAC*, AOAC International; *AACCI*, AACC International; *FDA BAM*, FDA's Bacteriological Analytical Manual; *Internal*, company method; *Hunter*, Hunter colorimeter; *Rotap*, machine to measure granulation

consumer complaints. Setting the specifications for raw ingredients is commonly the responsibility of product developers, but it is often the production and quality control staff at the processing facility that must deal with challenges that arise when there are problems associated with ingredient specifications. Companies must have in place a means to maintain control of the COAs and react to them. With careful control over the quality of raw ingredients/materials, less testing is required during processing and on the final product.

1.2.4.4 Properties Analyzed
Analyzing foods for the reasons described in Table 1.1 to manage product quality commonly involves testing the following: chemical composition/characteristics,

1.4 table Quality management tests for dried pasta

Component/property being measured	Name of test
Quality test done in-house on semolina	
Moisture content	Rapid moisture analyzer
Color (L*, a*, b* determined to calculate Linear E value)	Colorimeter
Quality tests done in-process	
Moisture	Rapid moisture analyzer
Dimensions (after extrusion)	Micrometer and tape measurements by trained personnel
Metal detection	In-line metal detection (ferrous, nonferrous, stainless steel)
Package weight	In-line check with weight scale
Quality tests done on final product	
Moisture	Rapid moisture analyzer
Color L*, a*, b* determined to calculate Linear E value)	Colorimeter
Dimensions (diameter and shape)	Micrometer and tape measurements by trained personnel
Cooking quality	Sensory test by trained personnel (descriptive test; biting of samples)
Label	Visual inspection (with probability sampling methods)

physical properties, sensory properties, and microbial quality. Table 1.4 shows the quality management tests typically done on dried pasta. Each test indicated would be done at some specified frequency and by a specified method. Note the relationship between the COA information for semolina in Table 1.3 and the quality tests done on in-house on semolina by the processor, as reported in Table 1.4. While the COA generally requires considerable testing, done by official methods, routine in-house testing of ingredients is typically limited in scope and often uses rapid methods. The nature of testing required is largely determined by the nature of the food ingredient/product, but not surprisingly sensory properties (including taste, smell, appearance) are commonly tested across all foods and ingredients. This book focuses only on methods of analysis for testing chemical composition/characteristics and physical properties, and not on sensory properties or microbial quality.

1.3 STEPS IN ANALYSIS

1.3.1 Select and Prepare Sample

In analyzing food samples of the types described previously, all results depend on obtaining a representative sample and converting the sample to a form that can be analyzed. Neither of these is as easy as it sounds! **Sampling** and **sample preparation** are covered in detail in Chap. 5.

Sampling is the initial point for sample identification. Analytical laboratories must keep track of incoming samples and be able to store the analytical data from the analyses. This analytical information often is stored on a **laboratory information management system** (LIMS), which is a computer database program. Especially beneficial to commercial analytical laboratories, this system can capture all data for any specific sample and make it accessible to customers so they can import, review, and analyze as they wish. Customers have real-time access to these data, which also enables them to review testing status. There is no need for hand typing data, which saves times and decreases chances for error, and the system settings can ensure standardized data entry for consistency and compliance to analytical requirements.

1.3.2 Perform the Assay

Performing the assay is unique for each component or characteristic to be analyzed and may be unique to a specific type of food product. Single chapters in this book address sampling and sample preparation (Chap. 5) and data handling (Chap. 4), while the remainder of the book addresses the step of actually performing the assay. Chapters 6, 7, 8, 9, 10, and 11 cover spectroscopy and spectrometry, and Chaps. 12, 13, and 14 cover chromatography. These major general topics are covered before chapters that cover specific methods for chemical composition and characterization, since many analytical methods utilize spectroscopy and chromatography techniques. The descriptions of the various specific procedures are meant to be overviews of the methods. For guidance in actually performing the assays, details regarding chemicals, reagents, apparatus, and step-by-step instructions are found in the books and articles referenced in each chapter. Numerous chapters in this book, and other books devoted to food analysis [8–12], make the point that for food analysis, we increasingly rely on expensive equipment, some of which requires considerable expertise. Also, it should be noted that numerous analytical methods utilize automated instrumentation, including autosamplers and robotics to speed the analyses.

1.3.3 Calculate and Interpret the Results

To make decisions and take action based on the results obtained from performing the assay that determined the composition or characteristics of a food product, one must make the appropriate calculations to interpret the data correctly. **Data handling**, covered in Chap. 4, includes important statistical principles.

1.4 METHOD SELECTION

1.4.1 Objective of the Assay

Selection of a method depends largely on the objective of the measurement (Fig. 1.1). For example, methods used for rapid online processing measurements may be less accurate than official methods (see Sect. 1.5) used for nutritional labeling purposes. Methods referred to as reference, definitive, official, or primary are most applicable in a well-equipped and staffed analytical lab. Quality control testing of raw ingredients at the processing facility, in-process testing, and final product testing often rely on secondary/rapid quality control methods. This is likely in contrast to primary/official methods used on raw ingredient specification and COAs and the testing for nutrition labeling. Both **primary methods** and **secondary methods** may be necessary and appropriate, with a secondary method calibrated against the primary method. The more rapid secondary or field methods may be more applicable on the manufacturing floor in a food processing facility. For example, refractive index may be used as a rapid, secondary method for sugar analysis (see Chaps. 15 and 19), with results correlated to those of the primary method, high-performance liquid chromatography (HPLC) (see Chaps. 13 and 19). Moisture content data for a product being developed in the pilot plant may be obtained quickly with a rapid moisture analyzer that has been calibrated using a more time-consuming forced air oven method (see Chap. 15). Many companies commonly use unofficial, rapid methods for testing in the processing plants but validate them against official methods. The calibration between these methods is critical.

1.4.2 Characteristics of the Method

Numerous methods often are available to assay food samples for a specific characteristic or component. To select or modify methods used to determine the chemical composition and characteristics of foods, one must be familiar with the principles underlying the procedures and the critical steps. Certain properties of methods and criteria described in Table 1.5 are useful to evaluate the appropriateness of a method in current use or a new method being considered. As highlighted in a review article on food analysis by Cifuentes [2], there is a continual need for the development of methods that can be characterized as being more robust, efficient, sensitive, and cost-effective. Many of the older "wet chemistry" methods have evolved into powerful and commonly used instrumental techniques. This has led to significant increases in accuracy, precision, detection limits, and sample throughput.

1.4.3 Validity of the Method

1.4.3.1 Overview

Numerous factors affect the usefulness and validity of the data obtained using a specific analytical method. One must consider certain characteristics of any method, such as specificity, precision, accuracy, and sensitivity (see Table 1.5 and Chap. 4). However, one also must consider how the variability of data from the method for a specific characteristic compares to differences detectable and acceptable to a consumer, and the variability of the specific characteristic inherent in processing of the food. One must consider the nature of the samples collected for the analysis, how representative the samples were of the whole, and the number of samples analyzed (Chap. 5). One must ask whether details of the analytical procedure were followed adequately, such that the results are accurate, repeatable, and comparable to data collected previously. For data to be valid, equipment to conduct the analysis must be standardized and appropriately used, and the performance limitations of the equipment recognized.

1.4.3.2 Standard Reference Materials

A major consideration for determining method validity is the analysis of materials used as controls, often referred to as **standard reference materials** or **check samples** [13]. Analyzing check samples concurrently with test samples is an important part of quality control [14]. Standard reference materials (SRMs) can be obtained in the United States from the National Institute of Standards and Technology (NIST) and from US Pharmacopeia (USP), in Europe from the Institute for Reference Materials and Measurements (IRMM), and from other specific organizations for other countries/regions. Besides government-related groups, numerous organizations offer check sample services that provide test samples to evaluate the reliability of a method [13]. For example, **AACC International** (AACCI) (formerly known as the American Association of Cereal Chemists, AACC) has a check sample service in which a subscribing

table 1.5 Criteria for choice of food analysis methods: characteristics of a method

Characteristic	Critical questions
Inherent properties	
Specificity/selectivity	Is the property being measured the same as that claimed to be measured, and is it the only property being measured?
	Are there interferences?
	What steps are being taken to ensure a high degree of specificity?
Precision	What is the precision of the method? Is there within-batch, batch-to-batch, or day-to-day variation?*
	What step in the procedure contributes the greatest variability?
Accuracy	How does the new method compare in accuracy to the old or a standard method?
	What is the percent recovery?
	Is it reproducible between labs?
Applicability of method to laboratory	
Reagents	Can you properly prepare them?
	What equipment is needed? Are they stable? For how long and under what conditions?
Equipment	Is the method very sensitive to slight or moderate changes in the reagents?
	Do you have the appropriate equipment?
	Are personnel competent to operate equipment?
Cost	What is the cost in terms of equipment, reagents, and personnel?
Applicability to food/sample	Destructive or nondestructive?
	Online or off-line?
	Official method/approval?
	Nature of food matrix?
Usefulness	
Time required	How fast is it? How fast does it need to be?
Reliability	How reliable is it from the standpoints of precision and stability?
Need	Does it meet a need or better meet a need?
	Simplicity of operation?
Personnel	
Safety	Is any change in method worth the trouble of the change?
Procedures	Who will do any required calculations?

*In-process samples may not accurately represent finished product; must understand what variation can and should be present

laboratory receives specifically prepared test samples from AACCI. The subscribing laboratory performs the specified analyses on the samples and returns the results to AACCI. The AACCI then provides a statistical evaluation of the analytical results and compares the subscribing laboratory's data with those of other laboratories to inform the subscribing laboratory of its degree of accuracy. The AACCI offers check samples such as flours, semolina, and other cereal-based samples, for analyses such as moisture, ash, protein, vitamins, minerals, sugars, total dietary fiber, and soluble and insoluble dietary fiber. Samples also are available for testing physical properties and for microbiological and sanitation analyses.

The **American Oil Chemists' Society** (AOCS) has a reference sample program that includes oilseeds, oils and fats, marine products, aflatoxins, cholesterol, trace metals, *trans*-fatty acids, and other samples.

Laboratories from many countries participate in the program to check the accuracy of their work, their reagents, and their laboratory apparatus against the statistical norm derived from the group data.

Standard reference materials are important tools to ensure reliable data. However, such materials need not necessarily be obtained from outside organizations. Control samples internal to the laboratory (i.e., **internal reference materials**) can be prepared by carefully selecting an appropriate type of sample, gathering a large quantity of the material, mixing and preparing to ensure homogeneity, packaging the sample in small quantities, storing the samples appropriately, and routinely analyzing the control sample when test samples are analyzed. Whatever the standard reference materials used, these should match closely the matrix of the samples to be analyzed by a specific method. AOAC International has a peer-review program of matching reference materials with respective official methods of analysis.

1.4.3.3 *ISO Certification*

To help ensure validity of results generated by methods of analysis, commercial, private, and government laboratories are increasingly using **ISO** (International Organization for Standardization) certification (currently 17025) to help assure their customers of the quality of their work (i.e., validity of data) [4, 15]. Certification can be for the entire laboratory and/or just individual assays (i.e., methods and procedures). Once certified, these laboratories are audited for recertification every two years. ISO certification involves review of standard operating procedures, forms, records, work instructions, document control process, and test methods. While ISO certification does not deal with safety or the business, it does address questions about the following: (1) results validity, (2) equipment, (3) training, (4) analysts operating the equipment, (5) reagents and chemicals, and (6) customer communication.

Method certification requires comparison to other laboratories and the use of control samples/standards. For a food company with a corporate laboratory and multiple plant operations, the corporate laboratory does proficiency testing, using check samples (e.g., USP, NIST), and makes control samples that can be tested at both the corporate and plant laboratories for comparison of results.

A benefit of certification can include increased business for the laboratories (i.e., some customers choose ISO-certified laboratories over noncertified laboratories, due to increased confidence in the data). Also, the certification tends to create a mind-set of continuous improvement within the laboratory. The greatest challenges for laboratories regarding certification are often updating procedures and balancing the need to get results to customers versus the need to check equipment and follow all required procedures. If a company fails an audit, the approach used is root cause analysis and corrective action investigation and implementation.

1.4.4 Consideration of Food Composition

Proximate analysis of foods refers to determining the major components of moisture (Chap. 15), ash (total minerals) (Chap. 16), lipids (Chap. 17), protein (Chap. 18), and carbohydrates (Chap. 19). The performance of many analytical methods is affected by the **food matrix** (i.e., its major chemical components, especially lipid, protein, and carbohydrate). In food analysis, it is usually the food matrix that presents the greatest challenge to the analyst [16]. For example, high-fat or high-sugar foods can cause different types of interferences than do low-fat or low-sugar foods. Digestion procedures and extraction steps necessary for accurate analytical results can be very dependent on the food matrix. The complexity of various food systems often requires having not just one technique available for a specific food component, but rather multiple techniques and procedures, as well as the knowledge about which to apply to a specific food matrix.

A task force of **AOAC International** (formerly known as the Association of Official Analytical Chemists, AOAC) suggested a "triangle scheme" for dividing foods into matrix categories [17–25] (Fig. 1.3). The apexes of the triangle contain food groups that

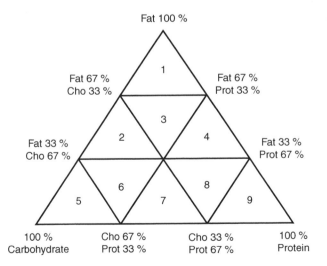
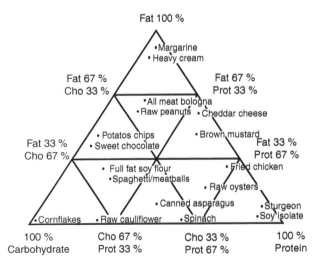

1.3 figure　Schematic layout of food matrixes based on protein, fat, and carbohydrate content, excluding moisture and ash (Reprinted with permission from [20], *Inside Laboratory Management*, September 1997, p. 33. Copyright 1997, by AOAC International])

were either 100% fat, 100% protein, or 100% carbohydrate. Foods were rated as "high," "low," or "medium" based on levels of fat, carbohydrate, and proteins, which are the three nutrients expected to have the strongest effect on analytical method performance. This created nine possible combinations of high, medium, and low levels of fat, carbohydrate, and protein. Complex foods were positioned spatially in the triangle according to their content of fat, carbohydrate, and protein, on a normalized basis (i.e., fat, carbohydrate, and protein normalized to total 100%). General analytical methods ideally would be geared to handle each of the nine combinations, replacing more numerous matrix-dependent methods developed for specific foods. For example, using matrix-dependent methods, one method might be applied to potato chips and chocolates, which are both low-protein, medium-fat, medium-carbohydrate foods, but another might be required for a high-protein, low-fat, high-carbohydrate food such as nonfat dry milk [19]. In contrast, a robust general method could be used for all of the food types.

The AACCI has approved a method studied using the triangle approach [21]. This approach also has been adopted by NIST for categorizing their SRMs [23–25] (see also Sect. 1.4.3.2). These SRMs can be used by laboratories to optimize methods for samples with specific matrices and to confirm that a method can be used reliably for a specific matrix. The food triangle categorization of both samples and reference materials enables analysts to choose the appropriate SRM for method development, for optimization, or for confidence in routine testing. NIST has, as a goal, the development of SRMs and other reference materials in all nine sections of the food matrix triangle [23].

1.5   OFFICIAL METHODS

The choice of method for a specific characteristic or component of a food sample is often made easier by the availability of **official methods**. Several nonprofit scientific organizations have compiled and published these methods of analysis for food products, which have been carefully developed and standardized. They allow for comparability of results between different laboratories that follow the same procedure, and for evaluating results obtained using new or more rapid procedures.

1.5.1   AOAC International

AOAC International, which dates back to 1884, is well known for its program to provide official methods of

analysis that fit their intended purpose (i.e., will perform with the necessary accuracy and precision under usual laboratory conditions). The official methods developed are intended for use by regulated industries, regulatory agencies, contract research organizations and testing laboratories, and academic institutions. AOAC International also has programs to develop standards, to test the performance of commercial and analytical methods, and for testing laboratory proficiencies.

AOAC International uses advisory panels, stakeholder panels, working groups, and expert review panels in their collaborative system of voluntary consensus building to develop fit-for-purpose methods and services to ensure quality measurements. The *Official Methods of Analysis* of AOAC International [26] is the widely used compilation of methods adopted over the years. Details of the method validation program are given online and in the front matter of that book. Methods validated and adopted by AOAC International and the data supporting the method validation are published in the *Journal of AOAC International*. AOAC International's *Official Methods of Analysis* was most recently published in 2016, but an online version of the book is available as a "continuous edition," with new and revised methods posted as soon as they are approved. The *Official Methods of Analysis* of AOAC International includes methods appropriate for a wide variety of products and other materials (Table 1.6). These methods often are specified by the FDA with regard to legal requirements for food products. They are generally the methods followed by the FDA and the **Food Safety and Inspection Service** (FSIS) of the **United States Department of Agriculture** (USDA) to check the nutritional labeling information on foods and to check foods for the presence or absence of undesirable residues or residue levels.

1.5.2   Other Endorsed Methods

The AACCI publishes a set of approved laboratory methods, applicable mostly to cereal products (e.g., baking quality, gluten, physical dough tests, staleness/texture). The AACCI process of adopting the *Approved Methods of Analysis* [27] is consistent with the process used by the AOACI and AOCS. Approved methods of the AACCI are continuously reviewed, critiqued, and updated (Table 1.7) and are now available online.

The AOCS publishes a set of official methods and recommended practices, applicable mostly to fat and oil analysis (e.g., vegetable oils, glycerin, lecithin) [28] (Table 1.8). AOCS is a widely used methodology source on the subjects of edible fats and oils, oilseeds and oilseed proteins, soaps and synthetic detergents,

1.6 table	Table of contents of 2016 *Official Methods of Analysis of AOAC International* [26]

Chapter	Title
1	Agriculture liming materials
2	Fertilizers
3	Plants
4	Animal feed
5	Drugs in feeds
6	Disinfectants
7	Pesticide formulations
8	Hazardous substances
9	Metals and other elements at trace levels in foods
10	Pesticide and industrial chemical residues
11	Waters; and salt
12	Microchemical methods
13	Radioactivity
14	Veterinary analytical toxicology
15	Cosmetics
16	Extraneous materials: isolation
17	Microbiological methods
18	Drugs: part I
19	Drugs: part II
20	Drugs: part III
21	Drugs: part IV
22	Drugs: part V
23	Drugs and feed additives in animal tissues
24	Forensic sciences
25	Baking powders and baking chemicals
26	Distilled liquors
27	Malt beverages and brewing materials
28	Wines
29	Nonalcoholic beverages and concentrates
30	Coffee and tea
31	Cacao bean and its products
32	Cereal foods
33	Dairy products
34	Eggs and egg products
35	Fish and other marine products
36	Flavors
37	Fruits and fruit products
38	Gelatin, dessert preparations, and mixes
39	Meat and meat products
40	Nuts and nut products
41	Oils and fats
42	Vegetable products, processed
43	Spices and other condiments
44	Sugars and sugar products
45	Vitamins and other nutrients
46	Color additives
47	Food additives: direct
48	Food additives: indirect
49	Natural toxins
50	Infant formulas, baby foods, and enteral products
51	Dietary supplements

1.7 table	Table of contents of 2010 *Approved Methods of AACC International* [27]

Chapter	Title
2	Acidity
4	Acids
6	Admixture of flours
7	Amino acids
8	Total ash
10	Baking quality
11	Biotechnology
12	Carbon dioxide
14	Color and pigments
20	Ingredients
22	Enzymes
26	Experimental milling
28	Extraneous matter
30	Crude fat
32	Fiber
33	Sensory analysis
38	Gluten
39	Infrared analysis
40	Inorganic constituents
42	Microorganisms
44	Moisture
45	Mycotoxins
46	Nitrogen
48	Oxidizing, bleaching, and maturing agents
54	Physical dough tests
55	Physical tests
56	Physicochemical tests
58	Special properties of fats, oils, and shortenings
61	Rice
62	Preparation of sample
64	Sampling
66	Semolina, pasta, and noodle quality
68	Solutions
74	Staleness/texture
76	Starch
78	Statistical principles
80	Sugars
82	Tables
86	Vitamins
89	Yeast

industrial fats and oils, fatty acids, oleochemicals, glycerin, and lecithin.

Standard Methods for the Examination of Dairy Products [29], published by the American Public Health Association, includes methods for the chemical analysis of milk and dairy products (e.g., acidity, fat, lactose, moisture/solids, added water) (Table 1.9). *Standard Methods for the Examination of Water and Wastewater* [30] is published jointly by the American Public Health Association, the American Water Works Association,

| **1.8** table | Table of contents of 2009 *Official Methods and Recommended Practices of the American Oil Chemists' Society* [28] |

Section	Title
A	Vegetable oil source materials
B	Oilseed by-products
C	Commercial fats and oils
D	Soap and synthetic detergents
E	Glycerin
F	Sulfonated and sulfated oils
G	Soapstocks
H	Specifications for reagents and solvents and apparatus
J	Lecithin
M	Evaluation and design of test methods
S	Official listings
T	Recommended practices for testing industrial oils and derivatives

| **1.9** table | Contents of Chap. 15 on chemical and physical methods in *Standard Methods for the Examination of Dairy Products* [29] |

15.010	Introduction
15.020	Acidity tests
15.030	Adulterant tests
15.040	Ash tests
15.050	Chloride tests
15.060	Contaminant tests
15.070	Extraneous material tests
15.080	Fat determination methods
15.090	Lactose and galactose tests
15.100	Minerals and food additives
15.110	Moisture and solids tests
15.120	Multicomponent tests
15.130	Protein/nitrogen tests
15.140	Rancidity tests
15.150	Sanitizer tests
15.160	Vitamins A, D2, and D3 in milk products, HPLC method
15.170	Functional tests
15.180	Cited references

and the Water Environment Federation. *Food Chemicals Codex* [31], published by USP, contains methods for the analysis of certain food additives. Some trade associations publish standard methods for the analysis of their specific products. The FDA refers analysts to manuals for general laboratory quality practices [32], elemental analysis methods [33], drug and chemical residue analysis methods [34], pesticide analysis methods [35, 36], and macroanalytical procedures (e.g., evaluation by sight, smell, or taste) [37]. The USDA refers analysts to the *Chemistry Laboratory Guidebook* [38], which contains test methods used by USDA laboratories for ensuring the safety and accurate labeling of foods regulated by the USDA (see also Chap. 3, Sect. 3.3).

1.6 SUMMARY

Food scientists and technologists determine the chemical composition and physical characteristics of foods routinely as part of their quality management, product development, or research activities. For example, the types of samples analyzed in a quality management program of a food company can include raw materials, process control samples, finished products, competitors' samples, and consumer complaint samples. Consumer, government, and food industry concern for food quality and safety has increased the importance of analyses that determine composition and critical product characteristics.

To successfully base decisions on results of any analysis, one must correctly conduct all three major steps in the analysis: (1) select and prepare samples, (2) perform the assay, and (3) calculate and interpret the results. The choice of analysis method is usually based on the objective of the analysis, the characteristics of the method itself (e.g., specificity, accuracy, precision, speed, cost of equipment, and training of personnel), and the food matrix involved. Validation of the method is important, as is the use of standard reference materials to ensure quality results. Rapid methods used for quality assessment in a production facility may be less accurate but much faster than official methods used for nutritional labeling. Endorsed methods for the chemical analyses of foods have been compiled and published by AOAC International, AACCI, AOCS, and certain other non-profit scientific organizations. These methods allow for comparison of results between different laboratories and for evaluation of new or more rapid procedures.

1.7 STUDY QUESTIONS

1. Identify six specific reasons you might need to determine certain chemical characteristics of a food product (or ingredient) as part of a quality management program.
2. You are considering the use of a new method to measure compound X in your food product. List six factors you will consider before adopting this new method in your quality assurance laboratory.
3. In your work at a food company, you mentioned to a coworker something about the *Official Methods of Analysis* published by AOAC International. The coworker asks you what AOAC International does and what the *Official Methods of Analysis* is. Answer your coworker's questions.
4. For each type of product listed below, identify a publication in which you can find standard

methods of analysis appropriate for the product:

(a) Ice cream
(b) Enriched flour
(c) Wastewater (from food processing plant)
(d) Margarine

Acknowledgments The author thanks the numerous former students and others associated with quality assurance in the food industry who reviewed this chapter and contributed ideas for its revision. Special thanks go to the following for their help with specific topics: Chris Fosse (ConAgra Foods) and Ryan Lane (TreeHouse Foods), specifications for semolina; Samantha Park (TreeHouse Foods) and Yuezhen He (former student at Purdue University), quality tests for pasta; Sandy Zinn (General Mills), ISO; Ryan Ellefson (Covance), LIMS; Karen Andrews (USDA), food matrix and standard reference materials; and Julie Culp (General Mills), food analysis linked to consumer demand and government regulations.

REFERENCES

1. McGorrin RJ (2009) One hundred years of progress in food analysis. J. Agric. Food Chem. 57(18):8076–8088
2. Cifuentes A (2012) Food analysis: Present, future, and foodomics. ISRN Analytical Chemistry. Article ID 801607
3. Spence JT (2006) Challenges related to the composition of functional foods. J Food Compost Anal 19 Suppl 1: S4–S6
4. Alli I (2016) Food quality assurance: principles and practices, 2nd edn. CRC, Boca Raton, FL
5. Vasconcellos JA (2004) Quality assurances for the food industry: a practical approach. CRC, Boca Raton, FL
6. Pearson D (1973) Introduction – some basic principles of quality control, Ch. 1. In: Laboratory techniques in food analysis. Wiley, New York, pp 1–26
7. Pomeranz Y, Meloan CE (1994) Food analysis: theory and practice, 3rd edn. Chapman & Hall, New York
8. Cruz RMS, Khmelinskii I, Vieira M (2014) Methods in food analysis. CRC, Boca Raton, FL
9. Nollett LML (2004) Handbook of Food Analysis, 2nd edn. vol 1: physical characterization and nutrient analysis, vol 2: residues and other food component analysis. CRC, Boca Raton, FL
10. Nollett LML, Toldra F (2012) Food analysis by HPLC, 3rd edn. CRC, Boca Raton, FL
11. Ötleş S (2008) Handbook of food analysis instruments. CRC, Boca Raton, FL
12. Ötleş S (2011) Methods of analysis of food components and additives, 2nd edn. CRC, Boca Raton, FL
13. Latimer GW, Jr (1997) Check sample programs keep laboratories in sync. Inside Lab Manage 1(4): 18–20
14. Ambrus A (2008) Quality assurance, Ch. 5. In: Tadeo JL (ed) Analysis of pesticides in food and environmental samples. CRC, New York, p 145
15. ISO/IEC (2010) ISO/IEC 17025:2005. General requirements for the competence of testing and calibration laboratories (last review in 2010). http://www.iso.org/iso/home/store/catalogue_tc/catalogue_detail.htm?csnumber=39883
16. Wetzel DLB, Charalambous G (eds) (1998) Instrumental methods in food and beverage analysis. Elsevier Science, Amsterdam, The Netherlands
17. Wolf W (1993) In: Methods of analysis for nutrition labeling. AOAC International, Ch 7, pp 115–120, as adapted from W. Ikins, DeVires J, Wolf W, Oles P, Carpenter D, Fraley N, Ngeh-Nwainbi J. The Referee. AOAC International 17(7), pp 1, 6–7
18. Wolf WR, Andrews KW (1995) A system for defining reference materials applicable to all food matrices. Fresenius J Anal Chem 352:73–76
19. Lovett RA (1997) U.S. food label law pushes fringes of analytical chemistry. Inside Lab Manage 1(4): 27–28
20. Ellis C, Hite D, van Egmond H (1997) Development of methods to test all food matrixes unrealistic, says OMB. Inside Lab Manage 1(8):33–35
21. DeVries JW, Silvera KR (2001) AACC collaborative study of a method for determining vitamins A and E in foods by HPLC (AACC Method 86–06). Cereal Foods World 46(5):211–215
22. Sharpless KE, Greenberg RR, Schantz MM, Welch MJ, Wise SA, Ihnat M (2004) Filling the AOAC triangle with food-matrix standard reference materials. Anal and Bioanal Chem 378:1161–1167
23. Wise SA, Sharpless KE, Sander LC, May WE (2004) Standard reference materials to support US regulations for nutrients and contaminants in food and dietary supplements. Accred Qual Assur 9:543–550
24. Sharpless KE, Lindstrom RM, Nelson BC, Phinney KW, Rimmer CA, Sander LC, Schantz MM, Spatz RO, Thomas JB, Turk GC, Wise SA, Wood LJ (2010) Preparation and characterization of standard reference material 1849 Infant/adult nutritional formula. 93(4):1262–1274
25. Thomas JB, Yen JH, Sharpless KE (2013) Characterization of NIST food-matrix standard reference materials for their vitamin C content. Anal Bioanal Chem 405: 4539–4548
26. AOAC International (2016) Official methods of analysis, 20th edn., 2016 (online). AOAC International, Rockville, MD
27. AACC International (2010) Approved methods of analysis, 11th edn (online), AACC International, St. Paul, MN
28. AOCS (2013) Official methods and recommended practices, 6th edn, 3rd printing, American Oil Chemists' Society, Champaign, IL
29. Wehr HM, Frank JF (eds) (2004) Standard methods for the examination of dairy products, 17th edn. American Public Health Association, Washington, DC
30. Rice EW, Baird RB, Eaton AD, Clesceri LS (eds) (2012) Standard methods for the examination of water and wastewater, 22nd edn. American Public Health Association, Washington, DC
31. U.S. Pharmacopeia (USP) (2016) Food chemicals codex, 10th edn. United Book Press, Baltimore, MD
32. FDA (2014) CFSAN laboratory quality manual (last updated 12/10/14) http://www.fda.gov/downloads/Food/FoodScienceResearch/UCM216952.pdf
33. FDA (2016) Elemental analysis manual for food and related products (last updated 1/08/16). http://www.fda.gov/food/foodscienceresearch/laboratory-methods/ucm2006954.htm
34. FDA (2015) Drug and chemical residue methods (last updated 6/29/15) http://www.fda.gov/Food/FoodScienceResearch/LaboratoryMethods/ucm2006950.htm
35. FDA (1999) Pesticide analytical manual volume I (PAM), 3rd edn (1994, updated October, 1999). http://www.fda.gov/Food/ScienceResearch/LaboratoryMethods/PesticideAnalysisManualPAM/ucm111455.htm

36. FDA (2002) Pesticide analytical manual volume II (PAM) (updated January, 2002). http://www.fda.gov/Food/ScienceResearch/LaboratoryMethods/PesticideAnalysisManualPAM/ucm113710.htm

37. FDA (2015) Macroanalytical Procedures Manual (last updated 6/05/15) http://www.fda.gov/Food/FoodScienceResearch/LaboratoryMethods/ucm2006953.htm

38. USDA (2016) Chemistry laboratory guidebook (last updated 6/26/15) http://www.fsis.usda.gov/wps/portal/fsis/topics/science/laboratories-and-procedures/guidebooks-and-methods/chemistry-laboratory-guidebook. Food Safety and Inspection Service, US Department of Agriculture, Washington, DC

US Government Regulations and International Standards Related to Food Analysis

S.Suzanne Nielsen

Department of Food Science,
Purdue University,
West Lafayette, IN 47907-2009, USA
e-mail: nielsens@purdue.edu

S. Nielsen (ed.), *Food Analysis*, Food Science Text Series,
DOI 10.1007/978-3-319-45776-5_2, © Springer International Publishing 2017

2.1 INTRODUCTION

Knowledge of government regulations relevant to the chemical analysis of foods is extremely important to persons working in the food industry. Federal laws and regulations reinforce the efforts of the food industry to provide wholesome foods, to inform consumers about the nutritional composition of foods, and to eliminate economic frauds. In some cases, they dictate what ingredients a food must contain, what must be tested, and the procedures used to analyze foods for safety factors and quality attributes. This chapter describes the US federal regulations related to the composition of foods. The reader is referred to Refs. [1–4] for a comprehensive coverage of US food laws and regulations. Many of the regulations referred to in this chapter are published in the various titles of the *Code of Federal Regulations* (CFR) [5]. This chapter also includes information about food standards and safety practices established by international organizations. Nutrition labeling regulations are not covered in this chapter, but Chap. 3 is devoted to this topic, as it is related to food analysis.

2.2 US FEDERAL REGULATIONS AFFECTING FOOD COMPOSITION

2.2.1 US Food and Drug Administration (FDA)

The FDA is a government agency within the **Department of Health and Human Services** (DHHS). The FDA is responsible for regulating, among other things, the safety of foods, cosmetics, drugs, medical devices, biologics, tobacco, and radiation-emitting products. It acts under laws passed by the US Congress to monitor the affected industries and ensure the consumer of the safety of such products. A comprehensive collection of federal laws, guidelines, and regulations relevant to foods and drugs has been published by the Food and Drug Law Institute [1, 2].

2.2.1.1 Legislative History

Since the original Food and Drug Act of 1906, various acts, amendments, and regulations have been put in place to regulate foods. Table 2.1 summarizes these acts, amendments, and regulations with food analysis implications. The FDA implements and enforces the **Federal Food, Drug, and Cosmetic (FD&C) Act** of 1938 as amended. Select aspects of this and other regulations are further described in the subsections that follow.

2.2.1.2 Food Definitions and Standards

The food definitions and standards established by the FDA are published in 21 CFR 100–169 and include standards of identity, quality, and fill. The **standards of identity**, which have been set for a wide variety of food products, are most relevant to the chemical analysis of foods because they specifically establish which ingredients a food must contain. They limit the amount of water permitted in certain products. The minimum levels for expensive ingredients are often set, and maximum levels for inexpensive ingredients are sometimes set. The kind and amount of certain vitamins and minerals that must be present in foods labeled "enriched" are specified. The standards of identity for some foods include a list of optional ingredients. As an example, the standard of identity for sour cream (21 CFR 131.160) is given in Fig. 2.1. Table 2.2 summarizes the standards of identity relevant to food analysis for a number of other foods. Note that the standard of identity often includes the recommended analytical method for determining chemical composition.

Although standards of quality and fill generally are less related to the chemical analysis of foods than are standards of identity, they are important for economic and quality control considerations. **Standards of quality**, established by the FDA for some canned fruits and vegetables, set minimum standards and specifications for factors such as color, tenderness, weight of units in the container, and freedom from defects. The FDA has also set a standard of quality for bottled water, which includes set allowable levels for the following: coliform, turbidity, color, odor, radium-226 and radium-228 activity, beta particle and photon radioactivity, and more than 70 chemical contaminants [21 CFR 165.110(b)]. The FDA has established **standards of fill** for some canned fruits and vegetables, tomato products, and seafood, stating how full a container must be to avoid consumer deception.

2.2.1.3 Inspection and Enforcement

The FDA has the broadest regulatory authority over most foods (generally, all foods other than meat, poultry, egg products; water supplies). However, the FDA shares responsibilities with other regulatory agencies for certain foods, as described in later sections of this chapter. Table 2.3 summarizes the food

2.1 table Food and Drug Administration acts, amendments, and regulations with food analysis implications

Act, amendment, or regulation (year)	Intent	Food analysis relevance
Federal Food, Drug, and Cosmetic Act (FD&C) (1938)	Intended to assure consumers that foods are safe, wholesome, produced under sanitary conditions, and packaged and labeled truthfully. Prohibits adulteration and misbranding of foods	Authorizes food definitions and standards of identify, many of which require testing food composition
Food Additive Amendment (1958)	Designed to protect the health of consumers by requiring a food additive to be proven safe before addition to a food, and used at only a safe level in foods. Delaney Clause (rider to amendment) prohibits FDA from setting any tolerance as a food additive for substances known to be a carcinogen	Requires testing food additives (type, amount, safety)
Color Additive Amendment (1960)	Defined color additives. Set rules for both certified and uncertified colors. Provides for approval of color additives. Allows for listing color additives for specific uses and sets quantity limitations. Contains Delaney Clause (see above)	Requires testing food additives (type, amount, safety)
Current Good Manufacturing Practices (cGMP) (1969) (21 CRS 110 – general; specific GMPs also exist for certain foods)	Designed to prevent adulterated foods in the marketplace	Requires testing for adulterants (including extraneous matter)
Hazard Analysis Critical Control Point (HACCP): 21 CFR Part 123 – seafood; 21 CRS Part 120 – juice; 9 CFR Part 417 – meat and poultry) (Refs. [6–9])	Intended to improve food safety and quality	Requires testing for microbial, chemical, and physical hazards
Nutrition Labeling and Education Act (NLEA) (1990)	Made nutrition labeling mandatory on most food products under FDA jurisdiction. Established definitions for health and nutrient claims	Requires testing for food composition (specific components and their levels)
Dietary Supplement Health and Education Act (DSHEA)(1994)	Defined supplements as "dietary ingredients," set criteria to regulate claims and labeling, and established government agencies to handle regulation. Changed the definition and regulations for dietary supplements, so no longer considered a food	Eliminated testing of dietary supplements as required when defined as a food
Food Safety Modernization Act (FSMA) (2011)	Aims to ensure the food supply is safe by shifting the focus from responding to contamination to preventing it	Requires testing for hazards and preventive controls, including physical and chemical hazards, food allergens, sanitation controls, and supply chain controls
Food Labeling: Revision of the Nutrition and Supplement Facts Label (2016)	Updated information on label to help consumers maintain health dietary practices	Changes the nutrients tested and reported on label

analysis-related scope of responsibility of the FDA, compared to and often in cooperation with other federal agencies, to ensure the quality and safety of foods in the USA. Table 2.4 summarizes which federal agency has responsibility for ensuring the quality and safety of specific types of foods.

When violations of the FD&C Act are discovered by the FDA through periodic inspections of facilities and products and through analysis of samples, the FDA can use warning letters, administrative detention, suspension of registration, seizures, injunctions, or recalls, depending on the circumstances. The

§131.160 Sour cream.

(a) *Description.* Sour cream results from the souring, by lactic acid producing bacteria, of pasteurized cream. Sour cream contains not less than 18 percent milkfat; except that when the food is characterized by the addition of nutritive sweeteners or bulky flavoring ingredients, the weight of the milkfat is not less than 18 percent of the remainder obtained by subtracting the weight of such optional ingredients from the weight of the food; but in no case does the food contain less than 14.4 percent milkfat. Sour cream has a titratable acidity of not less than 0.5 percent, calculated as lactic acid.

(b) *Optional ingredients.*
    (1) Safe and suitable ingredients that improve texture, prevent syneresis, or extend the shelf life of the product.
    (2) Sodium citrate in an amount not more than 0.1 percent may be added prior to culturing as a flavor precursor.
    (3) Rennet.
    (4) Safe and suitable nutritive sweeteners.
    (5) Salt.
    (6) Flavoring ingredients, with or without safe and suitable coloring, as follows:
        (i) Fruit and fruit juice (including concentrated fruit and fruit juice).
        (ii) Safe and suitable natural and artificial food flavoring.

(c) *Methods of analysis*: Referenced methods in paragraph (c) (1) and (2) of this section are from "Official Methods of Analysis of the Association of Official Analytical Chemists," 13th Ed. (1980), which is incorporated by reference. Copies may be obtained from the AOAC INTERNATIONAL, 481 North Frederick Ave., suite 500, Gaithersburg, MD 20877, or may be examined at the National Archives and Records Administration (NARA). For information on the availability of this material at NARA, call 202-741-6030, or go to: *http://www.archives.gov/federal_register/code_of_federal_regulations/ibr_location.html*
    (1) Milkfat content – "Fat-Official Final Action," section 16.172.
    (2) Titratable acidity – "Acidity-Official Final Action," section 16.023.

(d) *Nomenclature.* The name of the food is "Sour cream" or alternatively "Cultured sour cream". The full name of the food shall appear on the principal display panel of the label in type of uniform size, style, and color. The name of the food shall be accompanied by a declaration indicating the presence of any flavoring that characterizes the product, as specified in §101.22 of this chapter. If nutritive sweetener in an amount sufficient to characterize the food is added without addition of characterizing flavoring, the name of the food shall be preceded by the word "sweetened".

(e) *Label declaration.* Each of the ingredients used in the food shall be declared on the label as required by the applicable sections of parts 101 and 130 of this chapter. [42 FR 14360, Mar. 15, 1977, as amended at 47 FR 11824, Mar. 19, 1982; 49 FR 10092, Mar. 19, 1984; 54 FR 24893, June 12, 1989; 58 FR 2891, Jan. 6, 1993]

2.1 figure   Standard of identity for sour cream [From 21 CFR 131.160 (2015)]

FDA cannot file criminal charges, but rather recommends to the Justice Department that court action be taken that might result in fines or imprisonment of offenders. Details of these enforcement activities of the FDA are given in Refs. [1–4].

2.2.2 US Department of Agriculture (USDA)

The USDA administers several federal statutes relevant to food standards, composition, and analysis. These include standards of identity for meat products, grade standards, and inspection programs. Some programs for fresh and processed food commodities are mandatory, and others are voluntary.

2.2.2.1 *Standards of Identity for Meat Products*

Standards of identity have been established by the **Food Safety Inspection Service** (FSIS) of the USDA for many meat products (9 CFR 319). These commonly specify percentages of meat, fat, and water. Analyses are to be conducted using an AOAC method, if available.

2.2.2.2 *Grade Standards*

Grade standards developed for many foods by the USDA classify products in a range from substandard to excellent in quality. While most grade standards are not mandatory requirements (but they are mandatory for certain grains), they are widely used, voluntarily, by food processors and distributors as an aid in whole-

Selected chemical composition requirements of some foods with standards of identity

Section in 21 CFR[a]	Food product	Requirement	Number in 13th Edn.	AOAC method[b] Number in 18th Edn.	Name/description
131.110	Milk	Milk solids nonfat ≥8 1/4%	16.032	990.19	Total solids, by forced air oven after steam table
		Milk fat ≥3 1/4% Vitamin A (if added) ≥2,000 IU[c]/qt[d]	16.059	905.02	Roese-Gottlieb
		Vitamin D (if added) – 400 IU[c]/qt[d]	43.195–43.208	936.14	Bioassay line test with rats
133.113	Cheddar cheese	Milk fat ≥50% by wt. of solids	16.255 and calculation	933.05	Digest with HCl, Roese-Gottlieb
		Moisture ≤39% by wt.	16.233	926.08	Vacuum oven
		Phosphatase level ≤3 μg phenol equivalent/0.25 g[e]	16.275–16.277	946.03–946.03C	Residual phosphatase
137.165	Enriched flour	Moisture ≤15% Ascorbic acid ≤200 ppm (if added as dough conditioner)	14.002, 14.003	925.09, 925.09B	Vacuum oven
		Ash[f] ≤0.35 + (1/20 of the percent of protein, calculated on dwb[g])	14.006	923.03	Dry ashing
		(Protein) Thiamine, 2.9 mg/lb Riboflavin, 1.8 mg/lb Niacin, 24 mg/lb Iron, 20 mg/lb Calcium (if added), 960 mg/lb Folic acid 0.7 mg/lb	2.057	955.04C	Kjeldahl, for nitrate-free samples
146.185	Pineapple juice	Soluble solids ≥10.5°Brix[h] Total acidity ≤1.35 g/100 mL (as anhydrous citric acid)	31.009	932.14A	Hydrometer Titration with NaOH[i]
		Brix/acid ratio ≥12 Insoluble solids ≥5 and ≤30%			Calculated[j] Calculated from volume of sediment[k]
163.113	Cocoa	Cocoa fat ≤22% and ≥10%		963.15	Petroleum ether extraction with Soxhlet unit
164.150	Peanut butter	Fat ≤55%	27.006(a)	948.22	Ether extraction with Soxhlet unit

[a]CFR Code of Federal Regulations (2015)
[b]Official Methods of Analysis of AOAC International
[c]IU international units
[d]Within limits of good manufacturing practice
[e]If pasteurized dairy ingredients are used
[f]Excluding ash resulting from any added iron or salts of iron or calcium or wheat germ
[g]dwb, moisture-free or dry weight basis
[h]Exclusive of added sugars, without added water. As determined by refractometer at 20 °C uncorrected for acidity and read as °Brix on International Sucrose Scales. Exception stated for juice from concentrate
[i]Detailed titration method given in 21 CFR, 145.180 (b)(2)(ix)
[j]Calculated from °Brix and total acidity values, as described in 21 CFR 146.185 (b)(2)(ii)
[k]Detailed method given in 21 CFR 146.185 (b)(2)(iv)

2.3 table Food analysis-related scope of responsibility for federal agencies

US Food and Drug Administration (FDA)	US Department of Agriculture (USDA)	US Dept. Commerce, National Oceanic and Atmospheric Administration – National Marine Fisheries Service (NOAA-NMFS)	US Dept. Justice, Alcohol and Tobacco Tax and Trade Bureau (TTB)	US Environmental Protection Agency (EPA)	US Customs and Border Protection (CBP)	US Federal Trade Commission (FTC)
Standards of identity (most foods, except meats)	Standards of identity for meat products	Grading, standardization and inspection program for seafood (works with FDA and EPA)	Standards and labeling of most alcoholic beverages	Sets tolerances for pesticide residues (note: tolerance enforced by FDA)	Ensure imported products are taxed properly, are safe for human consumption, and are not economically deceptive (works with FDA)	Authority over advertising and sales promotion practices for foods
Standards of quality (canned fruits and vegetables)	Grade standards (voluntary; for many foods)			Sets standards for drinking water and allowed levels of contaminants		
Standards of fill (select foods)	Inspection programs (meats, poultry, eggs, grains)			Sets standards and guidelines for effluent from food processing plants		
Nutrition labeling (all foods but meats) (see Chap. 3 for details)	For select imported products, works with CBP					
Standards for some alcoholic beverages and cooking wines						
Enforces tolerances set on pesticides (note: tolerances set by EPA)						
For seafood, works with NMFS to ensure safety; Sets and enforces allowable levels of contaminants and pathogenic microorganisms in seafood						
For shellfish, through National Shellfish Sanitation Program (NSSP) (part of FDA): sanitation for products shipped interstate; Ensures shellfish-growing areas are free of sewage pollution and toxic industrial waste						
For imported foods, works with US Customs Service						
For dairy products, works with USDA and States for safety and wholesomeness						
Food additives (all foods)						
Color additives (all foods)						
Inspects food processing facilities for compliance with cGMP regulations and mandatory HACCP inspection programs						

| 2.4 table | Federal agencies with quality and safety responsibility for specific foods |

Water	Dairy products	Meats and meat products	Seafood	Shellfish	Alcoholic beverages	Fruits and vegetables	Harvested food or feed crops: pesticide residues	Infant formula
EPA (drinking; effluent) FDA (bottled)	FDA USDA States	USDA	Dept. Commerce: NOAA-NMFS FDA EPA	FDA: NSSP	Dept. Treasury: TTB (most products) FDA (select products)	FDA	EPA (sets tolerances) FDA (enforces tolerances)	FDA

sale trading, because the quality of a product affects its price. Such grade standards often are used as quality control tools. Consumers are familiar generally with grade standards for beef, butter, and eggs, but buyers for the retail market utilize grade standards for a wide variety of foods. Major users of standards include institutions such as schools, hospitals, restaurants, prisons, and the Department of Defense (see also Sect. 2.5).

The USDA has issued grade standards for more than 300 food products under authority of the Agricultural Marketing Act of 1946 and related statutes. Standards for grades are not required to be stated on the label, but if they are stated, the product must comply with the specifications of the declared grade. Official USDA grading services are provided, for a fee, to pickers, processors, distributors, and others who seek official certification of the grades of their products.

While complete information regarding the standards was published previously in the CFR, currently only some standards are published in the CFR because they are USDA Agricultural Marketing Service (AMS) Administrative Orders. All grade standards are available as pamphlets from USDA and also are accessible on the Internet.

Grade standards, issued by the AMS of the USDA for agricultural products and by the Department of Commerce for fishery products, must not be confused with standards of quality set by the FDA or standards of identity set by the FDA or FSIS of the USDA, as discussed previously. Grade standards exist for many types of meats, poultry, dairy products, fruits, vegetables, and grains, along with eggs, domestic rabbits, certain preserves, dry beans, rice, and peas. Additional information about grade standards for dairy products is given in Sect. 2.3, but examples of grade standards for several other types of foods follow here.

Standards for grades of processed fruits and vegetables often include factors such as color, texture or consistency, defects, size and shape, tenderness, matu-

| 2.5 table | USDA standards for grade determinants of frozen concentrated orange juice |

Quality	Analytical
Appearance	Concentrate:
Reconstitution	Brix
Color	Brix/acid ratio
Defects	Reconstituted juice:
Flavor	Brix
	Soluble orange solids
	Recoverable oil

rity, flavor, and a variety of chemical characteristics. Sampling procedures and methods of analysis are commonly given. As an example, the quality and analytical factors that determine the grade standards of frozen concentrated orange juice [10] are given in Table 2.5.

Grades for various grains (e.g., wheat, corn, soybeans, oats) are determined by factors such as test weight per bushel and percentages of heat-damaged kernels, broken kernels, and foreign material. Also, a grade limit is set commonly for moisture content. Grade standards for rice, beans, peas, and lentils are determined commonly by factors such as defects, the presence of foreign material, and insect infestation, and sometimes moisture content is specified.

2.2.2.3 Inspection Programs
The USDA administers some programs on inspection and certification that are mandatory, and some inspection programs are voluntary. Comprehensive inspection manuals specific to various types of foods have been developed to assist inspectors and industry personnel in interpreting and utilizing the regulations. Under the Federal Meat Inspection Act, the Poultry Products Inspection Act, and the Egg Products Inspection Act, the FSIS of the USDA inspects all meat, poultry, and egg products in interstate commerce (9 CFR 200–End). This

includes a review of foreign inspection systems and packing plants that export meat and poultry to the USA. Imported products are reinspected at ports of entry. Hazard Analysis Critical Control Point (HACCP) is a major component of FSIS rules for all slaughter and processing plants to improve safety of meat and poultry. A program within the **Grain Inspection, Packers and Stockyard Administration** (GIPSA) of the USDA administers the mandatory requirements of the US Grain Standards Act (7 CFR 800). Regulations to enforce this act provide for a national inspection system for grain and mandatory official grade standards of numerous types of grain. Another program of the USDA standardizes, grades, and inspects fruits and vegetables under various voluntary programs. The inspection programs rely heavily on the HACCP concept.

2.2.3 US Department of Commerce

2.2.3.1 Seafood Inspection Service

The **National Oceanic and Atmospheric Administration's** (NOAA) **National Marine Fisheries Service** (NMFS), a division of the **US Department of Commerce** (USDC), provides a seafood inspection service. The USDC Seafood Inspection Program ensures the safety and quality of seafoods consumed in the USA and certified for export through **voluntary grading, standardization**, and **inspection programs**, as described in 50 CFR 260. The inspection programs rely heavily on the HACCP concept. The US Standards for Grades of Fishery Products (50 CFR 261) are intended to help the fishing industry maintain and improve quality and to thereby increase consumer confidence in seafoods. Standards are based on attributes such as color, size, texture, flavor, odor, workmanship defects, and consistency.

2.2.3.2 Interaction Between FDA and Environmental Protection Agency

The FDA and the Environmental Protection Agency (EPA) work with the NMFS for the assurance of seafood safety. The FDA, under the FD&C Act, is responsible for ensuring that seafood shipped or received in interstate commerce is safe, wholesome, and not misbranded or deceptively packaged. The FDA has primary authority in setting and enforcing allowable levels of contaminants and pathogenic microorganisms in seafood. The EPA assists the FDA in identifying the range of chemical contaminants that pose a human health risk and are most likely to accumulate in seafood. A tolerance of 2.0 parts per million (ppm) for total polychlorinated biphenyls (PCBs) (21 CFR 109.30) is the only formal tolerance specified by the FDA to mitigate human health impacts in seafood. However, the EPA has established tolerances for certain pesticide residues, and the FDA has established guidance level for methyl mercury [11].

2.2.4 US Alcohol and Tobacco Tax and Trade Bureau

2.2.4.1 Regulatory Responsibility for Alcoholic Beverages

Beer, wines, liquors, and other alcoholic beverages are termed "food" according to the FD&C Act of 1938. However, regulatory control over their quality, standards, manufacture, and other related aspects is specified by the **Federal Alcohol Administration Act**, which is enforced by the **Alcohol and Tobacco Tax and Trade Bureau** (TTB) of the **US Department of Treasury**. Issues regarding the composition and labeling of most alcoholic beverages are handled by the Bureau. However, the FDA has jurisdiction over certain other alcoholic beverages and cooking wines. The FDA also deals with questions of sanitation, filth, and the presence of deleterious substances in alcoholic beverages.

2.2.4.2 Standards and Composition of Beer, Wine, and Distilled Beverage Spirits

Information related to **definitions, standards of identity**, and certain **labeling requirements** for beer, wine, and distilled beverage spirits is given in 27 CFR 1–30. Standards of identity for these types of beverages stipulate the need for analyses such as percent alcohol by volume, total solids content, volatile acidity, and calculated acidity. For example, the fruit juice used for the production of wine is often specified by its °Brix and total solids content. The maximum volatile acidity (calculated as acetic acid and exclusive of sulfur dioxide) for grape wine must not be more than 0.14 g/100 mL (20 °C) for natural red wine and 0.12 g/100 mL for other grape wine (27 CFR 4.21). The percent alcohol by volume is often used as a criterion for class or type designation of alcoholic beverages. For example, dessert wine is grape wine with an alcoholic content in excess of 14 % but not in excess of 24 % by volume, while table wines have an alcoholic content not in excess of 14 % alcohol by volume (27 CFR 4.21). No product with less than 0.5 % alcohol by volume is permitted to be labeled "beer," "lager beer," "lager," "ale," "porter," "stout," or any other class or type designation normally used for malt beverages with higher alcoholic content (27 CFR 7.24).

2.2.5 US Environmental Protection Agency (EPA)

The EPA regulatory activities most relevant to this book are control of pesticide residues in foods, drinking water safety, and the composition of effluent from food processing plants.

| 2.6 table | Tolerance for selected insecticides (I), fungicides (F), and herbicides (H) classified as food additives permitted in foods for human consumption |

Section	Food additive	Chemical classification	Food	Tolerance[a]
180.103	Captan (F)	Phthalimide	Apples	25
			Cattle, meat	0.2
			Milk	0.1
			Grapes	25
			Peach	15
			Strawberry	20
			Sunflower, seed	0.05
180.342	Chlorpyrifos[b](I)	Organophosphate	Apples	0.01
			Cattle, meat	0.05
			Corn oil	0.25
			Strawberry	0.2
180.435	Deltamethrin (I)	Pyrethroid	Cattle, meat	0.02
			Tomatoes	0.02
			Tomato products, conc.	1.0
180.292	Picloram (H)	Chloropyridine-carboxylic acid	Cattle, meat	0.02
			Milk	0.05
			Corn oil	2.5
			Wheat, grain	0.5

Adapted from 40 CFR 180 (2016)
[a]Parts per million
[b]Also known as Dursban™ and Lorsban™

2.2.5.1 Pesticide Registration and Tolerance Levels

Pesticides are chemicals intended to protect our food supply by controlling harmful insects, diseases, rodents, weeds, bacteria, and other pests. However, most pesticide chemicals can have harmful effects on people, animals, and the environment if they are improperly used. The three federal laws relevant to protection of food from pesticide residues are (1) certain provisions of the FD&C Act, (2) the Federal Insecticide, Fungicide, and Rodenticide Act (FIFRA), as amended, and (3) the **Food Quality Protection Act** of 1996. FIFRA, supplemented by the FD&C Act, authorizes a comprehensive program to regulate the manufacturing, distribution, and use of pesticides, along with a research effort to determine the effects of pesticides.

The Food Quality Protection Act amends both the FD&C Act and FIFRA to take pesticides out of the section of the FD&C Act that includes the Delaney Clause. This was done by changing the definition of a "food additive" to exclude pesticides. This redefinition leaves the Delaney Clause greatly reduced in scope and less relevant.

The EPA registers approved pesticides and sets tolerances for pesticide residues (see also Chap. 32, Sect. 32.3). The EPA is authorized to register approved pesticides and to establish an **allowable limit** or **tolerance level** for any detectable pesticide residues that

might remain in or on a harvested food or feed crop (see Chap. 31, Sect. 31.3). While the EPA establishes the tolerance levels, the FDA enforces the regulations by collecting and analyzing food samples, mostly agricultural commodities. Livestock and poultry samples are collected and analyzed by the USDA. Pesticide residue levels that exceed the established tolerances are considered in violation of the FD&C Act.

Regulations regarding pesticide tolerances in foods are given in 40 CFR 180, which specifies general categories of products and specific commodities with tolerances or exemptions, and in some cases which part of the agricultural product is to be examined. Unless otherwise noted, the specific tolerances established for the pesticide chemical apply to residues resulting from their application prior to harvest or slaughter. Tolerance levels for selected pesticides and insecticides permitted in foods as food additives are given in Table 2.6.

The analytical methods to be used for determining whether pesticide residues are in compliance with the tolerance established are identified among the methods contained or referenced in the *Pesticide Analytical Manual* [12] maintained by and available from the FDA. The methods must be sensitive and reliable at and above the tolerance level. Pesticides are generally detected and quantitated by gas chromatographic or high-performance liquid chromatographic methods (see Chaps. 13, 14, and 30).

| table | Effluent limitations for plants processing natural and processed cheese |

	Effluent characteristics					
	Metric units[a]			English units[b]		
Effluent limitations	BOD 5[c]	TSS[d]	pH	BOD 5	TSS	pH
Processing more than 100,000 lb/day of milk equivalent						
Maximum for any 1 day	0.716	1.088	(e)	0.073	0.109	(e)
Avg of daily values for 30 consecutive days shall not exceed	0.290	0.435	(e)	0.029	0.044	(e)
Processing less than 100,000 lb/day of milk equivalent						
Maximum for any 1 day	0.976	1.462	(e)	0.098	0.146	(e)
Avg of daily values for 30 consecutive days shall not exceed	0.488	0.731	(e)	0.049	0.073	(e)

Adapted from 40 CFR 405.62 (2016)
[a]Kilograms per 1,000 kg of BOD 5 input
[b]Pounds per 100 lbs of BOD 5 input
[c]BOD 5 refers to biochemical oxygen demand measurement after 5 days of incubation
[d]TSS refers to total soluble solids
[e]Within the range 6.0–9.0

2.2.5.2 Drinking Water Standards and Contaminants

While the FDA regulates bottled water with a standard of identify, a detailed standard of quality, and specific current Good Manufacturing Practices (cGMP) regulations, the EPA regulates drinking (i.e., tap) water. The EPA administers the **Safe Drinking Water Act** of 1974, which is to provide for the safety of drinking water supplies in the USA and to enforce national drinking water standards. The EPA has identified potential contaminants of concern and established their maximum acceptable levels in drinking water. The EPA has primary responsibility to establish the standards, while the states enforce them and otherwise supervise public water supply systems and sources of drinking water. **Primary and secondary drinking water regulations** (40 CFR 141 and 143, respectively) have been established. Concerns have been expressed regarding the special standardization of water used in the manufacturing of foods and beverages.

Maximum contaminant levels (MCL) for primary drinking water are set for certain **inorganic** and **organic chemicals, turbidity, certain types of radioactivity**, and **microorganisms**. Sampling procedures and analytical methods for the analysis of chemical contaminants are specified, with common reference to *Standard Methods for the Examination of Water and Wastewater* [13] published by the American Public Health Association; *Methods of Chemical Analysis of Water and Wastes* [14], published by the EPA; and *Annual Book of ASTM Standards* [15], published by the American Society for Testing Materials (ASTM). Methods commonly specified for the analysis of inorganic contaminants in water include atomic absorption (direct aspiration or furnace technique),

inductively coupled plasma (see Chap. 9), ion chromatography (see Chap. 13), and ion-selective electrode (see Chap. 21).

2.2.5.3 Effluent Composition from Food Processing Plants

In administering the **Federal Water Pollution and Control Act**, the EPA has developed effluent guidelines and standards that cover various types of food processing plants. Regulations prescribe effluent limitation guidelines for existing sources, standards of performance for new sources, and pretreatment standards for new and existing sources. Point sources of discharge of pollution are required to comply with these regulations, where applicable. Regulations are prescribed for specific foods under the appropriate point source category: dairy products processing (40 CFR 405), grain mills (40 CFR 406), canned and preserved fruits and vegetables processing (40 CFR 407), canned and preserved seafood processing (40 CFR 408), sugar processing (40 CFR 409), and meat and poultry products (40 CFR 432). **Effluent characteristics** commonly prescribed for food processing plants are **biochemical oxygen demand** (BOD) (see Chap. 28), **total soluble solids** (TSS) (see Chap. 15), and **pH** (see Chap. 22), as shown in Table 2.7 for effluent from a plant that makes natural and processed cheese. The test procedures for measurement of effluent characteristics are prescribed in 40 CFR 136.

2.2.6 US Customs and Border Protection (CBP)

Over 100 countries export food, beverages, and related edible products to the USA. The CBP assumes the central role in ensuring that imported products are taxed properly, safe for human consumption, and not economically deceptive. The CBP receives assistance from the FDA and USDA as it assumes these responsibilities. The major regulations promulgated by the CBP are given in Title 19 of the CFR.

All goods imported into the USA are subject to duty or duty-free entry according to their classification under applicable items in the **Harmonized Tariff Schedule of the United States** (TSUSA). The US tariff system has official tariff schedules for over 400 edible items exported into the USA [16]. The TSUSA specifies the food product in detail and gives the general rate of duty applicable to that product coming from most countries and any special higher or lower rates of duty for certain other countries.

The **rate of duty** for certain food products is determined by their chemical composition. For example, the rate of duty on some dairy products is determined in part by the fat content. The tariff for some syrups is determined by the fructose content, for some chocolate products by the sugar or butterfat content, for butter substitutes by the butterfat content, and for some wines by their alcohol content (percent by volume).

2.2.7   US Federal Trade Commission (FTC)

The FTC is the most influential of the federal agencies that have authority over various aspects of advertising and sales promotion practices for foods in the USA. The major role of the FTC is to keep business and trade competition free and fair.

2.2.7.1 *Enforcement Authority*

The **Federal Trade Commission Act** of 1914 authorizes the FTC to protect both the consumer and the business person from anticompetitive behavior and unfair or deceptive business and trade practices. The FTC periodically issues industry guides and trade regulations and rules that tell businesses what they can and cannot do. These issuances are supplemented with advisory opinions given to corporations and individuals upon request. The FTC not only has guidance and preventive functions but is also authorized to issue complaints or shut down orders and sue for civil penalties for violation of trade regulation rules. The **Bureau of Consumer Protection** is one of the FTC bureaus that enforce and develop trade regulation rules.

2.2.7.2 *Food Labels, Food Composition, and Deceptive Advertising*

While the **Fair Packaging and Labeling Act** of 1966 is administered by the FTC, that agency does not have specific authority over the packaging and labeling of foods. The FTC and FDA have agreed upon responsibilities: FTC has primary authority over advertising of foods, and FDA has primary authority over labeling of foods.

Grading, standards of identity, and labeling of foods regulated by several federal agencies as described previously have eliminated many potential problems in the advertising of foods. Such federal regulations and voluntary programs have reduced the scope of advertising and other forms of product differentiation. Misleading, deceptive advertising is less likely to be an issue and is more easily controlled. For example, foods such as ice cream, mayonnaise, and peanut butter have standards of identity that set minimum ingredient standards. If these standards are not met, the food must be given a different generic designation (e.g., salad dressing instead of mayonnaise) or be labeled "imitation." Grading, standards, and labeling of food aid consumers in making price-quality comparisons. Once again, analyses of chemical composition play an important role in developing and setting these grades, standards, and labels. In many cases in which the FTC intervenes, data from a chemical analysis become central evidence for all parties involved.

2.3   REGULATIONS AND RECOMMENDATIONS FOR MILK

The safety and quality of milk and dairy products in the USA are the responsibility of both federal (FDA and USDA) and state agencies. The FDA has regulatory authority over the dairy industry interstate commerce, while the USDA involvement with the dairy industry is voluntary and service oriented. Each state has its own regulatory office for the dairy industry within that state. The various regulations for milk involve several types of chemical analyses.

2.3.1   FDA Responsibilities

The FDA has responsibility under the FD&C Act, the Public Health Service Act, and the Import Milk Act to assure consumers that the US milk supply and imported dairy products are safe, wholesome, and not economically deceptive. As described in Sect. 2.2.1.2, the FDA promulgates standards of identity and labeling, quality, and fill-of-container requirements for milk and dairy products moving in interstate commerce.

For **Grade A milk and dairy products**, each state shares with the FDA the responsibility of ensuring safety, wholesomeness, and economic integrity. This is done through a **Memorandum of Understanding with the National Conference on Interstate Milk Shipments**, which is comprised of all 50 states. In cooperation with the states and the dairy industry, the FDA has also developed for state adoption model

2.8 table	Pasteurized milk ordinance standards for Grade A pasteurized milk and milk products and bulk-shipped heat-treated milk products

Criteria	Requirement
Temperature	Cooled to 7 °C (45 °F) or less and maintained thereat
Bacterial limits[a]	20,000 per mL
Coliform[b]	Not to exceed 10 per mL. Provided that in the case of bulk milk transport tank shipments shall not exceed 100 per mL
Phosphatase[b]	Less than 350 milliunits/L for fluid products and other milk products by the fluorometer or Charm ALP or equivalent
Drugs[c]	No positive results on drug residue detection methods

Adapted from US Department of Health and Human Services, Public Health Service, Food and Drug Administration [17]
[a]Not applicable to acidified or cultured products
[b]Not applicable to bulk-shipped heat-treated milk products
[c]Reference to specific laboratory techniques

2.9 table	US standards for grades of nonfat dry milk (spray process)

Laboratory tests[a]	US extra grade	US standard grade
Bacterial estimate, standard plate count per gram	10,000	75,000
Milk fat content, percent	1.25	1.50
Moisture content, percent	4.0	5.0
Scorched particle content, mg	15.0	22.5
Solubility index, mL	1.2	2.0
US high heat	2.0	2.5
Titratable acidity (lactic acid), percent	0.15	0.17

http://www.ams.usda.gov/sites/default/files/media/Nonfat_Dry_Milk_%28Spray_Process%29_Standard%5B1%5D.pdf
[a]All values are maximum allowed

regulations regarding sanitation and quality aspects of producing and handling Grade A milk. These regulations are contained in the **Grade A Pasteurized Milk Ordinance** (PMO) [17], which all states have adopted as minimum requirements.

The standards for Grade A pasteurized milk and milk products and bulk-shipped heat-treated milk products under the PMO are given in Tables 2.2, 2.3, 2.4, 2.5, 2.6, 2.7, and 2.8. The PMO specifies that "all sampling procedures, including the use of approved in-line samples, and required laboratory examinations shall be in substantial compliance with the most current edition of *Standard Methods for the Examination of Dairy Products* (SMEDP) of the American Public Health Association, and the most current edition of *Official Methods of Analysis* of the *AOAC INTERNATIONAL* (*OMA*)" [18–20]. The FDA monitors state programs for compliance with the PMO and trains state inspectors.

2.3.2 USDA Responsibilities

Under authority of the Agricultural Marketing Act of 1946, the **Dairy Quality Program** of the USDA offers **voluntary grading services** for manufactured or processed dairy products (7 CFR 58). If USDA inspection of a dairy manufacturing plant shows that good sanitation practices are being followed to meet the requirements in the *General Specifications for Dairy Plants Approved for USDA Inspection and Grading Service* [20], the plant qualifies for the USDA services of grading, sampling, testing, and certification of its products. A product such as nonfat dry milk is graded based on fla-

vor, physical appearance, and various laboratory analyses (Table 2.9).

The USDA, under an arrangement with the FDA, assists states in establishing safety and quality regulations for manufacturing grade milk. Much as described previously for the FDA with Grade A milk, the USDA has developed model regulations [21] for state adoption regarding the quality and sanitation aspects of producing and handling manufacturing grade milk.

2.3.3 State Responsibilities

As described previously, individual states have enacted safety and quality regulations for Grade A and manufacturing grade milk that are essentially identical to those in the PMO and the USDA Recommended Requirements, respectively. The department of health or agriculture in each state normally is responsible for enforcing these regulations. The states also establish their own standards of identity and labeling requirements for milk and dairy products, which are generally similar to the federal requirements.

2.4 REGULATIONS AND RECOMMENDATIONS FOR SHELLFISH

Shellfish include fresh or frozen oysters, clams, and mussels. They may transmit intestinal diseases such as typhoid fever or act as carriers of natural or chemical toxins. This makes it very important that they be obtained from unpolluted waters and handled and processed in a sanitary manner.

The growing, handling, and processing of shellfish must comply not only with the general requirements of the FD&C Act but also with the requirements of state health agencies cooperating in the **National Shellfish Sanitation Program** (NSSP), a federal, state, industry voluntary cooperative program, administered by the FDA [22]. The FDA has no regulatory power over shellfish sanitation unless the product is shipped interstate. However, the Public Health Service Act authorizes the FDA to make recommendations and to cooperate with state and local authorities to ensure the safety and wholesomeness of shellfish. Through the NSSP, state health personnel continually inspect and survey bacteriological conditions in shellfish-growing areas. Any contaminated location is supervised or patrolled so that shellfish cannot be harvested from the area.

A major concern is the ability of shellfish to concentrate radioactive material, insecticides, and other chemicals from their environment. Thus, one aspect of the NSSP is to ensure that shellfish-growing areas are free from sewage pollution and toxic industrial waste. **Pesticide residues** in shellfish are usually quantitated by gas chromatographic techniques, and **heavy metals** such as mercury are commonly quantitated by inductively coupled plasma-mass spectrometry (ICP-MS)

(Chap. 9, Sect. 9.6). Another safety problem with regard to shellfish is the control of **natural toxins**, which is a separate issue from sanitation. The naturally occurring toxins are produced by planktonic organisms, and testing is conducted using a variety of assays. Control of this toxicity is achieved by a careful survey followed by prohibition of harvesting from locations inhabited by toxic shellfish.

2.5 SPECIFICATIONS FOR FOODS PURCHASED BY GOVERNMENT AGENCIES

Large amounts of food products are purchased by federal agencies for use in domestic (e.g., school lunch) and foreign programs, prisons, veterans' hospitals, the armed forces, and other organizations. Specifications or descriptions developed for many food products are used by federal agencies in procurement of foods to ensure the safety and quality of the product specified. Such specifications or descriptions often include information that requires assurance of chemical composition, in addition to specified microbial quality. Many such documents are referred to as a **commercial item description** (CID). These specifications, with specific examples for foods and their content, are given in Table 2.10.

2.10 table

Specifications for foods purchased by government agencies

Specification type	Example product	Example content of specification
Commercial item description (CID)	Canned tuna [23]	Salt/sodium content, methylmercury, and histamine, with specified methods of analysis
Federal specification	Macaroni and cheese mix CID [24]	Fat and sodium contents and viscosity, with specified methods of analysis (AOAC International)
	Beans, precooked, dehydrated CID [25]	Moisture, fat, cholesterol, and sodium contents, with specified methods of analysis (AOAC International)
Department of Defense specification	Syrup CID [26] Instant tea mix CID [27] Nut butters CID [28]	Brix, ash content, color Moisture and sugar contents, titratable acidity Salt content, aflatoxin content
Commodity specification	Dried egg mix [29]	Vegetable oil composition/characteristics: free fatty acid value, peroxide value, linolenic acid, moisture, volatile matter, iodine value, Lovibond color, by specified methods of analysis (American Oil Chemists' Society)
	American cheese [30] and mozzarella cheese [31]	pH, milk fat, and moisture contents
USDA specification (e.g., Institutional Meat Purchase Specification)	Sausage products [32]	Fat content

2.6 INTERNATIONAL STANDARDS AND POLICIES

With the need to compete in the worldwide market, employees of food companies must be aware that allowed food ingredients, names of food ingredients, required and allowed label information, and standards for foods and food ingredients differ between countries. For example, colorings and preservatives allowed in foods differ widely between countries, and nutritional labeling is not universally required. To develop foods for, and market foods in, a global economy, one must seek such information from international organizations and from organizations in specific regions and countries.

2.6.1 Codex Alimentarius

The **Codex Alimentarius Commission** (Codex Alimentarius is Latin for "code concerned with nourishment") was established in 1962 by two United Nations organizations, the Food and Agriculture Organization (FAO) and the World Health Organization (WHO), to develop international standards and safety practices for foods and agricultural products [33]. The standards, published in the *Codex Alimentarius*, are intended to protect consumers' health, ensure fair business practices in food trade, and facilitate international trade of foods.

The *Codex Alimentarius* is published in 13 volumes: one on general requirements (includes labeling, food additives, contaminants, irradiated foods, import/export inspection, and food hygiene), nine on standards and codes of practice compiled on a commodity basis, two on residues of pesticides and veterinary drugs in foods, and one on methods of analysis and sampling (Table 2.11). Codex has efforts to validate and harmonize methods of food safety analysis among countries and regions, help maintain the smooth flow of international commerce, and ensure appropriate decisions on food exports and imports. The setting of international standards on food quality by Codex has been a high priority in world trade to minimize "nontariff" trade barriers. International trade of food and raw agricultural products has increased due to reduced economic trade restrictions and tariffs imposed.

2.6.2 Other Standards

Other international, regional, and country-specific organizations publish standards relevant to food composition and analysis. For example, the **Saudi Arabian Standards Organization** (SASO) publishes standards documents (e.g., labeling, testing methods) important in the Middle East (except Israel), and the **European Commission** sets standards for foods and

table | Content of the Codex Alimentarius [33]

Volume	Subject
1A	General requirements
1B	General requirements (food hygiene)
2A	Pesticide residues in foods (general text)
2B	Pesticide residues in foods (maximum residue limits)
3	Residues in veterinary drugs in foods
4	Foods for special dietary uses
5A	Processed and quick-frozen fruits and vegetables
5B	Fresh fruits and vegetables
6	Fruit juices
7	Cereals, pulses (legumes) and derived products, and vegetable proteins
8	Fats and oils and related products
9	Fish and fishery products
10	Meat and meat products, soups and broths
11	Sugars, cocoa products and chocolate, and misc. products
12	Milk and milk products
13	Methods of analysis and sampling

food additives for countries in the European Economic Community (EEC). In the USA, the Food Ingredients Expert Committee, which operates as part of the US Pharmacopeia, sets standards for the identification and purity of food additives and chemicals, published as the **Food Chemicals Codex** (FCC) [34]. For example, a company may specify in the purchase of a specific food ingredient that it be "FCC grade." Countries other than the USA adopt FCC standards (e.g., Australia, Canada). At an international level, the **Joint FAO/WHO Expert Committee on Food Additives** (JECFA) sets standards for purity of food additives [35]. The Codex Alimentarius Commission is encouraged to utilize the standards established by JECFA. Standards established by FCC and JECFA are used by many countries as they develop their own standards.

2.7 SUMMARY

Various kinds of standards set for certain food products by federal agencies make it possible to get essentially the same food product whenever and wherever purchased in the USA. The standards of identity set by the FDA and USDA define what certain food products must consist of. The USDA and NMFS of the Department of Commerce have specified grade standards to define attributes for certain foods. Grading programs are voluntary, while inspection programs may be either voluntary or mandatory, depending on the specific food product.

While the FDA has the broadest regulatory authority over most foods, responsibility is shared with other regulatory agencies for certain foods. The USDA has significant responsibilities for meat and poultry, the NOAA and the NMFS for seafood, and the TTB for alcoholic beverages. The FDA, the USDA, state agencies, and the dairy industry work together to ensure the safety, quality, and economic integrity of milk and milk products. The FDA, the EPA, and state agencies work together in the NSSP to ensure the safety and wholesomeness of shellfish. The EPA shares responsibility with the FDA for control of pesticide residues in foods and has responsibility for drinking water safety and the composition of effluent from food processing plants. The CBP receives assistance from the FDA and USDA in its role to ensure the safety and economic integrity of imported foods. The FTC works with the FDA to prevent deceptive advertising of food products, as affected by food composition and labels. The chemical composition of foods is often an important factor in determining the quality, grade, and price of a food. Government agencies that purchase foods for special programs often rely on detailed specifications that include information on food composition.

International organizations have developed food standards and safety practices to protect consumers, ensure fair business practices, and facilitate international trade. The Codex Alimentarius Commission is the major international standard-setting group for food safety and quality. Certain regional and country-specific organizations also publish standards related to food composition and analysis.

2.8   STUDY QUESTIONS

1. Define the abbreviations FDA, USDA, and EPA, and give two examples for each of what they do or regulate relevant to food analysis.
2. Differentiate "standards of identity," "standards of quality," and "grade standards" with regard to what they are and which federal agency establishes and regulates them.
3. Government regulations regarding the composition of foods often state the official or standard method by which the food is to be analyzed. Give the full name of three organizations that publish commonly referenced sources of such methods.
4. For each type of product listed below, identify the governmental agency (or agencies) that has regulatory or other responsibility for quality assurance. Specify the general nature of that responsibility and, if given, the specific types of analyses that would be associated with that responsibility.

   (a) Frozen fish sticks
   (b) Contaminants in drinking water
   (c) Dessert wine
   (d) Grade A milk
   (e) Frozen oysters
   (f) Imported chocolate products
   (g) Residual pesticide on wheat grain
   (h) Corned beef

5. Upon completing your college degree, you are employed by a major US food company that processes fruits and vegetables.

   (a) Where, specifically, would you look to find if a standard of identity exists for each of your processed products? What kind of information does such a standard include?
   (b) What US governmental agency sets the standards of identity for such products?
   (c) What are the minimum standards called that are set for some fruit and vegetable products?
   (d) What governmental agency sets the grade standards that you may want to use as a quality control tool and in marketing your products?
   (e) You are concerned about pesticide tolerances for the fruits and vegetables you process. What governmental agency sets those tolerances?
   (f) What governmental agency enforces the pesticide tolerances?
   (g) For nutrition labeling purposes for your products, you want to check on official methods of analysis. Where, specifically, should you look? (See Chap. 3)
   (h) You want to check the detailed rules on nutrition labeling that would apply to your products. Where, specifically, would you look to find those rules?
   (i) You are considering marketing some of your products internationally. What resource could you check to determine if there are international standards and safety practices specified for those products?

REFERENCES

1. Adams DG, Cooper RM, Hahn MJ, Kahan JS (2014) Food and drug law and regulation, 3rd edn. Food and Drug Law Institute, Washington, DC
2. Sanchez MC (2015) Food law and regulation for non-lawyers. A US perspective. Springer, New York
3. Piña W, Pines K (2014) A practical guide to FDA's food and drug law and regulation, 5th edn. Food and Drug Law Institute, Washington, DC
4. Curtis PA (2005) Guide to food laws and regulations. Wiley-Blackwell, San Francisco, CA

5. Anonymous (2016) Code of federal regulations. Titles 7, 9, 21, 27, 40, 50. US Government Printing Office, Washington, DC. Available on Internet.

6. Mortimore S, Wallace C (2013) HACCP A practical approach, 3rd edn. Spring, New York

7. Cramer MM (2006) Food plant sanitation: design, maintenance, and good manufacturing practices. CRC, Boca Raton, FL

8. Marriott NG, Gravani RB (2006) Principles of food sanitation. Springer, Berlin

9. Pierson MD, Corlett DA Jr (1992) HACCP principles and applications. Van Nostrand Reinhold, New York

10. USDA (1983) U.S. standards for grades of orange juice. 10 Jan 1983. Processed Products Branch, Fruit and Vegetable Division, Agricultural Marketing Service, US Department. of Agriculture. https://www.ams.usda.gov/sites/default/files/media/Canned_Orange_Juice_Standard%5B1%5D.pdf

11. FDA (2011) Fish and fisheries products hazards and controls guidance, 4th edn. Center for Food Safety and Applied Nutrition, Office of Food Safety, Food and Drug Administration, Washington, DC. Available on Internet.

12. FDA (1994) Pesticide analytical manual, vol 1 (PAMI) (updated Oct 1999) (Methods which detect multiple residues) and vol 2 (PAMII) (updated Jan 2002) (Methods for individual pesticide residues), 3rd edn. National Technical Information Service, Springfield, VA. Available on Internet.

13. Rice EW, Baird RB, Eaton AD, Clesceri LS (eds) (2012) Standard methods for the examination of water and wastewater, 22nd edn. American Public Health Association, Washington, DC

14. EPA (1983) Methods of chemical analysis of water and wastes. EPA-600/4-79-020, March 1979. Reprinted in 1983. EPA Environmental Monitoring and Support Laboratory, Cincinnati, OH. Available on Internet.

15. American Society for Testing Materials (ASTM) International (2009) Annual book of ASTM standards, section 11, water and environmental technology vol 11.02, water (II). ASTM International, West Conshohocken, PA. Available on Internet.

16. US International Trade Commission (USITC) (2016) Official harmonized tariff schedule of the United States. USITC. Available on Internet.

17. US Department of Health and Human Services, Public Health Service, Food and Drug Administration (2013) Grade A pasteurized milk ordinance. Available on Internet.

18. Wehr HM, Frank JF (eds) (2004) Standard methods for the examination of dairy products, 17th edn. American Public Health Association, Washington, DC. Available on Internet.

19. AOAC International (2016) Official methods of analysis, 20th edn, 2016 (On-line). AOAC International, Rockville, MD

20. USDA (2012) General specifications for dairy plants approved for USDA inspection and grading service. Dairy Program, Agricultural Marketing Service, US Department of Agriculture, Washington, DC. https://www.ams.usda.gov/sites/default/files/media/General%20Specifications%20for%20Dairy%20Plants%20Approved%20for%20USDA%20Inspection%20and%20Grading%20Service.pdf

21. USDA (2011) Milk for manufacturing purposes and its production and processing, recommended requirements. Dairy Program, Agricultural Marketing Service, US Department of Agriculture, Washington, DC. https://www.ams.usda.gov/sites/default/files/media/Milk%20for%20Manufacturing%20Purposes%20and%20its%20Production%20and%20Processing.pdf

22. FDA (1995) National shellfish sanitation program. Food and Drug Administration, Washington, DC. (last updated 06/08/2016) http://www.fda.gov/Food/GuidanceRegulation/FederalStateFoodPrograms/ucm2006754.htm

23. USDA (2009) Tuna, canned or in flexible pouches. A-A-20155D. July 8, 2009. Livestock and Seed Division, Agriculture Marketing Service, US Department of Agriculture, Washington, DC. https://www.ams.usda.gov/sites/default/files/media/CID%20Tuna,%20Canned%20or%20in%20Flexible%20Pouches.pdf

24. USDA (2002) Commercial item description. Macaroni and cheese mix. A-A-20308. 22 August 2002. General Services Administration, Specifications Section, Washington, DC. https://www.ams.usda.gov/sites/default/files/media/CID%20Macaroni%20and%20Cheese%20Mix%2C%20Dry.pdf

25. USDA (2002) Commercial item description. Beans, precooked, dehydrated. A-A-20337. 14 June 2002. General Services Administration, Specifications Section, Washington, DC. https://www.ams.usda.gov/sites/default/files/media/CID%20Beans%2C%20Precooked%2C%20Dehydrated.pdf0.35

26. USDA (2008) Commercial item description. Syrup. A-A-20124D. 17 April 2008. General Services Administration, Specifications Section, Washington, DC. https://www.ams.usda.gov/sites/default/files/media/CID%20Syrup.pdf

27. USDA (2014) Commercial item description. Tea mixes, Instant. A-A-220183D. 13 Aug 2014. General Services Administration, Specifications Unit, Washington, DC. https://www.ams.usda.gov/sites/default/files/media/CID%20Tea%2C%20Instant.pdf

28. USDA (2011) Commercial item description. Nut butters and nut spreads. A-A-20328B. 29 Sep 2011. General Service Administration Specification Unit, Washington, DC. https://www.ams.usda.gov/sites/default/files/media/CID%20Nut%20Butters%20and%20Nut%20Spreads.pdf

29. USDA (2013) Commodity specification of all-purpose egg mix. April 2013. Poultry Division, Agricultural Marketing Service, US Department of Agriculture, Washington, DC https://www.ams.usda.gov/sites/default/files/media/Commodity%20Specification%20for%20All%20Purpose%20Egg%20Mix%2C%20April%202013%20%28PDF%29.pdf

30. USDA (2007) USDA Commodity requirements, PCD6, Pasteurized process American Cheese for use in domestic programs. 5 Nov 2007. Kansas City Commodity Office, Commodity Credit Corporation, US Department of Agriculture, Kansas City, MO. http://www.fsa.usda.gov/Internet/FSA_File/pcd6.pdf

31. USDA (2007) USDA Commodity requirements, MCD4, Mozzarella cheese for use in domestic programs. 15 Oct 2007. Kansas City Commodity Office, Commodity Credit Corporation, US Department of Agriculture, Kansas City, MO. http://www.fsa.usda.gov/Internet/FSA_File/mcd4.pdf

32. USDA (1992) Institutional meat purchase specifications for sausage products. Series 800. Nov. 1992. Livestock

and Seed Division, Agricultural Marketing Service, US Department of Agriculture, Washington, DC. https://www.ams.usda.gov/sites/default/files/media/LSimps800.pdf

33. FAO/WHO. Codex Alimentarius. International Food Standards. Joint FAO/WHO food standards programme. Codex Alimentarius Commission, Food and Agriculture Organization of the United Nations/World Health Organization, Rome, Italy. http://www.fao.org/fao-who-codexalimentarius/standards/list-of-standards/en/

34. US Pharmacopeia (USP) (2016) Food chemicals codex, 10th edn. United Book Press, Baltimore, MD

35. JECFA (2006) Monograph 1: Combined compendium of food additive specifications. vol 4. Joint FAO/WHO Committee on Food Additives (JECFA). Web version updated Aug 2011. FAO, Rome, Italy. http://www.fao.org/3/a-a0691e.pdf

3
chapter

Nutrition Labeling

Lloyd E. Metzger (✉)
Department of Dairy Science,
South Dakota State University,
Brookings, SD 57007, USA
e-mail: lloyd.metzger@sdstate.edu

S. Suzanne Nielsen
Department of Food Science,
Purdue University,
West Lafayette, IN 47907, USA
e-mail: nielsens@purdue.edu

S. Nielsen (ed.), *Food Analysis*, Food Science Text Series,
DOI 10.1007/978-3-319-45776-5_3, © Springer International Publishing 2017

3.1　INTRODUCTION

Nutrition labeling regulations differ in countries around the world. The focus of this chapter is on nutrition labeling regulations in the USA, as specified by the **Food and Drug Administration** (FDA), with a brief summary of regulations for the **Food Safety and Inspection Service** (FSIS) of the **United States Department of Agriculture** (USDA). A major reason for analyzing the chemical components of foods in the USA is nutrition labeling regulations. Nutrition label information is not only legally required in many countries but also is of increasing importance to consumers as they focus more on health and wellness.

The FDA was authorized under the 1906 Federal Food and Drug Act and the 1938 **Federal Food**, **Drug**, *and* **Cosmetic** (FD&C) **Act** to require certain types of food labeling [1, 2]. This labeling information includes the amount of food in a package, its common or usual name, and its ingredients. In 1973, the FDA promulgated regulations that permitted, and in some cases required, food to be labeled with regard to their nutritional value. The **1990 Nutrition Labeling and Education Act** (NLEA) [2, 3] modified the 1938 FD&C Act to regulate nutrition labeling. Additionally, the 1997 **Food and Drug Administration Modernization Act** (FDAMA) [4] also amended the FD&C act and included provisions that sped up the process for approving health and nutrient content claims. The FDA amended in 2016 its nutrition labeling regulations for conventional foods and dietary supplements, with compliance dates of 2018 and 2019, depending on the sales level of the food manufacturer. These regulations, as they relate to food analysis, are the focus of this chapter.

The FDA and FSIS of the USDA have coordinated their regulations for nutrition labeling. The FDA regulations, as related to food analysis, will be described in some detail in the Sect. 3.2, focusing on the following:

1. What nutrients must be analyzed
2. How samples are to be collected
3. What methods of analysis are to be used
4. How data are to be reported
5. How data can be used to calculate caloric content
6. How data can be used for claims made on the food label

Following coverage of FDA regulations, Sect. 3.3 will give a general discussion of similarities and differences between FDA and USDA regulations.

Complete details of the current nutrition labeling regulations are available in the *Federal Register* and the *Code of Federal Regulations* (CFR) [5–8]. In developing a nutrition label for a food product, it is important to review the details of the regulations in the CFR and utilize other routinely updated resources available via the Internet. During the product development process, the effect of formulation changes on the nutritional label may be important. As an example, a small change in the amount of an ingredient may determine if a product can be labeled low fat. As a result, the ability to immediately approximate how a formulation change will impact the nutritional label can be valuable. The use of nutrient databases and computer programs designed for preparing and analyzing nutritional labels can be valuable and can simplify the process of preparing a nutritional label. The use of computer programs to prepare nutritional labels is beyond the scope of this chapter. However, an example computer program (TechWizard™, Owl Software) and a description of how this program can be used to prepare a nutrition label are found in the laboratory manual that accompanies this text.

3.2　US FOOD AND DRUG ADMINISTRATION FOOD LABELING REGULATIONS

For each aspect of nutrition labeling regulations related to food analysis described below, only FDA labeling requirements are covered. While the focus here is on **mandatory nutrition labeling**, it should be noted that the FDA has guidelines for **voluntary nutrition labeling** of raw fruit, vegetables, and fish (21 CFR 101.45).

3.2.1　Mandatory Nutrition Labeling

3.2.1.1　*Format*

The FDA regulations implementing the 1990 NLEA require nutrition labeling for most foods offered for sale and regulated by the FDA (21 CFR 101.9). Certain nutrient information is required on the label, and other information is voluntary. The standard vertical format label showing mandatory and voluntary nutrition information on food labels [21 CFR 101.9 (d)] is given in Fig. 3.1 (mandatory only) and Fig. 3.2 (includes mandatory and voluntary). Note that all nutrients, including vitamins and minerals, must be reported by weight, in addition to relevant rules about expressing as a percent of the Daily Value. A simplified format for nutrition information may be used under certain specific conditions. Also, certain foods are exempt from mandatory nutrition labeling requirements.

Standard Vertical

Nutrition Facts

8 servings per container
Serving size   2/3 cup (55g)

Amount per serving
Calories   230

	% Daily Value*
Total Fat 8g	**10%**
Saturated Fat 1g	**5%**
Trans Fat 0g	
Cholesterol 0mg	**0%**
Sodium 160mg	**7%**
Total Carbohydrate 37g	**13%**
Dietary Fiber 4g	**14%**
Total Sugars 12g	
Includes 10g Added Sugars	**20%**
Protein 3g	
Vitamin D 2mcg	10%
Calcium 260mg	20%
Iron 8mg	45%
Potassium 235mg	6%

* The % Daily Value (DV) tells you how much a nutrient in a serving of food contributes to a daily diet. 2,000 calories a day is used for general nutrition advice.

figure 3.1 Mandatory information shown on an example nutrition label, Nutrition Labeling and Education Act of 1990, amended 2016 (Courtesy of the Food and Drug Administration, Washington, DC)

3.2.1.2 *Daily Values and Serving Size*

Daily Value (DV) is a generic term used to describe two separate terms which are: (1) **Reference Daily Intake** (RDI) and (2) **Daily Reference Value** (DRV). The term RDI is used for essential vitamins and minerals (Table 3.1), while the term DRV is used for select other food components (Table 3.2). The DRVs are based on a 2000 reference Calorie intake. Nutrient content values and percent DV calculations for the nutrition label are based on serving size [21 CFR 101.12 (b), 101.9 (b)]. The labeled serving size and reference amount are important since the use of nutrient content claims (Sect. 3.2.3) is dependent on the serving size and the reference amount.

3.2.1.3 *Rounding Rules*

Increments for the numerical expression of quantity per serving are specified for all nutrients (Table 3.3) [21 CFR 101.9 (c)]. Values obtained from nutrient analysis are all rounded for reporting on the nutrition label, following very specific rules. For example, sodium con-

**Standard Vertical
(w/ Voluntary)**

Nutrition Facts

17 servings per container
Serving size   3/4 cup (28g)

Amount per serving
Calories   140

	% Daily Value*
Total Fat 1.5g	**2%**
Saturated Fat 0g	**0%**
Trans Fat 0g	
Polyunsaturated Fat 0.5g	
Monounsaturated Fat 0.5g	
Cholesterol 0mg	**0%**
Sodium 160mg	**7%**
Total Carbohydrate 22g	**8%**
Dietary Fiber 2g	**7%**
Soluble Fiber <1g	
Insoluble Fiber 1g	
Total Sugars 9g	
Includes 8g Added Sugars	**16%**
Protein 9g	**18%**
Vitamin D 2mcg (80 IU)	10%
Calcium 130mg	10%
Iron 4.5mg	25%
Potassium 115mg	2%
Vitamin A 90mcg	10%
Vitamin C 9mg	10%
Thiamin 0.3mg	25%
Riboflavin 0.3mg	25%
Niacin 4mg	25%
Vitamin B$_6$ 0.4mg	25%
Folate 200mcg DFE (120mcg folic acid)	50%
Vitamin B$_{12}$ 0.6mcg	25%
Phosphorus 100mg	8%
Magnesium 25mg	6%
Zinc 3mg	25%

* The % Daily Value (DV) tells you how much a nutrient in a serving of food contributes to a daily diet. 2,000 calories a day is used for general nutrition advice.

Calories per gram:
Fat 9   •   Carbohydrate 4   •   Protein 4

figure 3.2 Mandatory and voluntary information shown on an example nutrition label, Nutrition Labeling and Education Act of 1990, amended 2016 (Courtesy of the Food and Drug Administration, Washington, DC)

tent is to be reported to the nearest 5 mg amount up to and including 140 mg, and to the nearest 10 mg above 140 mg. Sodium content can be reported as zero if there are less than 5 mg per serving.

table 3.1 Reference daily intakes (RDIs) for vitamins and minerals essential in human nutrition

Nutrient	RDI
Vitamin A	900 µg
Vitamin C	90 mg
Calcium	1,300 mg
Iron	18 mg
Vitamin D	205 µg
Vitamin E	15 mg
Vitamin K	120 µg
Thiamin	1.2 mg
Riboflavin	1.3 mg
Niacin	16 mg
Vitamin B$_6$	1.7 mg
Folate	400 µg
Vitamin B$_{12}$	2.4 µg
Biotin	30 µg
Pantothenic acid	5 mg
Phosphorus	1,250 mg
Iodine	150 µg
Magnesium	420 mg
Zinc	11 mg
Selenium	55 µg
Copper	0.9 mg
Manganese	2.3 mg
Chromium	35 µg
Molybdenum	45 µg
Chloride	2,300 mg
Potassium	4,700 mg
Choline	550 mg

From [7]
Values are for adults and children 4 or more years of age. RDI values have also been established for infants, children under 4 years of age, and pregnant and lactating women

table 3.2 Daily reference values (DRVs) of food components[a]

Food component	DRV
Fat	78 g
Saturated fatty acids	20 g
Cholesterol	300 mg
Total carbohydrate	275 g
Fiber	28 g
Sodium	2,300 mg
Protein	50 g
Added sugars	50 g

From [7]
[a]Based on the reference calorie intake of 2,000 cal for adults and children ≥4 years

3.2.1.4 Caloric Content

Calories on the label can be expressed in numerous ways. A **calorie**, which is the standard for measurement of the energy value of substances and to express the body's energy requirement, is the amount of heat required to raise the temperature of 1 g of water 1 °C (1 cal = 4.184 J). The unit used in nutritional work is **Calorie** or "kilocalorie" (kcal), which equals 1,000 cal. In this chapter, the term Calorie is used to express caloric content. The FDA regulations specify multiple methods by which caloric content may be calculated, one of which uses bomb calorimetry [21 CFR 101.9 (c) (1)]:

1. Specific Atwater factors for Calories per gram of protein, total carbohydrate, and total fat
2. The general factors of 4, 4, and 9 Cal/g of protein, total carbohydrate, and total fat, respectively
3. The general factors of 4, 4, and 9 Cal/g of protein, total carbohydrate (less the amount of nondigestible carbohydrates and sugar alcohols), and total fat, respectively [Note: Regulations specify a general factor of 2 Cal/g for soluble nondigestible carbohydrates shall be used, and general factors for caloric value of sugar alcohols is provided in 21 CRF 101.9 (c)(1)(i)(F), i.e., 0–3.0, depending on the specific sugar alcohol.]
4. Data for specific food factors for particular foods or ingredients approved by the FDA
5. Bomb calorimetry data subtracting 1.25 Cal/g protein to correct for incomplete digestibility

3.2.1.5 Protein Quality

Reporting the amount of protein as a percent of its Daily Value on FDA-regulated foods is optional, except if a protein claim is made for the product, or if the product is represented or purported to be used by infants or children under 4 years of age, in which case the statement is required [21 CFR 101.9 (c) (7)]. For infant foods, the corrected amount of protein per serving is calculated by multiplying the actual amount of protein (g) per serving by the relative protein quality value. This relative quality value is the **protein efficiency ratio** (PER) value of the subject food product divided by the PER value for casein. For foods represented or purported for adults and children 1 year or older, the corrected amount of protein per serving is equal to the actual amount of protein (g) per serving multiplied by the **protein digestibility-corrected amino acid score** (PDCAAS). Both the PER and PDCAAS methods to assess protein quality are described in Chap. 24. The FDA allows use of the general factor 6.25 to calculate the protein content from the measured nitrogen content, except when official AOAC procedures require use of a different conversion factor (described in Chap. 18).

3.2.2 Compliance

3.2.2.1 Sample Collection

Random sampling techniques are used by the FDA to collect samples to be analyzed for compliance with nutrition labeling regulations. A "lot" is the basis for sample collection by the FDA, defined as "a collection of primary containers or units of the same size, type,

3.3 table Rounding rules for declaring nutrients on nutrition label

Nutrient/serving	Increment rounding [a,b]	Insignificant amount
Calories, calories from saturated fat	<5 Cal – express as zero ≤50 Cal – express to nearest 5 Cal increment >50 Cal – express to nearest 10 Cal increment	<5 Cal
Total fat, *trans* fat, polyunsaturated fat, monounsaturated, saturated fat	<0.5 g – express as zero <5 g – express to nearest 0.5 g increment ≥5 g – express to nearest 1 g increment	<0.5 g
Cholesterol	<2 mg – express as zero 2–5 mg – express as "less than 5 mg" >5 mg – express to nearest 5 mg increment	<2 mg
Sodium, potassium	<5 mg – express as zero 5–140 mg – express to nearest 5 mg increment >140 mg – express to nearest 10 mg increment	<5 mg
Total carbohydrate, total sugars, added sugars, sugar alcohols, dietary fiber, soluble fiber, insoluble fiber, protein	<0.5 g – express as zero <1 g – express as "Contains less than 1 g" or "less than 1 g" ≥1 g – express to nearest 1 g increment	<1 g

Vitamins and minerals	<2% of RDI – may be expressed as:
	  1. Zero
	  2. An asterisk that refers to statement "Contains less than 2% of the Daily Value of this (these) nutrient (nutrients)" (or use symbol < in place of "less than")
	  3. For vitamin D, calcium, iron, potassium: statement "Not a significant source of _____ (listing the vitamins or minerals omitted)"
	≤10% of RDI – express to nearest 2% increment
	>10–≤50% of RDI – express to nearest 5% increment
	>50% of RDI – express to nearest 10% increment
		<2% RDI

Fluoride	<0.1 mg – express as zero ≤0.8 mg – express to nearest 0.1 mg >0.8 mg – express to nearest 0.2 mg	

Summarized from [7]

Notes below taken from: 21 CRF 101.9 (c); Food Labeling Guide. Appendix H. FDA rounding rules. 2015. Center for Food Safety and Applied Nutrition, Food and Drug Administration, Washington, DC

[a]To express nutrient values to the nearest 1 g increment, for amounts falling exactly halfway between two whole numbers or higher (e.g., 2.5–2.99 g), round up (e.g., 3 g). For amounts less than halfway between two whole numbers (e.g., 2.01–2.49 g), round down (e.g., 2 g)

[b]The percent DV shall be calculated by diving either the amount declared on the label for each nutrient or the actual amount of each nutrient (i.e., before rounding) by the DRV for the nutrient, except that the percent for protein shall be calculated as specified in [21 CFR (c)(7)(ii)] (described in Sect. 3.2.1.5 of this textbook chapter)

When rounding % DV for nutrients other than vitamins and minerals, when the % CV values fall exactly halfway between whole numbers or higher (e.g., 2.5–2.99), the values round up (e.g., 3%). For values less than halfway between two whole numbers (e.g., 2.01–2.49), the values round down (e.g., 2%) (Note: Sodium %DV is rounded like for other nutrients that have a DRV, rather than like other minerals that have a RDI)

and style produced under conditions as nearly uniform as possible, and designated by a common container code or marking, or in the absence of any common container code or marking, a day's production." The sample used by the FDA for nutrient analysis consists of a "composite of 12 subsamples (consumer units), taken 1 from each of 12 different ran-domly chosen shipping cases, to be representative of a lot" [21 CFR 101.9 (g)].

3.2.2.2 *Methods of Analysis*

The FDA states that unless a particular method of analysis is specified in 21 CFR 101.9(c), appropriate methods of AOAC International published in the

| **3.4** table | Nutritional label components: commonly used AOAC International official methods and food analysis textbook coverage |

Nutrient	Name of method	AOAC official method number	AOAC official method locator number	Food analysis book chapter
Calories	One of multiple methods allowed by FDA: Calories = (g of carbohydrate × 4 Cal/g) + (g of protein × 4 Cal/g) + (g of fat × 9 Cal/g)			
Total carbohydrate	Calculation for proximate composition (100% = total fat + protein + total carbohydrate + moisture + ash)	Proximate analysis		
Dietary fiber	Enzymic-gravimetric method for total, soluble, and insoluble dietary fiber	2011.25	32.1.43	19
Total sugars	HPLC	977.20	44.4.13	19
Added sugar	No analytical method; Must be calculated based on formulation	–	–	
Protein	Dumas	968.06	4.2.04	18
Total fat	Gas chromatography	996.06	41.1.28A	17
Saturated fat	Gas chromatography	996.06	41.1.28A	17, 23
Trans fat	Gas chromatography	996.06	41.1.28A	17, 23
Cholesterol	Capillary gas chromatography	976.26	45.4.06	23
Moisture (to calculate total carbohydrate)	Forced draft oven at 105 ± 1 °C for 4 h (note: other methods would be official/more appropriate for certain foods)	925.10	32.1.03	15
Ash (to calculate total carbohydrate)	Dry ashing using muffle furnace at 585 °F	923.03	32.1.05	16
Sodium	Inductively coupled plasma – atomic emission spectroscopy	985.01	3.2.06	9
Vitamin D	Liquid chromatography – mass spectrometry	2002.05	45.1.22A	20
Calcium	Inductively coupled plasma – atomic emission spectroscopy	985.01	3.2.06	9
Iron	Inductively coupled plasma – atomic emission spectroscopy	985.01	3.2.06	9
Potassium	Inductively coupled plasma – atomic emission spectroscopy	985.01	3.2.06	9

AOAC official method listed are common use methods, based on information (in 2016) from commercial analytical laboratories that do nutrition label analyses. Other AOAC official methods are possible for many nutrients and in some cases may be more appropriate, depending on the nature of the food or ingredient

Official Methods of Analysis [9] are to be used. Other reliable and appropriate methods can be used if no AOAC method is available or appropriate. If scientific knowledge or reliable databases have established that a nutrient is not present in a specific product (e.g., dietary fiber in seafood, cholesterol in vegetables), the FDA does not require analyses for the nutrients.

Table 3.4 gives commonly used AOAC official methods associated with nutrition label components, as reported by analytical laboratories. Also listed for each nutrition label component is the book chapter in this textbook that describes the type of method identified, along with associated methods.

3.2.2.3 Levels for Compliance
The FDA monitors accuracy of nutrient content information for compliance based on two classes of nutrients

and an unnamed third group, as described in Table 3.5. Compliance regulations point to the importance of appropriate sample collection and sample preparation, and for accurate chemical analysis to ensure the nutrition label information is correct. For example, a product fortified with iron would be considered misbranded if it contained less than 100% of the label declaration. A product that naturally contains dietary fiber would be considered misbranded if it contained less than 80% of the label declaration. A product would be considered misbranded if it had a caloric content greater than 20% in excess of the label declaration. Reasonable excesses over labeled amounts (of a vitamin, mineral, protein, total carbohydrate, polyunsaturated or monounsaturated fat, or potassium) or deficiencies below label amounts (of Calories, sugars, total fat, saturated fat, cholesterol, or sodium) are acceptable within current

	Basis for compliance of nutrition labeling regulation by Food and Drug Administration and Food Safety and Inspection Service of the US Department of Agriculture		
3.5 table			
Class of nutrients	*Purposes of compliance*	*Nutrients regulated*	*% required[a]*
I	Added nutrients in fortified or fabricated foods	Vitamin, mineral, protein, dietary fiber	≥100%
II	Naturally occurring (indigenous) nutrients	Vitamin, mineral, protein, total carbohydrate, dietary fiber, soluble fiber, insoluble fiber, polyunsaturated or mono-unsaturated fat	≥80%
*[b]		Calories, total sugars, added sugar (when only source in food is added sugars), total fat, saturated fat, trans fat, cholesterol, sodium	≤120%

Summarized from [7]

[a]Amount of nutrient required in food sample as a percentage of the label declaration or else product is considered misbranded

[b]*Unnamed class

Good Manufacturing Practices (cGMP). Noncompliance with regard to a nutrition label can result in warning letters, recalls, seizures, and prosecution (21 CFR 1.21). Compliance with the regulations can be obtained by use of FDA-approved databases [7] [21 CFR 101.9 (g) (8)] that have been computed using FDA guidelines, and foods that have been handled under good manufacturing practice conditions to prevent nutritional losses. In certain instances, compliance includes record keeping of data for dietary fiber, added sugars, vitamin E, and folate [7] [21 CFR 101.0 (g) (10)].

3.2.3 Nutrient Content Claims

The FDA has defined **nutrient content claims** that characterize the level of a nutrient, according to specific definition (21 CFR 101.13, 101.54–101.67). The terms include the following: "free," "low," "lean," "light," "reduced," "less," "fewer," "added," "extra," "plus," "fortified," "enriched," "good source," "contains," "provides," "more," "high," "rich in," "excellent source of," and "high potency." Of these terms,

"less" (or "fewer"), "more," "reduced," "added" (or "extra," "plus," "fortified," and "enriched"), and "light" are relative terms and require label information about the food product that is the basis of the comparison. The percentage difference between the original food (reference food) and the food product being labeled must be listed on the label for comparison.

The use of nutrient content claims on food labels is typically based on the content of specific nutrients per reference amount or per serving. For example, the terms "free" or "low" for total fat, saturated fat, cholesterol, and sodium relate to specific amounts of those nutrients. Also, terms such as "high," "excellent source," and "enriched" require that the nutrient identified contain a specific percentage of the DV per reference amount. The term "healthy" or its derivatives may be used on the label or in labeling of foods under conditions defined by the FDA, replying on very specific levels of various nutrients. Note that the FDA requirements on nutrient content claims do not apply to infant formulas and medical foods.

3.2.4 Health Claims

The FDA has defined and will allow claims for certain relationships between a nutrient or a food and the risk of a disease or health-related condition (21 CFR 101.14). The FDA utilizes several types of oversight to determine which health claims may be used in labeling of food or dietary supplements, leading to multiple categories of health claims. One such category, the **NLEA-authorized health claims,** characterizes a relationship between a food, food component, dietary ingredient, or dietary supplement and risk of a disease (Table 3.6). Most of these NLEA-authorized health claims, and many in other health claim categories, are based on specific nutrients and therefore the chemical analysis of those food components.

3.3 US DEPARTMENT OF AGRICULTURE FOOD LABELING REGULATIONS

The FDA and FSIS of the USDA have coordinated their regulations for nutrition labeling. The USDA regulations require nutrition labeling of most meat or meat products (9 CFR 317.300–317.400) and poultry products (9 CFR 381.400–381.500). The differences that exist in the regulations are due principally to the inherent differences in the food products regulated by the FDA and USDA (USDA regulates only meat, poultry, and egg products). The two agencies maintain close harmony regarding interpretation of the regulations and changes made in regulations. Some general areas in which differences exist between FDA and USDA nutrition labeling requirements are the following: (1) nutrients

| 3.6 table | Nutrition Labeling Education Act (NLEA)-authorized health claims |

Claim	CFR[a] reference
Calcium and osteoporosis	21 CFR 101.72
Dietary fat and cancer	21 CFR 101.73
Sodium and hypertension	21 CFR 101.74
Dietary saturated fat and cholesterol and risk of coronary heart disease	21 CFR 101.75
Fiber-containing grain products, fruits, and vegetables and cancer	21 CFR 101.76
Fruits, vegetables, and grain products that contain fiber, particularly soluble fiber, and risk of coronary heart disease	21 CFR 101.77
Fruits and vegetables and cancer	21 CFR 101.78
Folate and neural tube defects	21 CFR 101.79
Dietary noncariogenic carbohydrate sweeteners and dental caries	21 CFR 101.80
Soluble fiber from certain foods and risk of coronary heart disease	21 CFR 101.81
Soy protein and risk of coronary heart disease	21 CFR 101.82
Plant sterol/stanol esters and risk of coronary heart disease	21 CFR 101.83

[a]*CFR* Code of Federal Regulations

allowed on label by voluntary declaration, (2) serving size regulations, (3) compliance procedures, and (4) nutrient content claims. One specific difference is that FSIS does not require *trans* fats as a mandatory nutrient, but permits it to be declared voluntairly. Two other specific areas different for USDA from FDA regulations are summarized below.

Regarding compliance, FSIS specifies for nutritional analysis the methods of the *USDA Chemistry Laboratory Guidebook* [14]. If no USDA method is available and appropriate for the nutrient, methods in the *Official Methods of Analysis* of AOAC International [9] are to be used. If no USDA, AOAC International, or specified method is available and appropriate, FSIS specifies the use of other reliable and appropriate analytical procedures as determined by the Agency. FSIS provides information on how it collects samples for compliance analysis is [9 CFR 317.309 (h), 381.409 (h)].

Regarding nutrient content claim differences between FDA- and USDA-regulated foods, the following are some examples:

1. "Enriched" and "Fortified" are not defined in the FSIS regulations.
2. "Lean" and Extra lean" are defined and approved for all USDA-regulated products but only for the FDA-regulated products of seafood, game meat, and meal products.
3. The term "___% Lean" is approved only for USDA-regulated products.

This section has only briefly summarized the USDA nutrition labeling regulations and compared them to the FDA regulations. The reader is referred to the CFR regulations listed above for all details of USDA nutrition labeling regulations.

3.4   SUMMARY

A major reason for analyzing the chemical components of food in the USA (and many other countries) is nutrition labeling regulation. The FDA and FSIS of the USDA have coordinated their regulations on nutrition labeling. Regulations that implement the NLEA of 1990 require nutrition labeling for most foods regulated by the FDA, and FSIS requires the same label on most meat and poultry products. The regulations were amended in 2016 regarding nutrition information on the label in an attempt to help consumers maintain healthy dietary practices. Nutrition labeling regulations define the format for the nutrition information and give the rules and methods to report specific information. Specifications include sample collection procedures, the method of analysis to be used, and the nutrient levels required to ensure compliance with nutrition labeling regulations. Specific nutrient content claims and health claims are allowed on the nutrition label. The nutrition labeling regulations covered in this chapter are only those closely linked to food analysis. Readers are referred to appropriate sections of the CFR for all details of FDA and USDA regulations.

3.5   STUDY QUESTIONS

1. Utilize the data in the table below that you obtained on the nutrient content of your cereal product (actual amount per serving) to help develop a nutrition label that meets FDA requirements under the NLEA, amended in 2016. Use appropriate rounding rules to

complete the blank columns. If you wanted to report the protein content as a percent of the Daily Value, what would you need to do?

	Actual amount per serving[a]	Amount per serving reported on label	% daily value reported on label
Calories	192		–
Total fat	1.1 g		
Saturated fat	0 g		
Trans fat	0 g		
Cholesterol	0 mg		
Sodium	267 mg		
Total carbohydrate	44.3 g		
Dietary fiber	3.8 g		
Total sugars	20.2 g		–
Incl. added sugars	6.6 g		
Protein	3.7 g		–
Vitamin D	2 µg		
Calcium	210 mg		
Iron	4.3 mg		
Potassium	217 mg		

[a]Serving size is 1 cup (55 g)

2. The FDA and FSIS of the USDA have very similar regulations for nutrition labeling.

   (a) Identify the differences in regulations between the FDA and FSIS regarding the first choice of methods for nutritional analysis.

   (b) Identify one difference between the agencies in nutrient content claims that is consistent with the statement that the differences in regulations between the agencies are primarily due to the inherent differences in the food products they regulate.

Acknowledgements The authors thank Drs. Ann Roland and Lance Phillips of Owl Software (Columbia, MO) for their review of this chapter and helpful comments to ensure consistency with the new nutrition labeling regulations.

REFERENCES

1. Adams DG, Cooper RM, Hahn HJ, Kahan JS (2014) Food and drug law and regulation. 3rd edn. Food and Drug Law Institute, Washington, DC
2. Piña KR, Pines WL (2014) A practical guide to food and drug law and regulation, 5th edn. Food and Drug Law Institute, Washington, DC
3. US Congress (1990) US public law 101–535. Nutrition labeling and education act of 1990. Nov 1990. US Congress, Washington, DC
4. US Congress (1997) US public law 105–115. Food and Drug Administration modernization act of 1997. 21 Nov 1997. US Congress, Washington, DC
5. Federal Register (1993) Department of Agriculture. Food Safety and Inspection Service. Part CFR Parts 317, 320, and 381. Nutrition labeling of meat and poultry products; final rule. 6 January 1993. 58(3): 631–685. Part III. 9 CFR Parts 317 and 381. Nutrition labeling: use of "healthy" and similar terms on meat and poultry products labeling; proposed rule. 6 January, 1993. 58(3): 687–691. Superintendent of documents. US Government Printing Office, Washington, DC
6. Federal Register (1993) 21 CFR Part 1, et al. Food labeling; general provisions; nutrition labeling; label format; nutrient content claims; health claims; ingredient labeling; state and local requirements; and exemptions; final rules. 6 January, 1993. 58(3):#2066–2941. Superintendent of documents. US Government Printing Office, Washington, DC
7. Federal Register (2016) Food labeling: Revision of the nutrition and supplemental facts labels (5/27/2016). https://www.federalregister.gov/documents/2016/05/2016/food-labeling-revision-of-the-nutrition-and-supplement-facts-label
8. Code of Federal Regulations (2015) (animal and animal products). 9 CFR 317 subpart B 317. 300–317.400; 9 CFR 381 subpart Y 381.400–381.500. US Government Printing Office, Washington, DC
9. AOAC International (2016) Official methods of analysis, 20th edn. (On-line). AOAC International, Rockville, MD
10. USDA (2016) Chemistry laboratory guidebook. http://www.fsis.usda.gov/wps/portal/fsis/topics/science/laboratories-and-procedures/guidebooks-and-methods/chemistry-laboratory-guidebook Food Safety and Inspection Service, US Department of Agriculture, Washington, DC

Evaluation of Analytical Data

J. Scott Smith

Department of Animal Sciences and Industry,
Kansas State University,
Manhattan, KS 66506-1600, USA
e-mail: jsschem@ksu.edu

S. Nielsen (ed.), *Food Analysis*, Food Science Text Series,
DOI 10.1007/978-3-319-45776-5_4, © Springer International Publishing 2017

4.1   INTRODUCTION

The field of food analysis, or any type of analysis, involves a considerable amount of time learning principles, methods, and instrument operations and perfecting various techniques. Although these areas are extremely important, much of our effort would be for naught if there were not some way for us to evaluate the data obtained from the various analytical assays. Several mathematical treatments are available that provide an idea of how well a particular assay was performed or how well we can reproduce an experiment. Fortunately, the statistics are not too involved and apply to most analytical determinations.

Whether analytical data are collected in a research laboratory or in the food industry, important decisions are made based on the data. Appropriate data collection and analysis help avoid bad decisions being made based on the data. Having a good understanding of the data and how to interpret the data (e.g., what numbers are statistically the same) are critical to good decision making. Talking with a statistician before designing experiments or testing products produced can help ensure appropriate data collection and analysis, for better decision making.

The focus in this chapter is primarily on how to evaluate replicate analyses of the same sample for accuracy and precision. In addition, considerable attention is given to the determination of best line fits for standard curve data. Keep in mind as you read and work through this chapter that there is a vast array of computer software to perform most types of data evaluation and calculations/plots.

Proper sampling and sample size are not covered in this chapter. Readers should refer to Chap. 5 and Garfield et al. [1] for sampling in general and statistical approaches to determine the appropriate sample size, and to Chap. 33, Sect. 33.4 for mycotoxin sampling.

4.2   MEASURES OF CENTRAL TENDENCY

To increase accuracy and precision, as well as to evaluate these parameters, the analysis of a sample is usually performed (repeated) several times. At least three assays are typically performed, though often the number can be much higher. Because we are not sure which value is closest to the true value, we determine the mean (or average) using all the values obtained and report the results of the **mean**. The mean is designated by the symbol $\bar{x}$ and calculated according to the equation below:

$$\bar{x} = \frac{x_1 + x_2 + x_3 + \ldots + x_n}{n} = \frac{\Sigma x_i}{n} \qquad (4.1)$$

where:

$$\bar{x} = \text{mean}$$

$$x_1, x_2, \text{etc.} = \text{individually measured values } (x_i)$$

$$n = \text{number of measurements}$$

For example, suppose we measured a sample of uncooked hamburger for percent moisture content four times and obtained the following results: 64.53 %, 64.45 %, 65.10 %, and 64.78 %:

$$\bar{x} = \frac{64.53 + 64.45 + 65.10 + 64.78}{4} = 64.72\% \qquad (4.2)$$

Thus, the result would be reported as 64.72 % moisture. When we report the mean value, we are indicating that this is the best experimental estimate of the value. We are not saying anything about how accurate or true this value is. Some of the individual values may be closer to the true value, but there is no way to make that determination, so we report only the mean.

Another determination that can be used is the **median**, which is the midpoint or middle number within a group of numbers. Basically, half of the experimental values will be less than the median and half will be greater. The median is not used often, because the mean is such a superior experimental estimator.

4.3   RELIABILITY OF ANALYSIS

Returning to our previous example, recall that we obtained a mean value for moisture. However, we did not have any indication of how repeatable the tests were or how close our results were to the true value. The next several sections will deal with these questions and some of the relatively simple ways to calculate the answers. More thorough coverage of these areas is found in references [2–4].

4.3.1   Accuracy and Precision

One of the most confusing aspects of data analysis for students is grasping the concepts of accuracy and precision. These terms are commonly used interchangeably in society, which only adds to this confusion. If we consider the purpose of the analysis, then these terms become much clearer. If we look at our experiments, we know that the first data obtained are the individual results and a mean value ($\bar{x}$). The next questions should be: "How close were our individual measurements?" and "How close were they to the true value?" Both questions involve accuracy and precision. Now, let us turn our attention to these terms.

Accuracy refers to how close a particular measure is to the true or correct value. In the moisture analysis for hamburger, recall that we obtained a mean of 64.72%. Let us say the true moisture value was actually 65.05%. By comparing these two numbers, you could probably make a guess that your results were fairly accurate because they were close to the correct value. (The calculations of accuracy will be discussed later.)

The problem in determining accuracy is that most of the time we are not sure what the true value is. For certain types of materials, we can purchase known samples from, for example, the National Institute of Standards and Technology and check our assays against these samples. Only then can we have an indication of the accuracy of the testing procedures. Another approach is to compare our results with those of other labs to determine how well they agree, assuming the other labs are accurate.

A term that is much easier to deal with and determine is **precision**. This parameter is a measure of how reproducible or how close replicate measurements become. If repetitive testing yields similar results, then we would say the precision of that test was good. From a true statistical view, the precision often is called **error**, when we are actually looking at experimental **variation**. So, the concepts of precision, error, and variation are closely related.

The difference between precision and accuracy can be illustrated best with Fig. 4.1. Imagine shooting a rifle at a target that represents experimental values. The bull's eye would be the true value, and where the bullets hit would represent the individual experimental values. As you can see in Fig. 4.1a, the values can be tightly spaced (good precision) and close to the bull's eye (good accuracy), or, in some cases, there can be situations with good precision but poor accuracy (Fig. 4.1b). The worst situation, as illustrated in Fig. 4.1d, is when both the accuracy and precision are poor. In this case, because of errors or variation in the determination, interpretation of the results becomes very difficult. Later, the practical aspects of the various types of error will be discussed.

When evaluating data, several tests are commonly used to give some appreciation of how much the experimental values would vary if we were to repeat the test (indicators of precision). An easy way to look at the variation or scattering is to report the range of the experimental values. The **range** is simply the difference between the largest and smallest observation. This measurement is not too useful and thus is seldom used in evaluating data.

Probably the best and most commonly used statistical evaluation of the precision of analytical data is the standard deviation. The **standard deviation** measures the spread of the experimental values and gives a good indication of how close the values are to each other. When evaluating the standard deviation, one has to remember that we are never able to analyze the entire food product. That would be difficult, if not impossible, and very time consuming. Thus, the calculations we use are only estimates of the unknown true value.

If we have many samples, then the standard deviation is designated by the Greek letter sigma (σ). It is calculated according to Eq. 4.3, assuming all of the food product was evaluated (which would be an infinite amount of assays):

$$\sigma = \sqrt{\frac{\Sigma\left(x_i - \mu\right)^2}{n}} \qquad (4.3)$$

where:

σ = standard deviation
xi = individual sample values
μ = true mean
n = total population of samples

Because we do not know the value for the true mean, the equation becomes somewhat simplified so that we can use it with real data. In this case, we now call the σ term the standard deviation of the sample and designate it by SD or σ. It is determined according to the calculation in Eq. 4.4, where $\bar{x}$ replaces the true mean term μ and n represents the number of samples:

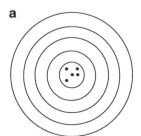
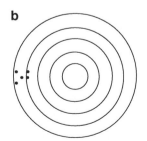
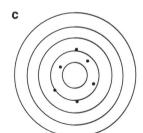
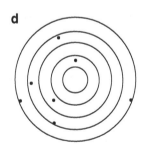

a   b   c   d

Comparison of accuracy and precision: (**a**) good accuracy and good precision, (**b**) good precision and poor accuracy, (**c**) good accuracy and poor precision, and (**d**) poor accuracy and poor precision

$$SD = \sqrt{\frac{\Sigma\left(x_i - \overline{x}\right)^2}{n}} \qquad (4.4)$$

If the number of replicate determinations is small (about 30 or less), which is common with most assays, the n is replaced by the $n-1$ term, and Eq. 4.5 is used. Unless you know otherwise, Eq. 4.5 is always used in calculating the standard deviation of a group of assays:

$$SD = \sqrt{\frac{\Sigma\left(x_i - \overline{x}\right)^2}{n-1}} \qquad (4.5)$$

Depending on which of the equations above is used, the standard deviation may be reported as SDn or σn and SDn_{-1} or σn_{-1}. (Different brands of software and scientific calculators sometimes use different labels for the keys, so one must be careful.) Table 4.1 shows an example of the determination of standard deviation. The sample results would be reported to average 64.72% moisture with a standard deviation of 0.293.

Once we have a mean and standard deviation, we must next determine how to interpret these numbers. One easy way to get a feel for the standard deviation is to calculate what is called the **coefficient of variation** (CV), also known as the **relative standard deviation**. This calculation is shown below for our example of the moisture determination of uncooked hamburger:

$$\% \text{ Coefficient of variation } (\%CV) = \frac{SD}{\overline{x}} \times 100 \qquad (4.6)$$

$$\%CV = \frac{0.293}{64.72} \times 100 = 0.453\% \qquad (4.7)$$

The CV tells us that our standard deviation is only 0.453% as large as the mean. For our example, that number is small, which indicates a high level of precision or reproducibility of the replicates. As a rule, a CV below 5% is considered acceptable, although it depends on the type of analysis.

4.1 table

Determination of the standard deviation of percent moisture in uncooked hamburger

Measurement	Observed % moisture	Deviation from the mean $\left(x_i - \overline{x}\right)$	$\left(x_i - \overline{x}\right)^2$
1	64.53	−0.19	0.0361
2	64.45	−0.27	0.0729
3	65.10	+0.38	0.1444
4	64.78	+0.06	0.0036
	$\Sigma x_i = 258.86$		$\Sigma\left(x_i - \overline{x}\right)^2 = 0.257$

$$\overline{x} = \frac{\Sigma x_i}{n} = \frac{258.86}{4} = 64.72$$

$$SD = \sqrt{\frac{\Sigma\left(x_i - \overline{x}\right)^2}{n-1}} = \sqrt{\frac{0.257}{3}} = 0.2927$$

Another way to evaluate the meaning of the standard deviation is to examine its origin in statistical theory. Many populations (in our case, sample values or means) that exist in nature are said to have a normal distribution. If we were to measure an infinite number of samples, we would get a distribution similar to that represented by Fig. 4.2. In a population with a **normal distribution**, 68% of those values would be within ±1 standard deviation from the mean, 95% would be within ±2 standard deviations, and 99.7% would be within ±3 standard deviations. In other words, there is a probability of less than 1% that a sample in a population would fall outside ±3 standard deviations from the mean value.

Another way of understanding the normal distribution curve is to realize that the probability of finding the true mean is within certain confidence intervals as defined by the standard deviation. For large numbers of samples, we can determine the **confidence limit** or **interval** around the mean using the statistical parameter called the **Z value**. We do this calculation by first looking up the Z value from statistical tables once we have decided the desired degree of certainty. Some Z values are listed in Table 4.2.

The confidence limit (or interval) for our moisture data, assuming a 95% probability, is calculated according to Eq. 4.8. Since this calculation is not valid for

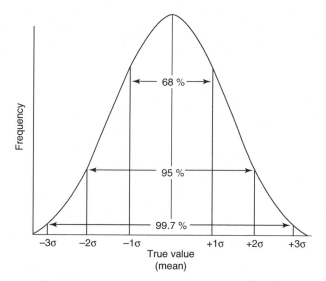

4.2 figure

A normal distribution curve for a population or a group of analyses

4.2 table

Values for Z for checking both upper and lower levels

Degree of certainty (confidence) (%)	Z value
80	1.29
90	1.64
95	1.96
99	2.58
99.9	3.29

small numbers, assume we ran 25 samples instead of four:

Confidence interval (CI)

$$= \bar{x} \pm Z \text{ value} \times \frac{\text{standard deviation (SD)}}{\sqrt{n}} \quad (4.8)$$

$$\text{CI (at 95\%)} = 64.72 \pm 1.96 \times \frac{0.2927}{\sqrt{25}}$$

$$= 64.72 \pm 0.115\% \quad (4.9)$$

Because our example had only four values for the moisture levels, the confidence interval should be calculated using statistical t-tables. In this case, we have to look up the t value from Table 4.3 based on the **degrees of freedom**, which is the sample size minus one ($n-1$), and the desired level of confidence.

The calculation for our moisture example with four samples (n) and three degrees of freedom ($n-1$) is given below:

$$\text{CI} = \bar{x} \pm t \text{ value} \times \frac{\text{standard deviation (SD)}}{\sqrt{n}} \quad (4.10)$$

$$\text{CI (at 95\%)} = 64.72 \pm 3.18 \times \frac{0.2927}{\sqrt{4}}$$

$$= 64.72 \pm 0.465\% \quad (4.11)$$

To interpret this number, we can say that, with 95% confidence, the true mean for our moisture will fall within $64.72 \pm 0.465\%$ or between 65.185% and 64.255%.

The expression $SD/\sqrt{n}$ is often reported as the **standard error of the mean**. It is then left to the reader to calculate the confidence interval based on the desired level of certainty.

Other quick tests of precision used are the relative deviation from the mean and the relative average deviation from the mean. The **relative deviation from the mean** is useful when only two replicates have been performed. It is calculated according to Eq. 4.12, with values below 2% considered acceptable:

4.3 table

Values of *t* for various levels of probability[a]

Degrees of freedom (n − 1)	Levels of certainty		
	95%	99%	99.9%
1	12.7	63.7	636
2	4.30	9.93	31.60
3	3.18	5.84	12.90
4	2.78	4.60	8.61
5	2.57	4.03	6.86
6	2.45	3.71	5.96
7	2.36	3.50	5.40
8	2.31	3.56	5.04
9	2.26	3.25	4.78
10	2.23	3.17	4.59

[a]More extensive t-tables can be found in statistics books

$$\text{Relative deviation from the mean} = \frac{x_i - \bar{x}}{\bar{x}} \times 100 \quad (4.12)$$

where:

x_i = individual sample value
$\bar{x}$ = mean

If there are several experimental values, then the **relative average deviation from the mean** becomes a useful indicator of precision. It is calculated similarly to the relative deviation from the mean, except the average deviation is used instead of the individual deviation. It is calculated according to Eq. 4.13:

Relative average deviation from the mean

$$= \frac{\frac{\Sigma |x_i - \bar{x}|}{n}}{\bar{x}} \times 1000$$

$$= \text{parts per thousand} \quad (4.13)$$

Using the moisture values discussed in Table 4.1, the $x_i - \bar{x}$ terms for each determination are −0.19, −0.27, +0.38, and +0.06. Thus, the calculation becomes:

$$\text{Rel. avg. dev.} = \frac{\frac{0.19 + 0.27 + 0.38 + 0.06}{4}}{64.72} \times 1000$$

$$= \frac{0.225}{64.72} \times 1000$$

$$= 3.47 \text{ parts per thousand} \quad (4.14)$$

Up to now, our discussions of calculations have involved ways to evaluate precision. If the true value is not known, we can calculate only precision. A low degree of precision would make it difficult to predict a realistic value for the sample.

However, we may occasionally have a sample for which we know the true value and can compare our results with the known value. In this case, we can calculate the error for our test, compare it to the known value, and determine the accuracy. One term that can be calculated is the **absolute error**, which is simply the difference between the experimental value and the true value:

$$\text{Absolute error} = E_{abs} = x - T \quad (4.15)$$

where:

x = experimentally determined value
T = true value

The absolute error term can have either a positive or negative value. If the experimentally determined value is from several replicates, then the mean (0) would be substituted for the x term. This is not a good test for error, because the value is not related to the magnitude of the true value. A more useful measurement of error is **relative error**:

$$\text{Relative error} = E_{\text{rel}} = \frac{E_{\text{abs}}}{T} = \frac{x - T}{T} \qquad (4.16)$$

The results are reported as a negative or positive value, which represents a fraction of the true value.

If desired, the relative error can be expressed as percent relative error by multiplying by 100%. Then the relationship becomes the following, where x can be either an individual determination or the mean (0) of several determinations:

$$\%\,E_{\text{rel}} = \frac{E_{\text{abs}}}{T} \times 100\% = \frac{x - T}{T} \times 100\% \qquad (4.17)$$

Using the data for the percent moisture of uncooked hamburger, suppose the true value of the sample is 65.05%. The percent relative error is calculated using our mean value of 64.72% and Eq. 4.17:

$$\%\,E_{\text{rel}} = \frac{\overline{x} - T}{T} \times 100\% = \frac{64.72 - 65.05}{65.05} \times 100\%$$
$$= -0.507\% \qquad (4.18)$$

Note that we keep the negative value, which indicates the direction of our error, that is, our results were 0.507% lower than the true value.

4.3.2 Sources of Errors [3]

As you may recall from the discussions of accuracy and precision, error (variation) can be quite important in analytical determinations. Although we strive to obtain correct results, it is unreasonable to expect an analytical technique to be entirely free of error. The best we can hope for is that the variation is small and, if possible, at least consistent. As long as we know about the error, the analytical method often will be satisfactory. There are several sources of error, which can be classified as: systematic error (determinate), random error (indeterminate), and gross error or blunders. Again, note that error and variation are used interchangeably in this section and essentially have the same meaning for these discussions.

Systematic or **determinate error** produces results that consistently deviate from the expected value in one direction or the other. As illustrated in Fig. 4.1b, the results are spaced closely together, but they are consistently off the target. Identifying the source of this serious type of error can be difficult and time consuming, because it often involves inaccurate instruments or measuring devices. For example, a pipette that consistently delivers the wrong volume of reagent will produce a high degree of precision yet inaccurate results. Sometimes impure chemicals or the analytical method itself is the cause. Generally, we can overcome systematic errors by proper calibration of instruments,

running blank determinations, or using a different analytical method.

Random or **indeterminate errors** are always present in any analytical measurement. This type of error is due to our natural limitations in measuring a particular system. These errors fluctuate in a random fashion and are essentially unavoidable. For example, reading an analytical balance, judging the endpoint change in a titration, and using a pipette all contribute to random error. Background instrument noise, which is always present to some extent, is a factor in random error. Both positive and negative errors are equally possible. Although this type of error is difficult to avoid, fortunately it is usually small.

Blunders are easy to eliminate, since they are so obvious. The experimental data are usually scattered, and the results are not close to an expected value. This type of error is a result of using the wrong reagent or instrument or of sloppy technique. Some people have called this type of error the "Monday morning syndrome" error. Fortunately, blunders are easily identified and corrected.

4.3.3 Specificity

Specificity of a particular analytical method means that it detects only the component of interest. Analytical methods can be very specific for a certain food component or, in many cases, can analyze a broad spectrum of components. Quite often, it is desirable for the method to be somewhat broad in its detection. For example, the determination of food lipid (fat) is actually the crude analysis of any compound that is soluble in an organic solvent. Some of these compounds are glycerides, phospholipids, carotenes, and free fatty acids. Since we are not concerned about each individual compound when considering the crude fat content of food, it is desirable that the method be broad in scope. On the other hand, determining the lactose content of ice cream would require a specific method. Because ice cream contains other types of simple sugars, without a specific method, we would overestimate the amount of lactose present.

There are no hard rules for what specificity is required. Each situation is different and depends on the desired results and type of assay used. However, it is something to keep in mind as the various analytical techniques are discussed.

4.3.4 Sensitivity and Limit of Detection [5]

Although often used interchangeably, the terms sensitivity and limit of detection should not be confused. They have different meanings, yet they are closely related. **Sensitivity** relates to the magnitude of change of a measuring device (instrument) with changes in compound concentration. It is an indicator of how little change can be made in the unknown material before we notice a difference on a needle gauge or a digital readout. We are all familiar with the process of tuning in a radio station on our stereo and know how, at some point, once the station is tuned in, we can move the dial without disturbing the reception. This is sensitivity. In many situations, we can adjust the sensitivity of an assay to fit our needs, that is, whether we desire more or less sensitivity. We even may desire a lower sensitivity so that samples with widely varying concentration can be analyzed at the same time.

Limit of detection (LOD), in contrast to sensitivity, is the lowest possible increment that we can detect with some degree of confidence (or statistical significance). With every assay, there is a lower limit at which point we are not sure if something is present or not. Obviously, the best choice would be to concentrate the sample so we are not working close to the detection limit. However, this may not be possible, and we may need to know the LOD so we can work away from that limit.

There are several ways to measure the LOD, depending on the apparatus that is used. If we are using something like a spectrophotometer, gas chromatograph, or high-performance liquid chromatography (HPLC), the LOD often is reached when the signal to noise ratio is 3 or greater [5]. In other words, when the sample gives a value that is three times the magnitude of the noise detection, the instrument is at the lowest limit possible. Noise is the random signal fluctuation that occurs with any instrument.

A more general way to define the limit of detection is to approach the problem from a statistical viewpoint, in which the variation between samples is considered. A common mathematical definition of limit of detection is given below [3]:

$$X_{LD} = X_{Blk} + \left(3 \times SD_{Blk}\right) \qquad (4.19)$$

where:

X_{LD} = minimum detectable concentration
X_{Blk} = signal of a blank
SD_{Blk} = standard deviation of the blank readings

In this equation, the variation of the blank values (or noise, if we are talking about instruments) determines the detection limit. High variability in the blank values decreases the limit of detection.

Another method that encompasses the entire assay method is the **method detection limit** (MDL). According to the US Environmental Protection Agency (EPA) [6], the MDL is defined as "the minimum concentration of a substance that can be measured and reported with 99 % confidence that the analyte concentration is greater than zero and is determined from

analysis of a sample in a given matrix containing the analyte." What differentiates the MDL from the LOD is that it includes the entire assay and various sample types thus correcting for variability throughout. The MDL is calculated based on values of samples within the assay matrix and thus is considered a more rigorous performance test. The procedures on how to set up the MDL are explained in Appendix B of Part 136 (40 CFR, Vol 22) of the EPA regulations on environmental testing.

Though the LOD or MDL is often sufficient to characterize an assay, a further evaluation to check is the **limit of quantitation** (LOQ). In this determination data are collected similar to the LOD except the value is determined as $X_{Blk} + (10 \times SD_{Blk})$ instead of $(X_{Blk} + 3 \times SD_{Blk})$.

4.3.5 Quality Control Measures [1–3]

Quality control/assurance is desirable to evaluate analysis performance of a method or process. To explain how analytical data and control charts can be used in the food industry for **statistical process control**, this section will briefly describe quality control from the perspective of monitoring a specific process in making a food product (e.g., drying of a product, thereby affecting final moisture content). If the process is well defined and has known variability, the analytical data gathered can be evaluated over time. This provides set control points to determine if the process is performing as intended. Since all processes are susceptible to changes or drift, a decision can be made to adjust the process.

The best way to evaluate quality control is by control charting. This entails sequential plotting of the mean observations (e.g., moisture content) obtained from the analysis along with a target value. The standard deviation then is used to determine acceptable limits at the 95 % or 99 % confidence level, and at what point the data are outside the range of acceptable values. Often the acceptable limits are set as two standard deviations on either side of the mean, with the action limits set at three standard deviations. The charts and limits are used to determine if variation has occurred that is outside the normal variation for the process. If this occurs, there is a need to determine the root cause of the variation and put in place corrective and preventive actions to further improve the process.

Two common types of **control charts** used are the Shewhart and CuSum charts described by Ellison et al. [2]. The CuSum chart is more involved and is better at highlighting small changes in the mean value. The Shewhart chart (Fig. 4.3) entails plots of the target value mean and both upper and lower limits for each measurement. An upper and lower warning limit and action limit are determined and added to the plot. The warning limit shows that the

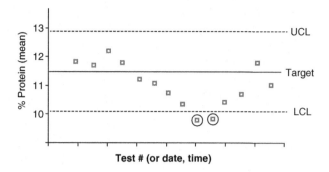

4.3 figure An example of a Shewhart control chart for protein analysis. The upper control limit (UCL) and lower control limit (LCL) are predetermined. Values which fall out of the limits (the circled values) indicate that the assay requires action and needs to be corrected or adjusted

measurements may be moving out of the desirable limit (e.g., upward trend). The action limit indicates that the measurements are past the acceptable limit so the process needs to be evaluated for the causes of the drift. Examples of the calculations and charts are provided in references [2, 3].

4.4 CURVE FITTING: REGRESSION ANALYSIS [2–4]

Curve fitting is a generic term used to describe the relationship and evaluation between two variables. Most scientific fields use curve fitting procedures to evaluate the relationship of two variables. Thus, curve fitting or curvilinear analysis of data is a vast area as evidenced by the volumes of material describing these procedures. In analytical determinations, we are usually concerned with only a small segment of curvilinear analysis, the standard curve, or regression line.

A **standard curve** or **calibration curve** is used to determine unknown concentrations based on a method that gives some type of measurable response that is proportional to a known amount of standard. It typically involves making a group of known standards in increasing concentration and then recording the particular measured analytical parameter (e.g., absorbance, area of a chromatography peak, etc.). What results when we graph the paired x and y values is a scatter plot of points that can be joined together to form a straight line relating concentration to observed response. Once we know how the observed values change with concentration, it is fairly easy to estimate the concentration of an unknown by interpolation from the standard curve.

As you read through the next three sections, keep in mind that not all correlations of observed values to

standard concentrations are linear (but most are). There are many examples of nonlinear curves, such as antibody binding, toxicity evaluations, and exponential growth and decay. Fortunately, with the vast array of computer software available today, it is relatively easy to analyze any group of data.

4.4.1 Linear Regression [2–4]

So how do we set up a standard curve once the data have been collected? First, a decision must be made regarding onto which axis to plot the paired sets of data. Traditionally, the concentration of the standards is represented on the x-axis, and the observed readings are on the y-axis. However, this protocol is used for reasons other than convention. The x-axis data are called the **independent variable** and are assumed to be essentially free of error, while the y-axis data (the **dependent variable**) may have error associated with them. This assumption may not be true because error could be incorporated as the standards are made. With modern-day instruments, the error can be very small. Although arguments can be made for making the y-axis data concentration, for all practical purposes, the end result is essentially the same. Unless there are some unusual data, the *concentration should be associated with the x-axis and the measured values with the y-axis*.

Figure 4.4 illustrates a typical standard curve used in the determination of caffeine in various foods. Caffeine is analyzed readily in foods by using HPLC coupled with an ultraviolet detector set at 272 nm. The area under the caffeine peak at 272 nm is directly proportional to the concentration. When an unknown sample (e.g., coffee) is run on the HPLC, a peak area is

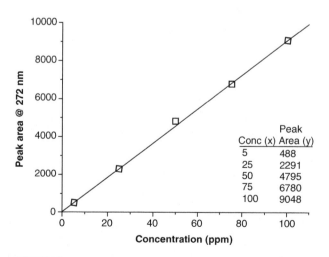

Peak	
Conc (x)	Area (y)
5	488
25	2291
50	4795
75	6780
100	9048

4.4 figure A typical standard curve plot showing the data points and the generated best fit line. The data used to plot the curve are presented on the graph

obtained that can be related back to the sample using the standard curve.

The plot in Fig. 4.4 shows all the data points and a straight line that appears to pass through most of the points. The line almost passes through the origin, which makes sense because zero concentration should produce no signal at 272 nm. However, the line is not perfectly straight (and never is) and does not quite pass through the origin.

To determine the caffeine concentration in a sample that gave an area of say 4,000, we could interpolate to the line and then draw a line down to the x-axis. Following a line to the x-axis (concentration), we can estimate the solution to be at about 42–43 ppm of caffeine.

We can mathematically determine the best fit of the line by using **linear regression**. Keep in mind the equation for a straight line, which is $y = ax + b$, where a is the slope and b is the y-intercept. To determine the slope and y-intercept, the regression equations shown below are used. We determine a and b and thus, for any value of y (measured), we can determine the concentration (x):

$$\text{slope } a = \frac{\Sigma(x_i - \overline{x})(y_i - \overline{y})}{\Sigma(x_i - \overline{x})^2} \quad (4.20)$$

$$y - \text{intercept } b = \overline{y} - a\overline{x} \quad (4.21)$$

where:

x_i and y_i = individual values

$\overline{x}$ and $\overline{y}$ = means of the individual values

Low-cost calculators and computer spreadsheet software can readily calculate regression equations, so no attempt is made to go through the mathematics in the formulas.

The formulas give what is known as the line of regression of y on x, which assumes that the error occurs in the y direction. The regression line represents the average relationship between all the data points and thus is a balanced line. These equations also assume that the straight-line fit does not have to go through the origin, which at first does not make much sense. However, there are often background interferences, so that even at zero concentration, a weak signal may be observed. In most situations, calculating the origin as going through zero will yield the same results.

Using the data from Fig. 4.4, calculate the concentration of caffeine in the unknown and compare with the graphing method. As you recall, the unknown had an area at 272 nm of 4,000. Linear regression analysis of the standard curve data gave the y-intercept (b) as 90.727 and the slope (a) as 89.994 ($r^2 = 0.9989$):

$$y = ax + b \quad (4.22)$$

or

$$x = \frac{y - b}{a} \quad (4.23)$$

$$x\,(\text{conc}) = \frac{4000 - 90.727}{89.994} = 43.4393 \text{ ppm caffeine} \quad (4.24)$$

The agreement is fairly close when comparing the calculated value to that estimated from the graph. Using high-quality graph paper with many lines could give us a line very close to the calculated one. However, as we will see in the next section, additional information can be obtained about the nature of the line when using computer software or calculators.

4.4.2 Correlation Coefficient

In observing any type of correlation, including linear ones, questions always surface concerning how to draw the line through the data points and how well the data fit to the straight line. The first thing that should be done with any group of data is to plot it to see if the points fit a straight line. By just looking at the plotted data, it is fairly easy to make a judgment on the linearity of the line. We also can pick out regions on the line where a linear relationship does not exist. The figures below illustrate differences in standard curves; Fig. 4.5a shows a good correlation of the data and Fig. 4.5b shows a poor correlation. In both cases, we can draw a straight line through the data points. Both curves yield the same straight line, but the precision is poorer for the latter.

There are other possibilities when working with standard curves. Figure 4.6a shows a good correlation between x and y, but in the negative direction, and Fig. 4.6b illustrates data that have no correlation at all.

The **correlation coefficient** defines how well the data fit to a straight line. For a standard curve, the ideal situation would be that all data points lie perfectly on a

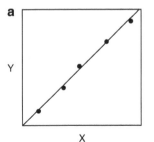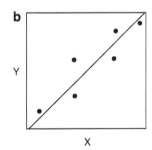

4.5 figure Examples of standard curves showing the relationship between the x and y variables when there are (**a**) a high amount of correlation and (**b**) a lower amount of correlation. Both lines have the same equation

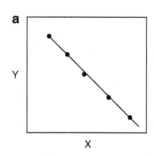 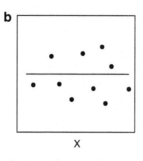

Examples of standard curves showing the relationship between the x and y variables when there are (**a**) a high amount of negative correlation and (**b**) no correlation between x and y values

straight line. However, this is never the case, because errors are introduced in making standards and measuring the physical values (observations).

The correlation coefficient and coefficient of determination are defined below. Essentially all spreadsheet and plotting software will calculate the values automatically:

correlation coefficient =

$$r = \frac{\Sigma(x_i - \bar{x})(y_i - \bar{y})}{\sqrt{\left[\Sigma(x_i - \bar{x})^2\right]\left[\Sigma(y_i - \bar{y})^2\right]}} \quad (4.25)$$

For our example of the caffeine standard curve from Fig. 4.3, $r = 0.99943$ (values are usually reported to at least four significant figures).

For standard curves, we want the value of r as close to +1.0000 or −1.000 as possible, because this value is a perfect correlation (perfect straight line). Generally, in analytical work, the r should be 0.9970 or better (this does not apply to biological studies).

The **coefficient of determination** (r^2) is used often because it gives a better perception of the straight line even though it does not indicate the direction of the correlation. The $r2$ for the example presented above is 0.99886, which represents the proportion of the variance of absorbance (y) that can be attributed to its linear regression on concentration (x). This means that about 0.114 % of the straight-line variation (1.0000–0.99886 = 0.00114 × 100 % = 0.114 %) does not vary with changes in x and y and thus is due to indeterminate variation. A small amount of variation is expected normally.

4.4.3 Errors in Regression Lines

While the correlation coefficient tells us something about the error or variation in linear curve fits, it does not always give the complete picture. Also, neither linear regression nor correlation coefficient will indicate that a particular set of data have a linear relationship. They only provide an estimate of the fit assuming the line is a linear one. As indicated before, plotting the data is critical when looking at how the data fit on the curve (actually, a line). One parameter that is used often is the **y-residuals**, which are simply the differences between the observed values and the calculated or computed values (from the regression line). Advanced computer graphics software can actually plot the residuals for each data point as a function of concentration. However, plotting the residuals is usually not necessary because data that do not fit on the line are usually quite obvious. If the residuals are large for the entire curve, then the entire method needs to be evaluated carefully. However, the presence of one point that is obviously off the line while the rest of the points fit very well probably indicates an improperly made standard.

One way to reduce the amount of error is to include more replicates of the data such as repeating the observations with a new set of standards. The replicate x and y values can be entered into the calculator or spreadsheet as separate points for the regression and coefficient determinations. Another, probably more desirable, option is to expand the concentrations at which the readings are taken. Collecting observations at more data points (concentrations) will produce a better standard curve. However, increasing the data beyond seven or eight points usually is not beneficial.

Plotting **confidence intervals**, or **bands** or **limits**, on the standard curve along with the regression line is another way to gain insight into the reliability of the standard curve. Confidence bands define the statistical uncertainty of the regression line at a chosen probability (such as 95 %) using the t-statistic and the calculated standard deviation of the fit. In some aspects, the confidence bands on the standard curve are similar to the confidence interval discussed in Sect. 4.3.1. However, in this case we are looking at a line rather than a confidence interval around a mean. Figure 4.7 shows the caffeine data from the standard curve presented before, except some of the numbers have been modified to enhance the confidence bands. The confidence bands (dashed lines) consist of both an upper limit and a lower limit that define the variation of the y-axis value. The upper and lower bands are narrowest at the center of the curve and get wider as the curve moves to the higher or lower standard concentrations.

Looking at Fig. 4.7 again, note that the confidence bands show what amount of variation we expect in a peak area at a particular concentration. At 60 ppm concentration, by going up from the x-axis to the bands and interpolating to the y-axis, we see that with our data the 95 % confidence interval of the observed peak area will be 4,000–6,000. In this case, the variation is large and would not be acceptable as a standard curve and is presented here only for illustration purposes.

Error bars also can be used to show the variation of y at each data point. Several types of error or variation statistics can be used such as standard error, standard deviation, or percentage of data (i.e., 5%). Any of these methods give a visual indication of experimental variation.

Even with good standard curve data, problems can arise if the standard curve is not used properly. One common mistake is to extrapolate beyond the data points used to construct the curve. Figure 4.8 illustrates some of the possible problems that might occur when extrapolation is used. As shown in Fig. 4.8, the curve or line may not be linear outside the area where the data were collected. This can occur in the region close to the origin or especially at the higher concentration level.

Usually a standard curve will go through the origin, but in some situations it may actually tail off as

zero concentration is approached. At the other end of the curve, at higher concentrations, it is fairly common for a plateau to be reached where the measured parameter does not change much with an increase in concentration. Care must be used at the upper limit of the curve to ensure that data for unknowns are not collected outside of the curve standards. Point Z on Fig. 4.7 should be evaluated carefully to determine if the point is an outlier or if the curve is actually tailing off. Collecting several sets of data at even higher concentrations should clarify this. Regardless, the unknowns should be measured only in the region of the curve that is linear.

4.5 REPORTING RESULTS

In dealing with experimental results, we are always confronted with reporting data in a way that indicates the sensitivity and precision of the assay. Ideally, we do not want to overstate or understate the sensitivity of the assay, and thus we strive to report a meaningful value, be it a mean, standard deviation, or some other number. The next three sections discuss how we can evaluate experimental values so as to be precise when reporting results.

4.5.1 Significant Figures

The term **significant figure** is used rather loosely to describe some judgment of the number of reportable digits in a result. Often, the judgment is not soundly based, and meaningful digits are lost or meaningless digits are retained. Exact rules are provided below to help determine the number of significant figures to report. However, it is important to keep some flexibility when working with significant figures.

Proper use of significant figures is meant to give an indication of the sensitivity and reliability of the analytical method. Thus, reported values should contain only significant figures. A value is made up of significant figures when it contains all digits known to be true and one last digit that is in doubt. For example, a value reported as 64.72 contains four significant figures, of which three digits are certain (64.7) and the last digit is uncertain. Thus, the 2 is somewhat uncertain and could be either 1 or 3. As a rule, numbers that are presented in a value represent the significant figures, regardless of the position of any decimal points. This also is true for values containing zeros, provided they are bounded on either side by a number. For example, 64.72, 6.472, 0.6472, and 6.407 all contain four significant figures. Note that the zero to the left of the decimal point is used only to indicate that there are no numbers above 1. We could have reported the value as .6472, but using the zero is better, since we know that a number was not inadvertently left off our value.

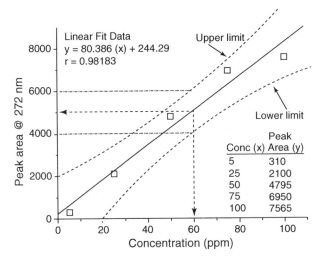

figure 4.7 A standard curve graph showing the confidence bands. The data used to plot the graph are presented on the graph as are the equation of the line and the correlation coefficient

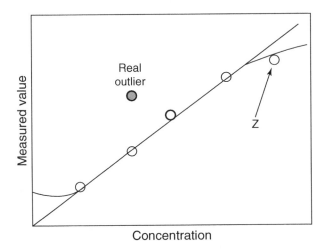

figure 4.8 A standard curve plot showing possible deviations in the curve in the upper and lower limits

Special considerations are necessary for zeros that may or may not be significant:

1. Zeros after a decimal point are always significant figures. For example, 64.720 and 64.700 both contain five significant figures.
2. Zeros before a decimal point with no other preceding digits are not significant. As indicated before, 0.6472 contains four significant figures.
3. Zeros after a decimal point are not significant if there are no digits before the decimal point. For example, 0.0072 has no digits before the decimal point; thus, this value contains two significant figures. In contrast, the value 1.0072 contains five significant figures.
4. Final zeros in a number are not significant unless indicated otherwise. Thus, the value 7,000 contains only one significant figure. However, adding a decimal point and another zero gives the number 7,000.0, which has five significant figures.

A good way to measure the significance of zeros, if the above rules become confusing, is to convert the number to the exponential form. If the zeros can be omitted, then they are not significant. For example, 7000 expressed in exponential form is 7×10^3 and contains one significant figure. With 7000.0, the zeros are retained and the number becomes 7.0000×10^3. If we were to convert 0.007 to exponential form, the value is 7×10^{-3} and only one significant figure is indicated. As a rule, determining significant figures in arithmetic operations is dictated by the value having the least number of significant figures. The easiest way to avoid any confusion is to perform all the calculations and then round off the final answer to the appropriate digits. For example, $36.54 \times 238 \times 1.1 = 9566.172$, and because 1.1 contains only two significant figures, the answer would be reported as 9600 (remember, the two zeros are not significant). This method works fine for most calculations, except when adding or subtracting numbers containing decimals. In those cases, the number of significant figures in the final value is determined by the numbers that follow the decimal point. Thus, when adding $7.45 + 8.725 = 16.175$, the sum is rounded to 16.18 because 7.45 has only two numbers after the decimal point. Likewise, $433.8 - 32.66$ gives 401.14, which rounds off to 401.1.

A word of caution is warranted when using the simple rule stated above, for there is a tendency to underestimate the significant figures in the final answer. For example, take the situation in which we determined the caffeine in an unknown solution to be 43.5 ppm (see Eq. 4.24). We had to dilute the sample 50-fold using a volumetric flask in order to fit the unknown within the range of our method. To calculate the caffeine in the original sample, we multiply our result by 50 or $43.5 \times 50 = 2,175\ \mu g/mL$ in the unknown. Based on our rule above, we then would round off the number to one significant figure (because 50 contains one significant figure) and report the value as 2,000. However, doing this actually underestimates the sensitivity of our procedure, because we ignore the accuracy of the volumetric flask used for the dilution. A Class-A volumetric flask has a tolerance of 0.05 mL; thus, a more reasonable way to express the dilution factor would be 50.0 instead of 50. We now have increased the significant figures in the answer by two, and the value becomes $2,180\ \mu/mL$.

As you can see, an awareness of significant figures and how they are adopted requires close inspection. The guidelines can be helpful, but they do not always work unless each individual value or number is closely inspected.

4.5.2 Outlier Data and Testing [2, 3]

Inevitably, during the course of working with experimental data, we will come across outlier values that do not match the others. Can you reject that value and thus not use it in calculating the final reported results?

The answer is "very rarely" and only after careful consideration. If you are routinely rejecting data to help make your assay look better, then you are misrepresenting the results and the precision of the assay. If the bad value resulted from an identifiable mistake in that particular test, then it is probably safe to drop the value. Again, caution is advised because you may be rejecting a value that is closer to the true value than some of the other values.

Consistently poor accuracy or precision indicates that an improper technique or incorrect reagent was used or that the test was not very good. It is best to make changes in the procedure or change methods rather than try to figure out ways to eliminate undesirable values.

There are several tests for rejecting outlier data. In addition, the use of more robust statistical estimators of the population mean can minimize effects of extreme outlier values [2]. The simplest test for rejecting outlier data is the Dixon Q test [7, 8], often called just the Q-test. The advantage is that this test can be easily calculated with a simple calculator and is useful for a small group of data. In this test, a **Q-value** is calculated as shown below and compared to values in a table. If the calculated value is larger than the table value, then the questionable measurement can be rejected at the 90% confidence level:

$$Q - \text{value} = x_2 - x_1 / W \qquad (4.26)$$

where:

x_1 = the questionable value

4.4 table

Q-values for the rejection of results

Number of observations	Q of rejection (90% level)
3	0.94
4	0.76
5	0.64
6	0.56
7	0.51
8	0.47
9	0.44
10	0.41

Adapted from Dean and Dixon [7]

x_2 = the next closest value to x_1

W = the total spread of all values, obtained by subtracting the lowest value from the highest value

Table 4.4 provides the rejection of Q-values for a 90% confidence level.

The example below shows how the test is used for the moisture level of uncooked hamburger for which four replicates were performed giving values of 64.53, 64.45, 64.78, and 55.31. The 55.31 value looks as if it is too low compared to the other results. Can that value be rejected? For our example, x_1 is the questionable value (55.31) and x_2 is the closest neighbor to x_1 (which is 64.45). The spread (W) is the high value minus the low measurement, which is 64.78–55.31:

$$Q - \text{value} = \frac{64.45 - 55.31}{64.78 - 55.31} = \frac{9.14}{9.47} = 0.97 \qquad (4.27)$$

From Table 4.4, we see that the calculated Q-value must be greater than 0.76 to reject the data. Thus, we make the decision to reject the 55.31% moisture value and do not use it in calculating the mean.

4.6 SUMMARY

This chapter focuses on statistical methods to measure data variability, precision, etc. and basic mathematical treatment that can be used in evaluating a group of data. For example, it should be almost second nature to determine a mean, standard deviation, and CV when evaluating replicate analyses of an individual sample. In evaluating linear standard curves, best line fits should always be determined along with the indicators of the degree of linearity (correlation coefficient or coefficient of determination). Fortunately, most computer spreadsheet and graphics software will readily perform the calculations for you. Guidelines are available to enable one to report analytical results in a way that tells something about the sensitivity and confidence of a particular test. A section is included which describes sensitivity and limit of detection as related to various analytical methods and regulatory agency policies. Additional information includes the proper use of significant figures, rules for rounding off numbers, and use of various tests to reject grossly aberrant individual values (outliers).

4.7 STUDY QUESTIONS

1. Method A to quantitate a particular food component was reported to be more specific and accurate than Method B, but Method A had lower precision. Explain what this means.
2. You are considering adopting a new analytical method in your lab to measure moisture content of cereal products. How would you determine the precision of the new method and compare it to the old method? Include any equations to be used for any needed calculations.
3. A sample known to contain 20 g/L glucose is analyzed by two methods. Ten determinations were made for each method and the following results were obtained:

Method A	Method B
Mean = 19.6	Mean = 20.2
Std. Dev. = 0.055	Std. Dev. = 0.134

(a) Precision and accuracy:

   (i) Which method is more precise? Why do you say this?

   (ii) Which method is more accurate? Why do you say this?

(b) In the equation to determine the standard deviation, $n-1$ was used rather than just n. Would the standard deviation have been smaller or larger for each of those values above if simply n had been used?

(c) You have determined that values obtained using Method B should not be accepted if outside the range of two standard deviations from the mean. What range of values will be acceptable?

(d) Do the data above tell you anything about the specificity of the method? Describe what "specificity" of the method means as you explain your answer.

4. Differentiate "standard deviation" from "coefficient of variation," "standard error of the mean," and "confidence interval."
5. Differentiate the terms "absolute error" versus "relative error." Which is more useful? Why?

6. For each of the errors described below in performing an analytical procedure, classify the error as random error, systematic error, or blunder, and describe a way to overcome the error:

   (a) Automatic pipettor consistently delivered 0.96 mL rather than 1.00 mL.
   (b) Substrate was not added to one tube in an enzyme assay.

7. Differentiate the terms "sensitivity" and "limit of detection."
8. The correlation coefficient for standard curve A is reported as 0.9970. The coefficient of determination for standard curve B is reported as 0.9950. In which case do the data better fit a straight line?

4.8   PRACTICE PROBLEMS

1. How many significant figures are in the following numbers: 0.0025, 4.50, 5.607?
2. What is the correct answer for the following calculation expressed in the proper amount of significant figures?

$$\frac{2.43 \times 0.01672}{1.83215} =$$

3. Given the following data on dry matter (88.62, 88.74, 89.20, 82.20), determine the mean, standard deviation, and CV. Is the precision for this set of data acceptable? Can you reject the value 82.20 since it seems to be different than the others? What is the 95% confidence level you would expect your values to fall within if the test were repeated? If the true value for dry matter is 89.40, what is the percent relative error?
4. Compare the two groups of standard curve data below for sodium determination by atomic emission spectroscopy. Draw the standard curves using graph paper or a computer software program. Which group of data provides a better standard curve? Note that the absorbance of the emitted radiation at 589 nm increases proportionally to sodium concentration. Calculate the amount of sodium in a sample with a value of 0.555 for emission at 589 nm. Use both standard curve groups and compare the results.

Sodium concentration (μg/mL)	Emission at 589 nm
Group A—sodium standard curve	
1.00	0.050
3.00	0.140
5.00	0.242
10.0	0.521
20.0	0.998

Sodium concentration (μg/mL)	Emission at 589 nm
Group B—sodium standard curve	
1.00	0.060
3.00	0.113
5.00	0.221
10.00	0.592
20.00	0.917

Answers

1. 2, 3, 4
2. 0.0222
3. Mean = 87.19, SD_{n-1} = 3.34:

$$CV = \frac{3.34}{87.18} \times 100\% = 3.83\%$$

Thus, the precision is acceptable because it is less than 5%:

$$Q-\text{calc value} = \frac{88.62 - 82.20}{89.20 - 82.20} = \frac{6.42}{7.00} = 0.917$$

Q_{calc} = 0.92; therefore, the value 82.20 can be rejected because it is more than 0.76 from Table 4.4, using 4 as number of observations:

$$CI\,(at\,95\%) = 87.19 \pm 3.18 \times \frac{3.34}{\sqrt{4}}$$
$$= 87.19 \pm 5.31$$

Relative error = $\%E_{rel}$ where mean is 87.19 and true value is 89.40:

$$\%E_{rel} = \frac{\bar{x} - T}{T} \times 100\% = \frac{87.19 - 89.40}{89.40} \times 100\%$$
$$= -2.47\%$$

4. Using linear regression, we get

   Group A: $y = 0.0504x - 0.0029$, $r^2 = 0.9990$.
   Group B: $y = 0.0473x + 0.0115$, $r^2 = 0.9708$.
   Group A r^2 is closer to 1.000 and is more linear and thus the better standard curve.

   Sodium in the sample using group A standard curve is

   $0.555 = 0.0504x - 0.0029$, $x = 11.1\mu g/mL$

   Sodium in the sample using group B standard curve is

   $0.555 = 0.0473x + 0.0115$, $x = 11.5\mu g/mL$

Acknowledgment The author wishes to thank Ryan Deeter for his contributions in preparation of the content on quality control measures.

REFERENCES

1. Garfield F M, Klestra E, and Hirsch J (2000) Quality assurance principles for analytical laboratories, AOAC International, Gaithersburg, MD. (This book covers mostly quality assurance issues, but Chap. 3 is a good chapter on sampling and on the basics of statistical applications to data.)

2. Ellison, SLR, Barwick, VJ, Farrant, TJD (2009) Practical Statistics for the Analytical Scientist: A Bench Guide, Chapter 5, 2nd ed. RSC Publishing, Cambridge, UK. (This an excellent text for beginner and advanced analytical chemists alike. It contains a fair amount of detail, sufficient for most analytical statistics, yet works through the material starting at a basic introductory level. The authors have an entire chapter, Chap. 5, devoted to the various test used to evaluating and rejecting data outliers.)

3. Miller JN, Miller JC (2010) Statistics and chemometrics for analytical chemistry, 6th edn. Pearson Prentice Hall, Upper Saddle River, NJ. (This is another excellent introductory text for beginner analytical chemists at the undergraduate and graduate level. It contains a fair amount of detail, sufficient for most analytical statistics, yet works through the material starting at a basic introductory level. Chapter 1 covers the types of experimental error. The assay evaluation parameters LOD, MDL, and LOQ are thoroughly discussed. The authors also cover outlier testing including the Q-test used for rejecting data.)

4. Skoog DA, West DM, Holler JF, Crouch SR (2014) Fundamentals of analytical chemistry, 9th edn. Cengage Brain, Independence, KY. (Part I, Chaps. 5–8, have thorough coverage of the statistics needed by an analytical chemist in an easy-to-read style.)

5. Shrivastava A, Gupta VB (2011) Methods for the determination of limit of detection and limit of quantitation of the analytical methods. Chron Young Sci 2(1):21–25

6. Code of Federal Regulations (2011) Environmental Protection Agency. 40 CFR, Vol. 22. Chapter I part 136 – Guidelines Establishing Test Procedures for the Analysis of Pollutants. Appendix B to Part 136-Definition and Procedure for the Determination of the Method Detection Limit-Revision 1.11

7. Dean RB, Dixon WJ (1951) Simplified statistics for small numbers of observations. Anal Chem 23(4): 636–638

8. Rorabacher DB (1991) Statistical treatment for rejection of deviant values: critical values of Dixon's "Q" parameter and related subrange ratios at the 95% confidence level. Anal Chem 63 (2):139–146

Sampling and Sample Preparation

Rubén O. Morawicki

Department of Food Science,
University of Arkansas,
Fayetteville, AR 72704, USA
e-mail: rmorawic@uark.edu

S. Nielsen (ed.), *Food Analysis*, Food Science Text Series,
DOI 10.1007/978-3-319-45776-5_5, © Springer International Publishing 2017

5.1 INTRODUCTION

Quality attributes in food products, raw materials, or ingredients are measurable characteristics that need monitoring to ensure specifications are met. Some quality attributes can be measured online by using specially designed sensors and results obtained in real time (e.g., color of vegetable oil in an oil extraction plant). However, in most cases quality attributes are measured on small portions of material that are taken periodically from continuous processes or on a certain number of small portions taken from a lot. The small portions taken for analysis are referred to as **samples**, and the entire lot or the entire production for a certain period of time, in the case of continuous processes, is called a **population**. The process of taking samples from a population is called **sampling**. If the procedure is done correctly, the measurable characteristics obtained for the samples become a very accurate estimation of the population.

By sampling only a fraction of the population, a quality estimate can be obtained accurately, quickly, and with less expense and personnel time than if the total population were measured. The reliability of sampling is dependent more on the sample size than on the population size [1]. The larger the sample size, the more reliable the sampling. However, sample size is limited by time, cost, sampling methods, and the logistics of sample handling, analysis, and data processing. Moreover, in the case of food products, analyzing a whole population would be practically impossible because of the destructive nature of most analytical methods. Paradoxically, estimated parameters using representative samples (discussed in Sects. 5.2 and 5.3) are normally more accurate than the same estimations done on the whole population (census).

The sample actually analyzed in the laboratory can be of any size or quantity [2]. This **laboratory sample** has generally undergone preparation such as homogenization or grinding to prepare it for analysis and is much smaller than the sample actually collected. Sampling and associated problems are discussed in Sects. 5.2, 5.3, and 5.4, while preparation of laboratory samples for testing is described in Sect. 5.5.

Sampling starts the series of steps needed to make decisions about data collected: sampling, sample preparation, laboratory analysis, data processing, and interpretation. In each step, there is a potential for error that would compromise the certainty, or reliability, of the final result. This final result depends on the cumulative errors at each stage that are usually described by the variance [3, 4]. **Variance** is an estimate of the uncertainty. The total variance of the whole testing procedure is equal to the sum of the variances associated with each step of the sampling procedure and represents the **precision** of the process. Precision is a measure of the reproducibility of the data. In contrast, **accuracy** is a measure of how close the data are to the true value. The most efficient way to improve accuracy is to improve the reliability of the step with the greatest variance, and that is frequently the initial sampling step. Attention is often given to the precision and accuracy of analytical methods, with less attention given to the validity of sampling and sample preparation. However, sampling can often be the greatest source of error in chemical analysis in general, but especially a problem in food analysis because of the food matrix [5–7] (Fig. 5.1). Sample homogenization and other aspects of sample preparation are additional sources of potential error, even prior to the actual analysis.

As you read each section of the chapter, consider application of the information to some specific examples of sampling needs in the food industry: sampling for nutrition labeling (see Study Question 7 in this chapter), pesticide analysis (see also Chap. 33, Sect. 33.3), mycotoxin analysis (see also Chap. 33, Sect. 33.4),

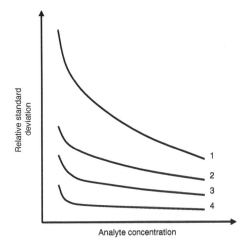

5.1 figure Typical relative standard deviation of error components for a nonhomogeneous food matrix. *1* Sampling; *2* homogenization; *3* sample preparation; *4* assay procedure (Used with permission from Lichon [6])

extraneous matter (see also Chap. 34), or rheological properties (see also Chap. 29). To consider sample collection and preparation for these and other applications subject to government regulations, you are referred also to the sample collection section of compliance procedures established by the Food and Drug Administration (FDA) and Food Safety and Inspection Service (FSIS) of the United States Department of Agriculture (USDA) (see Chap. 3, Sect. 3.2.2.1 for FDA and Sect. 3.3 for USDA). Additional information is available on how the FDA does sampling linked to the Food Safety Modernization Act [8] and how USDA collects samples for the National Nutrient Databank [9, 10].

It should be noted that sampling terminology and procedures used may vary between companies and between specific applications. However, the principles described in this chapter are intended to provide a basis for understanding, developing, and evaluating sampling plans and sample handling procedures for specific applications encountered.

5.2 SELECTION OF SAMPLING PROCEDURES

5.2.1 Definition and Purpose of Sampling Plan

The **International Union of Pure and Applied Chemistry** (IUPAC) defines a sampling plan as: "A predetermined procedure for the selection, withdrawal, preservation, transportation, and preparation of the portions to be removed from a lot as samples" [11]. A sampling plan should be a well-organized document that establishes the goals of the sampling plan, the factors to be measured, sampling point, sampling procedure, frequency, size, personnel, preservation of the samples, etc. The primary aim of sampling is to obtain a sample, subject to constraints of size that will satisfy the sampling plan specifications. A sampling plan should be selected on the basis of the sampling objective, the study population, the statistical unit, the sample selection criteria, and the analysis procedures. Depending on the purpose of the sampling plan, samples are taken at different points of the food production system, and the sampling plan may vary significantly for each point.

5.2.2 Factors Affecting the Choice of Sampling Plans

Each factor affecting the choice of sampling plans (Table 5.1) must be considered in the selection of a plan: (1) purpose of inspection, (2) nature of population, (3) nature of product, and (4) nature of test method. Once these are determined, a sampling plan that will provide the desired information can be developed.

5.1 table Factors that affect the choice of sampling plans

Factors to be considered	Questions
Purpose of the inspection	Is it to accept or reject the lot?
	Is it to measure the average quality of the lot?
	Is it to determine the variability of the product?
Nature of the population	Is the lot large but uniform?
	Does the lot consist of smaller, easily identifiable su bplots?
	What is the distribution of the units within the population?
Nature of the product	Is it homogeneous or heterogeneous?
	What is the unit size?
	How consistently have past populations met specifications?
	What is the cost of the material being sampled?
Nature of the test method	Is the test critical or minor?
	Will someone become sick or die if the population fails to pass the test?
	Is the test destructive or nondestructive?
	How much does the test cost to complete?

Adapted from Puri et al. [2]

5.2.2.1 Purpose of Inspection

Most sampling is done for a specific purpose and the purpose may dictate the nature of the sampling approach. The two primary objectives of sampling are often to estimate the average value of a characteristic and determine if the average value meets the specifications defined in the sampling plan. Sampling purposes vary widely among different food industries; however, the most important categories include the following:

1. Nutritional labeling
2. Detection of contaminants and foreign matter
3. Acceptance of raw materials, ingredients, or products (acceptance sampling)
4. Process control samples
5. Release of lots of finished product
6. Detection of adulterations
7. Microbiological safety
8. Authenticity of food ingredients.

5.2.2.2 *Nature of Population and Product*

One must clearly define the population and understand the nature of the product that is going to be sampled to select an appropriate sampling plan. Populations for sampling are often defined in terms of being homogeneous or heterogeneous and being either discrete or continuous. The population and product can also vary greatly in size.

The ideal population and product would be uniform throughout and identical at all locations. Such a population would be **homogeneous**. Sampling from such a population is simple, as a sample can be taken from any location, and the analytical data obtained will be representative of the whole. However, this occurs rarely, as even in an apparently uniform product, such as sugar syrup, suspended particles and sediments in a few places may render the population heterogeneous. In fact, most populations and products that are sampled are **heterogeneous**. Therefore, the location within a population where a sample is taken will affect the subsequent data obtained. However, sampling plans and sample preparation can make the sample representative of the population or take heterogeneity into account in some other way. If the sample is heterogeneous, some additional questions become important [6]: What is the nature of the variation? Should samples be pooled or replicated? Should different portions of the sample be analyzed separately? Should the surface of the sample be tested?

Sampling from **discrete** (or compartmentalized) populations is relatively easy, since the population is split into multiple separate subunits (e.g., cans in a palled of canned food, boxes of breakfast cereal in a truck, bottle of juice on a conveyer belt). The choice of the sampling plan is determined in part by the number and size of the individual subunits. Sampling is more difficult from a **continuous** population since different parts of the sample are not physically separated (e.g., potato chips on a conveyer belt, oranges in a semitruck) [6].

The population may vary in size from a production lot, a day's production, to the contents of a warehouse. Information obtained from a sample of a particular production lot in a warehouse must be used strictly to make inferences about that particular lot, but conclusions cannot be extended to other lots in the warehouse.

To use food analysis to solve problems in the food industry, it may be inadequate to focus just on the nature of the population and product to develop a sampling plan. For example, to collect samples to address a problem of variation of moisture content of packaged products coming off a production line, it would be important to examine each processing step. It would be necessary to understand where variation likely exists to then determine how to appropriately collect samples and data to analyze for variability.

One of the most challenging cases of accounting for the nature of the population, as it influences the sampling plan, is collecting samples to measure fungal toxins, named mycotoxins, in food systems. Mycotoxins are distributed broadly and randomly within a population and a normal distribution cannot be assumed [2]. Such distribution requires a combination of many randomly selected portions to obtain a reasonable estimate of mycotoxin levels. Methods of analysis that are extremely precise are not needed when determining mycotoxin levels, when sampling error is many times greater than analytical error [2]. In this case, sampling and good comminution and mixing prior to particle size reduction are more important than the chemical analysis itself. Additional information on sampling for mycotoxin analysis is provided in Chap. 33, Sect. 33.4.

5.2.2.3 *Nature of Test Method*

Procedures used to test samples collected vary in several characteristics that help determine the choice of sampling plan, e.g., cost, speed, accuracy, precision, and destructive vs. nondestructive. Low-cost, rapid, nondestructive tests that are accurate and precise make it more feasible to analyze many samples. However, limitations on any of these characteristics will make the nature of the test method a more important determinant of the sampling plan.

5.3 TYPES OF SAMPLING PLANS

5.3.1 Sampling by Attributes and Sampling by Variables

Sampling plans are designed for examination of either attributes or variables [4]. In **attribute sampling**, sampling is performed to decide on the acceptability of a population based on whether the sample possesses a certain characteristic or not. The result has a binary outcome of either conforming or nonconforming. Sampling plans by attributes are based on the hypergeometric, binomial, or Poisson statistical distributions. In the event of a binomial distribution (e.g., presence of *Clostridium botulinum*), the probability of a single occurrence of the event is directly proportional to the size of the sample, which should be at least ten times smaller than the population size. Computing binomial probabilities will allow the investigator to make inferences on the whole lot.

In **variable sampling**, sampling is performed to estimate quantitatively the amount of a substance (e.g., protein content, moisture content, etc.) or a characteristic (e.g., color) on a continuous scale. The estimate obtained from the sample is compared with an acceptable value (normally specified by the label, regulatory agencies, or the customer) and the deviation measured. This type of sampling usually produces data that have a *normal distribution* such as in the percent fill of a container and total solids of a food sample. In general, variable sampling requires smaller sample size than attribute sampling [1], and each characteristic should be sampled for separately when possible. However, when the FDA and the USDA's FSIS perform sampling for compliance of nutrition labeling, a composite of 12, and of at least six subsamples, respectively, is obtained and used for all nutrients to be analyzed.

5.3.2 Acceptance Sampling

Acceptance sampling is a procedure that serves a very specific role: to determine if a shipment of products or ingredients has enough quality to be accepted. **Acceptance sampling** can be performed by the food processor before receiving a lot of materials from a supplier or by a buyer who is evaluating the processor's output [6]. Acceptance sampling is a very broad topic that can be applied to any field; more specific literature can be consulted if needed.

Lot acceptance sampling plans that may be used for evaluation of attributes or variables, or a combination of both, fall into the following categories:

1. *Single sampling plans.* The decision of accepting or rejecting a lot for this type of plan is based just on one sample of items taken at random. These plans are usually denoted as (n, c) plans for a sample size n, where the lot is rejected if there are more than c defective samples [12]. If results are inconclusive, a second sample is taken, and the decision of accepting or rejecting is made based on the combined outcome of both samples.
2. *Double sampling plans.* Similar to single sampling plan, but two samples are taken (Fig. 5.2).
3. *Multiple sampling plans.* These are extensions of double sampling plans and use more than two samples to reach a conclusion.
4. *Sequential sampling plans.* Under this plan, which is an ultimate extension of multiple sampling, a sample is taken, and after analysis a decision of accepting, rejecting, or taking another sample is made. Therefore, the number of total samples to be taken depends exclusively on the sampling process. In this chart of Fig. 5.3, the cumulative observed number of defective samples is plotted against the number of samples taken. Two lines—the rejection and acceptance lines—are drawn, thus dividing the plot in three different regions: accept, reject, and continue sampling. An initial sample is taken and the results plotted in the graph. If the plotted point falls within the parallel lines, then a second sample is taken, and the process is repeated until the reject or accept zones are reached [12]. Details about the construction of this plot are beyond the scope of this book, and particulars can be found in more specialized literature.
5. *Skip lot sampling.* Only a fraction of the submitted lots is inspected with this type of plan. It is a

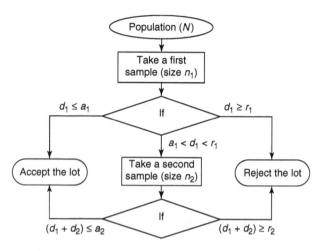

5.2 figure Example of a double sampling plan with two points where the decision of acceptance or rejection can be made (Adapted from NIST/SEMATECH [13]). N, population size; n_1 and n_2, sample size; a_1 and a_2, acceptance numbers; r_1 and r_2, rejection numbers; d_1 and d_2, number of non-conformities, *Subindices* 1 and 2 represent samples 1 and 2, respectively

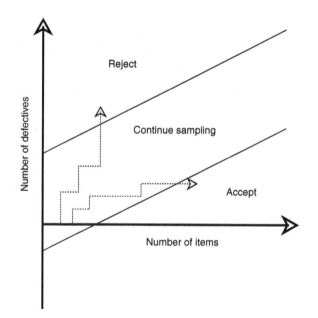

5.3 figure Sequential sampling plan (Adapted from NIST/SEMATECH [13])

money-saving sampling procedure, but it can be implemented only when there is enough proof that the quality of the lots is consistent.

5.3.3 Risks Associated with Acceptance Sampling

There are two types of risks associated with acceptance sampling: producer's and consumer's risks [13].

The **consumer's risk** describes the probability of accepting a poor-quality population. This should happen rarely (<5% of the lots), but the actual acceptable probability (β) of a consumer risk depends on the consequences associated with accepting an unacceptable lot. These may vary from major health hazards and subsequent fatalities to a lot being of slightly lower quality than standard lots. Obviously, the former demands a low or no probability of occurring whereas the latter would be allowed to occur more frequently. The **producer risk** is the probability of rejecting (α) an acceptable product. As with consumer's risk, the consequences of an error determine the acceptable probability of the risk. An acceptable probability of producer's risk is usually 5–10%. Further discussion of sampling plans can be found in the following section.

5.4 SAMPLING PROCEDURES

5.4.1 Introduction

The reliability of analytical data is compromised if sampling is not done properly. As shown in Table 5.1, the use of the data to be obtained is one major factor determining the sampling procedure. Details for the sampling of specific food products are described in the *Official Methods of Analysis* of AOAC International [14] and in the *Code of Federal Regulations* (CFR) [15]. Two such examples for specific foods follow.

5.4.2 Examples

The AOAC Method 925.08 [14] describes the method for sampling flour from sacks. The number of sacks to be sampled is determined by the square root of the number of sacks in the lot. The sacks to be sampled are chosen according to their exposure. The samples that are more frequently exposed are sampled more often than samples that are exposed less. Sampling is done by drawing a core from a corner at the top of the sack diagonally to the center. The sampling instrument is a cylindrical, polished trier with a pointed end. It is 13 mm in diameter with a slit at least one third of the circumference of the trier. A second sample is taken from the opposite corner in a similar manner. The cores are stored for analysis in a clean, dry, airtight container that has been opened near the lot to be sampled. The container should be sealed immediately after the sample is added. A separate container is used for each sack. Additional details regarding the container and the procedure also are described below.

Title 21 CFR specifies the sampling procedures required to ensure that specific foods conform to the standard of identity. In the case of canned fruits, 21 CFR 145.3 defines a sample unit as "container, a portion of the contents of the container, or a composite mixture of product from small containers that is sufficient for the testing of a single unit" [15]. Furthermore, a sampling plan is specified for containers of specific net weights. The container size is determined by the size of the lot. A specific number of containers must be filled for sampling of each lot size. The lot is rejected if the number of defective units exceeds the acceptable limit. For example, out of a lot containing 48001–84000 units, each weighing 1 kg or less, 48 samples should be selected. If six or more of these units fail to conform to the attribute of interest, the lot will be rejected. Based on statistical confidence intervals, this sampling plan will reject 95% of the defective lots examined, that is, 5% consumer risk [15].

5.4.3 Manual Versus Continuous Sampling

To obtain a **manual sample**, the person taking the sample must attempt to take a "random sample" to avoid human bias in the sampling method. Thus, the sample must be taken from a number of locations within the population to ensure it is representative of the whole population. For liquids in small containers, this can be done by shaking prior to sampling. When sampling from a large volume of liquid, such as that stored in silos, aeration ensures a homogeneous unit. Liquids may be sampled by pipetting, pumping, or dipping. However, when sampling grain from a rail car, mixing is impossible, and samples are obtained by probing from several points at random within the rail car. Such manual sampling of granular or powdered material is usually achieved with triers or probes that are inserted into the population at several locations. Errors may occur in sampling [10], as rounded particles may flow into the sampling compartments more easily than angular ones. Similarly, hygroscopic materials flow more readily into the sampling devices than does nonhygroscopic material. Horizontal core samples have been found to contain a larger proportion of smaller-sized particles than vertical ones [16].

Continuous sampling is performed mechanically. Figure 5.4 shows an automatic sampling device that is used to take liquid samples from a continuous production line. Continuous sampling should be less prone to human bias than manual sampling.

5.4.4 Statistical Considerations

5.4.4.1 *Probability Sampling*

Probability sampling plans prescribe the selection of a sample from a population based on chance. It provides a statistically sound basis for obtaining representative samples with elimination of human bias [2]. The probability of including any item in the sample is known and sampling error can be calculated. Several probability sampling methods are available to the researcher, and the most common ones are described in the next few paragraphs.

An automatic liquid sampling device that uses air under high pressure to collect multiple 1.5 mL samples. The control box (*left*) regulates the sampling frequency (Courtesy of Liquid Sampling Systems Inc., Cedar Rapids, IA)

Simple random sampling requires that the number of units in the population be known and each unit is assigned an identification number. Then using a random selection process, a certain number of identification numbers are selected according to the sample size. The sample size is determined according to the lot size and the potential impact of a consumer or vendor error. The random selection of the individuals units is done by using random number tables or computer-generated random numbers. Units selected randomly (sample) are analyzed and the results can be considered an unbiased estimate of the population.

Systematic sampling is used when a complete list of sample units is not available, but when samples are distributed evenly over time or space, such as on a production line. The first unit is selected at random (random start) and then units are taken every nth unit (sampling interval) after that.

Stratified sampling involves dividing the population (size N) into a certain number of mutually exclusive homogeneous subgroups (size $N1$, $N2$, $N3$, etc.) and then applying random or another sampling technique to each subgroup. Stratified sampling is used when subpopulations of similar characteristics can be observed within the whole population. An example of stratified sampling would be a company that produces tomato juice in different plants. If we need to study the residual activity of polygalacturonase in tomato juice, we can stratify on production plants and take samples on each plant.

Cluster sampling entails dividing the population into subgroups, or clusters, and then selecting randomly only a certain number of clusters for analysis. The main difference between cluster sampling and stratified sampling is that in the latter, samples are taken from every single subgroup, while in cluster sampling only some randomly clusters selected are sampled. The clusters selected for sampling may be either totally inspected or subsampled for analysis. This sampling method is more efficient and less expensive than simple random sampling, if populations can be divided into clusters. Going back to the tomato juice example, when using cluster sampling, we would consider all processing plants, but we would select randomly just a few for the purpose of the study.

Composite sampling is used to obtain samples from bagged products such as flour, seeds, and larger items in bulk. Small aliquots are taken from different bags, or containers, and combined in a simple sample (the composite sample) that is used for analysis. Composite sampling also can be used when a representative sample of a whole production day in a continuous process is needed. In this case, a systematic approach is used to take equal aliquots at different times, and then a representative sample is obtained by mixing the individual aliquots. A typical example of composite sampling is the sampling plan mandated by the FDA and FSIS for nutritional labeling. They require a composite of 12 samples with at least six subsamples taken and analyzed for compliance with nutrition labeling regulations [17].

5.4.4.2 Nonprobability Sampling

Randomization is always desired. However, it is not always feasible, or even practical, to take samples based on probability methods. Examples include in preliminary studies to generate hypothesis, in the estimation of the standard deviation so a more accurate sampling plan can be designed, or in cases for which the bulkiness of the material makes inaccessible the removal of samples. In these cases, nonprobability sampling plans may be more economical and practical than probability sampling. Moreover, in certain cases of adulteration such as rodent contamination, the objective of the sampling plan may be to highlight the adulteration rather than collect a representative sample of the population.

Nonprobability sampling can be done in many ways, but in each case the probability of including any specific portion of the population is not equal because the investigator selects the samples deliberately. Without the use of a methodology that gives every element of the population the same chance to be selected, it is not possible to estimate the sampling variability and possible bias.

Judgment sampling is solely at the discretion of the sampler and therefore is highly dependent on the person taking the sample. This method is used when it is the only practical way of obtaining the sample. It may result in a better estimate of the population than random sampling if sampling is done by an experienced individual, and the limitations of extrapolation from the results are understood [2]. **Convenience sampling** is performed when ease of sampling is the key factor. The first pallet in a lot or the sample that is most accessible is selected. This also is called "chunk sampling" or "grab sampling." Although this sampling requires little effort, the sample obtained will not be representative of the population and therefore is not recommended. **Restricted sampling** may be unavoidable when the entire population is not accessible. This is the case if sampling is from a loaded boxcar, but the sample will not be representative of the population. **Quota sampling** is the division of a lot into groups representing various categories, and samples are then taken from each group. This sampling method is less expensive than random sampling but also is less reliable.

5.4.4.3 *Mixed Sampling*
When the sampling plan is a mixture of two or more basic sampling methods that can be random or nonrandom, then the sampling plan is called **mixed sampling**.

5.4.4.4 *Estimating the Sample Size*
Sample size determination can be based on either **precision analysis** or **power analysis**. Precision and power analyses are done by controlling the confidence level (type I error) or the power (type II error). For the purpose of this section, the precision analysis will be used and will be based on the confidence interval approach and the assumptions that the population is normal.

The confidence interval for a sample mean is described by the following equation:

$$\bar{x} \pm z_{\alpha/2} \frac{\text{SD}}{\sqrt{n}} \qquad (5.1)$$

where:

$\bar{x}$ = sample mean
$z_{\alpha/2}$ = z-value corresponding to the level of confidence desired
SD = known, or estimated, standard deviation of the population
n = sample size

In Eq. 5.1, $z_{\alpha/2} \dfrac{\text{SD}}{\sqrt{n}}$ represents the maximum error (E) that is acceptable for a desired level of confidence. Therefore, we can set the equation $E = z_{\alpha/2} \dfrac{\text{SD}}{\sqrt{n}}$ and solve for n:

$$n = \left(\frac{z_{\alpha/2}\text{SD}}{E} \right)^2 \qquad (5.2)$$

The maximum error, E, in Eq. 5.2 can be expressed in terms of the accuracy (γ) as $E = \gamma \times \bar{x}$. Then Eq. 5.2 can be rearranged as

$$n = \left(\frac{z_{\alpha/2}\text{SD}}{\gamma \times \bar{x}} \right)^2 \qquad (5.3)$$

Now, we have an equation to calculate the sample size, but the equation is dependent on an unknown parameter: the standard deviation. To solve this problem, we can follow different approaches. One way is to take few samples using a nonstatistical plan and use the data to estimate the mean and standard deviation. A second approach is using data from the past or data from a similar study. A third method is to estimate the standard deviation as 1/6 of the range of data values [13]. A fourth method is to use typical coefficients of variation (defined as $100 \times [\text{standard deviation} / \text{population mean}]$) assuming we have an estimation of the population mean.

If the estimated sample is smaller than 30, then the Student's t distribution needs to be used instead of the normal distribution by replacing the $z_{\alpha/2}$ with the parameter t, with n-1 degrees of freedom. However, the use of the Student's t-test distribution comes with the additional cost or introducing another uncertainty into Eq. 5.3: the **degrees of freedom**. For the estimation of the t-score, we need to start somewhere by assuming the degrees of freedom, or assuming a t-score, and then calculating the number of samples, recalculating the t-score with n-1 degrees of freedom, and calculating the number of samples again. For a level of uncertainty of 95%, a conservative place to start would be assuming a t-score of 2.0 and then calculating the initial sample size. If we use a preliminary experiment to estimate the standard deviation, then we can use the sample size of the preliminary experiment minus one to calculate the t-score.

Example: We want to test the concentration of sodium in a lot of a ready-to-eat food product with a level of confidence of 95%. Some preliminary testing showed an average content of 1000 mg of sodium per tray with an estimated standard deviation of 500. Determine the sample size with an accuracy of 10%.

Data: Confidence level = 95% => $\alpha = 0.05$ => $z = 1.96$; $\gamma = 0.1$; $\bar{x} = 1000$; SD = 500

$$n = \left(\frac{z_{\alpha/2}\text{SD}}{\gamma \times \bar{x}} \right)^2 = \left(\frac{1.96 \times 500}{0.1 \times 1000} \right)^2 = 96 \text{ trays}$$

5.4.5 Problems in Sampling and Sample Storage

No matter how reliable our analytical technique is, our ability to make inferences on a population will always depend on the adequacy of sampling techniques. **Sampling bias**, due to nonstatistically viable convenience, may compromise reliability. Errors also may be introduced by not understanding the **population distribution** and subsequent selection of an inappropriate sampling plan.

Unreliable data also can be obtained by nonstatistical factors such as poor **sample storage** resulting in sample degradation. Samples should be stored in a container that protects the sample from moisture and other environmental factors that may affect the sample (e.g., heat, light, air). To protect against changes in moisture content, samples should be stored in an airtight container. Light-sensitive samples should be stored in containers made of opaque glass or the container wrapped in aluminum foil. Oxygen-sensitive samples should be stored under nitrogen or an inert gas. Refrigeration or freezing may be necessary to protect chemically unstable samples. However, freezing should be avoided when storing unstable emulsions. Preservatives (e.g., mercuric chloride, potassium dichromate, and chloroform) [1] can be used to stabilize certain food substances during storage. To help manage the issue of proper storage conditions for various samples, some laboratories use color-coded sample cups to ensure each sample is stored properly.

Mislabeling of samples causes mistaken sample identification. Samples should be clearly identified by markings on the sample container in a manner such that markings will not be removed or damaged during storage and transport. For example, plastic bags that are to be stored in ice water should be marked with water-insoluble ink.

If the sample is an official or **legal sample**, the container must be sealed to protect against tampering and the seal mark easily identified. Official samples also must include the date of sampling with the name and signature of the sampling agent. The chain of custody of such samples must be identified clearly.

5.5 PREPARATION OF SAMPLES

5.5.1 General Size Reduction Considerations

If the particle size or mass of the sample is too large for analysis, it must be reduced in bulk or particle size [1]. To obtain a smaller quantity for analysis, the sample can be spread on a clean surface and divided

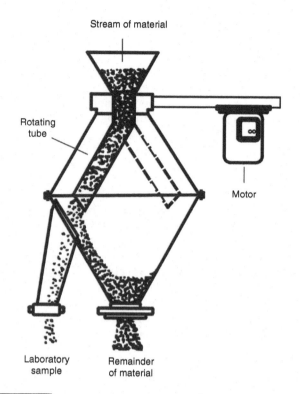

5.5 figure A rotating tube divider for reducing a large sample (ca. 880 kg) of dry, free flowing material to a laboratory size sample (ca. 0.2 kg) (Courtesy of Glen Mills, Inc., Clifton, NJ)

into quarters. The two opposite quarters are combined. If the mass is still too large for analysis, the process is repeated until an appropriate amount is obtained. This method can be modified for homogeneous liquids by pouring into four containers and can be automated (Fig. 5.5). The samples are thus homogenized to ensure negligible differences between each portion [2].

AOAC International [14] provides details on the preparation of specific food samples for analysis, which depends on the nature of the food and the analysis to be performed. For example, in the case of meat and meat products [14], it is specified in Method 983.18 that small samples should be avoided, as this results in significant moisture loss during preparation and subsequent handling. Ground meat samples should be stored in glass or similar containers, with air- and watertight lids. Fresh, dried, cured, and smoked meats are to be bone free and passed three times through a food chopper with plate openings no more than 3 mm wide. The sample then should be mixed thoroughly and analyzed immediately. If immediate analysis is not possible, samples should be chilled or dried for short-term and long-term storage, respectively.

A further example of size reduction is the preparation of solid sugar products for analysis as described in AOAC Method 920.175 [14]. The method prescribes that the sugar should be ground, if necessary, and mixed to uniformity. Raw sugars should be mixed thoroughly and rapidly with a spatula. Lumps are to be broken by a mortar and pestle or by crushing with a glass or iron rolling pin on a glass plate.

5.5.2 Grinding

5.5.2.1 Introduction

Grinding is important both for sample preparation prior to analysis and for food ingredient processing. Various mills are available for reducing particle size to achieve sample homogenization [17]. To homogenize moist samples, bowl cutters, meat mincers, tissue grinders, mortars and pestles, or blenders are used. However, mortars and pestles and mills are best for dry samples. Some foods are more easily ground after drying in a desiccator or vacuum oven. Grinding wet samples may cause significant losses of moisture and chemical changes. In contrast, grinding frozen samples reduces undesirable changes. The grinding process should not heat the sample, and therefore the grinder should not be overloaded because heat will be produced through friction. For especially heat-sensitive sample, grinders can be cooled with liquid nitrogen and then ground samples are stored at −80 °C. Contact of food with bare metal surfaces should be avoided if trace metal analysis is to be performed [18].

To break up moist tissues, a number of slicing devices are available: bowl cutters can be used for fleshy tubers and leafy vegetables, while meat mincers may be better suited for fruit, root, and meat [19]. Addition of sand as an abrasive can provide further subdivision of moist foods. Blenders are effective in grinding soft and flexible foods and suspensions. Rotating knives (25000 rpm) will disintegrate a sample in suspension. In colloidal mills, a dilute suspension is flowed under pressure through a gap between slightly serrated or smooth-surfaced blades until they are disintegrated by shear. Sonic and supersonic vibrations disperse foods in suspension and in aqueous and pressurized gas solution. The Mickle disintegrator sonically shakes suspensions with glass particles, and the sample is homogenized and centrifuged at the same time [19]. Alternatively, a low shear continuous tissue homogenizer is fast and handles large volumes of sample.

Another alternative is **cryogenic grinding** or **cryogrinding**. This method is ideal for biological samples and materials that are sensitive to oxygen or temperature. However, most materials are suitable for this technique. Cryogrinding can be performed manually with a mortar and pestle after freezing the sample with liquid nitrogen. The mortar and pestle have to be prechilled with liquid nitrogen before adding the material. Also, there are several brands of specialized grinding equipment with an integrated cooling system that perform the cryogenic freezing and grinding automatically.

5.5.2.2 Applications for Grinding Equipment

Mills differ according to their mode of action, being classified as a **burr, hammer, impeller, cyclone, impact, centrifugal, or roller mill** [19]. Methods for grinding dry materials range from a simple pestle and mortar to power-driven hammer mills. Hammer mills wear well and they reliably and effectively grind cereals and dry foods, while small samples can be finely ground by ball mills. A ball mill grinds by rotating the sample in a container that is half filled with ceramic balls. This impact grinding can take hours or days to complete. A chilled ball mill can be used to grind frozen foods without predrying and also reduces the likelihood of undesirable heat-initiated chemical reactions occurring during milling [19]. Alternatively, dry materials can be ground using an ultracentrifugal mill by beating, impacting, and shearing. The food is fed from an inlet to a grinding chamber and is reduced in size by rotors. When the desired particle size is obtained, the particles are delivered by centrifugal force into a collection pan [19]. Large quantities can be ground continuously with a cyclone mill.

5.5.2.3 Determination of Particle Size

Particle size is controlled in certain mills by adjusting the distance between burrs or blades or by screen mesh size/number. The **mesh number** is the number of square screen openings per linear inch of mesh. The final particles of dried foods should be 20 mesh for moisture, total protein, or mineral determinations. Particles of 40 mesh size are used for extraction assays such as lipid and carbohydrate estimation.

In addition to reducing particle size for analysis of samples, it also is important to reduce the particle size of many food ingredients for use in specific food products. For example, rolled oats for a grain-based snack bar may have a specified granulation size described as 15 % of the oats maximum passes through a #7 US standard sieve. A higher granulation (i.e., more smaller particles) would mean more fines and less whole oats in the finished bar. This would result in higher incidences of snack bar breakage.

There are a variety of methods for measuring particle size, each suited for different materials. The simplest way to measure particle sizes of dry materials of less than 50 µm in diameter is by passing the sample through a series of vertically stacked sieves with increasing mesh number. As the mesh number increases, the apertures between the mesh are smaller and only finer and finer particles pass through subsequent sieves (see Table 5.2). Sieve sizes have been specified for salt, sugar, wheat flour, cornmeal, semolina, and cocoa. The sieve method is inexpensive and fast,

| 5.2 table | US standard mesh with equivalents in inches and millimeters | | |

US standard mesh	Sieve opening	
	Inches	Millimeters
4	0.1870	4.760
6	0.1320	3.360
7	0.1110	2.830
8	0.0937	2.380
10	0.0787	2.000
12	0.0661	1.680
14	0.0555	1.410
16	0.0469	1.190
18	0.0394	1.000
20	0.0331	0.841
30	0.0232	0.595
40	0.0165	0.400
50	0.0117	0.297
60	0.0098	0.250
70	0.0083	0.210
80	0.0070	0.177
100	0.0059	0.149
120	0.0049	0.125
140	0.0041	0.105
170	0.0035	0.088
200	0.0029	0.074
230	0.0024	0.063
270	0.0021	0.053
325	0.0017	0.044
400	0.0015	0.037

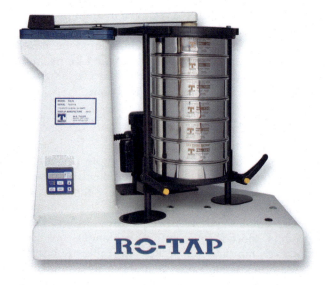

| 5.6 figure | W.S. Tyler® Ro-Tap® Sieve Shaker (Courtesy of W.S. Tyler®, Mentor, OH) |

but it is not suitable for emulsions or very fine powders [20]. A standard in the industry is the W.S. Tyler® **Ro-Tap® Sieve Shaker** (Fig. 5.6).

To obtain more accurate size data for smaller particles (<50 μm), characteristics that correlate to size are measured, and thus size is measured indirectly [21]. **Surface area** and **zeta potential** (electrical charge on a particle) are characteristics that are commonly used. Zeta potential is measured by an electroacoustic method whereby particles are oscillated in a high-frequency electrical field and generate a sound wave whose amplitude is proportional to the zeta potential. Optical and electron microscopes are routinely used to measure particle size. Optical microscopes are interfaced with video outputs and video-imaging software to estimate size and shape. The advantage of the visual approach is that a three-dimensional size and detailed particle structure can be observed.

A widespread technique to measure particle size distribution uses a principle known as **light scattering** or **laser diffraction**. When a coherent light source, such as laser beam, is pointed to a particle, four interactions of the beam with the particle can take place: reflection, refraction, absorption, and diffraction. **Reflection** is the portion of light that is rejected by the particle. **Refraction** is the light that goes through transparent or translucent materials and exits the particle

with an unchanged wavelength, at a different angle, and generally with a lower intensity. Some of the light can be **absorbed** by the particle and re-irradiated at a different frequency, which can display as fluorescence or heat. The last interaction, **diffraction**, is the result of light interacting with the edges of the particle. This interaction makes the light spread out, or scatter, and produce wave fronts around the particle that act as secondary spherical waves. This phenomenon is similar to Young's double-slit experiment, which readers can find in a simple Google search.

The secondary waves generated at the edge of the particles interact with each other, in some areas constructively and others destructively, thus producing interference patterns that correlate with the particle size. In general, larger particles scatter light at smaller angles and with a higher intensity. Smaller particles, on the other side, scatter light at wider angles and with less intensity. These scatter patterns can be correlated with particle size by using mathematical algorithms.

Besides size, shape and optical properties affect the angle and intensity of the scattered light. When particles are capable of transmitting light, the refractive index plays a role in particle size determination. Therefore, light scattering instruments take advantage of the refractive index to improve accuracy, especially for small particles.

A typical laser diffraction particle analyzer has a sample port, a medium for the sample transport, a flow cell, a laser beam, a detector, electronics, and software. The sample is introduced into the port, which contains some dispersion system to avoid particle agglomeration. The sample is then carried by a liquid, for wet samples, or compressed air, for dry samples, to a flow cell. As the sample goes through

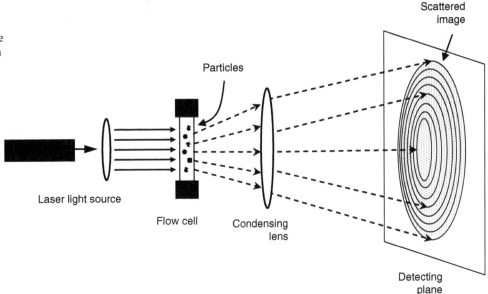

5.7 figure Basics of an instrument to measure particle size distribution by light scattering (laser diffraction) (Adapted from Shimadzu Corporation [22])

the flow cell, a laser beam irradiates the sample. This results in diffraction patterns that after passing through a condensing lens are projected on a detector where the diffraction angles and intensities are measured. This information is then transformed into a particle size distribution by the instrument electronics and a computer software algorithm (Fig. 5.7). Results are displayed as frequency distributions with particle size on the x-axis and frequency on the y-axis.

Determination of particle size by light scattering has many advantages, such as:

- Wide dynamic range—from the nanometer to the several millimeters range
- Suitable for wet and dry materials—such as solid in liquid suspensions, dry powders, liquid-liquid emulsions, and pastes
- Fast measurement—in the order of seconds
- No calibration needed
- Easy to use and interpret results
- Proficient at detecting mixes of large and small particles
- Technique covered by ISO standard 13320:2009

The most important disadvantages are the following:

- Assumption that particles are spherical, which introduced error in particles that deviates a lot from sphericity.
- Particles agglomeration.
- Results are expressed in volume basis, which gives more weight to large particles.
- Results cannot be compared with other size determination methods.

For particles in the nanoscale range, a widely used technique for particle analysis is **dynamic light scattering**. These instruments determine particle size and even the molecular weight of large molecules in solution. This is achieved by measurement of frequency shifts of light scattered by particles due to Brownian motion.

Understanding of the principles of the instrument used to obtain size data is vital to appreciate the limitations of each method. For example, data obtained from the same sample using sieves and light scattering will differ [21]. Sieves separate particles using square holes, and therefore they distinguish size in the smallest dimension, independent of shape. However, light scattering techniques assume that the particle is spherical, and data are derived from the average of all dimensions. Particle size measurement is useful to maintain sample quality, but care must be taken in choosing an appropriate method and interpreting the data.

5.5.3 Enzymatic Inactivation

Food materials often contain enzymes that may degrade the food components being analyzed. Enzyme activity therefore must be eliminated or controlled using methods that depend on the nature of the food. Heat denaturation to inactivate enzymes and freezer storage (−20 to −30 C) for limiting enzyme activity are common methods. However, some enzymes are more effectively controlled by changing the pH or by salting out [19]. Oxidative enzymes may be controlled by adding reducing agents.

5.5.4 Lipid Oxidation Protection

Lipids present particular problems in sample preparation. High-fat foods are difficult to grind and may need to be ground while frozen. Unsaturated lipids are sensitive to oxidative degradation and should be protected by storing under nitrogen or vacuum. Antioxidants may stabilize lipids and may be used if they do not interfere with the analysis. Light-initiated photooxidation of unsaturated lipids can be avoided by controlling storage conditions. In practice, lipids are more stable when frozen in intact tissues rather than as extracts [19]. Therefore, ideally, unsaturated lipids should be extracted just prior to analysis. Low-temperature storage is generally recommended to protect most foods.

5.5.5 Microbial Growth and Contamination

Microorganisms are present in almost all foods and can alter the sample composition. Likewise, microorganisms are present on all but sterilized surfaces, so sample cross-contamination can occur if samples are not handled carefully. The former is always a problem and the latter is particularly important in samples for microbiological examination. Freezing, drying, and chemical preservatives are effective controls and often a combination of these is used. The preservation methods used are determined by the probability of contamination, the storage conditions, storage time, and the analysis to be performed [19].

5.6 SUMMARY

Food quality is monitored at various processing stages but 100 % inspection is rarely possible or even desirable. To ensure a representative sample of the population is obtained for analysis, sampling and sample reduction methods must be developed and implemented. The selection of the sampling procedure is determined by the purpose of the inspection, the nature of the population and product, and the test method. Increasing the sample size will generally increase the reliability of the analytical results, and using t-test techniques will optimize the sample size necessary to obtain reliable data. Multiple sampling techniques also can be used to minimize the number of samples to be analyzed. Sampling is a vital process, as it is often the most variable step in the entire analytical procedure.

Sampling may be for attributes or variables. Attributes are monitored for their presence or absence, whereas variables are quantified on a continuous scale. Sampling plans are developed for either attributes or variables and may be single, double, or multiple. Multiple sampling plans reduce costs by rejecting low-quality lots or accepting high-quality lots quickly, while intermediate-quality lots require further sampling. There is no sampling plan that is risk-free. The consumer risk is the probability of accepting a poor-quality product, while the vendor risk is the probability of rejecting an acceptable product. An acceptable probability of risk depends on the seriousness of a negative consequence.

Sampling plans are determined by whether the population is homogeneous or heterogeneous. Although sampling from a homogeneous population is simple, it rarely is found in practical industrial situations. Sampling from heterogeneous populations is most common, and suitable sampling plans must be used to obtain a representative sample. Sampling methods may be manual or continuous. Ideally, the sampling method should be statistically sound. However, nonprobability sampling is sometimes unavoidable, even though there is not an equal probability that each member of the population will be selected due to the bias of the person sampling. Probability sampling is preferred because it ensures random sampling and is a statistically sound method that allows calculation of sampling error and the probability of any item of the population being included in the sample.

Each sample must be clearly marked for identification and preserved during storage until completion of the analysis. Official and legal samples must be sealed and a chain of custody maintained and identified. Often, only a portion of the sample is used for analysis and sample size reduction must ensure that the portion analyzed is representative of both the sample and population. Sample preparation and storage should account for factors that may cause sample changes. Samples can be preserved by limiting enzyme activity, preventing lipid oxidation, and inhibiting microbial growth/contamination.

5.7 STUDY QUESTIONS

1. As part of your job as supervisor in a quality assurance laboratory, you need to give a new employee instruction regarding choosing a sampling plan. Which general factors would you discuss with the new employee? Distinguish between sampling for attributes vs. sampling for variables. Differentiate the three basic sampling plans and the risks associated with selecting a plan.
2. Your supervisor wants you to develop and implement a multiple sampling plan. What would you take into account to define the acceptance and rejection lines? Why?
3. Distinguish probability sampling from nonprobability sampling. Which is preferable and why?
4. (a) Identify a piece of equipment that would be useful in *collecting* a representative sample for

analysis. Describe precautions to be taken to ensure that a representative sample is taken and a suitable food product that could be sampled with this device. (b) Identify a piece of equipment that would be useful for *preparing* a sample for analysis. What precautions should be taken to ensure that the sample composition is not changed during preparation?

5. For each of the problems identified below that can be associated with collection and preparation of samples for analysis, state one solution for how the problem can be overcome:

   (a) Sample bias
   (b) Change in composition during storage of sample prior to analysis
   (c) Metal contamination in grinding
   (d) Microbial growth during storage of product prior to analysis

6. The instructions you are following for cereal protein analysis specify grinding a cereal sample to 10 mesh before you remove protein by a series of solvent extractions:

   (a) What does 10 mesh mean?
   (b) Would you question the use of a 10-mesh screen for this analysis? Provide reasons for your answer.

7. You are to collect and prepare a sample of cereal produced by your company for the analyses required to create a standard nutritional label. Your product is considered "low fat" and "high fiber" (see information on nutrient claims and FDA compliance procedures in Chap. 3). What kind of sampling plan will you use? Will you do attribute or variable sampling? What are the risks associated with sampling in your specific case? Would you use probability or nonprobability sampling, and which specific type would you choose? What specific problems would you anticipate in sample collection, storage, and preparation? How would you avoid or minimize each of these problems?

REFERENCES

1. Horwitz W (1988) Sampling and preparation of samples for chemical examination. J Assoc Off Anal Chem 71: 241–245
2. Puri SC, Ennis D, Mullen K (1979) Statistical quality control for food and agricultural scientists. G.K. Hall and Co., Boston, MA
3. Harris DC (2015) Quantitative chemical analysis, 9th edn. W.H. Freeman and Co., New York
4. Miller JC (1988) Basic statistical methods for analytical chemistry. I. Statistics of repeated measurements. A review. Analyst 113: 1351–1355
5. Lichon, MJ (2000) Sample preparation for food analysis, general. In: Meyers RA (ed) Encyclopedia of analytical chemistry: applications, theory and instrumentation. John Wiley, New York
6. Lichon MJ (2004) Sample preparation, ch. 44. In: Nollet LML (ed) Handbook of food analysis, 2nd edn. Vol 3. Marcel Dekker, New York, pp. 1741–1755
7. Mälkki Y (1986) Collaborative testing of methods for food analysis. J Assoc Off Anal Chem 69(3):403–404
8. FDA (2016) Sampling to protect the food supply (last updated 5/5/2016) http://www.fda/gov/Food/ComplianceEnforcement/Sampling/ucm20041972.htm
9. Pehrsson PR, Haytowitz DB, Holden JM, Perry, CR, Beckler DG (2000) USDA's national food and nutrient analysis program: food sampling. J Food Comp Anal. 13:379–389
10. Pehrsson P, Perry C, Daniel M (2013) ARS, USDA updates food sampling strategies to keep pace with demographic shifts. Procedia Food Sci 2:52–59
11. IUPAC (1997) Compendium of chemical terminology, 2nd edn. (the "Gold Book"). Compiled by McNaught AD Wilkinson A. Blackwell Scientific, Oxford. XML online corrected version: http://goldbook.iupac.org (2006-) created by Nic M, Jirat J, Kosata B; updates compiled by Jenkins A. ISBN 0-9678550-9-8. doi:10.1351/goldbook
12. Weiers RM (2007) Introduction to business statistics, 6th edn. South-Western College, Cincinnati, OH
13. NIST/SEMATECH (2013) e-Handbook of statistical methods, chapter 6: process or product monitoring and control. http://www.itl.nist.gov/div898/handbook/
14. AOAC International (2016) Official methods of analysis, 20th edn., (On-line). AOAC International, Rockville, MD
15. Anonymous (2016) Code of federal regulations. Title 21. US Government Printing Office, Washington, DC
16. Baker WL, Gehrke CW, Krause GF (1967) Mechanism of sampler bias. J Assoc Off Anal Chem 50: 407–413
17. Anonymous (2016) Code of federal regulations. 21 CFR 101.9 (g), 9 CFR 317.309 (h), 9 CFR 381.409 (h). US Government Printing Office, Washington, DC
18. Pomeranz Y, Meloan CE (1994) Food analysis: theory and practice, 3rd edn. Chapman & Hall, New York
19. Cubadda F, Baldini M, Carcea M, Pasqui LA, Raggi A, Stacchini P (2001) Influence of laboratory homogenization procedures on trace element content of food samples: an ICP-MS study on soft and durum wheat. Food Addit Contam 18: 778–787
20. Kenkel JV (2003) Analytical chemistry for technicians, 3rd edn. CRC, Boca Raton, FL
21. Jordan JR (1999) Particle size analysis. Inside Lab Manage 3(7): 25–28
22. Shimadzu Corporation (2016) Particle Size Distribution Calculation Method. Lecture on Partical Analysis – Hands-on Course http://www.shimadzu.com/an/powder/support/practice/p01/lesson22.html [Accessed on June 29, 2016]

2
part

Spectroscopy and
Mass Spectrometry

Basic Principles of Spectroscopy

Michael H. Penner

Department of Food Science and Technology,
Oregon State University,
Corvallis, OR 97331-6602, USA
e-mail: mike.penner@oregonstate.edu

S. Nielsen (ed.), *Food Analysis*, Food Science Text Series,
DOI 10.1007/978-3-319-45776-5_6, © Springer International Publishing 2017

6.1   INTRODUCTION

Spectroscopy deals with the production, measurement, and interpretation of spectra arising from the *interaction of electromagnetic radiation with matter*. There are many different spectroscopic methods available for solving a wide range of analytical problems. The methods differ with respect to the species to be analyzed (such as molecular or atomic spectroscopy), the type of radiation-matter interaction to be monitored (such as absorption, emission, or diffraction), and the region of the electromagnetic spectrum used in the analysis. Spectroscopic methods are very informative and widely used for both quantitative and qualitative analyses. Spectroscopic methods based on the absorption or emission of radiation in the *ultraviolet* (UV), *visible* (Vis), *infrared* (IR), and radio (*nuclear magnetic resonance, NMR*) frequency ranges are most commonly encountered in traditional food analysis laboratories. Each of these methods is distinct in that it monitors different types of molecular or atomic transitions. The basis of these transitions is explained in the sections below.

6.2   LIGHT

6.2.1   Properties

Light may be thought of as particles of energy that move through space with wavelike properties. This image of light suggests that the energy associated with a ray of light is not distributed continuously through space along the wave's associated electric and magnetic fields but rather that it is concentrated in discrete packets. Light is therefore said to have a dual nature: *particulate* and *wavelike*. Phenomena associated with light propagation, such as interference, diffraction, and refraction, are most easily explained using the wave theory of electromagnetic radiation. However, the interaction of light with matter, which is the basis of absorption and emission spectroscopy, may be best understood in terms of the particulate nature of light. Light is not unique in possessing both wavelike and particulate properties. For example, fundamental particles of matter, such as electrons, protons, and neutrons, are known to exhibit wavelike behavior.

The wave properties of electromagnetic radiation are described in terms of the wave's frequency, wavelength, and amplitude. A graphical representation of a plane-polarized electromagnetic wave is given in Fig. 6.1. The wave is plane polarized in that the oscillating electric and magnetic fields making up

the wave are each limited to a single plane. The *frequency* (ν, the lower case Greek letter *nu*) of a wave is defined as the number of oscillations the wave will make at a given point per second. This is the reciprocal of the *period* (*p*) of a wave, which is the time in seconds required for successive maxima of the wave to pass a fixed point. The *wavelength* (λ) represents the distance between successive maxima on any given wave. The units used in reporting wavelengths will depend on the region of electromagnetic radiation used in the analysis. Spectroscopic data sometimes are reported with respect to *wave numbers* ($\bar{v}$), which are reciprocal wavelengths in units of cm^{-1}. Wave numbers are encountered most often in IR spectroscopy. The *velocity of propagation* (v_i) of an electromagnetic wave, in units of distance per second, in any given medium "i" can be calculated by taking the product of the frequency of the wave, in cycles per second, and its wavelength in that particular medium:

$$v_i = \nu \lambda_i \qquad (6.1)$$

where:

v_i = velocity of propagation in medium *i*
ν = frequency (of associated wave)
λ_i = wavelength in medium *i*

The frequency of an electromagnetic wave is determined by the source of the radiation, and it remains constant as the wave traverses different media. However, the velocity of propagation of a wave will vary slightly depending on the medium through which the light is propagated. The wavelength of the radiation will change in proportion to changes in wave velocity as defined by Eq. 6.1. The *amplitude of the wave* (*A*) represents the magnitude of the electric vector at the wave maxima. The *radiant power* (*P*) and *radiant intensity* (*I*) of

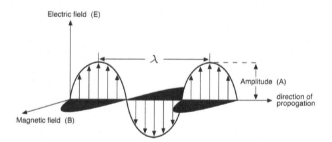

6.1
figure   Representation of plane-polarized electromagnetic radiation propagating along the *x*-axis. The electric and magnetic fields are in phase, perpendicular to each other and to the direction of propagation

a beam of radiation are proportional to the square of the amplitude of the associated waves making up that radiation. Figure 6.1 indicates that electromagnetic waves are composed of *oscillating magnetic* and *electric fields*, the two of which are mutually perpendicular, in phase with each other, and perpendicular to the direction of wave propagation. As drawn, the waves represent changes in the respective field strengths with time at a fixed location or changes in the respective field strengths over distance at a fixed time. The electrical and magnetic components of the waves are represented as a series of vectors whose lengths are proportional to the magnitude of the respective field. It is the oscillating electric field that is of most significance to spectroscopic phenomena such as absorption, transmission, and refraction. However, a purely electric field, without its associated magnetic field, is impossible.

6.2.2 Terminology

The propagation of electromagnetic waves is often described in terms of wave fronts or trains of waves (Fig. 6.2). A *wave front* represents the locus of a set of points, all of which are in phase. For a point source of light, a concentric ring that passes through the maxima of adjacent light rays will represent a wave front. The entire ring need not be drawn in all cases, such that wave fronts may represent planes of light in cases where the observation is sufficiently removed from the point source that the curved surface appears planar. Wave fronts are most typically drawn by connecting maxima, minima, or both for adjacent rays. If maxima are used for depicting wave fronts, then each of the wave fronts will be separated by one wavelength. A *train of waves*, or *wave train*, refers to a series of wave fronts, all of which are in phase, that is, each individual wave will have a maximum amplitude at the same location in space. A wave train also may be represented by a series of light rays. Rays of light are used

generally with reference to the corpuscular nature of light, representing the path of photons. A wave train would indicate that a series of photons, all in phase, followed the same path.

6.2.3 Interference

Interference is the term used to describe the observation that when two or more wave trains cross one another, they result in an instantaneous wave, at the point of intersection, whose amplitude is the algebraic sum of the amplitudes of the individual waves at the point of intersection. The law describing this wave behavior is known as the *principle of superposition*. Superposition of sinusoidal waves is illustrated in Fig. 6.3. Note that in all cases, the effective amplitude of the perceived wave at the point in question is the combined effect of each of the waves that crosses that point at any given instant. In spectroscopy, the amplitude of most general interest is that corresponding to the magnitude of the resulting electric field intensity. *Maximum constructive interference* of two waves occurs when the waves are completely in phase (i.e., the maxima of one wave align with the maxima of the other wave), while *maximum destructive interference* occurs when waves are 180° out of phase (the maxima of one wave align with the minima of the other wave). This concept of interference is fundamental to the interpretation of diffraction data, which represents a specialized segment of qualitative spectroscopy. Interference phenomena also are widely used in the design of spectroscopic instruments that require the dispersion or selection of radiation, such as those instruments employing grating monochromators or interference filters, as described in Chap. 7.

Interference phenomena are best rationalized by considering the wavelike nature of light. However,

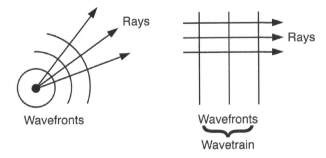

6.2 figure Wave fronts, wave trains, and rays (From Hugh D. Young, *University Physics* (8th. Ed.) (p. 947), © 1992 by Addison-Wesley, Reading, MA. Courtesy of the publisher)

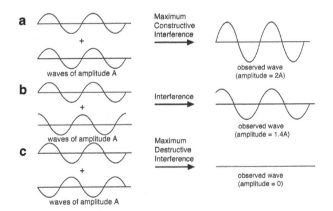

6.3 figure Interference of identical waves that are (**a**) in phase, (**b**) 90° out of phase, and (**c**) 180° out of phase

6.1 table Properties of light

Symbols/terms	Relationship	Frequently used units
λ = wavelength	$\lambda_i \nu = v_i$ ($v_i = c$ in a vacuum)	nm (nanometers, 10^{-9} m) Å (Ångstrom units, 10^{-10} m) μm (microns, 10^{-6} m) mμ (millimicrons, 10^{-9} m)
ν = frequency c = speed of light		Hz (hertz, 1 Hz = 1 oscillation per second) 2.9979×10^8 m s^{-1} in vacuum
$\bar{v}$ = wave number	$=1/\lambda$	cm^{-1} kK (kilokayser, 1 kK = 1,000 cm^{-1})
p = period E = energy	$p = 1/\nu$ $E = h\nu$ $= hc/\lambda$ $= hc\bar{v}$	s J (1 J = 1 kg m^2 s^{-2}) cal (calorie, 1 cal = 4.184 J) erg (1 erg = 10^{-7} J) eV (1 eV = 1.6022×10^{-19} J)
h = Planck's constant P = radiant power	Amount of energy striking a given unit area per unit time	6.6262×10^{-34} J s (Joules) (m^2)$^{-1}$(s)$^{-1}$

phenomena such as the absorption and emission of radiation are more easily understood by considering the particulate nature of light. The particles of energy that move through space with wavelike properties are called *photons*. The energy of a photon can be defined in terms of the frequency of the wave with which it is associated (Eq. 6.2):

$$E = h\nu \qquad (6.2)$$

where:

E = energy of a photon
h = Planck's constant
ν = frequency (of associated wave)

This relationship indicates that the photons making up *monochromatic light*, which is electromagnetic radiation composed of waves having a single frequency and wavelength, are all of equivalent energy. Furthermore, just as the frequency of a wave is a constant determined by the radiation source, the energy of associated photons also will be unchanging. The brightness of a beam of monochromatic light, when expressed in terms of the particulate nature of light, will be the product of the photon flux and the energy per photon. The *photon flux* refers to the number of photons flowing across a unit area perpendicular to the beam per unit time. It follows that to change the brightness of a beam of monochromatic light will require a change in the photon flux. In spectroscopy, the term *brightness* is generally not used, but rather one refers to the *radiant power* (P) or the *radiant intensity* (I) of a beam of light. Radiant power and radiant intensity often are used synonymously when referring to the amount of radiant energy striking a given area

per unit time. In terms of International System of Units (SI) (time, seconds; area, meters; energy, joules), radiant power equals the number of joules of radiant energy impinging on a 1 m^2 area of detector per second. The basic interrelationships of light-related properties and a general scheme of the electromagnetic spectrum are presented in Table 6.1 and Fig. 6.4, respectively.

6.3 ENERGY STATES OF MATTER

6.3.1 Quantum Nature of Matter

The energy content of matter is quantized. Consequently, the potential or internal energy content of an atom or molecule does not vary in a continuous manner but rather in a series of discrete steps. Atoms and molecules, under normal conditions, exist predominantly in the *ground state*, which is the state of lowest energy. Ground-state atoms and molecules can gain energy, in which case they will be elevated to one of their higher energy states, referred to as *excited states*. The quantum nature of atoms and molecules puts limitations on the energy levels that are available to these species. Consequently, there will be specific "allowed" *internal energy levels* for each atomic or molecular species. Internal energy levels not corresponding to an allowed value for that particular species are unattainable. The set of available energy levels for any given atom or molecule will be distinct for that species. Similarly, the potential energy spacings between allowed internal energy levels will be characteristic of a species. Therefore, the set of potential energy spacings for a species may be used qualitatively as a distinct fingerprint. Qualitative absorption

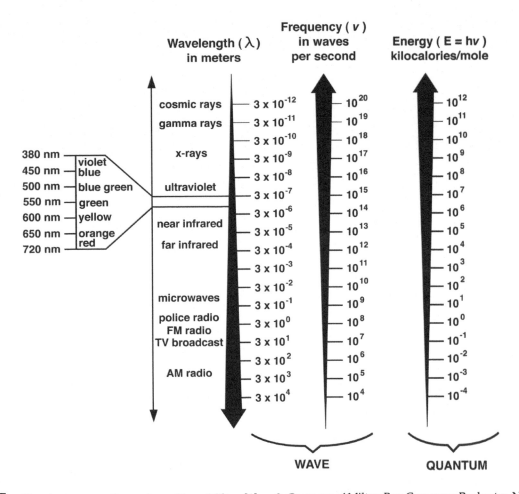

| Wavelength (λ) in meters | Frequency (v) in waves per second | Energy ($E = hv$) kilocalories/mole |

6.4 **figure** The electromagnetic spectrum (From Milton [1], p. 3. Courtesy of Milton Roy Company, Rochester, NY, a subsidiary of Sundstrand Corporation)

and emission spectroscopy make use of this phenomenon in that these techniques attempt to determine an unknown compound's relative energy spacings by measuring transitions between allowed energy levels.

6.3.2 Electronic, Vibrational, and Rotational Energy Levels

The relative *potential energy* of an atom or molecule corresponds to the energy difference between the energy state in which the species exists and that of the ground state. Figure 6.5 is a partial molecular energy-level diagram depicting potential energy levels for an organic molecule. The lowest energy state in the figure, bottom line in bold, represents the ground state. There are three *electronic energy states* depicted, each with its corresponding vibrational and rotational energy levels. Each of the electronic states corresponds to a given *electron orbital*. Electrons in different orbitals are of different potential energy. When an electron changes orbitals, such as when absorbing or emitting a photon of appropriate energy, it is termed an *electronic transition* since it is the electron that is

changing energy levels. However, any change in the potential energy of an electron will, by necessity, result in a corresponding change in the potential energy of the atom or molecule that the electron is associated with.

Atoms are like *molecules* in that only specific energy levels are allowed for atomic electrons. Consequently, an energy-level diagram of an atom would consist of a series of electronic energy levels. In contrast to molecules, the electronic energy levels of atoms have no corresponding vibrational and rotational levels and, hence, may appear less complicated. Atomic energy levels correspond to allowed electron shells (orbits) and corresponding subshells (i.e., 1s, 2s, 2p, etc.). The magnitude of the energy difference between the ground state and first excited states for valence electrons of atoms and bonding electrons of molecules is generally of the same range as the energy content of photons associated with UV and Vis radiation.

The wider lines within each electronic state of Fig. 6.5 depict the species' *vibrational energy levels*. The atoms that comprise a molecule are in constant motion,

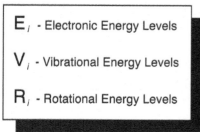

E_i - Electronic Energy Levels

V_i - Vibrational Energy Levels

R_i - Rotational Energy Levels

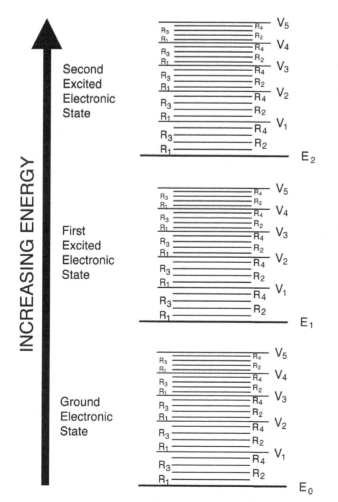

INCREASING ENERGY

Second
Excited
Electronic
State

First
Excited
Electronic
State

Ground
Electronic
State

6.5 figure Partial molecular energy-level diagram depicting three electronic states

vibrating in many ways. However, in all cases the energy associated with this vibrational motion corresponds to defined quantized energy levels. The energy differences between neighboring vibrational energy levels are much smaller than those between adjacent electronic energy levels. Therefore, it is common to consider that several vibrational energy levels are superimposed on each of the molecular electronic energy levels. Energy differences between allowed vibrational energy levels are of the same magnitude as the energy of photons associated with radiation in the

IR region. Vibrational energy levels would not be superimposed on an atomic potential energy-level diagram since this vibrational motion does not exist in a single atom. In this respect, the potential energy diagram for an atom is less complex than that for a molecule, the atomic energy-level diagram having fewer energy levels.

The potential energy of a molecule also is quantized in terms of the energy associated with the rotation of the molecule about its center of gravity. These *rotational energy levels* are yet more closely spaced than the corresponding vibrational levels, as depicted by the narrow lines within each electronic state shown in Fig. 6.5. Hence, it is customary to consider several rotational energy levels superimposed on each of the permitted vibrational energy levels. The energy spacings between rotational energy levels are of the same magnitude as the energy associated with photons of microwave radiation. Microwave spectroscopy is not commonly used in food analysis laboratories; however, the presence of these different energy levels will impact the spectrum observed in other forms of spectroscopy, as will be discussed later. Similar to the situation of vibrational energy levels, rotational energy levels are not of consequence to atomic spectroscopy.

In summation, the internal energy of an *atom* is described in terms of its *electronic energy levels*, while the internal energy of a *molecule* is dependent on its *electronic*, *vibrational*, and *rotational energies*. The algebraic form of these statements follows:

$$E_{\text{atom}} = E_{\text{electronic}} \tag{6.3}$$

$$E_{\text{molecule}} = E_{\text{electronic}} + E_{\text{vibrational}} + E_{\text{rotational}} \tag{6.4}$$

The spectroscopist makes use of the fact that each of these associated energies is quantized and that different species will have somewhat different energy spacings.

6.3.3 Nuclear Energy Levels in Applied Magnetic Fields

NMR spectroscopy makes use of yet another type of quantized energy level. The energy levels of importance to NMR spectroscopy differ with respect to those described above in that they are relevant only in the presence of an applied external *magnetic field*. The basis for the observed energy levels may be rationalized by considering that the nuclei of some atoms behave as tiny bar magnets. Hence, when the atoms are placed in a magnetic field, their nuclear magnetic moment will have a preferred orientation, just as a bar magnet would behave. The NMR-sensitive nuclei of general relevance to the food analyst have two permissible orientations. The energy difference between these allowed orientations depends on the effective magnetic field strength that the nuclei experience. The effective magnetic field strength will itself depend on the strength of the applied magnetic field and the chemical environment surrounding the nuclei in question. The applied magnetic field strength will be set by the spectroscopist, and it is essentially equivalent for each of the nuclei in the applied field. Hence, differences in energy spacings of NMR-sensitive nuclei will depend solely on the identity of the nucleus and its environment. In general, the energy spacings between permissible nuclear orientations, under usable external magnetic field strengths, are of the same magnitude as the energy associated with radiation in the radio frequency range.

6.4   ENERGY-LEVEL TRANSITIONS IN SPECTROSCOPY

6.4.1   Absorption of Radiation

The *absorption of radiation* by an atom or molecule is that process in which energy from a photon of electromagnetic radiation is transferred to the absorbing species. When an atom or molecule absorbs a photon of light, its internal energy increases by an amount equivalent to the amount of energy in that particular photon. Therefore, in the process of absorption, the species goes from a lower energy state to a more *excited state*. In most cases, the species is in the *ground state* prior to absorption. Since the absorption process may be considered quantitative (i.e., all of the photon's energy is transferred to the absorbing species), the photon being absorbed must have an energy content that exactly matches the energy difference between the energy levels across which the transition occurs. This must be the case due to the quantized energy levels of matter, as discussed previously. Consequently, if one plots photon energy versus the relative absorbance of radiation uniquely composed of photons of that energy, one observes a characteristic *absorption spectrum*, the shape of which is determined by the relative absorptivity of photons of different energy. The *absorptivity* of a compound is a wavelength-dependent proportionality constant that relates the absorbing species concentration to its experimentally measured absorbance under defined conditions. A representative absorption spectrum covering a portion of the UV radiation range is presented in Fig. 6.6. The independent variable of an absorption spectrum is most commonly expressed in terms of the wave properties (wavelength, frequency, or wave numbers) of the radiation, as in Fig. 6.6, rather than the energy of the associated photons.

Various molecular transitions resulting from the absorption of photons of different energy are shown schematically in Fig. 6.7. The transitions depicted represent those that may be induced by absorption of UV, Vis, IR, and microwave radiation. The figure also

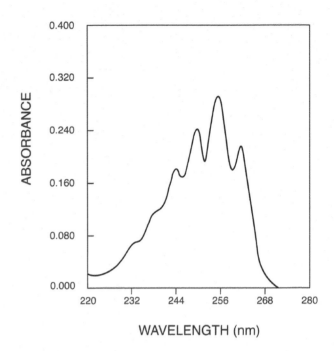

6.6 figure Absorption spectrum of a 0.005 M benzene in water solution

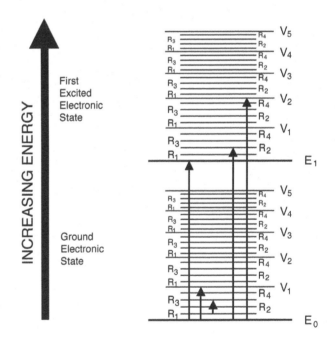

6.7 figure Partial molecular energy-level diagram including electronic, vibrational, and rotational transitions

includes transitions in which the molecule is excited from the ground state to an exited electronic state with a simultaneous change in its vibrational or rotational energy levels. Although not shown in the figure, the absorption of a photon of appropriate energy also may

6.2
table   Wavelength regions, spectroscopic methods, and associated transitions

Wavelength region	Wavelength limits	Type of spectroscopy	Usual wavelength range	Types of transitions in chemical systems with similar energies
Gamma rays	0.01–1 Å	Emission	<0.1 Å	Nuclear proton/neutron arrangements
X-rays	0.1–10 nm	Absorption, emission, fluorescence, and diffraction	0.1–100 Å	Inner-shell electrons
Ultraviolet	10–380 nm	Absorption, emission, and fluorescence	180–380 nm	Outer-shell electrons in atoms, bonding electrons in molecules
Visible	380–750 nm	Absorption, emission, and fluorescence	380–750 nm	Same as ultraviolet
Infrared	0.075–1000 μm	Absorption	0.78–300 μm	Vibrational position of atoms in molecular bonds
Microwave	0.1–100 cm	Absorption	0.75–3.75 mm	Rotational position in molecules
		Electron spin resonance	3 cm	Orientation of unpaired electrons in an applied magnetic field
Radio wave	1–1000 m	Nuclear magnetic resonance	0.6–10 m	Orientation of nuclei in an applied magnetic field

cause simultaneous changes in electronic, vibrational, and rotational energy levels. The ability of molecules to have simultaneous transitions between the different energy levels tends to broaden the peaks in the UV-Vis absorption spectrum of molecules relative to those peaks observed in the absorption spectrum of atoms. This would be expected when one considers that vibrational and rotational energy levels are absent in an atomic energy-level diagram. The depicted transitions between vibrational energy levels, without associated electronic transitions, are induced by radiation in the IR region. Independent transitions between allowed rotational energy levels also are depicted, these resulting from the absorption of photons of microwave radiation. A summary of transitions relevant to atomic and molecular absorption spectroscopy, including corresponding wavelength regions, is presented in Table 6.2.

6.4.2 Emission of Radiation

Emission is essentially the reverse of the absorption process, occurring when energy from an atom or molecule is released in the form of a photon of radiation. A molecule raised to an excited state will typically remain in the excited state for a very short period of time before relaxing back to the ground state. There are several *relaxation processes* through which an excited molecule may dissipate energy. The most common relaxation process is for the excited molecule to dissipate its energy through a series of small steps brought on by collisions with other molecules. The

energy is thus converted to kinetic energy, the net result being the dissipation of the energy as heat. Under normal conditions, the dissipated heat is not enough to measurably affect the system. In some cases, molecules excited by the absorption of UV or Vis light will lose a portion of their excess energy through the emission of a photon. This emission process is referred to as either *fluorescence* or *phosphorescence*, depending on the nature of the excited state. In molecular fluorescence spectroscopy, the photons emitted from the excited species generally will be of lower energy and longer wavelength than the corresponding photons that were absorbed in the excitation process. The reason is that, in most cases, only a fraction of the energy difference between the excited and ground states is lost in the emission process. The other fraction of the excess energy is dissipated as heat during vibrational relaxation. This process is depicted in Fig. 6.8, which illustrates that the excited species undergoes vibrational relaxation down to the lowest vibrational energy level within the excited electronic state, and then undergoes a transition to the ground electronic state through the emission of a photon. The photon emitted will have an energy that equals the energy difference between the lowest vibrational level of the excited electronic state and the ground electronic state level it descends to. The fluorescing molecule may descend to any of the vibrational levels within the ground electronic state. If the fluorescence transition is to an excited vibrational level within the ground electronic state, then it will quickly return to the ground state (lowest energy

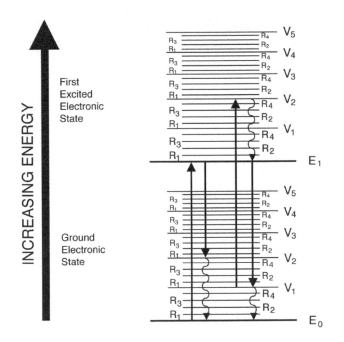

INCREASING ENERGY

First Excited Electronic State

E_1

Ground Electronic State

E_0

6.8 figure Partial molecular energy-level diagram including absorption, vibrational relaxation, and fluorescence relaxation

level) via vibrational relaxation. In yet other cases, an excited species may be of sufficient energy to initiate some type of photochemistry that ultimately leads to a decrease in the system's potential energy. In all cases, the relaxation process is driven by the tendency for a species to exist at its lowest permissible internal energy level. The relaxation process that dominates a system will be the one that minimizes the lifetime of the excited state. Under normal conditions, the relaxation process is so rapid that the population of molecules in the ground state is essentially unchanged.

6.4.3 Population of Energy Levels

The preceding text discussed the existence and quantum nature of molecular, atomic, and nuclear energy levels. It is now relevant to ask how a population of particles is likely to be distributed among their permissible energy levels, i.e., the relative population of the different energy levels. This is important to consider because the intensity of the signal generated in a spectroscopic method is a function of the number of particles in the energy level corresponding to the origin of the signal. For example, when doing traditional molecular absorption spectroscopy, one is measuring the excitation of molecules in the ground state (Chap. 7); when doing atomic emission spectroscopy, one is measuring photons emitted by atoms in an excited state (Chap. 9). In the former case, the observed signal will be a function of the number of molecules in the ground state; in the latter case the signal will be a function of

the number of molecules in the appropriate excited state. For a given collection of particles, the relative population of the different energy levels is described by the Boltzmann distribution. The *Boltzmann distribution* is a probability distribution or frequency distribution; when applied to a group of particles, it describes the average number of particles to be found in the different energy levels available to those particles (assuming the system is in thermal equilibrium). The Boltzmann distribution law may be expressed as follows:

$$(p' \, / \, p) = e^{-(E'-E)/kT} \qquad (6.5)$$

where:

p' = probability of finding particle in energy level E'

p = probability of finding particle in energy level E

E' = energy corresponding higher energy level

E = energy corresponding to lower energy level

k = Boltzmann constant

T = absolute temperature

The equation indicates that the fraction of molecules in the higher energy state (E') decreases exponentially with increasing ΔE (i.e., $E'-E$). In UV-Vis spectroscopy, the vast majority of the molecules are in the ground state since the energy difference between the ground state and the first excited state is relatively large. In nuclear magnetic resonance spectroscopy, the populations in the two energy states are nearly identical due to the relatively small energy difference between the lower and higher energy states (the lower energy state being slightly more populated). These relationships are consistent with UV-Vis spectroscopy being significantly more sensitive than nuclear magnetic resonance spectroscopy. The equation also illustrates that increasing the temperature of a system will increase the fraction of molecules in the higher energy state. This is relevant to atomic emission spectroscopy where molecules in the excited state are emitting the detected signal. At room temperature there are insufficient neutral atoms in the excited state to do atomic emission spectroscopy. However, as one continues to raise the temperature of the analyte mixture, the fraction of atoms in the excited state increases to the point that atomic emission spectroscopy becomes feasible.

6.5 SUMMARY

Spectroscopy deals with the interaction of electromagnetic radiation with matter. Spectrochemical analysis, a branch of spectroscopy, encompasses a wide range of techniques used in analytical laboratories for the qualitative and quantitative analysis of the chemical composition of foods. Common spectrochemical analysis methods include UV, Vis, and IR absorption spectroscopy, molecular fluorescence spectroscopy, and NMR spectroscopy. In each of these methods, the analyst attempts to measure the amount of radiation either absorbed or emitted by the analyte. All of these methods make use of the facts that the energy content of matter is quantized and that photons of radiation may be absorbed or emitted by matter if the energy associated with the photon equals the energy difference for allowed transitions of that given species. The above methods differ from each other with respect to the radiation wavelengths used in the analysis or the molecular vs. atomic nature of the analyte.

6.6 STUDY QUESTIONS

1. Which phenomena associated with light are most readily explained by considering the wave nature of light? Explain these phenomena based on your understanding of interference.

2. Which phenomena associated with light are most readily explained by considering the particulate nature of light? Explain these phenomena based on your understanding of the quantum nature of electromagnetic radiation.

3. What does it mean to say that the energy content of matter is quantized?

4. Molecular absorption of radiation in the UV-Vis range results in transitions between what types of energy levels?

5. Molecular absorption of radiation in the IR range results in transitions between what types of energy levels?

6. Why is an applied magnetic field necessary for NMR spectroscopy?

7. How do the allowed energy levels of molecules differ from those of atoms? Answer with respect to the energy-level diagram depicted in Fig. 6.5.

8. In fluorescence spectroscopy, why is the wavelength of the emitted radiation longer than the wavelength of the radiation used for excitation of the analyte?

RESOURCE MATERIALS

- Atkins P, de Paulo J (2012) Elements of physical chemistry 6th ed, W.H. Freeman, New York
- Ball DW (2001) The basics of spectroscopy. Society of Photo-optical Instrumentation Engineers, Bellingham, WA
- Currell G (2000) Analytical instrumentation – performance characteristics and quality. Wiley, New York, pp 67–91
- Duckett 2 (2000) Foundations of spectroscopy. Oxford University Press, New York
- Harris DC (2015) Quantitative chemical analysis, 9th edn. WH Freeman, New York
- Harris DC, Bertolucci MD (1989) Symmetry and spectroscopy. Dover, Mineola, NY
- Harwood LM, Claridge TDW (1997) Introduction to organic spectroscopy. Oxford University Press, New York
- Ingle JD Jr, Crouch SR (1988) Spectrochemical analysis. Prentice Hall, Englewood Cliffs, NJ
- Meyers RA (ed) (2000) Encyclopedia of analytical chemistry: applications, theory, and instrumentation. 5, 2857–4332
- Milton Roy educational manual for the SPECTRONIC® 20 & 20D spectrophotometers (1989) Milton Roy Co., Rochester, NY
- Pavia DL, Lampman GM, Kriz GS Vyvyan JA (2015) Introduction to spectroscopy. 5th Edition, Cengage Learning, Independence, KY
- Robinson JW, Frame EMS, Frame GM II (2014) Undergraduate instrumental analysis. 7th edn. Marcel Dekker, New York
- Young HD, Freedman RA (2011) Sears and Zemansky's university physics, vol. 2, 13th edn. Addison-Wesley, Boston
- Skoog DA, Holler FJ, Crouch SR (2007) Principles of instrumental analysis, 6th edn. Brooks/Cole, Pacific Grove, CA

7
chapter

Ultraviolet, Visible, and Fluorescence Spectroscopy

Michael H. Penner

Department of Food Science and Technology,
Oregon State University,
Corvallis, OR 97331-6602, USA
e-mail: mike.penner@oregonstate.edu

S. Nielsen (ed.), *Food Analysis*, Food Science Text Series,
DOI 10.1007/978-3-319-45776-5_7, © Springer International Publishing 2017

7.1 INTRODUCTION

Spectroscopy in the ultraviolet-visible (UV-Vis) range is one of the most commonly encountered laboratory techniques in food analysis. Diverse examples, such as the quantification of macrocomponents (total carbohydrate by the phenol-sulfuric acid method), quantification of microcomponents (thiamine by the thiochrome fluorometric procedure), estimates of rancidity (lipid oxidation status by the thiobarbituric acid test), and surveillance testing (enzyme-linked immunoassays), are presented in this text. In each of these cases, the analytical signal for which the assay is based is either the emission or absorption of radiation in the UV-Vis range. This signal may be inherent in the analyte, such as the absorbance of radiation in the visible range by pigments, or a result of a chemical reaction involving the analyte, such as the colorimetric copper-based Lowry method for the analysis of soluble protein.

Electromagnetic radiation in the UV-Vis portion of the spectrum ranges in wavelength from approximately 200–700 nm. The accessible **UV range** for common laboratory analyses runs from 200 to 350 nm and the **Vis range** from 350 to 700 nm (Table 7.1). The UV range is colorless to the human eye, while different wavelengths in the visible range each have a characteristic color, ranging from violet at the short wavelength end of the spectrum to red at the long wavelength end of the spectrum. Spectroscopy utilizing radiation in the UV-Vis range may be divided into two general categories, **absorbance** and **fluorescence** spectroscopies, based on the type of radiation-

matter interaction that is being monitored. Each of these two types of spectroscopy may be subdivided further into **qualitative** and **quantitative** techniques. In general, quantitative absorption spectroscopy is the most common of the subdivisions within UV-Vis spectroscopy.

7.2 ULTRAVIOLET AND VISIBLE ABSORPTION SPECTROSCOPY

7.2.1 Basis of Quantitative Absorption Spectroscopy

The objective of quantitative absorption spectroscopy is to determine the concentration of analyte in a given sample solution. The determination is based on the measurement of the amount of light absorbed from a reference beam as it passes through the sample solution. In some cases the analyte may naturally absorb radiation in the UV-Vis range, such that the chemical nature of the analyte is not modified during the analysis. In other cases analytes that do not absorb radiation in the UV-Vis range are chemically modified during the analysis, converting them to a species that absorbs radiation of the appropriate wavelength. In either case the presence of analyte in the solution will affect the amount of radiation transmitted through the solution, and, hence, the relative transmittance or absorbance of the solution may be used as an index of analyte concentration.

In actual practice, the solution to be analyzed is contained in an absorption cell and placed in the path of radiation of a selected wavelength(s). The amount of radiation passing through the sample is then measured relative to a reference sample. The relative amount of light passing through the sample is then used to estimate the analyte concentration. The process of absorption may be depicted as in Fig. 7.1. The radiation incident on the absorption cell, P_0, will have significantly greater radiant power than the radiation exiting the opposite side of the cell, P. The decrease in radiant power as the beam passes through the solution is due to the capture (absorption) of photons by the absorbing species. The relationship between the power of the incident and exiting beams typically is expressed in terms of either the transmittance or the absorbance of the solution. The **transmittance** (T) of a solution is defined as the ratio of P to P_0 as given in Eq. 7.1. Transmittance also may be expressed as a percentage as given in Eq. 7.2.

$$T = P / P_0 \qquad (7.1)$$

7.1
table | Spectrum of visible radiation

Wavelength (nm)	Color	Complementary hue[a]
<380	Ultraviolet	
380–420	Violet	Yellow-green
420–440	Violet-blue	Yellow
440–470	Blue	Orange
470–500	Blue-green	Red
500–520	Green	Purple
520–550	Yellow-green	Violet
550–580	Yellow	Violet-blue
580–620	Orange	Blue
620–680	Red	Blue-green
680–780	Purple	Green
>780	Near infrared	

[a]Complementary hue refers to the color observed for a solution that shows maximum absorbance at the designated wavelength assuming a continuous spectrum "white" light source

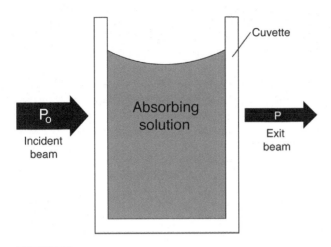

Incident beam P_0

Absorbing solution

Exit beam P

Cuvette

7.1 figure

Attenuation of a beam of radiation as it passes through a cuvette containing an absorbing solution

$$\%T = (P / P_0) \times 100 \qquad (7.2)$$

where:

T = transmittance
P_0 = radiant power of beam incident on absorption cell
P = radiant power of beam exiting the absorption cell
$\%T$ = percent transmittance

The terms T and $\%T$ are intuitively appealing, as they express the fraction of the incident light absorbed by the solution. However, T and $\%T$ are not directly proportional to the concentration of the absorbing analyte in the sample solution. The nonlinear relationship between transmittance and concentration is an inconvenience since analysts are generally interested in analyte concentrations. A second term used to describe the relationship between P and P_0 is **absorbance** (A). Absorbance is defined with respect to T as shown in Eq. 7.3.

$$A = \log (P_0 / P) = -\log T = 2 - \log \%T \qquad (7.3)$$

where:

A = absorbance
T and $\%T$ = as in Eqs. 7.1 and 7.2, respectively

Absorbance is a convenient expression in that, under appropriate conditions, it is directly proportional to the concentration of the absorbing species in the solution. Note that based on these definitions for A and T, the absorbance of a solution *is not* simply unity minus the transmittance. In quantitative spectroscopy, the fraction of the incident beam that is not transmitted does not equal the solution's absorbance (A).

The relationship between the absorbance of a solution and the concentration of the absorbing species is known as **Beer's law** (Eq. 7.4).

$$A = abc \qquad (7.4)$$

where:

A = absorbance
c = concentration of absorbing species
b = pathlength through solution (cm)
a = absorptivity

There are no units associated with absorbance, A, since it is the log of a ratio of beam powers. The concentration term, c, may be expressed in any appropriate units (M, mM, mg/mL, %). The pathlength, b, is in units of cm. The **absorptivity**, a, of a given species is a proportionality constant dependent on the molecular properties of the species. The absorptivity is wavelength dependent and may vary depending on the chemical environment (pH, ionic strength, solvent, etc.) the absorbing species is experiencing. The units of the absorptivity term are $(cm)^{-1}$ (concentration)$^{-1}$. In the special case where the concentration of the analyte is reported in units of molarity, the absorptivity term has units of $(cm)^{-1}$ $(M)^{-1}$. Under these conditions, it is designated by the symbol ε, which is referred to as the **molar absorption coefficient**. Beer's law expressed in terms of the molar absorption coefficient is given in Eq. 7.5. In this case, c refers specifically to the molar concentration of the analyte:

$$A = \varepsilon bc \qquad (7.5)$$

where:

A and b = as in Eq. 7.4
ε = molar absorption coefficient
c = concentration in units of molarity

Quantitative spectroscopy is dependent on the analyst being able to accurately measure the fraction of an incident light beam that is absorbed by the analyte in a given solution. This apparently simple task is somewhat complicated in actual practice due to processes other than analyte absorption that also result in significant decreases in the power of the incident beam. A pictorial summary of reflection and scattering processes that will decrease the power of an incident beam is given in Fig. 7.2. It is clear that these processes must be accounted for if a truly quantitative estimate of analyte absorption is necessary. In practice, a reference cell is used to correct for these processes. A **reference cell** is one that, in theory, exactly matches the sample absorption cell with the exception that it contains no analyte. Reference cells are often prepared by filling appropriate absorption cells with water. The reference cell is placed in the path of the light beam, and the power of the radiation exiting the reference cell is

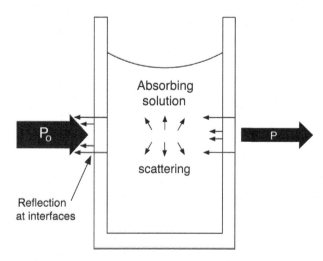

7.2
figure

Factors contributing to the attenuation of a beam of radiation as it passes through a cuvette containing an absorbing solution

measured and taken as P_0 for the sample cell. This procedure assumes that all processes except the selective absorption of radiation by the analyte are equivalent for the sample and reference cells. The absorbance actually measured in the laboratory approximates Eq. 7.6.

$$A = \log\left(P_{\text{solvent}} / P_{\text{analyte solution}}\right) \cong \log\left(P_0 / P\right) \quad (7.6)$$

where:

> P_{solvent} = radiant power of beam exiting cell containing solvent (blank)
> $P_{\text{analyte solution}}$ = radiant power of beam exiting cell containing analyte solution
> P_0 and P = as in Eq. 7.1.
> A as in Eq. 7.3.

7.2.2 Deviations from Beer's Law

It should never be assumed that Beer's law is strictly obeyed. Indeed, there are several reasons for which the predicted linear relationship between absorbance and concentration may not be observed. In general, Beer's law is applicable only to dilute solutions, up to approximately 10 mM for most analytes. The actual concentration at which the law becomes limiting will depend on the chemistry of the analyte. As analyte concentrations increase, the intermolecular distances in a given sample solution will decrease, eventually reaching a point at which neighboring molecules mutually affect the charge distribution of the other. This perturbation may significantly affect the ability of the analyte to capture photons of a given wavelength; that is, it may alter the analyte's absorptivity (a). This causes the linear relationship between concentration and absorption to break down since the absorptivity term is the constant of proportionality in Beer's law

(assuming a constant pathlength, b). Other chemical processes also may result in deviations from Beer's law, such as the reversible association-dissociation of analyte molecules or the ionization of a weak acid in an unbuffered solvent. In each of these cases, the predominant form of the analyte may change as the concentration is varied. If the different forms of the analyte (e.g., ionized versus neutral) have different absorptivities (a), then a linear relationship between concentration and absorbance will not be observed.

A further source of deviation from Beer's law may arise from limitations in the instrumentation used for absorbance measurements. Beer's law strictly applies to situations in which the radiation passing through the sample is monochromatic, since under these conditions a single absorptivity value describes the interaction of the analyte with all the radiation passing through the sample. If the radiation passing through a sample is polychromatic and there is variability in the absorptivity constants for the different constituent wavelengths, then Beer's law will not be obeyed. An extreme example of this behavior occurs when radiation of the ideal wavelength and stray radiation of a wavelength that is not absorbed at all by the analyte simultaneously pass through the sample to the detector. In this case, the observed transmittance will be defined as in Eq. 7.7. Note that a limiting absorbance value will be reached as $Ps \gg P$, which will occur at relatively high concentrations of the analyte:

$$A = \log\left(P_0 + P_s\right) / \left(P + P_s\right) \quad (7.7)$$

where:

> P_s = radiant power of stray light
> A = as in Eq. 7.3
> P and P_0 = as in Eq. 7.1.

7.2.3 Procedural Considerations

The goal of many quantitative measurements is to determine the concentration of an analyte with optimum precision and accuracy, in a minimal amount of time, and at minimal cost. To accomplish this, it is essential that the analyst consider potential errors associated with each step in a particular assay. Potential sources of error for spectroscopic assays include inappropriate sample preparation techniques, inappropriate controls, instrumental noise, and errors associated with inappropriate conditions for absorbance measurements (such as extreme absorbance/transmittance readings).

Sample preparation schemes for absorbance measurements vary considerably. In the simplest case, the analyte-containing solution may be measured directly following homogenization and clarification. Except for special cases, homogenization is required prior to any analysis to ensure a representative sample. Clarification of samples is essential prior to taking absorbance readings in order to avoid the apparent absorption due to scattering of light by turbid solutions. The **reference solution** for samples in this simplest case will be the sample solvent, the solvent being water or an aqueous buffer in many cases. In more complex situations, the analyte to be quantified may need to be chemically modified prior to making absorbance measurements. In these cases, the analyte that does not absorb radiation in an appropriate spectral range is specifically modified, resulting in a species with absorption characteristics compatible with a given spectrophotometric measurement. Specific reactions such as these are used in many colorimetric assays that are based on the absorption of radiation in the Vis range. The reference solution for these assays is prepared by treating the sample solvent in a manner identical with that of the sample. The reference solution therefore will help to correct for any absorbance due to the modifying reagents themselves and not the modified analyte.

A **sample-holding cell** or **cuvette** should be chosen after the general spectral region to be used in a spectrophotometric measurement has been determined. Sample-holding cells vary in composition and dimensions. The sample-holding cell should be composed of a material that does not absorb radiation in the spectral region being used. Cells meeting this requirement for measurements in the **UV range** may be composed of **quartz or fused silica**. For the **Vis range** cells made of **silicate glass** are appropriate, and inexpensive **plastic** cells also are available for some applications. The dimensions of the cell will be important with respect to the amount of solution required for a measurement and with regard to the pathlength term used in Beer's law. A typical absorption cell is 1 cm^2 and approximately 4.5 cm long. The pathlength for this traditional cell is 1 cm, and the minimum volume of solution needed for standard absorption measurements is approximately 1.5 mL. Absorption cells with pathlengths ranging from 1 to 100 mm are commercially available. Narrow cells, approximately 4 mm in width, with optical pathlengths of 1 cm, are also available. These narrow cells are convenient for absorbance measurements when limiting amounts of solution are available, e.g., less than 1 mL.

In many cases an analyst must **choose an appropriate wavelength** at which to make absorbance measurements. If possible, it is best to choose the wavelength at which the analyte demonstrates maximum absorbance and where the absorbance does not change rapidly with changes in wavelength (Fig. 7.3). This position usually corresponds to the apex of the highest absorption peak. Taking measurements at this apex has two advantages: (1) maximum sensitivity, defined as the absorbance change per unit change in analyte concentration, and (2) greater adherence to Beer's law since the spectral region making up the radiation beam is composed of wavelengths with rela-

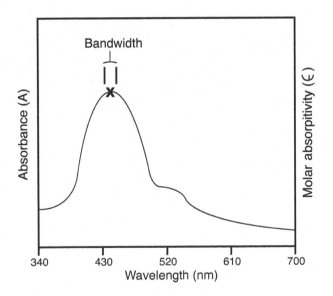

7.3 figure Hypothetical absorption spectrum between 340 and 700 nm. The effective bandwidth of the radiation used in obtaining the spectrum is assumed to be approximately 20 nm. Note that at the point indicated there is essentially no change in molar absorptivity over this wavelength range

tively small differences in their molar absorptivities for the analyte being measured (Fig. 7.3). The latter point is important in that the radiation beam used in the analysis will be composed of a small continuous band of wavelengths centered about the wavelength indicated on the instrument's wavelength selector.

The actual **absorbance measurement** is made by first calibrating the instrument for 0% and then 100% transmittance. The 0% transmittance adjustment is made while the photodetector is screened from the incident radiation by means of an occluding shutter, mimicking infinite absorption. This adjustment sets the base level current or "dark current" to the appropriate level, such that the readout indicates zero. The 100% transmittance adjustment then is made with the occluding shutter open and an appropriate reference cell/solution in the light path. The reference cell itself should be equivalent to the cell that contains the sample (i.e., a "matched" set of cells is used). In many cases, the same cell is used for both the sample and reference solutions. The reference cell generally is filled with solvent, that often being distilled/deionized water for aqueous systems. The 100% T adjustment effectively sets $T = 1$ for the reference cell, which is equivalent to defining P_0 in Eq. 7.1 as equivalent to the radiant power of the beam exiting the reference cell. The 0% T and 100% T settings should be confirmed as necessary throughout the assay. The sample cell that contains analyte then is measured without changing the adjustments. The adjustments made with the reference cell will effectively set the instrument to give a sample readout in terms of Eq. 7.6. The readout for the sample solution will be between 0 and 100% *T*. Most modern spectrophotometers allow the analyst to make readout measurements in either absorbance units or as percent transmittance. It is generally most convenient to make readings in absorbance units since, under optimum conditions, absorbance is directly proportional to concentration. When making measurements with an instrument that employs an analog swinging needle type of readout, it may be preferable to use the linear percent transmittance scale and then calculate the corresponding absorbance using Eq. 7.3. This is particularly true for measurements in which the percent transmittance is less than 20.

7.2.4 Calibration Curves

It is generally advisable to use calibration curves for quantitative measurements. Empirical assays that require the use of a calibration curve are common in food analyses. The calibration curve is used to establish the relationship between analyte concentration and absorbance. This relationship is established experimentally through the analysis of a series of samples of known analyte concentration. The standard solutions are best prepared with the same reagents and at the same time as the unknown. The concentration range covered by the standard solutions must include that expected for the unknown. Typical calibration curves are depicted in Fig. 7.4. **Linear calibration curves** are expected for those systems that obey Beer's law. **Nonlinear calibration curves** are used for some assays, but linear relationships generally are preferred due to the ease of processing the data. Nonlinear calibration curves may be due to concentration-dependent changes in the chemistry of the system or to limitations inherent in the instruments used for the assay. The nonlinear calibration curve in Fig. 7.4b reflects the fact that the **calibration sensitivity**, defined as change in absorbance per unit change in analyte concentration, is not constant. For the case depicted in Fig. 7.4b, the assay's concentration-dependent decrease in sensitivity obviously begins to limit its usefulness at analyte concentrations above 10 m*M*.

In many cases truly representative calibration standards cannot be prepared due to the complexity of the unknown sample. This scenario must be assumed when insufficient information is available on the extent of interfering compounds in the unknown. **Interfering compounds** include those that absorb radiation in the same spectral region as the analyte, those that influence the absorbance of the analyte, and those compounds that react with modifying reagents that are supposedly specific for the analyte. This means that calibration curves are potentially in error if the unknown and the standards differ with respect to pH, ionic strength, viscosity, types of impurities, and the like. In these cases, it is advisable to calibrate the assay system by

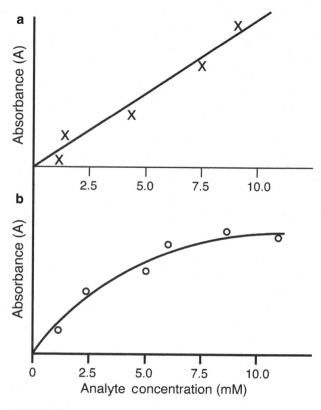

7.4 **figure** Linear (**a**) and nonlinear (**b**) calibration curves typically encountered in quantitative absorption spectroscopy

using a **standard addition protocol**. One such protocol goes as follows: to a series of flasks, add a constant volume of the unknown (V_u) for which you are trying to determine the analyte concentration (C_u). Next, to each individual flask, add a known volume (V_s) of a standard analyte solution of concentration C_s, such that each flask receives a unique volume of standard. The resulting series of flasks will contain identical volumes of the unknown and different volumes of the standard solution. Next, dilute all flasks to the same total volume, V_t. Each of the flasks is then assayed, with each flask treated identically. If Beer's law is obeyed, then the measured absorbance of each flask will be proportional to the total analyte concentration as defined in Eq. 7.8.

$$A = k\left[V_s C_s + V_u C_u\right)/(V_t)\right] \qquad (7.8)$$

where:

V_s = volume of standard
V_u = volume of unknown
V_t = total volume
C_s = concentration of standard
C_u = concentration of unknown
k = proportionality constant (pathlength x absorptivity)

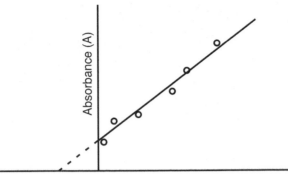

7.5
figure

Calibration curve for the determination of the analyte concentration in an unknown using a standard addition protocol. *A* absorbance, *Vs* volume of standard analyte solution; as discussed in text

The results from the assays are then plotted with the volume of standard added to each flask (V_s) as the independent variable and the resulting absorbance (A) as the dependent variable (Fig. 7.5). Assuming Beer's law, the line describing the relationship will be as in Eq. 7.9, in which all terms other than V_s and A are constants. Taking the ratio of the slope of the plotted line (Eq. 7.10) to the line's intercept (Eq. 7.11) and rearranging gives Eq. 7.12, from which the concentration of the unknown, C_u, can be calculated since C_s and V_u are experimentally defined constants:

$$A = kC_s V_s / V_T + V_u C_u k / V_t \qquad (7.9)$$

$$\text{Slope} = kC_s / V_t \qquad (7.10)$$

$$\text{Intercept} = V_u C_u k / V_t \qquad (7.11)$$

$$C_u = (\text{measured intercept} / \text{measured slope})(C_s / V_u) \qquad (7.12)$$

where:

V_s, V_u, V_t, C_s, C_u, and k = as in Eq. 7.8

7.2.5 Effect of Indiscriminant Instrumental Error on the Precision of Absorption Measurements

All spectrophotometric assays will have some level of **indiscriminant error** associated with the absorbance/transmittance measurement itself. Indiscriminant error of this type often is referred to as **instrument noise**. It is important that the assay be designed such that this source of error is minimized, the objective

being to keep this source of error low relative to the variability associated with other aspects of the assay, such as sample preparation, subsampling, reagent handling, and so on. Indiscriminant instrumental error is observed with repeated measurements of a single homogeneous sample. The relative concentration uncertainty resulting from this error is not constant over the entire percent transmittance range (0–100%). Measurements at intermediate transmittance values tend to have lower relative errors, thus greater relative precision, than measurements made at either very high or very low transmittance. **Relative concentration uncertainty** or relative error may be defined as S_c/C, where S_c is sample standard deviation and C is measured concentration. Relative concentration uncertainties of from 0.5% to 1.5% are to be expected for absorbance/transmittance measurements taken in the optimal range. The optimal range for absorbance measurements on simple, less expensive spectrophotometers is from approximately 0.2–0.8 absorbance units, or 15–65% transmittance. On more sophisticated instruments, the range for optimum absorbance readings may be extended up to 1.5 or greater. To be safe, it is prudent to always make absorbance readings under conditions at which the absorbance of the analyte solution is less than 1.0. If there is an anticipated need to make measurements at absorbance readings greater than 1.0, then the relative precision of the spectrophotometer should be established experimentally by repetitive measurements of appropriate samples. Absorbance readings outside the optimal range of the instrument may be used, but the analyst must be prepared to account for the higher relative error associated with these extreme readings. When absorbance readings approach the limits of the instrumentation, then relatively large differences in analyte concentrations may not be detected.

7.2.6 Instrumentation

There are many variations of spectrophotometers available for UV-Vis spectrophotometry. Some instruments are designed for operation in only the visible range, while others encompass both the UV and Vis ranges. Instruments may differ with respect to design, quality of components, and versatility. A basic spectrophotometer is composed of five essential components: the **light source**, the **monochromator**, the **sample/reference holder**, the **radiation detector**, and a **readout device**. A power supply is required for instrument operation. A schematic depicting component interrelationships is shown in Fig. 7.6.

7.2.6.1 Light Source

Light sources used in spectrophotometers must continuously emit a strong band of radiation encompassing the entire wavelength range for which the

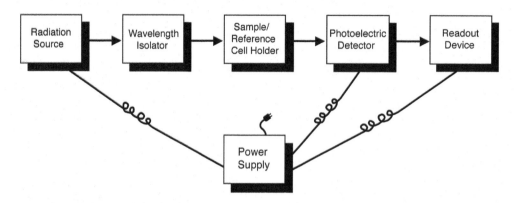

7.6
figure

Arrangement of components in a simple single-beam, UV-Vis absorption spectrophotometer

instrument is designed. The power of the emitted radiation must be sufficient for adequate detector response, and it should not vary sharply with changes in wavelength or drift significantly over the experimental time scale. The most common radiation source for Vis spectrophotometers is the **tungsten filament lamp**. These lamps emit adequate radiation covering the wavelength region from 350 to 2,500 nm. Consequently, tungsten filament lamps also are employed in near-infrared spectroscopy. The most common radiation sources for measurements in the UV range are **deuterium electrical-discharge lamps**. These sources provide a continuous radiation spectrum from approximately 160 nm through 375 nm. These lamps employ quartz windows and should be used in conjunction with quartz sample holders, since glass significantly absorbs radiation below 350 nm.

7.2.6.2 *Monochromator*

The component that functions to isolate the specific, narrow, continuous group of wavelengths to be used in the spectroscopic assay is the **monochromator**. The monochromator is so named because light of a single wavelength is termed **monochromatic**. Theoretically, **polychromatic radiation** from the source enters the monochromator and is dispersed according to wavelength, and **monochromatic radiation** of a selected wavelength exits the monochromator. In practice, light exiting the monochromator is not of a single wavelength, but rather it consists of a narrow continuous band of wavelengths. A representative monochromator is depicted in Fig. 7.7. As illustrated, a typical monochromator is composed of **entrance** and **exit slits, concave mirror(s)**, and a **dispersing element** (the grating in this particular example). Polychromatic light enters the monochromator through the entrance slit and is then culminated by a concave mirror. The culminated polychromatic radiation is then dispersed, dispersion being the physical separation in space of radiation of different wavelengths. The radiation of different wavelengths is then reflected from a concave

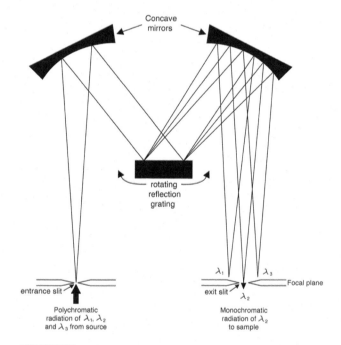

7.7
figure

Schematic of a monochromator employing a reflection grating as the dispersing element. The concave mirrors serve to culminate the radiation into a beam of parallel rays

mirror that focuses the different wavelengths of light sequentially along the focal plane. The radiation that aligns with the exit slit in the focal plane is emitted from the monochromator. The radiation emanating from the monochromator will consist of a narrow range of wavelengths presumably centered around the wavelength specified on the wavelength selection control of the instrument.

The size of the wavelength range passing out of the exit slit of the monochromator is termed the **bandwidth** of the emitted radiation. Many spectrophotometers allow the analyst to adjust the size of the monochromator exit slit (and entrance slit) and, consequently, the bandwidth of the emitted radiation.

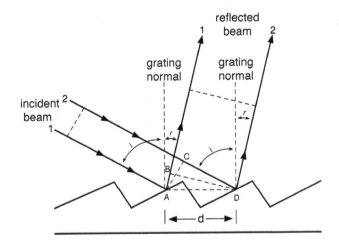

7.8 figure Schematic illustrating the property of diffraction from a reflection grating. Each reflected point source of radiation is separated by a distance *d*

Decreasing the exit slit width will decrease the associated bandwidth and the radiant power of the emitted beam. Conversely, further opening of the exit slit will result in a beam of greater radiant power but one that has a larger bandwidth. In some cases where resolution is critical, such as some qualitative work, the narrower slit width may be advised. However, in most quantitative work, a relatively open slit may be used since adsorption peaks in the UV-Vis range generally are broad relative to spectral bandwidths. Also, the signal-to-noise ratio associated with transmittance measurements is improved due to the higher radiant power of the measured beam.

The effective bandwidth of a monochromator is determined not only by the slit width but also by the quality of its dispersing element. The dispersing element functions to spread out the radiation according to wavelength. **Reflection gratings**, as depicted in Fig. 7.8, are the most commonly used dispersing elements in modern spectrophotometers. Gratings sometimes are referred to as **diffraction gratings** because the separation of component wavelengths is dependent on the different wavelengths being diffracted at different angles relative to the grating normal. A reflection grating incorporates a reflective surface in which a series of closely spaced grooves has been etched, typically between 1,200 and 1,400 grooves per millimeter. The grooves themselves serve to break up the reflective surface such that each point of reflection behaves as an independent point source of radiation.

Referring to Fig. 7.8, lines 1 and 2 represent rays of parallel monochromatic radiation that are in phase and that strike the grating surface at an angle *i* to the normal. Maximum constructive interference of this radiation is depicted as occurring at an angle *r* to the normal. At all other angles, the two rays will partially or completely cancel each other. Radiation of a differ-

ent wavelength would show maximum constructive interference at a different angle to the normal. The wavelength dependence of the diffraction angle can be rationalized by considering the relative distance the photons of rays 1 and 2 travel and assuming that maximum constructive interference occurs when the waves associated with the photons are completely in phase. Referring to Fig. 7.8, prior to reflection, photon 2 travels a distance CD greater than photon 1. After reflection, photon 1 travels a distance AB greater than photon 2. Hence, the waves associated with photons 1 and 2 will remain in phase after reflection only if the net difference in the distance traveled is an integral multiple of their wavelength. Note that for a different angle *r* the distance AB would change and, consequently, the net distance CD–AB would be an integral multiple of a different wavelength. The net result is that the component wavelengths are each diffracted at their own unique angles *r*.

7.2.6.3 *Detector*

In a spectroscopic measurement, the light transmitted through the reference or sample cell is quantified by means of a **detector**. The detector is designed to produce an electric signal when it is struck by photons. An ideal detector would give a signal directly proportional to the radiant power of the beam striking it, it would have a high signal-to-noise ratio, and it would have a relatively constant response to light of different wavelengths, such that it was applicable to a wide range of the radiation spectrum. There are several types and designs of radiation detectors currently in use. The most commonly encountered detectors are the **phototube**, the **photomultiplier tube**, and **photodiode detectors**. All of these detectors function by converting the energy associated with incoming photons into electrical current. The **phototube** consists of a semicylindrical cathode covered with a photoemissive surface and a wire anode, the electrodes being housed under vacuum in a transparent tube (Fig. 7.9a). When photons strike the photoemissive surface of the cathode, there is an emission of electrons; the freed electrons are collected at the anode. The net result of this process is that a measurable current is created. The number of electrons emitted from the cathode and the subsequent current through the system are directly proportional to the number of photons, or radiant power of the beam, impinging on the photoemissive surface. The **photomultiplier tube** is of similar design. However, in the photomultiplier tube, there is an amplification of the number of electrons collected at the anode per photon striking the photoemissive surface of the cathode (Fig. 7.9b). The electrons originally emitted from the cathode surface are attracted to a dynode with a relative positive charge. At the dynode, the electrons strike the surface, causing the emission of several

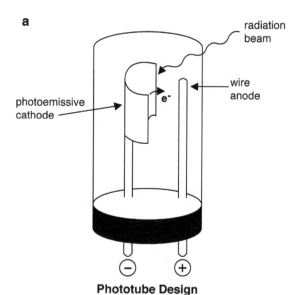

Phototube Design

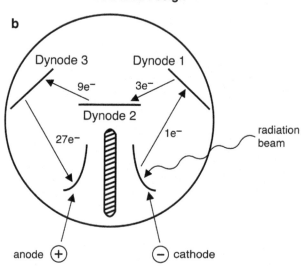

Photomultiplier Arrangement

7.9 **figure** Schematic diagram of a typical phototube design (**a**) and the cathode-dynode-anode arrangement of a representative photomultiplier tube (**b**)

more electrons per original electron, resulting in an amplification of the signal. Signal amplification continues in this manner, as photomultiplier tubes generally contain a series of such dynodes, with electron amplification occurring at each dynode. The cascade continues until the electrons emitted from the final dynode are collected at the anode of the photomultiplier tube. The final gain may be as many as 10^6–10^9 electrons collected per photon.

Photodiode detectors are now common in UV-Vis spectrophotometers. These are solid-state devices in which the light-induced electrical signal is

a result of photons exciting electrons in the semiconductor materials from which they are fabricated, most commonly silicon. Spectrophotometers using photodiode detectors may contain a single diode detector or a linear array of diodes (diode array spectrophotometers). If a single photodiode detector is used, then the arrangement of components is generally as depicted in Fig. 7.6. If an array of photodiode detectors is used, then the light originating from the source typically passes into the sample prior to it being dispersed. The light transmitted through the sample is subsequently dispersed onto the diode array, with each diode measuring a narrow band of the resulting spectrum. This design allows one to simultaneously measure multiple wavelengths, allowing nearly instantaneous collection of an entire absorption spectrum. Diode-based detectors are generally reported to be more sensitive than phototubes but less sensitive than photomultiplier tubes.

7.2.6.4 Readout Device

The signal from the detector generally is amplified and then displayed in a usable form to the analyst. The final form in which the signal is displayed will depend on the complexity of the system. In the simplest case, the analog signal from the detector is displayed on an **analog meter** through the position of a needle on a meter face calibrated in percent transmission or absorbance. Analog readouts are adequate for most routine analytical purposes; however, analog meters are somewhat more difficult to read, and, hence, the resulting data are expected to have somewhat lower precision than that obtained on a digital readout (assuming the digital readout is given to enough places). **Digital readouts** express the signal as numbers on the face of a meter. In these cases, there is an obvious requirement for signal processing between the analog output of the detector and the final digital display. In virtually all cases, the signal processor is capable of presenting the final readout in terms of either absorbance or transmittance. Many of the newer instruments include microprocessors capable of more extensive data manipulations on the digitized signal. For example, the readouts of some spectrophotometers may be in concentration units, provided the instrument has been correctly calibrated with appropriate reference standards.

7.2.7 Instrument Design

The optical systems of spectrophotometers fall into one of two general categories: they are either single-beam or double-beam instruments. In a **single-beam instrument**, the radiant beam follows only one path, that going from the source through the sample to the detector (Fig. 7.6). When using a single-beam

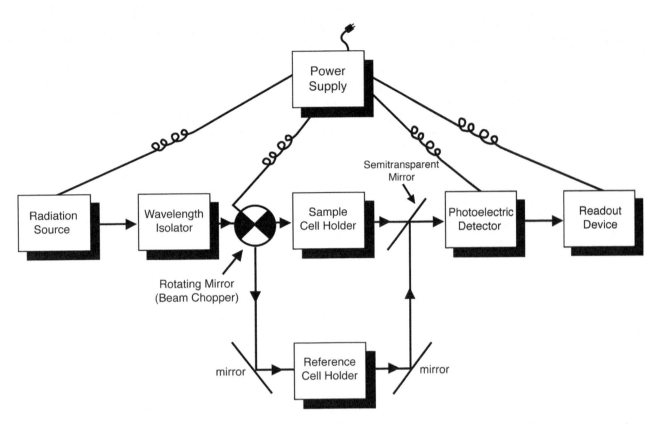

7.10 figure Arrangement of components in a representative double-beam UV-Vis absorption spectrophotometer. The incident beam is alternatively passed through the sample and reference cells by means of a rotating beam chopper

instrument, the analyst generally measures the transmittance of a sample after first establishing $100\% T$, or P_0, with a reference sample or blank. The blank and the sample are read sequentially since there is but a single light path going through a single cell-holding compartment. In a **double-beam instrument**, the beam is split such that one-half of the beam goes through one cell-holding compartment and the other half of the beam passes through a second. The schematic of Fig. 7.10 illustrates a double-beam optical system in which the beam is split in time between the sample and reference cell. In this design, the beam is alternately passed through the sample and reference cells by means of a rotating sector mirror with alternating reflective and transparent sectors. The double-beam design allows the analyst to simultaneously measure and compare the relative absorbance of a sample and a reference cell. The advantage of this design is that it will compensate for deviations or drifts in the radiant output of the source since the sample and reference cells are compared many times per second. The disadvantage of the double-beam design is that the radiant power of the incident beam is diminished because the beam is split. The lower energy throughput of the double-beam design is generally associated with

inferior signal-to-noise ratios. Computerized single-beam spectrophotometers now are available that claim to have the benefits of both the single- and double-beam designs. Their manufacturers report that previously troublesome source and detector drift and noise problems have been stabilized such that simultaneous reading of the reference and sample cell is not necessary. With these instruments, the reference and sample cells are read sequentially, and the data are stored, then processed, by the associated computer.

The Spectronic® 20 is a classic example of a simple single-beam visible spectrophotometer (Fig. 7.11). The white light emitted from the source passes into the monochromator via its entrance slit; the light is then dispersed into a spectrum by a diffraction grating, and a portion of the resulting spectrum then leaves the monochromator via the exit slit. The radiation emitted from the monochromator passes through a sample compartment and strikes the silicon photodiode detector, resulting in an electrical signal proportional to the intensity of impinging light. The lenses depicted in Fig. 7.11 function in series to focus the light image on the focal plane that contains the exit slit. To change the portion of the spectrum exiting the monochromator, one rotates the reflecting grating by means of the

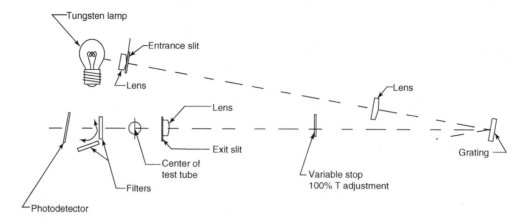

7.11 figure Optical system for the Spectronic® 20 spectrophotometer (Courtesy of Thermo Spectronic, Rochester, NY, Thermo Electron Spectroscopy)

wavelength cam. A shutter automatically blocks light from exiting the monochromator when no sample/reference cell is in the instrument; the zero percent T adjustment is made under these conditions. The light control occluder is used to adjust the radiant power of the beam exiting the monochromator. The occluder consists of an opaque strip with a V-shaped opening that can be physically moved in or out of the beam path. The occluder is used to make the $100\%\,T$ adjustment when an appropriate reference cell is in the instrument.

7.2.8 Characteristics of UV-Vis Absorbing Species

The absorbance of UV-Vis radiation is associated with electronic excitations within atoms and molecules. Commonly encountered analytical methods based on UV-Vis spectroscopy do not use UV radiation below 200 nm. Hence, the excitations of interest in traditional UV-Vis spectroscopy are the result of unsaturation and/or the presence of nonbonded electrons in the absorbing molecules. The UV-Vis absorption characteristics of several functional groups common to food constituents are tabulated in Table 7.2. The presented wavelengths of maximum absorbance and the associated molar absorption coefficients are only approximate since the environment to which the functional group is exposed, including neighboring constituents and solvents, will have an influence on the electronic properties of the functional group.

The type of information contained in Table 7.2 will likely be useful in determining the feasibility of UV-Vis spectroscopy for specific applications. For example, it is helpful to know the absorption characteristics of carboxyl groups if one is considering the feasibility of using UV-Vis absorption spectroscopy as a detection method to monitor non-derivatized organic acids eluting from liquid chromatography columns. With respect to this particular organic acids question, the table indicates that organic acids are likely to absorb radiation in the range accessible to most UV-Vis detectors (>200 nm). However, the table also indicates that sensitivity of such a detection method is likely to be limited due to the low molar absorption coefficient of carboxyl groups at such wavelengths. This explains why high-performance liquid chromatography methods for organic acid quantification sometimes make use of UV-Vis-detectors tuned to ~210 nm (e.g., Resource Material 3) and why there are research efforts aimed at developing derivatization methods to enhance the sensitivity of UV-Vis-based methods for the quantification of organic acids (Resource Material 11).

The data of Table 7.2 also illustrate the effect of conjugation on electronic transitions. Increased conjugation leads to absorption maxima at longer wavelengths due to the associated decrease in the electronic energy spacing within a conjugated system (i.e., lower energy difference between the ground and excited state). The aromatic compounds included in the table were chosen due to their relevance to protein quantification: benzene/phenylalanine, phenol/tyrosine, and indole/tryptophan (Chap. 18, Sect. 18.5.1). The table indicates that typical proteins will have an absorption maximum at approximately 278 nm (high molar absorption coefficient of the indole side chain of tryptophan), as well as another peak at around 220 nm. This latter peak corresponds to the amide/peptide bonds along the backbone of the protein, the rationale being deduced from the data for the simple amide included in the table (i.e., acetamide).

7.2 table

Representative absorption maxima above 200 nm for select functional groups

Chromophore	Example	λ_{max}[a]	ε_{max}[b]	Resource material
Nonconjugated systems				
R-CHO	Acetaldehyde	290	17	4
R_2-CO	Acetone	279	15	4
R-COOH	Acetic acid	208	32	4
R-CONH$_2$	Acetamide	220	63	24
R-SH	Mercaptoethane	210	1,200	19
Conjugated systems				
R_2C=CR$_2$	Ethylene	<200	–	24
R-CH=CH-CH=CH-R	1,3 Butadiene	217	21,000	24
R-CH=CH-CH=CH-CH=CH-R	1,3,5 Hexatriene	258	35,000	24
11 conjugated double bonds	β-Carotene	465	125,000	13
R_2C=CH-CH=O	Acrolein (2-propenal)	210	11,500	24
		315	14	24
HOOC-COOH	Oxalic acid	250	63	24
Aromatic compounds[c]				
C_6H_6	Benzene	256	200	24
C_6H_5OH	Phenol	270	1,450	24
C_8H_7N	Indole	278	2,500	NIST database[d]

[a] λ_{max}, wavelength (in nm) of a maximum absorbance greater than 200 nm

[b] ε_{max}, molar absorption coefficient, units of $(cm)^{-1}$ $(M)^{-1}$

[c] Spectra of the aromatic compounds generally contain an absorption band(s) of higher intensity at a lower wavelength (e.g., phenol has an absorption maxima of ~210 nm with a molar absorptivity of ~6,200 $(cm)^{-1}$ $(M)^{-1}$; values from reference [1]). Only the absorption maxima corresponding to the longer wavelengths are included in the table

[d] NIST Standard Reference Database 69: *NIST Chemistry WebBook*. (The presented values were estimated from the UV-Vis spectrum for indole presented online: http://webbook.nist.gov/cgi/cbook.cgi?Name=indole&Units=SI&cUV=on. The web site contains UV-Vis data for many compounds that are of potential interest to food scientists)

7.3 FLUORESCENCE SPECTROSCOPY

Fluorescence spectroscopy is generally one to three orders of magnitude more sensitive than corresponding absorption spectroscopy. In **fluorescence spectroscopy**, the signal being measured is the electromagnetic radiation that is emitted from the analyte as it relaxes from an excited electronic energy level to its corresponding ground state. The analyte is originally activated to the higher energy level by the absorption of radiation in the UV or Vis range. The processes of activation and deactivation occur simultaneously during a fluorescence measurement. For each unique molecular system, there will be an optimum radiation wavelength for sample excitation and another, of longer wavelength, for monitoring fluorescence emission. The respective wavelengths for excitation and emission will depend on the chemistry of the system under study.

The instrumentation used in fluorescence spectroscopy is composed of essentially the same components as the corresponding instrumentation used in UV-Vis absorption spectroscopy. However, there are

definite differences in the arrangement of the optical systems used for the two types of spectroscopy (compare Figs. 7.6 and 7.12). In fluorometers and spectrofluorometers, there is a need for two wavelength selectors, one for the **excitation beam** and one for the **emission beam**. In some simple fluorometers, both wavelength selectors are filters such that the excitation and emission wavelengths are fixed. In more sophisticated spectrofluorometers, the excitation and emission wavelengths are selected by means of grating monochromators. The photon detector of fluorescence instrumentation is generally arranged such that the emitted radiation that strikes the detector is traveling at an angle of 90° relative to the axis of the excitation beam. This detector placement minimizes signal interference due to transmitted source radiation and radiation scattered from the sample.

The **radiant power** of the fluorescence beam (P_F) emitted from a fluorescent sample is proportional to the change in the radiant power of the source beam as it passes through the sample cell (Eq. 7.13). Expressing this another way, the radiant power of the fluorescence

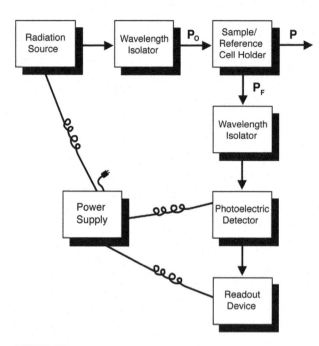

7.12
figure

Schematic diagram depicting the arrangement of the source, excitation and emission wavelength selectors, sample cell, photoelectric detector, and readout device for a representative fluorometer or spectrofluorometer

beam will be proportional to the number of photons absorbed by the sample:

$$P_F = \varphi(P_0 - P) \qquad (7.13)$$

where:

PF = radiant power of beam emitted from fluorescent cell
φ = constant of proportionality
$P_0\ and\ P$ = as in Eq. 7.1

The constant of proportionality used in Eq. 7.13 is termed the **quantum efficiency** (φ), which is specific for any given system. The quantum efficiency equals the ratio of the total number of photons emitted to the total number of photons absorbed. Combining Eqs. 7.3 and 7.5 allows one to define P in terms of the analyte concentration and P_0, as given in Eq. 7.14:

$$P = P_0 10^{-\varepsilon bc} \qquad (7.14)$$

where:

P_0 and P = as in Eq. 7.1
ε, b, and c = as in Eq. 7.5

Substitution of Eq. 7.14 into Eq. 7.13 gives an expression that relates the radiant power of the fluorescent beam to the analyte concentration and P_0, as shown in Eq. 7.15. At low analyte concentrations, $\varepsilon bc < 0.01$, Eq. 7.15 may be reduced to the expression of Eq. 7.16 (see Resource Material 20 for more on this). Further grouping of terms leads to the expression of Eq. 7.17, where k incorporates all terms other than P_0 and c:

$$P_F = \varphi P_0 \left(1 - 10^{-\varepsilon bc}\right) \qquad (7.15)$$

$$P_F = \varphi P_0\, 2.303\, \varepsilon bc \qquad (7.16)$$

$$P_F = kP_0 c \qquad (7.17)$$

where:

k = constant of proportionality
P_F = as in Eq. 7.13
c = as in Eq. 7.5

Equation 7.17 is particularly useful because it emphasizes two important points that are valid for the conditions assumed when deriving the equation, particularly the assumption that analyte concentrations are kept relatively low. First, the fluorescent signal will be directly proportional to the analyte concentration, assuming other parameters are kept constant. This is very useful because a linear relationship between signal and analyte concentration simplifies data processing and assay troubleshooting. Second, the sensitivity of a fluorescent assay is proportional to P_0, the power of the incident beam, the implication being that the sensitivity of a fluorescent assay may be modified by adjusting the source output.

Equations 7.16 and 7.17 will eventually break down if analyte concentrations are increased to relatively high values. Therefore, the **linear concentration range** for each assay should be determined experimentally. A representative calibration curve for a fluorescence assay is presented in Fig. 7.13. The nonlinear portion of the curve at relatively high analyte concentrations results from decreases in the fluorescence yield per unit concentration. The fluorescence yield for any given sample also is dependent on its environment. Temperature, solvent, impurities, and pH may influence this parameter. Consequently, it is imperative that these environmental parameters be accounted for in the experimental design of fluorescence assays. This may be particularly important in the preparation of appropriate reference standards for quantitative work.

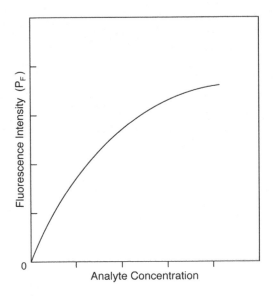

Relationship between the solution concentration of a fluorescent analyte and that solution's fluorescence intensity. Note that there is a linear relationship at relatively low analyte concentrations that eventually goes nonlinear as the analyte concentration increases

7.4 SUMMARY

UV and Vis absorption and fluorescence spectroscopy are used widely in food analysis. (See Chap. 8, Table 8.2, for a comparison of these types of spectroscopy, including their applications. The table also allows for comparison with other types of spectroscopy.) These techniques may be used for either qualitative or quantitative measurements. Qualitative measurements are based on the premise that each analyte has a unique set of energy spacings that will dictate its absorption/emission spectrum. Hence, qualitative assays generally are based on the analysis of the absorption or emission spectrum of the analyte. In contrast, quantitative assays most often are based on measuring the absorbance or fluorescence of the analyte solution at one wavelength. Quantitative absorption assays are based on the premise that the absorbance of the test solution will be a function of the solution's analyte concentration.

Under optimum conditions, there is a direct linear relationship between a solution's absorbance and its analyte concentration. The equation describing this linear relationship is known as Beer's law. The applicability of Beer's law to any given assay always should be verified experimentally by means of a calibration curve. The calibration curve should be established at the same time and under the same conditions that are used to measure the test solution. The analyte concentration of the test solution then should be estimated from the established calibration curve.

Molecular fluorescence methods are based on the measurement of radiation emitted from excited analyte molecules as they relax to lower energy levels. The analytes are raised to the excited state as a result of photon absorption. The processes of photon absorption and fluorescence emission occur simultaneously during the assay. Quantitative fluorescence assays are generally one to three orders of magnitude more sensitive than corresponding absorption assays. Like absorption assays, under optimal conditions there will be a direct linear relationship between the fluorescence intensity and the concentration of the analyte in the unknown solution. Most molecules do not fluoresce and, hence, cannot be assayed by fluorescence methods.

Instruments used for absorption and fluorescence methods have similar components, including a radiation source, wavelength selector(s), sample-holding cell(s), radiation detector(s), and a readout device.

7.5 STUDY QUESTIONS

1. Why is it common to use absorbance values rather than transmittance values when doing quantitative UV-Vis spectroscopy?
2. For a particular assay, the plot of absorbance vs. concentration is not linear; explain the possible reasons for this.
3. What criteria should be used to choose an appropriate wavelength at which to make absorbance measurements, and why is that choice so important?
4. In a particular assay, the absorbance reading on the spectrophotometer for one sample is 2.033 and for another sample 0.032. Would you trust these values? Why or why not?
5. Explain the difference between electromagnetic radiation in the UV and Vis ranges. How does quantitative spectroscopy using the UV range differ from that using the Vis range?
6. What is actually happening inside the spectrophotometer when the analyst "sets" the wavelength for a particular assay?
7. Considering a typical spectrophotometer, what is the effect of decreasing the exit slit width of the monochromator on the light incident to the sample?
8. Describe the similarities and differences between a phototube and a photomultiplier tube. What is the advantage of one over the other?
9. Your lab has been using an old single-beam spectrophotometer that must now be replaced by a new spectrophotometer. You obtain sales literature that describes single-beam and double-beam instruments. What are the basic differences between a single-beam and a double-beam

spectrophotometer, and what are the advantages and disadvantages of each?

10. Explain the similarities and differences between UV-Vis spectroscopy and fluorescence spectroscopy with regard to instrumentation and principles involved. What is the advantage of using fluorescence spectroscopy?

7.6 PRACTICE PROBLEMS

1. A particular food coloring has a molar absorption coefficient of 3.8×10^3 cm^{-1} M^{-1} at 510 nm.

   (a) What will be the absorbance of a $2 \times 10^{-4} M$ solution in a 1-cm cuvette at 510 nm?

   (b) What will be the percent transmittance of the solution in (a)?

2. (a) You measure the percent transmittance of a solution containing chromophore X at 400 nm in a 1-cm pathlength cuvette and find it to be 50%. What is the absorbance of this solution?

   (b) What is the molar absorption coefficient of chromophore X if the concentration of X in the solution measured in question 2a is 0.5 mM?

   (c) What is the concentration range of chromophore X that can be assayed if, when using a sample cell of pathlength 1, you are required to keep the absorbance between 0.2 and 0.8?

3. What is the concentration of compound Y in an unknown solution if the solution has an absorbance of 0.846 in a glass cuvette with a pathlength of 0.2 cm? The absorptivity of compound Y is 54.2 cm^{-1} (mg/mL)$^{-1}$ under the conditions used for the absorption measurement.

4. (a) What is the molar absorption coefficient of compound Z at 295 and 348 nm, given the absorption spectrum shown in Fig. 7.14 (which was obtained using a UV-Vis spectrophotometer and a 1 mM solution of compound Z in a sample cell with a pathlength of 1 cm)?

   (b) Assume you decide to make quantitative measurements of the amount of compound Z in different solutions. Based on the above spectrum, which wavelength will you use for your measurements? Give two reasons why this is the optimum wavelength.

Answers

1. (a)=0.76, (b)=17.4

   This problem requires a knowledge of the relationship between absorbance and transmittance and the ability to work with Beer's law.

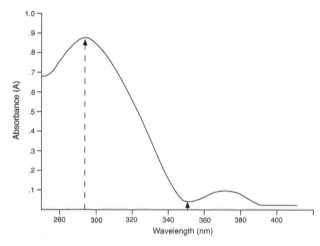

7.14 figure   Absorption spectrum of compound Z, to be used in conjunction with problems 4a and 4b

Given: molar absorption coefficient $=3.8 \times 10^3$ cm^{-1} M^{-1}

(a) Use Beer's law: A = εbc (see Eq. 7.5 of the text)

   where:

   $$\varepsilon = 3.8 \times 10^3 \, \text{cm}^{-1} M^{-1}$$
   $$b = 1 \text{cm}$$
   $$c = 2 \times 10^{-4} M$$

   Plugging into Beer's law gives the answer: Absorbance $=0.76$

(b) Use definition of absorbance: $A = -\log T$ (see Eq. 7.3 of text)

   where:

   $$T = P / P_0$$

   Rearranging Eq. 7.3:

   $$-A = \log T$$
   $$10^{-A} = T$$

   $A = 0.76$ [from part (a) of question]

   $$10^{-.76} = .1737 = T$$

   $$\%T = 100 \times T$$

   (combining Eqs. 7.1 and 7.2 of text)

   Answer: $\%T = 17.4$

2. (a)$=301$, (b)$=602$ cm^{-1} M^{-1}, (c)$=0.33 \times 10^{-3} M$ to $1.33 \times 10^{-3} M$

   This problem again requires knowledge of the relationship between absorbance and transmittance and the manipulation of Beer's law. Care must be taken in working with the appropriate concentration units.

(a) $T = 0.5$

Use $A = -\log T = -\log .5 = .301$

Answer: .301

(b) Given that the solution in part (a) is 0.5 mM (equivalent to $5 \times 10^{-4} M$)
Rearranging Beer's law: $\varepsilon = A/(bc)$

$$\varepsilon = .301 / \left[(1 \text{ cm}) \times \left(5 \times 10^{-4} M \right) \right]$$

Answer: $\varepsilon = 602 \text{ cm}^{-1} M^{-1}$

(c) To answer the problem, find the concentration that will give an absorbance of 0.200 (lower limit) and the concentration that will give an absorbance of 0.800 (upper limit). In both cases, use Beer's law to determine the appropriate concentrations:

where:

$$c = A / eb$$

Lowest concentration $= 0.2/[(602 \text{ cm}^{-1}M^{-1})(1 \text{ cm})]$
$\qquad = 3.3 \times 10^{-4} M$ (i.e., 0.33 mM)

Highest concentration $= 0.8/[(602 \text{ cm}^{-1}M^{-1})(1 \text{ cm})]$
$\qquad = 1.3 \times 10^{-3} M$ (i.e., 1.33 mM)

3. 0.078 mg/mL
This problem illustrates (1) that concentration need not be expressed in units of molarity and (2) that the pathlength of the cuvette must be considered when applying Beer's law. In the present problem the analyte concentration is given in mg/mL: thus, the absorptivity must be in analogous units:

$$\text{Apply}: c = A / \varepsilon b$$

where:

$A = 0.846$

$\varepsilon = 54.2 \text{ cm}^{-1} \left(\text{mg} / \text{mL} \right)^{-1}$

$b = 0.2 \text{ cm}$

Answer: $0.078 \text{ mg} / \text{mL}$

4. (a) $= 860$ at 295 nm, 60 at 348 nm; (b) $= 295$ nm; optimum sensitivity and more likely to adhere to Beer's law.

This problem presents the common situation in which one wants to use absorbance spectroscopy for quantitative measurements but is unsure what wavelength to choose for the measurements. Furthermore, the absorptivity of the analyte at the different wavelengths of interest is unknown. A relatively simple way to obtain the necessary information is to determine the absorption spectrum of the analyte at a known concentration.

(a) The arrows on the provided spectrum indicate the points on the spectrum corresponding to 295 and 348 nm. The problem notes that the absorption spectrum was obtained using a 1 mM solution (i.e., $1 \times 10^{-3} M$ solution) of the analyte and that the pathlength of the cuvette was 1 cm. The answer to the problem is thus determined by taking the absorbance of the analyte at the two wavelengths in question and then plugging the appropriate data into Beer's law. It is somewhat difficult to get an exact absorbance reading from the presented spectrum, but we can estimate that the absorbance of the 1 mM solution is ~0.86 at 295 nm and ~0.06 at 348 nm.

Using $\varepsilon = A / bc$

Answer:

At 295 nm $\varepsilon = 0.86/[(1 \text{ cm}) (.001 M)] = 860 \text{ cm}^{-1} M^{-1}$

At 348 nm $\varepsilon = 0.06/[(1 \text{ cm}) (.001 M)] = 60 \text{ cm}^{-1} M^{-1}$

(b) In general, analysts strive to obtain maximum sensitivity for their assays, where sensitivity refers to the change in assay signal per unit change in analyte concentration (the assay signal in this case is absorbance). The absorbance values for the analyte at the different wavelengths, taken from the absorption spectrum, and/or the relative absorptivity values for the analyte at the different wavelengths, provide a good approximation of the relative sensitivity of the assay at different wavelengths (it is an approximation because we have not determined the variability/precision of the measurements at the different wavelengths). It can be seen from the given spectrum that absorbance "peaks" were at ~298 and ~370 nm. The sensitivity of the assay, relative to neighboring wavelengths, is expected to be maximum at these absorbance peaks. The peak at 295 nm is significantly higher than that at 370 nm, so the sensitivity of the assay is expected to be significantly higher at 295 nm. Thus, this would be the optimum wavelength to use for the assay. A second reason to choose 295 nm is because it appears to be in the middle of the "peak," and, thus, small changes in wavelength due to instrumental/operator limitations are not expected to appreciably change the absorptivity values. Therefore, the assay is more likely to adhere to Beer's law.

There are situations in which an analyst may choose to not use the wavelength corresponding to an overall maximum absorbance. For example, if there are known to be interfering compounds that absorb at 295 nm, then an analyst may choose to do take absorbance measurements at 370 nm.

REFERENCE

1. Skoog DA, Holler FJ, Crouch SR (2007) Principles of instrumental analysis, 6th edn. Brooks/Cole, Pacific Grove, CA

RESOURCE MATERIALS

- Brown CW (2009) Ultraviolet, visible, near-infrared spectrophotometers. In: Cazes J (ed) Ewing's analytical instrumentation handbook, 3rd edn. Marcel Dekker, New York
- Currell G (2000) Analytical instrumentation – performance characteristics and quality. Wiley, New York, pp 67–91
- DeBolt S, Cook DR, Ford CM (2006) L-Tartaric acid synthesis from vitamin C in higher plants. Proc Natl Acad Sci 103: 5608–5613
- Feinstein K (1995) Guide to spectroscopic identification of organic compounds. CRC, Boca Raton, FL
- Hargis LG (1988) Analytical chemistry – principleses and techniques. Prentice-Hall, Englewood Cliffs, NJ
- Harris DC (2015) Quantitative chemical analysis, 9th edn. W.H. Freeman, New York
- Harris DC, Bertolucci MD (1989) Symmetry and spectroscopy. Dover, Mineola, NY
- Ingle JD Jr, Crouch SR (1988) Spectrochemical analysis. Prentice-Hall, Englewood Cliffs, NJ
- Lakowicz JR (2011) Principles of fluorescence spectroscopy, 3rd edn. Springer, New York
- Milton Roy educational manual for the SPECTRONIC® 20 & 20D spectrophotometers (1989) Milton Roy Co., Rochester, NY
- Miwa H (2000) High-performance liquid chromatography determination of mono-, poly- and hydroxycarboxylic acids in foods and beverages as their 2-nitrophenylhydrazides J Chromatogr A 881: 365–385
- Owen T (2000) Fundamentals of UV–visible spectroscopy, Agilent Technologies. https://www.agilent.com/cs/library/primers/Public/59801397_020660.pdf
- Pavia DL, Lampman GM, Kriz GS Jr (1979) Introduction to spectroscopy: a guide for students of organic chemistry. W. B. Saunders, New York
- Pavia DL, Lampman GM, Kriz GS Vyvyan JA (2015) Introduction to spectroscopy. 5th Edition, Cengage Learning, Independence, KY
- Perkampus H-H (1994) UV-Vis spectroscopy and its applications. Springer, Berlin, Germany
- Robinson JW, Frame EMS, Frame GM II (2014) Undergraduate instrumental analysis, 7th edn. CRC Press, Inc., Boca Raton, FL
- Robinson JW, Frame EMS, Frame GM II (2014) Undergraduate instrumental analysis, 7th edn. CRC Press, Inc., Boca Raton, FL
- Royal Society of Chemistry (2016) Learning Chemistry - Spectra School. http://www.rsc.org/learn-chemistry/collections/spectroscopy/
- Shriner RL, Fuson RC, Curtin DY, Morrill TC (1980) The systematic identification of organic compounds – a laboratory manual. 6th edn. Wiley, New York
- Smith BC (2003) Quantitative spectroscopy: theory and practice, Academic Press, Amsterdam
- Thomas MJK, Ando DJ (1996) Ultraviolet and visible spectroscopy, 2nd edn. Wiley, New York
- Valeur B, Berberan-Santos MN (2013) Molecular fluorescence: principles and applications, 2nd edn., Wiley-VCH, New York
- Yadav LDS (2005) Organic spectroscopy. Kluwer Academic, Boston, MA

Infrared and Raman Spectroscopy

Luis Rodriguez-Saona (✉)
Department of Food Science and Technology,
The Ohio State University,
Columbus, OH 43210, USA
e-mail: rodriguez-saona.1@osu.edu

Huseyin Ayvaz
Department of Food Engineering,
Canakkale Onsekiz Mart University,
Canakkale 17020, Turkey
e-mail: huseyinayvaz@comu.edu.tr

Randy L. Wehling
Department of Food Science and Technology,
University of Nebraska,
Lincoln, NE 68583-0919, USA
e-mail: rwehling1@unl.edu

S. Nielsen (ed.), *Food Analysis*, Food Science Text Series,
DOI 10.1007/978-3-319-45776-5_8, © Springer International Publishing 2017

8.1 INTRODUCTION

Infrared (IR) **spectroscopy** refers to measurement of the absorption of different frequencies of IR radiation by foods or other solids, liquids, or gases. IR spectroscopy began in 1800 with an experiment by Herschel [1]. When he used a prism to create a spectrum from white light and placed a thermometer at a point just beyond the red region of the spectrum, he noted an increase in temperature. This was the first observation of the effects of IR radiation. By the 1940s, IR spectroscopy had become an important tool used by chemists to identify functional groups in organic compounds. In the 1970s, commercial near-IR (NIR) reflectance instruments were introduced that provided rapid quantitative determinations of moisture, protein, and fat in cereal grains and other foods. Today, IR spectroscopy is used widely in the food industry for both qualitative and quantitative analyses of ingredients and finished foods.

In this chapter, the techniques of mid- and near-IR and Raman spectroscopy are described, including the principles by which molecules absorb IR radiation, the components and configuration of commercial IR spectrometers, sampling methods for IR spectroscopy, and qualitative and quantitative applications of these techniques to food analysis. Infrared and Raman microspectroscopy will not be covered in this chapter, but rather are covered in Chap. 32, Sects. 32.3.2 and 32.3.3.

8.2 PRINCIPLES OF IR SPECTROSCOPY

8.2.1 The IR Region of the Electromagnetic Spectrum

Infrared radiation is electromagnetic energy with **wavelengths** (λ) longer than visible light but shorter than microwaves. Generally, wavelengths from 0.8 to 100 micrometers (μm) can be used for IR spectroscopy and are divided into the **near-IR** (0.8–2.5 μm; 12,500–4000 cm^{-1}), the **mid-IR** (2.5–15.4 μm; 4000–650 cm^{-1}), and the **far-IR** (15.4–100 μm; 650–100 cm^{-1}) regions. One μm is equal to 1×10^{-6} m. The near- and mid-IR regions of the spectrum are most useful for quantitative and qualitative analysis of foods.

IR radiation also can be measured in terms of its **frequency**, which is useful because frequency is directly related to the energy of the radiation by the following relationship:

$$E = h\nu \qquad (8.1)$$

where:

$E =$ energy of the system
$h =$ Planck's constant
$\nu =$ frequency in hertz

Frequencies are commonly expressed as **wave numbers** ($\bar{\nu}$, in reciprocal centimeters, cm^{-1}). Wave numbers are calculated as follows:

$$\bar{\nu} = 1 / (\lambda \text{ in cm}) = 10^4 / (\lambda \text{ in } \mu m) \qquad (8.2)$$

8.2.2 Molecular Vibrations

A molecule can absorb IR radiation if it vibrates in such a way that its charge distribution, and therefore its electric dipole moment, changes during the vibration. Although there are many possible vibrations in a polyatomic molecule, the most important vibrations that produce a change in dipole moment are stretching (symmetric and asymmetric) and bending (scissoring, rocking, twisting, wagging) motions. Examples of these vibrations for the water molecule are shown in Fig. 8.1. Note that the stretching motions vibrate at higher frequencies than the scissoring motion. Also, asymmetric stretches are more likely to result in a change in dipole moment, with corresponding absorption of IR radiation, than are symmetric stretches.

8.2.3 Factors Affecting the Frequency of Vibration

The basic requirement for absorption of infrared radiation is that there must be a net change in dipole moment during the vibration of the molecule or functional group. A molecular vibration can be thought of as a **harmonic oscillator** (Fig. 8.2a), with the energy level for any molecular vibration given by the following equation:

$$E = \left(v + \frac{1}{2}\right)\left(\frac{h}{2\pi}\right)\sqrt{k / \frac{m_1 m_2}{m_1 + m_2}} \qquad (8.3)$$

where:

$v =$ vibrational quantum number (positive integer values, including zero, only)
$h =$ Planck's constant
$k =$ force constant of the bond
m_1 and $m_2 =$ masses of the individual atoms involved in the vibration

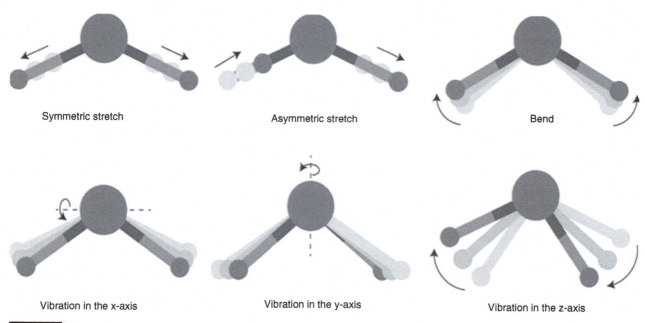

8.1 figure   Vibrational modes of the water molecule (Adapted from the SKC chemistry [67])

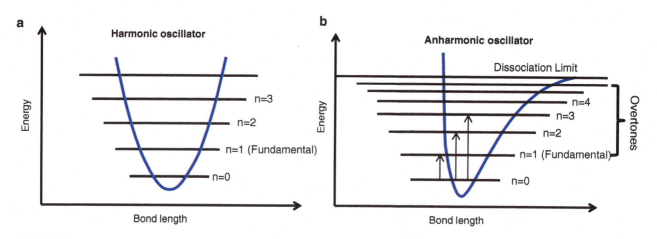

8.2 figure   Diagram of the differences in potential energy curves between the (**a**) harmonic and (**b**) anharmonic oscillator model

Note that the vibrational energy, and therefore the frequency of vibration, is directly proportional to the strength of the bond and inversely proportional to the mass of the molecular system. Thus, different chemical functional groups will vibrate at different frequencies. A vibrating molecular functional group can absorb radiant energy to move from the lowest ($v=0$) vibrational state to the first excited ($v=1$) state, and the frequency of radiation that will make this occur is identical to the initial frequency of vibration of the bond. This frequency is referred to as the **fundamental absorption**. The harmonic oscillator provides a good fit to explain bond stretching vibrations for fundamen-

tal vibrations. However, molecules also can absorb radiation to move to a higher ($v=2$ or 3) excited state, such that the frequency of the radiation absorbed is two or three times that of the fundamental frequency. These absorptions are referred to as **overtones**, and the intensities of these absorptions are much lower than the fundamental since these transitions are less favored. The **anharmonic oscillator model** (Fig. 8.2b) accounts for repulsion and attraction of the electron cloud and accommodates bond dissociation at higher energy levels. Overall, the fundamental vibrations are unaffected by the anharmonicity terms, but overtone transitions are influenced by anharmonicity, which must be taken into account when assessing the frequency of these higher frequency vibrations. Combination bands also can occur if two or more different vibrations interact to give bands that are sums of their fundamental frequencies. The model of the harmonic oscillator and its modification to account for anharmonicity allows explanation of the origin of many of the characteristic frequencies that can be assigned to particular combinations of atoms within a molecule [2]. The overall result is that each functional group within the molecule absorbs IR radiation in distinct wavelength bands rather than as a continuum.

8.3　MID-IR SPECTROSCOPY

Mid-IR spectroscopy measures a sample's ability to absorb light in the 2.5–15 μm (4000–650 cm^{-1}) region. Fundamental absorptions are primarily observed in this spectral region. Mid-IR spectroscopy is very useful in the study of organic compounds because the absorption bands are related to the vibrational modes of specific functional groups. The positioning of the band and its intensity are correlated with the energy of the bond, its environment, and its concentration in the matrix, making mid-IR spectroscopy ideal for both qualitative and quantitative applications.

8.3.1　Instrumentation

8.3.1.1　*Overview*
There are two types of spectrometers available for mid-IR analysis, dispersive and Fourier transform (FT) instruments. Dispersive systems have been available since the 1940s using prisms or gratings as dispersive elements. These systems contain components similar to ultraviolet–visible (UV–Vis) spectrometers, including a radiation source, a monochromator, a sample holder, and a detector connected to an amplifier system to record the spectra. In these systems, a filter, grating, or a prism is used to separate the IR radiation into its individual wavelengths. A major advance in the field of mid-IR spectroscopy was the development of Fourier transform infrared spectrometers (FTIR), which have mostly replaced the dispersive instruments due to dramatically improved quality of spectra and decreased time required to obtain data.

8.3.1.2　*Fourier Transform Instruments*
Compared to mid-IR dispersive instruments, FTIR spectrometers in food analysis allow for greater speed, higher sensitivity, superior wavelength resolution, and wavelength accuracy (details for advantages are in references [3, 4]). In **Fourier transform** (FT) instruments, the radiation is not dispersed, but rather all wavelengths arrive at the detector simultaneously,

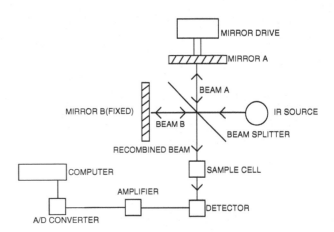

8.3 figure Block diagram of an interferometer and associated electronics typically used in an FTIR instrument

and a mathematical treatment is used to convert the results into a typical IR spectrum. Instead of a monochromator, the instrument uses an **interferometer**. A Michelson interferometer is the most commonly used design, and its mechanism is simple (Fig. 8.3). The infrared radiation from the source is split into two beams by a beam splitter, and each half of the beam goes to a mirror (either a fixed or moving mirror). The beams are reflected back and recombined at the beam splitter, resulting in interference that is directed to the sample (or reference) and then the detector. Motion of the moving mirror results in the change of optical path length between the two split beams so that constructive, destructive, and intermediate interference states occur (with destructive interference being dominant). The resulting output is referred to as an **interferogram**, which is the intensity measured by the detector as a function of the position of the moving mirror. When a sample interacts with the recombined beam ahead of the detector, molecules absorb at their characteristic frequencies, and thus the radiation reaching the detector is modified (Fig. 8.4). Once the data are collected, a mathematical transformation called a "Fourier transform" converts the interferogram from time domain (intensity versus time) to an IR spectrum in the frequency domain (intensity versus frequency). A computer allows the mathematical transformation to be completed rapidly.

　　The common **radiation sources** for mid-IR spectrometers are inert solids heated electrically to 1000–1800 °C. Three popular types of sources are the Nernst glower (constructed of rare-earth oxides), Globar (constructed of silicon carbide), and a Nichrome coil wrapped around a ceramic core that glows when an electrical current is passed through it. They all produce continuous radiation, but with different radiation energy profiles.

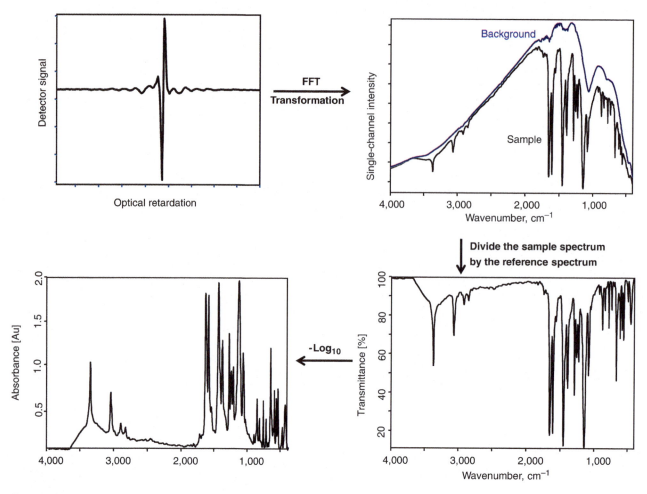

8.4 figure Illustration on the process to convert an interferogram into an infrared spectrum by using the Fourier transform algorithm

Detectors include **thermocouples** for which output voltage varies with temperature changes caused by varying levels of radiation striking the detector. In a **Golay detector**, the radiation strikes a sealed tube of xenon gas warming the gas and causing pressure changes within the tube. However, most modern instruments use either **pyroelectric detectors**, such as deuterated triglycine sulfate (DTGS) crystals, or solid-state **semiconductor detectors**. Variation in the amount of radiation striking a DTGS detector causes the temperature of the detector to change, which results in a change in the dielectric constant of the DTGS element. The resulting change in capacitance is measured. Semiconductor detectors, such as those made from a mercury–cadmium–telluride (MCT) alloy, have conductivities that vary according to the amount of radiation striking the detector surface. MCT detectors respond faster to smaller changes in radiation intensity than other detectors; however, they typically require

cryogenic cooling. DTGS and MCT detectors are the most commonly used detectors in Fourier transform instruments.

8.3.2 Sample Handling Techniques

Transmission mode is based on the IR beam passing through a sample that is placed in between two IR transparent windows. Liquids are often measured by **transmission IR spectroscopy**. Because absorptivity coefficients in the mid-IR are high, cells with path lengths of only 0.01–1.0 mm are commonly used. Quartz and glass absorb in the mid-IR region, so cell windows are made of non-absorbing materials such as halide or sulfide salts. Halide salts are soluble in water, and care must be taken when selecting cells for use with aqueous samples. Cells also are available with windows made from more durable and less soluble materials, such as zinc selenide, but are more expensive than those with halide salt windows. Liquid cells must be free of air bubbles and extra care needs to be taken when cleaning between samples.

Transmission spectra of solids can be obtained by finely grinding a small amount of the sample with potassium bromide (KBr), pressing the mixture into a pellet under high pressure and inserting the pellet into the IR beam. Limitations of this technique include difficulty of handling and storing the hygroscopic KBr and the complexity and time required to make a good KBr pellet. An alternative technique is to disperse a finely divided solid in Nujol mineral oil to form a mull.

Transmission spectra can be obtained from gas samples using a sealed 2–10 cm glass cell with IR transparent windows. For trace analysis, multiple-pass cells are available that reflect the IR beam back and forth through the cell many times to obtain path lengths as long as several meters. FTIR instruments also can be interfaced to a gas chromatograph, to obtain spectra of compounds eluting from the chromatography column.

Attenuated total reflectance (ATR) is a widely applied sampling technique in infrared spectroscopy because it requires little or no sample preparation, eliminates variation in cell path lengths, and provides consistent spectra collection. ATR allows for obtaining spectra from solid samples that are too thick for transmission measurements, e.g., pastes such as peanut butter and viscous liquids. ATR works based on the attenuation effect of infrared light (Fig. 8.5) when it is directed at an interface between an internal reflection element (crystal) with high refractive index properties (i.e., zinc selenide (ZnSe), thallium iodide–thallium bromide (KRS-5), germanium (Ge), silicon (Si), and diamond) and a low refractive index material (food sample) on its surface. Upon the interaction with the reflecting surface, radiation called an "evanescent wave" is formed, exits the high refractive index material, and slightly penetrates into the sample. The sample material selectively absorbs, the intensity of the reflected radiation is decreased at wavelengths for which the sample absorbs radiation, and the final attenuated radiation exiting the crystal is measured as being unique for the sample analyzed.

Radiation is not transmitted through the sample; therefore, there is no need for the sample to be thin enough to allow the transmission of the incident light. Since the penetration depth of the radiation is limited to a few micrometers (μm), the same spectrum is obtained regardless of the amount of the sample placed on the surface, and there is no need to dilute the samples.

The physical state of the sample is an important factor because it must be in intimate contact with the ATR crystal to obtain a good ATR spectrum. Liquids and pastes usually exhibit better ATR spectra than solid samples. A pressure clamping system is used

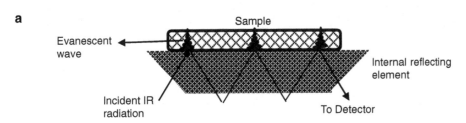

a

Sample

Evanescent wave

Internal reflecting element

Incident IR radiation

To Detector

Triple reflection ATR crystal

b   **Guidelines for selection of ATR crystals**

Material	Spectral range (cm^{-1})	Hardness (Kg/mm)	pH range	Refractive Index	Depth of penetration at 45°
Diamond	50,000–2,500	9,000	1–14	2.4	1.66
Germanium	5,000–550	780	1–14	4.0	0.65
Silicon	8,333–33	1,150	1–12	3.4	0.81
KRS-5	17,900–250	40	5–8	2.4	0.85
ZnSe	20,000–500	130	5–9	2.4	1.66
AMTIR	11,000–725	170	1–9	2.5	1.46

8.5 figure (**a**) Illustration of the reflection phenomena in a triple reflection attenuated total reflection accessory and formation of the evanescence wave into the sample. (**b**) Characteristics of common ATR crystals

with solid samples to deform the sample, increasing the extent of contact between the ATR crystal and the sample.

8.3.3 Applications of Mid-IR Spectroscopy

8.3.3.1 *Absorption Bands of Organic Functional Groups*

Infrared spectroscopy monitors the interaction of functional groups in chemical molecules with infrared light resulting in predictable vibrations that provide a "fingerprint" characteristic of chemical substances present in the sample. Spectra in the mid-IR region have well-resolved bands that can be assigned to functional groups of the components of foods. The positioning of the band facilitates structural characterization, and its intensity correlates with its concentration in the matrix, allowing for both qualitative and quantitative applications.

Spectra are commonly presented in wave numbers plotted on the x-axis and either percent transmittance or absorbance plotted on the y-axis. The mid-IR spectra of selected foods are shown in Fig. 8.6 displaying the major absorption bands that can be associated with functional groups (Table 8.1) in fat-, protein-, and carbohydrate-rich commodities.

The unique spectral profile can be used to identify specific functional groups present in an unknown substance. Comparing the mid-IR spectrum to a set of standard spectra and determining the closest match can accomplish identification of chemical compounds. Spectral libraries are available from several sources, but probably the largest collection of standards is the Sadtler Standard Spectra (Sadtler Division of Bio-Rad Inc., Philadelphia, PA) that contains over 225,000

infrared spectra. Algorithms are used to compare the unknown spectrum to each spectrum in the reference database, and the hit quality index (HQI) is determined, indicating the similarity between spectra. Several HQI values can be generated for the unknown compounds, and the software will sort and display the best matches in a search report. Noise and spectral artifacts can impact the HQI and lead to mistakes in identification; thus, it is imperative to perform visual comparisons to confirm a good match. Spectral searches will most commonly be done on purified substances, rather than foods or commodities.

8.3.3.2 *Applications*

Mid-IR spectroscopic measurements obey **Beer's law**, although deviations may be greater than in UV–Vis spectroscopy due to the low intensities of IR sources, the low sensitivities of IR detectors, and the relative narrowness of mid-IR absorption bands. One of the first and most extensive uses of this technique is the **infrared milk analyzer**, which has the ability to analyze hundreds of samples per hour. The fat, protein, and lactose contents of milk can be determined simultaneously with one of these instruments. The ester carbonyl groups of lipid absorb at 5.73 μm ($1742\ cm^{-1}$), the amide groups of protein absorb at 6.47 μm ($1545\ cm^{-1}$), and the hydroxyl groups of lactose absorb at 9.61 μm ($1045\ cm^{-1}$). These automated instruments homogenize the milk fat globules to minimize light scattering by the sample and then pump the milk into a flow-through cell through which the IR beam is passed. The instrument is calibrated using samples of known concentration to establish the slope and intercept of a Beer's law plot. Official methods have been adopted

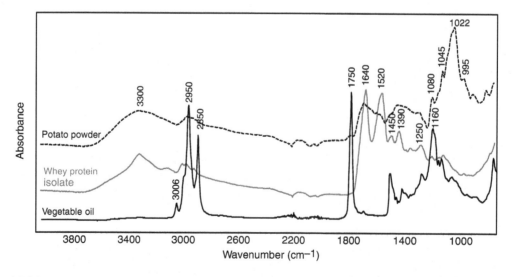

8.6 figure Mid-IR spectra of corn oil, whey protein isolate, and potato flour measured by triple-reflection ATR unit. Frequency in wave numbers is plotted on the x-axis, with intensity on the y-axis

8.1 table

Characteristic mid-IR and near-IR absorption frequencies of major food components

Food components	Infrared region	Frequency (cm⁻¹)	Assignment
Fats	Mid-IR	3006	cis-Olefinic groups
		3000–2800	C–H asym. and sym. str. of long-chain fatty acids
		1740	C=O str. of fatty acid esters
		1711	C=O str. of fatty acids
		1475–1435	Asym. bending defor. –CH₃ in branched alkanes
		1465	=C–H cis-bending
		1350–1395	C–H symmetric bending in –CH₃
		1350–1150	C–H bending
		1240 and 1163	–C–O and –CH₂– str. bending
		1150–1000	–C=O str.
		966	Isolated out of plane trans-C̲u̲C bending
		914	cis –C=C–H bending out of plane
	Near-IR	8700–8100 (1150–1235 nm)	C–H str., second overtone
		8563 (1168 nm)	cis Double bond
		7209 (1387 nm)	Combination C–H str. and C–H bending
		5807 (1722 nm), 5681 (1760 nm)	C–H str., first overtone
		4705 (2125 nm)	C–H vibrations in isolated C=C bonds
		4336 (2306 nm), 4269 (2342 nm)	Combination bands of C–H and C–O str.
Water	Mid-IR	3500	O–H str.
		1650	O–H bending
	Near-IR	6900 (1450 nm)	First overtones str. free OH
		5150 (1940 nm)	OH combination bands
Proteins	Mid-IR	3300	Amide A
		1640	Amide I
		1540	Amide II
		1330–1230	Amide III
	Near-IR	5000–4550	Amides
		4855 (2016 nm)	Combination of amide A and amide II
		4580 (2180 nm)	Combination of amide A and amide III
Carbohydrates	Mid-IR	1745	Ester, pectin
		1630–1605	Carboxylate, pectin
		Shifts at 1617 and 1420	Metal coordination by the pectate chains
		1460–1340	C–C–H and C–O–H deformations
		1250–950	Endo- and/or exocyclic C–C and C–O bonds
		1150	Characteristic of pyranose ring
		1110	CO ring, C–4–O, C–6–O
		1080	Typical bending of C–1–H
		1060	C–1–OH (fructose residue)
		1030	Typical C–4–OH vibration
		1050–1020	Starch retrogradation, increased crystallinity
		995	Glycosidic linkage
		950–750	α and β anomeric region of saccharide
	Near-IR	69405 (1440 nm)	OH str., first overtone, crystalline structure
		5924 (1688 nm)	C–H stretch first overtone
		5882 (1700 nm)	C–H str. methyl groups from cellulose or lignin
		4662 (2100 nm)	Combination C–O str./OH bend – carbohydrates
		4280 (2336 nm)	Cellulose and starch
		4386 (2280 nm)–4292 (2330 nm)	Combination C–H str. and CH₂ deformation
Aromatics	Mid-IR	3100–3000	Aromatic –CH stretch
		1600	–C=C– stretch
Alcohols	Mid-IR	3600–3200	–OH stretch
		1500–1300	–OH bend
		1220–1000	C–O stretch
	Near-IR	6850 (1460 nm)–6240 (1600 nm)	First overtone –OH
		6800 (1470 nm) and 7100 (1280 nm)	Typical first overtone bands of bonded OH–H bonding
		4550 (1800 nm)–5550 (2200 nm)	Combination OH str. and OH bending
Ethers	Mid-IR	1220–1000	C–O asymmetric stretch

for the IR milk analyzers, and specific procedures for operation of these instruments are given [5, 6].

There are many additional applications of mid-IR spectroscopy to food analysis. Examples of applications of mid-IR spectroscopy to foods are found in review articles [7–9]. Due to the complex nature of infrared spectra, multivariate statistical analysis techniques (chemometrics) must often be used to extract information from the infrared spectra, allowing for classification and quantitative analysis of multiple components in foods. Instrument calibration using chemometric techniques is discussed in more detail in Sect. 8.5.3.

8.4   NEAR-IR SPECTROSCOPY

Measurements in the **near-IR** (NIR) spectral region (0.7–2.5 µm, equal to 700–2500 nm) are more widely used for quantitative analysis of foods than are mid-IR measurements. Several commercial instruments are available for compositional analysis of foods using near-IR spectroscopy. A major advantage of near-IR spectroscopy is its ability to measure directly the composition of solid food products by the use of diffuse reflection techniques.

8.4.1   Principles

8.4.1.1 *Principles of Diffuse Reflection Measurements*

When radiation strikes a solid or granular material, part of the radiation is reflected from the sample surface. This mirrorlike reflection is called **specular reflection** and gives little useful information about the sample. Most of the specularly reflected radiation is directed back toward the energy source. Another portion of the radiation will penetrate through the surface of the sample and be reflected off several sample particles before it exits the sample. This is referred to as **diffuse reflection**, and this diffusely reflected radiation emerges from the surface at random angles through 180°. Each time the radiation interacts with a sample particle, the chemical constituents in the sample can absorb a portion of the radiation. Therefore, the diffusely reflected radiation contains information about the chemical composition of the sample, as indicated by the amount of energy absorbed at specific wavelengths.

The amount of radiation penetrating and leaving the sample surface is affected by the size and shape of the sample particles. Compensation for this effect may be achieved by grinding solid or granular materials with a sample preparation mill to a fine, uniform particle size, or by applying mathematical corrections when the instrument is calibrated [10].

8.4.1.2 *Absorption Bands in the Near-IR Region*

The absorption bands observed in the near-IR region are primarily overtones and combinations. Therefore, the absorptions tend to be weak in intensity. However, this is actually an advantage, since absorption bands that have sufficient intensity to be observed in the near-IR region arise primarily from functional groups that have a hydrogen atom attached to a carbon, nitrogen, or oxygen, which are common groups in the major constituents of food such as water, proteins, lipids, and carbohydrates. Table 8.1 lists the absorption bands associated with a number of important food constituents.

The absorption bands in the near-IR region tend to be broad and frequently overlap, yielding spectra that are quite complex. However, these broad bands are especially useful for quantitative analysis. Typical near-IR spectra of wheat, dried egg white, and cheese are shown in Fig. 8.7. Note that strong absorption bands associated with the -OH groups of water which are the dominant features in the spectrum of cheese, containing 30–40 % moisture, and they are still prominent even in the lower moisture wheat and egg white samples. Bands arising from protein (2060 and 2180 nm) in the wheat sample are partially obscured by a strong starch absorption band and centered at 2100 nm. Relatively sharp absorption bands arising from -CH groups in lipid can be observed at 2310, 2350 nm, and 1730 nm and are distinctly observable in the cheese spectrum.

8.4.2   Instrumentation

Two commercial near-IR spectrometers are shown in Fig. 8.8. The radiation source in most near-IR instruments is a tungsten–halogen lamp with a quartz enve-

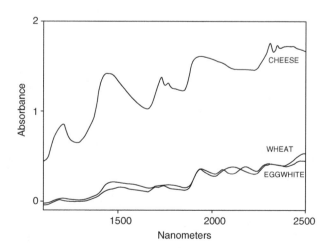

8.7 *figure*   Near-IR spectra of cheese, wheat, and dried egg white plotted as log (1/R) vs. wavelength in nm

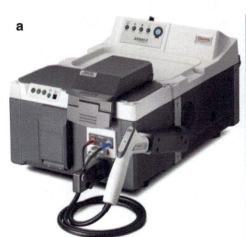
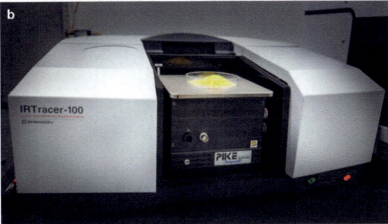

| 8.8 figure | Modern commercial near-IR instrument. (**a**) Thermo Scientific Antaris II. (**b**) Shimadzu IRTracer-100 equipped with an NIR integrating sphere accessory from PIKE Technologies (Photographs courtesy of Thermo Fisher Scientific and Shimadzu Scientific Instruments) |

lope, similar to a projector lamp. These lamps emit significant amounts of radiation in both the visible and near-IR spectral regions. Semiconductor detectors are most commonly used in near-IR instruments, with silicon detectors used in the 700–1100 nm range and lead sulfide used in the 1100–2500 nm region. In situations for which a rapid response to changing light intensity is needed, such as in online monitoring, indium–gallium–arsenide (InGaAs) detectors can be used. Many InGaAs detectors are limited to a maximum wavelength of 1700 nm, although commercial InGaAs detectors with a range extended to longer wavelengths are now available. Most commercial near-IR instruments use monochromators, rather than interferometers, although some commercial instruments are now using FT technology. Monochromator-based instruments may be of the scanning type, in which a grating is used to disperse the radiation by wavelength, and the grating is rotated to impinge a single wavelength (or more appropriately, a narrow band of wavelengths) onto a sample at any given time. Using this arrangement, it takes several seconds to collect a spectrum from a sample over the entire near-IR region. Some rapid scanning instruments impinge light over the entire near-IR region onto the sample. The reflected or transmitted light then is directed onto a fixed grating that disperses the light by wavelength and also focuses it onto a multichannel array detector that measures all wavelengths at once. These instruments can obtain a spectrum from a sample in less than 1 s.

Instruments dedicated to specific applications can use optical interference filters to select 6–20 discrete wavelengths that can be impinged on the sample. The filters are selected to obtain wavelengths that are known to be absorbed by the sample constituents. The instrument inserts filters one at a time into the light

beam to direct individual wavelengths of radiation onto the sample.

Either **reflection** or **transmission** measurements may be made in near-IR spectroscopy, depending on the type of sample. In the reflection mode, used primarily for solid or granular samples, it is desirable to measure only the diffusely reflected radiation that contains information about the sample. In many instruments, this is accomplished by positioning the detectors at a 45° angle with respect to the incoming IR beam, so that the specularly reflected radiation is not measured (Fig. 8.9a). Other instruments use an integrating sphere, which is a gold-coated metallic sphere with the detectors mounted inside (Fig. 8.9b). The sphere collects the diffusely reflected radiation coming at various angles from the sample and focuses it onto the detectors. The specular component escapes from the sphere through the same port by which the incident beam enters and strikes the sample.

Samples often are prepared by packing the food tightly into a cell against a quartz window, thereby providing a smooth, uniform surface from which reflection can occur. Quartz does not absorb in the near-IR region. At each wavelength, the intensity of light reflecting from the sample is compared to the intensity reflected from a non-absorbing reference, such as a ceramic or fluorocarbon material. Reflectance (*R*) is calculated by the following formula:

$$R = I / I_0 \qquad (8.4)$$

where:

I = intensity of radiation reflected from the sample at a given wavelength

I_0 = intensity of radiation reflected from the reference at the same wavelength

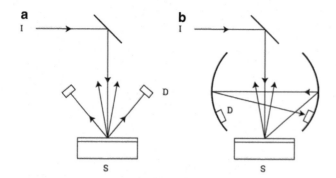

8.9
figure

Typical instrument geometries for measuring diffuse reflectance from solid food samples. Radiation from the monochromator (*I*) is directed by a mirror onto the sample (*S*). Diffusely reflected radiation is measured directly by detectors (*D*) placed at a 45° angle to the incident beam (**a**) or is collected by an integrating sphere and focused onto the detectors (**b**). In both cases, the specularly reflected radiation is not measured

Reflectance data are expressed most commonly as log (1/*R*), an expression analogous to absorbance in transmission spectroscopy. Reflectance measurements also are expressed sometimes as differences, or derivatives, of the reflectance values obtained from adjacent wavelengths:

$$\left(\log R_{\lambda 2} - \log R_{\lambda 1}\right) \qquad (8.5)$$

or

$$\left(2\log R_{\lambda 2} - \log R_{\lambda 1} - \log R_{\lambda 3}\right) \qquad (8.6)$$

These derivative values are measures of the changes in slope of the spectrum, where λ_1, λ_2, and λ_3 are adjacent wavelengths typically separated by 5–20 nm, with the higher numbers representing longer wavelengths.

Transmission measurements also can be made in the near-IR region, and this is usually the method of choice for liquid samples. A liquid is placed in a quartz cuvette and the absorbance measured at the wavelengths of interest. Transmission measurements also can be taken from solid samples, but generally only in the 700–1100 nm range. In this wavelength region, the absorption bands are higher overtones that are very weak, allowing the radiation to penetrate through several millimeters of a solid sample. The use of transmission measurements can minimize the degree of sample preparation needed. Since the IR beam passes through the entire sample, the need for a smooth, homogeneous sample surface is reduced.

Near-IR energy can be transmitted through a fiber optic cable some distance from the monochromator or interferometer, allowing reflection or transmission measurements to be made remotely from the

instrument. This is very useful to take measurements in a processing plant environment. Commercial probes are available that can be inserted directly into bulk granular materials, or inserted into a pipe carrying a liquid.

As with mid-IR, near-IR imaging instruments are now commercially available. These instruments use an array detector so that a digital image of a food sample can be obtained at various wavelengths or a spectrum can be obtained from a single pixel in a digital image. This technique is often referred to as hyperspectral imaging. It holds much potential for evaluating sample heterogeneity, or identifying small features or contaminants on an intact food sample.

8.4.3 Quantitative Methods Using Infrared Spectroscopy

Infrared instruments can be calibrated to measure various constituents in foods and agricultural commodities. Because of the overlapping nature of the near-IR absorption bands, it is usually necessary to take measurements at two or more wavelengths to quantify a food component reliably. Multivariate statistical techniques (**chemometrics**) are used to relate the spectral data collected at multiple wavelengths to the concentration of the component of interest in the food [11–13]. The simplest statistical technique used is **multiple linear regression** (MLR), which applies an equation of the following form to predict the amount of a constituent present in the food from the spectral measurements:

$$\% \text{ constituent} = z + a\log\left(1/R_{\lambda 1}\right) + b\log\left(1/R_{\lambda 2}\right)$$
$$+ c\log\left(1/R_{\lambda 3}\right) + \ldots \qquad (8.7)$$

where each term represents the spectral measurement at a different wavelength multiplied by a corresponding coefficient. Each coefficient and the intercept (*z*) are determined by multivariate regression analysis.

Absorbance or derivatized reflectance data also can be used in lieu of the log (1/*R*) format. The use of derivatized reflectance data has been found to provide improved results in some instances, particularly with samples that may not have uniform particle sizes. Other mathematical techniques also are available that can be applied to the reflectance data to correct for the effects of nonuniform particle size [10].

To **calibrate** an infrared instrument for quantitative measurement, a set of samples is obtained that represents the product to be measured and contains the component of interest over the expected range of interest. The samples are then analyzed using the conventional method of analysis (e.g., for protein analysis, use Kjeldahl or Dumas methods; Chap. 18, Sects. 18.2.1 and 18.2.2), and the infrared spectrum of each sample is collected. A computer-assisted MLR analysis is then performed to determine the combinations of wavelengths that best predict the concentration of the component of interest and the coefficients associated with each wavelength, as shown in Eq. (8.7). In chemometric techniques such as MLR, the wavelengths are chosen based on statistical correlation with the component being measured. However, the results should always be inspected to make sure that the wavelengths selected make sense from a spectroscopic perspective. Each calibration also should be tested using a second set of independent samples. Then, if the calibration yields satisfactory results, it can be used for routine analysis.

When using MLR, it may sometimes be difficult to include enough wavelengths to adequately define the relationship between the spectral and composition data. Adding too many wavelengths may "overfit" the calibration so that it does not apply well to samples that were not part of the original set. This can occur because the responses of individual wavelengths are highly intercorrelated. To overcome this problem and to obtain more robust predictions, multivariate calibration methods such as **partial least squares** (PLS) regression and **principal component regression** (PCR) can be used. PLS and PCR are often referred to as "data compression" techniques, since they take the spectral variation from the entire wavelength range and express most of that variation with a smaller number of variables that are not correlated. These variables then are used in to develop a regression equation. PLS and PCR often provide improved results compared to MLR because they can use information from the entire spectrum with less risk of "overfitting" the results. For this reason, these two techniques are now the most widely used methods for calibrating mid-IR and near-IR instruments for quantitative analysis. Readers interested in a more detailed explanation of these techniques should consult the references [11–16].

8.4.4 Qualitative Analysis by Infrared Spectroscopy

Infrared spectroscopy can be used to classify a sample into one of two or more groups, rather than to provide quantitative measurements. Classification techniques, such as **principal component analysis** (PCA), soft independent modeling of class analogy (SIMCA), or discriminant analysis, can be used to compare the infrared spectrum of an unknown sample to the spectra of samples from different groups. The unknown sample then is classified into the group to which its spectrum is most similar. While this technique has historically been used in the chemical and pharmaceutical industries for raw material identification, it is becoming more widely used for food applications, including the classification of wheat as hard red spring or hard red winter [17], the identification of orange juice samples from different sources [18, 19], authenti-

cation of the source of olive oils [20–22], and discrimination of beef [23–25]. Readers interested in a more detailed explanation of these classification techniques should consult the references [11, 26].

8.4.5 Applications of Near-IR Spectroscopy to Food Analysis

Theory and applications of near-IR spectroscopy to food analysis have been discussed in several publications [27–30]. The technique is widely used throughout the grain, cereal products, and oilseed processing industries. Near-IR techniques using measurements from ground or whole grain samples have been adopted as approved methods of analysis by AACC International [31] for measuring protein in barley, oats, rye, triticale, wheat, and wheat flour, as well as moisture, protein, and oil in soybeans. These approved methods describe the instruments available for making these measurements, including a list of current manufacturers with contact information in Method 39–30, as well as the proper techniques for preparing samples and calibrating the instruments. Near-IR instruments now are used by the official grain inspection agencies in both the US and Canada for measuring protein, moisture, and oil in cereals and oilseeds.

Food components such as protein and dietary fiber can be determined successfully in a number of cereal-based foods using near-IR spectroscopy [32–34]. Modern instruments and calibration techniques allow a wide variety of products, such as cookies, granola bars, and ready-to-eat breakfast foods, to be analyzed using the same calibration.

Near-IR spectroscopy also can be used for numerous other commodities and food products. The technique has been used successfully to evaluate composition and quality of red meats and processed meat products [35–37], poultry [38], and fish [39]. Near-IR spectroscopy is useful also for analyzing a number of dairy products and nondairy spreads, including measuring moisture in butter and margarine [40]; moisture, fat, and protein in cheese [41, 42]; and lactose, protein, and moisture in milk and whey powders [43]. Near-IR techniques also have shown promise for measuring total sugars and soluble solids in fruits, vegetables, and juices [44–46], are being used commercially for monitoring the sugar content in corn sweeteners [47], and can be used to quantitate sucrose and lactose in chocolate [48].

Near-IR spectroscopy also is showing potential for measuring specific chemical constituents in a food that affect its end-use quality, for monitoring changes that occur during processing or storage, and for directly predicting processing characteristics of a commodity that are related to its chemical composition. Examples include determining the amylose content in rice starch, an important determinant of rice quality [49, 50], monitoring peroxide value in vegetable oils [51], monitoring degradation of frying oils [52], and predicting corn-processing quality [53, 54].

These are only a few examples of current applications. If a substance absorbs in the near-IR region and is present at a level of a few tenths of a percent or greater, it has potential for being measured by this technique. The primary advantage of near-IR spectroscopy is that once the instrument has been calibrated, several constituents in a sample can be measured rapidly (from 30 s to 2 min) and simultaneously. To measure multiple constituents, a calibration equation for each constituent is stored into the memory of the instrument. Measurements are taken at all wavelengths needed by the calibrations, and each equation then is solved to predict the constituents of interest. No sample weighing is required, and no hazardous reagents are used or chemical waste generated. It also is adaptable for **online measurement systems** [55]. Disadvantages include the high initial cost of the instrumentation, which may require a large sample load to justify the expenditure, and the fact that specific calibrations may need to be developed for each product measured. Also, the results produced by the instrument can be no better than the data used to calibrate it, which makes careful analysis of the calibration samples of highest importance.

8.5 RAMAN SPECTROSCOPY

8.5.1 Principles

Raman spectroscopy is a vibrational spectroscopic technique that is complementary to IR measurements

[56]. When a photon of light collides with a molecule, the collision can result in the photon being scattered. Molecules in the sample can be excited and reach an unstable virtual energy state when they interact with the incident light as shown in Fig. 8.10. However, this transition to a high-energy state in the molecule is a short-lived process, and most of the molecules relax back to their initial low energy level resulting in the scattered photon having the same energy as the incident light. This is called elastic scattering or **Rayleigh scattering**. However, a few molecules relax to a higher vibrational state with a change in the vibrational and rotational energy of the molecule causing a shift in the wavelength of the scattered radiation. This is called as inelastic scattering or **Raman scattering**. For Raman scattering to occur, a molecule must undergo a change in polarizability of the electron cloud of the molecule, but does not need to undergo a change in dipole moment. Thus, Raman can observe symmetrical vibrations that cannot be detected by IR spectroscopy. Raman is complementary to IR spectroscopy in that some vibrations are only Raman active, some are only IR active, and some are both.

During Raman scattering, scattered photons (approximately 1 in 10^7 photons) will shift to a longer wavelength (lower frequency), and this shift in frequency is called a **Stokes lines** or Stokes shift. If the final energetic state is lower than the initial state, scattered photons will shift to a shorter wavelength (higher frequency), and this is called an **anti-Stokes lines** or anti-Stokes shift [57]. Intensities of the Stokes lines are higher than those of the anti-Stokes lines, and therefore the Stokes lines are usually measured as the Raman spectrum [58].

A typical Raman spectrum includes scattering intensity (photons per second) on the y-axis versus

Energy-level diagram of mid-IR, near-IR, Rayleigh, and Raman scattering (Adapted from Boyaci et al. [57])

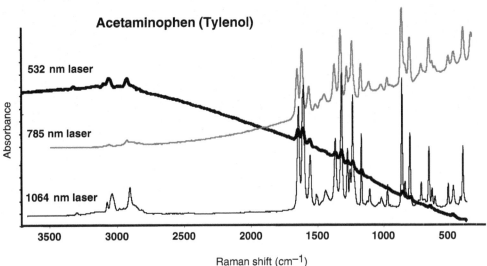

8.11 figure Raman spectrum of acetaminophen (Tylenol) generated using different laser sources

either increasing wavelength (nm) or Raman shift (cm^{-1}) on the x-axis. Each band in a Raman spectrum corresponds to a vibration of a chemical bond and/or functional group in the molecule as illustrated in a representative Raman spectrum of a pharmaceutical in Fig. 8.11. Similar to IR, a fingerprint of a molecule can be acquired, and by employing chemometric tools previously described in Sect. 8.5.3, both qualitative and quantitative analyses can be generated since the intensity of the Raman bands is proportional to the concentration of the analyte [57].

Raman spectroscopy is positioning as an attractive technique because: (1) it requires little or no sample preparation; (2) water and alcohols are weak Raman scatterers, allowing measurement of aqueous samples without any special accessories or sample preparation; and (3) it allows measurements through common transparent containers such as glass, quartz, and plastic eliminating the need to open containers to analyze the contents. For food applications of Raman spectroscopy, the challenge of sample fluorescence may be a limiting factor. In general, the intensity of the Raman scatter is proportional to $1/\lambda^4$, so shorter excitation wavelengths deliver a much stronger Raman signal. The caveat is that when using short excitation wavelengths, **fluorescence** is more likely to occur under these conditions.

8.5.2 Instrumentation

Raman spectrometers are based on **dispersive** and **Fourier transform** technologies (Fig. 8.12). Each technique has its unique advantages, and each is ideally suited to specific types of analysis.

A Raman dispersive system is composed of a laser source, sample, dispersing element (diffraction grating), detector, and a computer. During a typical Raman

spectrum collection process, a laser source gives a coherent beam of monochromatic light that is focused on the sample. The scattered light is passed through a notch filter that rejects the Rayleigh-scattered light, resulting in an important gain in sensitivity. Raman scattered photons enter the monochromator, where they are separated by wavelength and are collected by a detector that records the intensity of the Raman signal at each wavelength. Lasers are typically used as the radiation source, as the strength of the Raman signal is proportional to the intensity of the incident light. The use of the lasers as a source of radiation to generate the Raman scattering has been a crucial development in Raman instrumentation [59]. Typical wavelengths are 785, 633, 532, and 473 nm. Raman detectors are frequently photodiode arrays (PDA) or charge-coupled device (CCD) cameras. CCD detectors are extremely sensitive to light and contain thousands of picture elements (pixels) that acquire the whole spectrum at once in less than a second. CCD detectors allow the use of very low laser power, to prevent thermal or photochemical destruction of the sample.

FT-Raman spectrometers commonly use a near-IR laser, usually at 1030 nm or 1064 nm. Using lasers with excitation wavelengths in the near-IR region almost completely eliminates fluorescence; however, the Raman scattering intensity is weak. FT-Raman uses sensitive, single-element near-IR detectors such as **InGaAs** or liquid nitrogen-cooled germanium (**Ge**) detectors. An **interferometer** converts the Raman signal into an interferogram, permitting the detector to collect the entire spectrum simultaneously. Application of the Fourier transform algorithm to the **interferogram** converts the results into a conventional Raman spectrum. Besides removal of the fluorescence interference, FT-Raman spectroscopy provides excellent wave number accu-

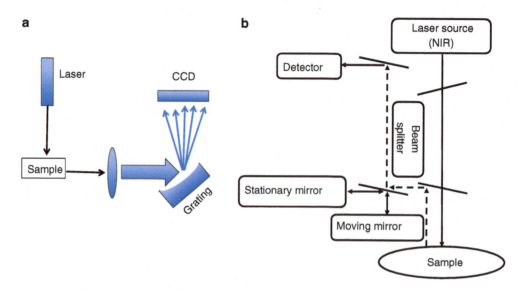

8.12 figure Basic illustration of the (**a**) dispersive Raman and (**b**) FT-Raman instrumentation (Adapted from Das and Agrawal [68])

racy as a result of the internal interferometer calibration by a built-in helium–neon laser.

8.5.3 Surface-Enhanced Raman Scattering (SERS)

As mentioned earlier, one of the drawbacks of traditional Raman spectroscopy is the very low level of Raman signals. One of the most popular techniques used to overcome this problem is surface-enhanced Raman scattering (SERS). Nanostructures on a metallic surface (typically gold or silver) can tremendously enhance the Raman signal of sample molecules on the order of 10^4–10^{11}, allowing detection in the ppb or single-molecule level [60, 61]. Simple mechanisms of traditional Raman and SERS are compared in Fig. 8.13. Variations in the magnitude of the signal enhancement depend on the morphology of the particle (roughened surfaces can provide much more enhancements compared to flat (smooth) metal surfaces) [62].

The SERS phenomena result from an enhanced electromagnetic field produced at the surface of the metal and a chemical enhancement due to a charge–transfer interaction between the metal and adsorbed molecules. When the wavelength of the incident light is close to a surface plasmon resonance (collective excitation of conductive electrons in small metallic structures), molecules adsorbed or in close proximity to the surface experience an exceptionally large electromagnetic field. In addition, the electronic transitions of many charge–transfer complexes between the metal surface and the molecule result in resonance.

The enormous signal enhancement achieved by SERS has positioned it as a very promising analytical tool for the biochemical, biomedical, and pharmaceutical fields. Food applications have been directed to food safety for the detection of food-borne pathogens as an alternative to current microorganism diagnostic tools, providing the possibility of developing portable pathogen sensors for on-site food inspection [60]. Other applications of SERS include the detection of food contaminants (pesticides and antibiotic residues) and adulteration (melamine, illegal food dyes, and mycotoxins) [60, 63].

8.6 HANDHELD AND PORTABLE TECHNOLOGY

Vibrational spectroscopy techniques are extremely well suited to be used as portable or handheld instruments. Their simplicity, speed, selectivity, and ability to operate without sample preparation make them ideal to be used outside the laboratory for process monitoring in challenging environments. A single spectrometer can be used to verify the identity of bulk materials, check for contamination, control processes, and confirm final product specifications. Field-based instruments have to tolerate harsh conditions, maintain reliability and accuracy, be easy to operate, battery powered, lightweight with an ergonomic design and intuitive user interface. Sample accessories must be robust, with limited or no sample preparation required and capable of fast analysis (Fig. 8.14).

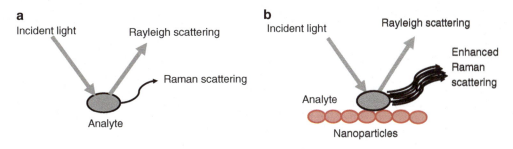

8.13 figure Comparison of (**a**) traditional Raman and (**b**) SERS mechanism (Adapted from Zheng and He [63])

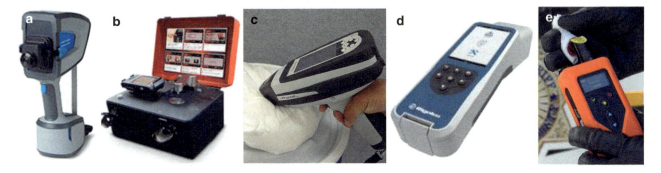

8.14 figure Portable/handheld vibrational spectrometers commercially available. (**a**) Agilent 4300 handheld FTIR, (**b**) Agilent 4500 Portable FTIR, (**c**) Thermo microPhazir Handheld Near-IR, (**d**) Progeny™ 1064 nm handheld Raman analyzer by Rigaku Analytical Devices, and (**e**) DeltaNu ReporteR handheld Raman spectrometer

The advantage of portable mid-IR spectrometers (compared to near-IR) is the higher fundamental signal, allowing detection of low analyte levels and its unique fingerprinting capabilities. However, if a sample contains water, its strong mid-IR absorption may swamp useful information. Near-IR allows analysis without any sample preparation. The most commonly used mode of sampling for solids is diffuse reflectance, while transflectance (combined transmission and reflectance) and transmission are suitable configurations for liquid analysis. Although the near-IR signal is 10–1000 times lower than mid-IR bands, the lower absorptivities permit the near-IR beam to penetrate deeper into the sample, resulting in more representative analysis. Advantages of portable Raman instruments for field deployment include little or no sample preparation, noncontact and nondestructive capabilities, and the relatively weak Raman response of water to allow measuring aqueous solutions. Near-IR and Raman analysis permit measurements through glass and plastic films.

Portable and handheld vibrational spectrometers are attractive fingerprinting techniques for various applications including food, pharmaceuticals, petrochemicals, and law enforcement [64]. Ellis et al. [65] summarized the applicability of various commercially available spectroscopy-based approaches for rapid on-site food fraud analysis. In addition, dos Santos et al. [66] reviewed the application of portable near-IR spectrometers in the agro-food industry.

8.7 SUMMARY

IR spectroscopy measures the absorption of radiation in the near-IR ($\lambda = 0.8$–$2.5\ \mu m$) or mid-IR ($\lambda = 2.5$–$15\ \mu m$) regions by molecules in food or other substances. IR radiation is absorbed as molecules change their vibrational energy levels. A summary of the most important characteristics of spectroscopy techniques is presented in Table 8.2. Mid-IR spectroscopy is especially useful for qualitative analysis, such as identifying specific functional groups present in a substance. Different functional groups absorb different frequencies of radiation, allowing the groups to be identified from the spectrum of a sample. Quantitative analysis also can be achieved by mid-IR spectroscopy, with milk analysis being a major application. Near-IR spectroscopy is used most extensively for quantitative applications, using either transmission or diffuse reflection measurements that can be taken directly from solid foods. By using multivariate statistical techniques, infrared instruments can be calibrated to measure the amounts

8.2 table Comparison of spectroscopy methods common in food analysis

Type of spectroscopy	Principles	Order of components	Application
Ultraviolet–visible (UV–Vis)	Based on absorption by molecules of radiant energy of specific wavelength from tungsten or UV lamp. Measure difference between amount of energy emitted by lamp and that reaching the detector. Absorption is proportional to concentration of molecule	UV or visible lamp, monochromator, sample, detector, readout	Quantitate molecules (by themselves or as the result of a chemical reaction) that will absorb radiant energy in the UV or vis region
Fluorescence	Based on emission, by molecules, of radiant energy of specific wavelength. Molecule is activated from its ground state to an excited electronic energy level by the absorption of radiation from a UV or vis lamp. Measure radiant emitted by the molecule as it relaxes from an excited electronic energy level to its corresponding ground state. Emitted energy is proportional to concentration of molecule	UV or visible lamp, monochromator, sample, monochromator, detector, readout	Quantitate molecules (by themselves or as the result of a chemical reaction) that will emit radiant energy in the UV or vis region
Mid-IR (MIR)	Based on absorption, by molecules, of radiant energy of specific wavelength (2.5–15 μm; 4000–670 cm^{-1}) from a mid-IR source. Molecules in the ground state absorb mid-IR radiation and make a transition to the first excitation state of vibration (fundamental vibration). When the frequency of the radiation matches the vibrational frequency of the molecule or functional group, then radiation causes vibrational and rotational changes and a net change in the dipole moment of the molecule. Measured by difference between amount of energy emitter by the mid-IR source and that reaching the detector. Absorption is proportional to concentration of the analyte	For dispersive systems: mid-IR source, grating, slit, sample, detector, readout For FT systems: mid-IR source, interferometer (He:Ne alignment laser, beam splitter, movable mirror, fixed mirror), sample, detector, readout	Qualitative and quantitative analysis of sample(s)
Near-IR (NIR)	Based on absorption, by molecules, of radiant energy of specific wavelength (0.8–2.5 μm; 12,500–4000 cm^{-1}) from a NIR source. Molecules absorb near-IR radiation and make transitions to higher excited states (v = 2, 3 or higher) resulting in overtones and combination vibrations of C–H, O–H, and N–H groups. A net change in the dipole moment and large mechanical anharmonicity of the vibrating atoms are required. Measured by the difference between amount of energy emitted by the near-IR source and that reaching the detector. Absorption is proportional to concentration of molecule	For dispersive systems: near-IR source, grating, slit, sample, detector, readout For FT systems: near-IR source, interferometer (He:Ne alignment laser, beam splitter, movable mirror, fixed mirror), sample, detector, readout	Qualitative and quantitative analysis of sample(s)
Raman	Based on inelastic scattering process of monochromatic laser beam through its interaction with vibrating molecule resulting in a change in polarizability of the electron cloud of the molecule. Scattered photon is shifted to higher (anti-Stokes shift) or lower (Stokes shift) frequencies than the incident photon, giving information about the vibrational modes in the system. Measure Raman scattered photons by a detector. Band intensities are proportional to concentration of molecule	For dispersive systems: UV or visible laser, sample, dispersing element (diffracting grating), detector, readout For FT systems: near-IR laser, interferometer (He:Ne alignment laser, beam splitter, movable mirror, fixed mirror), sample, detector, readout	Qualitative and quantitative analysis of sample(s)

(continued)

8.2
table   (Continued)

Type of spectroscopy	Principles	Order of components	Application
Atomic absorption (flame)	Based on absorption, by atoms, of energy of specific wavelength from hollow cathode lamp (HCL). Flame converts molecules to atoms. Energy from HCL activates atoms from ground state to excited state. Absorption is proportional to concentration of atoms	HCL, sample inserted in flame, monochromator, detector, readout	Quantitate minerals
Atomic emission (plasma)	Based on emission, by atoms, of energy of specific wavelength. Plasma converts molecules to atoms. Plasma activates atoms from ground state to excited state. Emission is proportional to concentration of atoms	Sample inserted in plasma, monochromator (or echelle optics), detector, readout	Quantitate minerals

of various constituents in a food sample based on the amount of IR radiation absorbed at specific wavelengths. Mid-IR, near-IR, and Raman spectroscopy requires much less time to perform quantitative analysis than do many conventional wet chemical or chromatographic techniques.

8.8   STUDY QUESTIONS

1. Describe the factors that affect the frequency of vibration of a molecular functional group and thus the frequencies of radiation that it absorbs. Also, explain how the fundamental absorption and overtone absorptions of a molecule are related.
2. Describe the essential components of an FT mid-IR spectrometer and their function, and compare the operation of the FT instrument to a dispersive instrument. What advantages do FT instruments have over dispersive IR spectrophotometers?
3. Describe the similarities and differences between mid-IR spectroscopy and Raman spectroscopy.
4. Of the three antioxidants, butylated hydroxytoluene (BHT), butylated hydroxyanisole (BHA), and propyl gallate, which would you expect to have a strong IR absorption band in the 1700–1750 cm^{-1} spectral region? Look up these compounds in a reference book if you are uncertain of their structure.
5. Describe the two ways in which radiation is reflected from a solid or granular material. Which type of reflected radiation is useful for making quantitative measurements on solid samples by near-IR spectroscopy? How are near-IR reflectance instruments designed to select for the desired component of reflected radiation?
6. Describe the steps involved in calibrating a near-IR reflectance instrument to measure the protein content of wheat flour. Why is it usually necessary to make measurements at more than one wavelength?

REFERENCES

1. Herschel W (1800) Investigation of the powers of the prismatic colours to heat and illuminate objects; with remarks, that prove the different refrangibility of radiant heat. to which is added, an inquiry into the method of viewing the sun advantageously, with telescopes of large apertures and high magnifying powers. Philos Trans R Soc London Ser A 90:255–283
2. Coates J (2000) Interpretation of Infrared Spectra, A Practical Approach. In: Meyers RA (Ed.) Encyclopedia of Analytical Chemistry, Vol. 12, Wiley, Chichester, UK, pp. 10 815
3. Subramanian A, Rodriguez-Saona LE (2009) Fourier transform infrared (FTIR) spectroscopy. In: Da-Wen Sun (Ed.) Infrared Spectroscopy for Food Quality Analysis and Control. Elsevier, London, UK, pp. 146
4. Griffiths PR, de Haseth JA (2007) Other Components of FT-IR Spectrometers, Ch. 6. In: Fourier Transform Infrared Spectrometry. 2nd edn. John Wiley & Sons Inc. Hoboken, NJ.
5. AOAC International (2016) Official methods of analysis, 20th edn., Method 972.12. AOAC International, Rockville, MD
6. Lahner BS (1996) Evaluation of Aegys MI 600 Fourier transform infrared milk analyzer for analysis of fat, protein, lactose, and solids nonfat: a compilation of eight independent studies. J AOAC Int 79:388
7. Rodriguez-Saona LE, Allendorf ME (2011) Use of FTIR for rapid authentication and detection of adulteration of food. Annu Rev Food Sci Technol 2:467
8. Nunes C (2014) Vibrational spectroscopy and chemometrics to assess authenticity, adulteration and intrinsic quality parameters of edible oils and fats. Food Res Intl 60:255
9. Cozzolino D (2015) The role of vibrational spectroscopy as a tool to assess economically motivated fraud and

counterfeit issues in agricultural products and foods. Anal. Methods 7:9390

10. Martens H, Naes T (2001) Multivariate calibration by data compression, Ch. 4. In: Williams PC, Norris KH (eds) Near infrared technology in the agricultural and food industries, 2nd edn. American Association of Cereal Chemists, St. Paul, MN, p 75

11. Roggo Y, Chalus P, Maurer L, Lema-Martinez C, Edmond A, Jent N (2007) A review of near infrared spectroscopy and chemometrics in pharmaceutical technologies. J Pharm Biomed Anal 44:683

12. Lavine BK, Workman J (2013) Chemometrics. Anal Chem 85:705

13. Martens H (2015) Quantitative Big Data: where chemometrics can contribute. J Chemom 29: 563

14. Wold S, Sjöström M, Eriksson L (2001) PLS-regression: a basic tool of chemometrics. Chemom Intell Lab Syst 58:109

15. Mevik BH, Wehrens R (2007) The pls package: principal component and partial least squares regression in R. J Stat Softw 18:1

16. Bjorsvik HR, Martens H (2008) Data analysis: calibration of NIR instruments by PLS regression, Ch. 8. In: Burns DA, Ciurczak EW (eds) Handbook of near-infrared analysis, 3rd edn. CRC Press, Boca Raton, FL

17. Delwiche SR, ChenYR, Hruschka WR (1995) Differentiation of hard red wheat by near-infrared analysis of bulk samples. Cereal Chem 72:243

18. Evans DG, Scotter CN, Day LZ, Hall MN (1993) Determination of the authenticity of orange juice by discriminant analysis of near infrared spectra. A study of pretreatment and transformation of spectral data. J Near Infrared Spectrosc 1:33

19. Goodner KL, Manthey JA (2005) Differentiating orange juices using Fourier transform infrared spectroscopy (FT-IR). P Fl St Hortic Soc 118:410

20. Bertran E, Blanco M, Coello J, Iturriaga H, Maspoch S, Montoliu I (2000) Near infrared spectrometry and pattern recognition as screening methods for the authentication of virgin olive oils of very close geographical origins. J Near Infrared Spectrosc 8:45

21. Gurdeniz G, Ozen B (2009) Detection of adulteration of extra-virgin olive oil by chemometric analysis of mid-infrared spectral data. Food Chem 116:519

22. Rohman A, Man YBC, Yusof FM (2014) The use of FTIR spectroscopy and chemometrics for rapid authentication of extra virgin olive oil. J Am Oil Chem Soc 91:207

23. Alomar D, Gallo C, Castaneda M, Fuchslocher R (2003) Chemical and discriminant analysis of bovine meat by near infrared reflectance spectroscopy (NIRS). Meat Sci 63:441

24. Al-Jowder O, Defernez M, Kemsley EK, Wilson, RH (1999) Mid-infrared spectroscopy and chemometrics for the authentication of meat products. J Agric Food Chem 47:3210

25. Meza-Márquez OG, Gallardo-Velázquez T, Osorio-Revilla G (2010) Application of mid-infrared spectroscopy with multivariate analysis and soft independent modeling of class analogies (SIMCA) for the detection of adulterants in minced beef. Meat Sci 86:511

26. Lavine BK (2000) Clustering and classification of analytical data. Robert A. Meyers (Ed). In "Encyclopedia of Analytical Chemistry". John Wiley & Sons Ltd, Chichester.

27. Osborne BG, Fearn T, Hindle PH (1993) Practical NIR spectroscopy with application in food and beverage analysis. Longman, Essex, UK.

28. Williams PC, Norris KH (eds) (2001) Near-infrared technology in the agricultural and food industries, 2nd edn. American Association of Cereal Chemists, St. Paul, MN.

29. Ozaki Y, McClure WF, Christy AA (2006) Near-infrared spectroscopy in food science and technology. Wiley, Hoboken, NJ.

30. Woodcock T, Downey G, O'Donnel CP (2008) Review: better quality food and beverages: the role of near infrared spectroscopy. J Near Infrared Spectrosc 16:1

31. AACC International (2010) Approved methods of analysis, 11th edn. (On-line). The American Association of Cereal Chemists, St. Paul, MN

32. Kays SE, Windham WR, Barton FE (1998) Prediction of total dietary fiber by near-infrared reflectance spectroscopy in high-fat and high-sugar-containing cereal products. J Agric and Food Chem 46:854.

33. Kays SE, Barton FE (2002) Near-infrared analysis of soluble and insoluble dietary fiber fractions of cereal food products. J Agric Food Chem 50:3024.

34. Kays WE, Barton FE, Windham WR (2000) Predicting protein content by near infrared reflectance spectroscopy in diverse cereal food products. J Near Infrared Spectrosc 8:35

35. Oh EK, Grossklaus D (1995) Measurement of the components in meat patties by near infrared reflectance spectroscopy. Meat Sci 41:157

36. Geesink GH, Schreutelkamp FH, Frankhuizen R, Vedder HW, Faber NM, Kranen RW, Gerritzen MA (2003) Prediction of pork quality attributes from near infrared reflectance spectra. Meat Sci 65:661

37. Naganathan GK, Grimes LM, Subbiah J, Calkins CR, Samal A, Meyer G (2008) Visible/near-infrared hyperspectral imaging for beef tenderness prediction. Comput Electron Agric 64:225

38. Windham WR, Lawrence KC, Feldner PW (2003) Prediction of fat content in poultry meat by near-infrared transmission analysis. J Appl Poultr Res 12:69

39. Solberg C, Fredriksen G (2001) Analysis of fat and dry matter in capelin by near infrared transmission spectroscopy. J Near Infrared Spectrosc 9:221

40. Isakkson T, Nærbo G, Rukke EO (2001) In-line determination of moisture in margarine, using near infrared diffuse transmittance. J Near Infrared Spectrosc 9:11

41. Pierce MM, Wehling RL (1994) Comparison of sample handling and data treatment methods for determining moisture and fat in Cheddar cheese by near-infrared spectroscopy. J Agric Food Chem 42:2830.

42. Rodriguez-Otero JL, Hermida M, Cepeda A (1995) Determination of fat, protein, and total solids in cheese by near infrared reflectance spectroscopy. J AOAC Int 78:802

43. Wu D, He Y, Feng S (2008) Short-wave near-infrared spectroscopy analysis of major compounds in milk powder and wavelength assignment. Anal Chim Acta 610:232

44. Tarkosova J, Copikova J (2000) Determination of carbohydrate content in bananas during ripening and storage by near infrared spectroscopy. J Near Infrared Spectrosc 8:21

45. Segtman VH, Isakkson T (2000) Evaluating near infrared techniques for quantitative analysis of carbohydrates in fruit juice model systems. J Near Infrared Spectrosc 8:109

46. Camps C, Christen D (2009) Non-destructive assessment of apricot fruit quality by portable visible-near infrared spectroscopy. LWT – Food Sci Technol 42:1125

47. Psotka J, Shadow W (1994) NIR analysis in the wet corn refining industry – A technology review of methods in use. Int Sugar J 96:358

48. Tarkosova J, Copikova J (2000) Fourier transform near infrared spectroscopy applied to analysis of chocolate. J Near Infrared Spectrosc 8:251.

49. Villareal CP, De la Cruz NM, Juliano BO (1994) Rice amylose analysis by near-infrared transmittance spectroscopy. Cereal Chem 71:292

50. Delwiche SR, Bean MM, Miller RE, Webb BD, Williams PC (1995) Apparent amylose content of milled rice by near-infrared reflectance spectrophotometry. Cereal Chem 72:182

51. Yildiz G, Wehling RL, Cuppett SL (2001) Method for determining oxidation of vegetable oils by near-infrared spectroscopy. J Am Oil Chem' Soc 78:495

52. Ng CL, Wehling RL, Cuppett SL (2007) Method for determining frying oil degradation by near-infrared spectroscopy. J Agric Food Chem 55:593

53. Wehling RL, Jackson DS, Hooper DG, Ghaedian AR (1993) Prediction of wet-milling starch yield from corn by near-infrared spectroscopy. Cereal Chem 70:720.

54. Paulsen MR, Singh M (2004) Calibration of a near-infrared transmission grain analyzer for extractable starch in maize. Biosyst Eng 89:79

55. Psotka J (2001) Challenges of making accurate on-line near-infrared measurements. Cereal Foods World 46:568

56. An introduction to Raman for the infrared spectroscopist. Inphotonics Technical Note No. 11, Norwood, MA (www.inphotonics.com/technote11.pdf)

57. Boyaci IH, Temiz HT, Geniş HE, Soykut EA, Yazgan NN, Güven B, Uysal RS, Bozkurt AG, İlaslan K, Torun O, Şeker FCD (2015) Dispersive and FT-Raman spectroscopic methods in food analysis. RSC Adv. 5:56606

58. Schrader B (Ed.) (2008) Infrared and Raman spectroscopy: methods and applications. John Wiley & Sons.

59. Li YS, Church JS (2014) Raman spectroscopy in the analysis of food and pharmaceutical nanomaterials. J Food Drug Anal 22:29

60. Craig AP, Franca AS, Irudayaraj J (2013) Surface-enhanced Raman spectroscopy applied to food safety. Annu Rev Food Sci Technol 4:369

61. Doering WE, Nie S (2002) Single-molecule and single-nanoparticle SERS: examining the roles of surface active sites and chemical enhancement. J Phys Chem B, 106:311

62. Vlckova B, Pavel I, Sladkova M, Siskova K, Slouf M (2007) Single molecule SERS: perspectives of analytical applications. J Mol Struct, 834:42

63. Zheng J, He L (2014) Surface-Enhanced Raman Spectroscopy for the Chemical Analysis of Food. Compr Rev Food Scie Food Saf 13:317

64. Volodina VA, Marina DV, Sachkovc VA, Gorokhova EB, Rinnertd H, Vergnatd M (2014) Applying an Improved Phonon Confinement Model to the Analysis of Raman Spectra of Germanium Nanocrystals. J Experiment Theoret Phys 118:65

65. Ellis DI, Muhamadali H, Haughey SA, Elliott CT, Goodacre, R (2015) Point-and-shoot: rapid quantitative detection methods for on-site food fraud analysis – moving out of the laboratory and into the food supply chain. Anal. Methods. 7:9401

66. dos Santos CAT, Lopo M, Páscoa RN, Lopes JA (2013) A review on the applications of portable near-infrared spectrometers in the agro-food industry. Appl Spectrosc 67:215

67. SKCchemistry, (2016) Infrared spectroscopy. Available from: https://skcchemistry.wikispaces.com/Infra+red+Spectroscopy. Accessed April 11, 2016

68. Das RS, Agrawal YK (2011) Raman spectroscopy: recent advancements, techniques and applications. Vib Spectrosc 57:163

9
chapter

Atomic Absorption Spectroscopy, Atomic Emission Spectroscopy, and Inductively Coupled Plasma-Mass Spectrometry

Vincent Yeung (✉)
Animal Science Department,
California Polytechnic State University,
San Luis Obispo, CA 93407, USA
e-mail: ckyeung@calpoly.edu

Dennis D. Miller
Department of Food Science,
Cornell University
Ithaca, NY 14853-7201, USA
e-mail: ddm2@cornell.edu

Michael A. Rutzke
School of Integrative Plant Science,
Cornell University,
Ithaca, NY 14853-7201, USA
e-mail: mar9@cornell.edu

S. Nielsen (ed.), *Food Analysis*, Food Science Text Series,
DOI 10.1007/978-3-319-45776-5_9, © Springer International Publishing 2017

9.1 INTRODUCTION

The development of accurate methods for measuring concentrations of mineral elements in foods and other biological samples has a long history. The major challenge is to accurately measure these elements in a food matrix that contains much higher concentrations of other components (i.e., carbohydrates, proteins, and fats) as well as other mineral elements that may interfere. Table 9.1 lists mineral elements of interest in foods [1, 2]. The USDA nutrient database [1] for calcium, iron, sodium, and potassium in foods is quite complete, but the database for the trace elements and toxic heavy metals is lacking for some food groups.

As the names imply, atomic absorption spectroscopy (AAS) quantifies the **absorption** of electromagnetic radiation by well-separated neutral atoms, while atomic emission spectroscopy (AES) measures **emission** of radiation from atoms in excited states. AAS and AES allow accurate measurements of mineral elements even in the presence of other components because the atomic absorption and emission spectra are unique for each individual element. The use of inductively coupled plasma (ICP), originally developed in the 1960s [3, 4], as an excitation source for emission spectroscopy has further expanded our ability to rapidly measure multiple elements in a single sample. In theory, virtually all of the elements in the periodic chart can be determined by AAS or AES. In practice, **atomic spectroscopy** is used primarily to determine mineral elements. Table 8.2 in Chap. 8 shows a comparison of different spectroscopy methods commonly available for food analysis, including AAS and AES.

More recently, ICP has been mated with mass spectrometry (MS) to form ICP-MS instruments that are capable of measuring mineral elements with extremely low detection limits. Moreover, mass spectrometers have the added advantage of being able to separate and quantify multiple isotopes of the same element. Taken together, these instrumental methods have largely replaced traditional wet chemistry methods for food mineral analysis, although traditional methods for calcium, chloride, fluoride, and phosphorus remain in use today (see Chap. 21).

This chapter deals with the basic principles that underlie analytical atomic spectroscopy and provides an overview of the instrumentation available for measuring atomic absorption and emission. A discussion of ICP-MS is also included. Readers interested in a more thorough treatment of the topic are referred to references 5–8.

9.1 table	Elements in foods classified according to nutritional essentiality, potential toxic risk, and inclusion in USDA Nutrient Database for Standard Reference

Essential nutrient[a]	Toxicity concern	USDA Nutrient Database[b]
Sodium	Lead	Calcium
Potassium	Mercury	Iron
Chloride	Cadmium	Magnesium
Calcium	Nickel	Phosphorous
Chromium	Arsenic	Potassium
Copper	Thallium	Sodium
Fluoride		Zinc
Iodine		Copper
Iron		Manganese
Magnesium		Selenium
Manganese		Fluoride
Molybdenum		
Phosphorous		
Selenium		
Zinc		
Arsenic		
Boron		
Nickel		
Silicon		
Vanadium		

Compiled based on U.S. Department of Agriculture, Agricultural Research Service [1] and Institute of Medicine (IOM) [2]

[a]Nutrients are considered essential if their removal from the diet causes some adverse change in physiological function. For arsenic, boron, nickel, silicon, and vanadium, there is evidence that trace amounts may have a beneficial role in some physiological processes in some species, but the available data is limited and often conflicting. Dietary Reference Intakes (DRIs) have been established by Institute of Medicine (IOM) for these minerals [2]

[b]Values for copper, manganese, selenium, and fluoride are not included for many foods in the USDA Nutrient Database for Standard Reference due to limited data

9.2 GENERAL PRINCIPLES

9.2.1 Energy Transitions in Atoms

Atomic absorption spectra are produced when ground-state atoms absorb energy from a radiation source. Atomic emission spectra are produced when neutral atoms in an excited state emit energy on returning to the ground state or a lower-energy state. As discussed in Chap. 6, atoms absorb or emit radiation of discrete wavelengths because the allowed energy levels of electrons in atoms are fixed and distinct. In other words, each element has a unique set of allowed electronic transitions and therefore a unique spectrum, enabling accurate identification and quantification even in the presence of other elements. The absorption and emission spectra of sodium are shown in Fig. 9.1. For absorption, transitions involve primarily the excitation of electrons in the ground state, so the number of transitions is relatively small. Emission, on the other hand, occurs

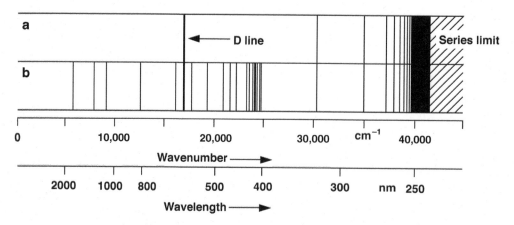

9.1 figure Spectra for sodium. The upper spectrum (**a**) is the absorption spectrum, and the lower (**b**) is the emission spectrum (From Welz [26], reprinted with permission of VCH Publishers (1985)

when electrons in various excited states fall to lower-energy levels including, but not limited to, the ground state. Therefore, the emission spectrum has more lines than the absorption spectrum of the same element as illustrated in Fig. 9.1. When a transition is from or to the ground state, it is termed a **resonance transition**, and the resulting spectral line is called a **resonance line**.

9.2.2 Atomization

Atomic spectroscopy requires that atoms of the element of interest to be in the atomic state (not combined with other elements in a compound) and to be well separated in space. In foods, virtually all elements are present as compounds or complexes and therefore must be converted to neutral atoms (i.e., atomization) before atomic absorption or emission measurements can be made. Atomization is usually accomplished by exposing a solution containing the analyte (the substance being measured) as a fine mist to high temperatures, typically in a flame or plasma. The solvent quickly evaporates, leaving solid particles of the analyte that vaporize and decompose to atoms that may absorb radiation (atomic absorption) or become excited and subsequently emit radiation (atomic emission). This process is shown schematically in Fig. 9.2. Three common methods for atomizing samples, including their atomization temperature ranges, are summarized in Table 9.2.

9.3 ATOMIC ABSORPTION SPECTROSCOPY

AAS is an analytical method based on the absorption of ultraviolet-visible (UV-Vis) radiation by free atoms in the gaseous state. It is a relatively simple method and was the most widely used form of atomic spectroscopy in food analysis for many years. It has been largely replaced by the more powerful ICP-based

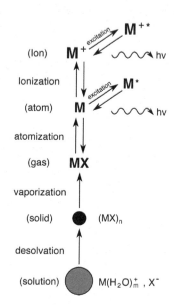

9.2 figure A schematic representation of the atomization of an element in a flame or plasma. The *large circle* at the bottom represents a tiny droplet of a solution containing the element (*M*) as part of a compound (From Boss and Fredeen [6], used with permission. Courtesy of the PerkinElmer Corporation, Shelton, CT)

 9.2 table Methods and temperature ranges for atomization of analytes

Source of energy for atomization	Approximate atomization temperature range (K)	Analytical method
Flame	2,000–3,400	AAS, AES
Electrothermal	1,500–3,300	AAS (graphite furnace)
Inductively coupled argon plasma	6,000–7,000	ICP-OES, ICP-MS

spectroscopy. Two types of atomization are commonly used in AAS: **flame atomization** and **electrothermal (graphite furnace) atomization**.

9.3.1 Principles of Flame Atomic Absorption Spectroscopy

A schematic diagram of a flame atomic absorption spectrometer is shown in Fig. 9.3. In flame AAS, a nebulizer-burner system is used to convert a sample solution into an atomic vapor. It is important to note that the sample must be in solution (usually an aqueous solution) before it can be analyzed by flame AAS. The sample solution is **nebulized** (dispersed into tiny droplets), mixed with fuel and an oxidant, and burned in a flame produced by oxidation of the fuel. Atoms and ions are formed within the hottest portion of the flame as the sample solution goes through the process of desolvation, vaporization, atomization, and ionization (Fig. 9.2). Atoms and ions of the same element absorb radiation of different wavelengths and produce different spectra. Therefore, it is desirable to choose a flame temperature that will maximize atomization and minimize ionization because atomic absorption spectrometers are tuned to measure *atomic* absorption, not *ionic* absorption.

Once the sample is atomized in the flame, the quantity of the analyte element is measured by determining the attenuation (decrease in intensity) of a beam of radiation passing through the flame, due to atomic absorption of incident radiation by the analyte element. For the measurement to be specific for the analyte element, the radiation source ideally should emit radiation of the exact discreet wavelengths that only the analyte element

is capable of absorbing. This can be accomplished by using lamps with cathodes fabricated with the element to be determined. Thus, the beam of radiation emitted from the lamp is the emission spectrum of the element. The spectral line of interest is isolated by passing the beam through a monochromator so that only radiation of a very narrow band width reaches the detector. Usually, one of the strongest spectral lines is chosen; for example, for sodium the monochromator is set to pass radiation with a wavelength of 589.0 nm. The principle of this process is illustrated in Fig. 9.4. Note that the intensity of the radiation leaving the flame is less than the intensity of radiation coming from the source. This is because sample atoms in the flame absorb some of the radiation. Note also that the line width of the radiation from the source is narrower than the corresponding line width in the absorption spectrum. This is because the higher temperature of the flame causes a broadening of the line width.

The amount of radiation absorbed by the analyte element is given by **Beer's law**:

$$A = \log(I_o / I) = abc \qquad (9.1)$$

where:

A = absorbance
I_o = intensity of radiation incident on the flame
I = intensity of radiation exiting the flame
a = molar absorptivity
b = path length through the flame
c = concentration of atoms in the flame

Clearly, absorbance is directly related to the concentration of atoms in the flame.

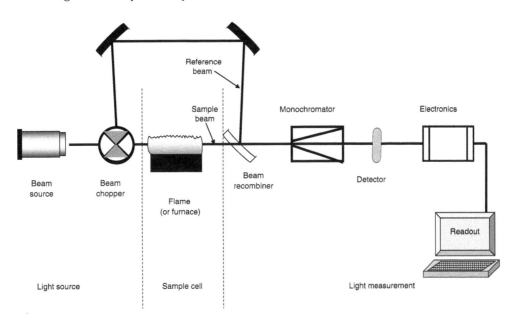

 Schematic representation of a double-beam atomic absorption spectrophotometer (Adapted from Beaty and Kerber [5])

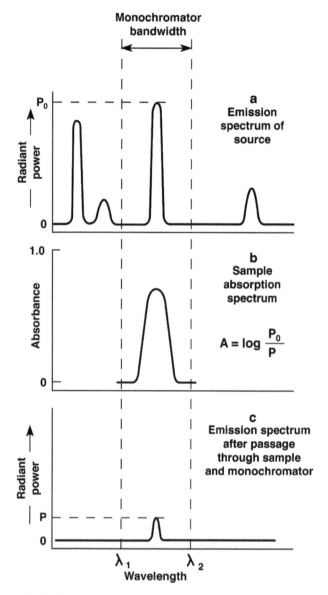

9.3.2 Principles of Electrothermal (Graphite Furnace) Atomic Absorption Spectroscopy

Electrothermal AAS is identical to flame AAS except for the atomization process. In electrothermal AAS, the sample is heated electrically in stages inside a graphite tube, commonly known as **graphite furnace**, to achieve atomization. The tube is aligned to the path of the radiation beam to be absorbed by the atomized sample and absorbance is determined. Electrothermal AAS requires smaller sample sizes and offers lower detection limits. Disadvantages are the added expense of the graphite furnace, lower sample throughput, higher matrix interferences, and lower precision.

9.3.3 Instrumentation for Atomic Absorption Spectroscopy

Atomic absorption spectrometers, typically with a double-beam design (see Chap. 7, Sect. 7.2.7), consist of the following components (Fig. 9.3):

1. **Radiation source**, a hollow cathode lamp (HCL) or an electrode-less discharge lamp (EDL)
2. **Atomizer**, usually a nebulizer-burner system or a graphite furnace
3. **Monochromator**, usually an UV-Vis grating monochromator
4. **Detector**, a photomultiplier tube (PMT) or a solid-state detector (SSD)
5. **Readout device**, an analog or a digital readout

(Radiation source and atomizers will be further discussed in the following paragraphs. See Chap. 7, Sects. 7.2.6.2, 7.2.6.3, and 7.2.6.4 for a more detailed discussion of monochromators, detectors, and readout devices, respectively.)

9.3.3.1 *Radiation Source*

A **hollow cathode lamp** (HCL) consists of a hollow tube filled with argon or neon gas, an anode made of tungsten, and a cathode made of the metallic form of the element being measured (Fig. 9.5). When voltage is applied across the electrodes, the lamp emits radiation

9.4 figure Schematic representation of the absorption of radiation by a sample during an atomic absorption measurement. The spectrum of the radiation source is shown in (**a**). As the radiation passes through the sample (**b**), it is partially absorbed by the element of interest. Absorbance is proportional to the concentration of the element in the flame. The radiant power of the radiation leaving the sample is reduced because of absorption by the sample (**c**) (From Skoog et al. [8], used with permission. Illustration from Principles of Instrumental Analysis, 6th ed., by DA Skoog, F. J. Holler, S.R. Crouch (2007). Reprinted with permission of Brooks/Cole, a division of Cengage Learning)

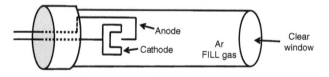

9.5 figure Schematic representation of a hollow cathode lamp (Adapted from Beaty and Kerber [5])

characteristic of the metal in the cathode. For example, if the cathode is made of iron, an iron spectrum will be emitted. As the radiation passes through the flame containing the atomized sample, only iron atoms (not atoms of other elements) will absorb this radiation because the emitted wavelengths from the HCL are specific for iron atoms. Of course, this means that it is necessary to use a different lamp for each element analyzed (there are a limited number of multielement lamps available that contain cathodes made of more than one element). HCLs for about 60 metallic elements are commercially available, suggesting that AAS may be used for the analysis of up to 60 elements.

An **electrode-less discharge lamp** (EDL) contains no electrodes but a hollow glass vessel filled with an inert gas plus the element of interest. The discharge is produced by a high-frequency generator coil rather than an electric current [9]. EDLs are suitable for volatile elements such as arsenic, mercury, and cadmium.

Radiation reaching the monochromator comes from three sources: (1) attenuated beam from the HCL (specific emission), (2) emission from sample atoms (including both analyte and non-analyte atoms) that were excited by the flame (nonspecific emission), and (3) radiation resulting from the combustion of the fuel to create the flame (nonspecific emission). Instruments are designed to eliminate nonspecific emissions from reaching the detector. This is accomplished by positioning a monochromator between the flame and the detector. The monochromator disperses wavelengths of light that are not specific to the analyte element and isolates a line that is specific. Thus, radiation reaching the detector is the sum of radiation from the attenuated HCL beam and radiation emitted by excited analyte atoms in the flame. Since we are interested only in the amount of HCL radiation absorbed by analyte atoms in the flame, it is necessary to correct for emission from excited analyte atoms in the flame. This is accomplished by positioning a **beam chopper** perpendicular to the light path between the HCL and the flame (Fig. 9.3). A beam chopper is a disk with segments removed. The disk rotates at a constant speed so that the light emitted from the HCL reaching the detector is either unimpeded or blocked at regular intervals, i.e., it is alternating. In contrast, emission from excited analyte atoms in the flame reaching the detector is continuous. The instrument electronics subtracts the continuous emission signal from the alternating emission signal so only the signal from the attenuated HCL beam is recorded in the readout.

9.3.3.2 *Atomizers*
Flame and **graphite furnace** atomizers are the two common types of atomizers used in AAS. When applicable, a **cold vapor technique** for mercury and **a**

hydride generation technique for a few elements are used to enhance sensitivity.

The **flame atomizer** consists of a **nebulizer** and a **burner**. The nebulizer is designed to convert the sample solution into a fine mist or aerosol. This is accomplished by aspirating the sample through a capillary into a chamber through which oxidant and fuel are flowing. The chamber contains baffles that remove larger droplets, leaving a very fine mist. Only about 1% of the total sample is carried into the flame by the oxidant-fuel mixture. The larger droplets fall to the bottom of the mixing chamber and are collected as waste. The burner head contains a long, narrow slot that produces a flame that may be 5–10 cm in length. This gives a long path length that increases the sensitivity of the measurement.

Flame characteristics may be manipulated by adjusting oxidant/fuel ratios and by choice of oxidant and fuel. **Air-acetylene** and **nitrous oxide-acetylene** are the most commonly used oxidant-fuel mixtures although other oxidants and fuels may be used for some elements. There are three types of flames:

1. **Stoichiometric.** This flame is produced from stoichiometric amounts (exact reaction ratios) of oxidant and fuel, so the fuel is completely burned and the oxidant is completely consumed. It is characterized by yellow fringes.
2. **Oxidizing.** This flame is produced from a fuel-lean (excess of oxygen) mixture. It is the hottest flame and has a clear blue appearance.
3. **Reducing.** This flame is produced from a fuel-rich mixture (excess of fuel compared to oxygen). It is a relatively cool flame and has a yellow color.

Analysts should follow manufacturer's guidelines or consult the literature for the proper type of flame for each element.

Flame atomizers have the advantage of being stable and easy to use. However, sensitivity is relatively low because much of the sample never reaches the flame and the residence time of the sample in the flame is short.

The **graphite furnace** is typically a cylindrical graphite tube connected to an electrical power supply. The sample is injected into the tube through an inlet using a microliter syringe with sample volumes ranging from 0.5 to 100 μL. During operation, the system is flushed with an inert gas to prevent the tube from burning and to exclude air from the sample compartment. The tube is heated electrically in stages: first the sample solvent is evaporated, then the sample is ashed, and finally the temperature is rapidly increased to ~2000–3000 K to quickly vaporize and atomize the sample.

The **cold vapor technique** works only for mercury, because mercury is the only mineral element that can exist as free atoms in the gaseous state at room temperature. Mercury compounds in a sample are first reduced to elemental mercury by the action of stannous chloride, a strong reducing agent. The elemental mercury is then carried in a stream of inert gas into an absorption cell without the need for further atomization. The **hydride generation technique** is limited to elements capable of forming volatile hydrides that include arsenic, lead, tin, bismuth, antimony, tellurium, germanium, and selenium. Samples containing these elements are reacted with sodium borohydride to generate volatile hydrides, which are carried into an absorption cell and decomposed by heat. Absorbance measurements with these two techniques are conducted in the same manner as with flame atomization, but sensitivity is greatly enhanced because there is very little sample loss [5].

9.3.4 General Procedure for Atomic Absorption Analysis

While the basic design of all atomic absorption spectrometers is similar, operation procedures do vary from one instrument to another. For any given method, it is always a good practice to carefully review standard operating procedures provided by the manufacturer before using the instrument. Most instruction manuals also provide pertinent information for the analysis of each particular element (wavelength and slit width requirements, interferences and corrections, flame characteristics, linear ranges, etc.).

9.3.4.1 *Safety Precautions*
General laboratory safety protocols and procedures as well as safety precautions recommended by the instrument manufacturers must be followed carefully to avoid personal injuries or costly accidents. The most commonly used fuel sources in flame AAS are mixtures of air-acetylene and nitrous oxide-acetylene. *ACETYLENE IS AN EXPLOSIVE GAS.* Proper ventilation must be in place before operation. The exhaust vent should be positioned directly above the burner to avoid the buildup of unburned fuel or any potentially hazardous toxic fumes. Flame atomic absorption spectrometers should never be left unattended while in operation.

9.3.4.2 *Calibration*
As illustrated in Fig. 9.6, a plot of absorbance versus concentration will deviate from linearity predicted by Beer's law when concentration exceeds a certain level. Therefore, properly constructed calibration curves using pure standards are essential for accurate quantitative measurements. If values for linear ranges are

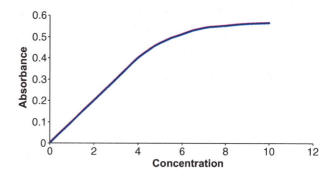

figure A plot of absorbance vs. concentration showing nonlinearity above a certain concentration

not provided by the manufacturer, the linear range of an element should be established by running a series of standards of increasing concentration and plotting absorbance versus concentration. The concentration of the unknown sample solution should be adjusted so that the measured absorbance always falls within the linear range of the calibration curve.

Instruction manuals from the manufacturers may provide values for characteristic concentrations for each element. For example, manuals for Perkin-Elmer atomic absorption spectrophotometers state that a 5.0 mg/L aqueous solution of iron "will give a reading of *approximately* 0.2 absorbance units." If the measured absorbance reading deviates significantly from this value, appropriate adjustments (e.g., flame characteristics, lamp alignment, etc.) should be made.

9.3.5 Interferences in Atomic Absorption Spectroscopy

Two types of interferences are encountered in AAS: spectral and nonspectral. **Spectral interferences** are caused by the absorption of radiation by other elemental or molecular species at wavelengths that overlap with the spectral regions of the analyte present in the sample. **Nonspectral interferences** are caused by sample matrices and conditions that affect the atomization efficiency and/or the ionization of neutral atoms in the atomizer.

9.3.5.1 *Spectral Interferences*
An element in the sample other than the element of interest may absorb at the wavelength of the spectral band being used. Such interference is rare because emission lines from the HCL are so narrow that only the element of interest is capable of absorbing the radiation in most cases. One example of when this problem does occur is with the interference of iron in zinc determinations. Zinc has a spectral line at 213.856 nm, which overlaps the iron line at 213.859 nm. The problem may be solved by choosing an alternative spectral

line for measuring zinc or by narrowing the monochromator slit width.

The presence of alkaline earth oxide and hydroxide molecules may also lead to several specific spectral interferences. Spectra of calcium oxide and magnesium hydroxide will appear as **background absorption** for atomic absorption measurements of sodium and chromium, respectively [10]. These interferences are weak but must be taken into account when working with a complex sample matrix.

9.3.5.2 *Nonspectral Interferences*

As mentioned above, quantitative results for an unknown sample are possible only through comparison with a series of standards of known concentrations. **Transport interferences** may occur when other components present in the sample matrix influence physical properties such as viscosity, surface tension, and vapor pressure of the sample solution, leading to differences in the rate of aspiration, nebulization, or transport between the sample solution and the standards during flame atomization. Such interferences often can be overcome by using the same solvent and by matching as closely as possible the physical properties of the sample solution and the standards. The standard addition protocol (see Chap. 7, Sect. 7.2.4) may also be used. Transport interferences are rarely a problem with graphite furnace instruments but matrix interferences are a common and serious problem.

Matrix composition of the sample solution also may affect the lateral migration of an analyte, resulting in **solute vaporization interferences**. For example, it has been observed in flame absorption and emission that alkaline earth metals are depressed by elevated levels of aluminum and phosphorus [11], and aluminum also suppresses the recovery of calcium [12]. **Chemical interferences** occur when an element forms thermally stable compounds that do not decompose during atomization. **Refractory metals** such as titanium and molybdenum may combine with oxygen to form stable oxides; higher-temperature flames are usually required for their dissociation. Also, phosphate in a sample matrix reacts with calcium to form calcium pyrophosphate which is not decomposed in the flame. A **releasing agent** such as lanthanum, which binds phosphate more strongly than calcium, may be added to the sample solution and the standards to free up calcium for atomization [12].

Ionization interferences are caused by ionization of analyte atoms in the flame and thereby reduce the concentration of neutral atoms for atomic absorption measurement. (Remember that atoms and ions of the same element absorb radiation of different wavelengths and produce different spectra.) Ionization increases with increasing flame temperature and normally is not a problem in cooler air-acetylene flames.

It can be a problem in hotter nitrous oxide-acetylene flames, especially with elements that have ionization potentials of 7.5 eV or less (e.g., alkali metals). The ionization of atoms results in an equilibrium situation:

$$M \leftrightarrows M^+ + e^-$$ (9.2)

Ionization interferences can be countered by spiking the sample solution and the standards with another **easily ionized element** (EIE) such as potassium or cesium, known as an **ionization suppressant**, to create a pool of free electrons in the flame, which shifts the above equilibrium to the left and suppresses the ionization of the analyte atoms.

9.4 ATOMIC EMISSION SPECTROSCOPY

In contrast to AAS, the source of the measured radiation in AES is the excited atoms in the sample, not radiation from a HCL. Figure 9.7 shows a simplified diagram of an atomic emission spectrometer. Sufficient energy is first applied to the sample to excite atoms to higher-energy levels; emissions of wavelengths characteristic of individual elements are then measured when electrons from excited atoms move back to the ground state or a lower-energy state. The ratio of the number of excited atoms to ground-state atoms occurring in a flame or plasma is described by the Maxwell-Boltzmann equation for resonance lines (Chap. 6, Sect. 6.4.3). This equation applies when there is thermal impact or collisions between atoms or molecules a portion of them will become excited.

Energy for excitation may be produced by **heat** (usually from a flame), **light** (from a laser), **electricity** (arcs or sparks), or **radio waves** (ICP). (*Note*: AES is also commonly called optical emission spectroscopy (OES), especially when combined with ICP. In this chapter, we will use ICP-OES rather than ICP-AES for our discussion, but the two terms are virtually interchangeable.)

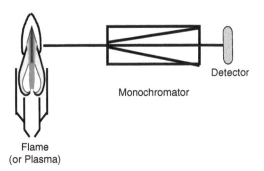

Detector

Monochromator

Flame
(or Plasma)

9.7 A simplified diagram of an atomic emission **figure** spectrometer (Adapted from Boss and Fredeen [6])

The two most common forms of AES used in food analysis are **flame emission spectroscopy** and **inductively coupled plasma-optical emission spectroscopy (ICP-OES)**.

9.4.1 Principles of Flame Emission Spectroscopy

Flame emission spectrometers employ a nebulizer-burner system to atomize and excite the atoms of the elements being measured. The flame with the excited atoms serves as the radiation source, so an external source (the HCL with the beam chopper) is not required. Otherwise instrumentation for flame emission spectroscopy is essentially identical to that for AAS. Many modern atomic absorption spectrometers can also be operated as flame emission spectrometers.

In some instruments, interference filters (instead of the more versatile monochromators typically found in absorption/emission spectrometers) are employed to isolate the spectral region of interest. **Flame photometers** are economical emission spectrometers equipped with interference filters and are specifically designed for the analysis of the alkali and alkaline earth metals in biological samples. Low flame temperatures are used so that only easily excited elements such as sodium, potassium, and calcium produce emissions. This results in a simpler spectrum and reduces interference from other elements present in the sample matrix.

9.4.2 Principles of Inductively Coupled Plasma-Optical Emission Spectroscopy

ICP-OES differs from flame emission spectroscopy in that it uses an argon plasma as the excitation source. A **plasma** is defined as a gaseous mixture containing significant concentrations of cations and electrons. Temperatures in argon plasmas could be as high as 10,000 K, with analyte excitation temperature typically ranging from 6,000 to 7,000 K.

The extremely high temperatures and the inert atmosphere of argon plasmas are ideal for the atomization, ionization, and excitation of the analyte atoms in the sample. The low oxygen content reduces the formation of oxides, which is sometimes a problem with flame methods. The nearly complete atomization of the sample minimizes chemical interferences. The relatively uniform temperatures in plasmas (compared to nonuniform temperatures in flames) and the relatively long residence time give good linear responses over a wide concentration range (up to 6 orders of magnitude).

9.4.3 Instrumentation for Inductively Coupled Plasma-Optical Emission Spectroscopy

Inductively coupled plasma-optical emission spectrometers typically consist of the following components (Fig. 9.8):

1. **Argon plasma torch**
2. **Monochromator, polychromator,** or **echelle optical system**
3. **Detector(s)**, a single or multiple PMT(s) or solid-state array detector(s)
4. **Computer** for data collection and treatment

9.4.3.1 Argon Plasma Torch

9.4.3.1.1 Characteristics of an Argon Plasma Torch

The plasma torch (Fig. 9.9) consists of three concentric quartz tubes centered in a copper coil, called the **load coil**. During operation of the torch, a stream of argon gas flows through the outer tube, and **radio frequency** (RF) power is applied to the load coil, creating a magnetic field oscillating at the frequency of the RF generator (usually 27 MHz or 40 MHz). The plasma is started by ionizing argon atoms with an electric spark to form argon ions and electrons. The oscillating magnetic field couples with the argon ions and electrons, forcing them to flow in an annular (ring-shaped) path. Heating does not involve burning fuel to directly heat and atomize the sample, as is the case with flame AAS (argon is a noble gas and will not combust). Rather, heating is accomplished by transferring RF energy to free electrons and argon ions in a manner similar to the transfer of microwave energy to water in a microwave oven. These high-energy electrons in turn collide with argon atoms, generating even more electrons and argon ions and causing a rapid increase in temperature to approximately 10,000 K. The process continues until about 1 % of the argon atoms are ionized. At this point the plasma is very stable and self-sustaining for as long as the RF field is applied at constant power. The transfer of energy to a system through the use of electromotive forces generated by magnetic fields is known as **inductive coupling** [13], hence the name ICP.

9.4.3.1.2 Sample Introduction and Analyte Excitation

Samples are nebulized and introduced as aerosols carried by another stream of argon gas in the inner tube inside the annulus of the plasma at the base of the RF load coil. The sample goes through the process

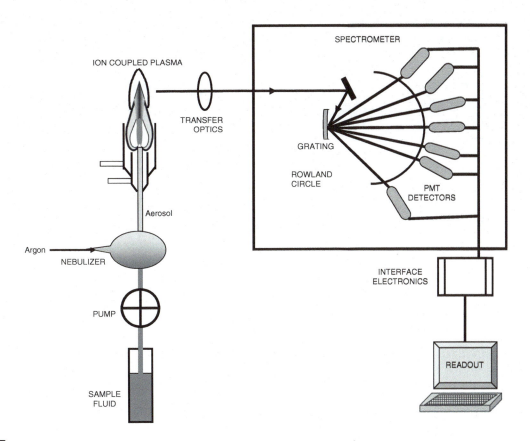

9.8 figure Schematic of an inductively coupled plasma-optical emission simultaneous spectrometer

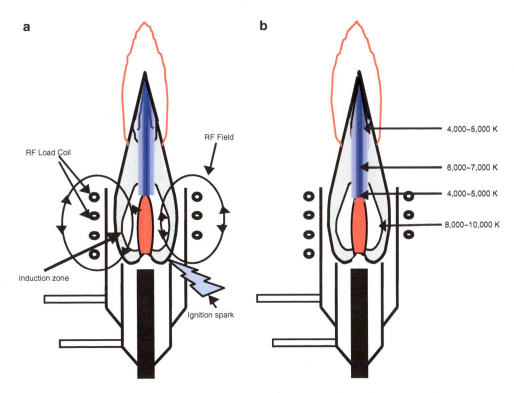

9.9 figure The ICP plasma: (**a**) the process by which the plasma is formed and sustained and (**b**) the temperature distribution of the plasma

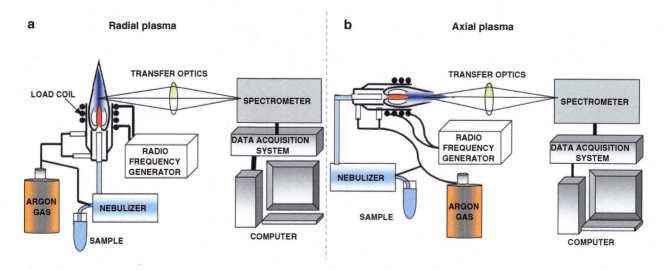

a   Radial plasma   **b**   Axial plasma

9.10 figure   Major components and typical layout of (**a**) a radially viewed ICP-OES instrument and (**b**) an axially viewed ICP-OES instrument

of desolvation, vaporization, atomization, ionization, and excitation as shown in Fig. 9.2. The exact mechanisms by which the analyte atoms and ions are excited in a plasma are not understood fully. Nevertheless, there is general agreement that excitation is dependent primarily on the number and temperature of the electrons in the plasma [14]. Presumably, when electrons are accelerated in a magnetic field, they acquire enough kinetic energy to excite analyte atoms and ions upon collision [14]. Exceptions to this mechanism include the excitation of neutral atoms of magnesium, copper, and a few other elements that are believed to be excited when an argon ion (Ar^+) extracts an electron from the analyte atom (M) to yield M^{+*} and Ar^0, where M is a generic abbreviation for a ground-state metal atom and M^{+*} is an excited metal ion. This mechanism is termed "charge transfer" [15, 16].

9.4.3.1.3 Radial and Axial Viewing

Emissions from ICP torches can be viewed either radially or axially. In the **radial view**, the optics are aligned perpendicular to the torch (Fig. 9.10a). In the **axial view**, the light is viewed by looking down the center of the torch (Fig. 9.10b). Axial viewing offers lower detection limits but is more prone to matrix interferences. Manufacturers of modern ICP-OES instruments have mostly combined the radial and axial configurations into single "dual-view" units, offering greater flexibility to the end user.

9.4.3.2 Detectors and Optical Systems

Older ICP-OES instruments were configured to focus emission lines from each of the analyte elements on separate PMTs arranged in a semicircle (the Rowland circle) as shown in Fig. 9.8. PMT-based instruments are relatively large and bulky, and while some are still in

use today, they have been mostly replaced by modern instruments equipped with an **echelle optical system** and a solid-state array detector (Fig. 9.11), capable of measuring continuous emission spectra over a wide wavelength range.

The echelle optical system employs two dispersing components in series, a prism and a diffraction grating. The prism first disperses the radiation from the plasma torch without any wavelength overlap (in the x-direction). The radiation then strikes a low-density ruled grating (about 53 groves per mm). This further separates the radiation in a direction perpendicular to the direction of radiation dispersed by the prism (in the y-direction), producing a two-dimensional spectrum with a wavelength range of 166–840 nm. When the radiation passing through the echelle optical system is focused onto the solid-state array detector, electrons are liberated proportionally to the intensity of the incident radiation and trapped in the silicon-based, light-sensitive elements called pixels for signal processing. ICP-OES instruments typically use one of three types of solid-state array detectors: **charge coupled device** (CCD), **complementary metal oxide semiconductor** (CMOS), or **charge injection device** (CID). A description of these detectors is beyond the scope of this chapter. Interested readers are referred to the comparisons between different types of solid-state array detectors provided in references [17, 18].

More recent development in ICP-OES instrumentation involves the use of the Optimized Rowland Circle Alignment (ORCA) design to further extend measurement over a wavelength range of 130–800 nm. This allows for the measurement of chlorine which emits a spectral line at 134 nm, below the range of an echelle system. In an ORCA system, detectors are positioned in a semicircular arrangement (Fig. 9.12). A holographic grating is used to separate wavelengths of

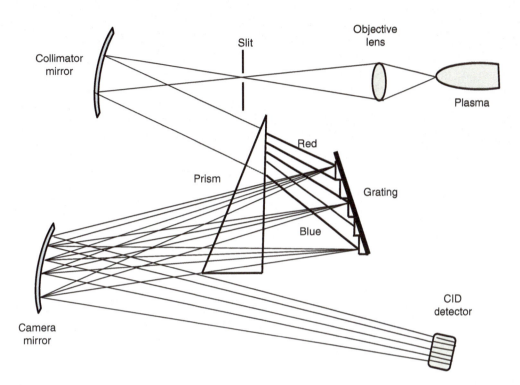

A simplified example of an echelle spectrometer. The echelle spectrometer combines two light dispersing elements in series. A prism first disperses the light in the x-direction without overlapping orders, followed by a grating which disperses the light in the y-direction, producing a two-dimensional spectrum on the detector

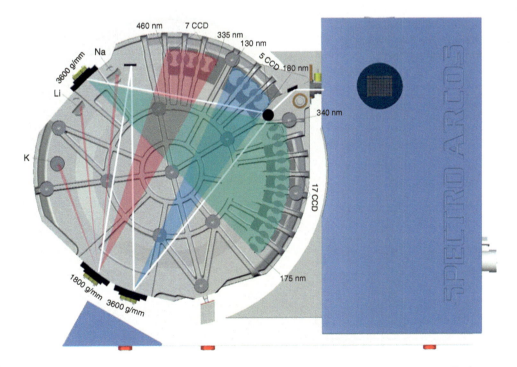

The Optimized Rowland Circle Alignment (ORCA) design using three concave gratings. Each concave grating has the same curvature of the Rowland circle and will disperse the light according to its wavelength. The dispersed light will then be focused back onto one of the detectors placed along the diameter of the Rowland circle. The system uses CCD detectors in place of photomultiplier tubes (PMTs) and allows for the determination of nearly all the elements listed in the periodic chart (Used with permission. Courtesy of SPECTRO Analytical Instruments GmbH Boschstr., Kleve, Germany)

the light coming from the plasma torch. The ORCA system has fewer light dispersing elements (gratings and prisms) than an echelle system. This reduces light loss in the system and therefore increases sensitivity. It also allows for a more uniform resolution and greater stability.

9.4.4 General Procedure for Inductively Coupled Plasma-Optical Emission Spectroscopy Analysis

As is the case with atomic absorption spectrometers, operating procedures for atomic emission spectrometers vary somewhat from instrument to instrument. ICP-OES instruments are almost always interfaced with computers. The software contains methods that specify instrument operating conditions. The computer may be programmed by the operator, or in some cases, default conditions may be used. Once the method is established, operation is highly automated.

9.4.5 Interferences in Inductively Coupled Plasma-Optical Emission Spectroscopy

Generally, interferences in ICP-OES analyses are less of a problem than with AAS, but they do exist and must be taken into account. **Spectral interferences** are the most common. Samples containing high concentrations of certain ions may cause an increase (shift) in background emissions at some wavelengths. This will cause a positive error in the measurement, referred to as **background shift interference** (Fig. 9.13). Correction is relatively simple. Two additional emission measurements are made at a wavelength above and below the

emission line of the analyte. The average of these two emissions is then subtracted from the emission of the analyte. (*Note*: In the example in Fig. 9.13, the intensity of the background shift from the aluminum is *increasing*, and this is why it is necessary to make measurements above and below the emission line for lead. If the intensity of the background shift is constant, then only one measurement near the wavelength of the analyte's emission line is required for correction.) Alternatively, another emission line could be chosen in a region where there is no background shift.

A more problematic type of spectral interference, called **spectral overlap interference** (Fig. 9.14), occurs when the resolution of the instrument is insufficient to prevent overlap of the emission line (i.e., a very narrow bandwidth) of one element with that of another. For example, when determining sulfur in a sample containing high concentrations of calcium, some of the emission from one of the calcium lines overlaps with the sulfur emission line at 180.731 nm. This will cause an apparent increase in the measured concentration of sulfur. This problem may be overcome by either choosing a different sulfur emission line or by calculating the **inter-element correction** (IEC) factor. In the above example, this would require first measuring the emission of calcium at a different wavelength to determine its concentration in the sample. A standard solution of pure calcium is then prepared at that same concentration, and the apparent sulfur concentration in the pure calcium standard solution, presumably containing no sulfur, is determined. Finally, the sample is analyzed for sulfur, and the contribution of calcium to the sulfur signal is subtracted, thereby giving an accurate estimate of the true sulfur concentration.

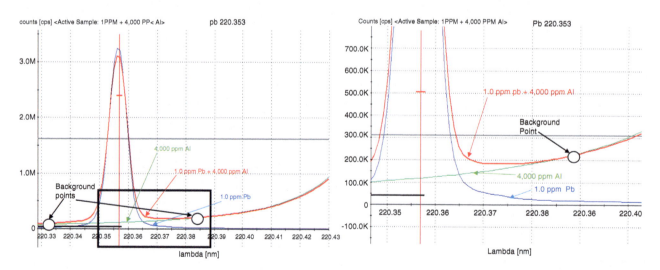

Background shift interference of aluminum on lead. The two graphs show that aluminum will increase the background signal near the lead line at 220.35 nm. This is corrected by placing background subtraction points on both sides of the lead peak. The graph on the right is a blowup of the area that shows the background shift interference

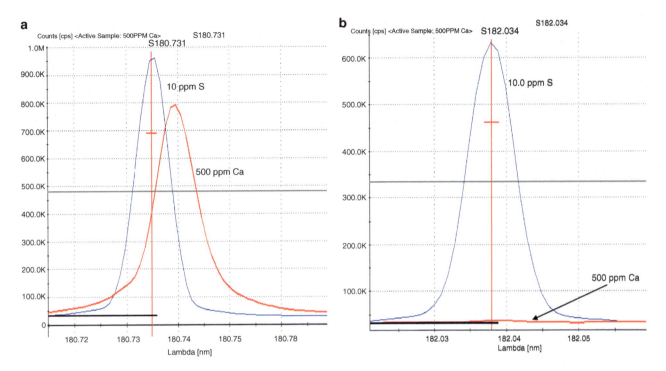

9.14 figure Spectral overlap interference of calcium on sulfur. (**a**) The overlapping spectra of sulfur and calcium indicate that the resolution of the instrument is insufficient to resolve the emission from sulfur at 180.731 nm and calcium at 180.734 nm. (**b**) If the sample is being analyzed for sulfur, it is best to use the sulfur line at 182.034 nm where there is no spectral overlap

9.5 APPLICATIONS OF ATOMIC ABSORPTION AND EMISSION SPECTROSCOPY

9.5.1 Uses

Atomic absorption and emission spectroscopy are widely used for the quantitative measurement of mineral elements in foods. In principle, any food may be analyzed with any of the atomic spectroscopy methods discussed. In most cases, it is necessary to ash the food to destroy organic matter and to dissolve the ash in a suitable solvent (usually water or dilute acid) prior to analysis (see Chap. 16 for details on ashing methodology). Proper ashing is critical to accuracy. Some elements may be volatile at temperatures used in **dry ashing** procedures. Volatilization is less of a problem in **wet ashing**, except for the determination of boron, which is recovered better using a dry ashing method. However, ashing reagents may be contaminated with the analyte. It is therefore wise to carry blanks through the ashing procedure.

Some liquid products may be analyzed without ashing, provided appropriate precautions are taken to avoid interferences. For example, vegetable oils may be analyzed by dissolving the oil in an organic solvent such as acetone or ethanol and aspirating the solution directly into a flame atomic absorption spectrometer. Milk samples may be treated with trichloroacetic acid to precipitate the protein; the resulting supernatant is analyzed directly. A disadvantage of this approach is that the sample is diluted in the process and the analyte can become entrapped or complexed to the precipitated proteins. This may be a problem when analytes are present in low concentrations. An alternative approach is to use a graphite furnace for atomization. For example, an aliquot of an oil may be introduced directly into a graphite furnace for atomization. The choice of method will depend on several factors, including instrument availability, cost, precision/sensitivity, and operator skill.

9.5.2 Practical Considerations

9.5.2.1 Reagents

Since concentrations of many mineral elements in foods are at the trace level, it is essential to use highly pure chemical reagents and water for preparation of samples and standard solutions. Only reagent grade chemicals should be used. Water may be purified by distillation, deionization, or a combination of the two. **Reagent blanks** should always be carried through the analysis.

9.5.2.2 Standards

Quantitative atomic spectroscopy depends on comparison of the sample measurement with appropriate standards. Ideally, standards should contain the analyte metal in known concentrations in a solution that closely approximates the sample solution in composition and physical properties. Because many factors can affect the measurement, such as flame temperature, aspiration rate, and the like, it is essential to run standards frequently, preferably right before and/or right after running the sample. Standard solutions may be purchased from commercial sources, or they may be prepared by the analyst. Obviously, standards must be prepared with extreme care since the accuracy of the analyte determination depends on the accuracy of the standard. Perhaps the best way to check the accuracy of a given assay procedure is to analyze a reference material of known composition and similar matrix. Standard reference materials may be purchased from the United States National Institute of Standards and Technology (NIST) [19].

9.5.2.3 *Labware*

Vessels used for sample preparation and storage must be clean and free of the elements of interest. Plastic containers are preferable because glass has a greater tendency to adsorb and later leach metal ions. All labware should be thoroughly washed with a detergent, carefully rinsed with distilled or deionized water, soaked in an acid solution (1 N HCl is sufficient for most applications), and rinsed again with distilled or deionized water.

9.6 INDUCTIVELY COUPLED PLASMA-MASS SPECTROMETRY

As described above, atomic absorption and emission spectrometers are designed to quantify mineral elements in a sample by measuring either the absorption or emission of radiation by the element of interest at a wavelength unique to that element. Another approach is to directly measure the atoms (as ions in this case) of the element in the sample. This is possible with an **inductively coupled plasma-mass spectrometry** (ICP-MS) instrument that combines ICP with a mass spectrometer (Fig. 9.15). This allows extremely low detection limits at the single part per trillion (ppt) levels, enhanced multielement capabilities, and the ability to quantify individual isotopes present in multi-isotope elements. Note that isotopic analysis is not possible with atomic absorption or emission spectroscopy since absorption and emission lines are the same for all isotopes of a given element.

9.6.1 Principles of Inductively Coupled Plasma-Mass Spectrometry

The principles of mass spectrometry and descriptions of instrumentation for different types of mass spectrometers are given in detail in Chap. 11. In ICP-MS

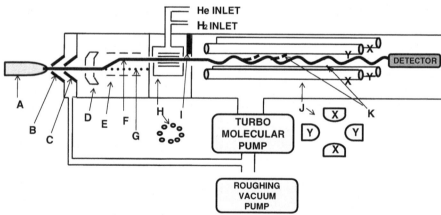

A Plasma
B Sample Cone
C Skimmer cone
D Extraction Lenses
E Omega and Focusing Lenses
F Ion Beam
G Photons and Neutrals being separated from the ion beam

H Collision/reaction Cell (CRC) or Octopole Reaction System (ORS); also showing a cross section of the octopole
I Gate valve used to maintain vacuum in the MS for maintenance
J Quadrupole rods; also showing a cross section of the 4 molybdenum hyperbolic rods
K Rejected ions of improper mass to charge ratio

A simplified diagram of an inductively coupled plasma-mass spectrometer. The ions, electrons, photons, and neutral species generated in the plasma (*A*) are guided through an interface made up of a water-cooled sampling cone (*B*) and a skimmer cone (*C*) with a partial vacuum in-between to remove argon gas and some neutral species. The remaining particles pass through the extraction lenses (*D*), which repel electrons and accelerate the positive ions further through to the off-axis ion omega lenses (*E*), bending the ion beam (*F*) as a result. The paths of the photons and neutral species (*G*) are unaffected and separated from the ion beam (see Chap. 11 for more details on the instrumentation of mass spectrometers)

mineral analyses, samples are prepared and aspirated into the ICP torch in the same manner as in ICP-OES, but instead of having an optical system and a device for separating and detecting radiation of specific wavelengths, the ICP-MS uses a mass spectrometer to separate and detect **ions** of the elements directly based on their unique **mass-to-charge (*m/z*) ratios**. As shown in Fig. 9.15, the interface between the high-temperature ICP operating at atmospheric pressure and the high vacuum mass spectrometer consists of: (1) two funnel-shaped water-cooled cones (the sampling cone and the skimmer cone) and (2) ion lenses (extraction lenses and omega lenses) to guide the analyte ions into the quadrupole while removing electrons, photons, and other neutral species from the ICP discharge.

9.6.2 Interferences in Inductively Coupled Plasma-Mass Spectrometry

Interferences in ICP-MS arise when different ionic species from the sample have the same *m/z* ratio, leading to overlapping of signals. For example, iron has four naturally occurring stable isotopes: ^{54}Fe, ^{56}Fe, ^{57}Fe, and ^{58}Fe, with natural abundances of 5.8%, 91.75%, 2.1%, and 0.28%, respectively. Nickel has five stable isotopes: ^{58}Ni, ^{60}Ni, ^{61}Ni, ^{62}Ni, and ^{64}Ni, with natural abundances of 68.04%, 26.22%, 1.14%, 3.63%, and 0.93%, respectively. The signals for ^{58}Fe and ^{58}Ni will overlap, resulting in an **isobaric interference** because *m/z* = 58 for both isotopes. In this case, the analyst would select ^{56}Fe for the determination of iron and ^{60}Ni for the determination of nickel in the sample. (It is best to select the isotope with the highest natural abundance because the measurement precision is higher for more abundant isotopes. The concentration of the isotope in the sample is equal to the concentration of the element multiplied by the % natural abundance.) Most elements have at least one isotope with a unique mass number, thus allowing identification and quantification of elements.

The **doubly charged interference** occurs when ions of a particular element exist with double positive charges (instead of the normal single positive charge). Ions with charges greater than +1 typically have negligible impact except for ^{138}Ba, which may lose one or two electrons to produce singly and doubly charged ions with *m/z* = 138 and *m/z* = 69, respectively. The presence of doubly charged ^{138}Ba in a sample would interfere with ^{69}Ga. Another possible source of interference, referred to as **polyatomic interference**, comes from the formation of molecular species in the plasma between argon and elements from acids (e.g., H, N, O, Cl, etc.) used in dissolving the sample. Table 9.3 shows some examples of polyatomic interferences. Modern ICP-MS

<table>
<tr><td>**9.3**
table</td><td colspan="3">Examples of polyatomic interference in ICP-MS</td></tr>
<tr><td colspan="2">*Polyatomic species*</td><td>*m/z*</td><td>*Interfered element/isotope*</td></tr>
<tr><td colspan="2">^{38}Ar^{1}H^{+}</td><td>39</td><td>^{39}K^{+}</td></tr>
<tr><td colspan="2">^{35}Cl^{16}O^{+}</td><td>51</td><td>^{51}V^{+}</td></tr>
<tr><td colspan="2">^{40}Ar^{12}C^{+}</td><td>52</td><td>^{52}Cr^{+}</td></tr>
<tr><td colspan="2">^{38}Ar^{16}O^{1}H^{+}</td><td>55</td><td>^{55}Mn^{+}</td></tr>
<tr><td colspan="2">^{40}Ar^{16}O^{+}</td><td>56</td><td>^{56}Fe^{+}</td></tr>
<tr><td colspan="2">^{40}Ar^{35}Cl^{+}</td><td>75</td><td>^{75}As^{+}</td></tr>
<tr><td colspan="2">^{40}Ar^{40}Ar^{+}</td><td>80</td><td>^{80}Se^{+}</td></tr>
</table>

Adapted from Thomas [7]

instruments can be equipped with devices called **collision/reaction cells** (CRC) through which gasses such as H_2, He, NH_3, or CH_4 may be introduced to remove doubly charged and polyatomic interferences based on differences in the physical size and kinetic energy between the analytes and the interfering species.

Another approach to reduce interferences from polyatomic species is to use a high-resolution ICP-MS, which utilizes a double focusing sector-field mass spectrometer to separate ions generated by the plasma [20]. For example, high-resolution ICP-MS has been shown to resolve the Fe peak at mass 55.935 from ArO at mass 55.957 [21, 22]. In this case the ions are measured sequentially, which is the case for most of the ICP-MS systems. However there is one ICP-MS system available that is capable of measuring all the elements from lithium to uranium simultaneously [23]. This system uses a double focusing sector-field mass spectrometer in a Mattauch-Herzog geometry, in which the ion beam energy band width is reduced using an electrostatic energy analyzer to achieve high resolution. The ion beam then passes through a magnetic field for mass separation and is focused on to one focal plane, enabling the entire spectrum to be measured with a flat surface detector simultaneously. This would be an ideal instrument for measuring isotope ratios from transient signals.

9.7 COMPARISON OF AAS, ICP-OES, AND ICP-MS

Table 9.4 provides a summary of advantages and disadvantages of AAS, ICP-OES, and ICP-MS. Flame AAS has enjoyed a long history of applications in mineral analysis. It is relatively inexpensive and easy to use, and is ideal for analyzing a single element in a given sample. The major disadvantages of flame AAS include relatively low sensitivity, narrow linear working range, the use of potentially dangerous fuel gas, and its limitations on multielement

9.4

9.4 table Advantages and disadvantages of AAS, ICP-OES and ICP-MS

	Flame AAS	Graphite furnace AAS	ICP-OES	ICP-MS
Detection limit[a]	Good detection limits with many elements at the part per billion (ppb) level	Better than flame AAS and better than ICP-OES for some elements	Better than flame AAS	Overall best detection limits compared to other techniques
Elemental analytical capability	Single	Single	Multiple	Multiple
Approximate analytical working range	3 orders of magnitude	2 orders of magnitude	6 orders of magnitude (could be higher with dual-view models)	9 orders of magnitude
Cost	Low	Low to medium	Medium	High
Use of explosive fuel gas	Yes (Flame AAS instruments should not be unattended while in operation.)	No	No	No
User-friendliness	Some skills required but relatively easy to use	Some skills required	Easy to use once the computer interface is set up and operation is automated	Method development requires more expertise compared to other techniques
Ideal application	Analyzing a limited number of elements in a given sample	Analyzing a limited number of elements, and requiring better detection limits than Flame AAS	Multiple elements in a large number of samples	Multiple elements at ultra-trace concentrations in a large number of samples
Isotopic analysis	N/A	N/A	N/A	Isotopic analysis possible because isotopes of the same element have different mass-to-charge ratios

[a]See Table 9.5 for more information on detection limit of each technique for different elements

analysis. Electrothermal AAS offers improved detection limits, but with the added cost of the graphite furnace, it somewhat negates the cost advantage of AAS. In fact, the cost of a high-end graphite furnace spectrometer overlaps with an entry-level ICP-OES system [24].

ICP-OES instruments are capable of determining concentrations of multiple elements in a single sample with a single aspiration. This offers a significant speed advantage and higher throughput over AAS when the objective is to quantify multiple elements (up to 70) in a given sample. The high temperature of the ICP torch also eliminates many nonspectral interferences (e.g., chemical interferences) encountered in AAS. Another advantage of ICP-OES over AAS is a much wider analytical working range. The analytical working range for ICP-OES is 4–6 orders of magnitude (i.e., 1 µg/L to 1 g/L without having to recalibrate the instrument), compared to about 3 orders of magnitude for AAS (i.e., 1 µg/L to 1 mg/L). All these advantages help explain the popularity of ICP-OES in commercial laboratories analyzing multiple elements in a large number of sam-

ples. ICP-OES is commonly used to obtain information for standard nutrition labeling.

ICP-MS retains the advantages offered by ICP-OES but, in conjunction with mass spectrometry, offers the lowest detection limits, enhanced multielement capabilities, a wider analytical working range (9 orders of magnitude, i.e., 1 ng/L to 1 g/L), and isotopic information potentially for tracking the geographical origins of food products [25]. Laboratories analyzing trace or ultra-trace concentrations of toxic heavy metals such as cadmium would be best served with an ICP-MS system. The major disadvantage for ICP-MS is perhaps the cost, which is about two to four times higher than its ICP-OES counterparts.

Table 9.5 lists the detection limits with different techniques for mineral elements of interest in foods. It should be noted that these are approximate values, and detection limits will vary depending on many factors such as the sample matrix and the stability of the instrument. A two- or threefold difference in detection limit is probably not meaningful, but an order of magnitude difference is noteworthy.

9.5 table	Approximate detection limits (µg/L) for selected elements analyzed with various instruments			

Element	Flame AAS	Graphite furnace AAS	ICP-OES	ICP-MS
Al	45	0.1	1	0.0004
As	150	0.05	1	0.0004
Ca	1.5	0.01	0.05	0.0003
Cd	0.8	0.002	0.1	0.00007
Cu	1.5	0.014	0.4	0.0002
Fe	5	0.06	0.1	0.0005
K	3	0.005	1	0.001
Hg	300	0.6	1	0.001
Mg	0.15	0.004	0.04	0.0001
Mn	1.5	0.005	0.1	0.0001
Na	0.3	0.005	0.5	0.0003
Ni	6	0.07	0.5	0.0002
P	75,000	130	4	0.04
Pb	15	0.05	1	0.00004
Se	100	0.05	2	0.0003
Tl	15	0.1	2	0.00001
Zn	1.5	0.02	0.2	0.0007

Adapted from Anonymous [24]
Detection limit is defined as the lowest concentration of the element in a solution that can be distinguished from the blank with 98 % confidence

9.8 SUMMARY

In comparison with traditional wet chemistry methods, AAS, AES, and ICP-MS methods are capable of measuring trace concentrations of elements in complex matrices rapidly and with excellent precision. For most applications, sample preparation involves the destruction of organic matter by dry or wet ashing, followed by dissolution of the ash in an aqueous solvent, usually a dilute acid. The sample solution is introduced as a fine mist into a flame atomizer or an ICP torch (or by injection into a graphite furnace) where it encounters very high temperatures (~2000–3000 K for flame or graphite furnace, and ~6000–7000 K for plasma). The sample goes through the process of desolvation, vaporization, atomization, and ionization. Analyte atoms, now in the gaseous state, are well separated and remain mostly neutral in a flame, but a significant fraction of them lose an electron and become charged in a plasma. The final step is to measure quantitatively the concentrations of the elements either by atomic spectroscopy or mass spectrometry.

Atomic spectroscopy depends on the absorption or emission of electromagnetic radiation (light) by the atoms in the gaseous state. Atoms absorb or emit radiation of discrete wavelengths because the allowed energy levels of electrons in atoms are fixed and distinct. In other words, each element has a unique set of allowed electronic transitions and therefore a unique spectrum. In AAS, light of a discrete wavelength from the element-specific hollow cathode lamp will only be absorbed by the atoms of that element in the sample. Furthermore, the amount of light absorbed is directly related to the concentration of the atoms in the sample. By measuring the absorbance of light of a particular wavelength by an atomized sample, analysts can determine the concentration of an element even in the presence of other elements. In emission spectroscopy, the optical approach involves exciting the electrons in an element to a higher-energy state by a flame or plasma, and then measuring the intensity of the light emitted when the electrons fall back to the ground state or a lower-energy state. Emission spectroscopy instruments are designed to separate the light emitted from excited atoms and to quantitatively measure the intensity of the emitted light.

In contrast to atomic spectroscopy, ICP-MS instruments are designed to measure ions of the element directly. This necessitates the ionization of atoms in the plasma. The ions of the element are then guided into a mass spectrometer which separates and detects ions according to their unique mass-to-charge (m/z) ratio. Quantification of elements with high sensitivity and specificity can be achieved because most elements have at least one isotope with a unique mass number.

Atomic spectroscopy is a powerful tool for the quantitative measurement of elements in foods. The development of these technologies over the past six decades has had a major impact on food analysis. Today, accurate and precise measurements of a large number of mineral nutrients and non-nutrients in foods can be made rapidly using commercially available instrumentation. Analysts contemplating what instruments to acquire could make a decision based on an assessment of the required cost, user-friendliness, analytical working range, detection limit, multielement capability, and the need for isotopic data.

9.9 STUDY QUESTIONS

1. AAS and AES instruments rely on energy transitions in atoms of elements being measured. What is an "energy transition" in this context and why can it be used to detect and quantify a given element in a sample containing multiple elements? What is the source of energy that produces this energy transition in an AAS instrument? In an AES instrument?

2. Describe the process of "atomization" as it pertains to AAS and AES analyses.

3. Your boss wants to purchase an AAS instrument for your analytical laboratory because it is cheaper but you want an ICP-OES instrument

because it is more versatile and will greatly increase your sample throughput. To convince your boss to go with the ICP-OES, you need to educate him on the capabilities and operating principles of the two instruments. Keep in mind that your boss is a food scientist who has not worked in a lab in 20 years.

(a) Explain the underlying principles of operation for an ICP-OES instrument in language your boss can understand. Describe the instrument you want to purchase (a simple drawing of the instrument might be helpful here).

(b) Explain how AAS differs in instrumentation and principle of operation from what you described previously for ICP-OES.

(c) Can you make the case that costs for an ICP-OES would be lower over the long term?

(d) For most types of food samples other than clear liquids, what type of sample preparation and treatment is generally required before using ICP-OES or AAS for analysis?

4. You are training a new technician in your analytical laboratory on mineral analysis by AAS and ICP-OES. Briefly describe the purpose of each of the following items

(a) HCL in AAS
(b) Plasma in ICP-OES
(c) Echelle optical system in ICP-OES
(d) Nebulizer in AAS and ICP-OES

5. In the quantitation of Na by atomic absorption, KCl or LiCl was not added to the sample. Would you likely over- or underestimate the true Na content? Explain why either KCl or LiCl is necessary to obtain accurate results.

6. Give five potential sources of error in sample preparation prior to atomic absorption analysis.

7. You are performing iron analysis on a milk sample using AAS. Your results for the blank are high. What could be causing this problem and what is a possible remedy?

8. The detection limit for calcium is lower for ICP-OES than it is for flame AAS. How is the detection limit determined, and what does it mean?

9. When analyzing a sample for mineral elements using ICP-MS, the instrument is programmed to count the number of ions with a specific m/z ratio striking the detector. You decide to determine the concentrations of potassium and calcium in a sample of wheat flour. What m/z ratio would you use for potassium? For calcium? Why? (Hint, study the masses of all the naturally occurring and stable isotopes for the two elements and for argon (see table) and select isotopes with no interferences.) Why is it important to know the masses of argon isotopes as well as potassium and calcium?

Isotope	Natural abundance (%)
^{36}Ar	0.34
^{38}Ar	0.063
^{40}Ar	99.6
^{39}K	93.2
^{40}K	0.012
^{41}K	6.73
^{40}Ca	96.95
^{42}Ca	0.65
^{43}Ca	0.14
^{44}Ca	2.086
^{46}Ca	0.004

9.10 PRACTICE PROBLEMS

1. Your company manufactures and markets an enriched all-purpose flour product. You purchase a premix containing elemental iron powder, riboflavin, niacin, thiamin, and folate which you mix with your flour during milling. To comply with the standard of identity (see Chap. 2) for enriched flour, you specified to the supplier that the premix be formulated so that when added to flour at a specified rate, the concentration of added iron is 20 mg/lb flour. However, you have reason to believe that the iron concentration in the premix is too low so you decide to analyze your enriched flour using your new atomic absorption spectrometer. You follow the following protocol to determine the iron concentration.

(a) Weigh out 10.00 g of flour, in triplicate (each replicate should be analyzed separately).

(b) Transfer the flour to an 800-mL Kjeldahl flask.

(c) Add 20 mL of deionized water, 5 mL of concentrated H_2SO_4, and 25 mL of concentrated HNO_3.

(d) Heat on a Kjeldahl burner in a hood until white SO_3 fumes form.

(e) Cool, add 25 mL of deionized water, and filter quantitatively into a 100-mL volumetric flask. Dilute to volume.

(f) Prepare iron standards with concentrations of 2, 4, 6, 8, and 10 mg/L.

(g) Install an iron hollow cathode lamp in your atomic absorption spectrometer and turn on the instrument and adjust it according to instructions in the operating manual.

(h) Run your standards and each of your ashed samples and record the absorbances.

Calculate the iron concentration in each of your replicates, express as mg Fe/lb flour.

Absorbance data for iron standards and flour samples

Sample	Fe Conc. (mg/L)	Absorbance	Corrected absorbance
Reagent blank	–	0.01	–
Standard 1	2.0	0.22	0.21
Standard 2	4.0	0.40	0.39
Standard 3	6.0	0.63	0.62
Standard 4	8.0	0.79	0.78
Standard 5	10.0	1.03	1.02
Flour sample 1	?	0.28	0.27
Flour sample 2	?	0.29	0.28
Flour sample 3	?	0.26	0.25

2. Describe a procedure for determining calcium, potassium, and sodium in infant formula using ICP-OES. *Note:* Concentrations of Ca, K, and Na in infant formula are around 700 mg/L, 730 mg/L, and 300 mg/L, respectively.

Answers

1. The following steps may be used to determine the iron concentration in the flour samples.

   (a) Enter the data for the standards into Excel. Using the scatter plot function, plot the standard curve and generate a trend line using linear regression. Include the equation for the line and the R^2 value. Your results should look like the standard curve shown.

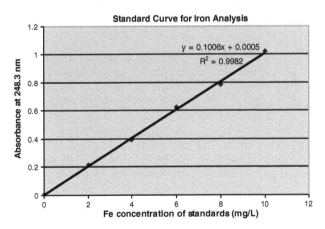

   (b) Using the equation, calculate the iron concentration in the solution in the volumetric flask for each of your samples. Your answers should be 2.68 mg/L, 2.79 mg/L, and

2.48 mg/L for samples 1, 2, and 3, respectively. The mean is 2.65 mg/L; the standard deviation is 0.16.

   (c) Now determine the iron concentration in the flour. Recall that you transferred the solution from the Kjeldahl flask quantitatively into the 100-mL volumetric flask. Therefore all of the iron in the flour sample should be in the volumetric flask. The mean concentration is 2.65 mg/L. The volume is 0.1 L. Therefore, the amount of iron in the 10 g of flour is 0.265 mg. To convert this to mg/lb, multiply by 454/10: 0.265 mg/10 g × 454 g/lb = 12 mg Fe/lb flour

   (d) Your suspicions are confirmed; your supplier shorted you on iron in the premix. You need to correct this as soon as possible because your flour does not conform to the FDA's standard of identity for enriched flour and you may be subject to legal action by the FDA.

2. Consult AOAC Method 984.27 (see Chap. 1 for a description of AOAC International), and the following approach may be used:

   (a) Shake can vigorously.
   (b) Transfer 15.0 mL of formula to a 100-mL Kjeldahl flask. (Carry two reagent blanks through with sample.)
   (c) Add 30 mL of HNO_3-$HClO_4$ (2:1).
   (d) Leave samples overnight.
   (e) Heat until ashing is complete (follow AOAC procedure carefully – mixture is potentially explosive.)
   (f) Transfer quantitatively to a 50-mL vol flask. Dilute to volume.
   (g) Calibrate instrument. Choose wavelengths of 317.9 nm, 766.5 nm, and 589.0 nm for Ca, K, and Na, respectively. Prepare calibration standards containing 200 μg/mL, 200 μg/mL, and 100 μg/mL for Ca, K, and Na, respectively.
   (h) The ICP-OES computer will calculate concentrations in the samples as analyzed. To convert to concentrations in the formula, use the following equation:

   Concentration in formula

   = Concentration measured by ICP

   $\times \left(50 \text{ mL} / 15 \text{mL}\right)$

REFERENCES

1. U.S. Department of Agriculture, Agricultural Research Service (2016) USDA Nutrient Database for Standard Reference, Release 28. Nutrient Data Laboratory Home Page, http://ndb.nal.usda.gov
2. Institute of Medicine (IOM) (2001) Dietary Reference Intakes for Vitamin A, Vitamin K, Arsenic, Boron, Chromium, Copper, Iodine, Iron, Manganese, Molybdenum, Nickel, Silicon, Vanadium, and Zinc. National Academy Press, Washington, D.C.
3. Greenfield S, Jones IL, Berry CT (1964) High-pressure plasma as spectroscopic emission sources. Analyst 89:713–720
4. Wendt RH, Fassel VA (1965) Induction-coupled plasma spectrometric excitation source. Anal Chem 37:920–922
5. Beaty RD, Kerber JD (1993) Concepts, instrumentation and techniques in atomic absorption spectrophotometry, 2nd edn. The Perkin-Elmer Corporation, Norwalk, CT
6. Boss CB, Fredeen KJ (2004) Concepts, instrumentation and techniques in inductively coupled plasma atomic emission spectrometry, 3rd edn. The Perkin-Elmer Corporation, Norwalk, CT
7. Thomas R (2013) Practical guide to ICP-MS: a tutorial for beginners, 3rd edn. CRC, Taylor and Francis Group, Boca Raton, FL
8. Skoog DA, Holler FJ, Crouch SR (2007) Principles of instrumental analysis, 6th edn. Thompson Brooks/Cole, Belmont, CA
9. Ganeev A, Gavare Z, Khutorshikov VI, Hhutorshikov SV, Revalde G, Skudra A, Smirnova GM, Stankov NR (2003) High-frequency electrodeless discharge lamps for atomic absorption spectrometry. Spectrochimica Acta Part B 58:879–889
10. Koirtyohann SR, Pickett EE (1966) Spectral interferences in atomic absorption spectrometry. Anal Chem 38:585–587
11. West AC, Fassel VA, Kniseley RN (1973) Lateral diffusion interferences in flame atomic absorption and emission spectrometry. Anal Chem 45:1586–1594
12. Dockery CR; Blew MJ; Goode SR (2008) Visualizing the Solute Vaporization Interference in Flame Atomic Absorption Spectroscopy. J Chem Educ 85:854–858
13. Hou X, Jones BT (2000) Inductively coupled plasma-optical emission spectrometry. In: Meyers RA (ed) Encyclopedia of Analytical Chemistry. John Wiley and Sons Ltd, Chichester, England
14. Hieftje GM, Rayson GD, Olesik JW (1985) A steady-state approach to excitation mechanisms in the ICP. Spectrochem Acta 40:167–176
15. Hasegawa T, Umemoto M, Haraguchi H, Hsiech C, Montaser A (1992) Fundamental properties of inductively coupled plasma. In: Montaser A, Golightly DW (eds) Inductively Coupled Plasmas in Analytical Atomic Spectroscopy. VCH Publishers, New York
16. Lazar AC, Farnsworth PB (1999) Matrix effect studies in the inductively coupled plasma with monodisperse droplets. Part 1: The influence of matrix on the vertical analyte emission profile. Appl Spectrosc 53:457–464
17. Litwiller D (2005) CMOS vs. CCD: Maturing Technologies, Maturing Markets-The factors determining which type of imager delivers better cost performance are becoming more refined. Photonic Spectra 39:54–61
18. Sweedler JV, Jalkian RD, Pomeroy RS, Denton MB (1989) A comparison of CCD and CID detection for atomic emission spectroscopy. Spectrochimica Acta Part B 44:683–692
19. National Institute of Standards and Technology (2016) http://www.nist.gov/srm/index.cfm. Accessed March, 2016
20. Jakubowski N, Moens L, Vanhaecke F (1998) Sector field mass spectrometers in ICP-MS. Spectrochimica Acta Part B 53:1739–1763
21. Thermo Scientific Application Note 30003 (2007) Determination of trace elements in clinical samples by high resolution ICP-MS. Thermo Fisher Scientific Inc., Waltham, MA
22. Thermo Scientific Application Note 30073 (2007) Determination of ultratrace elements in liquid crystal by high resolution ICP-MS. Thermo Fisher Scientific Inc., Waltham, MA
23. SPECTRO Analytical Instruments Inc. (2016). http://www.spectro.com/products/icp-ms-spectrometers/spectro-ms. Accessed March, 2016
24. Anonymous (2016) Atomic spectroscopy: guide to selecting the appropriate technique and system. PerkinElmer Inc. http://www.perkinelmer.com/PDFs/Downloads/BRO_WorldLeaderAAICPMSICPMS.pdf. Accessed March, 2016
25. Luykx DM, Van Ruth SM (2008) An overview of analytical methods for determining the geographical origin of food products. Food Chem 107:897–911
26. Welz, B. 1985. Atomic Absorption Spectrometry. VCH Weinheim, Germany

10
chapter

Nuclear Magnetic Resonance

Bradley L. Reuhs (✉)
Department of Food Science,
Purdue University,
West Lafayette, IN 47907-2009, USA
e-mail: breuhs@purdue.edu

Senay Simsek
Department of Plant Sciences,
North Dakota State University,
Fargo, ND 58108, USA
e-mail: senay.simsek@ndsu.edu

S. Nielsen (ed.), *Food Analysis*, Food Science Text Series,
DOI 10.1007/978-3-319-45776-5_10, © Springer International Publishing 2017

10.1 INTRODUCTION

Nuclear magnetic resonance (NMR) **spectroscopy** is a powerful analytical technique with a wide variety of applications. It may be used for complex structural studies, for protocol or process development, or as a simple quality assay for which structural information is important. It is nondestructive, and high-quality data may be obtained from milligram, even microgram, quantities of sample. Whereas other spectroscopy techniques may be used to determine the nature of the functional groups present in a sample, only NMR spectroscopy can provide the data necessary to determine the complete structure of a molecule. The applicability of NMR to food analysis has increased over the last three decades. In addition to improved instrumentation and much lower costs, very complex and specialized NMR techniques can now be routinely performed by a student or technician. These experiments can be set up with the click of a button/icon, as all the basic parameters are included in default experiment files listed in the data/work station software, and the results are obtained in a short time.

NMR instruments may be configured to analyze samples in solutions or in the solid state. In fact, these two types of analyses can be used in tandem to follow the fate of a given molecule within a specific system. For example, as a fruit ripens, many components will be released from the solid matrix around the plant cells into solution in the ripe fruit liquid. The development of this process can be followed by liquids versus solids analyses during the ripening time. As ripening progresses, some NMR signals will decrease in the solids NMR analyses and increase in the liquids NMR spectra.

Other food applications of NMR spectroscopy include structural analysis of food components, such as fiber, to correlate the structure to the rheological properties. Routine analyses are used to determine the quality of a product or to test the purity of ingredients. Related techniques, such as **NMR relaxometry**, can be used to assess processing operations; for example, relaxometry is used to follow the solubilization of powdered ingredients in water, to optimize processing parameters. **Magnetic resonance imaging** (MRI) is a nondestructive technique that can be used to image product quality and changes during processing and storage. For example, MRI is used to image the freezing process, with the goal of increasing shelf life. When combined with rheological analyses, sauce and paste flow in a processing system can be measured.

This chapter will cover the basic principles and applications of NMR spectroscopy, as well as a brief description of relaxometry, MRI, and a recent development in instrumentation that uses NMR as part of a rapid moisture and fat analysis system. Specific applications to food analysis will be highlighted.

10.2 PRINCIPLES OF NMR SPECTROSCOPY

10.2.1 Magnetic Field

NMR differs from most other forms of spectroscopy in that it is the **atomic nuclei** that are the subject of study, and the measured energy is in the radio-frequency range. Many nuclei possess an angular momentum, which means that they have a characteristic spin quantum number (I) and may be analyzed using NMR. The most common nuclei analyzed by NMR are the proton (H) and the ^{13}C isotope of carbon, as well as ^{19}F and ^{31}P, all of which have a spin $I = 1/2$. Nuclei with other spin quantum numbers will not be considered in this chapter, and the theoretical discussion will focus on the proton. These nuclei are charged, and a spinning charge generates a magnetic field. Simply put, the nuclei behave like tiny magnets that interact with an applied, external magnetic field.

Once the nuclei are placed within a strong **external magnetic field** (B_0), the spin of the nuclei will align with that field (Fig. 10.1). Because of quantum mechanical constraints (nuclei of spin I have $2I + 1$ possible orientations in the external magnetic field), there are only two orientations that the spin 1/2 nuclei can adopt: either aligned with the applied magnetic field (parallel or spin $+1/2$) or aligned against the field (antiparallel or spin $-1/2$). The parallel orientation has a slightly lower energy associated with it and, therefore, has a slightly higher population. It is this excess of nuclei in the spin $+1/2$ state that produces the net magnetization that is manipulated and measured during an NMR experiment. The spin of the nuclei is not around the center axis but comparable to a gyration (Fig. 10.1). The motion of a spinning charged particle in an external magnetic field is similar to that of a spinning gyroscope in a gravitational field. This type of motion is known as **precession**, and there is a specific precessional orbit and frequency, the **Larmor frequency**, which is related to the magnetic properties of the nuclei. The magnitude and direction of the local magnetic field describes the magnetic moment or **magnetic dipole** of the system, and due to the precession and the lower energy state excess of nuclei, there is a net vector parallel to the applied field (Fig. 10.1).

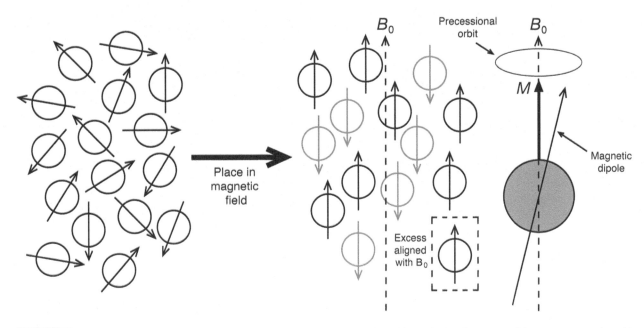

10.1 figure Nuclear spin and magnetic vectors are randomly ordered outside of the NMR magnet. However, once placed in an applied magnetic field, the NMR magnet and the nuclei align either with the applied field, B_0, (parallel) or against it (antiparallel). There is a slight excess in the population aligned parallel to B_0. Although the magnetic dipole tracks a precessional orbit, the net magnetization (M) is aligned with B_0

All nuclei of the same element, H, for example, will have a nearly identical Larmor frequency in a magnetic field. The specific frequency is dependent on the strength of the external magnetic field, and the Larmor frequency of H in this field defines the NMR instrument. For example, a proton has a Larmor frequency of 500 MHz in an 11.7 T magnet, so the instrument is termed a "500 MHz NMR spectrometer." The strength of the magnet not only determines the Larmor frequency of the nuclei but also the degree of excess nuclei in the parallel orientation. The excess of nuclei in the parallel orientation increases with an increase in the external magnetic field strength, and this in turn impacts the signal intensity of the NMR experiment in a higher field strength instrument. This is one reason researchers seek ever more powerful magnets for NMR (recent developments have yielded NMR magnets of greater than 23 T or 1000 MHz). Field strength impacts the signal-to-noise ratio, and, therefore, the sensitivity and resolution of the instrument and the information obtained from the NMR experiment.

The development of today's powerful NMR instruments was contingent on the advances in the production of cryo- or superconducting magnets, in which the magnet coil is held at the temperature of liquid helium (around 3 K). In addition to an increase in sensitivity, superconducting magnets also have the advantage that once charged, they maintain the magnetic field for years without the input of additional energy, due to the low temperature. The major disadvantage is the need for periodic addition of liquid N_2 and liquid He, which in the case of the latter can be quite expensive, particularly with the very large magnets associated with the high field strength instruments.

10.2.2 Radio-Frequency Pulse and Relaxation

Early NMR instruments relied on electromagnets and a simple **radio-frequency** (RF) transmitter, and the analyses were performed by a sweep through the frequency range of the instrument. The collected spectra contain the frequency information; hence, it is termed frequency-domain NMR. Although this enabled the development of NMR spectroscopy, it was not sufficient to facilitate the modern NMR experiment. One of the major developments in NMR technology was the **RF pulse**, in which a large range of frequencies is excited by a short pulse of RF energy around a centered carrier frequency, which is at the Larmor, or resonance frequency of the nuclei under study. This pulse simultaneously excites all of the protons in the sample, and the NMR data for all the protons is collected during a short time after the pulse is applied. The excitation of a range of radio frequencies by a pulse is similar to the excitation of a range of audio frequencies when a clapper strikes a bell, and the size and construction of the bell determines the range that is emitted. In NMR, the carrier

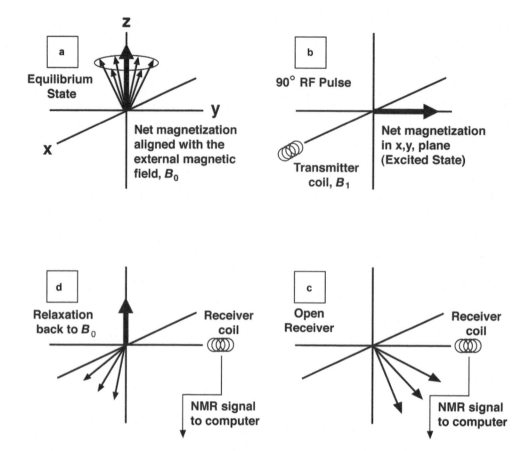

10.2 **figure** (a) Priyor to the RF energy pulse, the net magnetization (*M*) composed of all the component vectors is in the equilibrium state, aligned with B_0. (b) The 90° RF pulse, which covers the resonance frequencies of all relevant nuclei in the sample and originates perpendicular to the *z*-axis (B_1), causes the nuclei to move to a higher energy state, and the net magnetization rotates into the *xy*-plane. (c) Once in the *xy*-plane, the net magnetic vector separates into the component vectors for each unique population of nuclei. As these oscillate in the *xy*-plane, they emit RF signals that are detected by the NMR instrument after passing through the receiver coil, which is located perpendicular to both B_0 and the transmitter coil. (d) As the component vectors continue to oscillate in the *xy*-plane (and emit RF signals), the nuclei begin to relax back to the equilibrium state. The NMR instrument may be set up to repeat this process, with additional pulses, numerous times; the collected data are then added together to improve the signal/noise ratio and resolution

frequency, transmitter power, and duration of the RF pulse determine the frequency range of the pulse.

Once the sample is placed in the magnet, the protons align parallel or antiparallel to the applied, external magnetic field, $\mathbf{B_0}$, with an excess in parallel orientation. The net magnetization of the nuclei in the parallel orientation is aligned with the *z*-axis in an *xyz* graphical representation of the system (Fig. 10.2a). After a pulse of RF energy is applied to the system, the nuclei precess coherently and individual nuclei absorb energy and shift to a higher energy state. The pulse, which is applied by a transmitter coil perpendicular to the *z*-axis ($\mathbf{B_1}$), tilts the net magnetization vector away from the *z*-axis and toward the **xy-plane** (Fig. 10.2b).

Although the parameters that define a pulse include the transmitter power and the pulse duration, a specific pulse used in an NMR experiment is usually described by the degree to which the net magnetization is tilted. The most common pulse is the 90° pulse, which tilts the net magnetization exactly into the *xy*-plane, where the receiver coil is located, thereby maximizing the resulting signals. Many NMR experiments use a series of pulses, termed a **pulse sequence**, to manipulate the magnetization. Complex pulse sequences are essential for the two-dimensional (2D) NMR experiments (Sect. 10.2.5) that are required for structural analyses of complex molecules.

Once the net magnetization has been tilted into the *xy*-plane by a 90° pulse, the magnetization begins to decay back to the *z*-axis. This process is termed **NMR relaxation**, and it involves both **spin-lattice** (*T1*) and **spin-spin** (*T2*) relaxation. **T1 relaxation** is associated with the interaction of the magnetic fields of the excited-state nuclei with the magnetic fields of other nuclei within the "lattice" of the total sample. **T2 relaxation** involves the interactions of neighboring nuclei that lead to a diminishment in the energy state of the excited-state nuclei and the loss of phase coherence. The mechanisms behind relaxation are complex, but the process can be utilized for some specific NMR experiments that, among other things, take advantage of the fact that samples in different forms, liquids vs. solids, for example, relax at different rates (this is very important for the instruments and experiments that are utilized for food processing and content analyses).

10.2.3 Chemical Shift and Shielding

The total H population in the sample determines the net magnetization of the system in the external magnetic field, B_0. The exact frequency of a unique population of protons (i.e., all protons in a specific chemical configuration in the molecule), however, is also dependent on the immediate environment of the nucleus, principally the density of the electron cloud surrounding the nuclei, which determines the electronic environment of the nuclei. This is referred to as the "shielding" effect, because the electrons create a secondary, induced magnetic field that opposes the applied field, shielding the nuclei from the applied field. The resulting frequency differences are so small, relative to the Larmor frequency, that they are commonly reported in parts per million (ppm). However, they are large enough to be clearly detected and resolved during an NMR experiment. The frequency differences that result from the differences in the electronic environments yield the **chemical shift** of the

nuclei. Following the processing of the NMR data, these differences result in a series of resonance signals, each representing a unique proton population, along the x-axis in a one-dimensional (1D) NMR spectrum. Those protons that have a relatively dense electron cloud are considered shielded, since the electron cloud works in opposition to the external magnetic field, and the resonances will be found on the right, or upfield, side of the spectrum, at a lower chemical shift. As deshielding increases, the resonances are shifted further to the left, or downfield, at a progressively higher chemical shift (Fig. 10.3).

One of the most important determinants of the chemical shift for a specific population of protons (i.e., all the protons in the sample that are in an identical molecular environment) is the proximity to an electronegative group or atom, such as O (Fig. 10.3). For example, protons that are not located near (in a molecular context) any electronegative groups or atoms, such as those in the methyl group of a 6-deoxy sugar, are heavily shielded, and the resonances will be found on the far right side of a proton NMR spectrum (Fig. 10.3). In contrast, the proton on the C1 of a typical sugar (the anomeric proton) is near two O atoms, the O in the –OH group (or the O in the linkage to the next sugar in a polymer, such as starch) and the O that forms part of the hemiacetal ring structure of the sugar. Consequently, the resonances associated with the highly deshielded anomeric protons are found on the left side of the proton spectrum. Protons near one O, such as the ring protons in a typical sugar, will be partially shielded, and the resonances associated with these protons will be in the central region of the spectrum.

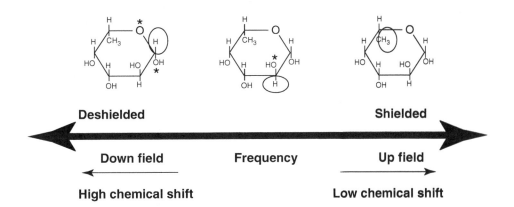

The shielding effect is responsible for the small, but detectable, differences in the resonance frequencies of nuclei such as protons. Protons that are not close to an electronegative group in the molecule, such as the protons in the methyl group in fucose (*top right*), a common 6-deoxy sugar, will be shielded by the electrons surrounding it, and it will have a low chemical shift, upfield in the NMR spectrum. Protons near one oxygen atom (indicated by an *asterisk*), such as the ring protons in sugars (*top middle*), will be intermediately shielded and have a chemical shift toward the middle of the spectrum. And a proton that is near two oxygen atoms, such as the anomeric proton in sugars (*top left*), will be relatively deshielded and have a high chemical shift downfield in the spectrum

10.2.4 1D NMR Experiment

For solution ^{1}H-NMR spectroscopy, the sample is dissolved in deuterated solvents produced for NMR analysis, such as D_2O (where D = deuterium or ^{2}H; this is to avoid overloading the NMR signal with solvent protons), and pipetted into a NMR tube, which is then capped and placed into the magnet. The net magnetization of the nuclei in the parallel orientation in the external magnetic field, B_0, is aligned with the z-axis; this is the equilibrium state. The 90° pulse, centered at the Larmor frequency of H, from the transmitter in the x-axis tilts the net magnetic vector into the xy-plane. There the component vectors (those representing each unique population of protons) will oscillate at their specific NMR resonance frequencies, separating the net vector into numerous component magnetic vectors, which then induce radio signals (the NMR signal) in the receiver coil located in the y-axis (Fig. 10.2c). At the same time, the magnitude of the vectors will decay in the xy-plane (relaxation) and return to the z-axis (Fig. 10.2d). This whole process is termed one **scan**.

The result of these actions is a combined signal for all protons in the sample, which rapidly decreases over time, yielding a **free induction decay** (FID) (Fig. 10.4a), which contains all the frequency and intensity information (as well as phase, which will not be considered) for each of the unique populations of nuclei in the sample. **Fourier transformation**, a mathematical operation that converts one function of a variable into another, is then applied to convert the time-domain FID to the frequency-domain **NMR spectrum** (an xy plot) (Fig. 10.4b). The frequency information for each unique proton population is presented on the x-axis of the NMR spectrum. The signal intensity is related to the y-axis; however, this is not labeled because it has no units, and, moreover, the linewidth differences of each resonance make the y-value imprecise for quantification in ^{1}H-NMR analyses. The best method to compare the signal intensity and, consequently, the relative abundance of specific protons in a 1D NMR spectrum is to compare the integration values of the resonances.

In practice, the NMR spectrometer is set up so that the pulse is applied numerous times, usually in increments of 16 scans. For samples that are present in a high concentration, 16 or 32 scans per experiment are common, and 256 or 512 scans are often used for more dilute samples. After each scan, the new data are added to the data already collected. The result of compiling the data from numerous scans is a significant improvement in the signal/noise ratio and resolution.

10.2.5 Coupling and 2D NMR

Another essential concept to consider is "**coupling**." Coupling is a result of the influence of electrons in covalent bonds on the local magnetic field of nearby

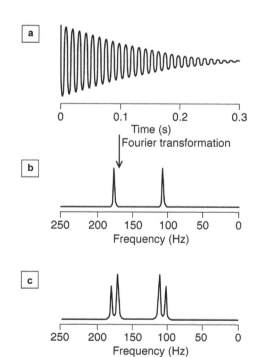

10.4 figures After the 90° RF pulse moves the magnetization to a higher energy state in the xy-plane, the receiver is turned on, and it collects the RF signals emitted by oscillating nuclei; the emitted RF signals rapidly decay over a short time as the magnetization relaxes back to the equilibrium state (see Fig. 10.2). (**a**) The result is NMR data that contains all of the frequency and intensity information for the nuclei under analysis and which diminishes as the signals decay; this data is termed a "free induction decay" (FID), and it represents the time-domain information obtained from the NMR experiment. (**b**) Once the FID has been processed by Fourier transformation, an NMR spectrum is obtained; this represents the frequency-domain NMR information. The resonance "peaks" found on the x-axis are due to unique populations of protons, two in this case. (**c**) If the two protons are coupled through the molecular bonds, they will each have the effect of splitting the resonance peaks into two distinct peaks. The degree of splitting, reported in Hz, is indicative of the strength of the coupling effect

nuclei. Through the intervening bonds, two nearby nuclei will affect the chemical shift of one another, resulting in the splitting of the resonances from each unique population of nuclei into two distinct resonances (Fig. 10.4c). The coupling strength is affected by both the proximity of the nuclei to one another and the geometry of the intervening bonds. For example, protons on a carbon backbone that have a *trans* relationship ("across" from each other) have a much stronger coupling than those that have a *cis* relationship ("on the same side" as one another). Thus, the use of coupling data yields information about the geometry of a specific molecule.

A more important impact of coupling is that complex 2D NMR experiments have been designed to take advantage of the coupling phenomenon, to produce the data necessary for the complete structure determination of a molecule. 2D NMR experiments are essentially a series of 1D experiments, in which the pulse sequence includes several pulses and a variable parameter, such as the delay time between two of the pulses. The computer collects all the spectra and plots them out as a 2D plot, in which "**cross peaks**" show the coupling correlations of nearby nuclei.

10.3   NMR SPECTROMETER

The typical research NMR spectrometer consists of a powerful cryomagnet, into which the sample is placed, a set of electronics for transmitting and collecting radio signals, and a data/work station (Fig. 10.5). Modern instruments use superconducting magnets that are cooled to a very low temperature by a jacket of liquid helium, which has a boiling point of 4.2 K. This jacket is, in turn, surrounded by an outer jacket of liquid nitrogen, which is cheaper and easier to work with than liquid helium. The core of the superconducting magnet consists of coil windings of thin wires made of superconducting alloys, such as niobium-titanium or niobium-tin. The coil (i.e., the magnet) and the coolants are contained in an insulated Dewar, termed a cryostat, which includes a vacuum chamber around the liquid jackets.

Once the magnet is cooled to the operating temperature, by the addition of the coolants, and energized by an external power supply, the magnet will maintain its charge and magnetic field for years. One

of the most important aspects of maintenance for NMR instruments is the routine filling, or topping off, of the coolants. A typical research instrument is filled with liquid nitrogen on a weekly basis, and liquid helium is added monthly. Failure to maintain the coolant levels will result in the **quenching** of the magnet, in which the coolants boil off violently and the magnet loses its charge. Should this happen, the magnet will need to be refilled and recharged, at the very least, and may also require expensive repairs.

Down the center of the magnet but external to the Dewar (i.e., at room temperature) is a tubular space, the **magnet bore**. A multifunction device, termed a **probe**, is placed inside the magnet bore from the bottom. Inside the probe, just inside of the main magnet coil, are small, secondary magnetic coils that receive power from the NMR instrument hardware. These small coils are manipulated by the operator to make fine adjustments to the magnetic field, to optimize the magnetic field; hence, they are called shims (in construction, a shim is a small piece of wood or metal that is placed between two layers of building material to obtain a better fit, such as when leveling a door frame). The probe also contains the coils for transmitting and receiving the RF energy. Thus, the probe is the central piece of hardware for the NMR experiment. Finally, at the top of the bore, and at the top of the whole system, is the sample insertion point. The sample tube, which is in a holder, is lowered down through the bore by a diminishing stream of forced air until it gently comes to rest at the top of the probe; there the sample is correctly aligned with the magnetic field and the probe hardware. One of the most common mistakes for an inexperienced operator is to drop the sample holder and sample into the magnet bore without the airstream flowing, which results in a rapid descent of the sample holder and sample, breaking the NMR tube and damaging the probe (as well as angering the instrument shop manager).

Connected to the magnet probe by several cables is an electronics console that includes the transmitter, the receiver, and other systems that control the NMR instrument, such as the sample temperature control unit. The transmitter includes systems to produce the pulse at the correct frequency for each nucleus that may be observed. For example, a 500-MHz NMR spectrometer requires a transmitter at 500 MHz for ^{1}H-NMR analyses and a second transmitter system at 125 MHz for ^{13}C-NMR. The console also houses the receiver electronics, which process the NMR signal from the probe receiver coils, the electronics that control the shim electromagnets, and the probe temperature control system. All of this is managed by a computer data/work station and some NMR-specific peripherals.

In contrast to the large (and expensive) research instruments are the benchtop **time-domain (TD)**

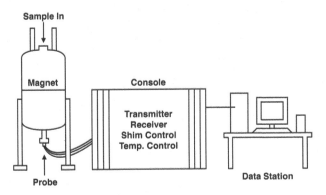

10.5 **figure** A diagram of an NMR spectrometer. The instrument consists of a superconducting cryomagnet (the NMR magnet), an electronics console, and a data/work station that also controls all the functions of the instrument

NMR **systems**, also termed **low-resolution NMR**. They do not provide frequency, or structural, information, but they can be applied to various types of content analyses. They are similar in size to a benchtop centrifuge and utilize a common magnet, with no need for coolants or a large Dewar flask. These instruments are relatively inexpensive and simple to use, and they have numerous food analysis applications, such as fat content.

10.4 APPLICATIONS

Table 10.1 lists recent applications of the NMR and related techniques to food analysis. The following discussion describes general applications and some specific applications to food research. See the references in Table 10.1 for detailed information on the various techniques.

10.4.1 NMR Techniques and General Applications

10.4.1.1 *Liquids*

Liquids NMR is used for relatively pure samples that readily go into solution in any of the many deuterated solvents produced for NMR analyses. Common samples include carbohydrates, proteins, lipids, phenolics, and many other classes of organic compounds. The experiment described in Sect. 10.2.4 is an example of a typical liquid 1D ^{1}H-NMR spectroscopy analysis. The result is a 1D NMR spectrum with the resonances plotted along the *x*-axis (the only plotted axis, hence, 1D). This is the simplest application of NMR spectroscopy, but it can be very informative. For example, plant- and yeast-derived β-glucans are currently an important topic of research, as they have many health-related benefits, and they are often discarded as waste from many food processing systems. NMR spectroscopy is the best analytical tool available to determine the

table Recent food applications of magnetic resonance spectroscopy and related techniques

NMR method	Food	Analysis	Ref.
MRI	Whole wheat bread	Water migration between arabinoxylan and gluten	[1]
^{1}H-MRI	Avocado	Nondestructive assessment of bruising	[2]
^{1}H-MRI	Honey	Authenticity screening	[3]
HR ^{1}H NMR	Pork meat	Quantitative fatty acid chain composition	[4]
qHNMR	Processed foods	Quantification of benzoic acid	[5]
^{1}H and spin-spin relaxation	Rice and potato starches	Impact of hydration levels on starch gelatinization	[6]
2D NMR	Green coffee bean extract	Analysis of organic compounds	[7]
^{1}H NMR, TOCSY, HSQC and HMBC	Hazelnut	Metabolic profiling of hazelnut cultivars	[8]
DOSY-NMR	Beverages	Sucrose quantification	[9]
^{1}H, 1H-DPFGSE, and F2-DPFGSE band-selective HSQC	Olive oil	Detection of aldehydes	[10]
^{13}C qNMR	Wine	In situ determination of fructose isomer concentration	[11]
Solid-state ^{13}C NMR	Milk protein concentrate	Change in molecular structure and dynamics of protein	[12]
UF iSQC NMR	Viscous liquid foods	Sugar content, quality testing, and determination of adulteration	[13]
TD-NMR	Mayonnaise and salad dressing	Through-package fat determination	[14]
TD-NMR	Biscuit dough	Influence of fiber on proton mobility	[15]
TD-NMR	Beef	Meat quality parameters	[16]
TD-NMR (SMART Trac™)	Organogels in cream cheese	Fat content	[17]
HR-MAS-NMR	Tomato	Metabolic profiling, tissue differentiation, and fruit ripening	[18]
^{1}H HR-MAS	Fish	Rapid assessment of freshness and quality	[19]
CP-MAS-NMR	Wheat bran	Hydration, plasticization, and disulfide bonds	[20]
CP-MAS-NMR	Starch	Chemicophysical properties	[20]

NMR nuclear magnetic resonance spectroscopy, *MRI* magnetic resonance imaging, *CP-MAS* cross polarization magic-angle-spinning NMR, *HR-MAS* high-resolution magic-angle spinning, *DOSY-NMR* diffusion-ordered ^{1}H NMR, *qHNMR*, quantitative proton NMR, *TD-NMR* time-domain NMR, *DPFGSE* double pulsed field gradient spin echo, *UF iSQC NMR* ultrafast intermolecular single-quantum coherence

purity and identity of the β-glucans as various food processors, particularly the cereal, baking, and brewing industries, work to extract these valuable byproducts from the waste stream in a cost-effective manner. Figure 10.6 shows a 1D NMR spectrum of 1,3–1,4 mixed linkage β-glucans from cereal (oat) processing waste. From this spectrum, both the purity and relative ratio of 1,3–1,4 linkages could be determined.

If additional structural information is needed, there are many powerful 2D and 3D NMR analyses available for the assignment of the chemical shifts of each ^1H and ^{13}C atom in an organic molecule. Once assigned, other experiments enable an assessment of the relative proximity of these nuclei through molecular bonds and through space. 2D NMR, therefore, can be applied to any sample for which structural information is required, such as a health-related fiber or a new sweetener. This information may be critical if a company or researcher wishes to file for a patent.

NMR spectroscopy also is a valuable assay tool in batch ingredient analysis for quality assurance. In such assays, the structural assignments of the spectra would not be as important as the consistency of the spectra compared to a spectrum of a high-quality control product. This application can be used with many types of ingredients, because NMR solvents are available for compounds with a range of solubility properties.

10.4.1.2 *Solids*

The principles that underlie solid-state NMR are similar to those discussed in Sect. 10.2; however, due to the fact that the sample is not freely tumbling about in solution, there is a "directional" aspect (anisotropic or orientation-dependent interactions) to the solid-state analysis. The anisotropic nature of solids results in very broad signals and yields spectra that lack the structural information obtained from samples in solu-

tion. One method to overcome this problem is **magic-angle spinning** (MAS), in which the line broadening due to the anisotropy is countered when the sample holder is spun at a specific "magic" angle relative to the external field, B_0, yielding much narrower lines. MAS is often combined with cross polarization enhancement (CP-MAS), in which the magnetization from more easily detected nuclei is transferred to those that are less easily detected (such as from ^1H to ^{13}C).

Solid-state NMR analyses can be applied to many types of samples, such as powders and fresh vegetable tissue. Solid-state ^{13}C-CPMAS-NMR techniques can be used to monitor the chemical composition and the physicochemical properties in the solid portion of an intact food sample. This has been applied to composition studies of different mushroom species, and solid-state ^{13}C-CPMAS-NMR spectroscopy showed significant differences in the ratio of carbohydrate to protein resonances between different species. Also, high-resolution ^1H-MAS-NMR techniques enabled food researchers to discriminate between durum wheat flours from Southern Italy, which differ in composition depending on the region of origin. A similar application was used to correlate composition with origin in a study of Parmesan cheese.

10.4.1.3 *Magnetic Resonance Imaging*

Magnetic resonance imaging (MRI) is unique in that the sample can be placed into the magnet in the native form, and 2D or 3D images of the sample can be generated. MRI involves variations in field strength and the center frequency of the pulses over time and space, along with the application of field gradients in different geometric positions relative to the magnet bore (B_0). The end result is a spatial "encoding" of the sample protons with different phase and frequency values. After multidimensional Fourier transformations of multiple FIDs from different spatial "slices" of the sample, an image of the sample is produced that contains information about the state of the tissue or other material under study.

The sample can be a medical patient, a small test animal, a diseased plant stem, a ripened fruit, or even a complex food product in various steps of processing or its final form. For example, a packaged product could be analyzed over time in the package to track water movement or loss. The MRI analyses would not affect the product. There are many potential applications for MRI in the food industry; for example, it can be used to image the freezing process in frozen food production, with the goal of increasing shelf life. MRI also may be applied to analyses of the composition and characteristics of pastes and sauces, to locate voids in products, or to examine the fat distribution in meats or the water/fat distribution in emulsions. It can also provide detailed information about the thickness of a

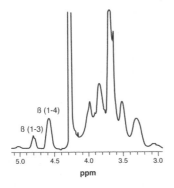

10.6 figure A 1D H-NMR spectrum of 1,3–1,4 mixed linkage β-glucans from oat processing waste. Both the purity of the sample and the ratio of 1,3–1,4 linkages could be determined from the spectrum

filling or coating; structural changes, including water loss, in a product via heat transfer (cooking); or changes associated with hydration of a food product during processing. When combined with rheological analyses, sauce and paste flow in a processing system may be monitored.

MRI images of clementine fruit are shown in Fig. 10.7. One image shows freeze damage to the interior pericarp region of the fruit. The other image shows the presence of an unwanted seed. Such problems often are undetected by a simple visual inspection of the fruit.

As the high costs of purchasing and maintaining a large-bore MRI instrument decrease over time, as they did for NMR spectroscopy, and as smaller bore instruments become more common, this important tool should become available to even small food companies and food science departments. These instruments may become a common sight in even modest research and development laboratories over the next decade.

10.4.1.4 *Relaxometry*

In the plastics industry, small molecules are mixed with the large polymers to make the system more fluid. These small molecules are termed plasticizers, and in food processing, the natural equivalent is water. The amount of water available to act as a plasticizer is a very important factor for food quality. An increase or a reduction in the amount of plasticizer can affect the glass transition process that, in turn, affects the quality of the final product. Water exists in several states in food, and the interaction of water molecules with food components can be investigated by the measurement of NMR relaxation. This includes both the spin-lattice ($T1$) and spin-spin ($T2$) relaxation times (Sect. 10.2.2) of the water protons. The relaxation times are related to the magnetic interactions of water protons with the surrounding environment, and the effective relaxation time is related to the extent of the association between the water molecules and immobilized or slowly moving macromolecules. In general, as the macromolecular content increases, the relaxation times of the water protons also increase.

10.4.1.5 *TD-NMR for Content Analyses*

A recent development in NMR spectroscopy is instrumentation consistent with AOAC Method 2008.06 (moisture and fat in meats), using the FAST Trac NMR system (CEM Corporation), which combines rapid drying (microwave oven) with low-resolution (time-domain) NMR in a benchtop instrument for moisture and fat analysis. The CEM Corporation (Matthew, NC) makes two such instruments that combine a microwave plus NMR for moisture/solids and fat contents (SMART TracII™ for wet products and HYBRID Trac™ for all products) and another instrument that only uses only an NMR to measure moisture /solids and fat for dry products (FAST Trac™). These relatively inexpen-

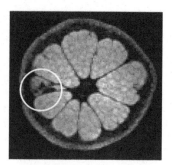

Freeze damage

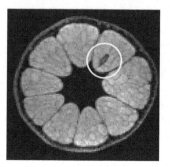

Seed

MRI images (18 mm slice thickness) of clementine citrus fruit with defects. Freeze damage is shown in the image on the top, and an unwanted seed is shown in the image on the bottom (Images courtesy of Michael McCarthy, Aspect AI Ltd., Netanya, Israel)

sive systems are operator friendly and easy to maintain. They provide rapid, reproducible information for quality control of food products, yielding information on the moisture and fat content of specific food items (e.g., FAST Trac™ for chocolate and potato chips).

10.4.2 Specific Food Application Examples

High-resolution NMR spectroscopy has been used for the analysis of complex systems such as food samples, biofluids, and biological tissues because it provides information on a wide range of compounds found in the food matrix in a single experiment. NMR spectrometry is nondestructive and offers advantages in the simplicity of sample preparation and rapidity of analysis. The short time frame needed to obtain NMR spectra (minutes), coupled with automation, enables the analysis of many samples with minimal operator input. There are two basic types of analysis in the application of NMR to the food industry: (1) identification of distinct resonances and, therefore, specific compounds and (2) use of chemometric profile analysis, in which spectral profiles are compared without assigning particular resonances.

10.4.2.1 Oil/Fat

10.4.2.1.1 Fatty Acid Profile

Physical and chemical properties of fats, oils, and their derivatives are mainly influenced by their fatty acid profile. Even though gas chromatography (GC) is usually used for determining the fatty acid profile (Chap. 17, Sect. 17.2.7; Chap. 23, Sect. 23.6.2), the common unsaturated fatty acids, such as oleic, linoleic, and linolenic acids in an oil or fat sample can be quantified using ^{1}H-NMR, by integration of select signals in the spectra. Although GC provides accurate information about complete fatty acid profile, it lacks information about the fatty acid distribution on the glycerol anchors, which is important to determine the functionality of the ingredient in processing, such as the crystallization point or how it plasticizes the dough in a baked product. For example, the correct type of fat is essential for quality pie crusts or croissants. The fatty acid distribution on the glycerol anchors can be obtained from ^{13}C-NMR analysis. There are two groups of resonances in the carbonyl region of the spectrum: the first is due to fatty acids in positions 1 and 3 and the second is from fatty acids in position 2 of the glycerol moiety.

10.4.2.1.2 Verification of Vegetable Oil Identity

Even though different oils or fats may be purposely mixed for specific reasons, the adulteration of high-value oils with oils of lesser value is an issue of economic and commercial importance. This is primarily a problem with olive oil, because it is expensive and has superior nutritional value. Accordingly, many studies from major olive oil-producing Mediterranean countries, such as Greece, Italy, and Spain, deal with identifying lower-value oils, such as hazelnut oil, used for adulterating olive oil. The adulteration problem is complicated by the fact that the lower-value oils usually have fatty acid profiles similar to olive oil. Among the methods used for analyzing potentially adulterated olive oil are ^{13}C-NMR and ^{1}H-NMR spectrometry. For example, NMR is utilized in conjunction with multivariate statistical analyses of specific resonances in NMR spectra of olive oil diluted with hazelnut or sunflower oil. These methods also can be used to identify the variety and geographical origin of the oil.

10.4.2.1.3 Monitoring of Oxidation

The oxidation of vegetable oils is a significant quality problem and can lead to further deterioration of the oil. Highly unsaturated fatty acids, with *bis*-allylic methylene groups, are particularly susceptible to oxidation. Primary and secondary oxidation products, such as hydroperoxides and aldehydes, are easily detected by ^{1}H-NMR analyses. ^{1}H-NMR is especially useful for such analyses because the samples do not require any additional treatments, such as derivatization, that could cause degradation.

10.4.2.1.4 Solid Fat Content (SFC)

While most analyses discussed in this chapter depend on high-resolution NMR instruments, a benchtop, low-resolution-pulsed NMR instrument can be used to determine the SFC of a sample (see also Chap. 23, Sect. 23.43.11). For example, the amount of solid triacylglycerols in the oil or fat at different temperatures can be determined. This method is based on the difference in relaxation times between solids and liquids, and after a delay, only the NMR signal of the liquid fat is measured. The solid content is then estimated. Crystallization mechanisms of fat blends also can be studied using SFC measurements.

10.4.2.2 Water

Glass transition is an important property of foods, and the glass transition temperature (Tg), which is dependent on water content, impacts both the processing and the storage of food products. Tg can be determined with an **NMR state diagram**, which is a curve relating NMR relaxation time to glass transition temperature at different moisture contents. This information is important because processing and storage temperatures above Tg at any point during production and distribution of a product are associated with more rapid deterioration. Spin-spin relaxation time (T2) is commonly used as an indication of proton mobility, which is different above and below the Tg of a given product. Although the differential scanning calorimeter (DSC) (Chap. 33, Sect. 33.3.2) is most commonly used for simple Tg analyses, the ability to generate NMR state diagrams increases the value of NMR for many applications.

10.4.2.3 Ingredient Assays

Adulteration in fruit juice is not easy to detect by taste or color. For example, orange juice can be blended with relatively inexpensive grapefruit juice, but the presence of the grapefruit juice in a commercially available orange juice product poses serious health risks for consumers with certain medical conditions. Grapefruit juice has a number of coumarin-like flavonoids and other powerful CYP450 inhibitors that negatively impact the metabolism of many prescribed drugs. Therefore, the detection and prevention of this kind of adulteration are especially important. NMR-based chemometric approaches using independent component analysis, a variant of principle component analysis, are now applied to this problem. Selected regions of the ^{1}H NMR spectra, which are known to contain distinguishing flavonoid glycoside signals, are accurately analyzed in a relatively short time. Another common issue with juice preparation is the differentiation between freshly squeezed juices and those produced from pulp washes, which can be added to fresh-squeezed orange juice to reduce production costs. ^{1}H NMR, in combination with principal component analyses, can easily and accurately distinguish the fresh-squeezed and pulp-wash orange juice.

NMR is also used in monitoring batch-to-batch quality and production site differences in beer. Large multinational breweries prepare their beers at many different geographic locations and require methods for quality control at a detailed molecular level. NMR can be used in conjunction with principal component analysis to distinguish beer from different production sites based on lactic acid, pyruvic acid, dextran, adenosine, inosine, uridine, tyrosine, and 2-phenylethanol content. Quantifying these compounds allows the producers to identify production sites where there is greater variability in these compounds (and therefore poorer quality control).

NMR methods are used by other producers to improve quality control in soft drink production, juice production, and vegetable oil manufacturing. Similar methods also are used to monitor the quality of functional foods and nutraceuticals (food extracts with positive medicinal effects) that are harvested from different geographic locations.

10.5 SUMMARY

Nuclear magnetic resonance technology provides powerful research instrumentation for a variety of applications, from structural elucidation of complex molecules, to 3D-imaging of fresh tissue, to simple ingredient assays for quality assurance. NMR differs from most other forms of spectroscopy because the nucleus is the subject of analysis, and the excitation step uses radiofrequency electromagnetic energy. The proton (H) and the ^{13}C isotope are the most commonly studied nuclei, and each has a characteristic charge and spin which results in a small, local magnetic field. NMR analyses require an external magnetic field, which causes the local magnetic fields of the nuclei to align in a parallel or antiparallel orientation. There is a slight excess in the parallel orientation (in the z-axis aligned with B_0), and it is the net magnetic vector of this population that is detected during an NMR experiment. A pulse of RF energy moves this net magnetism into the xy-plane, where a reemitted radio signal (the NMR signal) is detected. This signal, which decays quickly, contains the intensity and frequency information for all the nuclei in the sample, and the resulting FID is converted by Fourier transformation into the NMR spectrum, which shows the various resonances spread along the x-axis based on differences in frequency.

The NMR instrument consists of a cryomagnet with the transmitter and receiver antennae in the cen-

tral bore, an electronics console with the transmitter and receiver hardware, and a data/work station that controls all the functions of the instrument. In addition to NMR spectrometers, with both solids and liquids applications, there are other related instruments, such as MRI, that are based on the same principles, but yield different information.

Among the common applications of NMR to food science are structural studies that examine the correlation between chemical structure and health benefits or functionality of food ingredients, studies of the effects of processing on food properties and quality, composition studies of food ingredients or even fresh vegetable tissue, imaging of food products, and determination of SFC or ingredient purity.

10.6 STUDY QUESTIONS

1. Explain the basic principles associated with NMR spectroscopy, including the function of the magnet and the concept of nuclear spin.
2. Describe the interaction of the net magnetization with the RF pulse (90°) and the subsequent NMR signals.
3. Explain the concept of shielding and chemical shift.
4. Describe the FID and the NMR spectrum, including the concepts of time domain, frequency domain, and data transformation.
5. List the components of the NMR spectrometer and their functions.
6. What kinds of samples are analyzed by (a) liquid NMR and (b) solid-state NMR?
7. What kind of final data does one obtain with an MRI? List two applications of MRI.
8. What is the primary use of relaxometry in food analysis?
9. List the general types of food applications of NMR and give an example of each.

REFERENCES

1. Li J, Kang J, Wang L, Li Z, Wang R, Chen ZX, Hou GG (2012) Effect of water migration between arabinoxylans and gluten on baking quality of whole wheat bread detected by magnetic resonance imaging (MRI). J Agric Food Chem 60(26):6507–6514
2. Mazhar M, Joyce D, Cowin G, Brereton I, Hofman P, Collins R, Gupta M (2015) Non-destructive ^{1}H-MRI assessment of flesh bruising in avocado (Persea americana M.) cv. Hass. Postharvest Biol Tech 100:33–40
3. Spiteri M, Jamin E, Thomas F, Rebours A, Lees M, Rogers KM, Rutledge DN (2015) Fast and global authenticity screening of honey using ^{1}H-NMR profiling. Food Chem 189:60–66

4. Siciliano C, Belsito E, De Marco R, Di Gioia ML, Leggio A, Liguori A (2013) Quantitative determination of fatty acid chain composition in pork meat products by high resolution 1 H NMR spectroscopy. Food Chem 136(2):546–554

5. Ohtsuki T, Sato K, Sugimoto N, Akiyama H, Kawamura Y (2012) Absolute quantification for benzoic acid in processed foods using quantitative proton nuclear magnetic resonance spectroscopy. Talanta 99:342–348

6. Bosmans GM, Pareyt B, Delcour JA (2016) Non-additive response of blends of rice and potato starch during heating at intermediate water contents: A differential scanning calorimetry and proton nuclear magnetic resonance study. Food Chem 192:586–595

7. Wei F, Furihata K, Hu F, Miyakawa T, Tanokura M (2010) Complex mixture analysis of organic compounds in green coffee bean extract by two-dimensional NMR spectroscopy. Magnetic Resonance in Chemistry 48(11):857–865

8. Sciubba F, Di Cocco ME, Gianferri R, Impellizzeri D, Mannina L, De Salvador FR, Venditti A, Delfini M (2014) Metabolic profile of different Italian cultivars of hazelnut (Corylus avellana) by nuclear magnetic resonance spectroscopy. Natural Product Research. 28(14):1075–1081

9. Cao R, Nonaka A, Komura F, Matsui T (2015) Application of diffusion ordered-1 H-nuclear magnetic resonance spectroscopy to quantify sucrose in beverages. Food Chem 171:8–12

10. Dugo G, Rotondo A, Mallamace D, Cicero N, Salvo A, Rotondo E, Corsaro C (2015) Enhanced detection of aldehydes in Extra-Virgin Olive Oil by means of band selective NMR spectroscopy. Physica A: Statistical Mechanics and its Applications 420:258–264

11. Colombo C, Aupic C, Lewis AR, Pinto BM (2015) In situ determination of fructose isomer concentrations in wine using ^{13}C quantitative nuclear magnetic resonance spectroscopy. J Agric Food Chem 63(38):8551–9

12. Haque E, Bhandari BR, Gidley MJ, Deeth HC, Whittaker AK (2015) Change in molecular structure and dynamics of protein in milk protein concentrate powder upon ageing by solid-state carbon NMR. Food Hydrocolloids 44:66–70

13. Cai H-H, Chen H, Lin Y-L, Feng J-H, Cui X-H, Chen Z (2015) Feasibility of ultrafast intermolecular single-quantum coherence spectroscopy in analysis of viscous liquid foods. Food Anal Methods 8(7):1682–90

14. Pereira FMV, Pflanzer SB, Gomig T, Gomes CL, de Felicio PE, Colnago LA (2013) Fast determination of beef quality parameters with time-domain nuclear magnetic resonance spectroscopy and chemometrics. Talanta 108:88–91

15. Serial M, Canalis MB, Carpinella M, Valentinuzzi M, León A, Ribotta P, Acosta RH (2016) Influence of the incorporation of fibers in biscuit dough on proton mobility characterized by time domain NMR. Food Chem 192:950–957

16. Pereira FMV, Rebellato AP, Pallone JAL, Colnago LA (2015) Through-package fat determination in commercial samples of mayonnaise and salad dressing using time-domain nuclear magnetic resonance spectroscopy and chemometrics. Food Control 48:62–6

17. Bemer HL, Limbaugh M, Cramer ED, Harper WJ, Maleky F (2016) Vegetable organogels incorporation in cream cheese products. Food Res Intl 85:67–75

18. Pérez EMS, Iglesias MJ, Ortiz FL, Pérez IS, Galera MM (2010) Study of the suitability of HRMAS NMR for metabolic profiling of tomatoes: Application to tissue differentiation and fruit ripening. Food Chem 122(3):877–87

19. Heude C, Lemasson E, Elbayed K, Piotto M (2015) Rapid assessment of fish freshness and quality by ^{1}H HR-MAS NMR spectroscopy. Food Anal Methods 8(4):907–15

20. Bertocchi F, Paci M (2008) Applications of high-resolution solid-state NMR spectroscopy in food science. J Agric Food Chem 56(20):9317–27

RESOURCE MATERIALS

- Berger S, Braun S (2004) 200 and more NMR experiments: a practical course. Wiley–VCH, Weinheim, Germany
- Farhat IA, Belton PS, Webb GA (2007) Magnetic resonance in food science: from molecules to man. The Royal Society of Chemistry, Cambridge, England
- Günther H (1995) NMR spectroscopy, 2nd edn. Wiley, New York
- Hills B (1998) Magnetic resonance imaging in food science. Wiley, New York
- Jacobsen NE (2007) NMR spectroscopy explained: simplified theory, applications and examples for organic chemistry and structural biology. Wiley, New York
- McMurry JE (2007) Organic chemistry, 7th edn. Brooks-Cole, Salt Lake City, UT
- Skoog DA, Holler FJ, Crouch SR (2007) Principles of instrumental analysis, 6th edn. Brooks-Cole, Salt Lake City, UT
- Webb GA, Belton PS, Gill AM, Delgadillo I (eds) (2001) Magnetic resonance in food science: a view to the future. The Royal Society of Chemistry, Cambridge, England

Mass Spectrometry

J. Scott Smith (✉)

*Department of Animal Sciences and Industry,
Kansas State University,
Manhattan, KS 66506-1600, USA
e-mail: jsschem@k-state.edu*

Rohan A. Thakur

*Bruker Daltonics,
Billerica, MA 01821, USA
e-mail: Rohan.thakur@bruker.com*

S. Nielsen (ed.), *Food Analysis*, Food Science Text Series,
DOI 10.1007/978-3-319-45776-5_11, © Springer International Publishing 2017

11.1   INTRODUCTION

Over the past decade, **mass spectrometry** (MS) techniques have become indispensable for the identification, characterization, verification, and quantitation of small molecules (e.g., caffeine, 194 Da) to large complex biomolecules (e.g., immunoglobulin, 144,000 Da). Two important developments led to the rapid rise in popularity of MS as an analytical technique. First was the development of **hyphenated** MS techniques, which coupled the separation techniques of gas chromatography (GC) (Chap. 14) or liquid chromatography (LC) (Chap. 13) to MS. This coupling of chromatography and MS dramatically lowered the detection limits for quantitative analysis while simultaneously increasing the confidence of measurement through high specificity. Second was the development of **hybrid**, benchtop, MS instruments that made high-resolution, accurate mass, LC-MS analysis routine. Hyphenated, hybrid MS techniques deliver robust, highly sensitive, precise measurements that withstand the rigor of statistical analysis for the purposes of quantitative analysis, while significantly reducing sample preparation time and effort. These advantages made MS a "must-have" technique when faced with complex bioanalytical challenges such as pesticide screening in foods, trace analysis of environmental pollutants, characterization of natural products, or rapid identification of food-borne bacteria.

The power of the MS technique is due to its ability to place a charge on a molecule, thereby converting it to an **ion** in a process called **ionization**. The generated ions are then separated according to their **mass-to-charge ratio** (m/z) by subjecting them to a combination of radio-frequency (RF) and electrostatic fields in a **mass analyzer** and finally detected by highly sensitive **detectors**. The resulting signals from the detectors are digitized and processed by software to display the information as a mass spectrum, which reveals its molecular mass and its structural composition, leading to identification. An additional stage of **ion fragmentation** may be included before detection to elicit structural information in a technique known as **tandem MS**.

The most common MS technique remains GC-MS which was first used in the late 1960s, followed by the rapidly growing LC-MS technique which made ionization from liquids possible and started to gain adoption in the late 1980s, and matrix-assisted laser desorption ionization (MALDI) or MALDI time-of-flight (TOF) techniques which offers ionization from solid crystals discovered in 1988.

11.2   INSTRUMENTATION: THE MASS SPECTROMETER

11.2.1 Overview

Because there are so many acronyms associated with the MS instrumentation, a listing of the acronyms used in this chapter is given. Many of those acronyms are first used in Table 11.1, which summarizes mass spectrometer components and types of instruments. This table also helps introduce the three basic functions a MS performs. (1) There must be a way to **ionize** the molecules, which occurs in the ion source by a variety of techniques. (2) The charged molecular ion and its fragments must be **separated** according to their m/z, and this occurs in the **mass analyzer** section. (3) The separated, charged ions must be detected (electron multipliers, photomultipliers). The block diagram in Fig. 11.1 represents the various components of a mass spectrometer.

Sample introduction can be **static** (or **dynamic**, the latter of which involves interfacing with GC or LC instruments. Since all mass spectrometers work in high vacuum, regardless of the state of the sample (gas, liquid, or solid), all ions are introduced into the MS as a gas. The MS **interface** converts the samples into a form that is acceptable to introduction into the vacuum chamber. Common MS interfaces will be discussed in more detail in the sections on GC-MS and LC-MS.

Figure 11.2 depicts the interior of a typical GC-MS instrument that uses quadrupole mass analyzers. The region between ion generation and detection is maintained by different vacuum pumps. Each successive region from the ion source is kept at lower vacuum than the preceding region, with the mass analyzer/detector being in the region of strongest vacuum ($\approx 10^{-6}$–10^{-8} torr). A vacuum is necessary for two critical reasons: (1) to avoid ion-molecule reactions between the charged ions and other gaseous molecules before they reach the detector, and (2) for proper operation of ion lenses, mass analyzer electrodes, and ion-detectors that require the use of high voltages. Vacuum performance determines the sensitivity and resolution of mass spectrometers.

table   Summary of mass spectrometer components and types

	Types	Applications
Sample introduction		
Static method	Direct injection	Gas or volatile liquid
	Direct insertion probe	Solids
Dynamic	GC	Gas or volatile liquids
	LC	Nonvolatile solids or liquid
Ion source	Electron impact ionization (EI)	Primarily for GC-MS, for volatile compounds
	Electrospray ionization (ESI)	Most popular method for LC-MS, normally for polar or slightly polar compounds
	Atmospheric pressure chemical ionization (APCI)	Primarily for LC-MS, normally for compounds of low polarity and some volatility
	Atmospheric pressure photoionization (APPI)	Same uses as APCI but has advantages in signal-to-noise ratio and detection limit
	Matrix-assisted laser desorption ionization (MALDI)	A "soft ionization," ideal for large biopolymers and other fragile molecules
	Chemical ionization (CI)	A "soft ionization," ideal for large biopolymers and other fragile molecules
Mass analyzers	Quadrupoles mass analyzer/filter (Q)	Used in many types of instruments. Compact. Used in benchtop instruments
	Ion trap (IT)	LC-MS for MS/MS
	Time of flight (TOF)	Useful to analyze biopolymers and large molecules
	Fourier transform-based mass analyzer (FT-ion cyclotrons, FT-ICR; FS-orbitrap)	Allows for easy-to-use benchtop LC-MS
	Magnetic sector	Specialized applications requiring ultrahigh resolution, e.g., dioxin analysis
	Isotope ratio MS	Useful in geochemistry and nutrition science. Extreme specificity
	Accelerator mass spectrometer	Useful in geochemistry and nutrition science. Extreme specificity
Hybrid MS: common combinations of mass analyzers	Quadrupole TOF (e.g., Q-TOF, triple TOF)	Most LC-MS. Provides for MS/MS, benchtop instruments
	Triple quadrupole (e.g., TQ; tandem MS)	Common for LC-MS. Provides for MS/MS, benchtop instruments
	Ion trap (e.g., IT-FTMS, IT-orbitrap, Q-Trap)	Most LC-MS. Provides for MS/MS. Very high mass accuracy
Common MS instruments	Quadrupole MS (single quadrupole or TQ)	Quantitative and qualitative analysis
	ITMS	Qualitative analysis. Advantage of multistages of MS (MS^n)
	TOF/Q-TOF	High-resolution accurate mass needs
	FTMS	High-resolution accurate mass needs

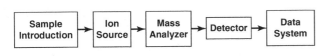

figure   A block diagram of the major components of a mass spectrometer

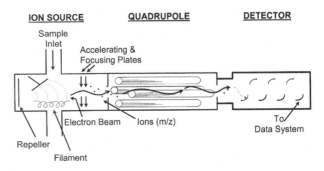

11.2.2 Sample Introduction

11.2.2.1 *Static Method*

The initial step in operating the MS is to get the sample into the ion source chamber. Pure compounds or sample extracts that are a gas or a volatile liquid are injected directly into the source region. This requires no special

figure   Schematic of a typical mass spectrometer. The sample inlets (interfaces) at the *top* and *bottom* can be used for direct injection or interfacing to a GC equipment or apparatus and is much the same as injecting a sample into a GC. Thus, this **static method** of introducing the sample to the source is called **direct injection**. With solids that are at least somewhat volatile, the **direct insertion probe method** is used, in which the sample is placed in a small cup at the end of a stainless steel rod or probe. The probe is inserted into the ion source through one of the sample inlets, and the source is heated until the solid vaporizes. The mass spectrum is then obtained on the vaporized solid material as with the direct injection method. Both direct injection and direct insertion probe methods work well with pure samples, but their use is very limited when analyzing complex mixtures of several compounds.

Direct analysis in real time (DART) is an example of a static sampling technique, where metastable He ions (19.8 eV energy) are used to initiate ionization of the analyses of interest via the Penning process (much like EI) resulting in radical cations ($M^{+\cdot}$). A mixture of heated He and nitrogen is used to initiate the metastable ionization process, essentially creating a plasmarich environment, wherein the metastable He reacts with ambient water, creating protonated water clusters, resulting ultimately in a charge transfer to the analyte of interest. The process has been well described by Hajslova et al. [3] for food QC and safety analysis.

11.2.2.2 Dynamic Method

For mixtures, sample introduction is a **dynamic method** in which the sample must be separated into the individual compounds and then analyzed by the MS. This is done typically by GC or HPLC units connected to an MS by an **interface** (see Sects. 11.4 and 11.5). The interface removes excess GC carrier gas or HPLC solvent that would otherwise overwhelm the vacuum pumps of the MS.

11.2.3 Ionization

There are many methods used to produce ions for the compounds, depending on the type of chromatographic interface and nature of the compounds (Table 11.1). The major types of ion sources are briefly described in subsections that follow.

11.2.3.1 Electron Impact Ionization (EI)

In GC-MS techniques, once the compound(s) coming from the GC enters the ion source, it is exposed to a beam of electrons emitted from a filament composed of rhenium or tungsten metal. When a direct current is applied to the filament (usually 70 electron volts, eV), it heats and emits electrons that move across the ion chamber toward a positive electrode. As the electrons pass through the source region, they come in close proximity to the sample molecule and extract an electron, forming an ionized molecule. Once ionized, the molecules contain such high internal energies they can further

fragment into smaller molecular fragments. This entire process is called **electron impact** (EI) **ionization**, although the emitted electrons rarely hit a molecule.

11.2.3.2 Electrospray Ionization (ESI)

Electrospray ionization, the most popular LC-MS technique in use today, functions at atmospheric pressure and is a highly sensitive technique. Normally, polar compounds are amenable to ESI analysis, with the type of ion produced depending on the initial charge. That is, positively charged compounds yield positive ions, while negatively charged compounds such as those containing free carboxylic acid functional groups will produce negative ions.

The ESI source as depicted in Fig. 11.3 consists of a nozzle that contains a fused-silica capillary sample tube (serves to transfer the LC effluent) coaxially positioned within a metal capillary tube to which a variable electrical potential can be applied against a counter-electrode, which is usually the entrance to the MS. Compressed nitrogen gas at high velocity is coaxially introduced to aid in the nebulization of the LC effluent as it exits the tip of the metal capillary tube. The relative velocity difference between the streams of nitrogen gas (fast moving) and LC effluent (slow moving) at the ESI tip results in producing a fine spray of highly charged droplets. At nanoflow rates (<1 µL/min), the force of the electrical field is strong enough to break up the LC effluent into fine droplets without the use of nebulizing gas, in a process known as nanospray. For conventional HPLC flow rates (1–1000 µL/min), the sheer volume of liquid requires an initial droplet size reduction through the use of nebulizing nitrogen gas, creating the required microdroplets, which can now be influenced by the prevailing electrical field.

At this point, the repulsive forces due to the accumulation of "like" charges inside the rapidly reducing microdroplet volume create an imbalance with the forces of surface tension that are trying to conserve the

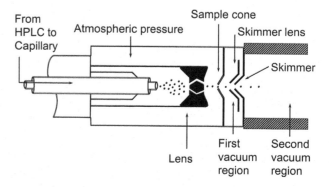

11.3 figure　Schematic of an electrospray LC-MS interface

spherical structure of the microdroplet. The positive charge is drawn out, but cannot escape the surface of the liquid, and forms what is known as a **Taylor cone**. Further reduction of the diameter of the droplets causes the Taylor cone to stretch to a critical point, at which the charge escapes the liquid surface and is emitted as a gas-phase ion in a process known as a **coulombic explosion**.

One of the many advantages of the ESI process is its ability to generate multiple-charged ions and tolerate conventional HPLC flow rates. Proteins and other large polymers (e.g., between 2000 and 70,000 Da) can be easily analyzed on LC-MS systems having a mass limit of m/z 2000, due to this multiple charging phenomenon. Powerful software can process in excess of +50 charge states, to yield the molecular ion information for larger proteins. A limitation of the ESI process is the phenomenon of ion suppression/enhancement or matrix effects, which usually causes a variation in response for the analyte signal intensity in presence of matrix components. Matrix factor corrections are used to account for ion suppression/enhancement effects, including the use of stable-labeled internal standards or matrix-assisted calibration curves for quantitative analysis.

11.2.3.3 Atmospheric Pressure Chemical Ionization (APCI)

The **APCI interface**, which like ESI operates at atmospheric pressure, is normally used for compounds of low polarity and some volatility. It is harsher than ESI and is a gas-phase ionization technique. Therefore gas-phase chemistries of the analyte and solvent vapor play an important part in the APCI process.

Figure 11.4 shows the schematic diagram of an APCI interface. The LC effluent-carrying fused-silica capillary tube protrudes about halfway inside a silicon-carbide (ceramic) vaporizer tube. The vaporizer tube is maintained at approximately 400–500 °C and serves to vaporize the LC effluent. High voltage is applied to a corona needle positioned near the exit of the vaporizer tube. The high voltage creates a **corona discharge** that forms reagent ions from the mobile phase and nitrogen nebulizing gas. These ions react with the sample molecules (M) and convert them to ions. A common cascade of reactions occurring in the presence of water, nitrogen gas, and the high-voltage corona discharge is as follows:

$$e^- + N_2 \rightarrow N_2^{+\bullet} + 2e^- \tag{11.1}$$

$$N_2^{+\bullet} + H_2O \rightarrow N_2 + H_2O^{+\bullet} \tag{11.2}$$

$$H_2O^{+\bullet} + H_2O \rightarrow H_3O^+ + OH^\bullet \tag{11.3}$$

$$M + H_3O^+ \rightarrow (M+H)^+ + H_2O \tag{11.4}$$

The APCI interface is a robust interface and can handle high flow rates of up to 2 mL min. It is unaffected by minor changes in buffer strength or composition and is typically used to analyze molecules less than 2000 Da. It does not facilitate multiple charges and hence cannot be used to analyze large biomolecules/polymers. In terms of matrix effects, APCI usually shows "ion enhancement" rather than "ion suppression." This is due to the matrix components enriching the plasma generation process, thereby enhancing the efficiency of the ionization process. As a result, there is an increase in response for the analyte signal in the presence of matrix components, requiring matrix factor correction through the appropriate use of stable-labeled internal standards or matrix-assisted calibration curves for quantitative analysis.

11.2.3.4 Atmospheric Pressure Photoionization (APPI)

APPI is an ionization technique that improves on the interface possible with APCI. The APPI interface, which uses a krypton or xenon light source to generate a beam of photons instead of a corona discharge-generated plasma as in APCI. Compounds having ionization potentials lower than the wavelength of the light source will be ionized. Since most HPLC solvents do not ionize at the wavelengths generated by the commonly used photon sources, APPI improves in the signal-to-noise ratio and hence detection limits.

11.2.3.5 Matrix-Associated Laser Desorption Ionization (MALDI)

In MALDI, the sample is dissolved in a matrix and ionized using an UV laser. The matrix plays an important role in ionization, acting both as the absorber of the laser energy, which causes it to

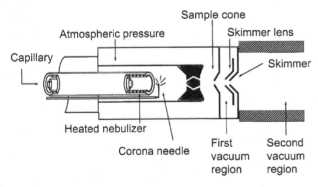

 Schematic of an atmospheric pressure chemical ionization LC-MS interface

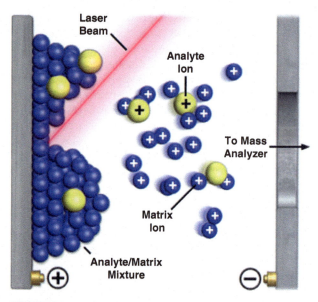

11.5 **figure** Diagram of the MALDI desorption and ionization process used in some TOF instruments (From Chughtai and Heeren [2], used with permission of National High Magnetic Field Laboratory, www.magnet.fsu.edu)

vaporize, and as a proton donor and acceptor to initiate charge transfer to the analyte (Fig. 11.5). Since the sample is not directly ionized, MALDI is considered a "soft ionization" technique and amenable to ionization of large biopolymers and other fragile molecules such as nucleic acids or carbohydrates [2]. The matrix used in MALDI is usually a weak organic acid with UV-absorbing properties [e.g., 2,5-dihydroxy benzoic acid (DHB), sinapinic acid (3,5-dimethoxy-4-hyroxycinnamic acid), gentisic acid (DHDA, 2,5-dihydroxybenzoic acid), or α-cyano-4-hydroxycinnamic acid (CHCA)]. Sensitivity is usually dependent on critical pairing of the chemistries of matrix with the sample, especially for samples that are inherently nonvolatile or insoluble in most aqueous solvents.

The typical laser used for MALDI applications is neodymium-doped yttrium aluminum garnet (Nd-YAG) nitrogen laser operating at 337 or 355 nm (3.7–3.5 eV photon energies) in vacuum and pulsed at a repetition rate between 1 and 10 KHz. The laser beam size can be attenuated between 5 and 100 um, which allows hundreds of laser shots to raster through a single sample spot. While the ionization mechanism is not fully understood, it is believed that a two-step process occurs; in step one the matrix absorbs the UV energy from the laser and is consequently ionized $(M+H)^+$; in step two a charge transfer to the sample $(S+H)^+$ is completed, allowing the charged sample to be focused into the mass analyzer. Infrared lasers also are used but are less popular, as is the case with atmospheric pressure-based MALDI ion sources.

11.2.3.6 *Matrix Effects on Ionization*

One key issue with all types of ionization is a phenomenon called matrix effect. This is when ionization of a molecule is either suppressed or enhanced by coeluting endogenous interferences contained in the matrix after sample cleanup. This effect has a direct impact on sensitivity; for the same level (e.g., 1 ng/mL), the ion intensity of the compound of interest will change in response to the coeluting matrix interferences such as salts, fatty acids, phospholipids, etc. For high-sensitivity quantitative MS analysis, a study of matrix effects is essential before quantitation can be performed. For example, if the matrix is spinach extract versus corn extract, each matrix will have to be individually studied for matrix effects. A set of test pesticide standards at known levels are spiked in a clean matrix (spinach or corn not exposed to pesticides) and their peak intensities compared to the same level spiked in pure solvent. The difference in peak intensities between the pure standard and matrix-spiked standard will determine the matrix effect during the final analysis of the samples to be tested. While all modes of ionization are susceptible to matrix effects, ESI seems to be most prone to ion suppression, while techniques such as APCI can be prone to ion enhancement, wherein the peak intensity increased in the presence of matrix as compared to the pure standard.

11.2.3.7 *Transition from Ion Source to Mass Analyzer*

The eventual outcome of the ionization process, by any of the methods described in sections above, is both negatively and positively charged molecules of various sizes unique to each compound. When the repeller plate at the back of the ion source is positively charged, it repels the positive fragments toward the quadrupole mass analyzer. Thus, we look only at the **positive fragments**, although negative fragments are sometimes analyzed. As the positively charged fragments leave the ion source, they pass through holes in the accelerating and focusing plates. These plates serve to increase the energy of the charged molecules and to focus the beam of ions, so that a maximum amount reaches the mass analyzer.

11.2.4 **Mass Analyzers**

11.2.4.1 *Overview*

The heart of an MS is the **mass analyzer**. It performs the fundamental task of separating the charged molecules or their fragments based on their m/z, and it dictates the mass range, accuracy, resolution, and sensitivity. Listed in Table 11.1 are the basic types of mass analyzers, the common combinations of basic mass analyzers (call **hybrid MS**), and the most common types of MS instruments, along with their typical applications. Described in the subsections below are only the four types of mass analyzers most commonly applied to food analysis.

11.2.4.2 Quadrupole Mass Analyzers (Q)

The word "quadrupole" of the quadrupole mass analyzer is derived from the Latin words for "fourfold" (quadruplus), and "pole," to describe the array of four rods that are used (Fig. 11.2). The four rods are used to generate two equal but out-of-phase electric potentials: one is alternating current (AC) frequency of applied voltage that falls in radio-frequency (RF) range, and one is direct current (DC). The potential difference can be varied to create an oscillating electrical field between two of the opposite rods, resulting in their having equal but opposite charges.

When, for example, a positive-charged ion enters the quadrupole field, it will be instantly attracted toward a rod maintained at a negative potential, and if the potential of that rod changes before the ion impacts, it will be deflected (i.e., change direction). Thus, every stable ion (i.e., ion with stable flight path) entering the quadrupolar region traces a sine wave-type pattern on its way to the detector. By adjusting the potentials on the rods, selected ions, a mass range, or only a single ion can be made stable and detected. The unstable ions impact one of the four rods, releasing them from the influence of the oscillating field, and they are pumped away by the vacuum pumps. A quadrupole mass analyzer is commonly referred to as a **mass filter**, because, in principle, the device filters ions that achieve stability from those that do not.

11.2.4.3 Ion Trap (IT) Mass Analyzers

Ion traps are essentially multidimensional quadrupole mass analyzers that store ions (trap) and then eject these trapped ions according to their m/z ratios. Once the ions are trapped, multiple stages of MS (MS^n) can be achieved, mass resolution can be increased, and sensitivity can be improved. The major difference between an ion trap and a quadrupole mass analyzer is that in an ion trap, the unstable ions are ejected and detected while the stable ions are trapped (referred to as MS in time); whereas in a quadrupole, the ions with a stable flight path reach the detector, and the unstable ions hit the rods and are pumped away (MS in space).

Figure 11.6 shows the cross-sectional view of a 3-D ion trap mass analyzer. It consists of a ring electrode sandwiched between a perforated entrance, end-cap electrode, and a perforated exit, end-cap electrode. An AC (RF) voltage and variable amplitude is applied to the ring electrode, producing a 3-D quadrupole field within the mass analyzer cavity.

Ions formed in the source are electronically injected into the ion trap, where they come under the influence of a time-varying RF field. The ions are trapped within the mass analyzer cavity, and the applied RF voltage drives ion motion in a wure eight toward the end-caps. Thus, for an ion to be trapped, it must have a stable trajectory in both the axial and radial directions. To detect the ions, the frequency applied to the ring electrode is changed, and the ion trajectories are made unstable. Helium is continuously infused into the ion trap cavity and primarily serves as a dampening gas. Recent developments in ion trap technology have resulted in **2-D ion traps**, which substantially increase ion trapping volume by spreading the ion cloud in a quadrupole-like assembly [1].

11.2.4.4 Time-of-Flight Mass Analyzer (TOF)

Time-of-flight mass analyzers separate ions according to the time required to reach the detector while traveling over a known distance (Fig. 11.7). Ions are pulsed from the source with the same kinetic energy, which causes ions of different m/z ratios to acquire different velocities (lighter ions travel faster while heavier ions travel slower). The difference in velocities translates to

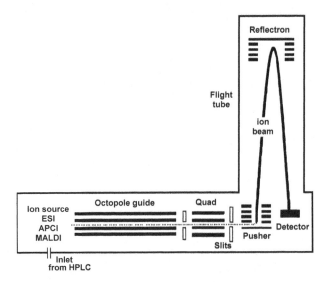

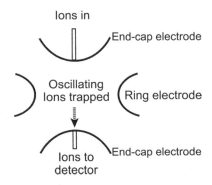

 Diagram of an ion trap mass analyzer

 Diagram of a single-stage time-of-flight instrument

difference in time reaching the detector, upon which the mass spectrum is generated. Theoretically, TOF instruments have no upper mass range, which makes them useful for the analysis of biopolymers and large molecules, and have fast cycles since they technically transmit all m/z ions (full scan mode). The use of reflectrons (ion mirrors) can quickly increase mass resolution of TOF instruments by increasing ion drift path length by bouncing ions in a V or W pattern without drastically increasing the instrument footprint.

11.2.4.5 *Fourier Transform-Based Mass Spectrometry (FTMS)*

Fourier transform-based mass spectrometers deconvolute image currents produced by ion motion (harmonic oscillations or cyclotron motion) into mass spectra. A Fourier transform ion cyclotron resonance mass analyzer traps ions in a magnetic field (Penning traps), while a Fourier transform orbitrap mass analyzer traps ions in an electric field. Both analyzer types are unique from the previously listed mass analyzers because the ions themselves are not detected by impinging upon a detector, but rather the frequency (cyclotron motion) is measured as a function of the applied electric (orbitrap) or magnetic field (ICR). Commercially launched in 2005, the orbitrap brought high resolution (400 K resolution @ m/z 200) to the benchtop by using electrostatic fields, resulting in a

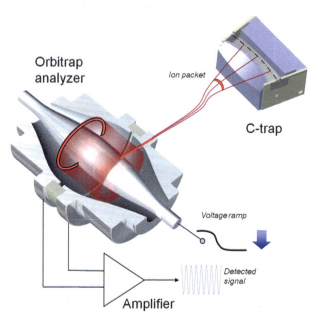

11.8
figure
Diagram of the orbitrap analyzer. Ions are captured in the C-trap after ionization during typical GC or LC-MS. They are then sent to the trap analyzer where the detected signal is then converted to mass (Used with permission of Thermo Fisher Scientific (Bremen), Waltham, MA USA – Artwork by Thermo Fisher Scientific, CC BY-SA 3.0)

simpler-to-operate, LC-MS instrument (Fig. 11.8). While traditional FTMS delivers significantly higher resolution (7 M resolution@ m/z 600), they require liquid helium-cooled superconducting magnets making them large in size, significantly more expensive, and require specialist operators, which limit their widespread adoption. This results in sub-part-per-million (ppm) mass accuracy measurements allowing determination of elemental composition. Extremely high resolution can be achieved with this type of mass analyzer which gives the ability to determine fine isotope structure.

11.3 INTERPRETATION OF MASS SPECTRA

As previously indicated, a **mass spectrum** is a plot (or table) of the intensity of various mass fragments (m/z) produced when a molecule is subjected to one of the many types of ionization techniques. In classis GC-MS, the electron beam generated by a heated filament (used to ionize the molecules) is usually kept at a constant potential of 70 eV because this produces sufficient ions without too much fragmentation, which would result in a loss of the higher-molecular-weight ions. Another advantage of using 70 eV for ionization is that the resulting mass spectra are usually very similar regardless of the make and model of the instrument. This allows for computer-assisted mass spectral matching to database libraries that help in unknown compound identification. In fact, most MSs now come with a MS spectral database and the required matching software.

Typical mass spectra include only positive fragments that usually have a charge of +1. Thus, the **mass-to-charge ratio** is the molecular mass of the fragment divided by +1, which equals the mass of the fragment. As yet, the mass-to-charge ratio unit has no name and is currently abbreviated by the symbol m/z (older books use m/e).

A mass spectrum for butane is illustrated in Fig. 11.9. The relative abundance is plotted on the y-axis and the m/z is plotted on the x-axis. Each line on the bar graph represents a m/z fragment with the abundance unique to a specific compound. The spectrum always contains what is called the **base peak** or **base ion**. This is the fragment (m/z) that has the highest abundance or intensity. When the signal detector is processed by the computer, the m/z with the highest intensity is taken to be 100%, and the abundance of all the other m/z ions is adjusted relative to the base peak. The base peak always will be presented as 100% relative abundance. Butane has the base peak at a m/z of 43.

Another important fragment is the **precursor ion** (often called the **molecular ion** or **parent ion**), designated by the symbol M$^{+\bullet}$. This peak has the highest

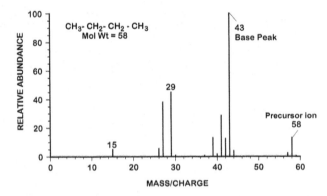

11.9 figure Mass spectrum of butane obtained by electron impact ionization

mass number and represents the positively charged intact molecule with a m/z equal to the molecular mass. The harsher ionizing techniques such as the EI shown here (Fig. 11.9) produces an ion (radical cation) at m/z 58 by stripping an electron. Because the mass of a single electron can be considered insignificant, the molecular ion produced by EI-type ionization is indicative of the molecular weight of that compound. All other molecular fragments originate from this charged species, so it is easy to see why it is called the precursor (molecular) ion. It is not always present because, sometimes, the precursor ion decomposes before it has a chance to traverse the mass analyzer. However, a mass spectrum is still obtained, and this becomes a problem only when determining the molecular mass of an unknown. The remainder of the mass spectrum is a consequence of the stepwise cleavage of large fragments to yield smaller ones termed **product ions** (**daughter ions**). The process is relatively straightforward for alkanes, such as butane, making possible identification of many of the fragments.

As indicated previously, the initial step in EI ionization is the abstraction of an electron from the molecule as electrons from the beam pass in close proximity. The equation below illustrates the first reaction that produces the positively charged product ion.

$$M + e \text{ (from electron beam)} \Rightarrow \begin{array}{l} M^{+\bullet} \text{ (molecular ion)} \\ + \\ 2e \text{ (one electron from the electron beam and one from the molecular ion, M)} \end{array}$$

$$(11.5)$$

The M symbolizes the unionized molecule as it reacts with the electron beam and forms a radical cation. The cation will have a m/z equal to the molecular weight.

The precursor (molecular) ion then sequentially fragments in a unimolecular fashion. (Note that the ion is often written as M^+, for which the free electron, symbolized by the dot, is assumed. Regardless, the molecule has lost one electron and still retains all the protons; thus, the net charge must be positive.) The reactions of butane as it forms several of the predominant product ion (daughter) fragments are shown below:

$$CH_3 - CH_2 - CH_2 - CH_3 + e \Rightarrow CH_2 - CH_2 - CH_2 - CH_3{}^{+\bullet}$$
$$(m/z = 58) + 2e \quad (11.6)$$

$$CH_3 - CH_2 - CH_2 - CH_3{}^{+\bullet} \Rightarrow CH_3 - CH_2 - CH_2 - CH_2{}^{+}$$
$$(m/z = 57) + {}^{\bullet}H \quad (11.7)$$

$$CH_3 - CH_2 - CH_2 - CH_3{}^{+\bullet} \Rightarrow CH_3 - CH_2 - CH_2{}^{+}$$
$$(m/z = 43) + \bullet CH_3 \quad (11.8)$$

$$CH_3 - CH_2 - CH_2 - CH_3{}^{+\bullet} \Rightarrow CH_3 - CH_2{}^{+\bullet}$$
$$(m/z = 29) + \bullet CH_3 - CH_3$$
$$(11.9)$$

$$CH_3 - CH_2 \Rightarrow CH_3{}^{+\bullet} (m/z = 15) + \bullet CH_2 \quad (11.10)$$

Many of the fragments for butane result from direct cleavage of the methylene groups. With alkanes, you will always see fragments in the mass spectrum that are produced by the sequential loss of CH_2 or CH_3 groups.

Close examination of the butane mass spectrum in Fig. 11.9 reveals a peak that is $1\, m/z$ unit larger than the molecular ion at $m/z = 58$. This peak is designated by the symbol $M + 1$ and is due to the naturally occurring isotopes. The most abundant isotope of carbon has a mass of 12; however, a small amount of ^{13}C is also present (1.11 %). Any ions that contained a ^{13}C or a deuterium isotope would be $1\, m/z$ unit larger, although the relative abundance would be low.

Another example of MS fragmentation patterns is shown for methanol in Fig. 11.10. Again, the fragmentation pattern is straightforward. The precursor ion ($CH_3 - OH^{+\bullet}$) is at a m/z of 32, which is the molecular weight. Other fragments include the base peak at a m/z of 31 due to $CH_2 - OH^+$, the CHO^+ fragment at a m/z of 29, and the $CH_3{}^+$ fragment at a m/z of 15.

The ionization method of **chemical ionization** (CI) (see Table 11.1) is classified as a **soft ionization** because only a few fragments are produced. In this technique, a gas is ionized, such as methane (CH_4), which then directly ionizes the molecule. The most important use of CI is in the determination of the molecular ion since there is usually a fragment that is

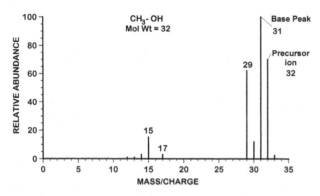

11.10 **figure** Mass spectrum of methanol obtained by electron impact ionization

$1\,m/z$ unit larger than that obtained with EI. Thus, a mass spectrum of butane taken by the CI method would have a quasimolecular (parent) ion at $m/z = 59$ (M + H). Many LC-MS interfaces use CI or electrospray ionization methods so it is common to see the (M + H)$^+$ precursor ion. As can be seen in Eqs. 11.6, 11.7, 11.8, 11.9, and 11.10, the reactions of the cleavage process can be quite involved. Many of the reactions are covered in detail in the book by McLafferty and Turecek listed in resource materials.

11.4 GAS CHROMATOGRAPHY-MASS SPECTROMETRY

Although samples can be introduced directly into the MS ion source, many applications require chromatographic separation before analysis. The rapid development of **gas chromatography-mass spectrometry** (GC-MS) has allowed for the coupling of the two methods for routine separation problems (see Chap. 14). A MS coupled to GC allows the peaks to be identified or confirmed, and, if an unknown is present, it can be identified using a computer-assisted search of a library containing known MS spectra. Another critical function of GC-MS is to ascertain the purity of each peak as it elutes from the column. Does the material eluting in a peak contain one compound, or is it a mixture of several that just happen to coelute with the same retention time?

In most cases a **capillary GC column** is connected directly to the MS source via a **heated capillary transfer line**. The transfer line is kept hot enough so as to avoid condensation of the volatile component eluting from the GC column on its way into the low-pressure MS source. The sample flows through the GC column into the interface and then on to be processed by the MS. A computer is used to store and process the data from the MS.

An example of the power of GC-MS is shown below in the separation of the methyl esters of several long-chain fatty acids (Fig. 11.11). Long-chain fatty acids must have the carboxylic acid group converted or blocked with a methyl group to make them volatile. Methyl esters of palmitic (16:0), oleic (18:1), linoleic (18:2), linolenic (18:3), stearic (18:0), and arachidic (20:0) acids were injected onto a column that was supposed to be able to separate all the naturally occurring fatty acids. However, the GC tracing showed only four peaks, when it was known that six different methyl esters were in the sample. The logical explanation is that one or two of the peaks contain a mixture of methyl esters resulting from poor resolution on the GC column.

The purity of the peaks is determined by running the GC-MS and taking mass spectra at very short increments of time (1 s or less). If a peak is pure, then the mass spectra taken throughout the peak should be the same. In addition, the mass spectrum can be compared with the library of spectra stored in the computer.

The **total ion current** (TIC) **chromatogram** of the separation of the fatty acid methyl esters is shown in Fig. 11.11. There are four peaks eluting off the column between 15.5 and 28 min. The first peak at 15.5 min has the same mass spectrum throughout, indicating that only one compound is eluting. A computer search of the MS library gives an identification of the peak to the methyl ester of palmitic acid. The mass spectra shown in Fig. 11.12 compare the material eluting from the column to the library mass spectrum.

Most of the fragments match, although the GC-MS scan does have many small fragments not present on the library mass spectrum. This is a common background noise and usually does not present a problem. The data from the rest of the chromatogram indicate that the peaks at both 20 and 27 min contain only one component. The computer match identifies the peak at 20 min as stearic acid, methyl ester, and the peak at 27 min as arachidic acid, methyl ester. However, the peak located at 19.5 min is shown to have several different mass spectra, indicating impurity or coeluting compounds.

In Fig. 11.13, the region around 19 min has been enlarged. The arrows indicate where different mass spectra were obtained. The computer identified the material in the peak at 19.5 min as linoleic acid, methyl ester; the material at 19.7 min as oleic acid, methyl ester; and the material at 19.8 min as linolenic acid, methyl ester. Thus, as we originally suspected, several of the methyl esters were coeluting off the GC column. This example illustrates the tremendous power of GC-MS used in both a quantitative and a qualitative manner.

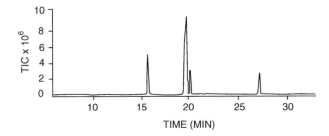

11.11
figure

Total ion current GC chromatogram of the separation of the methyl esters of six fatty acids. Detection is by electron impact ionization using a direct capillary interface

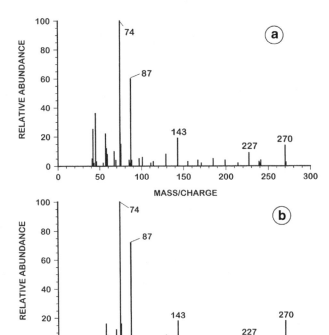

11.12
figure

Mass spectra of (**a**) the peak at 15.5 min in the TIC chromatogram shown in Fig. 11.11 and (**b**) the methyl ester of palmitic acid from a computerized MS library

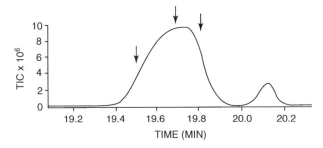

11.13
figure

Enlargement of the region 19.2–20.2 min from the TIC chromatogram shown in Fig. 11.11. *Arrows* indicate where mass spectra were obtained

11.5 LIQUID CHROMATOGRAPHY-MASS SPECTROMETRY

For a **high-performance liquid chromatography-mass spectrometry** (LC-MS) interface, the same overall requirements must be met as for GC-MS. There must be a way to remove the excess solvent, while converting a fraction of the liquid effluent into the gas phase, making it amenable for MS analysis. Furthermore most compounds analyzed by HPLC are either nonvolatile or thermally labile, making the task of liquid-to-gas phase transition even more challenging, especially while maintaining compound integrity.

How does LC-MS work? A modern LC-MS ionization interface converts liquid (LC eluent) into gas-phase ions (sampled by the MS) by a process of desolvation in the presence of a highly charged electrical field at atmospheric pressure. The energy applied to evaporate the solvent (thermal and electrical) is almost completely used in the desolvation process, and it does not contribute to degradation (usually thermal) of any labile species present in the LC eluant. Of the many different types of LC-MS ionization interfaces developed over the years, it was the development of the atmosphere pressure-based ionization interfaces, ESI (Sect. 11.2.3.2) and APCI (Sect. 11.2.3.3) that made LC-MS a routine technique. More recently, an APPI (Sect. 11.2.3.4) has been developed as a complementary technique to APCI.

11.6 TANDEM MASS SPECTROMETRY

Tandem MS (MS/MS, MS^n) is used in both GC-MS and LC-MS but is especially helpful in LC-MS since it allows for characterization, verification, and quantitation at ultrahigh sensitivity. There are two basic types of tandem MS, one which is a result of **collision-induced dissociation** (CID) typically observed on beam-type instruments (triple quadrupoles, TQ), and the other is a result of **collision-activated dissociation** (CAD) or MS^n, typically observed in ion trap MS (MS in time type instruments). Other fragmentation modes such as electron transfer dissociation (ETD), infrared multiphoton dissociation (IRMPD), and electron capture dissociation (ECD) are used for the analysis of compounds difficult to fragment.

Tandem MS using LC **triple quadrupoles** (TQ) (Fig. 11.14) operated in the selected reaction monitoring (SRM) or multiple reaction monitoring (MRM) scan mode provides a factor of 100–1000× more sensitivity than ultraviolet (UV) or diode array detectors, making them indispensable for high-sensitivity quantitative LC-MS analysis. Using the mechanism of CID, the precursor molecule filtered by Q1 energetically collides with argon or nitrogen gas, buffered in the collision cell (Q2), and a specific production is

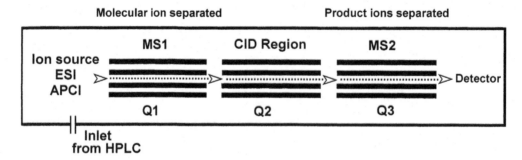

Molecular ion separated   **Product ions separated**

11.14 Diagram of a triple quadrupole mass spectrometer capable of doing MS/MS. Q1 and Q2 are used to separate
figure ions, and Q3 is the collision-induced dissociation area (CID). Once the compounds are ionized, they can be
allowed to pass through without the CID activated, or the molecular ion can be further fragmented in the Q2
area via CID to yield product ions (fragments)

transmitted through the third quadrupole (Q3). This mode offers the highest sensitivity critical for quantitative analysis that has been the mainstay of bioanalytical quantitative analysis for the past two decades. The high sensitivity of the SRM/MRM modes are due to two factors: (1) a dramatic improvement in signal-to-noise ratio by eliminating noise, and (2) TQ operation at nearly 100 %, allowing seven scans or more to ensure capturing the top of the eluting chromatographic peak to accurately determine the area under the peak. The other advantages are simplicity of data analysis and the high specificity resulting from the MRM mode. The triple quadrupole remains the instrument of choice for high-sensitive, routine LC-MS quantitative analysis, even though it was first introduced in the early 1990s.

Multiple MS or "MS-to-the-n[th]" made ion traps popular for structural analysis and verification of molecules. Subtle changes can be identified by piecing together the various fragments, which result in the different stages of the MS/MS process. The difference between tandem MS on a TQ and the MS[n] mode on the IT is how the ion undergoes fragmentation. In the CID process on a TQ or Q-TOF, the precursor ion selected in Q1 is accelerated into the collision cell (Q2) filled with argon or nitrogen and smashed into its component pieces in one highly energetic step. In contrast, the CAD process occurring in an ion trap is energetically much gentler involving a gradual ramping of voltages so that the internal energy of the specified trapped ion increases as it collides with the helium buffer gas, until the most fragile chemical bond breaks. As soon as this occurs, the m/z changes, and the product ions no longer experience the gradual ramping of the energy but remain trapped. This process allows for multiple steps of MS/MS analysis at high sensitivity. The most abundant fragment (typically) is then isolated and re-trapped, and the CAD process is repeated. In most cases, CAD can easily perform MS[4], but there have been cases where MS[10] could be accomplished. Such a method of MS[n] analysis makes it significantly easier to piece together the intact structure from its key

fragments and determine or verify changes. Since ITMS is tandem MS in time, it has a significant advantage over TQ's operating Q3 in the "full scan MS/MS" mode. The main advantage of MS[n] on ITMS is the high sensitivity of the MS[n] scan mode, especially when connected to ultra-HPLC (UHPLC) (Chap. 13, Sect. 13.2.3.1).

11.7 HIGH-RESOLUTION MASS SPECTROMETRY (HRMS)

The widespread adoption of orbitrap and Q-TOF MS instruments has made high-resolution, accurate mass applications a growing trend. Resolution in a mass spectrometer is defined in terms of full width half maxima (FWHM), which is the mass spectrum peak width at half height for a known m/z [4]. At unit mass resolution (nominal mass instruments such as quadrupoles and ion traps), FWHM is typically around 0.6 Da, so at m/z 300, the resolution would be 500 (300 ÷ 0.6). An orbitrap or a Q-TOF, for example, can deliver FWHM of 0.01 Da, so it would deliver a resolution of 30,000 (300 ÷ 0.01) for m/z 300. An FTMS can deliver FWHM of 0.0001 Da, enabling it to deliver a resolution of 3,000,000 (300 ÷ 0.0001). High-resolution analysis has significant advantages in terms of S/N, fine isotopic structure analysis, and determining elemental composition from highly precise accurate mass assignment.

Accurate mass is important in mass spectrometry because it can give you elemental composition and enable the identification of unknowns. With accurate mass and MS/MS fragmentation high-resolution data, uncertainty is significantly reduced in identification of "known" unknowns (compounds that are in a database, but not known to the analyst) and invaluable for screening type qualitative analysis. For mass accuracy below 5 ppm, elemental composition determination becomes straightforward through the application of sophisticated software algorithms.

The listing below for the compound caffeine ($C_8H_{10}N_4O_2$) illustrates the value of accurate mass:

- **Nominal mass:** 194 (used synonymously with molecular mass and is the sum of the integer mass of the most abundant isotope for each element, $C = 12$, $H = 1$, $N = 14$, $O = 16$)
- **Monoisotopic mass:** 194.0804 (sum of the most abundant isotopic mass for each of the constituent elements, $C = 12.0000$, $H = 1.007825$, $N = 14.003074$, $O = 15.994915$)
- **Average mass:** 194.1906 (sum of the average atomic masses or sum of isotopes taking into account the relative abundances for the constitute elements, e.g., $O_{16} = 15.994915$, but the average mass is 15.999405, which takes into account the O_{17} and O_{18} isotopes and their relative abundances)

Accurate mass is measured in terms of parts per million (ppm) and calculated by dividing the mass error by the theoretical mass. In the example above, assume that the measured mass on the high-resolution MS was 194.0811. Given the theoretical monoisotopic mass of 194.0804, this results in a mass error of 0.0007 (194.0811–194.0804), which would result in a mass accuracy of 3.6 ppm for that measurement ($[194.0804/0.0007]*10^6$). For mass accuracy below 5 ppm, elemental composition determination becomes straightforward through the application of sophisticated software algorithms.

The advancement of high-resolution MS coupled with availability of more databases has enhanced the determination of unknown compounds, a process termed nontargeted or "scouting" analysis. With accurate mass determinations, an unknown can be compared against LC-MS libraries for identification. The elemental composition can be determined and further verified by observing the fragment ions for definitive identification. There are several databases of compounds and their ionization patterns (i.e., ions produced). Most notably is the METLIN metabolite database by the Scripps Center for Metabolomics where the ion spectra are produced by ESI Q-TOF. In the case of caffeine (mass 194.080376), the spectrum contains a total of seven major ions, more than enough for positive identification. Another database that contains some mass spectra is ChemSpider. Unfortunately the databases only cover a limited amount of molecules, and when present, the mass spectrum depends on the type of ionization interface/method (e.g., ESI, IT, TOF), which is not the case with EI GC-MS libraries. However, the limitations of these databases will only improve as more scientists provide additional data. Both Milman [5] and Lehotay et al. [6] discuss MS libraries, screening of molecules, and nontarget identification.

11.8   APPLICATIONS

The use of MS in the field of food science is well established and growing rapidly as food exports from Asia increase yearly to the USA and Europe. While GC-MS has been used for years, LC-MS/MS instrumentation has become indispensible for the analysis of compounds such as chloramphenicol, nitrofurans, sulfonamides, tetracyclines, melamine, acrylamide, and malachite green in foods such as honey, fish, shrimp, and milk (see Chap. 33). Agencies such as the Food and Drug Administration (FDA), Center for Food Safety and Applied Nutrition (CFSAN), European Food Safety Authority (EFSA), Health Canada, and Japan Food Safety Commission use MS-based techniques to drive regulatory standards for banned substances, safeguarding the food supply for human consumption.

To give an appreciation of the usefulness of LC-MS, several examples are provided below. It is important to remember that there are a wide variety of methods now available to analyze just about any type of sample in a variety of matrices.

Due to the prevalence of consumption of caffeine-containing drinks throughout the world, the analysis of this small bioactive compound has been of interest for many years. Over 20 years ago HPLC methods were published showing that caffeine and other alkaloids, theobromine and theophylline, could be analyzed by HPLC using an ultraviolet detector. While HPLC-UV analysis is quite acceptable, the use of LC-MS can verify and enhance identification in a variety of complex food systems.

Figure 11.15 illustrates a reversed-phase HPLC column separation and MS spectrum obtained using the ESI interface coupled with MS/MS [7]. An aqueous coffee extract was filtered and separated by HPLC using an acetic acid-acetonitrile mobile phase. For comparative purposes a separate HPLC separation was achieved with the same HPLC column except detection was with a UV detector. The HPLC chromatogram in Fig. 11.15a shows the TIC, an indicator of total ions and thus compounds eluting, and matches the HPLC-UV chromatogram (not shown). Figure 11.15b is the selected ion trace of ions $m/z = 180.7$–181.7, which would correspond to the protonated molecular ions $(M+H)^+$ of theobromine and theophylline at 181.2 (both compounds are isomers and have identical masses of 180.2). The chromatogram in Fig. 11.15c is the selected ion trace of $m/z = 194.2$–196.2, which corresponds to the protonated molecular ion of caffeine (195.2). The MS/MS of caffeine and theobromine is presented in Fig. 11.15d, e and shows the protonated molecular ions for both caffeine and theobromine, and several ion fragment m/z.

The **melamine** food contamination issue of 2007–2008 is another excellent example of the use of LC-MS and MS/MS in both a detective role and later as an official analytical method. Melamine is a six-membered cyclic nitrogenous ring compound with three amines attached to the carbons in the ring and contains 67 % nitrogen (Fig. 11.16). Since a common test for protein measures nitrogen (Chap. 18, Sect. 18.2), the compound was used as an economic

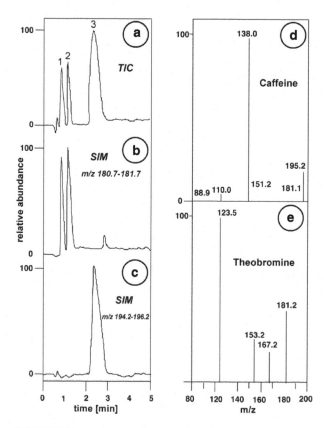

11.15 figure Reversed-phase LC-MS (parts **a–c**) and MS/MS separation of an aqueous coffee extract: (**a**) TIC, total ion current of extract; (1) theobromine, (2) theophylline, and (3) caffeine; (**b**) selected ion trace, $m/z = 80.7–181.7$; (**c**) selected ion trace, $m/z = 194.2–196.2$; (**d**) MS/MS ionization of caffeine; (**e**) MS/MS of theobromine (From Huck et al. [7], used with permission of Elsevier, New York)

11.16 figure Chemical structure of melamine

adulterant in wheat gluten and dried milk powder to artificially increase the apparent protein content. Due to the polar nature of the amine groups, melamine is not volatile and thus cannot be analyzed with GC unless a derivative is synthesized, thus LC-MS is a preferred method. Methods have been presented that entail the separation of melamine by HPLC with detection by ESI-MS/MS. ESI produces a strong protonated molecular ion at $m/z = 127.1$ with

a MS/MS precursor ion (molecular) of 127.1 and a product ion of 85.1. One of the analytical methods suggested by the FDA entails monitoring of all these ions, which produces very good specificity and sensitivity.

The area of bioactive food components has grown dramatically over the last 10 years in part due to the availability of LC-MS. Many bioactive compounds in fruits, vegetables, and spices are polar and are not volatile. Thus, identification and evaluations were very difficult without HPLC methods coupled with MS. A good example of the use of LC-MS is in the measurement of the polyphenolic flavonoids called the catechins. These antioxidant compounds present in catechu, green tea, cocoas, and chocolates appear to have several beneficial biological effects including enhanced heart and blood vessel health.

Figure 11.17 shows a separation of the major green tea catechins by HPLC with ESI-MS detection [8]. At the bottom of the figure depicted is the chromatogram showing the separation of the catechins by HPLC. The ESI-MS detector was used to monitor all ions from m/z 120–2200 (TIC mode). The top left panel shows the MS obtained for the epicatechin peak eluting at about 12.1 min. As typical with ESI, there are few fragments, though the M + H molecular ion at 291.3 m/z is predominant. Further fragmentation (MS/MS) of the epicatechin yields two major protonated fragments, m/z 273.3 (loss of OH from the C ring) and 139.3 (oxidation and cleavage of the A ring). With these data it is possible to elucidate isomers and also possible degradation pathways.

High-resolution analysis on orbitrap technology has found wide application in food analysis, with LC, GC, and supercritical fluid chromatography (SCF) as hyphenated front-end separation techniques before the mass analysis is performed. Rajski et al. [9] performed a large multi-residue screening study (>250 pesticides) in fruits and vegetables using three different resolution settings on the orbitrap ($R = 17.5$ K, 35 K, and 70 K). The study revealed that using a ±0.2 min retention time window, $R = 70,000$, and mass tolerance of 5 ppm gave less than 5 % false-positive results at the 10 ug/kg level. Ishibashi et al. [10] used supercritical fluid (SFC) coupled to an orbitrap to simultaneously analyze 373 pesticides (10 ug/kg level which is the provisional MRL in Japan) in a QuEChERS spinach extract using an $R = 70,000$ and a mass tolerance of 5 ppm. The high-throughput advantage of SFC separation allowed for 72 samples in approximately 45 min using SFC-OT technology, for compounds whose molecular weights ranged from m/z 99 to m/z 900.

Recently, a GC-capable version of the orbitrap was launched [11], making high-resolution GC-MS analysis possible at resolutions exceeding $R > 50,000$. In contrast to GC-TOF instruments currently available, which focused mainly on MS acquisition speed at resolutions around 5000 (higher than quadrupole based instruments), the

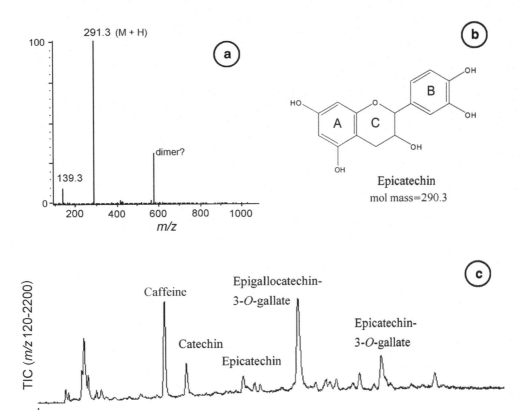

100 ┬ 291.3 (M + H)

(a)

139.3

dimer?

0 ┤

200 400 600 800 1000

m/z

(b)

Epicatechin
mol mass=290.3

(c)

TIC (*m/z* 120–2200)

Caffeine

Epigallocatechin-
3-*O*-gallate

Catechin

Epicatechin-
3-*O*-gallate

Epicatechin

0   5   10   15   20   25

Time (min)

11.17 LC-ESI-MS of green tea leaf extracts. (**a**) ESI mass spectra of epicatechin. (**b**) Chemical structure of epicatechin.
figure (**c**) TIC scanned from *m/z* 120–2,200 (From Shen et al. [8], used with permission of American Chemical Society, Washington, DC)

GC coupling to the orbitrap truly allows for high-resolution, accurate mass analysis in both MS and MS/MS modes. The superior chromatographic separation power of GC, coupled to high resolution ($R=60,000$ @ m/z 200), allows for unambiguous analysis of 132 pesticides in even the most complex matrices at the 10 ng/g level, approaching sensitivity performance (IDL 10 ng/g) of routine GC-triple quadrupoles. These orbitrap-based techniques are currently used extensively for screening of pesticide residues and highly sensitive quantitative analysis of banned substances.

FTMS provides ultrahigh-resolution MS at the isotope fine-structure level, i.e., resolution of isobaric species of different elemental composition, which can yield elemental composition analysis. Essentially, isotope fine structure is the fingerprint of a small molecule because the m/z value and their intensities in the fingerprint for a specific molecule exactly reflect the atomic composition of the molecule. Because FTMS delivers resolutions in excess of $R >1,000.000$ (million), this allows for isotopologue analysis of the A+1 and A+2 ion signals. Using this capability, bioactive sulfur-containing compounds were identified in asparagus, purported to have angiotensin-converting enzyme (ACE) inhibitory

function [12]. One such compound identified was aspartame (no UV-chromophores) detected at m/z 307.0893 (and confirmed by high-resolution MS/MS) that required acquisition at $R = 1,000,000$ to determine the elemental composition of $C_{10} H_{18} N_4 O_3 S_2$. Such discoveries are only possible with FTMS where isotope fine-structure analysis is possible.

Recently, MALDI-TOF has found key applications in food microbiology, with rapid identification of bacteria and fungi through their protein signatures. In this technique, bacteria or fungi from the culture plates are directly spotted onto the MALDI target plate, sprayed with matrix, and then directly analyzed on the MALDI-TOF instrument. The resulting spectrum representative of the microorganisms proteomic fingerprint is matched against a known, verified spectrum in the library, and, if there is positive hit, the bacteria or fungi is rapidly identified. Wieme et al. [13] used MALDI-TOF to catalogue 4200 mass spectra from 273 acetic acid bacteria and lactic acid bacteria covering 52 species responsible for beer spoilage and then used the resulting library for routine quality control in the brewing industry. MALDI-TOF has been used for the rapid confirmatory identification (within 24 h) of more than 120 strains for *S. aureus* in milk, a pathogen that causes toxic shock syndrome toxin-1, a deadly form of food poisoning. This food-borne pathogen is usually a result of subclinical and clinical mastitis-affected dairy cattle [14].

11.9 SUMMARY

MS is a powerful analytical technique that can solve most complex problems faced by the food analytical chemist, both in a qualitative and quantitative manner. Its principles are fairly simple when examined closely. The basic requirements are to: (1) get the sample into an ionizing chamber where ions are produced; (2) separate the ions formed by magnets, quadrupoles, drift tubes, and electric fields; (3) detect the m/z of the precursor ion; (4) fragment the m/z selectively to derive more information if required; and (5) output the data to a computer for software evaluation.

Since the qualitative and quantitative aspects of MSs are so powerful, they are routinely coupled to a GC or HPLC, and find growing use with static sample introduction techniques. The interface for GCs is versatile and easy to use; however, the extensive sample preparation required for GC-MS analysis makes its utility cumbersome. The adoption of LC-MS as an analytical technique has greatly increased because of far simpler sample preparation procedures, wider ionization ranges for different classes of compounds, faster analysis times, routine high sensitivity, access to accurate mass capability, and the advent of UHPLC.

Acronyms

AMS	Accelerator mass spectrometer
APCI	Atmospheric pressure chemical ionization
API	Atmospheric pressure ionization
APPI	Atmospheric pressure photoionization
CAD	Collision-activated dissociation
CI	Chemical ionization
CID	Collision-induced dissociation
DART	Direct analysis in real time
ECD	Electron capture dissociation
EI	Electron impact ionization
ES	Electrospray
ESI	Electrospray ionization
ETD	Electron transfer dissociation
FT	Fourier transform
FT-ICR	Fourier transform-based ion cyclotrons
FTMS	Fourier transform mass spectrometry
GC	Gas chromatography
GC-MS	Gas chromatography-mass spectrometry
HRMS	High-resolution mass spectrometry
ICP-MS	Inductively coupled plasma-mass spectrometry
IMS	Ion mobility mass spectrometry
IT	Ion trap
ITMS	Ion trap mass spectrometry
LC	Liquid chromatography
LC-MS	Liquid chromatography-mass spectrometry
m/z	Mass-to-charge ratio
MALDI	Matrix-assisted laser desorption ionization
MALDI-TOF	Matrix-assisted laser desorption ionization time of flight
MRM	Multiple reaction monitoring
MS	Mass spectrometry
MS/MS	Tandem mass spectrometry
MS^n	Multiple stages of MS (tandem mass spectrometry)
OT	Orbitrap
Q	Quadrupoles mass filters
Q-TOF	Quadrupole time of flight
SFC	Supercritical fluid chromatography
SRM	Selected reaction monitoring
TIC	Total ion current
TOF	Time of flight
TQ	Triple quadrupole
TWIM	Traveling wave
UHPLC	Ultrahigh-performance liquid chromatography

11.10 STUDY QUESTIONS

1. What are the basic components of a MS?
2. What are the unique aspects of data that a MSs provide? How is this useful in the analysis of foods?
3. What is EI ionization? What is CI ionization?
4. What is the base peak on a mass spectrum? What is the precursor ion peak?
5. What is the difference between nominal mass and monoisotopic mass?
6. What are the major ions (fragments) expected in the EI mass spectrum of ethanol (CH_3-CH_2-OH)?
7. What are the major differences in how ionization occurs in the electrospray versus the APCI interface? What is ion suppression?
8. What does MALDI stand for and how does it differ from ESI?
9. What are the major differences between the quadrupole, ion trap, time of flight, and Fourier transform mass analyzer? What are the advantages of using each analyzer? What is especially unique about a Fourier transform-based mass analyzer?
10. What is the working principle behind the MALDI-TOF-based microbiology identification?
11. Which MS type is popular for quantitative analysis
12. What is the difference between CAD and CID?

REFERENCES

1. Silveira JA, Ridgeway ME, Park MA (2014) High Resolution Trapped Ion Mobility Spectrometery of Peptides. Anal. Chem. 86(12):5624–5627
2. Chughtai, K, Heeren, RMA (2010) Mass spectrometric imaging for biomedical tissue analysis. Chem. Rev. 110(5):3237–3277
3. Hajslova J, Cajka T, Vaclavik L (2011) Challenging applications offered by direct analysis in real time (DART) in food-quality and safety analysis. Trends Anal. Chem. 30(2):204–218
4. Balogh, MP (2004) Debating resolution and mass accuracy. LC-GC Europe 17(3):152–159
5. Milman BL, (2015) General principles of identification by mass spectrometry. Trends Anal. Chem. 69:24–33
6. Lehotay SJ, Sapozhnikova Y, Hans G.J. Mol, HGJ (2015) Current issues involving screening and identification of chemical contaminants in foods by mass spectrometry. Trends Anal. Chem. 69:62–75
7. Huck CW, Guggenbichler W, Bonn GK (2005) Analysis of caffeine, theobromine and theophylline in coffee by near infrared spectroscopy (NIRS) compared to high-performance liquid chromatography (HPLC) coupled to mass spectrometry. Analytica Chimica Acta 538(1–2):195–203
8. Shen D, Wu Q, Wang M, Yang Y, Lavoie EJ, Simon JE (2006) Determination of the predominant catechins in Acacia catechu by liquid chromatography/electrospray ionization-mass spectrometry. J Agric Food Chem 54(9):3219–3224
9. Rajski L, Gomez-Ramos MDM, Fernandez-Alba, AR (2014) Large pesticide multiresidue screening method by liquid chromatography-Orbitrap mass spectrometry in full scan mode applied to fruit and vegetables. J. Chromatogr. A 1360:119–127
10. Ishibashi M, Izumi Y, Sakai M, Ando T, Fukusaki E, Bamba T. (2015) High-throughput simultaneous analysis of pesticides by supercritical fluid chromatography coupled with high-resolution mass spectrometry. J. Agric. Food Chem. 63(18):4457–4463
11. Peterson AC, Hauschild J-P, Quarmby ST, Krumwiede D, Lange O, Lemke RAS, Grosse-Coosmann F, Horning S, Donohue TJ, Westphall MS, Coon JJ, Griep-Raming J (2014) Development of a GC/Quadrupole-Orbitrap Mass Spectrometer, Part I: Design and Characterization. Anal. Chem. 86:10036–10043
12. Nakabayashi R, Yang Z, Nishizawa T, Mori T, Saito K (2015) Top-down Targeted Metabolomics Reveals a Sulfur-Containing Metabolite with Inhibitory Activity against Angiotensin-Converting Enzyme in Asparagus officinalis. J. Nat. Prod. 78(5):1179–1183
13. Wieme AD, Spitaels F, Vandamme P, Landschoot AV, (2014) Application of matrix-assisted laser desorption/ionization time-of-flight mass spectrometry as a monitoring tool for in-house brewer's yeast contamination: a proof of concept. J. Inst. Brewing. 120(4):438–443
14. El Behiry A, Zahran RN, Tarabees R, Marzouk E, Al-Dubaib M. Phenotypical and Genotypical Assessment Techniques for Identification of Some Contagious Mastitis Pathogens. Int J. Med Health Biomed Bioeng Pharm Eng. 8(5):236–242

RESOURCE MATERIALS

- Balogh, MP (2009) The mass spectrometer primer. Waters, Milford, MA. A very good introduction to modern mass spectrometry including newer developments in LC-MS technologies.
- Barker J (1999) Mass spectrometry: Analytical Chemistry by Open Learning, 2nd edn. Wiley, New York. One of the best introductory texts on mass spectrometry in its second edition. The author starts at a very basic level and slowly works through all aspects of MS, including ionization, fragmentation patterns, GC-MS, and LC-MS.
- Ho C-T, Lin J-K, Shahidi F (eds) (2009) Tea and tea products. CRC, Boca Raton, FL. A good review of current literature on the chemistry and health-promoting properties of tea. Includes several chapters that discuss analytical methods for analyzing bioactive compounds and flavonoids in teas.
- Lee TA (1998) A beginner's guide to mass spectral interpretation. Wiley, New York. A good basic introduction to Mass Spectrometry with many practical examples.
- McLafferty FW, Turecek F (1993) Interpretation of mass spectra, 4th edn. University Science Books, Sausalito, CA. The fourth edition of an essential classic book on how molecules fragment in the ion source. Contains many examples of different types of molecules.
- Niessen WMA (2006) Liquid chromatography-mass spectrometry. 3rd edn. Taylor and Francis, New York. A thorough, though somewhat dated, review of LC-MS methods and interfaces for a variety of types of biological compounds.

Chromatography

Basic Principles of Chromatography

Baraem P. Ismail

Department of Food Science and Nutrition,
University of Minnesota,
St. Paul, MN 55108-6099, USA
e-mail: bismailm@umn.edu

S. Nielsen (ed.), *Food Analysis*, Food Science Text Series,
DOI 10.1007/978-3-319-45776-5_12, © Springer International Publishing 2017

12.1 INTRODUCTION

Chromatography has a great impact on all areas of analysis and, therefore, on the progress of science in general. Chromatography differs from other methods of separation in that a wide variety of materials, equipment, and techniques can be used. [Readers are referred to references [1–29] for general and specific information on chromatography.] This chapter will focus on the principles of chromatography, mainly **liquid chromatography** (LC). Detailed principles and applications of **gas chromatography** (GC) will be discussed in Chap. 14. In view of its widespread use and applications, high-performance liquid chromatography (HPLC) will be discussed in a separate chapter (Chap. 13). The general principles of extraction are first described as a basis for understanding chromatography

12.2 EXTRACTION

In its simplest form, extraction refers to the transfer of a solute from one liquid phase to another. Extraction in myriad forms is integral to food analysis – whether used for preliminary sample cleanup, concentration of the component of interest, or as the actual means of analysis. Extractions may be categorized as **batch**, **continuous**, or **countercurrent** processes. (Various extraction procedures are discussed in detail in other chapters: traditional solvent extraction in Chaps. 14, 17, and 33; accelerated solvent extraction in Chap. 33; solid-phase extraction in Chaps. 14 and 33; and solid-phase microextraction and microwave-assisted solvent extraction in Chap. 33).

12.2.1 Batch Extraction

In **batch extraction** the solute is extracted from one solvent by shaking it with a second immiscible solvent. The solute **partitions**, or distributes, itself between the two phases, and, when equilibrium has been reached, the **partition coefficient**, K, is a constant:

$$K = \frac{\text{Concentration of solute in phase 1}}{\text{Concentration of solute in phase 2}} \quad (12.1)$$

After shaking, the phases are allowed to separate, and the layer containing the desired constituent is removed, for example, in a separatory funnel. In batch extraction, it is often difficult to obtain a clean separation of phases, owing to emulsion formation. Moreover, partition implies that a single batch extraction is usually incomplete.

12.2.2 Continuous Extraction

Continuous extraction requires special apparatus, but is more efficient than batch separation. One example is the use of a Soxhlet extractor (Chap. 17, Sect. 17.2.5) for extracting fat from solids using organic solvents. Solvent is recycled so that the solid is repeatedly extracted with fresh solvent. Other types of equipment have been designed for the continuous extraction of substances from liquids and/or solids, and different extractors are used for solvents that are heavier or lighter than water.

12.2.3 Countercurrent Extraction

Countercurrent distribution refers to a serial extraction process. It separates two or more solutes with different partition coefficients from each other by a series of partitions between two immiscible liquid phases. Liquid-liquid partition chromatography (Sect. 12.4.2), also known as countercurrent chromatography, is a direct extension of countercurrent extraction. Years ago the countercurrent extraction was done with a "Craig apparatus" consisting of a series of glass tubes designed such that the lighter liquid phase (**mobile phase**) was transferred from one tube to the next, while the heavy phase (**stationary phase**) remained in the first tube [5]. The liquid-liquid extractions took place simultaneously in all tubes of the apparatus, which was usually driven electromechanically. Each tube in which a complete equilibration took place corresponded to one theoretical plate of the chromatographic column (refer to Sect. 12.5.1.2.1). The greater the difference in the **partition coefficients** of various substances, the better was the separation. A much larger number of tubes were required to separate mixtures of substances with close partition coefficients, which made this type of countercurrent extraction very tedious. Modern **liquid-liquid partition chromatography** (Sect. 12.4.2) is much more efficient and convenient.

12.3 CHROMATOGRAPHY

12.3.1 Historical Perspective

Modern chromatography originated in the late nineteenth and early twentieth centuries from independent work by David T. Day, a distinguished American geologist and mining engineer, and Mikhail Tsvet, a Russian botanist. Day developed procedures for fractionating crude petroleum by passing it through Fuller's earth, and Tsvet used a column packed with chalk to separate leaf pigments into colored bands. Because Tsvet recognized and correctly interpreted the chromatographic processes and named the phenomenon **chromatography**, he is generally credited with its discovery.

After languishing in oblivion for years, chromatography began to evolve in the 1940s due to the development of column partition chromatography by Martin and Synge and the invention of paper chromatography. The first publication on GC appeared in 1952. By the late 1960s, GC, because of its importance to the petroleum industry, had developed into a sophisticated instrumental technique, which was the first instrumental chromatography to be available commercially. Since early applications in the mid-1960s, HPLC, profiting from the theoretical and instrumental advances of GC, has extended the area of LC into an equally sophisticated and useful method. Supercritical fluid chromatography (SFC), first demonstrated in 1962, has been gaining popularity in food analysis [7]. Efficient chromatographic techniques, including automated systems, continue to be developed for utilization in the characterization and quality control of food ingredients and products [4, 7–13].

12.3.2 General Terminology

Chromatography is a general term applied to a wide variety of separation techniques based on the partitioning or distribution of a sample (**solute**) between a moving or mobile phase and a fixed or stationary phase. Chromatography may be viewed as a series of equilibrations between the mobile and stationary phase. The relative interaction of a solute with these two phases is described by the **partition** (K) or **distribution** (D) **coefficient** (ratio of concentration of solute in stationary phase to concentration of solute in mobile phase). The mobile phase may be either a gas (for GC) or liquid (for LC) or a supercritical fluid (for SFC). The stationary phase may be a liquid or a solid. The field of chromatography can be subdivided according to the various techniques applied (Fig. 12.1) or according to the physicochemical principles involved in the separation. Table 12.1 summarizes some of the chromatographic procedures or methods that have been developed on the basis of different mobile-stationary phase combinations. Inasmuch as the nature of interactions between solute molecules and the mobile or stationary phases differ, these methods have the ability to separate different kinds of molecules. (The reader is urged to review Table 12.1 again after having read this chapter.)

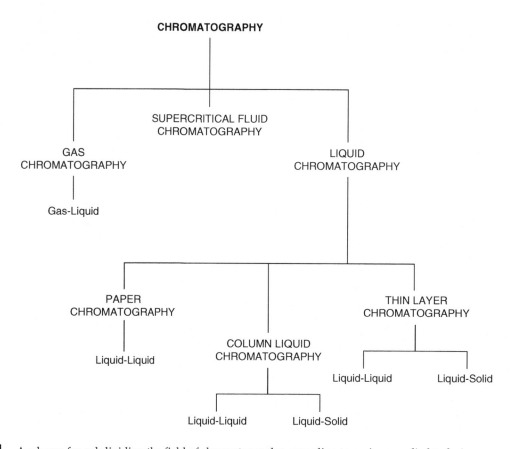

 A scheme for subdividing the field of chromatography, according to various applied techniques

12.1 table Characteristics of different chromatographic methods

Method	Mobile/phase	Stationary phase	Retention varies with
Gas-liquid chromatography	Gas	Liquid	Molecular size/polarity
Gas-solid chromatography	Gas	Solid	Molecular size/polarity
Supercritical fluid chromatography	Supercritical fluid	Solid	Molecular size/polarity
Reversed-phase chromatography	Polar liquid	Nonpolar liquid or solid	Molecular size/polarity
Normal-phase chromatography	Less polar liquid	More polar liquid or solid	Molecular size/polarity
Ion-exchange chromatography	Polar liquid-Ionic solid	Ionic solid	Molecular charge
Size-exclusion chromatography	Liquid	Solid	Molecular size
Hydrophobic interaction chromatography	Polar liquid	Nonpolar liquid or solid	Molecular size/polarity
Affinity chromatography	Water	Binding sites	Specific structure

Reprinted from Heftmann [1], p. A21, with kind permission from Elsevier Science-NL, Sara Burgerhartstraat 25, 1055 KV Amsterdam, The Netherlands

12.3.3 Gas Chromatography

Gas chromatography is a column chromatography technique, in which the mobile phase is gas and the stationary phase is mostly an immobilized liquid on an inert solid support in either a packed or capillary-type column. GC is used to separate thermally stable volatile components of a mixture. Gas chromatography, specifically gas-liquid chromatography, involves vaporizing a sample and injecting it onto the head of the column. Under a controlled temperature gradient, the sample is transported through the column by the flow of an inert, gaseous mobile phase. Volatiles are then separated based on several properties, including boiling point, molecular size, and polarity. Physiochemical principles of separation are covered in Sect. 12.4. However, details of the chromatographic theory of separation as it applies specifically to GC, as well as detection and instrumentation of GC, are detailed in Chap. 14.

12.3.4 Liquid Chromatography

There are several **liquid chromatography** techniques applied in food analysis, namely, **planar chromatography** (both paper and thin-layer chromatography) and column liquid chromatography, all of which involve a liquid mobile phase and either a solid or a liquid stationary phase. However, the physical form of the stationary phase is quite different in each case. Separation of the solutes is based on their physicochemical interactions with the two phases, which is discussed in Sect. 12.4.

12.3.4.1 *Planar Chromatography*

12.3.4.1.1 *Paper Chromatography*
Paper chromatography was introduced in 1944, and today it is mostly used as a teaching tool. In paper chromatography the stationary phase (water) and the mobile phase (organic solvent) are both liquid (**partition chromatography**, see Sect. 12.4.2), with

paper (usually cellulose) serving as a support for the liquid stationary phase. The support also may be impregnated with a nonpolar organic solvent and developed with water or other polar solvents (reversed-phase paper chromatography). The dissolved sample is applied as a small spot or streak about 1.5 cm from the edge of a strip or square of the paper, which is then allowed to dry. The dry strip is suspended in a closed container in which the atmosphere is saturated with the **developing solvent** (mobile phase) and the paper chromatogram is **developed**. The end closer to the sample is placed in contact with the solvent, which then travels up or down the paper by capillary action (depending on whether **ascending** or **descending** development is used), separating the sample components in the process. When the solvent front has traveled the length of the paper, the strip is removed from the developing chamber, and the separated zones are detected by an appropriate method.

In the case of complex sample mixtures, a two-dimensional technique may be used. The sample is spotted in one corner of a square sheet of paper, and one solvent is used to develop the paper in one direction. The chromatogram is then dried, turned 90°, and developed again, using a second solvent of different polarity. Another means of improving resolution is the use of **ion-exchange** (Sect. 12.4.4) papers, i.e., paper that has been impregnated with ion-exchange resin or paper, with derivatized cellulose hydroxyl groups (with acidic or basic moieties).

In planar chromatography, components of a mixture are often characterized by their relative mobility (R_f) value, where:

$$R_f = \frac{\text{Distance moved by component}}{\text{Distance moved by solvent}} \quad (12.2)$$

Unfortunately, R_f values are not always constant for a given solute/sorbent/solvent but depend on many factors, such as the quality of the stationary phase, layer thickness, humidity, development distance, and temperature.

12.3.4.1.2 Thin-Layer Chromatography

Thin-layer chromatography (TLC), first described in 1938, has largely replaced paper chromatography because it is faster, more sensitive, and more reproducible. The resolution in TLC is greater than in paper chromatography because the particles on the plate are smaller and more regular than paper fibers. Experimental conditions can be easily varied to achieve separation and can be scaled up for use in column chromatography, although thin-layer and column procedures are not necessarily interchangeable, due to differences such as the use of binders with TLC plates, vapor phase equilibria in a TLC tank, etc. There are several distinct advantages to TLC over paper chromatography and in some instances over column chromatography: high sample throughput, separations of complex mixtures, low cost, analysis of several samples and standards simultaneously, minimal sample preparation, and possibility to store the plate for later identification and quantification. Advances in TLC led to the development of **high-performance thin-layer chromatography** (HPTLC), which simply refers to TLC performed using plates coated with smaller, more uniform particles. This permits better separations in shorter times.

TLC, more so HPTLC, is applied in many fields, including environmental, clinical, forensic, pharmaceutical, food, flavors, and cosmetics. Within the food industry, TLC may be used for quality control. For example, corn and peanuts are tested for aflatoxins/mycotoxins prior to their processing into corn meal and peanut butter, respectively. Applications of TLC to the analysis of a variety of compounds, including lipids, carbohydrates, vitamins, amino acids, natural pigments, and sugar substitutes, are discussed in references [14, 17, 18].

1. **TLC General Procedures.** TLC utilizes a thin (ca. 250 µm thick) layer of **sorbent** or **stationary phase** bound to an **inert support**. The support is often a glass plate (traditionally, 20 × 20 cm), but plastic sheets and aluminum foil also are used. Pre-coated plates, of different layer thicknesses, are commercially available in a wide variety of sorbents, including chemically modified silica. Four frequently used TLC sorbents are silica gel, alumina, diatomaceous earth, and cellulose. Modified silica for TLC may contain polar or nonpolar groups, so both normal and reversed-phase (see Sect. 12.4.2.1) thin-layer separations may be carried out.

   If **adsorption** TLC is to be performed, the sorbent is first **activated** by drying for a specified time

and temperature. As in paper chromatography, the sample (in carrier solvent) is applied as a spot or streak about 1.5 cm from one end of the plate. After evaporation of the carrier solvent, the TLC plate is placed in a closed **developing chamber**, solvent migrates up the plate (**ascending development**) by capillary action, and sample components are separated. After the TLC plate has been removed from the chamber and solvent allowed to evaporate, the separated bands are made visible or detected by other means. Specific **chemical reactions (derivatization)**, which may be carried out either before or after chromatography, often are used for this purpose. Two examples are reaction with sulfuric acid to produce a dark charred area (a **destructive chemical method**) and the use of iodine vapor to form a colored complex (a **nondestructive method** inasmuch as the colored complex is usually not permanent). Common **physical detection methods** include the measurement of absorbed or emitted electromagnetic radiation, such as measuring fluorescence when stained with 2,7-dichlorofluorescein, and measurement of β-radiation from radioactively labeled compounds. Different reagents that can react selectively to generate colored products also are used [17]. **Biological methods** or biochemical inhibition tests can be used to detect toxicologically active substances. An example is measuring the inhibition of cholinesterase activity by organophosphate pesticides.

   Quantitative evaluation of thin-layer chromatograms may be performed [17]: (1) in situ (directly on the layer) by using a **densitometer** [18], or (2) scraping a zone off the plate, eluting compound from the sorbent, and then analyzing the resultant solution (e.g., by liquid scintillation counting).

2. **Factors Affecting Thin-Layer Separations.** In both planar and column liquid chromatography, the nature of the compounds to be separated determines what type of stationary phase is used. Separation can occur by adsorption, partition, ion-exchange, size-exclusion, or multiple mechanisms (Sect. 12.4). Table 12.2 lists the separation mechanisms involved in some typical applications on common TLC sorbents.

   Solvents for TLC separations are selected for specific chemical characteristics and **solvent strength** (a measure of interaction between solvent and sorbent; see Sect. 12.4.1). In simple adsorption TLC, the higher the solvent strength, the greater the R_f value of the solute. An R_f value of 0.3–0.7 is typical. Mobile phases have been developed for the separation of various compound classes on the different sorbents (see Table 7.1 in reference [19]).

In addition to the sorbent and solvent, several other factors must be considered when performing planar chromatography. These include the **type of developing chamber** used, **vapor phase conditions** (saturated vs. unsaturated), **development mode** (ascending, descending, horizontal, radial, etc.), and **development distance**. For additional reading refer to references [14–18].

12.3.4.2 Column Liquid Chromatography

Column liquid chromatography generally has enhanced resolution of solutes in a mixture and enables precise analysis compared to planar chromatography. Fractionation of solutes occurs as a result of differential migration through a closed tube of station-

ary phase, and analytes can be monitored while the separation is in progress. In column liquid chromatography, the mobile phase is liquid, and the stationary phase can be either solid or liquid supported by an inert solid. A system for **low-pressure** (i.e., performed at or near atmospheric pressure) column liquid chromatography is illustrated in Fig. 12.2.

Having selected a stationary and mobile phase suitable for the separation problem at hand, the analyst must first prepare the **stationary phase** (**resin, gel**, or **packing material**) for use according to the supplier's instructions (e.g., the stationary phase often must be **hydrated** or **preswelled** in the mobile phase). The prepared stationary phase then is **packed** into a column

table Thin-layer chromatography sorbents and mode of separation

Sorbent	Chromatographic mechanism	Typical application
Silica gel	Adsorption	Steroids, amino acids, alcohols, hydrocarbons, lipids, aflatoxins, bile acids, vitamins, alkaloids
Silica gel RP	Reversed phase	Fatty acids, vitamins, steroids, hormones, carotenoids
Cellulose, kieselguhr	Partition	Carbohydrates, sugars, alcohols, amino acids, carboxylic acids, fatty acids
Aluminum oxide	Adsorption	Amines, alcohols, steroids, lipids, aflatoxins, bile acids, vitamins, alkaloids
PEI cellulose[a]	Ion exchange	Nucleic acids, nucleotides, nucleosides, purines, pyrimidines
Magnesium silicate	Adsorption	Steroids, pesticides, lipids, alkaloids

Reprinted from Touchstone [19] by permission of Wiley, New York
[a]PEI cellulose refers to cellulose derivatized with polyethyleneimine (PEI)

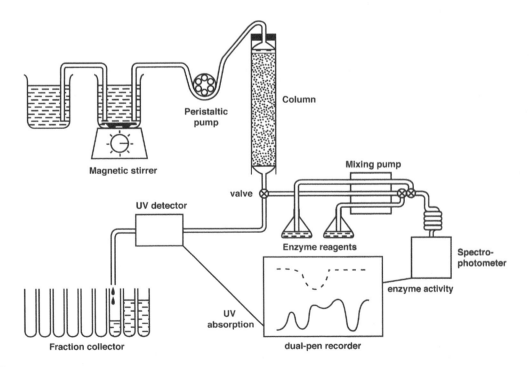

figure A system for low-pressure column liquid chromatography. In this diagram, the column effluent is being split between two detectors in order to monitor both enzyme activity (*at right*) and UV absorption (*at left*). The two tracings can be recorded simultaneously by using a dual-pen recorder (Adapted from Scopes [28], with permission)

(usually glass), the length and diameter of which are determined by the amount of sample to be loaded, the separation mode to be used, and the degree of resolution required. Longer and narrower columns usually enhance resolution and separation (Sect. 12.5.1). Adsorption columns may be either dry or wet packed; other types of columns are wet packed. The most common technique for wet packing involves making a slurry of the adsorbent with the solvent and pouring it into the column to the desired bed height. Pouring uniform columns is an art that is mastered with practice. If the packing solvent is different from the initial eluting solvent, the column must be thoroughly washed (**equilibrated** with 2–3 column volumes) with the starting mobile phase.

The sample to be fractionated is dissolved in a minimum volume of the starting mobile phase, injected through a sample injection port, and carried by the mobile phase onto the column. Low-pressure chromatography utilizes only **gravity** flow or a high-precision **peristaltic pump** to maintain a constant flow of **mobile phase** (**eluent** or **eluting solvent**) through the column. If eluent is fed to the column by a peristaltic pump (see Fig. 12.2), then the flow rate is determined by the pump speed. Depending on the dimensions of the column, the flow rate is adjusted not to exceed the max pressure sustained by the pump.

The process of passing the mobile phase through the column is called **elution**, and the portion that emerges from the outlet end of the column is called the **eluate** (or effluent). Elution may be **isocratic** (constant mobile phase composition) or a **gradient** (changing the mobile phase, e.g., increasing solvent strength or pH). Gradient elution enhances resolution and decreases analysis time (see also Sect. 12.5.1). As elution proceeds, components of the sample are selectively retarded by the stationary phase based on the strength of interaction with the stationary phase and thus are eluted at different times.

The column eluate may be directed through a detector and then into tubes, changed at intervals by a fraction collector. The detector response, in the form of an electrical signal, may be recorded (the **chromatogram**) using a computerized software. Signals are then integrated for either qualitative or quantitative analysis (Sects. 12.5.2 and 12.5.3). The fraction collector may be set to collect eluate at specified time intervals or after a certain volume or number of drops have been collected. Components of the sample that have been chromatographically separated and collected can be further analyzed as needed.

12.3.5 Supercritical Fluid Chromatography

Supercritical fluid chromatography is performed above the **critical pressure** (P_c) and **critical temperature** (T_c) of the mobile phase. A supercritical fluid (or compressed gas) is neither a liquid nor a typical gas. The combination of P_c and T_c is known as the **critical point**. A supercritical fluid can be formed from a conventional gas by increasing the pressure or from a conventional liquid by raising the temperature. Carbon dioxide frequently is used as a mobile phase for SFC; however, it is not a good solvent for polar and high-molecular-weight compounds. A small amount of a polar, organic solvent such as methanol can be added to a nonpolar supercritical fluid to enhance solute solubility, improve peak shape, and alter selectivity. Other supercritical fluids that have been used in food applications include nitrous oxide, trifluoromethane, sulfur hexafluoride, pentane, and ammonia.

Supercritical fluids confer chromatographic properties intermediate to LC and GC. The **high diffusivity** and **low viscosity** of supercritical fluids mean decreased analysis times and improved resolution compared to LC. An additional benefit of short analysis time is the reduced solvent consumption. SFC offers a wide range of **selectivity** (Sect. 12.5.2) adjustment, by changes in **pressure** and **temperature** as well as changes in **mobile phase composition** and the **stationary phase**. Compared to HPLC (Chap. 13), SFC is better for separating compounds with a broader range of polarities. In addition, SFC makes possible the separation of **nonvolatile**, **thermally labile compounds**, which cannot be analyzed by GC without derivatization. In fact, SFC today is mainly used for the analysis of nonvolatiles.

Either **packed** or **capillary** columns can be used in SFC. Packed column materials are similar to those used for HPLC. Small particle, porous, high surface area, hydrated silica, either bare or bonded silica, is commonly used as the column packing material (Sect. 12.4.2.3 and Chap. 13). Capillary columns are generally coated with a polysiloxane (-Si-O-Si) film containing different functional groups, which is then cross-linked to form a polymeric stationary phase that cannot be washed off by the mobile phase. Polysiloxanes containing different functional groups, such as methyl, phenyl, pyridine, or cyano, may be used to vary the polarity of this stationary phase.

The development of SFC benefited from advancement in the instrumentation for HPLC and GC. One major difference in instrumentation is the presence of a back pressure regulator to control the **outlet pressure** of the system. Without this device, the fluid would expand to a low-pressure, low-density gas. Detectors used in GC and HPLC also can be used with SFC, including coupling with mass spectrometry (MS) (Chap. 11).

SFC has been used for the analysis of a wide range of compounds. Fats, oils, and other lipids are compounds to which SFC is increasingly applied. For example, the noncaloric fat substitute, Olestra®, was characterized by SFC-MS. Other researchers have used SFC to detect pesticide residues, study thermally labile compounds from members of the *Allium* genus, fractionate citrus essential oils, and characterize compounds extracted from microwave packaging [20]. Bernal et al. [11] highlighted the use of packed column and capillary SFC for the analysis of food and natural products, namely, lipids and their derivatives, carotenoids, fat-soluble vitamins, polyphenols, carbohydrates, and food adulterants such as Sudan dyes.

12.4 PHYSICOCHEMICAL PRINCIPLES OF CHROMATOGRAPHIC SEPARATION

Several physicochemical principles (illustrated in Fig. 12.3) are involved in chromatography mechanisms employed to separate or fractionate various compounds of interest, regardless of the specific techniques applied (discussed in Sect. 12.3). The mechanisms described below apply mainly to liquid chromatography; GC mechanisms will be detailed in Chap. 14. Although it is more convenient to describe each of these phenomena separately, it must be emphasized that more than one mechanism may be involved in a given fractionation. For example, many cases of partition chromatography also involve adsorption. Table 12.3 summarizes the different separation modes and associated stationary phases, mobile phases, and types of interactions.

12.4.1 Adsorption (Liquid-Solid) Chromatography

Adsorption chromatography is the oldest form of chromatography and originated with Tsvet in 1903 in the experiments that spawned modern chromatography. In this chromatographic mode, the stationary phase is a finely divided solid to maximize the surface area. The stationary phase (**adsorbent**) is chosen to permit differential interaction with the components of the sample to be resolved. The intermolecular forces thought to be primarily responsible for chromatographic adsorption include the following:

- Van der Waals forces
- Electrostatic forces
- Hydrogen bonds
- Hydrophobic interactions

Sites available for interaction with any given substance are heterogeneous. Binding sites with greater affinities, the most active sites, tend to be populated first, so that additional solutes are less firmly bound. The net result is that adsorption is a concentration-dependent process, and the **adsorption coefficient** is *not* a constant (in contrast to the **partition coefficient**). Sample loads exceeding the adsorptive capacity of the stationary phase will result in relatively poor separation (Sect. 12.5.1).

Classic adsorption chromatography utilizes mostly **silica** (slightly acidic), **alumina** (slightly basic), or charcoal (nonpolar). Both silica and alumina are polar adsorbents, possessing surface hydroxyl groups, and Lewis acid-type interactions determine their adsorption characteristics. The elution order of compounds from these adsorptive stationary phases can often be predicted on the basis of their relative **polarities** (Table 12.4). Compounds with the most polar functional groups are retained most strongly on polar adsorbents and, therefore, are eluted last. Nonpolar solutes are eluted first.

One model proposed to explain the mechanism of liquid-solid chromatography is that **solute** and **solvent** molecules are competing for active sites on the adsorbent. Thus, as relative adsorption of the mobile phase increases, adsorption of the solute must decrease. Solvents can be rated in order of their strength of adsorption on a particular adsorbent, such as silica. Such a **solvent strength** (or solvent polarity in this case) **scale** is called an **eluotropic series**. An eluotropic series for alumina is listed in Table 12.5. Silica has a similar rank ordering. Once an adsorbent has been chosen, solvents can be selected from the eluotropic series for that adsorbent. Mobile phase strength (polarity in this case) can be increased (often by admixture of more polar solvents) until elution of the compound(s) of interest has been achieved.

Adsorption chromatography can be used to separate aromatic or aliphatic nonpolar compounds, based primarily on the type and number of functional groups present. The labile, fat-soluble chlorophyll and carotenoid pigments from plants have been studied extensively by adsorption column chromatography. Adsorption chromatography also has been used for the analysis of fat-solu is also another ble vitamins. Frequently, it is used as a batch procedure for the removal of impurities from samples prior to other analyses. For example, disposable solid-phase extraction cartridges (see Chap. 14) containing silica have been used for analyses of lipids in soybean oil, carotenoids in citrus fruit, and vitamin E in grain. Adsorption chromatography is also applied in specialized forms for the analysis of a wide range of compounds. Several of the chromatographic separation techniques described in the following sections are forms of specialized adsorption chromatography.

12.4.2 Partition (Liquid-Liquid) Chromatography

12.4.2.1 Introduction

In 1941, Martin and Synge undertook an investigation of the amino acid composition of wool, using a countercurrent extractor (Sect. 12.2.3) consisting of 40 tubes

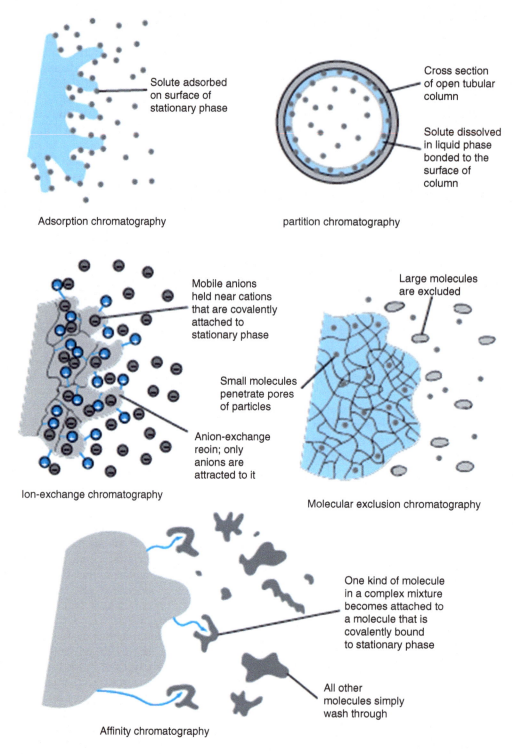

Adsorption chromatography

partition chromatography

Ion-exchange chromatography

Molecular exclusion chromatography

Affinity chromatography

Physicochemical principles of chromatography (From Harris [2] *Quantitative Chemical Analysis* 9e, by Daniel C. Harris, copyright 2016 by W.H. Freeman and Company. Used with permission of publisher)

12.3 table Summary of different chromatographic separation modes

Separation mode	Stationary phase	Mobile phase	Increasing mobile phase strength	Compounds eluting first/ eluting last	Type of interactions between solutes and stationary phase
Normal phase (can be in the form of adsorption or partition chromatography)	Polar sorbent (e.g., silica, alumina, water)	Nonpolar solvent (e.g., methanol, acetonitrile)	Decreasing concentration of organic solvent (i.e., increasing polarity, making the mobile phase more like the stationary phase)	Least polar/most polar	H-bonding mostly
Reversed-phase (can be in the form of adsorption or partition chromatography)	Nonpolar sorbent (e.g., bonded silica, C8 or C18)	Polar solvent (e.g., water)	Increasing concentration of organic solvent (i.e., decreasing polarity, making the mobile phase more like the stationary phase)	Most polar/least polar	H-bonding; Van der Waals, hydrophobic interactions
Hydrophobic interaction	Nonpolar sorbent (e.g., butyl-sepharose and phenyl-sepharose)	Salt solution/buffer (e.g., 1 M ammonium sulfate; Phosphate buffer)	Decreasing concentration of salt (result in reduced interaction of solute with the sorbent)	Least hydrophobic surface/most hydrophobic surface	Hydrophobic interactions
Cation exchange	Negatively charged functional groups (e.g., RSO_3^-, RCO_2^-)	Buffers of specific pH and ionic strength	Increasing the pH (in case of weak cation exchanger) or increasing salt concentration (e.g., increasing the pH will result in deprotonation, i.e., loss of positive charge of the solute so it no longer interacts with the functional group of the stationary phase, and increasing the salt concentration will provide counter ions that will displace the solute on the functional groups of the stationary phase)	Solutes with the least positive charge density/most positive charge density	Electrostatic
Anion exchange	Positively charged functional groups (e.g., RNR'_3^+, $R\text{-}NHR'_2^+$)	Buffers of specific pH and ionic strength	Decreasing the pH (in case of weak anion exchanger) or increasing salt concentration (e.g., decreasing the pH will result in protonation, i.e., loss of negative charge of the solute so it no longer interacts with the functional group of the stationary phase, and increasing the salt concentration will provide counter ions that will displace the solute on the functional groups of the stationary phase)	Solutes with the least negative charge density/most negative charge density	Electrostatic
Affinity	Highly specific ligand bound to inert surface (e.g., antibodies, enzyme inhibitors, lectins)	Buffer	Changing the pH or ionic strength, or adding a ligand similar to the bound ligand of the stationary phase	Solutes with least affinity to bound ligands/most affinity to bound	H-bonding; Van der Waals, hydrophobic interactions, electrostatic
Size exclusion	Porous inert material (e.g., Sephadex, a cross-linked dextran)	Mostly water or buffer	Not applicable	Largest in size/smallest in size	None

with chloroform and water flowing in opposite directions. The efficiency of the extraction process was improved enormously when a column of finely divided inert support material was used to hold one liquid phase (stationary phase) immobile, while the second liquid, an immiscible solvent (mobile phase), flowed over it, thus providing intimate contact between the two phases. Solutes partitioned between the two liquid phases according to their partition coefficients, hence the name **partition chromatography**.

In partition chromatography, depending on the characteristics of the compounds to be separated, the nature of the two liquid phases can be varied, usually by combination of solvents or pH adjustment of buffers. Often, the more polar of the two liquids is held stationary on the inert support, and the less polar solvent is used to elute the sample components (**normal-phase chromatography**). Reversal of this arrangement, using a **nonpolar stationary phase** and a **polar mobile phase**, has come to be known as **reversed-phase chromatography** (see Sect. 12.4.2.3).

Polar **hydrophilic** substances, such as amino acids, carbohydrates, and water-soluble plant pigments, are often separable by **normal-phase** partition chromatography. **Lipophilic** compounds, such as lipids, fat-soluble pigments, and **polyphenols**, may be resolved better with **reversed-phase** systems. Liquid-liquid partition chromatography has been invaluable to carbohydrate chemistry. Column liquid chromatography on finely divided cellulose has been used extensively in preparative chromatography of sugars and their derivatives.

12.4.2.2 *Coated Supports*

In its simplest form, the stationary phase for partition chromatography consists of a liquid coating on a solid matrix. The solid support should be as inert as possible and have a large surface area in order to maximize the amount of liquid held. Some examples of solid supports that have been used are silica, starch, cellulose powder, and glass beads. All are capable of holding a thin film of water, which serves as the stationary phase. It is important to note that materials prepared for adsorption chromatography must be **activated** by drying them to remove surface water. Conversely, some of these materials, such as silica gel, may be used for partition chromatography if they are deactivated by impregnation with water or the desired stationary phase. One disadvantage of liquid-liquid chromatographic systems is that the liquid stationary phase is often stripped off. This problem can be overcome by chemically bonding the stationary phase to the support material.

12.4.2.3 *Bonded Supports*

The liquid stationary phase may be covalently attached to a support by a chemical reaction. These **bonded** **phases** have become very popular for HPLC use, and a wide variety of both polar and nonpolar stationary phases is now available. It is important to note that mechanisms other than partition may be involved in the separation using bonded supports. Reversed-phase HPLC (see Chap. 13), with a nonpolar bonded stationary phase (e.g., silica bonded with C8 or C18 groups) and a polar solvent (e.g., water-acetonitrile), is widely used. Bonded-phase HPLC columns have greatly facilitated the analysis of vitamins in foods and feeds, as discussed in Chap. 3 of reference [21]. Additionally, bonded-phase HPLC is widely used for the separation and identification of polyphenols such as phenolic acids (e.g., *p*-coumaric, caffeic, ferulic, and sinapic acids) and flavonoids (e.g., flavonols, flavones, isoflavones, anthocyanidins, catechins, and proanthocyanidins).

12.4.3 Hydrophobic Interaction Chromatography

Hydrophobic interaction chromatography (HIC) has gained popularity over recent years for the purification of compounds on a preparative and semi-analytical scale. In HIC biomolecules adsorb to a weak hydrophobic surface at high salt concentration. Elution of adsorbed molecules is achieved by decreasing the salt concentration of the mobile phase over time. This technique takes advantage of hydrophobic moieties on the surface of a compound. Accordingly, HIC is very commonly used for the purification of food proteins, enzymes, and antibodies while offering high resolution. The high salt concentration allows biomolecules with high surface charge to adsorb to the hydrophobic **ligands** by shielding the charges. Salt precipitation of proteins, in particular, onto the column does not cause denaturation, since the interaction with the hydrophobic ligands is weak.

The stationary phase in HIC consists of hydrophilic support bonded to hydrophobic ligands. Several polymeric materials can be used as support, including cellulose, agarose, chitosan, polymethacrylate, or silica. The support must be hydrophilic so as not to contribute additional hydrophobicity and hence strong interactions that may cause denaturation of the protein. Often the polymer used has a high degree of cross-linking to provide rigidity and high surface area. Most commonly used ligands, chemically bonded to the support polymer, are linear chain alkanes. The size of the alkyl chains used depends on the surface hydrophobicity of the biomolecules to be separated, with the length of the chain being higher for more hydrophobic biomolecules. Sometimes phenyl and other aromatic groups are also used. Often times though, ligands with intermediate hydrophobicity are used to avoid strong interactions. Butyl-Sepharose® and phenyl-Sepharose® are among the most commonly used stationary phases.

12.4
table   Compounds class polarity scale

Fluorocarbons

Saturated hydrocarbons

Olefins

Aromatics

Halogenated compounds

Ethers

Nitro compounds

Esters ≈ ketones ≈ aldehydes

Alcohols ≈ amines

Amides

Carboxylic acids

From Johnson and Stevenson [5], used with permission
Listed in order of increasing polarity

12.5
table   Eluotropic series for alumina

Solvent

1-Pentane

Isooctane

Cyclohexane

Carbon tetrachloride

Xylene

Toluene

Benzene

Ethyl ether

Chloroform

Methylene chloride

Tetrahydrofuran

Acetone

Ethyl acetate

Aniline

Acetonitrile

2-Propanol

Ethanol

Methanol

Acetic acid

From Johnson and Stevenson [5], used with permission
Listed in order of increasing polarity

Different salts are used in HIC depending on their effects on protein precipitation. The most commonly used salt is ammonium sulfate. Concentration of the salt is also a determining factor for the precipitation of the protein on the column. Often 1 M ammonium sulfate is used for initial screening. It is important to prepare the sample in the same salt solution/buffer as the mobile phase. This, however, necessitates care in loading the sample to avoid precipitation of the protein

prior to reaching the column. A wash step generally precedes loading the sample to allow for washing out impurities. Depending on the sample characteristics, elution can be performed gradually by decreasing salt concentration over time, which may allow for isolation of different proteins in the sample in less time. Changing of salt concentration over time requires optimization for best resolution and shortest analysis time. Switching directly to water after the salt wash also may be performed to elute all bound protein with minimal separation. This is dependent on the level of fractionation and purification required. If a compound resists elution even after reducing salt concentration to zero, other HIC ligands should be tried. Cleaning and regeneration of the column are required after several runs. Often 0.1–1 M NaOH is used to prevent fouling of the column, and sometimes detergents and alcohol washes are used.

The pH of the mobile phase also can influence retention on the column and elution. However, oftentimes buffers with pH 7 are used. Additives, such as water-miscible alcohols and detergents, are sometimes used to help elute the protein faster. The hydrophobic parts of these additives will compete with the protein for binding to the ligand, causing desorption of the protein. Finally, temperature may play a role in HIC. As temperature increases, hydrophobic interaction increases allowing for better retention on the column; lowering the temperature aids in elution. Control of temperature during a run may enhance resolution and reduce analysis time (see also Sect. 2.5.1). For further details on HIC and applications, readers are directed to reference [22].

12.4.4 Ion-Exchange Chromatography

Separation by ion exchange can be categorized into three types: (1) ionic from nonionic, (2) cationic from anionic, and (3) mixtures of similarly charged species. In the first two cases, one substance binds to the ion-exchange medium, whereas the other substance does not. Batch extraction methods can be used for these two separations; however, chromatography is needed for the third category.

Ion-exchange chromatography is viewed as a specialized type of adsorption chromatography in which interactions between solute and stationary phase are primarily **electrostatic** in nature. The stationary phase (ion exchanger) contains fixed functional groups that are either negatively or positively charged (Fig. 12.4a). Exchangeable counterions preserve charge neutrality. A sample ion (or charged sites on large molecules) can exchange with the counterion for the interaction with the charged functional group. Ionic equilibrium is established as depicted in Fig. 12.4b. The functional group of the stationary phase determines whether cations or anions are

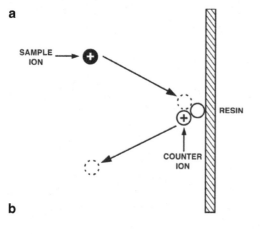

a

SAMPLE
ION

RESIN

COUNTER
ION

b

$M^+ + (Na^+ \, {}^-O_3S - \boxed{Resin}) \rightleftharpoons (M^+ \, {}^-O_3S - \boxed{Resin}) + Na^+$
$M^+ = Cation$

$X^- + (Cl^- \, {}^+R_4N - \boxed{Resin}) \rightleftharpoons (X^- \, {}^+R_4N - \boxed{Resin}) + Cl^-$
$X^- = Anion$

The basis of ion-exchange chromatography. (**a**) Schematic diagram of the ion-exchange process; (**b**) ionic equilibria for cation- and anion-exchange processes (From Johnson and Stevenson [5], used with permission)

exchanged. **Cation exchangers** possess covalently bound negatively charged functional groups, whereas **anion exchangers** possess bound positively charged groups. The chemical nature of these acidic or basic residues determines how stationary phase ionization is affected by the mobile phase pH.

The strongly acidic sulfonic acid moieties (RSO3-) of "strong" cation exchangers are completely ionized at all pH values above 2. Strongly basic quaternary amine groups (RNR'3+) on "strong" anion exchangers are ionized at all pH values below 10. Since maximum negative or positive charge is maintained over a broad pH range, the exchange or binding capacity of these stationary phases is essentially constant, regardless of mobile phase pH. "Weak" cation exchangers contain weakly acidic carboxylic acid functional groups (RCO2−); consequently, their exchange capacity varies considerably between ca. pH 4 and 10. Weakly basic anion exchangers possess primary, secondary, or tertiary amine residues (R-NHR'2+) that are deprotonated in moderately basic solution, thereby losing their positive charge and the ability to bind anions. Thus, one way of eluting solutes bound to weak ion-exchange medium is to change the mobile phase pH. Changing of the pH in the case of weak ion exchangers allows for enhanced separation and better selectivity compared to strong ion exchangers when separating compounds with very similar charge densities. While strong ion exchangers are generally used for initial screening and optimization of the separation, oftentimes, weak ion

exchangers are used to separate compounds that have similar adsorption coefficient, making use of a pH gradient. Changing the pH, however, may present a challenge when separating proteins. One must avoid the isoelectric point of the proteins during chromatographic separation; otherwise they will fall out of solution. A second way to elute bound solutes from either strong or weak ion exchangers is to increase the ionic strength (e.g., use NaCl) of the mobile phase, to weaken the electrostatic interactions.

Chromatographic separations by ion exchange are based upon differences in affinity of the exchangers for the ions (or charged species) to be separated. The factors that govern **selectivity** of an exchanger for a particular ion include: (1) ionic valence, radius, and concentration, (2) nature of the exchanger (including its displaceable counterion), and (3) composition and pH of the mobile phase. To be useful as an ion exchanger, a material must be both ionic in nature and highly permeable. Synthetic ion exchangers are thus cross-linked polyelectrolytes, and they may be inorganic (e.g., aluminosilicates) or, more commonly, organic compounds. Two commonly used types of organic compound-based synthetic ion exchangers are polystyrene and polysaccharide.

Polystyrene, made by cross-linking styrene with divinylbenzene (DVB), may be modified to produce either anion- or cation-exchange resins (Fig. 12.5). Polymeric resins such as these are commercially available in a wide range of particle sizes and with different degrees of cross-linking (expressed as weight percent of DVB in the mixture). The extent of cross-linking controls the rigidity and porosity of the resin, which, in turn, determines its optimal use. Lightly cross-linked resins permit rapid equilibration of solute, but particles swell in water, thereby decreasing charge density and selectivity (relative affinity) of the resin for different ions. More highly cross-linked resins exhibit less swelling, higher exchange capacity, and selectivity, but longer equilibration times. The small pore size, high charge density, and inherent hydrophobicity of the older ion-exchange resins have limited their use to small molecules [molecular weight (MW) <500].

Ion exchangers based on **polysaccharides**, such as cellulose, dextran, or agarose, have proven very useful for the separation and purification of large molecules, such as proteins and nucleic acids. These materials, called **gels**, are much softer than polystyrene resins, thus may be derivatized with strong or with weak acidic or basic groups via OH moieties on the polysaccharide backbone (Fig. 12.6). They have much larger pore sizes and lower charge densities than the older synthetic resins.

Food-related applications of ion-exchange chromatography include the separation of amino acids, sugars, alkaloids, and proteins. Fractionation of amino

CH = CH₂

Styrene

CH = CH₂

CH = CH₂

Divinylbenzene

Crosslinked styrene-divinylbenzene copolymer

R = H, Plain polystyrene

R = CH₂N⁺(CH₃)₃Cl⁻, Anion-exchanger

R = SO₃⁻H⁺, Cation-exchanger

Chemical structure of polystyrene-based ion-exchange resins

a

Derivatization sites

b

— OCH₂COO⁻   **Carboxymethyl - (CM)**
                                 (weak acid)

— OCH₂CH₂⁺NH(CH₂CH₃)(CH₂CH₃)   **Diethylaminoethyl - (DEAE)**
                                 (weak base)

— O — P(=O)(O⁻) — O⁻   **Phospho - (P)**
                               (intermediate acid)

— OCH₂CH₂⁺NCH₂CH(OH)(CH₃)(CH₂CH₃)   **Quaternaryaminoethyl- (QAE)**
                               (strong base)

— OCH₂CH₂SO₃⁻   **Sulfoethyl - (SE)**
                               (strong acid)

12.6 figure   Chemical structure of one polysaccharide-based ion-exchange resin. (**a**) Matrix of cross-linked dextran ("Sephadex," Pharmacia Biotech Inc, Piscataway NJ); (**b**) functional groups that may be used to impart ion-exchange properties to the matrix

12.4.5 Affinity Chromatography

Affinity chromatography is also another specialized form of adsorption chromatography, with the separation being based on very specific, reversible interaction between a solute molecule and a ligand immobilized on the chromatographic stationary phase (types of interactions are listed in Sect. 12.4.1). Affinity chromatography involves immobilized biological ligands as the stationary phase. These ligands can be antibodies, enzyme inhibitors, lectins, or other molecules that selectively and reversibly bind to complementary analyte molecules in the sample. Although both ligands and the species to be isolated are usually biological macromolecules, the term affinity chromatography also encompasses other systems, such as separation of small molecules containing *cis*-diol

acids in protein hydrolysates was initially carried out by ion-exchange chromatography; automation of this process led to the development of commercially produced amino acid analyzers (see Chap. 24). Many drugs, fatty acids, and organic acids, being ionizable compounds, may be chromatographed in the ion-exchange mode. For additional details on the principles and applications of ion chromatography, please refer to reference [23].

groups via phenylboronic acid moieties on the stationary phase.

The principles of affinity chromatography are illustrated in Fig. 12.7. A ligand, chosen based on its specificity/selectivity and strength of interaction with the analyte, is immobilized on a suitable support material. As the sample is passed through the column, analytes that are complementary to the bound ligand are adsorbed, while other sample components are eluted. Bound analytes are subsequently eluted via a change in the mobile phase composition as will be discussed below. After reequilibration with the initial mobile phase, the stationary phase is ready to be used again. The ideal support for affinity chromatography should be a porous, stable, high-surface-area material that does not adsorb anything itself. Thus, polymers such as agarose, cellulose, dextran, and polyacrylamide are used, as well as controlled pore glass.

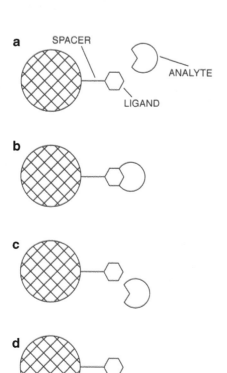

 Principles of bioselective affinity chromatography. (a) The support presents the immobilized ligand to the analyte to be isolated. (b) The analyte makes contact with the ligand and attaches itself. (c) The analyte is recovered by the introduction of an eluent, which dissociates the complex holding the analyte to the ligand. (d) The support is regenerated, ready for the next isolation (Reprinted from Lesellier and West [8], p. A311, with kind permission from Elsevier Science-NL, Sara Burgerhartstraat 25, 1055 KV Amsterdam, The Netherlands)

Affinity ligands are usually attached to the support or matrix by covalent bond formation, and optimum reaction conditions often must be found empirically. Immobilization generally consists of two steps: **activation** and **coupling**. During the activation step, a reagent reacts with functional groups on the support, such as hydroxyl moieties, to produce an activated matrix. After removal of excess reagent, the ligand is coupled to the activated matrix. (Preactivated supports are commercially available, and their availability has greatly increased the use of affinity chromatography.) The coupling reaction most often involves free amino groups on the ligand, although other functional groups can be used. When small molecules such as phenylboronic acid are immobilized, a **spacer arm** (containing at least four to six methylene groups) is used to hold the ligand away from the support surface, enabling it to reach into the binding site of the analyte.

Ligands for affinity chromatography may be either **specific** or **general** (i.e., group specific). Specific ligands, such as antibodies, bind only one particular solute. General ligands, such as nucleotide analogs and lectins, bind to certain classes of solutes. For example, the lectin concanavalin A binds to all molecules that contain terminal glucosyl and mannosyl residues. Bound solutes then can be separated as a group or individually, depending upon the elution technique used. Some of the more common general ligands are listed in Table 12.6. Although less selective, general ligands provide greater convenience.

Elution methods for affinity chromatography may be divided into **nonspecific** and **(bio)specific** methods. Nonspecific elution involves disrupting ligand analyte binding by changing the mobile phase pH, ionic strength, dielectric constant, or temperature. If additional selectivity in elution is desired, for example, in the case of immobilized general ligands, a biospecific elution technique is used. Free ligand, either identical to or different from the matrix-bound ligand, is added to the mobile phase. This free ligand competes for binding sites on the analyte. For example, glycoproteins bound to a concanavalin A (lectin) column can be eluted by using buffer containing an excess of lectin. In general, the eluent ligand should display greater affinity for the analyte of interest than the immobilized ligand.

In addition to protein purification, affinity chromatography may be used to separate supramolecular structures such as cells, organelles, and viruses; concentrate dilute protein solutions; investigate binding mechanisms; and determine equilibrium constants. Affinity chromatography has been useful especially in the separation and purification of enzymes and glycoproteins. In the case of the latter, carbohydrate-derivatized adsorbents are used to isolate specific lectins, such as concanavalin A, and lentil

12.6 table
General affinity ligands and their specificities

Ligand	Specificity
Cibacron Blue F3G-A dye, derivatives of AMP, NADH, and NADPH	Certain dehydrogenases via binding at the nucleotide binding site
Concanavalin A, lentil lectin, wheat germ lectin	Polysaccharides, glycoproteins, glycolipids, and membrane proteins containing sugar residues of certain configurations
Soybean trypsin inhibitor, methyl esters of various amino acids, D-amino acids	Various proteases
Phenylboronic acid	Glycosylated hemoglobins, sugars, nucleic acids, and other *cis*-diol-containing substances
Protein A	Many immunoglobulin classes and subclasses via binding to the F_c region
DNA, RNA, nucleosides, nucleotides	Nucleases, polymerases, nucleic acids

Reprinted with permission from Walters [26]. Copyright 1985 American Chemical Society

or wheat germ lectin. The lectin then agarose may be coupled to agarose, such as concanavalin A-agarose or lentil lectin-agarose, to provide a stationary phase for the purification of specific glycoproteins, glycolipids, or polysaccharides. Other applications of affinity chromatography include purification and quantification of mycotoxins [24] and pesticide residues/metabolites [25] in food/crops. For additional details on affinity chromatography, refer to references [26, 27].

12.4.6 Size-Exclusion Chromatography

Size-exclusion chromatography (SEC), also known as molecular exclusion, gel permeation (GPC), and gel filtration chromatography (GFC), is probably the easiest mode of chromatography to perform and to understand. It is widely used in the biological sciences for the resolution of macromolecules, such as proteins and carbohydrates, and also is used for the fractionation and characterization of synthetic polymers. Unfortunately, nomenclature associated with this separation mode developed independently in the literature of the life sciences and in the field of polymer chemistry, resulting in inconsistencies.

In the ideal SEC system, molecules are separated solely on the basis of their size; no interaction

occurs between solutes and the stationary phase. In the event that solute/support interactions do occur, the separation mode is termed nonideal SEC. The stationary phase in SEC consists of a column packing material that contains pores comparable in size to the molecules to be fractionated. Solutes too large to enter the pores travel with the mobile phase in the interstitial space (between particles) outside the pores. Thus, the largest molecules are eluted first from an SEC column. The volume of the mobile phase in the column, termed the column **void volume**, V_o, can be measured by chromatographing a very large (totally excluded) species, such as Blue Dextran, a dye of molecular weight (MW) two million.

As solute dimensions decrease, approaching those of the packing pores, molecules begin to diffuse into the packing particles and, consequently, are slowed down. Solutes of low MW (e.g., glycyltyrosine) that have free access to all the available pore volume are eluted in the volume referred to as V_t, the **total permeation volume** of the column. The V_t is equal to the column void volume, V_o, plus the volume of liquid inside the sorbent pores (**internal pore volume**), V_i ($V_t = V_o + V_i$). These relationships are illustrated in Fig. 12.8. Solutes are ideally eluted between the void volume and the total permeation volume of the column. Because this volume is limited, only a relatively small number of solutes (ca. 10), of a particular size range, can be completely resolved by SEC under ordinary conditions.

The behavior of a molecule in a size-exclusion column may be characterized in several different ways. Each solute exhibits an **elution volume**, V_e, as illustrated in Fig. 12.9. However, V_e depends on

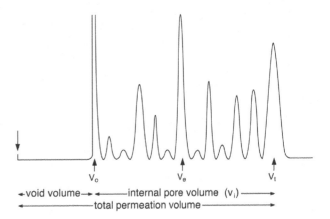

12.8 figure
Schematic elution profile illustrating some of the terms used in size-exclusion chromatography (Adapted from Lesellier and West [8], p. A271, with kind permission from Elsevier Science – NL, Sara Burgerhartstraat 25, 1055 KV Amsterdam, The Netherlands)

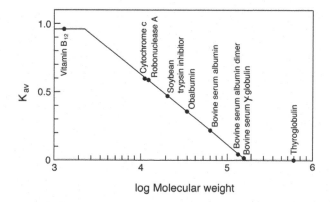

 figure

Relationship between K_{av} and log (molecular weight) for globular proteins chromatographed on a column of Sephadex G-150 Superfine (Reproduced by permission of Pharmacia Biotech, Inc., Piscataway, NJ)

column dimensions, the way in which the column was packed, and the sorbent pore size. The **available partition coefficient**, Kav, is used to define solute behavior independent of these variables:

$$Kav = (Ve - Vo)/(Vt - Vo) \qquad (12.3)$$

where:

Kav = available partition coefficient
Ve = elution volume of solute
Vo = column void volume
Vt = total permeation volume of column

The value of Kav calculated from experimental data for a solute chromatographed on a given SEC column defines the proportion of pores that can be occupied by that molecule. For a large, totally excluded species, such as Blue Dextran or DNA, Ve = Vo and Kav=0. For a small molecule with complete access to the internal pore volume, such as glycyltyrosine, Ve = Vt and Kav=1.

For each size-exclusion packing material, a plot of Kav vs. the logarithm of the MW for a series of solutes, similar in molecular shape and density, will give an S-shaped curve (Fig. 12.9). In the case of proteins, Kav is actually better related to the Stokes radius, the average radius of the protein in solution. The central, linear portion of this curve describes the **fractionation range** of the matrix, wherein maximum separation among solutes of similar MW is achieved. This correlation between solute elution behavior and MW (or size) forms the basis for a widely used method for characterizing large molecules such as proteins and polysaccharides. A size-exclusion column is calibrated with a series of solutes of known MW (or Stokes radius) to obtain a curve similar to that shown in Fig. 12.9. The value of Kav for the unknown is then determined, and

an estimate of MW (or size) of the unknown is made by interpolation of the calibration curve.

Column packing materials for SEC can be divided into two groups: semirigid, hydrophobic media and soft, hydrophilic gels. The former are usually derived from polystyrene and are used with organic mobile phases (GPC or nonaqueous SEC) for the separation of polymers, such as rubbers and plastics. Soft gels, polysaccharide-based packings, are typified by Sephadex, a cross-linked dextran (see Fig. 12.6a). These materials are available in a wide range of pore sizes and are useful for the separation of water-soluble substances in the MW range $1–2.5 \times 10^7$ Da. In selecting an SEC column packing, both the purpose of the experiment and size of the molecules to be separated must be considered. If the purpose of the experiment is group separation, for which molecules of widely different molecular sizes need to be separated, a matrix is chosen such that the larger molecules, e.g., proteins, are eluted in the void volume of the column, whereas small molecules are retained in the total volume. When SEC is used for separation of macromolecules of different sizes, the molecular sizes of all the components must fall within the fractionation range of the gel.

As discussed previously, SEC can be used, directly, to fractionate mixtures, to purify, or, indirectly, to obtain MW/size information about a dissolved species. In addition to MW estimations, SEC is used to determine the MW distribution of natural and synthetic polymers, such as dextrans and gelatin preparations. Fractionation of biopolymer mixtures is probably the most widespread use of SEC, since the mild elution conditions employed rarely cause denaturation or degradation. Often times SEC is used as an early chromatographic separation step in a multidimensional chromatographic approach toward purification. It is also a fast, efficient alternative to dialysis for desalting solutions of macromolecules, such as proteins.

12.5 ANALYSIS OF CHROMATOGRAPHIC PEAKS

Once the chromatographic technique (Sect. 12.3) and chromatographic mechanism (Sect. 12.4) have been chosen, the analyst has to ensure adequate separation of constituents of interests from a mixture, in a reasonable amount of time. After separation is achieved and chromatographic peaks are obtained, qualitative as well as quantitative analysis can be carried out. Basic principles of separation and resolution will be discussed in the subsequent sections. Understanding these principles allows the analyst to optimize separation and perform qualitative and quantitative analysis.

12.5.1 Separation and Resolution

This section will discuss separation and resolution as it pertains mainly to LC; separation and resolution optimization as it pertains specifically to GC will be discussed in Chap. 14.

12.5.1.1 *Developing a Separation*

There may be numerous ways to accomplish a chromatographic separation for a particular compound. In many cases, the analyst will follow a standard laboratory procedure or published methods. In the case of a sample that has not been previously analyzed, the analyst begins by evaluating what is known about the sample and defines the goals of the separation. How many components need to be resolved? What degree of resolution is needed? Is qualitative or quantitative information needed? Molecular weight (or MW range)/size, polarity, and ionic character of the sample will guide the choice of chromatographic separation mechanism (**separation mode**). Figure 12.10 shows that more than one correct choice may be possible. For example, small ionic compounds may be separated by ion-exchange, ion-pair reversed-phase

(see Sect. 13.3.2.1), or reversed-phase LC. In this case, the analyst's choice may be based on convenience, experience, and personal preference.

Having chosen a separation mode for the sample at hand, one must select an appropriate stationary phase, elution conditions, and a detection method. Trial experimental conditions may be based on the results of a literature search, the analyst's previous experience with similar samples, or the general recommendations from chromatography experts.

To achieve separation of sample components by all modes except SEC, one may utilize either isocratic or gradient elution. **Isocratic elution** is the simplest technique, in which solvent composition and flow rate are held constant throughout the run. **Gradient elution** involves reproducibly varying mobile phase composition or flow rate (flow programming) during the LC analysis. Gradient elution is used when sample components vary a lot in inherent characteristics such as polarity and/or charge density, so that an isocratic mobile phase does not elute all components within a reasonable time. The change may be continuous or stepwise. Gradients of increasing ionic strength, or

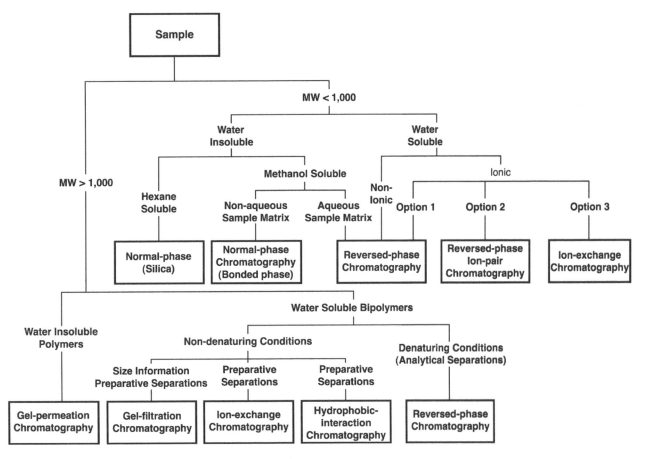

12.10 A schematic diagram for choosing a chromatographic separation mode based on sample molecular weight and

figure solubility (From Lough and Wainer [29], used with permission)

changing pH, are extremely valuable in ion-exchange chromatography (see Sect. 12.4.4), whereas gradients of increasing or decreasing polarity are valuable in normal or reversed-phase chromatography, respectively (Sect. 12.4.2). Increasing the "strength" of the mobile phase (Sect. 12.4.1), either gradually (continuously) or in a stepwise fashion, shortens the analysis time.

Method development may begin with an isocratic mobile phase, possibly of intermediate solvent strength; however, using gradient elution for the initial separation may ensure that some level of separation is achieved within a reasonable time period and nothing is likely to remain on the column. Data from this initial run allows one to determine if isocratic or gradient elution is needed, and to estimate optimal isocratic mobile phase composition or gradient range.

Once an initial separation has been achieved, the analyst can proceed to optimize resolution (separation of closely related analytes). This generally involves changing mobile phase variables, including the following: (1) nature and percentage of organic solvents (and type of organic solvents), (2) pH, (3) ionic strength, (4) nature and concentration of additives (such as ion-pairing agents), (5) flow rate, and (6) temperature. In the case of gradient elution, gradient steepness (slope, i.e., rate of change) is another variable to be optimized. However, the analyst must be aware of the principles of chromatographic resolution as will be discussed in the following section.

12.5.1.2 Chromatographic Resolution

12.5.1.2.1 Introduction

The main goal of chromatography is to segregate components of a sample into separate bands or peaks as they migrate through the column. A **chromatographic peak** is defined by several parameters including **retention time** (Fig. 12.11), **peak width**, and **peak height** (Fig. 12.12). The volume of the mobile phase required to elute a compound from an LC column is called the **retention volume**, V_R. The associated time is the **retention time**, t_R. Shifts in retention time and changes in peak width greatly influence **chromatographic resolution**.

Differences in column dimensions, loading, temperature, mobile phase flow rate, system dead volume, and detector geometry may lead to discrepancies in retention time. By subtracting the time required for the mobile phase or a non-retained solute (t_m or t_o) to travel through the column to the detector, one obtains an **adjusted retention time**, t'_R (or volume), as depicted in Fig. 12.11. The adjusted retention time (or volume) corrects for differences in system dead volume and signifies the time the analyte spends adsorbed on the stationary phase.

The **resolution** of two peaks from each other is related to the **separation factor**, α. Values for α (Fig. 12.11)

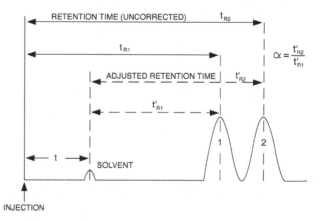

12.11 figure Measurement of chromatographic retention (Adapted from Johnson and Stevenson [5], with permission)

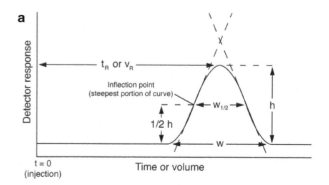

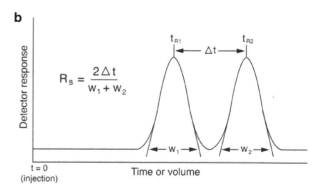

12.12 figure Measurement of peak width and its contribution to resolution. (**a**) Idealized Gaussian chromatogram, illustrating the measurement of w and $w_{1/2}$; (**b**) the resolution of two bands is a function of both their relative retentions and peak widths (Adapted from Johnson and Stevenson [5], with permission)

depend on temperature, flow rate, stationary phase, and mobile phase used. Resolution is defined as follows:

$$R_s = \frac{2\Delta t}{w_2 + w_1} \qquad (12.4)$$

where:

R_s = resolution
Δt = difference between retention times of peaks 1 and 2
w_2 = width of peak 2 at baseline
w_1 = width of peak 1 at baseline

Figure 12.12 illustrates the measurement of peak width [part (a)] and the values necessary for calculating resolution [part (b)]. (Retention and peak or band width must be expressed in the same units, i.e., time or volume).

Chromatographic resolution is a function of column **efficiency**, **selectivity**, and the **capacity factor**. Mathematically, this relationship is expressed as follows:

$$R_s = \underbrace{\frac{1}{4}\sqrt{N}}_{a}\underbrace{\left(\frac{\alpha - 1}{\alpha}\right)}_{b}\underbrace{\left(\frac{k'}{k'+1}\right)}_{c} \qquad (12.5)$$

where:

a = column **efficiency** term
b = column **selectivity** term
c = **capacity** term

These terms, and factors that contribute to them, will be discussed in the following sections.

12.5.1.2.2 Column Efficiency

If faced with the problem of improving resolution, a chromatographer should first examine the **efficiency** of the column. An efficient column keeps the solutes from spreading widely on the column and hence gives narrow peaks. Column efficiency can be calculated by:

$$N = \left(\frac{t_R}{\sigma}\right)^2 = 16\left(\frac{t_R}{w}\right)^2 = 5.5\left(\frac{t_R}{w\frac{1}{2}}\right)^2 \qquad (12.6)$$

where:

N = number of theoretical plates
t_R = retention time
σ = standard deviation for a Gaussian peak
w = peak width at baseline ($w = 4\sigma$)
$w_{\frac{1}{2}}$ = peak width at half height

The measurement of t_R, w, and $w_{\frac{1}{2}}$ is illustrated in Fig. 12.12. (Retention volume may be used instead of t_R; in this case, peak width is also measured in units of volume.) Although some peaks are not actually Gaussian in shape, normal practice is to treat them as if they were. In the case of peaks that are incompletely resolved or slightly asymmetric, peak width at half height is more accurate than peak width at baseline.

The value N calculated from the above equation is called the number of **theoretical plates**. The theoretical plate concept, borrowed from distillation theory, can best be understood by viewing chromatography as a series of equilibrations of solutes between mobile and stationary phases, analogous to countercurrent distribution. Thus, a column would consist of N segments (theoretical plates) with one equilibration occurring in each. As a first approximation, N is independent of retention time and is therefore a useful measure of column performance. One method of monitoring column performance over time is to chromatograph a standard compound periodically, under constant conditions, and to compare the values of N obtained. It is important to note that columns often behave as if they have a different number of plates for different solutes in a mixture. Different solutes have different partition coefficient and thus have distinctive series of equilibrations between mobile and stationary phases. Peak broadening due to column deterioration will result in a decrease of N for a particular solute. Peak broadening is a result of an extended time for a solute to reach equilibrium between mobile and stationary phases (causing them to spread over a wider area on the column).

The number of theoretical plates is generally proportional to column length. The longer the column, the higher the number of theoretical plates, and hence the better is the resolution. Because columns are available in various lengths, it is useful to have a measure of column efficiency that is independent of column length. This may be expressed as:

$$HETP = \frac{L}{N} \qquad (12.7)$$

where:

HETP = height equivalent to a theoretical plate
L = column length
N = number of theoretical plates

The so-called **HETP, height equivalent to a theoretical plate**, is sometimes more simply described as **plate height** (H). If a column consisted of discrete segments, HETP would be the height of each imaginary segment. Small plate height values (a large number of plates) indicate good efficiency of separation (minimal spread of solute within the column, resulting in sharp peaks). Conversely, reduced number of plates results in poor separation due to the extended equilibrium time (spread of solutes on the column, resulting in wide peaks) in a deteriorating column.

In reality, columns are not divided into discrete segments and equilibration is not infinitely fast. The plate theory is used to simplify the equilibration concept. The movement of solutes through a chromatography column takes into account the finite rate at which a solute can equilibrate itself between stationary and mobile phases. Thus, peak shape depends on the rate of elution and is affected by solute diffusion. Any mechanism that causes a peak of solute to broaden will increase HETP and decrease column efficiency. The various factors that contribute to plate height are expressed by the **Van Deemter equation**:

$$\text{HETP} = A + \frac{B}{u} + Cu \qquad (12.8)$$

where:

HETP = height equivalent to a theoretical plate
A, B, C = constants
u = mobile phase rate

The constants A, B, and C are characteristic for a given column, mobile phase, and temperature. The A term represents the **eddy diffusion** or multiple flow-paths. Eddy diffusion refers to the different microscopic flowstreams that the mobile phase can take between particles in the column (analogous to eddy streams around rocks in a brook). Sample molecules can thus take different paths as well, depending on which flowstreams they follow. As a result, solute molecules spread from an initially narrow band to a broader area within the column. The larger is the particle size of the packing material, the more paths a solute might take. Eddy diffusion may be minimized by good column packing techniques and the use of small diameter particles of narrow particle size distribution.

The B term of the Van Deemter equation, sometimes called the **longitudinal diffusion** term, exists because all solutes diffuse from an area of high concentration (the center of a chromatographic peak) to one of low concentration (the leading or trailing edge of a chromatographic peak). In LC, the contribution of this term to HETP is small except at low flow rate of the mobile phase. With slow flow rates there will be more time for a solute to spend on the column; thus, its diffusion will be greater.

The C (mass transfer) term arises from the finite time required for solute to equilibrate between the mobile and stationary phases. **Mass transfer** is practically the partitioning of the solute into the stationary phase, which does not occur instantaneously and depends on the solute's partition and diffusion coefficients. If the stationary phase consists of porous particles (see Chap. 13, Sect. 13.2.3.2, Fig. 13.3), a sample molecule entering a pore ceases to be transported by the solvent flow and moves by diffusion only. Subsequently, this solute molecule may diffuse back to the mobile phase flow or it may interact with the stationary phase. In either case, solute molecules inside the pores are slowed down relative to those outside the pores, and peak broadening occurs. Contributions to HETP from the C term can be minimized by using porous particles of small diameter or pellicular packing materials (Chap. 13, Sect. 13.2.3.2.2). Also, using a narrower column with a smaller inner diameter reduces the C value, given

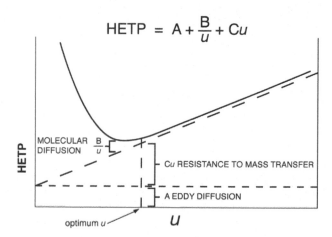

12.13 figure Van Deemter plot of column efficiency (HETP) vs. mobile phase rate (u). Optimum u is noted (Courtesy of Hewlett-Packard Co., Analytical Customer Training, Atlanta, GA)

that equilibrium time will be reduced since there is less packing material.

As expressed by the Van Deemter equation, **mobile phase flow rate**, u, contributes to plate height in opposing ways – increasing the flow rate increases the equilibration point (Cu), but decreases longitudinal diffusion of the solute particles (B/u). A Van Deemter plot (Fig. 12.13) may be used to determine the mobile phase flow rate at which plate height is minimized and column efficiency is maximized. Flow rates above the optimum may be used to decrease analysis time if adequate resolution is still obtained. However, at very high flow rates, there will be less time to approach equilibrium, which will lead to a shorter retention time and co-elution of closely related solutes.

In addition to flow rate, temperature can affect the longitudinal diffusion and the mass transfer. Increasing the temperature causes enhanced movement of the solute between the mobile phase and the stationary phase and within the column, thus leading to faster elution and narrower peaks.

12.5.1.2.3 Column Selectivity

Chromatographic resolution depends on column selectivity as well as efficiency. **Column selectivity** refers to the distance, or relative separation, between two peaks and is given by:

$$\alpha = \frac{t_{R2} - t_o}{t_{R1} - t_o} = \frac{t'_{R2}}{t'_{R1}} = \frac{K_2}{K_1} \qquad (12.9)$$

where:

α = separation factor
t_{R1} and t_{R2} = retention times of components 1 and 2, respectively
t_o (or t_m) = retention time of unretained components (solvent front)

t'_{R1} and t'_{R2} = adjusted retention times of components 1 and 2, respectively

K_1 and K_2 = distribution coefficients of components 1 and 2, respectively

Retention times (or volumes) are measured as shown in Fig. 12.11. The time, t_o, can be measured by chromatographing a solute that is not retained under the separation conditions (i.e., travels with the solvent front). When this parameter is expressed in units of volume, V_o or V_m, it is known as the **dead volume** of the system. Selectivity is a function of the stationary and/or mobile phase. For example, selectivity in ion-exchange chromatography is influenced by the nature and number of ionic groups on the matrix but also can be affected by pH and ionic strength of the mobile phase. Changing the pH or the ionic strength of the mobile phase will impact the extent of interaction of the solutes with the column. Selectivity can be either static as in isocratic elution (the choice of the mobile phase from the beginning impact selectivity) or dynamic as in gradient elution (selectivity changing over time). Good selectivity is probably more important to a given separation than high efficiency (Fig. 12.14), since resolution is directly related to selectivity but is quadratically related to efficiency; thus, a fourfold increase in N is needed to double R_s (Eq. 12.5).

12.5.1.2.4 Column Capacity Factor

The **capacity** or **retention factor**, k', is a measure of the amount of time a chromatographed species (solute) spends in/on the stationary phase relative to the mobile phase. The relationship between capacity factors and chromatographic retention (which may be expressed in units of either volume or time) is shown below:

$$k' = \frac{KV_s}{V_m} = \frac{V_R - V_m}{V_m} = \frac{t_R - t_o}{t_o} \qquad (12.10)$$

where:

k' = capacity factor
K = distribution coefficient of the solute
V_s = volume of stationary phase in column
V_m = volume of mobile phase
V_R = retention volume of solute
t_R = retention time of solute
t_o = retention time of unretained components (solvent front)

Small values of k' indicate little retention, and components will be eluted close to the solvent front, resulting in poor separations. Overuse or misuse of the column may lead to the loss of some functional groups, thus resulting in small k' values. Large values of k' result in improved separation but also can lead to broad peaks and long analysis times. On a practical basis, k' values within the range of 1–15 are generally used. (In the equation for R_s, k' is actually the average of k_1' and k_2' for the two components separated.)

12.5.2 Qualitative Analysis

Once separation and resolution have been optimized, identification of the detected compounds can be achieved. (Various detection methods are outlined in Chaps. 13 and 14). Comparing V_R or t_R to that of standards chromatographed under identical conditions often enables one to identify an unknown compound. When it is necessary to compare chromatograms obtained from two different systems or slightly different columns, it is better to compare **adjusted retention time**, t'_R (see Sect. 12.5.1.2). Oftentimes comparison across different systems and columns is not possible. Different compounds may have identical retention times using the same system. In other words, even if the retention time of an unknown and a standard are equivalent, the two compounds might not be identical. Therefore, other techniques are needed to confirm peak identity. For example:

1. Spike the unknown sample with a known compound and compare chromatograms of the original and spiked samples to see which peak has increased. Only the height of the peak of interest should increase, with no change in retention time, peak width, or shape.

2. A diode array detector can provide absorption spectra of designated peaks (see Sects. 13.2.6 and 13.2.4.1). Although identical spectra do not prove identity, a spectral difference confirms

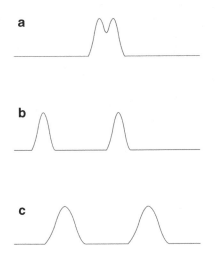

figure Chromatographic resolution: efficiency vs. selectivity. (**a**) Poor resolution; (**b**) good resolution due to high column efficiency; (**c**) good resolution due to column selectivity (From Johnson and Stevenson [5], used with permission) that sample and standard peaks are different compounds.

3. In the absence of spectral scanning capability, other detectors, such as absorption or fluorescence, may be used in a ratioing procedure. Chromatograms of sample and standard are monitored at each of two different wavelengths. The ratio of peak areas at these wavelengths should be the same if sample and standard are identical.

4. Peaks of interest can be collected and subjected to additional chromatographic separation using a different separation mode.

5. Collect the peak(s) of interest and establish their identity by another analytical method (e.g., mass spectrometry, which can give a mass spectrum that is characteristic of a particular compound; see Chap. 11).

12.5.3 Quantitative Analysis

Assuming that good chromatographic resolution and identification of sample components have been achieved, quantification involves mostly measuring peak area and comparing the data with those for standards of known concentration. Nowadays all chromatography systems use data analysis software that recognizes the start, maximum, and end of each chromatographic peak, even when not fully resolved from other peaks. These values then are used to determine retention times, peak height, and peak areas. At the end of each run, a report is generated that lists these data and post-run calculations, such as relative peak areas, areas as percentages of the total area, and relative retention times. If the system has been standardized, data from external or internal standards can be used to calculate analyte concentrations (discussed below). In low-pressure, preparative chromatography, post-chromatography analysis of collected fractions is sometimes used to identify samples eluted. Examples of post-chromatography analysis include the BCA (bicinchoninic acid) protein assay (Chap. 18, Sect. 18.4.2.3) and the phenol-sulfuric acid assay for carbohydrate (Chap. 19, Sect. 19.3). After obtaining the absorbance reading on a spectrophotometer for such assays, the results are plotted as fraction number on the x-axis and absorbance on the y-axis to determine which fractions contain protein and/or carbohydrate.

Having determined peak areas, one must compare these data with appropriate standards of known concentration to determine sample concentrations. Comparisons may be by means of **external** or **internal** standards. Comparison of peak area of unknown samples with standards injected separately (i.e., **external standards**) is common practice. Standard solutions covering the desired concentration range (preferably diluted from one stock solution) are chromatographed,

and the peak area is plotted versus concentration to obtain a standard curve. An identical volume of sample is then chromatographed, and an area of the sample peak is used to determine sample concentration via the standard curve (Fig. 12.15a). This absolute calibration method requires precise analytical technique and requires that detector sensitivity be constant from day to day if the calibration curve is to remain valid.

The use of an **internal standard** can minimize errors due to losses incurred during sample preparation, apparatus noise (inherent apparatus error), and systematic errors in the operator's technique (including the volume of sample injected). In other words, the internal standard, upon addition to the sample, will be subjected to the same sequence of events that may introduce changes to the actual content of the compounds of interest and the standard. In this technique, compared to the compounds of interest in the sample, the internal standard compound must: (1) be chemically/structurally related to the compounds of interest, (2) elute at a different time, and (3) not be naturally present in the sample. Basically, the amount of each component in the sample is determined by comparing the area of that component peak to the area of the internal standard peak. However, variation in detector response between compounds of different chemical

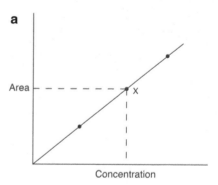

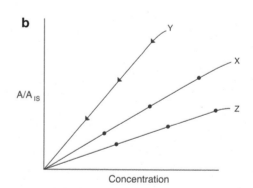

 Calibration curves for quantification of a sample component, x. (**a**) External standard technique; (**b**) internal standard technique (Adapted from Johnson and Stevenson [5], with permission)

structure must be taken into account. One way to do this is by first preparing a set of standard solutions containing varying concentrations of the compound(s) of interest. Each of these solutions is made to contain a known and constant amount of the internal standard. These standard solutions are chromatographed, and peak area is measured. Ratios of peak area (compound of interest/internal standard) are calculated and plotted against concentration to obtain calibration curves such as those shown in Fig. 12.15b. A separate response curve must be plotted for each sample component to be quantified. Next, a known amount of internal standard is added to the unknown sample, and the sample is chromatographed. Peak area ratios (compound of interest/internal standard) are calculated and used to calculate the concentration of each relevant component using the appropriate calibration curve. The advantages of using internal standards are that injection volumes need not be accurately measured and the detector response need not remain constant since any change will not alter ratios.

Standards should be included during each analytical session, since detector response may vary from day to day. Analyte recovery should be checked periodically. This involves addition of a known quantity of standard to a sample (usually before extraction) and determination of how much is recovered during subsequent analysis. During routine analyses, it is highly desirable to include a control or check sample, a material of known composition. This material is analyzed parallel to unknown samples. When the concentration of analyte measured in the control falls outside an acceptable range, data from other samples analyzed during the same period should be considered suspect. Carefully analyzed food samples and other substances are available from the National Institute of Standards and Technology (formerly the National Bureau of Standards) for use in this manner.

12.6 SUMMARY

Chromatography is a separation method based on the partitioning of a solute between a mobile phase and a stationary phase. The mobile phase may be liquid, gas, or a supercritical fluid. The stationary phase may be an immobilized liquid or a solid, in either a planar or column form. Based on the physicochemical characteristics of the analyte, and the availability of instrumentation, a chromatographic system is chosen to separate, identify, and quantify the analyte. Chromatographic modes include adsorption, partition, hydrophobic interaction, ion exchange, affinity, and size-exclusion chromatography. Factors to be considered when developing a separation include mobile phase variables (polarity, ionic strength, pH, temperature, and/or flow

rate) and column efficiency, selectivity, and capacity. Following detection, a chromatogram provides both qualitative and quantitative information via retention time and peak area data.

For an introduction to the techniques of HPLC and GC, the reader is referred again to Chaps. 13 and 14 in this text or to the excellent 5th edition of *Quantitative Chemical Analysis* by D.C. Harris [2]. The book by R.M. Smith [3] also contains information on basic concepts of chromatography and chapters devoted to TLC, LC, and HPLC, as well as an extensive discussion of GC. References [14–16] contain a wealth of information on TLC, and references [8] and [9] cover SFC. *Chromatography* [1], the standard work edited by E. Heftmann (2004 and earlier editions), is an excellent source of information on both fundamentals (Part A) and applications (Part B) of chromatography. Part B includes chapters on the chromatographic analysis of amino acids, proteins, lipids, carbohydrates, and phenolic compounds. In addition, *Fundamental and Applications Reviews* published (in alternating years) by the journal *Analytical Chemistry* relate new developments in all branches of chromatography, as well as their application to specific areas, such as food. Recent books and general review papers are referenced, along with research articles published during the specified review period.

12.7 STUDY QUESTIONS

1. Explain the principle of countercurrent extraction and how it developed into partition chromatography.
2. For each set of two (or three) terms used in chromatography, give a brief explanation as indicated to distinguish between the terms:

   (a) Adsorption vs. partition chromatography

	Adsorption	Partition
Nature of stationary phase		
Nature of mobile phase		
How solute interacts with the phases		

   (b) Normal-phase vs. reversed-phase chromatography

	Normal-phase chromatography	Reversed-phase chromatography
Nature of stationary phase		
Nature of mobile phase		
What elutes last		

(c) Cation vs. anion exchangers

	Cation exchanger	Anion exchanger
Charge on column		
Nature of compounds bound		

(d) Internal standards vs. external standards

	Nature of stds.	How stdsq. are handled in relation to samples	What is plotted on std. curve
Internal standard			
External standard			

(e) TLC vs. column liquid chromatography

	Thin-layer chromatography	Column liquid chromatography
Nature and location of stationary phase		
Nature and location of mobile phase		
How sample is applied		
Identification of solutes separated		

   (f) HETP vs. N vs. L (from the equation HETP = L/N)

3. State the advantages of TLC as compared to paper chromatography.
4. State the advantages of column liquid chromatography as compared to planar chromatography.
5. Explain how SFC differs from LC and GC, including the advantages of SFC.
6. What is the advantage of bonded supports over coated supports for partition chromatography?
7. You are performing LC using a stationary phase that contains a polar nonionic functional group. What type of chromatography is this, and what could you do to increase the retention time of an analyte?
8. You applied a mixture of proteins, in a buffer at pH 8.0, to an anion-exchange column. On the basis of some assays you performed, you know that the protein of interest adsorbed to the column:

   (a) Does the anion-exchange stationary phase have a positive or negative charge?

   (b) What is the overall charge of the protein of interest that adsorbed to the stationary phase?
   (c) Is the isoelectric point of the protein of interest (adsorbed to the column) higher or lower than pH 8.0?
   (d) What are the two most common methods you could use to elute the protein of interest from the anion-exchange column? Explain how each method works. (See also Chap. 24).

9. Would you use a polystyrene- or a polysaccharide-based stationary phase for work with proteins? Explain your answer.
10. Explain the principle of affinity chromatography, why a spacer arm is used, and how the solute can be eluted.
11. Explain how you would use SEC to estimate the molecular weight of a protein molecule. Include an explanation of what information must be collected and how it is used.
12. What is gradient elution from a column, and why is it often advantageous over isocratic elution?
13. A sample containing compounds A, B, and C is analyzed via LC using a column packed with a silica-based C_{18} bonded phase. A 1:5 solution of ethanol and H_2O was used as the mobile phase. The following chromatogram was obtained:

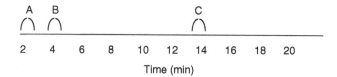

Assuming that the separation of compounds is based on their polarity:

   (a) Is this normal- or reversed-phase chromatography? Explain your answer.
   (b) Which compound is the most polar?
   (c) How would you change the mobile phase so compound C would elute sooner, without changing the relative positions of compounds A and B? Explain why this would work.
   (d) What could possibly happen if you maintained an isocratic elution mode at low solvent strength?

14. Using the Van Deemter equation, HETP, and N, as appropriate, explain why the following changes may increase the efficiency of separation in column chromatography:

(a) Changing the flow rate of the mobile phase
(b) Increasing the length of the column
(c) Reducing the inner diameter of the column
(d) Decreasing the particle size of the packing material

15. State the factors and conditions that lead to poor resolution of two peaks.
16. How can chromatographic data be used to quantify sample components?
17. Why would you choose to use an internal standard rather than an external standard? Describe how you would select an internal standard for use.
18. To describe how using internal standards works, answer the following questions:

    (a) What specifically will you do with the standards?
    (b) What do you actually measure and plot?
    (c) How do you use the plot?

REFERENCES

1. Heftmann E (ed) (2004) Chromatography, 6th edn. Fundamentals and applications of chromatography and related differential migration methods. Part A: fundamentals and techniques. Part B: applications. J Chromatog Library Ser vols 69A and 69B. Elsevier, Amsterdam
2. Harris DC (2016) Quantitative chemical analysis, 9th edn. W.H. Freeman, New York
3. Smith RM (1988) Gas and liquid chromatography in analytical chemistry, Wiley, Chichester, England
4. Fanali S, Haddad PR, Poole C, Schoenmakers P, Lloyd DK (2013) Liquid chromatography: fundamentals and instrumentation. Elsevier, Amsterdam.
5. Johnson EL, Stevenson R (1978) Basic liquid chromatography. Varian Associates, Palo Alto, CA
6. Craig LC (1943) Identification of small amounts of organic compounds by distribution studies. Application to Atabrine. J Biol Chem 150: 33–45
7. Bernal J, Martín M, Toribio L (2013) Supercritical fluid chromatography in food analysis. J Chromatogr A 1313:24–36
8. Lesellier E, West C (2015) The many faces of packed column supercritical fluid chromatography – A critical review. J Chromatogr A, 1382: 2–46
9. Caude M, Thiébaut D (1999) Practical supercritical fluid chromatography and extraction. Harwood Academic, Amsterdam
10. Radke W (2014) Polymer separations by liquid interaction chromatography: Principles – prospects – limitations. J Chromatogr A, 1335:62–79
11. Bernal J, Ares A, Pól J, Wiedmer S (2011) Hydrophilic interaction liquid chromatography in food analysis. J Chromatogr A 1218:7438–7452
12. Núñez O, Gallart-Ayala H, Martins C, Lucci P (2015) Fast Liquid Chromatography-Mass Spectrometry Methods in Food Analysis. Imperial College Press, London
13. Cordero C, Kiefl J, Schieberle P (2015) Comprehensive two-dimensional gas chromatography and food sensory properties: potential and challenges. Anal Bioanal Chem, 407: 169–191
14. Fried B, Sherma J (1999) Thin-layer chromatography, 4th edn. Marcel Dekker, New York
15. Hahn-Deinstrop E (2007) Applied thin-layer chromatography: best practice and avoidance of mistakes, 2nd edn. Wiley-VCH, Weinheim, Germany
16. Wall PE (2005) Thin-layer chromatography: a modern practical approach. The Royal Society of Chemistry, Cambridge, UK
17. Fuchs B, Süß R, Teuber K, Eibisch M, Schiller J (2011) Lipid analysis by thin-layer chromatography-A review of the current state. J Chromatogr A 1218:2754–2774
18. Morlock G, Meyer S, Zimmermann B, Roussel J (2014) High-performance thin-layer chromatography analysis of steviol glycosides in Stevia formulations and sugar-free food products, and benchmarking with (ultra) high-performance liquid chromatography. J Chromatogr A 1350:102–111
19. Touchstone JC (1992) Practice of thin layer chromatography. Wiley, New York
20. Chester TL, Pinkston JD, Raynie DE (1996) Supercritical fluid chromatography and extraction (fundamental review). Anal Chem 68: 487R–514R
21. Snyder LR, Kirkland JJ (eds) (1979) Introduction to modern liquid chromatography, 2nd ed. Wiley, New York
22. Tomaz CT, Queiroz JA (2013) Hydrophobic interaction chromatography, In Liquid chromatography: fundamentals and instrumentation, Eds. Fanali S, Haddad PR, Poole, Schoenmakers P, Lloyd DK. Elsevier, Amsterdam, p 122–141.
23. Weiss J (2004) Handbook of ion chromatography, 3rd edn. Wiley-VCH Verlag GmbH & Co. KGaA, Weinheim
24. Wacoo AP, Wendiro D, Vuzi, PC, Hawumba JF (2014) Methods for detection of aflatoxins in agricultural food crops. Applied Chem 2014, 1–15.
25. Rollag JG, Beck-Westermeyer M, Hage DS (1996) Analysis of pesticide degradation products by tandem high-performance immunoaffinity chromatography and reversed-phase liquid chromatography. Anal Chem 68,, 3631–3637.
26. Walters RR (1985) Report on affinity chromatography. Anal Chem 57: 1099A–1113A
27. Zachariou M (2008) Affinity chromatography: methods and protocols. Humana, Totowa, NJ
28. Scopes RK (1994) Protein purification: principles and practice, 3rd edn. Springer-Verlag, New York
29. Lough WJ, Wainer IW (eds) (1995) High performance liquid chromatography: fundamental principles and practice. Blackie Academic & Professional, Glasgow, Scotland

High-Performance Liquid Chromatography

Bradley L. Reuhs

Department of Food Science,
Purdue University,
West Lafayette, IN 47907-2009, USA
e-mail: breuhs@purdue.edu

S. Nielsen (ed.), *Food Analysis*, Food Science Text Series,
DOI 10.1007/978-3-319-45776-5_13, © Springer International Publishing 2017

13.1 INTRODUCTION

High-performance liquid chromatography (HPLC) was developed during the 1960s as a direct offshoot of classic column liquid chromatography through improvements in the technology of columns and instrumental components (pumps, injection valves, and detectors). Originally, HPLC was the acronym for *high-pressure liquid chromatography*, reflecting the relatively high operating pressures generated by early columns. By the late 1970s, however, *high-performance liquid chromatography* had become the preferred term, emphasizing the effective separations achieved. In fact, newer columns and packing materials offer high performance at moderate pressure (although much higher than gravity-flow systems). HPLC can be applied to the analysis of any compound with solubility in a liquid that can be used as the mobile phase. Although most often employed as an **analytical** technique, HPLC also may be used in **preparative** applications. There are many *advantages* of HPLC over traditional low-pressure column liquid chromatography:

1. Speed (many analyses can be accomplished in 30 min or less)
2. A wide variety of stationary phases
3. Improved resolution
4. Greater sensitivity (many different detectors can be employed)
5. Easy sample recovery (less eluent volume to remove)

Application of HPLC to the analysis of food began in the late 1960s, and its use increased with the development of column packing materials that would separate sugars. Using HPLC to analyze sugars was justified economically as a result of sugar price increases in the mid-1970s, which motivated soft drink manufacturers to substitute high-fructose corn syrup for sugar. Monitoring sweetener content by HPLC assured a good quality product. Other early food applications included the analysis of pesticide residues in fruits and vegetables, organic acids, lipids, amino acids, toxins (such as aflatoxins in peanuts), and vitamins [1]. HPLC continues to be applied to these, and many more, food-related analyses today [2–4].

Driven by the modern market for drug development and analysis, microbore and nanobore HPLC technology has improved drastically in the last few years. Along with improved detectors and coupled mass spectrometry (MS) detection systems, these HPLC systems have had a significant impact on the use of HPLC in the food and drug industries.

13.2 COMPONENTS OF AN HPLC SYSTEM

A schematic diagram of a basic HPLC system is shown in Fig. 13.1. The main components of this system—**pump, injector, column, detector,** and **data system**—are discussed briefly in the sections below. Also important are the mobile phase (**eluent**) reservoirs and a fraction collector, the latter of which is used if further analysis of separated components is needed. Connecting tubing, tube fittings, and the materials out of which components are constructed also influence system performance and lifetime. References [1, 5–9] include detailed discussions of HPLC equipment, with the book by Bidlingmeyer [1] especially appropriate for beginners. The unique organization of reference [7] is intended for those who may need to learn chromatography quickly in an industrial environment. Two useful books on HPLC troubleshooting are those written by Gertz [10] and Snyder et al. [11]. In addition, much information on HPLC equipment, hardware, and troubleshooting hints may be found in publications such as *LC•GC, American Laboratory, Chemical & Engineering News*, and similar periodicals. Manufacturers are also a source of practical information on HPLC instrumentation and columns/stationary phase material.

13.2.1 Pump

The **HPLC pump** delivers the mobile phase through the system, typically at a flow rate of 0.4–1 mL/min, in a controlled, accurate, and precise manner. The majority of pumps currently used in HPLC (>90 %) are reciprocating, piston-type pumps. The dual piston pump systems with ball check valves are the most efficient pumps available. One disadvantage of reciprocating pumps is that they produce a pulsating flow, requiring the addition of pulse dampers to suppress fluctuations. A mechanical **pulse damper** or **dampener** consists of a device (such as a deformable metal component or tubing filled with compressible liquid) that can change its volume in response to changes in pressure.

Gradient elution systems are used to vary the mobile phase concentration during the run, by mixing mobile phase from two or more reservoirs. This is accomplished with low-pressure mixing, in which mobile phase components are mixed before entering the high-pressure pump, or high-pressure mixing, in which two or more independent, programmable pumps are used. For low-pressure gradient systems, a computer-controlled proportioning valve, followed by a mixing chamber at the inlet to the pumps is used, which results in extremely accurate and reproducible gradients. Gradient HPLC is extremely important for the effective elution of all components of a sample and for optimal

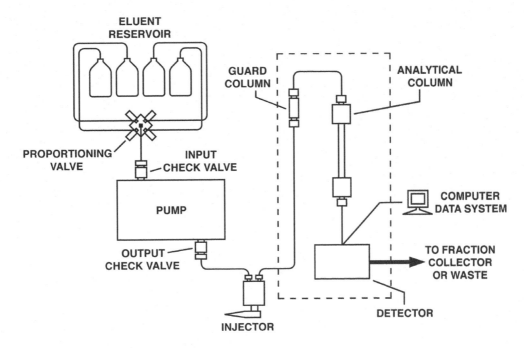

 Schematic representation of a system for high-performance liquid chromatography (not drawn to scale). Column(s) and detector may be thermostated, as indicated by the *dashed line*, for operation at elevated temperature

resolution. It is routinely applied to all modes of HPLC except size-exclusion chromatography.

Many commercially available HPLC pumping systems and connecting lines are made of high-grade ANSI 316 stainless steel, which can withstand high pressure, and is resistant to corrosion by oxidizing agents, acids, bases, and organic solvents (although mineral acids and halide ions will damage stainless steel). In other systems, all components that come into contact with the eluent are made of sturdy, inert polymers, and even employ sapphire pistons, which are resistant to extreme pH and high salt concentration. The latter systems can be used for all applications except normal phase, which uses organic solvents as the mobile phase. The polymer-based systems have facilitated a wider application of ion-exchange HPLC.

All HPLC pumps contain moving parts such as check valves and pistons and are quite sensitive to dust and particulate matter in the eluent. Therefore, it is advisable to filter the mobile phase using 0.45- or 0.22-μm filters prior to use. Degassing HPLC eluents, by the application of a vacuum or by sparging with helium, also is recommended to prevent the problems caused by air bubbles in a pump or detector.

13.2.2 Injector

The role of the injector is to place the sample into the flowing mobile phase for introduction onto the column.

Virtually all HPLC systems use **valve injectors**, which separate sample introduction from the high-pressure eluent system. With the injection valve in the *load* position (Fig. 13.2a), the sample is loaded into an **external, fixed-volume loop** using a syringe. Eluent, meanwhile, flows directly from the pump to the column at high pressure. When the valve is rotated to the *inject* position (Fig. 13.2b), the loop becomes part of the eluent flow stream, and sample is carried onto the column. Such injectors are generally trouble free and precise.

Changing the loop allows different volumes to be injected. Although injection volumes of 10–100 μL are typical, both larger (e.g., 1–10-mL) and smaller (e.g., ≤2-μL) sample volumes can be loaded by utilizing special hardware. An important advantage of the loop valve design is that it is readily adapted to automated operation. Thus, automated sample injectors, or **autosamplers**, may be used to store and inject large numbers of samples. Samples are placed in uniform-size vials, sealed with a septum, and held in a (possibly refrigerated) tray. A computer-actuated syringe needle penetrates the septum to withdraw solution from the vial, and a mechanical system introduces it onto the column. Autosamplers can reduce the tedium and labor costs associated with routine HPLC analyses and improve assay precision. However, because samples may remain unattended for 12–24 h prior to automatic injection, sample stability is a limiting factor for using this accessory.

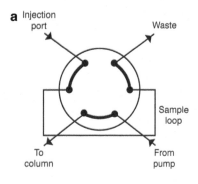

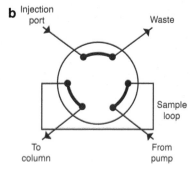

Valve-type injector. The valve allows the sample loop to be (**a**) isolated from the pump eluent stream (LOAD position) or (**b**) positioned in it (INJECT position) (From Gertz [10], used with permission)

13.2.3 Column

The first two sections below describe the basic column hardware and packing material for HPLC, followed by a third section to describe how those components, along with operating pressure, differ for ultra-HPLC.

13.2.3.1 *Column Hardware*

An **HPLC column** is usually constructed of stainless steel tubing with terminators that allow it to be connected between the injector and detector systems (Fig. 13.1). Columns also are made from glass, fused silica, titanium, and polyether ether ketone (PEEK) resin; the PEEK columns are essential for the high pH, high salt concentrations necessary for the powerful ion-exchange HPLC systems. Many types and sizes of columns are commercially available, ranging from 5 × 50-cm (or larger) preparative columns down to wall-coated capillary columns.

13.2.3.1.1 *Precolumns*

Auxiliary columns that precede the analytical HPLC column are termed **precolumns**. Short (≤5-cm) expendable columns, called **guard columns**, often are used to protect the analytical column from strongly adsorbed sample components. A guard column (or

cartridge) is installed between the injector and analytical column via short lengths of capillary tubing (or a cartridge holder). They may be filled with either **pellicular** media (see Sect. 13.2.3.2.2) of the same bonded phase as the analytical column or with **microparticulate** (≤10-μm) packing material identical to that of the analytical column. Microparticulate guard columns are usually purchased as prepacked, disposable inserts for use in a special holder and cost much less than replacing an analytical column. A guard column (or cartridge) should be repacked or replaced before its binding capacity is exceeded and contaminants pollute the expensive analytical column.

13.2.3.1.2 *Analytical Columns*

The most commonly used **analytical** HPLC columns are 10, 15, or 25-cm long with an internal diameter of 4.6 or 5 mm [7]. Short (3-cm) columns, packed with ≤3-μm particles, are gaining popularity for fast separations; for example, in method development or process monitoring. In recent years, the use of columns with smaller internal diameters (<0.5–2.0 mm), including wall-coated capillary columns, has increased. The advantages of using smaller diameter columns include a decreased use of mobile phase, an increased peak concentration, increased resolution, and the ability to couple HPLC with MS [12].

Various names have been used for the reduced-volume columns. Dorsey et al. [13] refer to columns with internal diameters of 0.5–2.0 mm as **microbore**, while packed or open tubular columns having internal diameters of <0.5 mm are termed **microcolumns** or **capillary columns** (a capillary column is a narrow-bore open tubular column, in which the inner surface is coated with a thin layer of stationary phase). In the case of the packed columns, the microbore or microcolumns contain very small particle size packing material (≤2 μm).

In general, to achieve good performance from microcolumns, it is essential to have an HPLC system with very low dead volume, so that peak broadening (sample diffusion) outside the column does not destroy resolution achieved within the column. Systems designed specifically for use with these columns are available from commercial suppliers (see Sect. 13.2.3.3).

13.2.3.2 *HPLC Column Packing Materials*

The development of a wide variety of column packing materials has contributed substantially to the success and widespread use of HPLC.

13.2.3.2.1 *General Requirements*

In most modes of chromatography, the column **packing material** serves as both the **support matrix** and the **stationary phase**. Requirements for HPLC column packing materials are good chemical stability, sufficient mechanical strength to withstand the high pressure generated during use, and a well-defined particle size with a narrow particle size distribution [11]. Two materials that meet the above criteria are porous silica and synthetic organic resins (see below Sects. 13.2.3.2.2 and 13.2.3.2.3, respectively).

13.2.3.2.2 Silica-Based Column Packings

Porous silica can be prepared in a wide range of particle and pore sizes, with a narrow particle size distribution. Both **particle size** and **pore diameter** are important: Small particles reduce the distance a solute must travel between stationary and mobile phases, which facilitates equilibration and results in good column efficiencies (Chap. 12, Sect. 12.5.2.2.2). However, small particles also yield greater flow resistance and higher pressure at equivalent flow rates. Spherical particles of 3-, 5-, or 10-μm diameter are utilized in analytical columns. One-half or more of the volume of porous silica consists of the **pores** [8]. Use of the smallest possible pore diameter will maximize **surface area** and **sample capacity**, which is the amount of sample that can be effectively separated on a given column. Packing materials with a pore diameter of 50–100 Å and surface area of 200–400 m²/g are used for low-molecular-weight (<500-Da) solutes. For increasingly larger molecules, such as proteins and polysaccharides, it is necessary to use materials with a wider pore (≥300 Å), so that internal surface is accessible to the solute [8].

Bonded phases (Fig. 13.3a) are made by covalently bonding hydrocarbon moieties to -OH groups (silanols) on the surface of silica particles [8, 14]. Often, the silica is reacted with an organochlorosilane:

$$\begin{matrix} & & R_1 & & & R_1 \\ & & | & & & | \\ \diagdown & & | & & & | \\ -Si-OH + Cl-Si-R_3 & \longrightarrow & Si-O-Si-R_3 + HCl & \quad [13.1] \\ \diagup & & | & & & | \\ & & R_2 & & & R_2 \end{matrix}$$

Substituents R_1 and R_2 may be halides or methyl groups. The nature of R_3 determines whether the resulting bonded phase will exhibit normal-phase, reversed-phase, or ion-exchange chromatographic behavior. The main disadvantage of silica and silica-based bonded-phase column packings is that the silica skeleton slowly dissolves in aqueous solutions, and the rate of this process becomes prohibitive at pH <2 and >8.

A **pellicular packing material** (Fig. 13.3b) is made by depositing a thin layer or coating onto the surface of an inert, usually nonporous, microparticulate **core**. Functional groups, such as ion-exchange sites, are present at the surface only. Core material may be either inorganic, such as silica, or organic, such as poly(styrene-divinylbenzene) or latex. The rigid core ensures good physical strength, and the thin stationary phase provides for rapid mass transfer and favorable column efficiency.

13.2.3.2.3 Porous Polymeric Column Packings

Synthetic organic resins offer the advantages of good chemical stability and the possibility to vary interactive properties through direct chemical modification. Two major categories of porous polymeric packing materials exist.

Microporous or gel-type resins (Fig. 13.3c) are comprised of crosslinked copolymers in which the apparent porosity, evident only when the gel is in its swollen state, is determined by the degree of crosslinking. These gel-type packings undergo swelling and contraction with changes in the chromatographic mobile phase. Microporous polymers of less than ca. 8% crosslinking are not sufficiently rigid for HPLC use.

Macroporous resins are highly crosslinked (e.g., ≥50%) and consist of a network of microspheric gel beads joined together to form a larger bead (Fig. 13.3d). Large, permanent pores, ranging from 100 to 4000 Å or more in diameter, and large surface areas (≥100 m²/g)

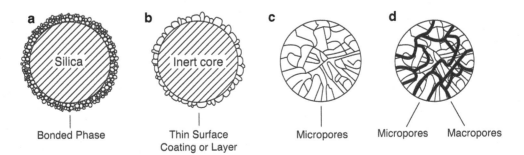

 13.3 figure Some types of packing materials utilized in HPLC. (**a**) Bonded-phase silica; (**b**) pellicular packing; (**c**) microporous polymeric resin; (**d**) macroporous polymeric resin (Adapted from Tomaz and Queiroz [19], p. 621, by courtesy of Marcel Dekker, Inc.)

are the result of interstitial spaces between the micro-beads [14]. Rigid microparticulate poly(styrene-divinylbenzene) packing materials of the macroporous type are popular for HPLC use. They are stable from pH 1 to 14 and are available in a variety of particle and pore sizes. These resins can be used in unmodified form for reversed-phase chromatography or functionalized for use in other HPLC modes.

13.2.3.3 Ultra-HPLC

Ultra-HPLC (UHPLC) and **ultra-performance liquid chromatography** (UPLC) are based on the same technique. The term "UPLC" was trademarked by the Waters Corporation in 2004, based on using ≤2-μm diameter porous particles and operating pressures much higher than that of regular HPLC. When other vendors entered the market with similar technology, the more general term "UHPLC" was used by manufacturers other than the Waters Corporation. UHPLC and UPLC instruments both use microbore or microcolumns with packing material of ≤2 μm and use much higher pressure than regular HPLC (HPLC max. of ~5800 psi; UHPLC/UPLC max. of ~8700–15000 psi, depending on the manufacturer). The flow rate for UHPLC/UPLC is generally lower than for regular HPLC, but this is dependent on the column dimensions. It is the combination of small packing material and increased pressure that results in reduced on-column dispersion (i.e., diffusion) of sample molecules [15]. With UPLC/UHPLC, efficiency of separation is increased, overall separation time is reduced, and less solvent is needed, compared to HPLC. The resolution and speed of both UPLC and UHPLC instruments are well suited to linkage analysis when coupled to MS. UPLC/UHPLC is the basis for newly approved AOAC official methods for numerous vitamins (i.e., A, B_6, B_{12}, C, D, and folate). Besides vitamins, UPLC/UHPLC is referred to in this textbook as it applies to the separation of proteins (Chap. 24, Sect. 24.2.3.1), phenolic compounds (Chap. 25, Sect. 25.2.3.1), and antibiotic residues (Chap. 33, Sect. 33.5.2.2).

13.2.4 Detector

A **detector** translates sample concentration changes in the HPLC column effluent into electrical signals. Spectrochemical, electrochemical, or other properties of solutes may be measured by a variety of instruments, each of which has advantages and disadvantages. The choice of which to use depends on solute type and concentration, and on detector sensitivity, linear range, and compatibility with the solvent and elution mode to be used. Cost also may influence detector selection. One common feature of most HPLC detectors is the presence of a flow cell, through which the eluent flows as it is analyzed by the detector system. These flow cells are often delicate and easily polluted or damaged, so care must be taken when handling them.

The most widely used HPLC detectors are based on ultraviolet-visible (UV-Vis) and fluorescence spectrophotometry, refractive index determination, and electrochemical analysis (see Chap. 7 for detailed discussion of UV-Vis and fluorescence spectrophotometry). Many other methods, such as light scattering or mass spectrometry, also can be applied to the detection of analytes in HPLC eluents. More than one type of HPLC detector may be used in series, to provide increased **specificity** and **sensitivity** for multiple types of analytes. In one food-related application, a multidetector HPLC system equipped with a diode array absorption detector coupled to fluorescence and electrochemical detectors was used to monitor a wide variety of Maillard reaction products.

13.2.4.1 UV-Vis Absorption Detectors

Many HPLC analyses are carried out using a **UV-Vis absorption** detector, which can measure the absorption of radiation by chromophore-containing compounds. The three main types of UV-Vis absorption detectors are **fixed wavelength**, **variable-wavelength**, and **diode array spectrophotometers** [8]. As its name implies, the simplest design operates at a single, **fixed wavelength**. A filter is used to isolate a single emission line (e.g., at 254 nm) from a source such as a mercury lamp. This type of detector is easy to operate and inexpensive but of limited utility.

The most popular general purpose HPLC detector today is the **variable-wavelength** detector, in which deuterium and tungsten lamps serve as sources of ultraviolet and visible radiation, respectively. Wavelength selection is provided by a **monochromator**, a device that acts to deflect light, and an exit slit that allows light from a limited range of wavelengths to pass through to the sample. Rotating the monochromator via a selector switch allows one to change the operating wavelength.

Diode array spectrophotometric detectors can provide much more information about sample composition than is possible with monochromatic detection. In this instrument, the light from a deuterium lamp is spread into a spectrum across an array of photodiodes mounted on a silicon chip. These are read almost simultaneously by a microprocessor to provide the full absorption spectrum from 200 to 700 nm every 0.1 s, which may enable the components of a mixture to be identified. Although considerably more expensive than variable-wavelength detectors, they are useful in method development and in routine analyses in which additional evidence of peak identity, without further analysis, is needed.

13.2.4.2 *Fluorescence Detectors*

Some organic compounds can reemit a portion of absorbed UV-Vis radiation at a longer wavelength (lower energy). This is known as **fluorescence**, and measurement of the **emitted light** provides another useful detection method. Fluorescence detection is both selective and very sensitive, providing up to 1000-fold lower detection limits than for the same compound in absorbance spectrophotometry. Although relatively few compounds are inherently fluorescent, analytes often are converted into fluorescent derivatives (see Sect. 13.2.4.7). Ideal for trace analysis, fluorescence detection has been used for the determination of various vitamins in foods and supplements, monitoring aflatoxins in stored cereal products, and the detection of aromatic hydrocarbons in wastewater.

13.2.4.3 *Refractive Index Detectors*

Refractive index (RI) detectors measure change in the **RI** of the mobile phase due to dissolved analytes, which provides a nearly universal method of detection. However, because a bulk property of the eluent is being measured, RI detectors are less sensitive than other types. Another disadvantage is that they cannot be used with gradient elution, as any change in eluent composition will alter its RI, thereby changing the baseline signal. RI detectors are widely used for analytes that do not contain UV-absorbing chromophores, such as carbohydrates and lipids, when the analytes are present at relatively high concentration.

13.2.4.4 *Electrochemical Detectors*

Electroanalytical methods used for HPLC detection are based either on electrochemical oxidation-reduction of the analyte or on changes in conductivity of the eluent. **Amperometric detectors** measure the change in current as the analyte is oxidized or reduced by the application of voltage across electrodes in the flow cell. This method is highly selective (nonreactive compounds give no response) and very sensitive. A major application of electrochemical detection has been for the routine determination of catecholamines, which are phenolic compounds of clinical importance that are present in blood and tissues at very low levels. The development of a triple-pulsed amperometric detector, which overcame the problem of electrode poisoning (accumulation of oxidized product on the electrode surface), has allowed electrochemical detection to be applied to the analysis of carbohydrates (see Sect. 13.3.4.2.2). Pulsed electrochemical detection also has excellent sensitivity for the quantification of flavor-active alcohols, particularly terpenols.

Analytes that are ionized and carry a charge can be detected by measuring the change in eluent **conductivity** between two electrodes. **Conductivity** detection has been used mainly to detect inorganic anions and cations and organic acids upon elution from weak ion-exchange columns. Its principal application has been as the basis of **ion chromatography** (Sect. 13.3.3.2.1). An excellent overview of electrochemical detection is provided by Swedesh [7].

13.2.4.5 *Other HPLC Detectors*

Unfortunately, there is no *truly universal* HPLC detector with *high sensitivity*. Thus, there have been many attempts to find new principles that could lead to improved instrumentation. One interesting concept is the **evaporative light scattering detector**. The mobile phase is sprayed into a heated air stream, evaporating volatile solvents and leaving nonvolatile analytes as aerosols. These droplets or particles can be detected because they will scatter a beam of light [7]. HPLC with light scattering detection has been applied to the analysis of wheat flour lipids. Also, light scattering detectors are quite useful for the characterization of polymers by size-exclusion chromatography. Improvements in laser applications brought about the development of **low-angle laser light scattering** (LALLS) and **multi-angle laser light scattering** (MALLS) detectors. With these detectors, there is no need to evaporate the mobile phase, as the laser beam is directed at the flow cell, and scattered laser light is then monitored by photo detectors set at specific angles to the cell. In MALLS there may be many different photo detectors at discrete angles, each continuously collecting and analyzing the scattered light; from this data, the computer can determine the molecular weight of the eluting sample.

Radioactive detectors are widely used for pharmacokinetic and metabolism studies with radiolabeled drugs. Decay of a radioactive nucleus leads to excitation of a scintillator, which subsequently loses its excess energy by photon emission. Photons are counted by a photomultiplier tube, and the number of counts per second is proportional to radiolabeled analyte concentration [8].

A **chemiluminescent nitrogen detector** (CLND) allows nitrogen-containing compounds, such as amino acids, to be detected without using chemical derivatization (Sect. 13.2.4.7). This nitrogen-specific detection system has been used to quantify caffeine in coffee and soft drink beverages and to analyze capsaicin in hot peppers [2].

13.2.4.6 *Coupled Analytical Techniques*

To obtain more information about the analyte(s), eluent from an HPLC system can be passed on to a second analytical instrument, such as infrared (IR), nuclear magnetic resonance (NMR), or MS (see Chaps. 8, 10, and 11, respectively, or reference [5]).

The coupling of spectrometers with liquid chromatography (LC) was initially slow to gain application, due to many practical problems. For example, in the case of **HPLC with mass spectrometric detection** (LC-MS), the liquid mobile phase affected the vacuum in the MS. This problem was addressed by the development of commercial interfaces that allow the solvent to be evaporated, so that only analyte is transferred to the spectrometer. Two commonly used interface techniques are discussed in detail by Harris [5]. The use of microbore or capillary HPLC columns with a low flow volume also facilitates direct coupling of the two instruments [12]. LC-MS systems continue to improve, and the applications are expanding to nearly every class of relatively low-molecular-weight compounds, including bioactives and contaminants.

In addition to the coupled techniques described above, LC can be coupled to itself, to create two-dimensional LC (2D-LC), just like two GC columns can be coupled to create multidimensional GC (Chap. 14, Sect. 14.3.5.9). In both cases, the two different separation stages are used to separate the injected sample, with the eluent from the first column being injected into the second column. Bands that may not have been completely resolved on the first column may be completely separated on a second column, which typically has a different separation mechanism. A single detector can be used after the second column, or a different detector can be used after each of the two LC columns [17].

13.2.4.7 *Chemical Reactions*

Detection sensitivity or specificity may sometimes be enhanced by converting the analyte to a **chemical derivative** with different or additional characteristics. An appropriate reagent can be added to the sample prior to injection (i.e., **precolumn derivatization**) or combined with column effluent before it enters the detector (i.e., **postcolumn derivatization**). Automated amino acid analyzers utilize postcolumn derivatization, usually with ninhydrin, for reliable and reproducible analysis of amino acids. Precolumn or postcolumn derivatization of amino acids with *o*-phthalaldehyde or similar reagents permits highly sensitive HPLC determination of amino acids using fluorescence detection (Chap. 24, Sect. 24.3.1.2). In addition, fractions may be collected after passing through the detector and aliquots of each fraction analyzed by various means, including chemical/colorimetric assays, such as the bicinchoninic acid method for protein (Chap. 18, Sect. 18.4.2.3) or a total carbohydrate assay (Chap. 19, Sect. 19.3). The results can then be plotted and overlaid with the detector plot, yielding very important information about the compounds eluting in various peaks.

13.2.5 Data Station Systems

As a detector provides an electronic signal related to the composition of the HPLC column effluent, it is the function of the last element of the HPLC system to display the chromatogram and integrate the peak areas. **Data stations** and **software packages** are nearly ubiquitous with modern HPLC, and all come with very powerful tools for sample identification and quantitation. As an HPLC analysis progresses, the data from the HPLC detector(s) are digitized and saved to a computer hard drive. The data then can be manipulated (annotated) by assigning and integrating the peaks, for example, and then printed out, as plots and tables, for further assessment. Importantly, the software programs can be set up prior to the analysis to perform nearly all these functions, without further input from the operator. For example, retention times can be calculated relative to an internal standard in pesticide residue analysis and the results compared to a stored database of standards that the software automatically accesses when the analysis is complete. The software will then assign and integrate the peaks, and construct a complete report that is displayed when the file is opened, even on a remote computer (e.g., in the office).

The current data stations are more comprehensive: the software packages include all the parameters needed to run the HPLC, including start and stop, autosampler injection of the sample, and developing the gradient via control of the proportioning pump systems. The data station can carry out the entire operation, on hundreds of samples, in the absence of an operator, and, with networking, deliver the analysis file/report to any connected computer.

13.3 APPLICATIONS IN HPLC

The basic physicochemical principles underlying all liquid chromatographic separations—adsorption, partition, hydrophobic interaction, ion exchange, affinity, and size exclusion—are discussed in Chap. 12, and details will not be repeated here. The number of separation modes utilized in HPLC, however, is greater than that available in classic chromatography [18]. Examples of HPLC applications in food analysis are given in Table 13.1. This is attributable to the success of bonded phases, initially developed to facilitate liquid-liquid partition chromatography (Chap. 12, Sect. 12.4.2). In fact, reversed-phase chromatography is the most widely used separation mode in modern HPLC.

13.3.1 Normal Phase

13.3.1.1 *Stationary and Mobile Phases*

In **normal-phase** HPLC, the **stationary phase** is a **polar adsorbent** such as bare silica or silica to which polar nonionic functional groups—hydroxyl, nitro, cyano (nitrile), or amino—have been chemically linked. These bonded phases are moderately polar and the surface is more uniform, resulting in better elution profiles. The **mobile phase** for this mode consists of a **nonpolar solvent**, such as hexane, to which is added a more polar modifier, such as methylene chloride, to control solvent strength and selectivity. Solvent strength refers to solvent effects on the migration rate of the sample: Relatively **weak solvents** increase retention values (large k') and **strong solvents** decrease retention values (small k').

13.3.1.2 *Applications of Normal-Phase HPLC*

In the past, normal-phase HPLC was used for the analysis of fat-soluble vitamins, although reverse phase is currently applied more frequently for these analyses (see Table 13.1). Normal phase is currently used for the analyses of biologically active polyphenols from natural plant sources, such as grape and cocoa. It is also used for the analysis of relatively polar vitamins, such as vitamins A, D, E, and K (see Chap. 20), and also natural carotenoid pigments, which impart both color and health benefits to foods. Highly hydrophilic species, such as carbohydrates (see Chap. 19, Sect. 19.4.2.1), also may be resolved by normal-phase chromatography, using amino bonded-phase HPLC columns. Other applications include the analysis of antioxidants, such as butylated hydroxytoluene (BHT), butylated hydroxyanisole (BHA), and tert-butylhydroquinone (TBHQ, tertiary butylhydroquinone), and vitamin E compounds, such as tocopherols (TCP) [3]. Analysis of these compounds is increasingly necessary for the proper assessment of food products, as well as additives.

13.3.2 Reversed Phase

13.3.2.1 *Stationary and Mobile Phases*

More than 70% of all HPLC separations are carried out in the reversed-phase mode, which utilizes a **nonpolar stationary phase** and a **polar mobile phase**. Octadecylsilyl (ODS) **bonded phases**, with an octadecyl (C_{18}) chain $[-(CH_2)_{17}CH_3]$, are the most popular reversed-phase packing materials, although shorter chain hydrocarbons [e.g., octyl (C_8) or butyl (C_4)] or phenyl groups are also used. Many silica-based, reversed-phase columns are commercially available. Differences in their chromatographic behavior result from variation in the type of organic group bonded to the silica matrix or the chain length of organic moiety.

Reversed-phase HPLC utilizes **polar mobile phases**, usually water mixed with methanol, acetonitrile, or tetrahydrofuran. Solutes are retained due to **hydrophobic interactions** with the **nonpolar stationary phase** and are eluted in order of increasing hydrophobicity (decreasing polarity). Increasing the polar (aqueous) component of the mobile phase increases solute retention (larger k' values) (see Chap. 12, Sect. 12.5.2.2.4), whereas increasing the organic solvent content of the mobile phase decreases retention (smaller k' values). Various additives can serve additional functions. For example, although ionic compounds often can be resolved without them, **ion-pair reagents** may be used to facilitate chromatography of ionic species on reversed-phase columns. These reagents are ionic surfactants, such as octanesulfonic acid, which can neutralize charged solutes and make them more lipophilic. This type of chromatography is referred to as **ion-pair reversed phase**.

A modification of reverse phase systems has led to the development of hydrophobic interaction

13.1 table Example applications of HPLC in the analysis of various food constituents

Analyte	Separation mode	Method of detection	Chapters	Sections
Mono- and oligosaccharides	Ion exchange; normal or reversed phase	Electrochemical; refractive index; postcolumn analysis	19	19.4.2.1
Vitamins	Normal or reversed phase	Fluorescence; electrochemical; UV	20	20.4.1
Amino acids	Ion exchange; reversed phase	Post- or precolumn derivatization	24	24.3.1
Protein separation	Ion exchange, reversed-phase, affinity, hydrophobic interaction	UV	24	24.2.3
Phenolics	Reversed phase	UV	25	25.2.3.1
Pesticides	Normal or reversed phase	UV; fluorescence; mass spectrometry	33	33.3.3.2.3; 18.3.3.3.2
Mycotoxins	Reversed phase; Immunoaffinity	UV; fluorescence	33	33.4.3.2.1
Antibiotics	Reversed phase	UV	33	33.5.2.2
Various food contaminants (e.g., acrylamide, melamine)	Reversed phase, ion exchange	UV; mass spectrometry	33	33.8
Sulfites	Ion exchange	UV; electrochemical	33	33.8.2

chromatography (HIC), which utilizes column matrices and elution conditions that minimize or eliminate protein denaturation.

13.3.2.2 *Applications of Reversed-Phase HPLC*

Reversed phase has been the HPLC mode most used for analysis of plant proteins. Cereal proteins, among the most difficult of these proteins to isolate and characterize, are now routinely analyzed by this method. Both water- and fat-soluble vitamins (Chap. 20) can be analyzed by reversed-phase HPLC [2–5], and the availability of fluorescence detectors has enabled researchers to quantitate very small amounts of the different forms of vitamin B_6 (vitamins) in foods and biological samples. Figure 13.4 shows the separation of several of these vitamins in a rice bran extract achieved by reversed-phase ion-pair HPLC [16].

Reversed-phase ion-pair HPLC can be used to resolve carbohydrates on C_{18} bonded-phase columns [10], and the constituents of soft drinks (caffeine, aspartame, etc.) can be rapidly separated. Reversed-phase HPLC using a variety of detection methods, including RI, UV, and light scattering, has been applied to the analysis of lipids [2–5, 11]. Antioxidants, such as BHA and BHT, can be extracted from dry foods and analyzed with simultaneous UV and fluorescence detection [2, 3]. Phenolic flavor compounds (such as vanillin) and pigments (such as chlorophylls, carotenoids, and anthocyanins) are also easily analyzed [2–5, 11]. A typical chromatogram of carotenoids present in a carrot extract is shown in Fig. 13.5. Reversed-phase ion-pair chromatography also is used for the separation of synthetic food colors (e.g., FD&C Red No. 40 and FD&C Blue No. 1) [4].

Chlorogenic acid, an antioxidant with antidiabetic effects, is also analyzed by RP-HPLC, as is the nutrient quality of sweet potatoes, which contain ß-carotene, in addition to many other beneficial components. The Scoville scale for hot peppers is based on capsicum content, which may also be measured by HPLC. This is particularly relevant to the modern food industry, as hot food has gained popularity around the world.

13.3.3 Hydrophobic Interaction

With the use of special matrices, hydrophobic interaction chromatography (HIC) offers the possibility to separate many compounds, including many non-denatured proteins, by less harsh eluents and conditions. The technique is based on the presence of hydrophobic groups on or near the surface of the sample molecules that can interact with the hydrophobic column matrix under high saline (but non-denaturing) conditions. The sample is then eluted by lowering the salt concentration of the eluent. This allows for the collection of active proteins, such as enzymes, which can then be studied further in the lab [19].

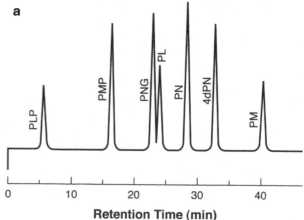

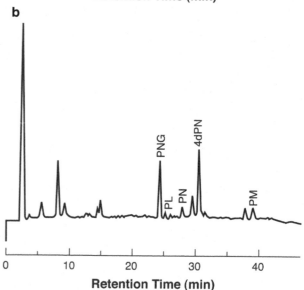

13.4 figure Analysis of vitamin B_6 compounds by reversed-phase HPLC with fluorescence detection. Some of the standard compounds (**a**) are present in a sample of rice bran extract (**b**). Sample preparation and analytical procedures are described in reference Stoll et al. [17]. Abbreviations: *PL* pyridoxal, *PLP* pyridoxal phosphate, *PM* pyridoxamine, *PMP* pyridoxamine phosphate, *PN* pyridoxine, *PNG* pyridoxine β-D-glucoside (Reprinted in part with permission from Tomaz and Queiroz [19]. Copyright 1991 American Chemical Society)

13.3.4 Ion Exchange

13.3.4.1 *Stationary and Mobile Phases*

Packing materials for **ion-exchange HPLC** are usually **functionalized organic resins**, such as sulfonated or aminated poly(styrene-divinylbenzene) (Chap. 12, Sect. 12.4.3). **Macroporous resins** are most effective for HPLC columns due to their rigidity and permanent pore structure. **Pellicular packings** also are utilized, particularly in the CarboPac™ (Dionex) series, in

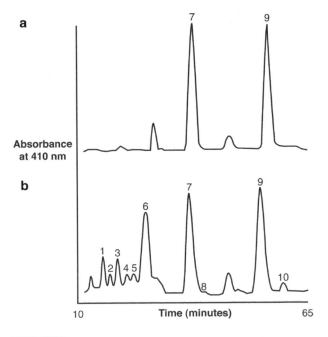

13.5
figure

Reversed-phase HPLC separation of α-carotene (AC) and β-carotene (BC) isomers in (**a**) fresh and (**b**) canned carrots using a 5-μm C_{30} stationary phase. Peak 1, 13-*cis* AC; 2, unidentified *cis* AC; 3, 13' –*cis* AC; 4, 15-*cis* BC; 5, unidentified *cis* AC; 6, 13-*cis* BC; 7, all-*trans* AC; 8, 9-*cis* AC; 9, all-*trans* BC; 10, 9-*cis* BC (Reprinted with permission from Lessin et al. [20]. Copyright 1997 American Chemical Society)

which the nonporous, latex resin beads are coated with functionalized microbeads. The **mobile phase** in ion-exchange HPLC is usually an **aqueous buffer**, and solute retention is controlled by changing mobile phase ionic strength and/or pH. **Gradient elution** (gradually increasing ionic strength) is frequently employed.

13.3.4.2 Applications of Ion-Exchange HPLC

Ion-exchange HPLC has many applications, ranging from the detection of simple inorganic ions, to analysis of carbohydrates and amino acids, to the preparative purification of proteins oligosaccharides.

13.3.4.2.1 Ion Chromatography

Ion chromatography is simply high-performance ion-exchange chromatography using a relatively **low-capacity stationary phase** (either anion- or cation-exchange) and, usually, a **conductivity detector**. All ions conduct an electric current; thus, measurement of electrical conductivity is an obvious way to detect ionic species. Because the mobile phase also contains ions, however, background conductivity can be relatively high. One step toward solving this prob-

lem is to use much lower capacity ion-exchange packing materials, so that more dilute eluents may be employed. In **non-suppressed** or **single-column ion chromatography**, the detector cell is placed directly after the column outlet, and eluents are carefully chosen to maximize changes in conductivity as sample components elute from the column. **Suppressed ion chromatography** utilizes an eluent that can be selectively removed by the use of ion-exchange membranes [11]. Suppressed ion chromatography permits the use of more concentrated mobile phases and gradient elution. Ion chromatography can be used to determine inorganic anions and cations, transition metals, organic acids, amines, phenols, surfactants, and sugars. Some specific examples of ion chromatography applied to food matrices include the determination of organic and inorganic ions in milk, organic acids in coffee extract and wine, chlorine in infant formula, and trace metals, phosphates, and sulfites in foods. Figure 13.6 illustrates the simultaneous determination of organic acids and inorganic anions in coffee by ion chromatography.

13.3.4.2.2 Ion-Exchange Chromatography of Carbohydrates and Proteins

Both cation- and anion-exchange stationary phases have been applied to HPLC of carbohydrates. The advantage of separating carbohydrates by **anion exchange** is that retention and selectivity may be altered by changes in eluent composition. Carbohydrate analysis has benefited greatly by the development of a system that involves **anion-exchange HPLC** at high pH (≥12) and detection by a **pulsed amperometric detector** (PAD). Pellicular column packings (see Sect. 13.2.3.2.2), consisting of nonporous latex beads coated with a thin film of strong anion exchanger, provide the necessary fast exchange, high efficiency, and resistance to strong alkali. These systems may be used in a variety of applications, from routine quality control to basic research. One common application is the determination of oligosaccharide distributions in corn syrups and other starch hydrolysates (Fig. 13.7).

Amino acids have been resolved on polymeric ion exchangers for more than 40 years (see Chap. 24, Sect. 24.3.1.2). Ion exchange is one of the most effective modes for HPLC of proteins and, recently, has been recognized as valuable for the fractionation of peptides.

13.3.5 Size Exclusion

Size-exclusion chromatography (SEC) fractionates solutes solely on the basis of size, with larger molecules eluting first. Due to the limited separation volume available in this chromatographic mode, as explained in Chap. 12 (Sect. 12.4.4), the resolution capacity of a size-exclusion column is relatively small;

Anions in Coffee

1. Acetate
2. Glycolate
3. Quinate
4. Formate
5. Chloride
6. Tartrate
7. Oxalate
8. Fumarate
9. Phosphate
10. Citrate

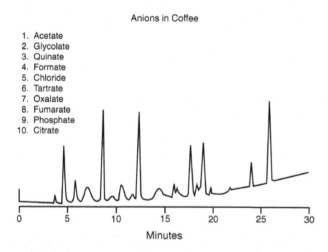

13.6 figure Ion-chromatographic analysis of organic acids and inorganic anions in coffee. Ten anions (listed) were resolved on an IonPac AS5A column (Dionex) using a sodium hydroxide gradient and suppressed conductivity detection (Courtesy of Dionex Corp., Sunnyvale, CA)

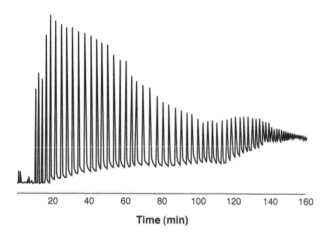

13.7 figure Anion exchange analysis of isoamylase-treated waxy corn starch. The enzyme de-branches the amylopectin, and the chromatogram represents the branch chain-length distribution, from four sugars in length and up. The analysis was performed with anion-exchange HPLC (Dionex™), with a pulsed amperometric detector

thus, the "high-performance" aspect of HPLC is not really applicable. The main advantage gained from use of small particle packing materials is speed. Relatively small amounts of sample can be analyzed or separated and collected in ≤60 min compared to ≤24-h separations using conventional low-pressure systems. A second advantage is that the sample

concentration is higher and the relative volume is lower, so there is much less eluent to remove.

13.3.5.1 *Column Packings and Mobile Phases*

Size-exclusion packing materials or columns are selected so that matrix pore size matches the molecular weight range of the species to be resolved. Prepacked columns of microparticulate media are available in a wide variety of pore sizes. **Hydrophilic packings**, for use with water-soluble samples and aqueous mobile phases, may be surface-modified silica or methacrylate resins. **Poly(styrene-divinylbenzene) resins** are useful for nonaqueous size-exclusion chromatography of synthetic polymers.

The mobile phase in this mode is chosen for sample solubility, column compatibility, and minimal solute-stationary phase interaction. Otherwise, it has little effect on the separation. Aqueous buffers are used for biopolymers, such as proteins and polysaccharides, both to preserve biological activity and to prevent adsorptive interactions. Tetrahydrofuran or dimethylformamide is generally used for size-exclusion chromatography of other polymer samples, to ensure sample solubility.

13.3.5.2 *Applications of High-Performance SEC*

Hydrophilic polymeric size-exclusion packings are used for the rapid determination of **average molecular weight** and **molecular weight range** of polysaccharides, including amylose, amylopectin, and other soluble gums such as xanthan, pullulan, guar, and water-soluble cellulose derivatives. **Molecular weight distribution** can be determined directly from high-performance size-exclusion chromatography, if LALLS or MALLS is used for detection [7]. The application of aqueous size-exclusion chromatography to two commercially important polysaccharides, xanthan and carboxymethyl cellulose, is discussed in detail in reference [7].

SEC analysis is useful to better understand various food components and systems. SEC analysis of tomato cell wall pectin from hot- and cold-break tomato preparations (Fig. 13.8) showed that the cell wall pectin was not differentially degraded by the different processing procedures. Size-exclusion HPLC has been shown to be a rapid, one-step method for assessing soybean cultivars on the basis of protein content (proteins in the extracts of non-defatted flours from five soybean cultivars were separated into six common peaks, and cultivars could be identified by the percent total area of the fifth peak). A size-exclusion liquid chromatographic method also has been applied to the determination of polymerized triacylglycerols in oils and fats [7].

13.3.6 Affinity

Affinity chromatography is based on the principle that the molecules to be purified can form a selective but reversible interaction with another molecular species that has been immobilized on a chromatographic support. Although almost any material can be immobilized on a suitably activated support, the major ligands are proteins, including lectins (Chap. 24, Sect. 24.2.3.2.2), nucleic acids, and dyes (Chap. 12, Sect. 12.4.5). Affinity chromatography is used to purify many glycoproteins. Affinity chromatography using immobilized folate-binding protein is an effective tool in purifying sample extracts for HPLC analysis of folates in foods (e.g., reference [21]).

13.4 SUMMARY

HPLC is a chromatographic technique of great versatility and analytical power. A basic HPLC system consists of a pump, injector, column, detector, and data system. The pump delivers mobile phase through the system. An injector allows sample to be placed into the flowing mobile phase for introduction onto the column. The HPLC column consists of stainless steel or polymer hardware filled with a separation packing material. Various auxiliary columns, particularly guard columns, may be used prior to the analytical column. Detectors used in HPLC include UV-Vis absorption, fluorescence, RI, electrochemical, and light scattering, as well as coupled analytical systems, such as a mass spectrometer. Detection sensitivity or specificity sometimes can be enhanced by chemical derivatization of the analyte. Computer-controlled data station systems offer data collection and processing capabilities and can run the instrument when an automated system is needed. A broad variety of column packing materials have contributed greatly to the widespread use of HPLC. These column packing materials may be categorized as silica-based (porous silica, bonded phases, pellicular packings) or polymeric (microporous, macroporous, or pellicular/nonporous). The success of silica-based bonded phases has expanded the applications of normal-phase and reversed-phase modes of separation in HPLC. Separations also are achieved with ion-exchange, size-exclusion, and affinity chromatography. HPLC is widely used for the analysis of small molecules and ions, such as sugars, vitamins, and amino acids, and is applied to the separation and purification of macromolecules, such as proteins and polysaccharides.

13.5 STUDY QUESTIONS

1. Why might you choose to use HPLC rather than traditional low-pressure column chromatography?
2. What is a guard column and why is it used?
3. Give three general requirements for HPLC column packing materials. Describe and distinguish among porous silica, bonded phases, pellicular, and polymeric column packings, including the advantages and disadvantages of each type.
4. What is the primary function of an HPLC detector (regardless of type)? What factors would you consider in choosing an HPLC detector? Describe three different types of detectors and explain the principles of operation for each.
5. You are performing HPLC using a stationary phase that contains a polar nonionic functional group. What type of chromatography is this, and what could you do to increase the retention time of an analyte?
6. Why are external standards commonly used for HPLC (unlike in GC, for which internal standards are more commonly used)?
7. Ion chromatography has recently become a widely promoted chromatographic technique in food analysis. Describe ion chromatography and give at least two examples of its use.
8. Describe one application each for ion-exchange and size-exclusion HPLC.

Acknowledgment Dr. Baraem Ismail is acknowledged for her preparation of Table 13.1.

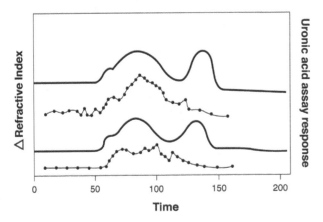

13.8 figure Analysis of tomato cell wall pectin from hot- and cold-break tomato preparations by size-exclusion chromatography. The *solid lines* are from a refractive index detector response. The *lines* with markers result from a post chromatography analysis. Aliquots of collected fractions were analyzed by a colorimetric chemical assay that is specific for pectic sugars

REFERENCES

1. Bidlingmeyer BA (1993) Practical HPLC methodology and applications. Wiley, New York
2. Matissek R, Wittkowski R (eds) (1993) High performance liquid chromatography in food control and research. Behr's Verlag, Hamburg, Germany
3. Nollet LML, Toldra F (eds) (2012) Food analysis by HPLC, 3rd edn. Marcel Dekker, New York
4. Macrae R (ed) (1988) HPLC in food analysis, 2nd edn. Academic, New York, NY
5. Harris DC (2015) Quantitative chemical analysis, 9th edn. W.H. Freeman and Co., New York
6. Hanai T (2004) HPLC: a practical guide. The Royal Society of Chemistry, Cambridge
7. Swadesh J (ed) (2000) HPLC: practical and industrial applications, 2nd edn. CRC, Boca Raton, FL
8. Lough WJ, Wainer IW (eds) (2008) High performance liquid chromatography: fundamental principles and practice. Springer, New York
9. LaCourse WR (2000) Column liquid chromatography: equipment and instrumentation (fundamental review). Anal Chem 72: 37R–51R
10. Gertz C (1990) HPLC tips and tricks. LDC Analytical, Riviera Beach, FL
11. Snyder LR, Kirkland JJ, Glajch JL (2012) Practical HPLC Method Development. John Wiley & Sons, Hoboken, New Jersey
12. Ishii D (ed) (1988) Introduction to microscale high-performance liquid chromatography. VCH Publishers, New York
13. Dorsey JG, Cooper WT, Siles BA, Foley JP, Barth HG (1996) Liquid chromatography: theory and methodology (fundamental review). Anal Chem 68:515R–568R
14. Waters Corporation (2014) Beginners Guide to UPLC: Ultra-Performance Liquid Chromatography (Waters Series) 1st edn. Milford MA
15. Unger KK (1990) Packings and stationary phases in chromatographic techniques. Marcel Dekker, New York
16. Gregory JF, Sartain DB (1991) Improved chromatographic determination of free and glycosylated forms of vitamin B_6 in foods. J Agric Food Chem 39:899–905
17. Stoll DR, Wang X, Carr PW (2008) Comparison of the practical resolving power of one-and two-dimensional high-performance liquid chromatography analysis of metabolomics samples. Anal Chem 80(1):268–278
18. Synder LR, Glajch JL, Kirkland JJ (1997) Practical HPLC method development, 2nd edn. Wiley, New York
19. Tomaz CT, Queiroz JA (2013) Hydrophobic interaction chromatography, In Liquid chromatography: fundamentals and instrumentation, Eds. Fanali S, Haddad PR, Poole, Schoenmakers P, Lloyd DK. Elsevier, Amsterdam, p 122–141
20. Lessin WJ, Catignani GL, Schwartz SJ (1997) Quantification of *cis-trans* isomers of provitamin A carotenoids in fresh and processed fruits and vegetables. J Agric Food Chem 45:3728–3732
21. Pfeiffer C, Rogers LM, Gregory JF (1997) Determination of folate in cereal-grain food products using trienzyme extraction and combined affinity and reverse-phase liquid chromatography. J Agric Food Chem 45: 407–413

Gas Chromatography

Michael C. Qian (✉)
Department of Food Science and Technology,
Oregon State University,
Corvallis, OR 97331-6602, USA
e-mail: michael.qian@oregonstate.edu

Devin G. Peterson
Department of Food Science and Technology,
The Ohio State University,
Columbus, OH 43210, USA
e-mail: peterson.892@osu.edu

Gary A. Reineccius
Department of Food Science and Nutrition,
University of Minnesota,
St. Paul, MN 55108-6099, USA
e-mail: greinecc@umn.edu

S. Nielsen (ed.), *Food Analysis*, Food Science Text Series,
DOI 10.1007/978-3-319-45776-5_14, © Springer International Publishing 2017

14.1 INTRODUCTION

The first publication on gas chromatography (GC) was in 1952 [1], while the first commercial instruments were manufactured in 1956. James and Martin [1] separated fatty acids by GC, collected the column effluent, and titrated the individual fatty acids for quantitation. GC has advanced greatly since that early work and is now considered to be a mature field that is approaching theoretical limitations.

The types of analysis that can be done by GC are very broad. GC has been used for the determination of fatty acids, triglycerides, cholesterol and other sterols, gases, solvent analysis, water, alcohols, and simple sugars, as well as oligosaccharides, amino acids and peptides, vitamins, pesticides, herbicides, food additives, antioxidants, nitrosamines, polychlorinated biphenyls (PCBs), drugs, flavor compounds, and many more. The fact that GC has been used for these various applications does not necessarily mean that it is the best method—often better choices exist. GC is ideally suited to the analysis of **thermally stable volatile substances**. Substances that do not meet these requirements (e.g., sugars, oligosaccharides, amino acids, peptides, and vitamins) are more suited to analysis by a technique such as high-performance liquid chromatography (HPLC) or supercritical fluid chromatography (SFC). Yet gas chromatographic methods appear in the literature for these substances after derivatization.

This chapter will discuss sample preparation for GC, GC hardware, columns, and chromatographic theory as it is uniquely applied to GC. Texts devoted to GC in general [2–4] and food applications in particular [5, 6] should be consulted for more detail.

14.2 SAMPLE PREPARATION FOR GAS CHROMATOGRAPHY

14.2.1 Introduction

One cannot generally directly inject a food product into a GC without some sample preparation. The high temperatures of the injection port will result in the degradation of nonvolatile constituents and create a number of false GC peaks corresponding to the volatile degradation products formed. In addition, very often the constituent of interest must be isolated from the food matrix simply to permit concentration such that it is at detectable limits for the GC or to isolate it from the bulk of the food. Thus, one must generally do some type of sample preparation, component isolation, and concentration prior to GC analysis.

Sample preparation often involves grinding, homogenization, or otherwise reducing particle size. There is substantial documentation in the literature showing that foods may undergo changes during sample storage and preparation. Many foods contain active enzyme systems that will alter the composition of the food product. This is very evident in the area of flavor work [7–9]. Inactivation of enzyme systems via high-temperature short-time thermal processing, sample storage under frozen conditions, drying the sample, or homogenization with alcohol may be necessary.

Microbial growth or chemical reactions may occur in the food during sample preparation. Chemical reactions often will result in false peaks on the GC. Thus, the sample must be maintained under conditions such that degradation does not occur. Microorganisms often are inhibited by certain chemicals (e.g., sodium fluoride), thermal processing, drying, or frozen storage.

14.2.2 Isolation of Analytes from Foods

14.2.2.1 Introduction

The isolation procedure may be quite complicated depending upon the constituent to be analyzed. For example, if one were to analyze the triglyceride-bound fatty acids in a food, one would first have to extract the lipids (free fatty acids; mono-, di-, and triglycerides; sterols; fat-soluble vitamins, etc.) from the food (e.g., by solvent extraction) and then isolate only the triglyceride fraction (e.g., by adsorption chromatography on silica). The isolated triglycerides then would have to be treated to first hydrolyze the fatty acids from the triglycerides and subsequently to form esters to improve gas chromatographic properties. The two latter steps might be accomplished in one reaction by transesterification (e.g., boron trifluoride in methanol) as described in Chap. 17, Sect. 17.3.6.2, and Chap. 23, Sect. 23.6.2. Thus, many steps involving several types of chromatography may be used in sample preparation for GC analysis.

The analysis of volatiles in foods (e.g., packaging or environmental contaminants, alcohols, and flavors or off-flavors) can be achieved by GC. These materials for GC analysis may be isolated by headspace analysis (static or dynamic), simple solvent extraction, distillation, preparative chromatography (e.g., solid-phase extraction, column chromatography on silica gel), or some combination of these basic methods. Table 14.1 summarizes various methods of isolation described in some detail in Sect. 14.2.2. The procedure used for a particular food will depend on the food matrix as well as the compounds to be analyzed. The primary considerations are to isolate the compounds of interest from nonvolatile food constituents (e.g., carbohydrates, proteins, vitamins) or those that would interfere with GC (e.g., lipids). Some of the chromatographic methods that might be applied to this task have been discussed in the basic chromatography chapter (Chap. 12) of this text. Methods for the isolation of volatile substances will be covered briefly as they pertain to the isolation of components for gas chromatographic analysis.

It should be emphasized that the isolation procedure used is critical in determining the results obtained. An improper choice of method or poor technique at this step negates the best gas chromatographic analysis of the isolated analytes. The influence of isolation technique on gas chromatographic analysis of aroma compounds has been demonstrated [10]. These biases are discussed in the sections that follow and in more detail in books edited by Marsili [11] and Mussinan and Morello [12]. While these books relate to the analysis of aroma compounds in foods, the techniques for the isolation of these volatiles are the same as used in the analysis of other volatiles in foods.

14.2.2.2 Headspace Methods

One of the simplest methods of isolating volatile compounds from foods is by direct injection of the headspace vapors above a food product. There are two types of headspace sampling: static headspace sampling and dynamic headspace sampling.

Static (i.e., **direct**) **headspace sampling** has been used extensively when rapid analysis is necessary and major component analysis is satisfactory. In static headspace sampling, a food sample is placed into a vessel, and the vessel is closed with an inert septum. At equilibrium, the headspace of the sample is taken using a gas-tight syringe and then directly injected onto the GC. Examples of method applications include measure-

14.1 table

Methods to isolate analytes from food for gas chromatography analysis

General types	Specific methods	Description	Advantages/applications	Disadvantages/limitations
Headspace sampling	Direct headspace sampling	Headspace taken from sample in closed vessel for direct injection into GC	Good for rapid analysis of very low-boiling point compounds	Low sensitivity
	Dynamic headspace sampling (purge and trap)	Sample is purged with inert gas to strip volatiles from the sample and retained by absorbent or cryogenic traps. The volatiles are then desorbed from the trap thermally or by organic solvent for GC analysis	Collects volatiles, no matter the polarity or boiling point. With adsorbent trap, no further extraction is required, and the system is readily automated	With cryogenic trap, usually must further extract with organic solvent and dry to concentrate. With adsorbent trap, get differential adsorption affinity and limited capacity
Distillations methods	Steam distillation (at normal pressure or in vacuum)	Use steam at atmospheric pressure or in vacuum to heat and codistill volatiles from the sample	Convenient and efficient	Solvent extraction is required on dilute aqueous solution collected
	Simultaneous distillation and extraction (SDE)	Product steam and solvent vapors are intermixed and condensed. The solvent extracts organic volatiles from condensed steam	Time savings due to one-step isolation and concentration. Less solvent use vs. what is needed to condense the sample in regular steam distillation	Get artifact formation due to elevated temperature
Solvent extraction	Simple batch extraction	Use typically organic solvent with, e.g., separatory funnel, to extract analytes of interest into the solvent	Efficient if multiple extractions and excessive shaking used	Need to remove nonvolatiles from the extracted sample
	Continuous extraction	Use organic solvent or CO_2 with equipment to make continuous	More efficient than batch extraction	Need to remove nonvolatiles from extracted sample. Requires elaborate equipment
	Solvent-assisted flavor evaporation (SAFE)	Sample with extracting solvent are added slowly to the SAFE device. Volatiles and solvent are distilled and collected in the cold trap	High efficiency. Operate at low temperature. Minimum thermal artifact formation	Need special device and high vacuum system

14.1 table (Continued)

General types	Specific methods	Description	Advantages/applications	Disadvantages/limitations
Solid-phase extraction (SPE)			Compared to traditional liquid-to-liquid extraction: less solvent, glassware, and time required, better precision and accuracy, minimum solvent evaporation for further analysis, readily automated	
	Simple solid-phase extraction (SPE)	The liquid sample is passed through the column of filter disk with chromatography stationary phase. Solutes with affinity are retained on the phase. After rinsing the column with water or weak solvent, a strong eluent is used to elute analytes of interest		
	Solid-phase microextraction (SPME)	Stationary phase is bound to fine fused silica filament. The filament is immersed in the sample or headspace of the sample and then pulled into the metal sheath that is forced through the septum of GC. Volatiles are thermally desorbed from the filament	Simple operation. No solvent contamination or disposal. Many phases of fiber available, so analytes with wide range of polarity and volatility can be analyzed	Poor reproducibility. Lot-to-lot variation: Short life of filament. Fiber is highly selective, sample saturation, competitive absorption. Highly influenced by other volatiles in the sample
	Solid-phase dynamic extraction (SPDE)	Similar to SPME, but the polymer is coated inside a special needle used to draw headspace of food. Volatiles are absorbed to the phase and then released when the needle is injected into GC	Similar to SPME, but less issues with analyte saturation and competition	Lack of commercial device
	Stir bar sorptive extraction (SBSE)	Magnetic stir bar is enclosed in a glass coated with absorbent film. The bar spins in the sample and absorbs analytes and then the stir bar is transferred to thermal desorption unit and delivered to GC column	Higher sensitivity and accuracy than SPME. Highly automated. Can extract less volatile compounds	Limited phase available, instrumentation is expensive, high operation cost, not good for solid samples (use HS-SBSE)
Direct injection		Inject 2–3-μl sample directly into the GC column	Simple, if the sample and column allow for this	Thermal degradation of nonvolatiles. Damage to the column. Decreased separation if water in the sample. Contamination of the column and injection port by nonvolatiles

ment of hexanal as an indicator of oxidation [13, 14] and 2-methylpropanal, 2-methylbutanal, and 3-methylbutanal as indicators of nonenzymatic browning [15]. The determination of residual solvents in packaging materials also may be approached by this method. Unfortunately, this method does not provide the sensitivity needed for trace analysis. Instrumental constraints typically limit headspace injection volumes to 5 mL or less. Therefore, only volatiles present in the headspace at concentrations greater than 10^{-7}-g/l headspace would be at detectable levels (using a flame ionization detector (FID)). Furthermore, the static headspace only allows the investigation of very low-boiling compounds.

Dynamic headspace sampling (i.e., **purge and trap**) has found wide usage in recent years (Fig. 14.1). In dynamic headspace method, the sample is purged with an inert gas, such as nitrogen or helium, which strips volatile constituents from the sample. This method may involve simply passing large volumes of headspace vapors through a cryogenic trap or, alternatively, an adsorption trap. A **cryogenic trap** (if properly designed and operated) will collect headspace vapors irrespective of compound polarity and boiling point. However, water is typically the most abundant volatile in a food product, and, therefore, this distillate must be extracted with an organic solvent, dried, and then concentrated for analysis. These additional steps add analysis time and provide opportunity for sample contamination and loss. A more commonly used technique is to use adsorbent traps. **Adsorbent traps** offer the advantages of providing a water-free volatile isolate (trap material typically has little affinity for water) and are readily automated. Tenax, charcoal, or synthetic porous polymers (Porapaks® and Chromosorbs®) are frequently used trapping materials. These polymers exhibit good thermal stability and reasonable capacity. The trapped volatiles are then recovered from the trap with a suitable solvent or by thermal desorption. For automated purge-trap system, the adsorbent traps are generally placed in a closed system and loaded, desorbed, and so on via the use of automated multiport valving systems. The automated closed system approach provides reproducible GC retention times and quantitative precision necessary for some studies. The primary disadvantage of adsorbent traps is their differential adsorption affinity and limited capacity. Therefore, the GC profile may only poorly represent the actual food composition due to biases introduced by the purging and trapping steps.

14.2.2.3 Distillation Methods

Distillation processes are quite effective at isolating volatile compounds from foods for GC analysis. Product moisture or outside steam is used to heat and codistill the volatiles from a food product. The most frequently used distillation method is steam distillation at atmospheric pressure or in vacuum. Steam

14.1 figure Purge and trap accessory for headspace sampling

distillation at normal pressure is a common method for isolating essential oils from plant materials such as hop oil. For most food analysis, this means that a very dilute aqueous solution of volatiles results, and a solvent extraction must be performed on the distillate to permit concentration for analysis. The distillation method most commonly used today is **simultaneous distillation and extraction** (SDE), modified from original Nickerson-Likens distillation head (Fig. 14.2) In this apparatus, a sample is boiled in one side flask and a small amount of extracting solvent in another. The product steam and solvent vapors are intermixed and condensed; the solvent extracts the organic volatiles from the condensed steam. The solvent and extracted distillate return to their respective flasks and are distilled to again extract the volatiles from the food. Its one-step isolation-concentration of flavor constituents allows a dramatic time saving over the separated operation and, because of their continuous recycling, a great reduction of organic solvents used. The drawback inherent in SDE is artifact formation. The elevated temperature applied during distillation may lead to lipid oxidation, Maillard browning, or Strecker reaction, which introduces errors. Also, the SDE system can be operated under vacuum, but it is not easy to regulate. Furthermore, some food aroma compounds, such as furaneol, may have low recovery by the SDE method [16]. While the distillation method is convenient and efficient, the volatile isolate can be contaminated by: (1) artifacts formed from solvents used in extraction, antifoam agents, and steam supply (contaminated water), (2) thermally-induced chemical changes, and (3) leakage of contaminated laboratory air into the system.

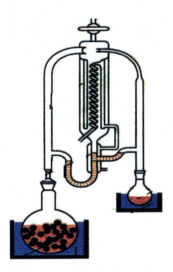

 Simultaneous distillation and extraction system for sample distillation

14.2.2.4 *Solvent Extraction*

Solvent extraction is often the preferred method for the recovery of volatiles from foods. Recovery of volatiles will depend upon solvent choice and the solubility of the analytes being extracted. Solvent extraction typically involves the use of an organic solvent (unless sugars, amino acids, or some other water-soluble components are of interest).

Solvent extractions may be carried out in quite elaborate equipment, such as supercritical CO_2 extractors, or can be as simple as a batch process in a separatory funnel. Batch extractions can be quite efficient if multiple extractions and extensive shaking are used [17]. The continuous extractors (liquid-liquid) are more efficient but require more costly and elaborate equipment.

Extraction with organic solvents limits the method to the isolation of volatiles from fat-free foods (e.g., wines; some breads, fruits, and berries; some vegetables; and alcoholic beverages). Even for the fat-free food systems, a small amount of fats, waxes, or other nonvolatile compounds can be extracted. These nonvolatile materials need to be removed; otherwise, they will interfere with subsequent concentration and GC analysis.

The nonvolatile compounds can be removed using **high vacuum distillation, molecular distillation, or solvent-assisted flavor evaporation** (SAFE) distillation. SAFE is a compact and versatile distillation apparatus (Fig. 14.3) which can offer fast and reliable isolation of volatiles from complex matrices [18]. During SAFE distillation, samples with extracting solvent are introduced slowly into the flask (top left) at low temperature under high vacuum, and volatiles and solvents are evaporated instantly and condensed in a cold trap (right, typically liquid nitrogen), leaving nonvolatiles in another flask. SAFE is highly efficient and can be used to isolate volatiles from solvent extracts or even directly from foods or beverages. The drawback is that compounds with high boiling points may not be completely recovered.

14.2.2.5 *Solid-Phase Extraction*

The extractions discussed above involve the use of two immiscible phases (water and an organic solvent). However, a newer and very rapidly growing alternative to such extractions is **solid-phase extraction** (SPE) [19, 20]. In one version of this technique, a liquid sample (most often aqueous based) is passed through a column (2–10 mL vol) filled with chromatographic packing or a filter disk (25–90 mm in diameter) that has the chromatographic packing embedded in it. The chromatographic packing (i.e., stationary phase coated on silica solid supports) may be any of a number of different materials (e.g., ion-exchange resins or a host of different reversed- or normal-phase HPLC column packings).

When a sample is passed through the cartridge or filter, analytes that have an affinity for the chromatographic phase will be retained on the phase, while those with little or no affinity will pass through. The phase is next rinsed with water, perhaps a weak solvent (e.g., pentane), and then a stronger solvent (e.g., dichloromethane). The strong eluent is chosen such that it will remove the analytes of interest.

SPE has been a very popular method for sample extraction and cleanup. Many SPE columns are commercially available in different phases and formats. C18, amino, and silica gel are commonly used phases for food component analysis. C18 cartridges have high retention for nonpolar compounds, whereas highly cross-linked styrene-divinylbenzene (DVB) copolymers can extract a wide range of nonpolar and polar compounds. The new generation of polymers (Oasis® HLB, copolymer of divinylbenzene and N-vinylpyrrolidone) can even extract lipophilic, hydrophilic, acidic, and basic compounds with a single cartridge and perform fractionation based on the functional group [21].

Overall, solid-phase extraction has numerous advantages over traditional liquid-liquid extractions including: (1) less solvent required, (2) speed, (3) less glassware needed (less cost and potential for contamination), (4) better precision and accuracy, (5) minimum solvent evaporation before GC analysis, and (6) being readily automated. However, SPE

 Solvent-assisted flavor evaporation (SAFE) distillation system

method development requires tedious empirical procedures to achieve the best separation and recovery for an analyte.

Another version of this method is called **solid-phase microextraction** (SPME). This method was developed originally for environmental work [22, 23]. In this adaptation, the phase is bound onto a fine fused silica or metal filament (e.g., fiber) (approximately the size of a 10-µl syringe needle, Fig. 14.4). The fiber is immersed in a sample or in the headspace above a sample. After the desired extraction time, the fiber is pulled into a protective metal sheath, removed from the sample, and forced through the septum of a gas chromatograph where the adsorbed volatiles are thermally desorbed from the fiber (Fig. 14.5).

Many different phases of fiber are commercially available, and compounds with a wide range of polarity or volatility can be analyzed. PDMS (polydimethylsiloxane) is a nonpolar phase coating and can be used to extract nonpolar compounds. Polar analytes can be extracted with polar phases (e.g., polyacrylate and Carbowax® coatings). Divinylbenzene (DVB) coating is good for many volatile compounds. The coating has various film thicknesses. Thicker film fibers (100 µm) are better for volatiles, whereas thinner film fibers (7 and 30 µm) are better for larger molecules. Multiphase fibers (such as Carboxen®/PDMS, Carboxen®/DVB/PDMS) are also available to extract both polar and nonpolar compounds. A Carboxen®/PDMS fiber is good for highly volatile compounds,

particularly for volatile sulfur compounds [24]. A 2-cm 50/30 DVB/Carboxen®/PDMS fiber is frequently used for volatile and semi-volatile flavor analysis in food system.

SPME has been widely used to extract volatile and semi-volatile organic compounds from environmental, biological, and food samples [25–29]. The main advantages of this technique are the simplicity of operation and no solvent contamination or disposal. With the autosampler, high precision and sample throughput can be achieved. However, depending on the fiber used, the compounds extracted can be highly selective. Furthermore, SPME has limited absorption capacity, and the fiber can be easily saturated. Thus, other volatile compounds in the sample can compete for the active site and cause competitive absorption. Other drawbacks are relatively poor reproducibility, fiber lot-to-lot variation, sensitivity to solvents, life time of the fiber, and difficulty in quantification.

Solid-phase dynamic extraction (SPDE) is another technique for volatile extraction. SPDE is similar to SPME, except the phase is coated inside a special needle. A gas-tight syringe is used for SPDE to draw the headspace of food and the volatiles are absorbed by the phase. The process can be repeated many times by moving the plunger up and down to achieve maximum absorption. The needle then can be injected into GC where the volatiles are thermally desorbed for analysis. Different phases are available. The volume of the phase is about 4.5 µl compared with only 0.6 µl for SPME, so the SPDE has less issue with analyte saturation and competition.

Stir bar sorptive extraction (SBSE) is a relatively new technique for volatile extraction. In SBSE (Fig. 14.6), a magnet stir bar is enclosed in a glass coated with a thick film of polymers such as polydimethylsiloxane (PDMS). The bar spins in the sample solution and absorb the analytes. Subsequently, the stir bar is transferred to a compact thermal desorption unit (TDU) mounted on a gas chromatograph (GC). The analytes are thermally desorbed in the TDU and delivered to a GC column (Fig. 14.7). The stir bar also can just hang in the headspace for volatile extraction the same way as the SPME.

The stir bar has 50–250 times more volume of absorbent than SPME. The PDMS volume is about 0.5 µl with SPME compared to 24–126 µl with SBSE. Due to the increased volume of absorbent phase, SBSE has much higher sensitivity than SPME and has minimum competition and saturation effects [30, 31]. The high sensitivity (ppt to ppg) and flexibility of PDMS-based SBSE for nonpolar and medium polar compounds make it an effective and time-saving method for extracting trace volatile compounds from complex matrices [31]. Food samples even containing fat (<3%) or alcohol (<10%) can be efficiently extracted with this technique.

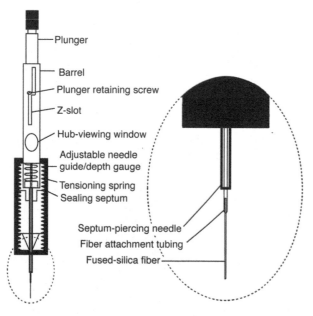

Plunger
Barrel
Plunger retaining screw
Z-slot
Hub-viewing window
Adjustable needle guide/depth gauge
Tensioning spring
Sealing septum
Septum-piercing needle
Fiber attachment tubing
Fused-silica fiber

14.4 **figure** Schematic of a solid-phase microextraction (SPME) device [21] (Courtesy of Dr. Janusz Pawliszyn, Dept. of Chemistry, University of Waterloo, Waterloo, Ontario, Canada)

Extraction Procedure

Pierce Sample
Septum Expose Fiber/
Extract Retract Fiber/
Remove

Desorption Procedure

Pierce GC Inlet
Septum Expose Fiber/
Desorb Retract Fiber/
Remove

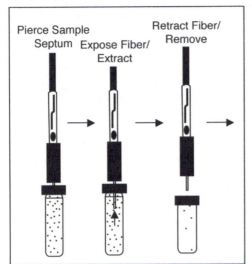
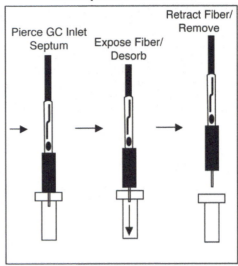

14.5
figure

Schematic showing the steps involved in the use of a solid-phase microextraction (SPME) device (Reprinted with permission of Supelco, Bellefonte, PA.)

SBSE is considered to be superior to SPME in terms of sensitivity and accuracy for determinations at trace level in difficult matrices. The PDMS phase is robust; it does not absorb water, alcohol, or pigment; and it is very good for flavor extraction of alcoholic beverages [32, 33]. PDMS-based SBSE had been successful applications in trace analysis in environmental, food, and biomedical fields [34–37]. However, the PDMS phase is not selective for polar compounds.

In addition to PDMS nonpolar phase, polyacrylate (PA) stir bar and the ethylene glycol (EG)-silicone stir bar are also commercially available. These phases, particularly EG-silicone, extract polar compounds more efficiently than the PDMS due to their polar nature. In addition, the EG-silicone can efficiently extract nonpolar compounds due to the properties of silicone materials. The use of new phase stir bars has been applied to different analytical fields including food and wine [38].

14.2.2.6 *Direct Injection*
It is theoretically possible to analyze some foods by direct injection of the food into a gas chromatograph. Assuming one can inject a 2–3-μl sample into a GC and the GC has a detection limit of 0.1 ng (0.1 ng/2 μl), one could detect volatiles in the sample at concentrations greater than 50 ppb. Problems with direct injection arise due to: (1) thermal degradation of any nonvolatile food constituents in the injector, (2) decreased separation efficiency due to water in the food sample, and (3) contamination of the column and injection port by nonvolatile materials. Despite these concerns, direct injection is sometimes used to determine oxidation in vegetable oils [39, 40]. A relatively large volume of oil (50–100 μl) can be directly injected into an injection

14.6
figure

Diagram of stir bar sorptive extraction (SBSE) device (Courtesy of Gerstel, Inc., Linthicum, MD)

port of a GC that has been packed with glass wool. Since vegetable oils are reasonably thermally stable and free of water, this method is particularly well suited to oil analysis.

There are numerous other approaches for the isolation of volatiles from foods. Some are simple variations of these methods, while others are unique. Several review articles are available that provide a more complete view of methodology [11, 12, 41].

14.2.3 Sample Derivatization

The compounds one wishes to determine by GC must be thermally stable under the GC conditions employed. Thus, for some compounds (e.g., pesticides, aroma compounds, polychlorinated biphenyls (PCBs), and volatile contaminants) the analyst can simply isolate the components of interest from a food as discussed above and directly inject them into the GC. For compounds that are thermally unstable, too low in volatility (e.g., sugars and amino acids), or yield poor chromatographic separation due to polarity (e.g., phenols or acids), a derivatization step must be included prior to GC analysis (see also Chaps. 19 and 23). A listing of some of the reagents used in preparing volatile derivatives for GC is given in Table 14.2. Most commonly used derivatizations are: (1) silylation for alcohols, cholesterol, and carbohydrates, (2) esterification for fatty acids, and (3) oxime formation and derivatization for aldehydes and ketones. The conditions of use for these reagents are often specified by the supplier or can be found in the

literature [42]. Derivatization can be performed in liquid extract, or on an SPE extraction column, on an SPME fiber, or on SBSE, to allow automated extraction, derivatization, and GC analysis.

14.3 GAS CHROMATOGRAPHIC HARDWARE AND COLUMNS

The major parts of a GC are the **gas supply** system, **injection port, oven, column, detector, electronics,** and **recorder/data handling system** (Fig. 14.8). The hardware as well as operating parameters used in any GC analysis must be accurately and completely recorded. The information that must be included is presented in Table 14.3.

14.3.1 Gas Supply System

The gas chromatograph will require at least a supply of carrier gas and, most likely, gases for the detector (e.g., hydrogen and air for a FID). The gases used must be of high purity and all regulators, gas lines, and fittings of good quality. High-quality pressure regulators must be used to provide stable and continuous gas supply. The regulators should have stainless steel rather than polymer diaphragms since polymers will give off volatiles that may contribute peaks to the analytical run. All gas lines must be clean and contain no residual drawing oil. Nitrogen, helium, and hydrogen gases are typically used as the **carrier gas** (i.e., mobile phase) to transport the analytes in the GC column. The carry gas line should have traps (moisture trap, oxy-

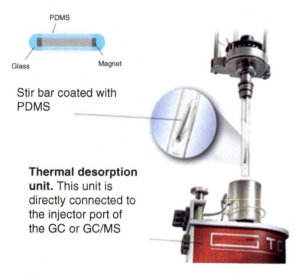

Stir bar coated with PDMS

Thermal desorption unit. This unit is directly connected to the injector port of the GC or GC/MS

Design of a commercial stir bar sorptive extraction device (Courtesy of Gerstel, Inc., Linthicum, MD) (www.gerstel.com)

14.2 table

Reagents used for making volatile derivatives of food components for GC analysis

Reagent	Chemical group	Food constituent

Silyl reagents
  Trimethylchlorosilane/hexamethyldisilazane
  BSA [N,O-bis(trimethylsilyl) acetamide
  BSTFA (N,O-bis (trimethylsilyl) trifluoroacetamide)
  t-BuDMCS (t-butyldimethylchlorosilyl/imidazole)
  TMSI (N-trimethylsilylimidazole)
	Hydroxy, amino carboxylic acids	Sugars, sterols, amino acids

Esterifying reagents
  Methanolic HCl
  Methanolic sodium methoxide
  N,N-Dimethylformamide dimethyl acetal
  Boron trifluoride methanol
	Carboxylic acids	Fatty acids, amines, amino acids, triglycerides, wax esters, phospholipids, cholesteryl esters

Miscellaneous		
Acetic anhydride/pyridine	Alcoholic and phenolic	Phenols, aromatic hydroxyl groups, alcohols
N-Trifluoroacetylimidizole/N-heptafluorobutyrlimidizole	Hydroxy and amines	Same as above
Alkylboronic acids	Polar groups on neighboring atoms	
O-Alkylhydroxylamine	Compounds containing both hydroxyl and carbonyl groups	Ketosteroids, prostaglandins

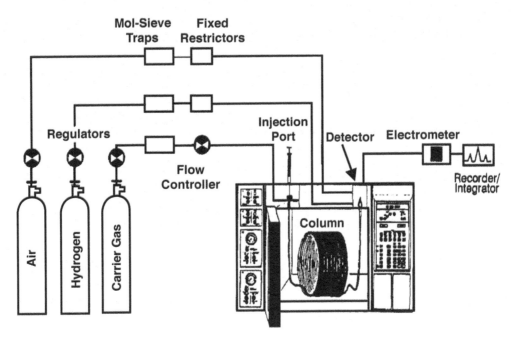

Mol-Sieve Traps   Fixed Restrictors

Regulators

Injection Port   Detector   Electrometer

Flow Controller

Air   Hydrogen   Carrier Gas

Column

Recorder/ Integrator

14.8 figure Diagram of a gas chromatographic system (Courtesy of Agilent Technologies, Inc., Santa Clara, CA)

gen trap, and hydrocarbon trap) in line to remove any moisture and contaminants from the incoming gas. These traps must be periodically replaced to maintain effectiveness.

14.3.2 Injection Port

14.3.2.1 *Hardware*

The injection port serves the purpose of providing a place for sample introduction, its vaporization, and possibly some dilution and splitting. Liquid samples make up the bulk of materials analyzed by GC, and they are always done by syringe injection (manual or automated). The injection port contains a soft septum that provides a gas-tight seal but can be penetrated by a syringe needle for sample introduction.

Samples may be introduced into the injection port using a **manual syringe technique** or an **automated sampling system**. Manual sample injection is generally the largest single source of poor precision in GC analysis. Ten-microliter syringes are usually chosen since they are more durable than the microsyringes, and sample injection volumes typically range from 1 to 3 μl. These syringes will hold about 0.6 μl in the needle and barrel (this is in addition to that measured on the barrel). Thus, the amount of sample that is injected into the GC depends upon the proportion of this 0.6 μl that is included in the injection and the ability of the analyst to accurately read the desired sample volume

14.3 table Gas chromatographic hardware and operating conditions to be recorded for all GC separations

Parameter	Description
Sample	Name and injection volume
Injection	Type of injection (e.g., split versus splitless and conditions (injection port flow rates))
Capillary column	Phase, length, diameter, film thickness, and manufacturer
Packed columns	Solid support, size mesh, coating, loading (%)
Temperatures	Injector, detector, oven, and any programming information
Carrier gas	Flow rate (velocity) and type
Detector	Type

on the syringe barrel. This can be quite variable for the same analyst and be grossly different between analysts. This variability between injections and the small sample volumes injected are the reasons internal (vs. external) standards are common for GC (see Chap. 12, Sect. 12.5.3).

14.3.2.2 *Sample Injection Techniques*

The sample must be vaporized in the injection port to pass through the column for separation. This vaporization can occur quickly by flash evaporation (standard injection ports) or slowly in a gentler manner (temperature-programmed injection port or on-column injection). The choice depends upon the thermal stability of the analytes. Due to the various sample as well as instrumental requirements, there are several different designs of injection ports available.

14.3.2.2.1 Split Injection

Capillary columns have limited capacity, and the injection volume may have to be reduced to permit efficient chromatography. The injection port may serve the additional function of splitting the injection so that only a portion of the analyte goes on the column (i.e., **split injection**) (Fig. 14.9, split vent valve open). The injection port is operated about 20 °C warmer than the maximum column oven temperature (commonly at 250 °C). The sample may be diluted with carrier gas to accomplish a split (1: 50–1: 100 preferred, split ratio = (column flow)/(column flow + venting flow), whereby only a small portion (1 part) of the analyte (more exactly, 1 part of gas flow) goes on the column, and the majority (49/50 or 99/100 parts) of the analytes are vented to the split vent. High split ratio typically gives a sharp, narrow peak.

14.3.2.2.2 Splitless Injection

To increase the sensitivity, a splitless injection mode can be used. In **splitless injection**, the split vent valve is closed and all of the analyte goes on the column (Fig. 14.5). Similar to the split injection, the temperature of the injector is operated at 20 °C higher than the maximum column oven temperature. Splitless injection requires to set up the initial column temperature 10–20 °C lower than the boiling point of the sample solvent, so the solvent can recondense in the column for acceptable chromatography of early eluting compounds (called solvent effect).

14.3.2.2.3 Programmed Temperature Vaporization Injection

For **programmed temperature vaporization injection** (PTV) ports, the sample is introduced into an ambient temperature port and then it is temperature programmed to some desired temperature. Since the sample is not introduced to the hot injector, the technique is desired for temperature-sensitive analytes. In addition, this technique is very useful to inject large amount sample when it is used together with split/splitless injection mode (solvent vent) to increase the sensitivity. For example, 10 μl of liquid sample can be injected at low temperature using a high split ratio to let the solvent to vent out, and then the injection mode can be changed to "splitless" as the injector is heated up to evaporate and transfer analytes onto the column.

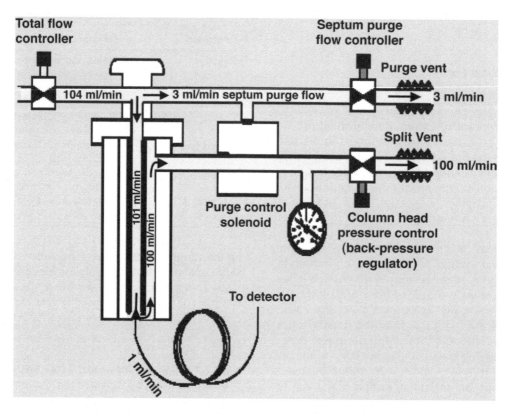

 Schematic of a GC injection port (Courtesy of Agilent Technologies, Inc., Santa Clara, CA)

14.3.2.2.4 *On-Column Injection*

On-column injection is a technique whereby the sample is directly introduced into the column for which the temperature is at that of the GC oven. The sample is then slowly volatized as the oven heats up. The initial oven temperature needs to be below the boiling point of the solvent. This technique is good for thermally labile analytes.

14.3.2.2.5 *Thermal Desorption Injection*

The volatiles can be introduced onto the head of a GC column for chromatographic separation directly from food samples through **thermal desorption**. The sample is heated in a thermal desorption unit, and the volatiles are carried through to a split/splitless injector. Cryofocusing with liquid nitrogen or CO_2 either in the injector or column is needed to attain sharp peaks. Alternatively, the volatiles can be retained using an absorbent such as a Tenax® trap during the purge stage and then thermally desorbed onto the column. The samples can be extracted with SPME or SBSE techniques described previously (Sect. 14.2.2.5) and then thermally desorbed onto the column for analysis. This technique has gained popularity to analyze volatile aroma compounds in foods including friuts [36, 43] and wine [45].

14.3.3 Oven

The oven controls the temperature of the column. In GC, one takes advantage of both an interaction of the analyte with the stationary phase and the boiling point for separation of compounds. Thus, the injection is often made at a lower oven temperature and is then temperature programmed to some elevated temperature. While analyses may be done isothermally, temperature-programmed runs are most common. It should be obvious that higher temperatures will cause the sample to elute faster and, therefore, be at a cost of resolution.

Oven temperature program rates can range from as little as 0.1 °C/min to the maximum temperature heating rate that the GC can provide. A rate of 2–10 °C/min is most common.

The capillary column (Sect. 14.3.4.2) also can be directly heated with an insulated heating wire based on **low thermal mass** (LTM) **technology**. A temperature sensor is mounted on the column. The column, the heating wire, and the sensor are all coiled together and wrapped with alumina foil. The column can be uniformly heated very rapidly to improve the separation and efficiency. Since the system does not have much void volume and other insulation materials, it cools very quickly. The total heating and cooling cycle is much shorter than the traditional standard GC oven, which makes it ideal for fast GC analysis. The module is available with almost any standard capillary GC column.

14.3.4 Column and Stationary Phases

The GC column may be either **packed** or **capillary**. Early chromatography was done on packed columns, but the advantages of capillary chromatography with regard to separation efficiency (see Sect. 14.4.2) so greatly outweigh those of packed column chromatography that few packed column instruments are sold any longer. While some use high-resolution gas chromatography (HRGC) to designate capillary GC, GC today means capillary chromatography to most individuals.

14.3.4.1 *Packed Columns*

The packed column is most commonly made of stainless steel or glass and may range from 1.6 to 12.7 mm in outer diameter and be 0.5–5.0 m long (generally 2–3 m). It is packed with a granular material consisting of a "liquid" coated on an allegedly inert solid support. The **solid support** is most often diatomaceous earth (skeletons of algae) that has been purified, possibly chemically modified (e.g., silane treated), and then sieved to provide a definite mesh size (60/80, 80/100, or 100/120).

The liquid loading is usually applied to the solid support at 1–10% by weight of the solid support. While the liquid coating can be any one of the approximately 200 available, the most common are silicone-based phases (methyl-, phenyl-, or cyano-substituted) and Carbowax™ (ester based).

The liquid phase and the percent loading are determined by the analysis desired. The choice of liquid is typically such that it is of similar polarity to the analytes to be separated. Loading influences time of analysis (retention time is proportional to loading), resolution (generally improved by increasing phase loading, within limits), and "bleed." The liquid coatings are somewhat volatile and will be lost from the column at high temperatures (this is dependent upon the phase itself). This results in an increasing baseline (**column bleeding**) during temperature programming.

14.3.4.2 *Capillary Columns*

The capillary column is a hollow fused silica glass (<100-ppm impurities) tube ranging in length from 5 to 100 m. The walls are so thin, ca. 25 µm, that they are flexible. The column outer walls are coated with a polyamide material to enhance strength and reduce breakage. Column inner diameters are typically 0.1 mm (**microbore**), 0.2–0.32 mm (**normal capillary**), or 0.53 mm (**megabore**).

Megabore columns (0.53 mm i.d.) were initially designed to replace packed columns without modification of instrumentation hardware. The most commonly used capillary columns are now 0.32 and 0.25 mm i.d. columns. Smaller diameters (0.10 mm and 0.18 mm i.d.) columns are used for fast GC analysis. The most common lengths of the GC column are 15, 30, and 60 m although special column can be over

100 m. Longer column will require longer analysis time. Although longer column gives improved resolution, this benefit of better separation is not particularly obvious due to already high-resolution power of capillary GC column.

As many as 200 different stationary phases have been developed for GC. As GC has changed from packed to capillary columns, fewer stationary phases are now in use since column efficiency has substituted for phase selectivity (i.e., high efficiency has resulted in better separations even though the stationary phase is less suited for the separation). Now, we find fewer than a dozen phases in common use (Table 14.3). The most durable and efficient phases are those based on polysiloxane (-Si-O-Si-).

Stationary phase selection involves some intuition, knowledge of chemistry, and help from the column manufacturer and the literature. There are general rules, such as choosing polar phases to separate polar compounds and the converse or phenyl-based column phase to separate aromatic compounds. However, the high efficiency of capillary columns often results in separation even though the phase is not optimal. For example, a 5% phenyl-substituted methyl silicone phase applied to a capillary column will separate most polar and nonpolar compounds (Table 14.4).

Liquid coating is chemically bonded to the glass walls of capillary columns and internally cross-linked at phase thicknesses ranging from 0.1 to 5 μm. Film thickness directly affects separation. Thicker films retain compounds longer in the stationary phase, thus the analytes will have longer interaction with the stationary phase to achieve separation. Generally, a thick-filmed column should be used to separate very volatile compounds. For example, an FFAP (polyethylene glycol treated with nitroterephthalic acid) column with 1-μm film thickness can effectively hold and separate hydrogen sulfide and other highly volatile sulfur compounds [24]. However, a thick film also will give higher baseline due to bleeding. A thin-film (0.25 μm) column is usually used to separate high-molecular-weight compounds; the analytes will stay in the stationary phase less time. A thin-film column also has less bleeding at high temperature, so it is used frequently for GC-MS.

Most compounds can be separated using nonpolar 5% phenyl 95% dimethylpolysiloxane-based columns (e.g., DB-5, Agilent-5, RTX-5). This type of column has a very wide temperature range (−60 °C to 325 °C) and is very stable. However, to separate very polar compounds such as alcohols and free fatty acids, a polar column is needed such as WAX (polyethylene glycol) or FFAP. A WAX type of column has superior

14.4 table   Common stationary phases

Composition	Polarity	Applications[a]	Phases with similar McReynolds constants[b]	Temperature limits[c]
100% Dimethyl polysiloxane (gum)	Nonpolar	Phenols, hydrocarbons, amines, sulfur compounds, pesticides, PCBs	OV-1 SE-30	−60 °C–325 °C
100% Dimethyl polysiloxane (fluid)	Nonpolar	Amino acid derivatives, essential oils	OV-101, SP-2100	0–280 °C
5% Phenyl 95% dimethyl polysiloxane	Nonpolar	Fatty acids, methyl esters, alkaloids, drugs, halogenated compounds	SE-52 OV-23 SE-54	−60 °C–325 °C
14% Cyanopropylphenyl-methyl polysiloxane	Intermediate	Drugs, steroids, pesticides	OV-1701	−200 °C–280 °C
50% Phenyl, 50% methyl methyl polysiloxane	Intermediate	Drugs, steroids, pesticides, glycols	OV-17	60–240 °C
50% Cyanopropylmethyl, 50% phenylmethyl polysiloxane	Intermediate	Fatty acids, methyl esters, alditol acetates	OV-225	60–240 °C
50% Trifluoropropyl polysiloxane	Intermediate	Halogenated compounds, aromatics	OV-210	45–240 °C
Polyethylene glycol-TPA modified	Polar	Acids, alcohols, aldehydes, acrylates, nitrites, ketones	OV-351 SP-1000	60–240 °C
Polyethylene glycol	Polar	Free acids, alcohols, esters, essential oils, glycols, solvents	Carbowax 20 M	60–220 °C

[a]Specific application notes from column suppliers provide information for choosing a specific column
[b]McReynolds constants are used to group stationary phases together on the basis of separation properties
[c]Stationary phases have both upper and lower temperature limits. Lower temperature limit is often due to a phase change (liquid to solid) and upper temperature limit to a volatilization of phase

separation power; however, it has a narrow usable temperature range (40–240 °C). It becomes solid (lost separation power) at low temperature and bleeds highly at high temperature. It is also sensitive to residue oxygen, so it deteriorates quickly if oxygen is not removed from the carrier gas. Cyanopropyl-based columns (SP-2560, CP-Sil 88) are good for *trans*-fatty acid esters. Other specialty phase columns have been developed to improve specific resolution. Ionic liquid-based GC columns [45, 46] have very good thermal stability and can be used to analyze very polar compounds. Ionic liquid column has been used to separate fatty acid methyl esters [47]. However, their applications in food systems are still limited. The ionic liquid column cannot easily replace WAX-based column due to the strong interactions with acids (octanoic and decanoic acids), resulting in poor chromatography for these compounds. The β-cyclodextrin-based column is useful to separate chiral isomers of essential oil and other volatile compounds [48].

14.3.4.3 Gas-Solid (PLOT) Chromatography

Gas-solid chromatography is a very specialized area of chromatography accomplished without using a liquid phase – the analyte interaction is with a porous material. This material has been applied both to packed and capillary columns. For the capillary column, the porous material is chemically or physically (by deposition) coated on the inner wall of the capillary, and the column is called a **porous-layer open-tabular** (PLOT) column. The most popular porous materials are alumina oxide, carbon, molecular sieve, and synthetic polymer such as Poropak® or Chromosorb® (trade names of polymers based on vinyl benzene). Separations usually involve water or other very volatile compounds such as head-

space gas composition (N_2, O_2, CO_2, CO) in packaged food and ethylene during fruit ripening and storage.

14.3.5 Detectors

There are numerous detectors available for GC, each offering certain advantages in either sensitivity (e.g., electron capture) or selectivity (e.g., atomic emission detector). The most common detectors are the **FID, thermal conductivity** (TCD), **electron capture** (ECD), **flame photometric** (FPD), **pulsed flame photometric detector** (PFPD), **photoionization** (PID) detectors, and **mass spectrometry** (MS) (see Chap. 11 on MS). The operating principles and food applications of these detectors are discussed below. The characteristics of these detectors are summarized in Table 14.5.

Not described in detail below with traditional GC detectors, but mentioned here is **GC-olfactory** (GC-O). In GC systems with an olfactory detector outlet, the column effluent is split so that a portion of the effluent goes to a "sniffing port" and the remainder goes to a GC detector. The sniffing port typically consists of a glass cone for the operator to use their nose to identify "aroma-active" components eluting from the column (see Chap. 35, Sect. 35.5.2.1).

14.3.5.1 Thermal Conductivity Detector

14.3.5.1.1 Operating Principles

As the carrier gas passes over a hot filament (tungsten), it cools the filament at a certain rate depending on carrier gas velocity and composition. The temperature of the filament determines its resistance to electrical current. As a compound elutes with the carrier gas, the cooling effect on the filament is typically less, resulting in a temperature increase in the

14.5 table

Characteristics of most common detectors for gas chromatography

Characteristic	Thermal conductivity detector	Flame ionization detector	Electron capture detector	Flame photometric detector	Photoionization detector
Specificity	Very little. Detects almost anything, including H_2O. Called the "universal detector"	Most organics	Halogenated compounds and those with nitro- or conjugated double bonds	Organic compounds with S or P (determined by which filter is used)	Depends on ionization energy of lamp relative to bond energy of analytes
Sensitivity limits	ca. 400 pg. Relatively poor. Varies with thermal properties of compound	10–100 pg for most organics. Very good	0.05–1 pg. Excellent	2 pg for S and 0.9 pg for P compounds. Excellent	1–10 pg depending on compound and lamp energy. Excellent
Linear range	10^4, poor. Response easily becomes nonlinear	10^6–10^7. Excellent	10^4. Poor	10^4 for P. 10^3 for S	10^7. Excellent

filament and an increase in resistance that is monitored by the GC electronics. Older style TCDs used two detectors and two matching columns; one system served as a reference and the other as the analytical system. Newer designs use only one detector (and column), which employs a carrier gas switching value to pass alternately carrier gas or column effluent though the detector (Fig. 14.10). The signal is then a change in cooling of the detector as a function of which gas is passing through the detector from the analytical column or carrier gas supply (reference gas flow).

The choice of carrier gas is important since differences between its thermal properties and the analytes determine response. While hydrogen is the best choice, He is most commonly used since H_2 is flammable.

14.3.5.1.2 Applications
The most valuable properties of this detector are that it is *universal* in response and nondestructive to the sample. Thus, it is used in food applications for which there is no other detector that will adequately respond to the analytes (e.g., water, permanent gases, CO) or when the analyst wishes to recover the separated compounds for further analysis (e.g., trap the column effluent for infrared, nuclear magnetic resonance (NMR), or sensory analysis). It does not find broad use because it is relatively insensitive, and often the analyst desires specificity in detector response to remove interfering compounds from the chromatogram. The detector is most useful for gas composition analysis (CO_2, CO, O_2, N_2) for food packaging and fruit ripening and self-life.

14.3.5.2 Flame Ionization Detector

14.3.5.2.1 Operating Principles
As compounds elute from the analytical column, they are burned in a hydrogen flame (Fig. 14.11). A potential (often 300 V) is applied across the flame. The flame will carry a current across the potential which is proportional to the organic ions present in the flame from the burning of an organic compound. The current flowing across the flame is amplified and recorded. The FID responds to organics on a weight basis. It gives virtually no response to H_2O, NO_2, CO_2, and H_2S and limited response to many other compounds. Response is best with compounds containing C-C or C-H bonds. The FID can be modified to include a methanizer to convert CO and CO_2 to methane using Ni catalyst and then detected by FID. An alternative is to oxidize all compounds to CO_2 first and then reduce to methane before FID. This approach increases the sensitivity and detectability for many carbon-containing compounds. In addition, since all compounds are converted to methane and have the same response on FID, the detector allows for the analysis of compounds without using standards.

14.3.5.2.2 Applications
The food analyst is most often working with organic compounds, to which this detector responds well. Its very good sensitivity, wide linear range in response (necessary in quantitation), and dependability make this detector the choice for most food work. Thus, this detector is used for virtually all food analyses for which a specific detector is not desired or sample destruction is acceptable (column eluant is burned in flame). This includes flavor studies, fatty acid analysis, carbohydrate analysis, sterols, contaminants in foods, and antioxidants.

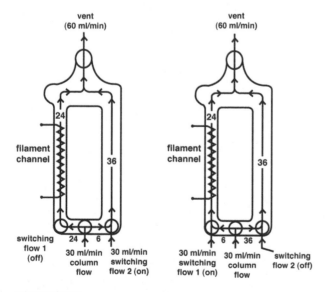

14.10 figure Schematic of the thermal conductivity detector (Courtesy of Agilent Technologies, Inc., Santa Clara, CA)

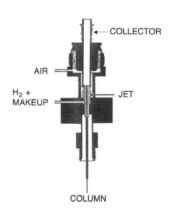

14.11 figure Schematic of the flame ionization detector designed for use with capillary columns (Courtesy of Agilent Technologies, Inc., Santa Clara, CA)

14.3.5.3 *Electron Capture Detector*

14.3.5.3.1 *Operating Principles*

The ECD contains a radioactive foil coating that emits electrons as it undergoes decay (Fig. 14.12). The electrons are collected on an anode, and the standing current is monitored by instrument electronics. As an analyte elutes from the GC column, it passes between the radioactive foil and the anode. Compounds that capture electrons reduce the standing current and thereby give a measurable response. Halogenated compounds or those with high electrophilic compounds (diketones) give the greatest detector response. Unfortunately this detector becomes saturated quite easily and thus has a very limited linear response range.

14.3.5.3.2 *Applications*

In food applications, the ECD has found its greatest use in determining PCBs and pesticide residues (see Chap. 33). The specificity and sensitivity of this detector make it ideal for this application. It also is used to analyze diacetyl and other vicinal diketone in beer because ECD is very sensitive for vicinal diketones.

14.3.5.4 *Flame Photometric Detector and Pulsed Flame Photometric Detector*

14.3.5.4.1 *Operating Principles*

The FPD works by burning all analytes eluting from the analytical column and then measuring specific wavelengths of light that are emitted from the flame using a filter and photometer (Fig. 14.13). The wavelengths of light that are suitable in terms of intensity and uniqueness are characteristic of sulfur (S) and phosphorus (P). Thus, this detector gives a greatly enhanced signal for these two elements (several thousand folds for S- or P-containing organic molecules versus non-S or P-containing organic molecules). Detector response to S-containing molecules is nonlinear and thus quantification must be done with care.

The PFPD is very similar to the FPD. Unlike traditional flame photometric detection (FPD), which uses a continuous flame, the PFPD ignites, propagates, and self-terminates two to four times per second (Fig. 14.14). Specific elements have their own emission profile: hydrocarbons will complete emission early, while sulfur emissions begin at a relatively later time after combustion. Therefore, a timed "gate delay" can selectively allow for only emissions due to sulfur to be integrated, producing a clean chromatogram. This timed "gate delay" greatly improves the sensitivity. The PFPD can detect sulfur-containing compounds at a much lower detection limit than nearly all other methods of detection [49].

14.3.5.4.2 *Applications*

Both the FPD and PFPD have found major food applications in the determination of organophosphorus pesticides and volatile sulfur compounds in general. The determination of sulfur compounds has typically been in relation to flavor studies.

14.3.5.5 *Photoionization Detector*

14.3.5.5.1 *Operating Principles*

The PID uses ultraviolet (UV) irradiation (usually 10.2 eV) to ionize analytes eluting from the analytical column (Fig. 14.15). The ions are accelerated by a polarizing electrode to a collecting electrode. The small current formed is magnified by the electrometer of the GC to provide a measurable signal.

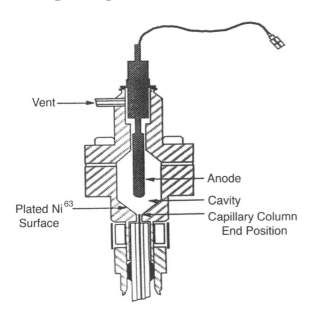

Vent

Anode

Cavity

Capillary Column End Position

Plated Ni 63 Surface

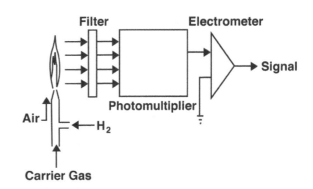

Filter

Electrometer

Signal

Photomultiplier

Air

H₂

Carrier Gas

14.12
figure

Schematic of the electron capture detector (Courtesy of Agilent Technologies, Inc., Santa Clara, CA)

14.13
figure

Schematic of the flame photometric detector (Courtesy of Agilent Technologies, Inc., Santa Clara, CA)

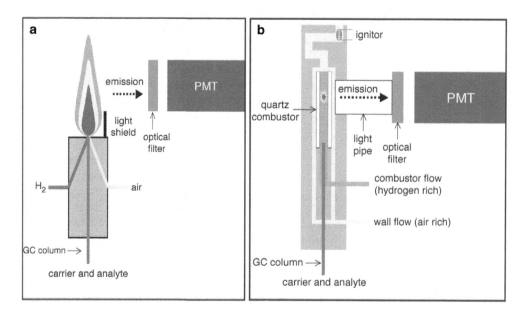

 14.14 Comparison of flame photometric detector (**a**) and pulsed flame photometric detector (**b**) (Courtesy of Varian Inc., Palo Alto, CA)

This detector offers the advantages of being quite sensitive and nondestructive and may be operated in a selective response mode. The selectivity comes from being able to control the energy of ionization, which will determine the classes of compounds that are ionized and thus detected.

14.3.5.5.2 Applications

The PID finds primary use in analyses for which excellent sensitivity is required from a nondestructive detector. This is most often a flavor application in which the analyst wishes to smell the GC effluent to determine the sensory character of the individual GC peaks. While this detector might find broader use, the widespread availability of the FID (which is suitable for most of the same applications) meets most of these needs.

14.3.5.6 Electrolytic Conductivity Detector

14.3.5.6.1 Operating Principles

Compounds entering the **electrolytic conductivity detector** (ELCD) are mixed with a reagent gas (oxidizing or reducing depending on the analysis) in a nickel reaction tube producing ionic species. These products are mixed with a deionized solvent, interfering ions are scrubbed from the effluent, and the ionic analyte-transformation product is detected within the electrolyte conductivity cell. This detector can be used for the specific detection of sulfur-, nitrogen-, or halogen-containing molecules. For example, when operated in the nitrogen mode, analyte is mixed with H_2 gas and hydro-

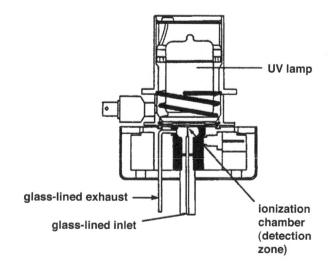

14.15 Schematic of the photoionization detector (Courtesy of Agilent Technologies, Inc., Santa Clara, CA)

genated over a nickel catalyst at 850 °C. Acidic hydrogenation products are removed from the effluent by passage through an $Sr(OH)_2$ trap, and the NH_3 from the analyte passes to the conductivity cell where it is measured [50].

14.3.5.6.2 Applications

This detector can be used in many applications for which element specificity is desired. Examples would be pesticide, herbicide, nitrosamine, or flavor analysis. The ELCD is very selective and quite sensitive, having detection limits of 0.1–1 pg of chlorinated compounds, 2 pg for sulfur, and 4 pg for nitrogen.

14.3.5.7 *Thermionic Detector*

14.3.5.7.1 *Operating Principles*
The thermionic detector (also called the nitrogen phosphorus detector, NPD) is a modified FID in which a nonvolatile ceramic bead is used to suppress the ionization of hydrocarbons as they pass through a low-temperature fuel-poor hydrogen plasma. The ceramic bead is typically composed of rubidium that is heated to 300–350 °C. Most commonly this detector is used for the selective detection of nitrogen- or phosphorus-containing compounds. It does not detect inorganic nitrogen or ammonia.

14.3.5.7.2 *Applications*
This detector is primarily used for the measurement of specific classes of flavor compounds, nitrosamines, amines, and pesticides.

14.3.5.8 *Hyphenated Gas Chromatographic Techniques*
Hyphenated gas chromatographic techniques are those that combine GC with another major technique. Examples are **GC-AED** (atomic emission detector), **GC-FTIR** (Fourier transform infrared), and **GC-MS** (mass spectrometry). While all of the techniques are established methods of analysis in themselves, they become powerful tools when combined with a technique such as GC. GC provides the separation and the hyphenated technique provides the detector. GC-MS has long been known to be a most valuable tool for the identification of volatile compounds (see Chap. 11). The MS, however, may perform the task of serving as a specific detector for the GC by selectively focusing on ion fragments unique to the analytes of interest. The analyst can detect and quantify components without their gas chromatographic resolution in this manner. The same statements can be made about GC-FTIR (see Chap. 9). The FTIR can readily serve as a GC detector.

In GC-AED, the GC column effluent enters a microwave-generated helium plasma that excites the atoms present in the analytes. The atoms emit light at their characteristic wavelengths. This results in a very sensitive and specific elemental detector.

14.3.5.9 *Multidimensional Gas Chromatography*
Multidimensional gas chromatography (MDGC) greatly increases the separation ability of gas chromatography [51]. By simply coupling two GC columns, each of opposite polarity, an overall improvement in separation can be accomplished [52]. However, this tandem operation of GC columns does not actually represent multidimensionality, but rather resembles the use of a mixed-stationary phase column [51]. True MDGC involves a process known as orthogonal separation for which a sample is first dispersed by one column, and the simplified subsamples are then applied onto another column for further separation. MDGC techniques can be generally divided into two classes: (1) conventional, or "heart-cut," MDGC and (2) comprehensive two-dimensional gas chromatography (GC×GC).

14.3.5.9.1 *Conventional Two-Dimensional GC*
Conventional two-dimensional GC is achieved by using coupled capillary columns for which a small portion, or heart-cut, of the effluent from the first ("pre-separation") column is transferred to the second ("analytical") column. The concept of conventional MDGC is almost identical to that of preparative GC operations, for which one column is used to obtain a partially separated fraction of a complex aroma mixture, which is then reinjected onto another GC column, usually with an opposite stationary phase, for further separation. The only difference is that with MDGC there are no requirements for manual collection of the effluent obtained from the pre-separation column since the two columns are directly connected.

Because the second column in the MDGC system is only injected with a small portion of the total sample at one time, a large quantity of the sample can be injected onto the first column without the worry of chromatographic band smearing during analytical separations [53]. Therefore, trace compounds can be easily enriched for more successful detection and identification.

The MDGC technique is particularly useful to study enantiomers of flavor compounds. The interested compound can be "heart-cut" and transferred to an analytical column with an enantioselective stationary phase for good separation of targeted chiral compounds.

14.3.5.9.2 *Comprehensive Two-Dimensional GC*
Comprehensive two-dimensional GC is among the most powerful two-dimensional gas chromatographic techniques that have been developed to date (Fig. 14.16). Unlike conventional MDGC in which only particular segments are transferred from the pre-separation column onto the analytical column, comprehensive MDGC, or GC×GC, involves the transfer of the entire effluent from the first column onto a second column by way of a modulation interface so that complete two-dimensional data can be obtained for the entire run of the first column. The operation of the modulator involves the generation of narrow injection bands from the first column, which are continuously, but individually, sent to the secondary column for final separation.

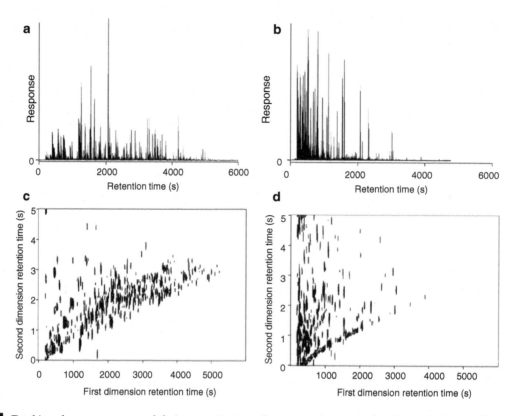

14.16 figure Total ion chromatograms and their respective two-dimensional contour plots for an Arabica coffee extract separated by GC×GC using two different column sets: polar × nonpolar (**a**, **c**) along with nonpolar×polar (**b**, **d**) (Reprinted from Ref. [54], used with permission)

GC×GC requires that the second column can operate quickly enough to generate a complete set of data during the time that a single peak elutes from the first GC column, generally within 5 s [51, 55]. The data from both time axes are combined to create a set of coordinates for each peak so that the resultant chromatogram is actually a two-dimensional (2D) plane rather than a straight line. Peak area information can be obtained by summing the integration over both dimensions.

In GC×GC, the two columns perform independently of each other; therefore, the overall peak capacity becomes the product of the capacities for each column. Because analytes elute from the second column so quickly, data acquisition must be adequately fast enough for proper detection. Time-of-flight mass spectrometry (TOF-MS) and rapid-scanning quadrupole mass spectrometry (qMS) have both been used as effective detection methods for GC×GC to obtain mass spectral information [56, 57].

GC×GC has been used in the field of foods and beverages [58–60]. Although the instrumentation can be quite expensive, the use of comprehensive GC×GC for volatile aroma analysis has exponentially increased over the past few years as methodologies have become more established and systems have become commercially available. Overall, the application of MDGC,

both conventional and comprehensive, has allowed for advanced separations of complex aromas to occur by using state-of-the-art instrumentation.

14.4 CHROMATOGRAPHIC THEORY

14.4.1 Introduction

GC may depend on several types (or principles) of chromatography for separation. The principles of chromatographic separations and chromatographic theory are discussed in Chap. 12, Sect. 12.4. For example, **size-exclusion chromatography** is used in the separation of permanent gases such as N_2, O_2, and H_2. A variation of size exclusion is used to separate chiral compounds on cyclodextrin-based columns; one enantiomorphic form will fit better into the cavity of the cyclodextrin than will the other form, resulting in separation. **Adsorption chromatography** is used to separate very volatile polar compounds (e.g., alcohols, water, and aldehydes) on porous polymer columns (e.g., Tenax[R] phase). **Partition chromatography** is the workhorse for gas chromatographic separations. There are over 200 different liquid phases that have been developed for gas chromatographic use over time. Fortunately, the vast majority of separations can be accomplished with only a few of these phases, and the other phases have fallen into disuse. GC depends not only upon adsorption, partition, and/or size exclusion for separation but also upon **analyte boiling point** for additional resolving powers. Thus, the separations accomplished are based on several properties of the analytes. This gives GC virtually unequaled resolution powers as compared to most other types of chromatography (e.g., HPLC, paper, or thin-layer chromatography).

A brief discussion of chromatographic theory will follow. The purpose of this additional discussion is to apply this theory to GC to optimize separation efficiency so that analyses can be done faster, less expensively, or with greater precision and accuracy. If one understands the factors influencing resolution in GC, one can optimize the process and gain in efficiency of operation.

14.4.2 Separation Efficiency

A good separation has narrow-based peaks and ideally, but not essential to quality of data, baseline separation of compounds. This is not always achieved. Peaks broaden as they pass through the column – the more they broaden, the poorer the separation and efficiency. As discussed in Chap. 12 (Sect. 12.5.2.2) a measure of this broadening is **height equivalent to a theoretical plate** (HETP). This term is derived from N, the number of plates in the column, and L, the length of the column. A good packed column might have $N = 5000$, while a good capillary column should have about 3000–4000 plates per meter for a total of 100000–500000 plates depending on column length. HETP will range from about 0.1 to 1 mm for good columns.

14.4.2.1 Carrier Gas Flow Rates and Column Parameters

Several factors influence column efficiency (peak broadening). As presented in Chap. 12, these are related by the **Van Deemter equation (Eq. 14.1)** (HETP values should be small):

$$\text{HETP} = Au^{1/3} + B/u + Cu \qquad (14.1)$$

where:

HETP = height equivalent to a theoretical plate
A = eddy diffusion
B = band broadening due to diffusion
u = velocity of the mobile phase
C = resistance to mass transfer

A is eddy diffusion; this is a spreading of the analytes in the column due to the carrier gas having various pathways or nonuniform flow (Fig. 14.17). In capillary chromatography, the A term is relatively very small compared to packed column chromatography.

However, as the diameter of the capillary column increases, the flow properties deteriorate, and band spreading occurs. The most efficient capillary columns have small diameters (0.1 mm), and efficiency decreases rapidly as one goes to megabore columns (Fig. 14.18). Megabore columns are only slightly more efficient than packed columns. While column efficiency increases as one goes to smaller columns, column capacity decreases rapidly. Microbore columns are easily overloaded (capacity may be 1–5 ng per analyte), resulting again in poor chromatography. Thus, column diameter is generally chosen as 0.2–0.32 mm to compromise efficiency with capacity.

B is band broadening due to diffusion; analytes will go from a high to a low concentration. The term u is velocity of the mobile phase. Thus, very slow flow rates result in large amounts of diffusion band broadening, and faster flow rates minimize this term. The term u is influenced by the carrier gas choice. Larger-molecular-weight carrier gases (e.g., nitrogen) are more viscous than the lighter-molecular-weight gases (e.g., helium or hydrogen), and thus peak spreading is less for nitrogen than for helium or hydrogen carrier gases. This results in nitrogen having the lowest HETP of the carrier gases and theoretically being the best choice for a carrier gas. However, other considerations that will be discussed in Sect. 14.4.2.2 make nitrogen a very poor choice for a carrier gas.

C is resistance to mass transfer. If the flow (u) is too fast, the equilibrium between the phases is not established, and poor efficiency results. This can be visualized in the following way: if one molecule of solute is dissolved in the stationary phase and another is not,

14.17
figure

Illustration of flow properties that lead to large eddy diffusion (Term A)

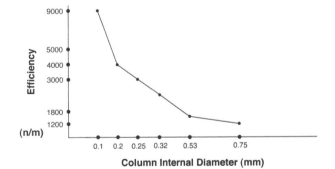

14.18
figure

The influence of column diameter on column efficiency (plates/meter) (Courtesy of Agilent Technologies, Inc., Santa Clara, CA)

the undissolved molecule continues to move through the column, while the other is retained. This results in band spreading within the column. Another factor that influences this term is thickness of the stationary phase. Thick films give greater capacity (ability to handle larger amounts of a solute) but at a cost in terms of band spreading (efficiency of separation) since thick films provide more variation in diffusion properties in and out of the stationary phase. Thus, phase thickness is a compromise between maximizing separation efficiency and sample capacity (with too much sample, the column is overloaded and separation ability destroyed). Capillary columns with phase thicknesses of 0.25–1 μm are commonly used for most applications.

If the Van Deemter equation is plotted, giving the figure discussed in Chap. 12 (Fig. 13), we see an optimum flow rate due to the opposing effects of the B and the C terms. It should be noted that the GC may not be operated at a carrier flow velocity yielding maximum efficiency (lowest HETP). Analysis time is directly proportional to carrier gas flow velocity. If the analysis time can be significantly shortened by operating above the optimum flow velocity and adequate resolution is still obtained, velocities well in excess of optimum should be used.

14.4.2.2 *Carrier Gas Type*
The relationship between HETP and carrier gas flow velocity is strongly influenced by carrier gas choice (Fig. 14.16). Nitrogen is the most efficient (lowest HETP) carrier gas, as discussed in Sect. 14.4.2.1, but its minimum HETP occurs at a very low flow velocity. This low mobile phase velocity results in unnecessarily long analysis times. Considering the data plotted in Fig. 14.19, nitrogen has an HETP of about 0.25 at an optimum flow velocity of 10 cm/s. The HETP of He is only about 0.35 at 40 cm/s flow velocity. This is a small loss in resolution to reduce the analysis time fourfold (10 cm/s for nitrogen versus 40 cm/s for helium). One can potentially even push the flow velocity up to 60 or 70 cm/s and accomplish separation in even shorter times.

The plots in Fig. 14.19 suggest that hydrogen is an even better choice for a carrier gas than He (i.e., has a flatter relationship between carrier gas flow velocity and HETP). However, there are some concerns about hydrogen being flammable and reports in the literature that some compounds may be hydrogenated in the GC system. Precautions should be taken when hydrogen is used as a carrier gas.

14.4.2.3 *Summary of Separation Efficiency*
In summary, an important goal of analysis is to achieve the necessary separation in the minimum amount of time. The following factors should be considered:

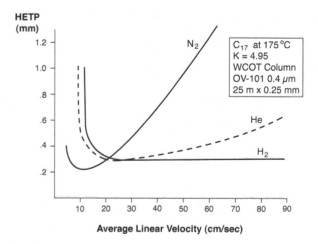

14.19 figure Influence of carrier gas type and flow rate on column efficiency (Courtesy of Agilent Technologies, Inc., Santa Clara, CA)

1. Column diameter: In general, small diameter columns should be used since separation efficiency is strongly dependent on column diameter. While small diameter columns will limit column capacity, limited capacity often can be compensated for by increasing phase thickness. Increased phase thickness also will decrease column efficiency but to a lesser extent than increasing column diameter.

2. Column operating temperature: Lower column operating temperatures should be used—if elevated column temperatures are required for the compounds of interest to elute, use a shorter column if resolution is adequate.

3. Column length: One should keep columns as short as possible (analysis time is directly proportional to column length—resolution is proportional to the square root of length).

4. Type of carrier gas: Use hydrogen as the carrier gas if the detector permits. Some detectors have specific carrier gas requirements.

5. Flow rate: Operate the GC at the maximum carrier gas velocity that provides resolution.

The pyramid shown in Fig. 14.20 summarizes the compromises that must be made in choosing the analytical column and gas chromatographic operating conditions. One cannot optimize any given operating conditions and column choices to get one of these properties without compromising another property. For example, optimizing chromatographic resolution (small bore capillary diameter, thin phase coating, long column lengths, and slow or optimum carrier gas flow rate) will be at the cost of capacity (large bore columns and thick phase coating) and speed (thin film coating, high carrier gas flow velocities, and short columns). Capacity will be at a cost of resolution and speed. The choice of column and operating parameters must consider the needs of the analyst and the compromises involved in these choices.

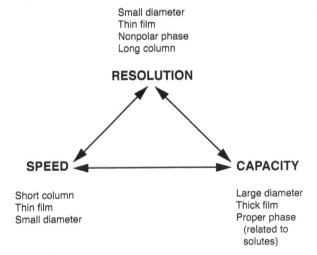

Small diameter
Thin film
Nonpolar phase
Long column

RESOLUTION

SPEED   **CAPACITY**

Short column   Large diameter
Thin film   Thick film
Small diameter   Proper phase
                                             (related to
                                             solutes)

14.20 figure

Relationships among column capacity, efficiency, resolution, and analysis speed

14.5   APPLICATIONS OF GC

While some detail on the application of GC to food analyses has been presented in Chaps. 19, 23, and 33, a few additional examples will be presented below to illustrate separations and chromatographic conditions.

14.5.1   Residual Volatiles in Packaging Materials

Residual volatiles in packaging materials can be a problem both from health (if they are toxic) and quality standpoints (produce off-flavors in the food). As the industry has turned from glass to polymeric materials, there have been more problems in this respect. GC is most commonly used to determine the residual volatiles in these materials [61].

The GC chromatogram presented in Fig. 14.21 illustrates the analysis of styrene and ethylbenzene monomers, from polystyrene (PS) packaging, which migrate into foodstuffs (in the current example from yogurt). Peak 6 and 9 represent ethylbenzene and styrene, respectively. The sample was extracted via HS-SPME with a divinylbenzene/Carboxen/polydimethylsiloxane fiber and analyzed by fast GC to determine the targeted VOC by headspace isolation [62].

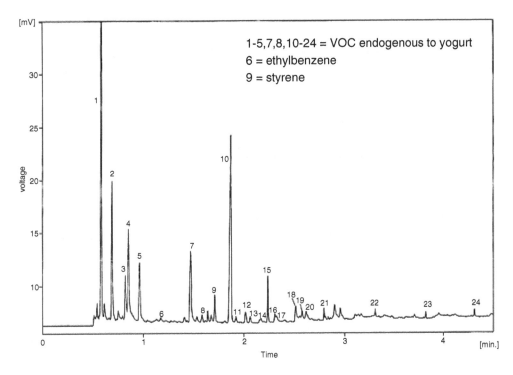

1-5,7,8,10-24 = VOC endogenous to yogurt
6 = ethylbenzene
9 = styrene

14.21 figure

SPME-Fast GC determination of packaging monomers in yogurt (From Verzera et al. [62], used with permission)

14.5.2 Multidimensional GC×GC-MS for the Generation of Reference Compounds

Volatile flavor analysis is a challenging analytical task, requiring the isolation of compounds of diverse chemical species from complex matrices often present at trace concentrations. Standard reference compounds are necessary to positively identify and quantify targeted flavor compounds and furthermore define the flavor significance by subsequent sensory analysis. However, not all compounds are readily commercially available without custom synthesis, which is often an expensive and time-consuming task. Furthermore, synthetic procedures can complicate human evaluation due to food grade protocols. Multidimensional GC×GC-MS methods provide an alternative approach to develop standards needed for flavor or food analysis. For example, the potent flavor compound 2-acetyl-pyrroline (2AP), which is an important ubiquitous

odorant in foodstuffs usually present in trace concentrations (ppb or less), described as having a cracker/popcorn-like odor character, is not available without custom synthesis. Figure 14.22 illustrates the application of GC×GC-MS for the isolation and purification of 2AP from a known botanical source of this compound, pandan leaf (*Pandanus amaryllifolius*):

Extraction and analysis details: An ether extract of the leaf material was purified using two-dimensional GC-MS system coupled with a fraction collector to yield a high purity compound standard. The analytical system consisted of an Agilent 6890 N GC coupled with an Agilent 5973 N MSD equipped with two Dean switch devices, two Gerstel modular accelerated column heaters (MACH), an Agilent 7683 autosampler, injector and a Gerstel preparative fraction collector (PFC). An RTX-5S ILMS (30 m×0.25 mm×0.25 μm) was used as a primary column, and a DB-Wax (30 m×0.25 mm×0.25 μm) was used as a secondary

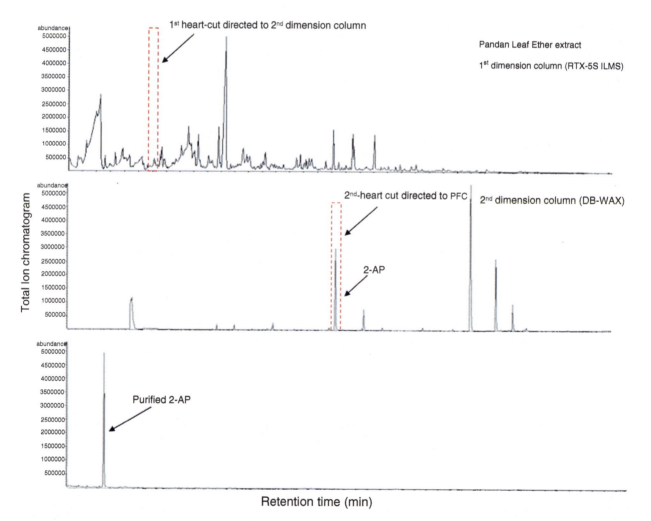

14.22 figure Application of multidimensional GC×GC-MS for the generation of standard compounds. Isolation of the flavor compound 2-acetyl-pyrroline (2AP) from pandan leaf extract

column. The valve-switching program for heart-cutting and isolation of 2AP was as follows: valve 1 switched on at 6.29 min, directing the flow to the secondary column, and switched off at 7.10 min; then valve 2 switched on at 15.4 min directing the flow to the cryotrap ($-70\,°C$) of the PFC for 2AP collection and switched off at 15.75 min.

In Fig. 14.22, the top chromatogram illustrates the first dimension column separation of the pandan leaf ether extract and the first "heart-cut" region containing 2AP, which was redirected to the second dimension column. The middle chromatogram is the resultant first dimension heart-cut that was reinjected on the second dimension column; the second dimension heart-cut region containing 2AP is also shown, which was again redirected to the fraction collector. The bottom chromatogram (analyzed by GC-MS/TOF) illustrates the high purity of the final isolate of 2AP obtained.

14.6 SUMMARY

GC has found broad application in both the food industry and academia. It is exceptionally well suited to the analysis of volatile thermally stable compounds. This is due to the outstanding resolving properties of the method and the wide variety of detectors that can provide either sensitivity or selectivity in analysis.

Sample preparation generally involves the isolation of solutes from foods, which may be accomplished by headspace analysis, distillation, preparative chromatography (including solid-phase extraction), or extraction (liquid-liquid). Some analytes can then be directly analyzed, while others must be derivatized prior to analysis.

The gas chromatograph consists of a gas supply and regulators (pressure and flow control), injection port, column and column oven, detector, electronics, and a data recording and processing system. The analyst must be knowledgeable about each of these GC components: carrier and detector gases; injection port temperatures and operation in split, splitless, temperature-programmed, or on-column modes; column choices and optimization (gas flows and temperature profile during separation); and detectors (TCD, FID, NPD, ECD, FPD, and PID). The characteristics of these GC components and an understanding of basic chromatographic theory are essential to balancing the properties of resolution, capacity, speed, and sensitivity.

Unlike most of the other chromatographic techniques, traditional GC has reached the theoretical limits in terms of both resolution and sensitivity. Thus, this method will not change significantly in the future other than for minor innovations in hardware or associated computer software. New development and applications will be more related to multidimensional GC including GC X GC.

GC as a separation technique has been combined with AED, FTIR, and MS as detection techniques to make GC an even more powerful tool. Such hyphenated techniques are likely to continue to be developed and refined, especially in the area of GC X GC-TOF-MS.

14.7 STUDY QUESTIONS

1. For each of the following methods to isolate analytes from food prior to GC analysis, describe the procedure, the applications, and the cautions in use of the method:

   (a) Headspace methods
   (b) Distillation methods
   (c) Solvent extraction

2. What is solid-phase extraction and why is it advantageous over traditional liquid-liquid extractions?
3. Why must sugars and fatty acids be derivatized before GC analysis, while pesticides and aroma compounds need not be derivatized?
4. Why is the injection port of a GC at a higher temperature than the oven temperature?
5. You are doing GC with a WAX column and notice that the baseline rises from the beginning to the end of each run. Explain a likely cause for this increase.
6. The most common detectors for GC are TCD, FID, ECD, FPD, and PID. Differentiate each of these with regard to the operating principles. Also, indicate below which detector(s) fits the description given:

   (a) Least sensitive
   (b) Most sensitive
   (c) Least specific
   (d) Greatest linear range
   (e) Nondestructive to sample
   (f) Commonly used for pesticides
   (g) Commonly used for volatile sulfur compounds

7. What types of chromatography does GC rely upon for separation of compounds?
8. In GC, explain why a balance has to be maintained between efficiency and capacity. Also, give an example situation in which you would sacrifice capacity for efficiency.
9. You plan to use GC to achieve good chromatographic separation of Compounds A, B, and C in your food sample. You plan to use an internal standard to quantitate each compound. By

answering the following questions, describe how using an internal standard works for this purpose (see also Chap. 12, Sect. 12.5.3):

(a) How do you choose the internal standard for your application?
(b) What do you do with the internal standard, relative to the standard solutions for Compounds A, B, and C and relative to the food sample? Be specific in your answer.
(c) What do you measure?
(d) If you were to prepare a standard curve, what would you plot?
(e) Why are internal standards commonly used for GC?

10. A fellow lab worker is familiar with HPLC for food analysis but not with GC. As you consider each component of a typical chromatographic system (and specifically the components and conditions for GC and HPLC systems), explain GC to the fellow worker by comparing and contrasting it to HPLC. Following that, state in general terms the differences among the types of samples appropriate for analysis by GC versus HPLC, and give several examples of food constituents appropriate for analysis by each (see also Chap. 13).

REFERENCES

1. James AT, Martin AJP (1952) Gas-liquid chromatography: the separation and microestimation of volatile fatty acids from formic acid to dodecanoic acid. Biochem J 50:679
2. Niessen WMA (2001) Current practice of gas chromatography – mass spectrometry. Marcel Dekker, New York
3. Rood D (1999) A practical guide to the care, maintenance, and troubleshooting of capillary gas chromatographic systems, 3rd edn. Weinheim, New York
4. Schomburg G (1990) Gas chromatography: a practical course. Weinheim: New York.
5. Gordon MH (1990) Principles and applications of gas chromatography in food analysis. E. Horwood, New York
6. O'Keeffe M (2000) Residue analysis in food: principles and applications. Harwood Academic, Amsterdam
7. Drawert F, Heimann W, Enberger R, Tressl R (1965) Enzymatische verandrung des naturlichen apfelaromass bei der aurfarbeitung. Naturwissenschaften 52:304
8. Fleming HP, Fore SP, Goldblatt LA (1968) The formation of carbonyl compounds in cucumbers. J Food Sci 33: 572
9. Kazeniak SJ, Hall RM (1970) Flavor chemistry of tomato volatiles. J Food Sci 35:519
10. Leahy MM, Reineccius GA (1984) Comparison of methods for the analysis of volatile compounds from aqueous model systems. In: Schreier P (ed) Analysis of volatiles: New methods and their application. DeGruyter, Berlin
11. Marsili R (1997) Techniques for analyzing food aroma. Marcel Dekker, New York
12. Mussinan CJ, Morello MJ (1998) Flavor analysis. American Chemical Society, Washington DC
13. Sapers GM, Panasiuk O, Talley FB (1973) Flavor quality and stability of potato flakes: Effects of raw material and processing. J Food Sci 38:586
14. Seo EW, Joel DL (1980) Pentane production as an index of rancidity in freeze-dried pork. J Food Sci 45:26
15. Buttery RG, Teranishi R (1963) Measurement of fat oxidation and browning aldehydes in food vapors by direct injection gas-liquid chromatography. J Agric Food Chem 11:504
16. Engel W, Bahr W, Schieberle P (1999) Solvent assisted flavor evaporation-a new and versatile technique for the careful and direct isolation of aroma compounds from complex food matrices. Z Lebensm Unters Forsch 209:237–241
17. Reineccius GA, Keeney PA, Weiseberger W (1972) Factors affecting the concentration of pyrazines in cocoa beans. J Agric Food Chem 20:202
18. Engel W, Bahr W, Schieberle P (1999) Solvent assisted flavour evaporation – a new and versatile technique for the careful and direct isolation of aroma compounds from complex food matrices. Eur Food Res Tech 209: 237–241
19. Majors RE (1986) Sample preparation for HPLC and GC using solid-phase extraction. LC – GC 4:972
20. Markel C, Hagen DF, Bunnelle VA (1991) New technologies in solid-phase extraction. LC – GC 9:332
21. Li KM, Rivory LP, Clarke SJ (2006) Solid-phase extraction (SPE) techniques for sample preparation in clinical and pharmaceutical analysis: a brief overview. Curr Pharma Anal 2: 95–102
22. Pawliszyn J (1997) Solid phase microextraction: theory and practice. VCH Publishers, New York
23. Zhang Z, Yang ML, Pawliszyn J (1994) Solid phase-microextraction: a solvent-free alternative for sample preparation. Anal Chem 66:844A–857
24. Fang Y, Qian MC (2005) Sensitive quantification of sulfur compounds in wine by headspace solid-phase microextraction technique. J Chromatogr A 1080:177–185
25. Verhoeven H, Beuerle T, Schwab W (1997) Solid-phase microextraction: artifact formation and its avoidance. J Chromatogr 46:63–66
26. Song J, Fan L, Beaudry RM (1998) Application of solid phase microextraction and gas chromatography/time-of-flight mass spectrometry for rapid analysis of flavor volatiles in tomato and strawberry fruits. J Agric Food Chem 46:3721–3726
27. Jetti RR, Yang EN, Kurnianta A, Finn C, Qian MC (2007) Quantification of selected aroma-active compounds in strawberries by headspace solid-phase microextraction gas chromatography and correlation with sensory descriptive analysis. J Food Sci 72:S487–S496
28. Kataoka H, Lord HL, Pawliszyn J (2000) Application of solid-phase microextraction in food analysis. J Chromatogr A 880:35–62
29. Vas G, Vekey K (2004) Solid-phase microextraction: A powerful sample preparation tool prior to mass spectrometric analysis. J Mass Spectrom 39:233–254
30. Pfanncoch E, Whitecavage J (2002) Stir bar sorptive extraction capacity and competition effects. Gerstel Global, Baltimore, MD, pp 1–8
31. David F, Tienpont B, Sandra P (2003) Stir-bar sorptive extraction of trace organic compounds from aqueous matrices. LC-GC Europe 16:410
32. Ou C, Du X, Shellie K, Ross C, Qian MC (2010) Volatile compounds and sensory attributes of wine from cv. Merlot (Vitis vinifera l.) grown under differential levels of

water deficit with or without a kaolin-based, foliar reflectant particle film. J Agric and Food Chem 58:12890–12898

33. Song J, Smart RE, Dambergs RG, Sparrow AM, Wells RB, Wang H, Qian MC (2014) Pinot noir wine composition from different vine vigour zones classified by remote imaging technology. Food Chem 153:52–59

34. David F, Sandra P (2007) Stir bar sorptive extraction for trace analysis. J Chromatogr A 1152:54–69

35. Kreck M, Scharrer A, Bilke S, Mosandl A (2001) Stir bar sorptive extraction (SBSE)-enantio-mdgc-ms - a rapid method for the enantioselective analysis of chiral flavour compounds in strawberries. Eur Food Res Technol 213:389–394

36. Malowicki SMM, Martin R,Qian MC (2008) Volatile composition in raspberry cultivars grown in the pacific northwest determined by stir bar sorptive extraction-gas chromatography-mass spectrometry. J Agric Food Chem 56:4128–4133

37. Malowicki SMM, Martin R, Qian MC (2008) Comparison of sugar, acids, and volatile composition in raspberry bushy dwarf virus-resistant transgenic raspberries and the wild type 'meeker' (Rubus idaeus l.). J Agric Food Chem 56:6648–6655

38. Zhou Q, Qian Y, Qian MC (2015) Analysis of volatile phenols in alcoholic beverage by ethylene glycol-polydimethylsiloxane based stir bar sorptive extraction and gas chromatography–mass spectrometry. J Chromatogr A 1390:22–27

39. Dupuy HP, Fore SP, Goldblatt LA (1971) Elution and analysis of volatiles in vegetable oils by gas chromatography. J Amer Oil Chem Soc 48:876

40. Legendre MG, Fisher GS,Fuller WH, Dupuy HP,Rayner ET (1979) Novel technique for the analysis of volatiles in aqueous and nonaqueous systems. J Amer Oil Chem Soc 56:552

41. Widmer HM (1990) Recent developments in instrumental analysis. In Flavor science and technology, Bessiere Y, Thomas AF, Eds. John Wiley & Sons: Chichester, England, 1990; p 181.

42. Blau K, King GS (1993) Handbook of derivatives for chromatography, Vol 2. Wiley, New York.

43. Du X, Qian M (2008) Quantification of 2,5-dimethyl-4-hydroxy-3(2h)-furanone using solid-phase extraction and direct microvial insert thermal desorption gas chromatography-mass spectrometry. J Chromatogr. A 1208:197–201

44. Fang Y, Qian MC (2006) Quantification of selected aroma-active compounds in pinot noir wines from different grape maturities. J Agric Food Chem 54:8567–8573

45. Armstrong DW, He L, Liu Y-S (1999) Examination of ionic liquids and their interaction with molecules, when used as stationary phases in gas chromatography. Anal Chem 71:3873–3876

46. Berthod A, Ruiz-Angel M, Carda-Broch S (2008) Ionic liquids in separation techniques. J Chromatogr A 1184:6–18

47. Delmonte P, Kia A-RF, Kramer JK, Mossoba MM, Sidisky L, Rader JI (2011) Separation characteristics of fatty acid methyl esters using slb-il111, a new ionic liquid coated capillary gas chromatographic column. J ChromatogrA 1218:545–554

48. Bicchi C, Manzin V, D'Amato A, Rubiolo P (1995) Cyclodextrin derivatives in gc separation of enantiomers of essential oil, aroma and flavour compounds. Flavour and Fragrance J 10:127–137

49. Amirav A, Jing H (1995) Pulsed flame photometer detector for gas chromatography. Anal Chem 67:3305–3318

50. Buffington R, Wilson MK (1987) Detectors for gas chromatography. Hewlett-Packard Corp., Avondale, PA

51. Shellie R, Marriott P (2003) Opportunities for ultra-high resolution analysis of essential oils using comprehensive two-dimensional gas chromatography: a review. Flavour Fragrance J 18:179–191

52. Merritt C (1971) Application in flavor research. In: Zlatkis A, Pretorius V, (Eds) Preparative gas chromatography, Wiley-Interscience, New York, 1971; pp 235–276

53. Kempfert KD (1989) Evaluation of apparent sensitivity enhancement in gc/ftir using multidimensional gc techniques. J Chromatogr Sci 27:63–70

54. Ryan D, Shellie R, Tranchida P, Cassilli A, Mondello L, Marriott P (2004) Analysis of roasted coffee bean volatiles by using comprehensive two-dimensional gas chromatography-time-of-flight mass spectrometry. J ChromatographyA 1054:57–65

55. Phillips JB, Xu J (1995) Comprehensive multi-dimensional gas chromatography (review). J Chromatogr A 703:327–334

56. Shellie R, Mondello L, Marriott P, Dugo G (2002) Characterisation of lavender essential oils by using gas chromatography-mass spectrometry with correlation of linear retention indices and comparison with comprehensive two-dimensional gas chromatography. J Chromatogr 970:225–234

57. Adahchour M, Brandt M, Baier H-U, Vreuls RJJ, Batenburg AM, Brinkman UAT (2005) Comprehensive two-dimensional gas chromatography coupled to a rapid-scanning quadrupole mass spectrometer: Principles and applications. J Chromatogr A 1067:245–254

58. Campo E, Cacho J, Ferreira V (2007) Solid phase extraction, multidimensional gas chromatography mass spectrometry determination of four novel aroma powerful ethyl esters. J Chromatogr A 1140:180–188

59. Campo E, Ferreira V, Lopez R, Escudero A, Cacho J (2006) Identification of three novel compounds in wine by means of a laboratory-constructed multidimensional gas chromatographic system. J Chromatogr A 1122:202–208

60. Adahchour M, Wiewel J, Verdel R, Vreuls RJJ, Brinkman UAT (2005) Improved determination of flavour compounds in butter by solid-phase (micro)extraction and comprehensive two-dimensional gas chromatography. J Chromatogr A 1086:99–106

61. Hodges K (1991) Sensory-directed analytical concentration techniques for aroma-flavor characterization and quantification. In: Risch SJ, Hotchkiss JH (Eds), Food packaging interactions ii, American Chemical Society, Washington, DC. p 174

62. Verzera A, Condurso C, Romeo V, Tripodi G, Ziino M. (2010). Solid-phase microextraction coupled to fast gas chromatography for the determination of migrants from polystyrene-packaging materials into yoghurt. Food Anal Methods 3(2):80–84

part

Compositional Analysis of Foods

Moisture and Total Solids Analysis

Lisa J. Mauer (⊠)
Department of Food Science,
Purdue University,
West Lafayette, IN 47907, USA
e-mail: mauerl@purdue.edu

Robert L. Bradley Jr
Department of Food Science,
University of Wisconsin,
Madison, WI 53706, USA
e-mail: rbradley@wisc.edu

S. Nielsen (ed.), *Food Analysis*, Food Science Text Series,
DOI 10.1007/978-3-319-45776-5_15, © Springer International Publishing 2017

15.1 INTRODUCTION

Moisture assays can be one of the most important analyses performed on a food product and yet one of the most difficult from which to obtain accurate and precise data. Water molecules are small and ubiquitous in the environment in which foods are produced, stored, and used. Moisture exchange between foods and the environment can lead to under- or overestimation from moisture assays, and water can be difficult to completely remove from foods. The first sections of this chapter describe various methods for moisture content analysis – their principles, procedures, applications, cautions, advantages, and disadvantages. Water activity measurement also is described later in this chapter, since it parallels the measurement of total moisture as an important stability and quality factor. Determining both the water content and the water activity of a food provides a complete moisture analysis. With an understanding of the techniques described, one can apply appropriate moisture analyses to a wide variety of food products.

15.1.1 Importance of Moisture Assays

One of the most fundamental and important analytical procedures that can be performed on a food product is an assay for the amount of moisture, referred to as the **moisture or water content** of the food [1–4]. In this context, the words "water" and "moisture" are generally used interchangeably. The dry matter that remains after moisture removal is commonly referred to as **total solids**. This analytical value is of great economic importance to a food manufacturer, and there are legal limits as to how much water must or can be present in some foods. Some examples in which moisture content is important to the food processor are provided in Table 15.1.

In addition to quantifying the amount of water in foods, it is also important to document the energy status of the water in the food by determining the **water activity** [5]. It is the water activity, more so than the moisture content, that influences microbial growth, physical properties, and chemical and enzymatic reactions in foods. Additionally, it is differences in water activity, not moisture content, that drive moisture migration between different food components (such as between a crust and a filling) or between a food and the environment. Water molecules move from regions of high water activity to regions of low water activity until equilibrium water activity is reached.

table 15.1 Importance of moisture content in the food industry

Moisture content is important for	Food examples
Preservation and stability	Dehydrated vegetables, potatoes, and fruits
	Dried milks and infant formulas
	Powdered eggs, coffees, and teas
	Spices and herbs
	Crispy fried or baked chips and crackers
	Cotton candy
Quality factors	Jams and jellies, to inhibit sugar crystallization
	Sugar syrups
	Prepared cereals: conventional 4–8% H_2O, puffed 7–8% H_2O
Convenience in packaging or shipping (usually reduced moisture)	Concentrated milks and fruit juices
	Liquid cane sugar (67% solids) and liquid corn sweetener (80% solids)
	Dehydrated products (which clump or become sticky and difficult to package if moisture content is too high)
Meeting compositional standards and standards of identity	Cheddar cheese must be ≤39% H_2O
	Enriched flour must be ≤15% H_2O
	Pineapple juice must have soluble solids of ≥10.5°Brix (conditions specified)
	Glucose syrup must have ≥70% total solids
	% H_2O in processed meats is commonly specified
Accurate computation of nutritional value	
Expressing results of other analytical determinations on a uniform basis	

15.1.2 Water in Foods

The amount, physical state, and location of water in foods will affect the types of analyses best suited to a particular food product, the ease and rate of water removal, the time required for assay equilibration, and the sample handling.

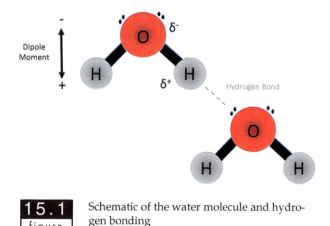

15.1 **figure**   Schematic of the water molecule and hydrogen bonding

15.2 **table**   Properties of water

Property	Value
Molecular weight	18.0153 g
Melting point (1 atm)	0.000 °C
Boiling point (1 atm)	100.000 °C
Enthalpy of vaporization (1 atm)	40.647 kJ/mol
Glass transition temperature (1 atm)	136 K
Critical temperature	373.99 °C
Density (20 °C)	0.99821 g/cm^3
Vapor pressure (20 °C)	2.3388 kPa
Heat capacity (20 °C)	4.1818 J/gK
Thermal conductivity (20 °C)	0.5984 W/mK
Thermal diffusivity (20 °C)	1.4×10^{-7} m^2/s
Dielectric constant (20 °C)	80.20
Water activity (20 °C)	1.000
Water activity (−20 °C)	0.82

Adapted from [5, 6, 38]

15.1.2.1 Structure of the Water Molecule

Water molecules are comprised of two hydrogen atoms covalently bound to an oxygen atom in a distorted tetrahedral arrangement (Fig. 15.1) [6]. There is a slight partially negative charge on the oxygen atom and a slight partially positive charge on the hydrogen atoms. Thus, water molecules are small and highly polar, have two hydrogen bond donor sites (the hydrogen atoms), and have two hydrogen bond acceptor sites (the two nonbonding electron pairs on the oxygen atom). Hydrogen bonds are relatively weak attractive interactions with short life spans (picoseconds), readily breaking and reforming as water molecules are constantly moving, particularly when temperature or humidity fluctuations occur [7, 8]. This poses challenges for accurately quantifying the amount or activity of water in a sample since the target can be moving from or escaping to the environment during sample handling.

15.1.2.2 Physical States and Properties of Water

The temperature, pressure, and extent of intermolecular hydrogen bonding between water molecules result in water existing in different physical states: as a solid (ice crystals), liquid (water), and gas (water vapor, humidity). Determining the total amount of water as well as the water activity in foods may involve analyzing water in all three of these states. Many direct moisture assays are based on the weight lost from evaporating all water from a sample, often by application of heat to overcome the enthalpy of vaporization. Other properties of water may be exploited for the indirect determination of moisture contents, including the dielectric constant, density, and freezing point. These and other properties of water are summarized in Table 15.2.

15.1.2.3 Water Interactions with Food Ingredients

Although water does not covalently bond with food ingredients, it can be difficult to remove from foods. Having some understanding of the water intermolecular interactions with food ingredients and the locations of water in foods can improve the application of moisture assays. The hydrogen bond donor and acceptor sites on water molecules interact with food ingredients via hydrogen bonds, dipole-dipole interactions, ionic attractions, and van der Waals forces [6]. Water molecules tend to cluster around ions and charged groups, as well as other hydrogen bond donor and acceptor groups such as hydroxyl groups (-OH), carbonyl oxygens (=O), and amine groups (-NH$_2$), adopting a more ordered structure in these hydration shell clusters than is found in bulk water [9]. To remove water from food, enough energy must be applied to overcome these intermolecular interactions. There are also different mechanisms by which water interacts with solids, including surface interactions (adsorption), condensed water (capillary condensation and deliquescence), and internalized water (absorption and crystal hydrate formation) [10]. Water also can be physically entrapped in food matrices, such as in gel structures or dense dehydrated or fried products, making it more difficult to remove than surface water. It is nearly impossible to remove all water from foods, particularly the internalized water for which heat and mass transfer rates and the properties of the food material can complicate complete moisture egress.

15.1.3 Sample Collection and Handling

General procedures for sampling, sample handling and storage, and sample preparation are given in Chap. 5. These procedures are perhaps the greatest potential source of error in any analysis. Precautions must be taken to minimize inadvertent **moisture losses** or **gains** that occur during these steps. Obviously, any exposure of a sample to the open atmosphere should be as short as possible. Any heating of a sample by friction during grinding should be minimized. Headspace in the sample storage container should be minimal because moisture transfer between the sample and the container environment will likely occur. It is critical to control temperature fluctuations since moisture will migrate in a sample to the colder part. To control this potential error, the entire sample should be removed from the container and reblended quickly, and a new test portion should then be sampled [11, 12].

To illustrate the need for optimum efficiency and speed in weighing samples for analysis, Bradley and Vanderwarn [13] showed, using shredded Cheddar cheese (2–3 g in a 5.5 cm aluminum foil pan), that moisture loss within an analytical balance was a straight line function. The rate of loss was related to the relative humidity (RH). At 50 % RH, it required only 5 s to lose 0.01 % moisture. This time doubled at 70 % RH or 0.01 % moisture loss in 10 s. While one might expect a curvilinear loss, the moisture loss was actually linear over a 5 min study interval. Samples with lower a_w than the environmental RH, such as many powders and crispy fried products, may pick up moisture during handling leading to overestimation of moisture contents. These examples demonstrate the necessity of absolute control during collection of samples through weighing, before drying or other analysis.

15.2 MOISTURE/WATER CONTENT

15.2.1 Overview

The moisture content of foods varies greatly, as shown in Table 15.3 [3]. Water is a major constituent of many food products. The approximate, expected moisture content of a food can affect the choice of the method of measurement. It also can guide the analyst in determining the practical level of accuracy required when measuring moisture content, relative to other food constituents. The method used for determining moisture content may measure more or less of the water present. This is the reason for official methods with stated procedures [14–16]. However, several official methods may exist for a particular product. For example, the AOAC International methods for cheese include Method 926.08, vacuum oven; 948.12, forced draft oven; 977.11, microwave oven; and 969.19, distillation [14]. Usually, the first method listed by AOAC International is preferred over others in any section.

15.3 table Moisture content of selected foods

Food item	Approximate percent moisture (wet weight basis)
Cereals, bread, and pasta	
Wheat flour, whole grain	10.3
White bread, enriched (wheat flour)	13.4
Corn flakes cereal	3.5
Crackers saltines	4.0
Macaroni, dry, enriched	9.9
Dairy products	
Milk, reduced fat, fluid, 2 %	89.3
Yogurt, plain, low fat	85.1
Cottage cheese, low fat or 2 % milk fat	80.7
Cheddar cheese	36.8
Ice cream, vanilla	61.0
Fats and oils	
Margarine, regular, hard, corn, hydrogenated	15.7
Butter, with salt	15.9
Oil – soybean, salad, or cooking	0
Fruits and vegetables	
Watermelon, raw	91.5
Oranges, raw, California navels	86.3
Apples, raw, with skin	85.6
Grapes, American type, raw	81.3
Raisins	15.3
Cucumbers, with peel, raw	95.2
Potatoes, microwaved, cooked in skin, flesh and skin	72.4
Snap beans, green, raw	90.3
Meat, poultry, and fish	
Beef, ground, raw, 95 % lean	73.3
Chicken, broilers and fryers, light meat, meat and skin, raw	68.6
Finfish, flatfish (flounder and sole species), raw	79.1
Egg, whole, raw, fresh	75.8
Nuts	
Walnuts, black, dried	4.6
Peanuts, all types, dry roasted with salt	1.6
Peanut butter, smooth style, with salt	1.8
Sweeteners	
Sugar, granulated	0
Sugar, brown	1.3
Honey, strained or extracted	17.1

From the US Department of Agriculture, Agricultural Research Service (2016) USDA National Nutrient Database for Standard Reference. Release 28. Nutrient Data Laboratory Home Page, http://ndb.nal.usda.gov [3]

The different types of moisture content assays can be generally categorized into direct and indirect methods. **Direct methods** for moisture content are often done by removing water, although the method for moisture removal may vary. Drying, distillation, and extraction are commonly used for moisture removal followed by weighing, volumetry, or titration to determine moisture content. **Indirect methods** are based on properties of the food that are related to the presence of water, such as capacitance, specific gravity, density, refractive index, freezing point, and electromagnetic absorption.

15.2.2 Oven Drying Methods

In **oven drying methods**, which are direct methods, the sample is heated under specified conditions, and the loss of weight is used to calculate the moisture content of the sample. The amount of moisture determined is highly dependent on the type of oven used, conditions within the oven, and the time and temperature of drying. Various oven methods are approved by AOAC International for determining the amount of moisture in many food products. The methods are simple, and many ovens allow for simultaneous analysis of large numbers of samples. The time required may be from ~1 h to over 24 h.

15.2.2.1 General Information

15.2.2.1.1 Removal of Moisture
Water evaporates more quickly at higher temperatures. Any oven method used to evaporate moisture has as its foundation the fact that the boiling point of water is 100 °C; however, this considers only pure water at sea level. According to Raoult's law, if 1 molecular weight (1 mol) of a solute is dissolved in 1.0 L of water, the boiling point would be raised by 0.512 °C. This boiling point elevation continues throughout the drying process as solute concentrations increase.

Moisture removal is sometimes best achieved in a two-stage process. Liquid products (e.g., juices, milk) are commonly pre-dried over a **steam bath** before drying in an oven. Products such as bread and field-dried grain are often air-dried and then ground and oven dried, with the moisture content calculated from moisture loss at both air and oven drying steps. Particle size, particle size distribution, sample sizes, hygroscopicity, and surface area during drying influence the rate and efficiency of moisture removal.

15.2.2.1.2 Decomposition of Other Food Constituents
Moisture loss from a sample during analysis is a function of time and temperature. Decomposition enters the picture when time is extended too much or temperature is too high. Thus, most methods for food moisture content analysis involve a compromise between

time and a particular temperature at which limited decomposition might be a factor. One major problem exists in that the physical process must separate all the moisture without decomposing any of the constituents that could release water. For example, carbohydrates decompose at elevated temperatures and release water to form dehydrated hydrocarbon compounds. The moisture generated in this decomposition leads to overestimation of the moisture content and is not the moisture that we want to measure. Certain other chemical reactions (e.g., sucrose hydrolysis) can result in utilization of moisture, which would reduce the moisture for measurement. The loss of **volatile constituents**, such as acetic, propionic, and butyric acids, and alcohols, esters, and aldehydes among flavor compounds, can also lead to errors. While weight changes in oven drying methods are assumed to be due to moisture loss, weight gains also can occur due to oxidation of unsaturated fatty acids and certain other compounds.

Nelson and Hulett [17] determined that moisture was retained in biological products to at least 365 °C, which is coincidentally near the critical temperature for water. Their data indicate that among the decomposition products at elevated temperatures were CO, CO_2, CH_4, and H_2O. These were not given off at any one particular temperature but at all temperatures and at different rates at the respective temperature in question.

By plotting moisture liberated against temperature, curves were obtained that show the amount of moisture liberated at each temperature (Fig. 15.2). Distinct breaks were shown that indicated the temperature at which decomposition became measurable.

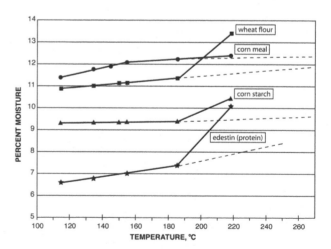

15.2 figure Moisture content of several foods held at various temperatures in an oven. The hyphenated line extrapolates data to 275 °C, the true moisture content (Reprinted with permission from [17]. Copyright 1920, American Chemical Society)

None of these curves showed any break before 184 °C. Generally, proteins decompose at temperatures somewhat lower than required for starches and celluloses. Extrapolation of the flat portion of each curve to 250 °C gave a true moisture content based on the assumption that there was no adsorbed water present at the temperature in question.

15.2.2.1.3 Temperature Control

Drying methods utilize specified drying temperatures and times, which must be carefully controlled. Moreover, there may be considerable variability of temperature, depending on the type of oven used for moisture analysis. One should determine the extent of variation within an oven before relying on data collected from its use.

Consider the temperature variation in three types of ovens: **convection (atmospheric)**, **forced draft**, and **vacuum**. The greatest temperature variation exists in a convection oven. This is because hot air slowly circulates without the aid of a fan. Air movement is obstructed further by pans placed in the oven. When the oven door is closed, the rate of temperature recovery is generally slow. This is dependent also upon the load placed in the oven and upon the ambient temperature. A 10 °C temperature differential across a convection oven is not unusual. This must be considered in view of anticipated analytical accuracy and precision. A convection oven should not be used when precise and accurate measurements are needed.

Forced draft ovens have the least temperature differential across the interior of all ovens, usually not greater than 1 °C. Air is circulated by a fan that forces air movement throughout the oven cavity. Forced draft ovens with air distribution manifolds have the added benefit of horizontal air movement across shelving. Thus, no matter whether the oven shelves are filled completely with moisture pans or only half filled, the result would be the same for a particular sample [13].

Two features of some **vacuum ovens** contribute to a wider temperature spread across the oven. One feature is a glass panel in the door. Although from an educational point of view it may be fascinating to observe some samples in the drying mode, the glass is a heat sink. The second feature is the way by which air is bled into the oven. If the air inlet and discharge are on opposite sides, conduct of air is virtually straight across the oven. Some newer models have air inlet and discharge manifolds mounted top and bottom. Air movement in this style of vacuum oven is upward from the front and then backward to the discharge in a broad sweep. The effect is to minimize cold spots as well as to exhaust moisture in the interior air.

15.2.2.1.4 Types of Pans for Oven Drying Methods

Pans used for moisture content determinations are varied in shape and may or may not have a cover. The AOAC International [14] moisture pan is about 5.5 cm in diameter with an insert cover. Other pans have covers that slip over the outside edge of the pan. These pans, while reusable, are expensive, in terms of labor costs to clean appropriately to allow reuse.

Pan covers are necessary to control loss of sample by spattering during the heating process. If the cover is metal, it must be slipped to one side during drying to allow for moisture evaporation. However, this slipping of the cover also creates an area where spattering will result in product loss. Examine the interior of most moisture ovens and you will detect odor and deposits of burned-on residue, which, although undetected at the time of occurrence, produce erroneous results and large standard deviations [13].

Consider the use of **disposable pans** whenever possible; then purchase **glass fiber discs** for covers. At 5.5 cm in diameter, these covers fit perfectly inside disposable aluminum foil pans and prevent spattering while allowing the surface to breathe. Paper filter discs foul with fat and thus do not breathe effectively. Drying studies done on cheese using various pans and covers have shown that fat does spatter from pans with slipped covers and fiberglass is the most satisfactory cover.

15.2.2.1.5 Handling and Preparation of Pans

The preparation and handling of pans before use require consideration. Use only **tongs** to handle any pan or wear gloves. Even the oils in fingerprints have weight. All pans must be oven treated to prepare them for use. This is a factor of major importance unless disproved by the technologist doing moisture determinations with a particular type of pan. Disposable aluminum pans must be vacuum oven dried for 3 h before use. At 3 and 15 h in either a vacuum or forced draft oven at 100 °C, pans varied in their weight within the error of the balance or 0.0001 g, and therefore longer drying times are not needed [13]. Store dried moisture pans in a functioning **desiccator**. The glass fiber covers should be dried for 1 h before use.

15.2.2.1.6 Control of Surface Crust Formation (Sand Pan Technique)

Some food materials tend to form a semipermeable crust or lump together during drying, which will contribute to erratic and erroneous results. To control this problem, analysts use the **sand pan technique**. Clean, dry sand and a short glass stirring rod are pre-weighed into a moisture pan. Subsequently, after weighing in a sample, the sand and sample are admixed with the stirring rod left in the pan. The remainder of the procedure follows a standardized method if available; otherwise the sample is dried to constant weight. The purpose of the sand is twofold: to prevent **surface crust** from forming and to disperse the sample so evaporation of moisture is less impeded. The amount of sand used is a function of sample size. Consider 20–30 g sand/3 g sample to obtain desired distribution in the pan. Similar to the procedure, applications, and advantages of using sand, other heat-stable inert materials such as diatomaceous earth can be used in moisture content determinations, especially for sticky fruits.

The inert matrices such as sand and diatomaceous earth function to disperse the food constituents and minimize retention of moisture in the food products. However, the analyst must ascertain that the inert matrix used does not give erroneous results for the assay because of decomposition or entrapped moisture loss. Test the sand or other inert matrix for weight loss before using in any method. Add approximately 25 g of sand into a moisture pan and heat at 100 °C for 2 h and weigh to 0.1 mg. Add 5 mL water and mix with the matrix using a glass rod. Heat dish, matrix, cover, and glass rod for at least 4 h at 100 °C and reweigh. The difference between weighing must be less than 0.5 mg for any suitable matrix [18].

15.2.2.1.7 Calculations

Moisture (wwb and dwb) and total solids contents of foods can be calculated as follows using oven drying procedures:

$$\% \text{ Moisture(wwb)}$$
$$= \frac{\text{wt of wet sample} - \text{wt of dry sample}}{\text{wt of wet sample}} \times 100$$
$$= \frac{\text{wt } H_2O \text{ in sample}}{\text{wt of wet sample}} \times 100 \tag{15.1}$$

$$\% \text{ Moisture(dwb)} = \frac{\text{wt } H_2O \text{ in sample}}{\text{wt of dry sample}} \times 100 \tag{15.2}$$

$$\% \text{ Total solids(wt/wt)}$$
$$= \frac{\text{wt of dry sample}}{\text{wt of wet sample}} \times 100$$
$$= 100 - \% \text{ Moisture (wwb)} \tag{15.3}$$

15.2.2.2 Forced Draft Oven

When using a forced draft oven, the sample is rapidly weighed into a pre-dried moisture pan, covered, and placed in the oven for an arbitrarily selected time if no standardized method exists. Drying time periods for this method are 0.75–24 h (Table 15.4), depending on the food sample and its pretreatment; some liquid samples are dried initially on a steam bath at 100 °C to minimize spattering. In these cases, drying times are shortened to 0.75–3 h. A forced draft oven is used with or without a steam table pre-drying treatment to determine the solids content of fluid milks (AOAC Method 990.19, 990.20).

An alternative to selecting a time period for drying is to weigh and reweigh the dried sample and pan until two successive weighings taken 30 min apart agree within a specified limit, for example, 0.1–0.2 mg for a 5 g sample. The user of this second method must be aware of sample transformation, such as browning which suggests moisture loss of the wrong form. Lipid oxidation and a resulting sample weight gain can occur at high temperatures in a forced draft oven. Samples high in carbohydrates should not be dried in a forced draft oven but rather in a vacuum oven at a temperature no higher than 70 °C.

15.2.2.3 Vacuum Oven

By drying under reduced pressure (25–100 mm Hg), the rate of evaporation is faster, and one is able to obtain a more complete removal of water without decomposition within a 3–6 h drying time. Vacuum

15.4 table Forced draft oven temperature and times for selected foods

Product	Dry on steam bath	Oven temperature (°C ± 2)	Time in oven (h)
Buttermilk, liquid	X[a]	100	3
Cheese, natural type only		100	16.5 ± 0.5
Chocolate and cocoa		100	3
Cottage cheese		100	3
Cream, liquid and frozen	X	100	3
Egg albumin, liquid	X	130	0.75
Egg albumin, dried	X	100	0.75
Ice cream and frozen desserts	X	100	3.5
Milk	X	100	3
Whole, low fat, and skim		100	3
Condensed skim		100	3
Nuts: almonds, peanuts, walnuts		130	3
Fruit, dried		70	6
Coffee, roasted		70	16

From Wehr and Frank [15] p. 492, with permission
[a]X = samples must be partially dried on steam bath before being placed in oven

ovens need a dry air purge in addition to temperature and vacuum controls to operate within method definition. In older methods, a vacuum flask is used, partially filled with concentrated sulfuric acid as the desiccant. One or two air bubbles per second are passed through the acid. Recent changes now stipulate an air trap that is filled with calcium sulfate containing an indicator to show water saturation (such as DrieRite™). Between the trap and the vacuum oven is an appropriately sized rotameter to measure air flow (100–120 ml/min) into the oven.

The following are important points in the use of a vacuum drying oven:

1. **Temperature** used depends on the product, such as 95–102 °C for some foods and lower temperatures (60–70 °C) for fruits and other high-sugar products. Even with reduced temperature, there can be some decomposition.
2. If the product to be assayed has a high concentration of **volatiles**, you should consider the use of a correction factor to compensate for the loss.
3. Analysts should remember that in a **vacuum**, heat is not conducted well. Thus pans must be placed directly on the metal shelves to conduct heat.
4. **Evaporation** is an endothermic process; thus, a pronounced cooling is observed. Because of the cooling effect of evaporation, when several samples are placed in an oven of this type, you will note that the temperature will drop. Do not attempt to compensate for the cooling effect by increasing the temperature; otherwise samples during the last stages of drying will be overheated.
5. The **drying time** is a function of the total moisture present, nature of the food, surface area per unit weight of sample, whether sand is used as a dispersant, and the relative concentration of sugars and other substances capable of retaining moisture or decomposing. The drying interval is determined experimentally to give reproducible results.

15.2.2.4 Microwave Analyzer

Determination of moisture contents in food products has traditionally been done using a standard oven, which, though accurate, can take many hours to dry a sample. **Microwave moisture analysis**, often called **microwave drying**, was the first precise and rapid technique that allowed some segments of the food industry to make in-process adjustment of the moisture content in food products before final packaging. For example, processed cheese could be analyzed, and the composition adjusted before the blend was dumped from the cooker. The ability to adjust the composition of a product in-process helps food manufacturers reduce production costs, meet regulatory requirements, and ensure product consistency. Such control could effectively pay for the microwave analyzer within a few months.

A particular microwave moisture/solids analyzer (CEM Corporation, Matthews, NC), or equivalent, is specified in the AOAC International procedures for total solids analysis of processed tomato products (AOAC Method 985.26) and moisture content analysis of meat and poultry products (AOAC Method 985.14).

The general procedure for use of a microwave moisture/solids analyzer has been to set the microprocessor controller to a percentage of full power to control the microwave output. Power settings are dependent upon the type of sample and the recommendations of the manufacturer of the microwave moisture analyzer. Next, the internal balance is tared with two sample pads on the balance. As rapidly as possible, a sample is placed between the two pads, and then pads are centered on the pedestal and weighed against the tare weight. Time for the drying operation is set by the operator and "start" is activated. The microprocessor controls the drying procedure, with percentage moisture indicated in the controller window. Some newer models of microwave moisture analyzers have a temperature control feature to precisely control the drying process, removing the need to guess appropriate time and power settings for specific applications. These new models also have a smaller cavity that allows the microwave energy to be focused directly on the sample.

There are some considerations when using a microwave analyzer for moisture determination: (1) the sample must be of a uniform, appropriate size to provide for complete drying under the conditions specified; (2) the sample must be centrally located and evenly distributed, so some portions are not burned and other areas under-processed; and (3) the amount of time used to place an appropriate sample weight between the pads must be minimized to prevent moisture loss or gain before weight determination. Sample pads also should be considered. There are several different types, including fiberglass and quartz fiber pads. For optimum results, the pads should not absorb microwave energy, as this can cause the sample to burn, nor should they fray easily, as this causes them to lose weight and can affect the analysis. In addition, they should absorb liquids well.

Another style of microwave oven that includes a vacuum system is used in some food plants. This vacuum microwave oven will accommodate one sample in triplicate or three different samples at one time. In 10 min, the results are reported to be similar to 5 h in a vacuum oven at 100 °C. The vacuum microwave oven is not nearly as widely used as conventional microwave analyzers, but can be beneficial in some applications.

Microwave drying provides a fast (4–8 min), accurate method to analyze many foods for moisture content. The method is sufficiently accurate for routine assay. The distinct advantage of rapid analysis far outweighs its limitation of testing only single samples [19].

15.2.2.5 Infrared Drying

Infrared drying involves penetration of heat into the sample being dried, as compared to heat conductivity and convection with conventional ovens. Such heat penetration to evaporate moisture from the sample can significantly shorten the required drying time to 10–25 min. The infrared lamp used to supply heat to the sample results in a filament temperature of 2000–2500 K. Factors that must be controlled include distance of the infrared source from the dried material and thickness of the sample. The analyst must be careful that the sample does not burn or case harden while drying. Infrared drying ovens may be equipped with forced ventilation to remove moist air and an analytical balance to read moisture content directly. No infrared drying moisture analysis techniques are approved by AOAC International currently. However, because of the speed of analysis, this technique is suited for qualitative in-process use.

15.2.2.6 Rapid Moisture Analyzer Technology

Many rapid moisture/solids analyzers based on thermogravimetric principles are used by the food industry. In addition to those based on infrared and microwave drying as described previously, compact instruments that depend on high heat created by different types of heaters are available. Two main categories of heating elements include halogen heaters (e.g., Halogen Moisture Analyzers, Mettler Toledo, Columbus, OH) and ceramic heaters (e.g., Computrac®, Arizona Instrument LLC, Chandler, AZ). These analyzers detect moisture levels from 50 ppm to 100 % using sample weights of 150 mg to 54 g. Smaller samples tend to dry more quickly, but it is important to use enough sample to be representative of the product. The test sample is placed on an aluminum pan or filter paper, and the heat control program (with a heating range of 25 °C to >200 °C) elevates the test sample to a constant temperature. As the moisture is driven from the sample, the instrument automatically weighs and calculates the percentage moisture or solids. The samples are not removed from the oven which minimizes weighing errors, and accurate results are obtained within minutes. These analyzers are utilized for both production

and laboratory assays with results comparable to reference methods.

15.2.2.7 Thermogravimetric Analyzer

In a thermogravimetric analyzer (TGA), the mass of a sample is continuously measured as it is heated at a controlled rate in a controlled atmosphere. The sample (often 10–50 mg) is loaded into a pan that is then placed into the TGA instrument that contains a furnace and a precision balance. Foods can lose mass during heating through release of adsorbed compounds (such as water), chemical reactions, and decomposition. To determine moisture content, the sample chamber is purged with an inert gas (e.g., nitrogen) so the sample only reacts with temperature during heating and mass changes due to oxidation are minimized. The furnaces in TGA instruments cover a wider temperature range than many other moisture content determination techniques and may range from sub-ambient temperatures to >1500 °C. This enables not only moisture assessment (moisture content, temperature of dehydration, stoichiometry of a hydrate, and dehydration kinetics) but also at higher temperatures measurement of pyrolysis, decomposition, and weight % ash. More detailed information on TGA is provided in the chapter on thermal analysis (Chap. 30). To use TGA, a survey scan is often run on a sample from approximately 100 °C below to 100 °C above the transition of interest at a heating rate of 20 °C per minute. For water content determination of many foods, scans from ambient to 200 °C are generally sufficient, and water content is determined by the mass lost up to ~100 °C. A derivative plot of the rate of mass loss will generate a peak showing the onset, midpoint, and endpoint temperatures for water loss. The precision balance and *in situ* mass monitoring result in precise and accurate moisture content determinations; however, decomposition or volatile loss at temperatures overlapping water evaporation can lead to errors in moisture content determination.

15.2.3 Distillation Procedures

15.2.3.1 Overview

Distillation is used as another direct measure of moisture content. Distillation techniques involve codistilling the moisture in a food sample with a high boiling point solvent that is immiscible in water, collecting the mixture that distills off, and then measuring the volume of water. Two distillation procedures are in use today: **direct** and **reflux distillations**, with a variety of solvents. For example, in direct distillation with immiscible solvents of higher boiling point than water, the sample is heated in mineral oil or liquid with a flash point well above the boiling point for water. Other immiscible liquids with boiling point only slightly above water can be used (e.g., toluene, xylene, and benzene). However, reflux distillation with the immiscible solvent toluene is the most widely used method.

Distillation techniques were originally developed as rapid methods for quality control work, but they are not adaptable to routine testing. The distillation method is an AOAC-approved technique for moisture content analysis of spices (AOAC Method 986.21), cheese (AOAC Method 969.19), and animal feeds (AOAC Method 925.04). It also can give good accuracy and precision for nuts, oils, soaps, and waxes.

Distillation methods cause less thermal decomposition of some foods than oven drying at high temperatures. Adverse chemical reactions are not eliminated but can be minimized by using a solvent with a lower boiling point. This, however, will increase distillation times. Water is measured directly in the distillation procedure (rather than by weight loss), but reading the volume of water in a receiving tube may be less accurate than using a weight measurement.

15.2.3.2 *Reflux Distillation with Immiscible Solvent*

Reflux distillation uses either a solvent less dense than water (e.g., toluene, with a boiling point of 110.6 °C, or xylene, with a boiling range of 137–140 °C) or a solvent more dense than water (e.g., tetrachlorethylene, with a boiling point of 121 °C). The advantage of using this last solvent is that the material to be dried floats; therefore it will not char or burn. In addition, there is no fire hazard with this solvent.

A **Bidwell-Sterling moisture trap** (Fig. 15.3) is commonly used as part of the apparatus for reflux distillation with a solvent less dense than water. The distillation procedure using such a trap requires about 1 h and involves using a brush to dislodge adhering water drops from the glassware, thereby minimizing error.

Three potential sources of error with distillation should be eliminated if observed:

1. Formation of water-solvent emulsions that will not break. Usually this can be controlled by allowing the apparatus to cool after distillation is completed and before reading the amount of moisture in the trap.
2. Clinging of water droplets to dirty apparatus. Clean glassware is essential, but even with this, a burette brush is needed to dislodge water droplets.
3. Decomposition of the sample with production of water. If this is a measurable problem, discontinue method use and find an alternative procedure.

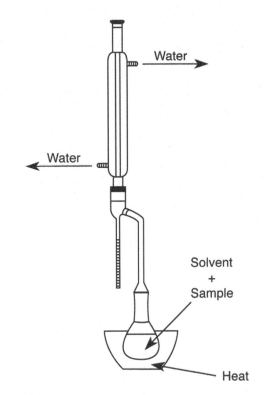

15.3 figure Apparatus for reflux distillation of moisture from a food. Key to this setup is the Bidwell-Sterling moisture trap. This style can be used only where the solvent is less dense than water

15.2.4 Chemical Method: Karl Fischer Titration

The **Karl Fischer titration**, a direct measure of moisture content, is particularly adaptable to food products that show erratic results when heated or submitted to a vacuum. This is the method of choice for determination of water content in many low-moisture foods such as dried fruits and vegetables (AOAC Method 967.19 E-G), candies, chocolate (AOAC Method 977.10), roasted coffee, oils and fats (AOAC Method 984.20), or any low-moisture food high in sugar or protein. The method is quite rapid and accurate and uses no heat. This method is based on the fundamental reaction described by Bunsen in 1853 [20] involving the reduction of iodine by SO_2 in the presence of water:

$$2H_2O + SO_2 + I_2 \rightarrow H_2SO_4 + 2HI \quad (15.4)$$

This was modified to include methanol and pyridine in a four-component system to dissolve the iodine and SO_2:

$$C_5H_5N \bullet I_2 + C_5H_5N \bullet SO_2 + C_5H_5N + H_2O$$
$$\rightarrow 2C_5H_5N \bullet HI + C_5H_5N \bullet SO_3 \quad (15.5)$$

$$C_5H_5N \bullet SO_3 + CH_3OH$$
$$\rightarrow C_5H_5N(H)SO_4 \bullet CH_3 \quad (15.6)$$

These reactions show that for each mole of water, 1 mol of iodine, 1 mol of SO_2, 3 mol of pyridine, and 1 mol of methanol are used. For general work, a methanolic solution is used that contains these components in the ratio of 1 iodine:3 SO_2:10 pyridine, and at a concentration so that 3.5 mg of water = 1 ml of reagent. A procedure for standardizing this reagent is given below.

In a **volumetric titration** procedure (Fig. 15.4 is manual titration unit; Fig. 15.5 is an example of automated titration unit), iodine and SO_2 in the appropriate form are added to the sample in a closed chamber protected from atmospheric moisture. The excess of I_2 that cannot react with the water can be determined **visually**. The endpoint color is dark red-brown. Some instrumental systems are improved by the inclusion of a potentiometer (i.e., **conductometric method**) to electronically determine the endpoint, which increases the sensitivity and accuracy. The automated volumetric titration, units (used for 100 ppm water to very high concentrations) use a pump for mechanical addition of titrant, and use the conductometric method for endpoint, determination (i.e., detection of excess iodine is by applying a current and measuring the potential).

The volumetric titration procedure described above is appropriate for samples with a moisture content greater than ~0.03 %. A second type of titration, referred to as **coulometric titration**, is ideal for prod-

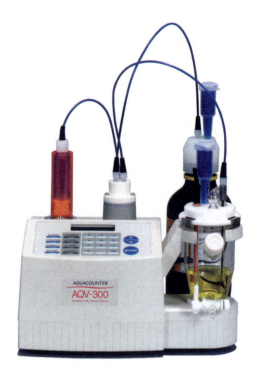

15.5 figure Automated Karl Fischer volumetric titration unit (Courtesy of Mettler Toledo, Columbus, OH)

ucts with very low levels of moisture, from 0.03 % down to parts per million (ppm) levels. In this method, iodine is electrolytically generated ($2I \rightarrow I_2 + 2e^-$) to titrate the water. The amount of iodine required to titrate the water is determined by the current needed to generate the iodine. Just like for volumetric titration, automated coulometric titration units are available commercially.

In a Karl Fischer volumetric titration, the **Karl Fischer reagent** (KFR) is added directly as the titrant if the moisture in the sample is accessible. However, if moisture in a solid sample is inaccessible to the reagent, the moisture is extracted from the food with an appropriate solvent (e.g., methanol). (Particle size affects efficiency of extraction directly.) Then the methanol extract is titrated with KFR.

The noxious odor of pyridine makes it an undesirable reagent. Therefore, researchers have experimented with other amines capable of dissolving iodine and sulfur dioxide. Some aliphatic amines and several other heterocyclic compounds were found suitable. On the basis of these amines, **one-component reagents** (solvent and titrant components together) and **two-component reagents** (solvent and titrant components separate) have been prepared. The one-component reagent may be more convenient to use, but the two-component reagent has greater storage stability.

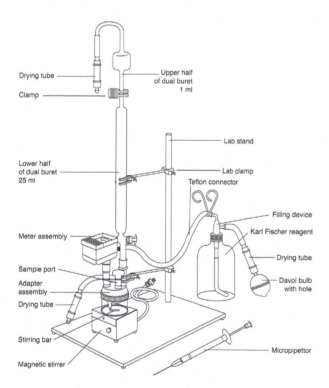

15.4 figure Manual Karl Fischer titration unit (Courtesy of Lab Industries, Inc., Berkeley, CA)

Before the amount of water found in a food sample can be determined, a **KFR water (moisture) equivalence** (KFReq) must be determined. The KFReq value represents the equivalent amount of water that reacts with 1 ml of KFR. Standardization must be checked before each use because the KFReq will change with time.

The KFReq can be established with **pure water**, a **water-in-methanol standard**, or **sodium tartrate dihydrate**. Pure water is a difficult standard to use because of inaccuracy in measuring the small amounts required. The water-in-methanol standard is premixed by the manufacturer and generally contains 1 mg of water/mL of solution. This standard can change over prolonged storage periods by absorbing atmospheric moisture. Sodium tartrate dihydrate ($Na_2C_4H_4O_6 \cdot 2H_2O$) is a primary standard for determining KFReq. This compound is very stable, contains 15.66 % water under all conditions expected in the laboratory, and is the material of choice to use.

The KFReq is calculated as follows using sodium tartrate dihydrate:

$$KFReq(mg\,H_2O/mL)$$
$$= \frac{36\,g\,H_2O/mol\,Na_2C_4H_4O_6 \cdot 2H_2O \times S \times 1000}{230.08\,g/mol \times A} \quad (15.7)$$

where:

> KFReq = Karl Fischer reagent water (moisture) equivalence
> S = weight of sodium tartrate dihydrate (g)
> A = mL of KFR required for titration of sodium tartrate dihydrate

Once the KFReq is known, the moisture content of the sample is determined as follows:

$$\%\,H_2O = \frac{KFReq \times Ks}{S} \times 100 \quad (15.8)$$

where:

> KFReq = Karl Fischer reagent water (moisture) equivalence
> Ks = mL of KFR used to titrate sample
> S = weight of sample (mg)

The major difficulties and sources of error in the Karl Fischer titration methods are as follows:

1. **Incomplete moisture extraction**. For this reason, fineness of grind (i.e., particle size) is important in preparation of cereal grains and some foods.
2. **Atmospheric moisture**. External air must not be allowed to infiltrate the reaction chamber.
3. **Moisture adhering** to walls of unit. All glassware and utensils must be carefully dried.
4. **Interferences** from certain food constituents. **Ascorbic acid** is oxidized by KFR to dehydroascorbic acid to overestimate moisture content; **carbonyl compounds** react with methanol to form acetals and release water, to overestimate moisture content (this reaction also may result in fading endpoints); **unsaturated fatty acids** will react with iodine, so moisture content will be overestimated.

15.2.5 Physical Methods

Most physical methods are indirect measures of moisture content and do not separate the water from the sample for analysis. These techniques can be rapid and nondestructive, which has led to their widespread use in food production and quality control; however, they must be calibrated against data collected by a direct method to quantify the amount of water in samples.

15.2.5.1 *Dielectric Method*

The electrical properties of water are used in the **dielectric method** to determine the moisture content of certain foods, by measuring the change in **capacitance** or **resistance to an electric current** passed through a sample. These instruments require calibration against samples of known moisture content as determined by standard direct methods. Sample density or weight/volume relationships and sample temperature are important factors to control in making reliable and repeatable measurements by dielectric methods. These techniques can be very useful for process control measurement applications, where continuous measurement is required. These methods are limited to food systems that contain no more than 30–35 % moisture.

The moisture content determination in dielectric-type meters is based on the fact that the dielectric constant of water (80.37 at 20 °C) is higher than that of most solvents. The **dielectric constant** is measured as an index of capacitance. As an example, the dielectric method is used widely for cereal grains. Its use is based on the fact that water has a dielectric constant of 80.37, whereas starches and proteins found in cereals have dielectric constants of 10. By determining this property on samples in standard metal condensers, dial readings may be obtained and the percentage of moisture determined from a previously constructed standard curve for a particular cereal grain.

15.2.5.2 *Hydrometry*

Hydrometry is the science of measuring **specific gravity** or **density**, which can be done using several different principles and instruments. While hydrometry is considered archaic in some analytical circles, it is still widely used and, with proper technique, is highly accurate. Specific gravity measurements with various types of **hydrometers** or with a **pycnometer** are commonly used for routine testing of moisture (or solids) content of numerous food products. These include beverages, salt brines, and sugar solutions. Specific gravity measurements are best applied to the analysis of solutions consisting of only one solute in a medium of water.

15.2.5.2.1 Hydrometer

One approach to measuring specific gravity is based on **Archimedes' principle**, which states that a solid suspended in a liquid will be buoyed by a force equal to the weight of the liquid displaced. The weight per unit volume of a liquid is determined by measuring the volume displaced by an object of standard weight. A hydrometer is a standard weight on the end of a spindle, and it displaces a weight of liquid equal to its own weight (Fig. 15.6). For example, in a liquid of low density, the hydrometer will sink to a greater depth, whereas in a liquid of high density, the hydrometer will not sink as far. Hydrometers are available in narrow and wide ranges of specific gravity. The spindle of the hydrometer is calibrated to read specific gravity directly at 15.5 °C or 20 °C. A **hydrometer** is not as accurate as a pycnometer, but the speed with which you can do an analysis is a decisive factor. The accu-

racy of specific gravity measurements can be improved by using a hydrometer calibrated in the desired range of specific gravities.

The rudimentary but surprisingly accurate hydrometer comes equipped with various modifications depending on the fluid to be measured:

1. The Quevenne and New York Board of Health **lactometer** is used to determine the density of milk. The Quevenne lactometer reads from 15 to 40 lactometer units and corresponds to 1.015–1.040 specific gravity. For every degree above 60 °F, 0.1 lactometer unit is added to the reading, and 0.1 lactometer unit is subtracted for every degree below 60 °F.
2. The **Baumé hydrometer** was used originally to determine the density of salt solutions (originally 10 % salt), but it has come into much wider use. From the value obtained in the Baumé scale, you can convert to specific gravity of liquids heavier than water. For example, it is used to determine the specific gravity of milk being condensed in a vacuum pan.
3. The **Brix hydrometer** is a type of **saccharometer** used for sugar solutions such as fruit juices and syrups, and one usually reads directly the percentage of sucrose at 20 °C. **Balling saccharometers** are graduated to indicate percentage of sugar by weight at 60 °F. The terms **Brix** and **Balling** are interpreted as the weight percentage of pure sucrose.
4. **Alcoholometers** are used to estimate the alcohol content of beverages. Such hydrometers are calibrated in 0.1° or 0.2° proof to determine the percentage of alcohol in distilled liquors (AOAC Method 957.03).
5. The **Twaddell hydrometer** is only for liquids heavier than water.

15.2.5.2.2 Pycnometer

Another approach to measuring specific gravity is a comparison of the weights of equal volumes of a liquid and water in standardized glassware, a **pycnometer** (Fig. 15.7). This will yield density of the liquid compared to water. In some texts and reference books, *20/20* is given after the specific gravity number. This indicates that the temperature of both fluids was 20 °C when the weights were measured. Using a clean, dry pycnometer at 20 °C, the analyst weighs it empty, fills it to the full point with distilled water at 20 °C, inserts the thermometer to seal the fill opening, and then touches off the last drops of water and puts on the cap for the overflow tube. The pycnometer is wiped dry in case of any spillage from filling and is reweighed. The density of the sample is calculated as follows:

15.6 figure Hydrometers (Courtesy of Cole-Parmer Instrument Company, Vernon Hills, IL)

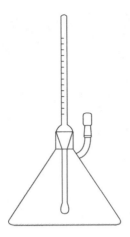

Pycnometer

$$\frac{\begin{array}{c}\text{weight of sample} - \text{filled pycnometer} \\ -\text{weight of empty pycnometer}\end{array}}{\begin{array}{c}\text{weight of water} - \text{filled pycnometer} \\ -\text{weight of empty pycnometer}\end{array}}$$

$$= \text{density of sample} \qquad (15.9)$$

This method is used for determining alcohol content in alcoholic beverages (e.g., distilled liquor, AOAC Method 930.17), solids in sugar syrups (AOAC Method 932.14B), and solids in milk (AOAC Method 925.22).

15.2.5.3 *Refractometry*

Moisture in liquid sugar products and condensed milks can be determined using a Baumé hydrometer (solids), a Brix hydrometer (sugar content), gravimetric means, or a **refractometer**. If it is performed correctly and no crystalline solids are evident, the refractometer procedure is rapid and surprisingly accurate (AOAC Method 932.14C, for solids in syrups). The refractometer has been valuable in determining the soluble solids in fruits and fruit products (AOAC Method 932.12; 976.20; 983.17).

The **refractive index** (RI) of an oil, syrup, or other liquid is a dimensionless constant that can be used to describe the nature of the food. While some refractometers are designed only to provide results as refractive indices, others, particularly handheld, quick-to-use units, are equipped with scales calibrated to read the percentage of solids, percentage of sugars, and the like, depending on the products for which they are intended. Tables are provided with the instruments to convert values and adjust for temperature differences. Refractometers are used not just on the laboratory bench or as handheld units. Refractometers can be installed in a liquid processing line to monitor the °Brix of products such as carbonated soft drinks, dissolved solids in orange juice, and the percentage of solids in milk [21].

When a beam of light is passed from one medium to another and the density of the two differs, then the beam of light is bent or refracted. Bending of the light beam is a function of the media and the sines of the angles of incidence and refraction at any given temperature and pressure, and is thus a constant (Fig. 15.8). The (RI) (η) is a ratio of the sines of the angles:

$$\eta = \frac{\text{sine incident ray angle}}{\text{sine refracted ray angle}} \qquad (15.10)$$

All chemical compounds have an index of refraction. Therefore, this measurement can be used for the qualitative identification of an unknown compound by comparing its RI with literature values. RI varies with **concentration** of the compound, **temperature**, and **wavelength of light**. Instruments are designed to give a reading by passing a light beam of a specific wavelength through a glass prism into a liquid, the sample. Benchtop or handheld units use **Amici prisms** to obtain the **D line of the sodium spectrum** or 589 nm from white light. Whenever refractive indices of standard fluids are given, these are prefaced with $\eta_D^{20} = $ a value from 1.3000 to 1.7000. The Greek letter η is the symbol for RI; the 20 refers to temperature in °C; and D is the wavelength of the light beam, the D line of the sodium spectrum.

Benchtop instruments are more accurate compared to handheld units mainly because of tempera-

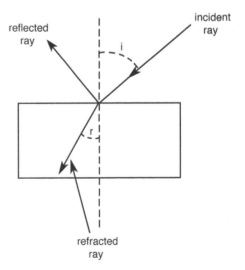

Reflection and refraction concepts of refractometry

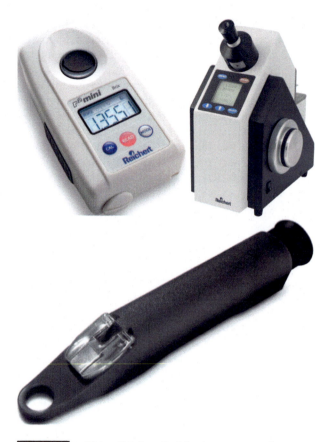

15.9
figure

Rhino Brix handheld refractometer, R² mini digital handheld refractometer, and Mark III Abbe refractometer (Courtesy of Reichert Analytical Instruments, Depew, NY)

ture control (Fig. 15.9). These former units have provisions for water circulation through the head where the prism and sample meet. Digital and **Abbe refractometers** are common for laboratory use. Care must be taken when cleaning the prism surface following use. The contact surface should be wiped clean with lens paper and rinsed with distilled water and then ethanol. The prism chamber should be closed and the instrument covered with a bag when not in use to protect the delicate prism surface from dust or other debris that might lead to scratches and inaccuracy.

The fact that the RI of a solution increases with concentration has been exploited in the analysis of total soluble solids of carbohydrate-based foods such as sugar syrups, fruit products, and tomato products. Because of this use, these refractometers are calibrated in °**Brix** (g of sucrose/100 g of sample), which is equivalent to percentage sucrose on a wt/wt basis. Refractive index measurements are used widely to approximate sugar concentration in foods, even though values are accurate only for pure sucrose solutions.

15.2.5.4 *Infrared Analysis*

Infrared spectroscopy (see Chap. 8) has attained a primary position in monitoring the composition of food products before, during, and following processing [22]. It has a wide range of food applications and has proven successful in the laboratory and online. Similar to the use of ultraviolet (UV) or visible (Vis) light in UV-Vis spectroscopy, in infrared spectroscopy a sample is exposed to IR radiation (near IR 700–2400 nm or mid IR 2500–25,000 nm), specific wavelengths are absorbed, and an IR spectrum is measured by calculating the intensity of the IR radiation before and after passing through the sample. The absorbance peaks are related to the type and amount of functional groups present. However, infrared spectrometers must be calibrated for each analyte to be measured, and the analyte must be uniformly distributed in the sample.

For water, near-infrared (NIR) bands (1400–1450; 1920–1950 nm) are characteristic of the -OH stretch of the water molecule and can be used to determine the moisture content of a food. NIR has been applied to moisture content analysis of a wide variety of food commodities and is an official method for moisture content determination in dried vegetables (AOAC Method 967.19).

The use of mid-infrared milk analyzers to determine fat, protein, lactose, and total solids in milk (AOAC Method 972.16) is covered in Chap. 8 of this text. The instrument must be calibrated using a minimum of eight milk samples that were previously analyzed for fat (F), protein (P), lactose (L), and total solids (TS) by standard methods. Then, a mean difference value, *a*, is calculated for all samples used in calibration:

$$a = \sum (TS - F - P - L)/n \qquad (15.11)$$

where:

  a = solids not measurable by the F, P, and L methods
  n = number of samples
  F = fat percentage
  P = protein percentage
  L = lactose percentage
  TS = total solids percentage

Total solids then can be determined from any infrared milk analyzer results by using the formula

$$TS = a + F + P + L \qquad (15.12)$$

The *a* value is thus a standard value mathematically derived. Newer instruments have the algorithm in their computer software to ascertain this value automatically. Although not directly measured, the moisture content can be calculated by subtracting the total solids content from 100.

15.2.5.5 *Microwave Absorption*

The absorption of microwaves, which are electromagnetic waves with wavelength of 0.001–1 m and frequency of 0.3–300 GHz, can be used to determine the moisture contents of a wide variety of food products. The approach is based on **dielectric constant** or **permittivity value** differences between water and dry materials. The permittivity of most dry materials is much lower than that of water, and small changes in the amount of water in a sample lead to measurable changes in permittivity. As microwave energy is passed through a sample (<40 mm thick) placed between a microwave transmitter and receiver, the absorption at 2.450 GHz (the most widely used frequency in microwaves) is linearly related to moisture content [23]. After calibration against laboratory standards, moisture contents can be determined in ~2 s using the microwave absorption method. Power absorption and attenuation may be affected by the dimensions, surface area, temperature, and dielectric properties of some food ingredients; however, the technique is rapid, nondestructive, and useful for some heterogeneous, powdered, multilayered, and frozen foods [4, 24].

15.2.5.6 *Freezing Point*

When water is added to a food product, many of the physical constants are altered. Some properties of solutions depend on the number of solute particles as ions or molecules present. These properties are vapor pressure, freezing point, boiling point, and osmotic pressure. Measurement of any of these properties can be used to determine the concentration of solutes in a solution. However, the most commonly practiced assay for milk is the change of the freezing point value. It has economic importance with regard to both raw and pasteurized milk. The **freezing point** of milk is its most constant physical property. While termed a physical constant, the freezing point varies within narrow limits, and the vast majority of samples from individual cows fall between −0.503 and −0.541 °C, with the average very close to −0.521 °C. Herd or bulk milk will exhibit a narrower range unless the supply was watered intentionally or accidentally or if the milk is from an area where severe drought has existed.

The AOAC Method 961.07 for water added to milk uses a **cryoscope** to test for freezing points, and assumes a freezing point for normal milk of −0.527 °C. The Food and Drug Administration will reject all milk with freezing points above −0.507 °C. Since the difference between the freezing points of milk and water is slight and since the freezing point can be used to calculate the amount of water added, it is essential that the method be as pre-

cise as possible. The thermistor used can sense temperature change to 0.001 °C. The general technique is to supercool the solution and then induce crystallization by a vibrating reed. The temperature will rise rapidly to the freezing point or eutectic temperature as the water freezes. In the case of pure water, the temperature remains constant until all the water is frozen. In the case of milk, the temperature is read when there is no further temperature rise. Time required for the automated instruments is 1–2 min per prechilled sample.

15.2.6 Comparison of Moisture Content Determination Methods

15.2.6.1 *Principles*

Characteristics of the various moisture content analysis methods described in Sects. 15.2.2, 15.2.3, 15.2.4, and 15.2.5 are summarized in Table 15.5. Direct moisture content determination methods often remove water from a sample and determine moisture contents by mass, volumetry, or titration. Oven drying methods involve the removal of moisture from the sample and then a weight determination of the solids remaining to calculate the moisture content (and consequently also the total solids content). Non-water volatiles can be lost during drying, but their loss is generally a negligible percentage of the amount of water lost. Distillation procedures also involve a separation of the moisture from the solids, and the moisture is quantitated directly by volume. Karl Fischer titration is based on chemical reactions of the moisture present, reflected as the amount of titrant used.

Indirect moisture content determination methods analyze some property of the food that is related to the presence of water. The dielectric method is based on electrical properties of water. Hydrometric methods are based on the relationship between specific gravity and moisture content. The refractive index method is based on how water in a sample affects the refraction of light. Near-infrared analysis of water in foods is based on measuring the absorption at wavelengths characteristic of the molecular vibration in water. Microwave absorption is based on the dielectric properties of water. Freezing point is a physical property of milk that is changed by a change in solute concentration.

15.2.6.2 *Nature of Sample*

While many foods will tolerate oven drying at high temperatures, some foods contain volatiles that are lost at such temperatures. Some foods have constituents that undergo chemical reactions at high temperatures to generate or utilize water or other compounds, and these reactions affect the calculated moisture content. Vacuum oven drying at reduced temperatures may overcome such problems for some foods. However, a distillation technique is necessary for some food to minimize vola-

15.5
table

Comparison of moisture analysis methods

Method	Principle	What is actually measured?	How is water removed/reacted/ identified/etc.?	Cautions/things to control	Advantages	Disadvantages	Typical applications
Forced draft oven	Sample is heated in oven to evaporate water. Weight loss equals moisture content	Weight change	Heat evaporates water when it boils at 100 °C	Control time, temperature; control sample particle size. Must pre-dry some samples to avoid splattering	Easy to handle many samples at one time	Takes long time to get results. High temperature can cause loss of volatiles, lipid oxidation, Maillard browning, sucrose hydrolysis, so not suitable for some types of foods	Official method for many types of samples. Not suitable of rapid quality control results. Not suitable for samples subject to loss of volatiles, lipid oxidation, Maillard browning, or sucrose hydrolysis
Vacuum oven	Sample is heated in oven under reduced pressure, so water evaporates at a lower temperature. Weight loss equals moisture content	Weight change	Heat sample under reduced pressure to evaporate water at ~70 °C	Control time and temperature. Pull and release vacuum slowly	Easy to handle many samples at one time. Lower temperature for evaporating water reduced problems with high-sugar products	Takes long time to get results (though usually less time than with forced draft oven). More expensive than forced draft oven	An official method for many types of products. Not suitable for rapid quality control. Not suitable for powdered products, since they can blow around when vacuum is pulled and released
Microwave drying oven	Sample is heated with microwave energy to evaporate water. Weight loss equals moisture content	Weight change	Heat from microwave energy causes water evaporation	Control power and time to prevent sample decomposition. Spread sample evenly. Check calibration of analytical balance	Rapid	More expensive than other drying methods listed. Can only run one sample at a time	Suitable for rapid quality control, especially for liquid products, since use of pads avoids splattering
Infrared drying oven	Infrared lamp supplies heat that penetrates sample to evaporate water. Weight loss equals moisture content	Weight change	Heat from infrared lamp evaporates water	Control time and temperature. Spread sample evenly	Rapid	Expensive. Can only run one sample at a time	Suitable for rapid quality control, but not for high moisture products (would get splattering)
Rapid moisture analyzer	Sample is heated with heating elements to evaporate water. Weight loss equals moisture content	Weight change	Heat evaporates water when it boils at 100 °C	Control time and temperature. Spread sample evenly. Regular calibration of analytical balance	Rapid	Expensive. Can only run one sample at a time	Suitable for rapid quality control, but not for high moisture products (would get splattering)

(continued)

Method	Principle	Measures	Basis	Precautions	Advantages	Disadvantages	Applications/comments
Reflux distillation (with toluene)	When sample is heated to toluene (an immiscible liquid), the toluene and water are co-distilled. Collected moisture distills off, is condensed and collected, and volume of water is measured	Volume of water from sample collected after distillation and condensation	Co-distill water from sample with toluene. Collect water and measure	Any emulsion formed must break to read volume of water. Need very clean glassware with no water. Use caution with solvents (fire hazards; toxic)	Causes less thermal decomposition of some foods than oven drying. Solvent protects sample from losing volatiles and minimized oxidation. Water is measured directly	Can only run one sample at a time. Solvent is likely flammable and toxic. Reading volume of water in receiving tube may be less accurate than by gravimetric method	AOAC method for spices
Karl Fischer	In titration of sample with Karl Fischer reagent, water in sample reacts with sulfur dioxide to cause reduction of iodine. Endpoint of titration is detected when excess iodine cannot react with water. Volume of titrant is used to calculate % moisture	Volume of Karl Fischer Reagent titrated	Water in sample reacts with iodine and sulfur dioxide to cause reduction of iodine	Control particle size of sample and humidity of room. Prevent any water in glassware. Must standardize KRF. Choose another method if interferences from certain food constituents (e.g., ascorbic acid, carbonyl compounds, unsaturated fatty acids)	No heat, so no thermal decomposition; rapid. Higher accuracy than many other methods for low-moisture foods	Can only run one sample at a time. Expensive, if using automated unit	Method of choice for many low-moisture foods (e.g., dried fruits and vegetables, candies, chocolate, roasted coffee, oils and fats, and many low-moisture foods that are high in sugar or protein). Good method to try if method with heating and/or vacuum gives erratic results
Hydrometer	Archimedes' principle. Compare relative density (specific gravity) of sample to that of water at same temperature	Volume displaced by hydrometer. Read specific gravity directly from hydrometer. Measuring solids content	Based on solids content of the solution, to determine specific gravity compared to pure water	Control temperature. Need clean hydrometer	Rapid. Easy. Enexpensive	Limited applications. Measures only solids content	Commonly used as rapid method to measure solids content of beverages, salt brines, and sugar solutions. Best applied to solutions with only one solute in a medium of water

(continued)

table 15.5 (continued)

Method	Principle	What is actually measured?	How is water removed/reacted/identified/etc.?	Cautions/things to control	Advantages	Disadvantages	Typical applications
Refractometer	Based on bending of light (i.e., refraction. Measure refractive index) as it hits surface of product. Refractive index can be used to determine concentration of compound of interest if nature of compound, temperature of sample, and wavelength of light are constant	Refractive index. Measuring solids content. Commonly calibrated in degree Brix (g of sucrose/100 g sample)	Based on solids content of solution, to determine refractive index	Control temperature. Need clean contact surface	Rapid. Easy. inexpensive	Limited applications; measures only solids content	Commonly used as rapid method to measure solids content of beverages and milk, and soluble solids of fruits and fruit products and of tomato products
Infrared analyzer	Measure absorption of infrared radiation at wavelength characteristic of the -OH stretch of the water molecule. Concentration of water is determined by energy that is reflected or transmitted, which is inversely proportional to energy absorbed	Amount of NIR light reflected from sample	Molecular vibration of functional groups of water determines absorption, which is inversely related to reflected light what is measured	Must calibrate instrument for each type of product and each analyte being analyzed. Control sample particle size. Prevent scratches on glass container for sample; remember that values obtained are only a prediction	Rapid. Easy. Can be used to estimate content of various food constituents	Expensive. Can only run one sample at a time. Value obtained are only estimates/predictions. Must calibrate the instrument for each analyte for each type of sample	Has wide range of food applications, in the laboratory, at-line, and on-line. NIR is heavily used in the grain/cereal industry for moisture, protein, and fat. NIR is heavily used in the dairy industry for total solids, fat, protein, and lactose in milk

15.6 table	Common food industry uses of moisture content and water activity assays

Technique		Where used: Production, quality control	Product development	Basic research
Water content measurement				
Direct methods	Forced draft oven drying	X	X	X
	Vacuum oven drying	X	X	X
	Microwave analyzer	X	X	
	Infrared drying	X	X	
	Rapid moisture analyzer technology	X	X	
	Thermogravimetric analyzer		X	X
	Lyophilization		X	X
	Chemical desiccation		X	X
	Karl Fischer titration	X	X	X
Indirect methods	Dielectric capacitance	X		
	Hydrometer	X		
	Pycnometer	X		
	Refractometer	X	X	X
	NIR spectroscopy (absorbance or reflectance)	X	X	X
	Freezing point, cryoscope		X	X
	Microwave absorption	X	X	X
	Conductivity	X	X	
Water activity measurement	Dewpoint analyzer	X	X	X
	Electric (capacitance or electrolyte) hygrometer	X	X	X
	Freezing point depression			X
	Tunable diode laser sensor		X	X

tilization and decomposition. For foods very low in moisture or high in fats and sugars, Karl Fischer titration is often the method of choice. The use of a pycnometer, hydrometer, and refractometer requires liquid samples, ideally with limited constituents.

15.2.6.3 *Intended Purposes*

Moisture content analysis data may be needed quickly for quality control purposes, in which high accuracy may not be necessary. Of the oven drying methods, microwave drying, infrared drying, and the rapid moisture analyzer techniques are fastest. Some forced draft oven procedures require less than 1 h drying, but most forced draft oven and vacuum oven procedures require a much longer time. The electrical, hydrometric, refractive index, and microwave absorption methods are very rapid but often require correlation to less empirical methods. Oven drying procedures are official methods for a variety of food products. Reflux distillation is an AOAC method for chocolate, dried vegetables, dried milk, and oils and fats. Such official methods are used for regulatory and nutrition labeling purposes. A survey of food industry usage of the moisture content assays is summarized in Table 15.6, wherein the most commonly used techniques across a variety of compa-

nies for production, product development, and basic research applications are identified.

15.3 WATER ACTIVITY

15.3.1 Overview

Water content alone is not a reliable indicator of food stability, since foods with the same water content differ in their perishability [6]. It is the water activity (a_w) of foods that has been correlated to microbial growth, physical properties, and chemical and enzymatic reactions.

The water activity (a_w) of foods varies greatly, as shown in Table 15.7. For referring to water activity, the a is lower case and the w is a subscript because an activity coefficient of water is being used to describe its energy state. Generally, foods with higher moisture contents also have higher a_ws, although the relationship between moisture content and a_w is not linear. **Water activity** is a thermodynamic property of water in foods defined as the ratio of the fugacity (or escaping tendency) of water in the food to the fugacity of pure water at the same temperature and pressure [4, 6, 25]. Because fugacity cannot be directly measured, a_w is more commonly determined

15.7

Table   Water activity (25 °C) of foods and saturated solutions of food ingredients and salts

| Foods | | Saturated solutions | | | |
| | | Food ingredients | | a_w control salts | |
Type	a_w	Type	a_w	Type	a_w
Potato chips	0.07	Single ingredients		LiCl	0.11
Hard candy	0.12	Malic acid	0.58	CH_3CO_2K	0.23
Crisp crackers	0.13–0.20	Fructose	0.62	$MgCl_2$	0.33
Sugar-free hard candy	0.25	Sorbitol	0.67	K_2CO_3	0.43
Crisp cookie	0.25	Glucose	0.74	$Mg(NO_3)_2$	0.53
Chewy cookie	0.55	Citric acid	0.78	$CoCl_2$	0.65
Honey	0.56	Xylitol	0.79	NaCl	0.75
Beef jerky	0.61	Sucrose	0.85	KCl	0.84
Gummy candy	0.66	Lactose	0.97	K_2SO_4	0.97
High fructose corn syrup	0.75	Maltose	0.97		
Condensed milk, strawberry preserves	0.84	Ingredient blends			
Soy sauce	0.87	NaCl +fructose	0.45		
Salted butter	0.90	Fructose + citric acid	0.50		
Bread	0.94	NaCl+sucrose	0.65		
Reduced sugar ketchup, unsalted butter	0.97	NaCl+glucose	0.71		
Juices, milk, fruits, vegetables	0.98–0.99				

All measurements were done using an AquaLab 4TE instrument (Decagon Devices, Pullman, WA)

as the ratio of the vapor pressure of water in a food (p) to the vapor pressure of water (p_o) at the same temperature and barometric pressure, as shown in Eq. 15.13.

$$a_w = \frac{p}{p_o} \qquad (15.13)$$

This concept is similar to how the relative humidity (RH) of the atmosphere is determined. The relationship between the a_w and equilibrium relative humidity (ERH) can be expressed as $a_w = ERH / 100$. Water activity is a dimensionless number between 0 (absolute no water) and 1 (pure water).

There are fewer official methods for determining the a_w of foods than there are for determining the moisture content [4, 14]. AOAC Method 978.18 describes techniques to determine the a_w of canned vegetables: in the regulation of acidified foods, an a_w of 0.85 is used as the cutoff for pathogen growth. In 21CFR Part 114 [26], low-acid foods are defined as foods with an equilibrium pH of >4.6 and $a_w > 0.85$, and acidified foods are defined as low-acid foods to which acid has been added to create a finished equilibrium pH of ≤4.6 and that have an $a_w > 0.85$. Foods with $a_w < 0.85$ are not covered by 21CFR Part 108, 113, or 114 [26]; however, it is important to recognize that lowering the a_w is not a kill step.

15.3.2 Importance of Water Activity

The a_w of foods affects important quality and safety factors, and therefore a_w measurements are commonly used in food development and production. The a_w influ-

ences all aspects of microbial growth, and each microorganism has a threshold a_w below which it will not grow, although microorganisms may not be dead below this threshold a_w [27]. Understanding which spoilage or pathogenic microorganisms are a concern for a specific food and their a_w limits for growth can provide a foundation for formulation efforts to reduce the a_w of the food below that level. Because of the direct correlation between a_w and microbial growth, there are many food safety regulations that incorporate a_w guidelines (21CFR 110, 113, and 114 [26]; ANSI/NSF Standard 75 [28]; ISO 21807:2004(E) [29]; AOAC Method 978.18 [14]), and a_w can be a critical control point in HACCP plans.

Food stability maps have been created to display the relationships between water activity, microbial growth, chemical and biochemical reaction rates, and physical properties [30]. An example of a general map for an amorphous food is provided in Fig. 15.10, although the reaction rates and properties will vary across different a_w ranges for different foods. In addition to controlling microbial growth, maintaining the desired a_w of a product is a critical aspect in maintaining its texture and quality throughout shelf-life, and there will be critical a_ws beyond which unwanted physical and/or chemical changes will occur in most products. If a crispy fried product (potato chip) is exposed to an environment with a higher RH than its a_w, moisture will migrate into the potato chip, increasing its a_w and leading to a softening of its texture once the critical a_w is exceeded. Likewise, if a food product contains multiple components with different a_ws (cheese and cracker, pizza sauce on crust, pastry with filling, etc.), water migrates from

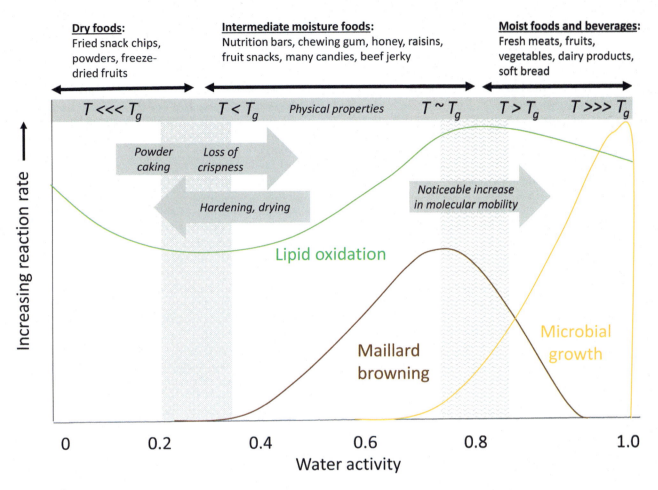

Dry foods:
Fried snack chips, powders, freeze-dried fruits

Intermediate moisture foods:
Nutrition bars, chewing gum, honey, raisins, fruit snacks, many candies, beef jerky

Moist foods and beverages:
Fresh meats, fruits, vegetables, dairy products, soft bread

$T \lll T_g$ $T < T_g$ *Physical properties* $T \sim T_g$ $T > T_g$ $T \ggg T_g$

Powder caking   Loss of crispness

Hardening, drying

Noticeable increase in molecular mobility

Lipid oxidation

Maillard browning

Microbial growth

Increasing reaction rate →

0   0.2   0.4   0.6   0.8   1.0

Water activity

15.10 figure   Food stability map (Adapted from Labuza et al. [30] and overlaid with data from Table 15.7)

the region of high *aw* to the region of low a_w until a_w equilibration is attained, which can result in a hardening of one phase and a softening of the other. If these changes are unwanted, then reformulation of the phases to the same a_w (if possible) or physical separation of the phases in barrier packaging could extend shelf-life. Increasing the a_w above a critical value can lead to clumping and caking of powders, recrystallization of sugars or salts in some products, a glassy-to-rubbery transition in amorphous solids, or even deliquescence of water-soluble deliquescent crystalline compounds.

15.3.3 Water Activity Measurement

15.3.3.1 *Principles*

Since $a_w = p/p_o$, determining the a_w often involves the direct measurement of vapor pressures, although there are also indirect measurements for a_w. For the majority of a_w measurement techniques, a representative food sample is sealed in a container and allowed to reach equilibrium (i.e., the water will migrate out of the sample into the container headspace, or vice versa, until the a_w of the food and the RH of the headspace are

15.11 figure   Dewpoint water activity meter (Courtesy of Decagon Devices, Pullman, WA)

equal) (Fig. 15.11). Enough sample must be used to assure that moisture migration during equilibration will not dry the sample but not so much sample that the sensor is compromised. A general recommendation is to fill half of the container with sample. More sample might be warranted for dense, dry products, while less sample could be used for wet products or liquids (AOAC Method 978.18 recommends canned vegetables fill >1/20 of the sample container volume).

The time required to attain headspace equilibrium can vary between samples and between assay types. Slow water-emitting samples (dense, dry, dehydrated, high fat, high viscosity) may require several hours for precise a_w measurement, while samples containing more water or those with water as the continuous phase may equilibrate much faster. For establishing the equilibrium criterion for a sample, the a_w could be monitored over time, such as at 15, 30, 60, 120 min (AOAC Method 978.18), extending out in 60 min intervals until consecutive readings vary by <0.01 a_w (or other designated equilibrium criteria for a particular product). While some samples may require 24 h for equilibration, many foods (barring the slow water emitters) may reach equilibrium within 5 min for dewpoint analyses or 30–90 min for electric hygrometers [31].

Temperature is an important factor in equilibration and a_w – higher temperatures often lead to higher a_w readings for many foods if the moisture content is kept constant (water molecules move faster as temperature is increased). This is why temperature is reported along with a_w values. Additionally, inequalities in temperature between samples and containers or equipment can lead to unwanted water condensation and/or errors in determining the a_w. For example, a 1 °C temperature difference between sample and dew point leads to an a_w error of 0.06 in dewpoint instruments [4, 32]. The majority of reported a_w values were measured at 25 °C (as stipulated in AOAC Method 978.18) or 20 °C, although the temperatures of interest for a particular product may vary (particularly for refrigerated or frozen products).

It is important to use a calibrated sensor for a_w measurements. Calibration is often done using salt solution standards of known a_w (Table 15.7). It is important to comply with instrument manufacturer guidelines for cleaning, calibrating, and maintaining a_w instruments. Calibration may involve at least three measurement points (ISO 21807) or ≥5 salts (AOAC Method 978.18) that encompass the a_w range of interest for the sample and sensor to generate a standard curve and/or determine an offset adjustment. Once the instrument is calibrated and temperature and headspace equilibrium are reached, the headspace is analyzed to determine the a_w. There are a variety of sensors and methods to directly or indirectly determine the equilibrium water vapor pressure in the headspace, although only the most common used by the food industry (Tables 15.6 and 15.8) will be further described. Examples of a_w methods include dewpoint measurement, electric hygrometer sensors (including capacitance and electrolyte sensors), direct measurement of manometric pressure, change in length of a hair/thread, increase in sorbent mass in an isopiestic method, tunable diode laser sensor, thermocouple psychrometry, and freezing point determination [5, 33].

15.8 table Comparison of water activity measurement methods

Method	Instrument	aw range	Accuracy	Repeatability	Resolution	Internal temperature control	Read time	Volatile interference
Dewpoint	AquaLab4 (Decagon Devices)	0.030–1.000	0.003	0.001	0.0001	15–50 °C	2–5 min	Yes
Electric (resistive electrolytic based on electrolyte)	LabMaster-aw (Novasina)	0.03–1.000	0.003	0.001	0.001	0–50 °C	10–15 min	Yes[a]
Electric (capacitance)	HygroLab C1 (Rotronic)	0.0–1.000	0.008	0.001	0.001	No	4–6 min	Minimal[b]
Electric (capacitance)	Aw Therm (Rotronic)	0.005–1.000	0.005	0.001	0.001	10–60 °C	4–6 min	Minimal[b]
Tunable diode laser	TDL (Decagon Devices)	0.03–1.000	0.005	0.001	0.0001	15–50 °C	2–5 min	No

Adapted from Fontana and Campbell [4] and personal communications with Brady Carter (Decagon Devices), Markus Bernasconi (Novasina), and Harry Trainor and Rico Hasler (Rotronic)
[a]For the LabMaster-aw, filters can be used to block out low to medium levels of volatiles, or use the CM-3 sensor that is resilient to alcohol
[b]The sensors in the HygroLab C1 and Aw Therm instruments are affected by volatiles when subjected to high concentrations and long-term exposure. A single a_w measurement in 4–6 min is not enough exposure in most cases for volatiles to affect the analysis

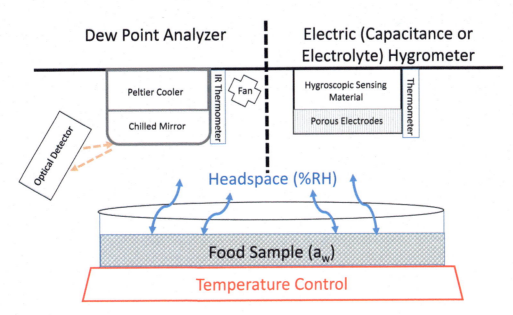

 Schematic of water activity measurement chambers for the two most common techniques used by the food industry: dewpoint and capacitance/electrolyte sensors

15.3.3.2 *Chilled Mirror Dew Point*

In a chilled mirror dewpoint instrument, such as a Decagon Devices AquaLab (Fig. 15.12), a sample is equilibrated in a temperature-controlled sealed chamber that contains a fan to circulate air in the headspace, an infrared thermometer to measure sample temperature, a temperature-controlled mirror, and a sensor that detects condensation on the mirror [5]. The mirror temperature is controlled by a thermoelectric Peltier cooler, and a photodetector monitors the reflectance of light off of the mirror. The a_w of the sample is the same as the RH of the headspace at equilibrium (a_w = ERH/100), and it is the headspace RH that is analyzed. The temperature of the mirror is cooled until condensation from the headspace first appears, which is determined by a change in the reflectance. When condensation occurs, the temperatures of the sample and the mirror are recorded. The temperature of the sample is used to determine the water vapor pressure (p_o), and the dewpoint temperature is used to determine the vapor pressure of water in the headspace (p). The a_w of the sample is determined by applying Eq. 15.13, $a_w = p/p_o$. Volatiles that may condense on the mirror, such as alcohols, acetic acid, or propylene glycol, interfere with accurate a_w measurements using dewpoint techniques.

15.3.3.3 *Electric Hygrometer*

In an electric hygrometer, which for our purposes will include capacitance and electrolyte sensors such as Novasina and Rotronic instruments, a sample container is equilibrated in a chamber with a potentiometer and a sensor (such as an electrolytic moisture sensor) that measures the headspace RH and tem-

perature [31]. Some models have internal temperature control features, while others may require an incubator or water-jacketed sample chamber for temperature control. The sensors monitor the electrical conductivity of their immobilized salt solutions or a hygroscopic polymer (the sensor composition is proprietary) that sorb or desorb water in response to the headspace RH. Changes in electric conductivity are calibrated to reflect changes in the headspace RH, and a_w is calculated as a_w = ERH/100. Volatile compounds may interfere with a_w measurements by electric hygrometers, and tunable diode laser techniques may be a better option for accurately determining the a_w of samples that contain alcohols, acetic acid, or propylene glycol.

15.4 MOISTURE SORPTION ISOTHERMS

15.4.1 Overview

A **moisture sorption isotherm** is a plot of the equilibrium relationship between a sample's moisture content and a_w at a constant temperature and pressure. The **hygrocapacity**, or water-holding capacity, of a sample is the amount of water it contains when equilibrated at a set a_w. Hygrocapacity depends on the affinity between water and the sample, temperature, RH, surface area, and water-solid interaction mechanism [34]. A key to establishing moisture sorption isotherms is to ensure that the samples have equilibrated at the set a_ws. Different samples exhibit different kinetics of moisture sorption and desorption. In the absence of equilibrium, a plot of moisture content versus a_w

would simply be called a **sorption profile** (not an isotherm).

Generally as a_w increases, moisture content also increases although the relationship is most certainly not linear. Different types of solids and different foods have different shapes of moisture sorption profiles, and these shapes have been organized into six major types (I–IV) [35]. Types II (sigmoid) and III (_/-shaped) are most common for foods. The inflection points and slopes of the different isotherms provide useful information about hygroscopicity, potential for reformulating or drying to alter a_w, how a sample might respond to different environments or formulations, and stability (microbial, physical, and chemical). Inflection points in sorption profiles occur during phase changes, such as a glassy-to-rubbery transition, hydrate formation, or deliquescence. Recrystallization events are often accompanied by expulsion of water from the matrix and a corresponding weight loss. Overlaying isotherms from different food types is a useful practice when considering co-formulation and multicomponent foods, keeping in mind that water does not move if the a_ws of the different components are equal. If differences in a_w are initially present between the different food components, water will migrate until a_w equilibrium is reached.

There are several approaches and instruments used to generate moisture sorption isotherms (or profiles), although there are no government regulations or AOAC official methods dictating techniques and parameters. A description of two common approaches for moisture sorption isotherm measurement is provided below.

15.4.2 Isopiestic Desiccator Method

In desiccator methods, samples are equilibrated in RH-controlled desiccators at set temperatures, and, upon equilibrium, their moisture contents are determined by either a mass change from a known starting amount or any of the moisture content methods described above. The RH in the desiccators is most often controlled by saturated salt solutions (Table 15.7) with the exception of the 0%RH condition, which is created using calcium sulfate (DrieRite™) or phosphorous pentoxide (P_2O_5), the latter of which is more effective [36]. Enough saturated salt solution and headspace must be present to maintain a constant RH as moisture is exchanged with stored samples. General recommendations are to use ≥10% of the container volume filled with salt solution, >10:1 ratio of salt solution surface area to sample surface area, and a 20:1 ratio of headspace volume to sample volume. Vacuum may or may not be applied to the desiccators. Six to nine different RH chambers are recommended to generate a moisture sorption isotherm for a product [37]. Temperature is controlled by placing the desiccators in temperature-controlled incubators, ideally with temperature varia-

tion <1 °C. Samples should be stored and analyzed in triplicate over time, weighing to ±0.0001 g until some equilibrium criterion is reached. For example, equilibrium could be defined as <0.01% mass change across three consecutive days of weighings. The amount of time it takes for samples to reach equilibrium varies by sample type and RH condition, often from 10 to 21 days. Desiccator methods are time and labor intensive, there is potential for RH fluctuation as samples are removed from the desiccators for weighing, and challenges with chemical change or microbial growth in samples during extended equilibration may be encountered. There are a limited number of RHs generated by saturated salt solutions, and therefore equations (such as the GAB and BET models) must be used to calculate sorption profiles from the measured data points [37]. However, the relative simplicity of the technique and ability to store multiple samples are attractive features of using the desiccator method to generate moisture sorption isotherms.

15.4.3 Automated Gravimetric Moisture Sorption Balance

In automated gravimetric moisture sorption balances, sometimes referred to as static dynamic vapor sorption (DVS) instruments, the mass of a sample is monitored as a function of time, while the sample is exposed to a series of tightly controlled RH step changes at constant temperature. An example of data collected using this technique is provided in Fig. 15.13. The initial moisture content of the sample must be known, or a sample can be dried in the instrument prior to analysis. The parameters used in generating automated moisture sorption isotherms (or profiles), including equilibrium criterion (% weight change in a set time period, e.g., 0.001% or 0.01%), step time (how long will a set RH be held before the next programmed RH step is initiated, in the absence of attaining the equilibrium criterion), and RH step size (1% RH, 5% RH,

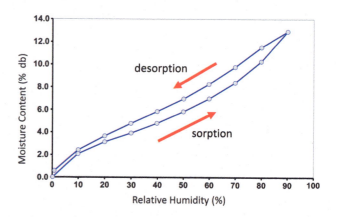

15.13 **figure** Moisture sorption isotherm (type II) of an amorphous powder

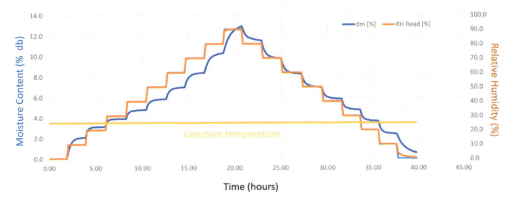

15.14
figure

Moisture sorption data for an amorphous powder collected on a DVS instrument (SPS, ProUmid GmbH, Germany) and used to generate the moisture sorption isotherm shown in Fig. 15.13

10 % RH, etc.), have a profound influence on the curves generated (Fig. 15.14).

When a DVS instrument moves to a programmed RH step, air with the programmed RH is moved into the sample chamber, and the mass of the sample is monitored as it sorbs (or desorbs) water. Usually the initial sorption is more rapid than later stages of sorption, as seen in Fig. 15.14. If true equilibrium is reached between the sample and the surrounding environmental RH, then no more moisture uptake would occur and the equilibrium criterion would be 0.00 % mass change over any time scale. However, due to the labile nature of some foods and the need to collect data for foods over a period of days not months, an equilibrium criterion of 0.01 % weight change over some time scale (1–5 h) is more common. It is important to recognize that the shorter the step time, the more likely the sample is not in equilibrium with the surrounding RH resulting in sorption profiles and not isotherms.

Careful control and reporting of test parameters are important for reproducibility of moisture sorption isotherms. Advantages of DVS methods compared to desiccator methods include precise RH control and mass determinations, a smaller amount of sample (5 mg–1.5 g), smaller RH step sizes, and possibly a reduced time to generate a moisture sorption isotherm for a single sample. Disadvantages of DVS methods include high instrumentation costs, the upper RH limit being 95 % (or lower RH at higher temperatures), and a more limited sample size (many DVS instruments only analyze one sample at a time, although there is currently one that analyzes up to 23 samples simultaneously (SPS, ProUmid GmbH & Co., Germany)).

A modified approach to the DVS instrumentation is found in the AquaLab Vapor Sorption Analyzer (Decagon Devices, Pullman, WA) operated in DDI mode. In the DDI mode, a small amount of saturated (100 % RH) or dry (0 % RH) air is introduced into the sample chamber, the chamber is sealed, the RH is continuously monitored by a capacitance RH sensor

until a constant RH is reached (on average 5–6 min), and at this point both the mass and the water activity of the sample (based on the chilled mirror dewpoint technique) are recorded. While it is not possible to program RH steps or timing between measurements into the DDI technique, the DDI sorption profiles are very high resolution with numerous data points collected and may be collected in less time than the static DVS moisture sorption isotherms.

15.4.4 Phase Diagrams Containing a_w, Moisture Content, and T_g Relationships

In addition to routine moisture analyses for quality control, the food industry is increasingly taking a fundamental materials science approach to designing foods and processes and controlling food quality. An important practice here is to establish not only moisture sorption isotherm relationships for foods but also to establish the relationship between moisture content and glass transition temperature (T_g) as well as the relationship between water activity and glass transition temperature (T_g). More detailed information about T_g measurements is provided in Chap. 31.

15.5 SUMMARY

The moisture content and water activity of foods are important to food processors and consumers for a variety of reasons. While moisture determination may seem simplistic, it is often one of the most difficult assays in obtaining accurate and precise results. Direct moisture content analysis methods involve a separation of moisture in the sample from the solids and then quantitation by weight or volume. Indirect methods do not involve such a separation but instead are based on some physical or chemical property of the sample that varies with respect to its moisture content. A major difficulty with many methods is attempting to remove or otherwise quantitate all water present. This often is complicated by decomposition or interference by other food constituents. For each moisture analysis method, there are factors that must be controlled or precautions that must be taken to ensure accurate and precise results. Careful sample collection and handling procedures are extremely important and cannot be overemphasized. The choice of moisture analysis method is often determined by the expected moisture content, nature of other food constituents (e.g., highly volatile, heat sensitive), equipment available, speed necessary, accuracy and precision required, and intended purpose (e.g., regulatory or in-plant quality control).

15.6 STUDY QUESTIONS

1. Identify five factors that one would need to consider when choosing a moisture content analysis method for a specific food product.
2. Why is standardized methodology needed for moisture content determinations?
3. What are the potential advantages of using a vacuum oven rather than a forced draft oven for moisture content determination?
4. In each case specified below, would you likely overestimate or underestimate the moisture content of a food product being tested? Explain your answer.

   (a) Forced draft oven:
   - Particle size too large
   - High concentration of volatile flavor compounds present
   - Lipid oxidation
   - Sample very hygroscopic
   - Alteration of carbohydrates (e.g., Maillard browning)
   - Sucrose hydrolysis
   - Surface crust formation
   - Splattering
   - Desiccator containing DrieRite™ or P_2O_5 desiccant (~0%RH) with dried sample not sealed properly

   (b) Toluene distillation:
   - Emulsion between water in sample and solvent not broken
   - Water clinging to condenser

   (c) Karl Fischer:
   - Very humid day when weighing original samples
   - Glassware not dry
   - Sample ground coarsely

- Food high in vitamin C
- Food high in unsaturated fatty acids

5. The procedure for an analysis for moisture in a liquid food product requires the addition of 1–2 ml of deionized water to the weighed sample in the moisture pan. Why should you add water to an analysis in which moisture is being determined?
6. A new instrument based on near-infrared principles has been received in your laboratory to be used in moisture analysis. Briefly describe the way you would ascertain if the new instrument would meet your satisfaction and company standards.
7. A technician you supervise is to determine the moisture content of a food product by the Karl Fischer method. Your technician wants to know what is this "Karl Fischer reagent water equivalence" that is used in the equation to calculate percentage of moisture in the sample, why is it necessary, and how it is determined. Give the technician your answer.
8. You are fortunate to have available in your laboratory the equipment for doing moisture content analysis by essentially all methods – both official and rapid quality control methods. For each of the food products listed below (with the purpose specified as rapid quality control or official), indicate (a) the name of the method you would use, (b) the principle (not procedure) for the method, (c) a justification for use of that method (as compared to using a hot air-drying oven), and (d) two cautions in use of the method to ensure accurate results:

   (a) Ice cream mix (liquid) – quality control
   (b) Milk chocolate – official
   (c) Spices – official
   (d) Syrup for canned peaches – quality control
   (e) Oat flour – quality control

9. You are a manufacturer of processed cheese. The maximum allowed moisture content for your product is 40%. Your current product has a mean moisture content of 38%, with a standard deviation of 0.7. It would be possible to increase your mean moisture content to 39.5% if you could reduce your standard deviation to 0.25. This would result in a saving of $3.4 million per year. You can accomplish this by rapidly analyzing the moisture content of the cheese blend prior to the cooking step of manufacture. The cheese blend is prepared in a batch process, and you have 10 min to adjust the moisture content of each batch:

(a) Describe the rapid moisture analysis method you would use. Include your rationale for selecting the method.

(b) How would you ensure the accuracy and precision of this method (you need to be sure your standard deviation is below 0.25)?

10. You work in a milk drying plant. As part of the production process, you need to rapidly analyze the moisture content of condensed milk:

(a) What rapid secondary method would you use, and what primary method would you use to calibrate the secondary method? Additionally, how would you ensure the accuracy and precision of your secondary method?

(b) Your results with the secondary method are consistently high (about 1%), based on the secondary method you chose. What are some potential problems and how would your correct them?

11. During a 12 h period, 1000 blocks (40 lbs each) from ten different vats (100 blocks per vat) of Cheddar cheese were produced. It was later realized that the cooking temperature was too low during cheesemaking. You are concerned that this might increase the moisture content of the cheese above the legal requirement. Describe the sampling plan and method of analysis you would use to determine the moisture content of the cheese. You want the results within 48 h so you can determine what to do with the cheese.

12. Compare and contrast moisture content and water activity in terms of measurement approaches, effects on food texture and moisture migration with the environment, influence on microbial growth, and roles in vitamin stability, lipid oxidation, and the Maillard reaction.

13. Identify three factors that would lead to underestimation of the water activity of a food sample. Then identify three factors that would lead to overestimation of the water activity of a food sample.

14. You are put in charge of creating moisture sorption isotherms of several food products. Describe what method you will use and what approach you will take for assuring you have created moisture sorption isotherms and not moisture sorption profiles of:

(a) Potato chips
(b) An intermediate moisture nutrition bar containing oats, chocolate pieces, and a binder syrup
(c) A gummy candy
(d) Cheddar cheese

15.7 PRACTICE PROBLEMS

1. As an analyst, you are given a sample of condensed soup to analyze to determine if it is reduced to the correct concentration. By gravimetric means, you find that the concentration is 26.54% solids. The company standard reads 28.63%. If the starting volume were 1000 gal at 8.67% solids and the weight is 8.5 lb per gallon, how much more water must be removed?

2. Your laboratory just received several sample containers of peas to analyze for moisture content. There is a visible condensate on the inside of the container. What is your procedure to obtain a result?

3. You have the following gravimetric results: weight of dried pan and glass disk is 1.0376 g, weight of pan and liquid sample is 4.6274 g, and weight of the pan and dried sample is 1.7321 g. What was the moisture content of the sample and what is the percent solids?

Answers

1. The weight of the soup initially is superfluous information. By condensing the soup to 26.54% solids from 8.67% solids, the volume is reduced to 326.7 gal [(8.67%/26.54%) × 1000 gal]. You need to reduce the volume further to obtain 28.63% solids [(8.67%/28.63%) × 1000 gal], or 302.8 gal. The difference in the gallons obtained is 23.9 gal (326.7−302.8 gal), or the volume of water that must be removed from the partially condensed soup to comply with company standards.

2. This problem focuses on a real issue in the food processing industry – when do you analyze a sample and when don't you? It would appear that the peas have lost moisture that should be within the vegetable for correct results. You will need to grind the peas in a food mill or blender. If the peas are in a Mason jar or one that fits a blender head, no transfer is needed. Blend the peas to a creamy texture. If a container transfer was made, then put the blended peas back into the original container. Mix with the residual moisture to a uniform blend. Collect a sample for moisture analysis. You should note on the report form containing the results of the analysis that the pea samples had free moisture on container walls when they arrived.

3. Note Eqs. 15.2, 15.3, and 15.4 in Sect. 15.2.2.1.7. To use any of the equations, you must subtract the weight of the dried pan and glass disk. Then you obtain 3.5898 g of original sample and 0.6945 g when dried. By subtracting these results, you have removed water (2.8953 g).

Then (0.6945 g/3.5898 g) × 100 = 19.35% solids
and (2.8953 g/3.5898 g) × 100 = 80.65% water.

REFERENCES

1. Pomeranz Y, Meloan C (1994) Food analysis: theory and practice, 3rd edn. Chapman & Hall, New York
2. Josyln MA (1970) Methods in food analysis, 2nd edn. Academic, New York
3. US Department of Agriculture, Agricultural Research Service (2016). USDA National Nutrient Database for Standard References, Release 28. Nutrient Data Laboratory Home Page. http://ndb.nal.usda.gov
4. Fontana Jr AJ, Campbell CS (2004) Water activity (Chapter 3). In: Nollet L (ed.) Handbook of Food Analysis, 2nd edn. Marcel Dekker, Inc., New York p. 39–54.
5. Fontana Jr AJ (2007) Measurement of water activity, moisture sorption isotherms, and moisture content of foods (Chapter 7). In: Barbosa-Canovas BV, Fontana AJ, Schmidt SJ, Labuza TP (eds). Water activity in foods: Fundamentals and applications. Blackwell Publishing, Ames, IA, p. 155–71.
6. Reid DS, Fennema OR (2007) Water and ice (Chapter 2). In: Damodaran S, Parkin K, Fennema OR (eds). Fennema's food chemistry, 4th edn. CRC, Boca Raton, FL
7. Arunan EDG, Klein RA, Sadlej J, Scheiner S, Alkorta I, Clary DC, Crabtree RH, Dannenbery JJ, Hobza P, Kjaergaard HG, Legon AC, Mennucci B, Nesbitt DJ (2011) Definition of the hydrogen bond (IUPAC recommendations 2011). Pure Applied Chemistry 83(8):1637–41
8. Martiniano H, Galamba, N (2013) Insights on hydrogen-bond lifetimes in liquid and supercooled water. J Physical Chem B. 117:16188–95
9. Schmidt S (2004)Water and solids mobility in foods. Adv Food and Nutr Res. 48:1–101
10. Zografi G (1988) States of water associated with solids. Drug Development And Industrial Pharmacy. 14(14):1905–26
11. Emmons DB, Bradley RL Jr, Sauvé J P, Campbell C, Lacroix C, Jimenez - Marquez SA (2001) Variations of moisture measurements in cheese. J AOAC Int. 84:593–604
12. Emmons DB, Bradley RL Jr, Campbell C, Sauve JP (2001) Movement of moisture in refrigeration cheese samples transferred to room temperature. J AOAC. 84:620–622
13. Bradley RL Jr, Vanderwarn MA (2001) Determination of moisture in cheese and cheese products. J AOAC Int. 84:570–592
14. AOAC International (2016) Official methods of analysis, 20th edn. (online). AOAC International, Rockville, MD
15. Wehr HM, Frank JF (eds) (2004) Standard methods for the examination of dairy products, 17th edn. American Public Health Association, Washington, DC
16. AACC International (2010) Approved methods of analysis, 11th edn. AACC International, St. Paul, MN
17. Nelson OA, Hulett GA (1920) The moisture content of cereals. J Ind Eng Chem. 12:40–45
18. International Dairy Federation (1982) Provisional Standard 4A. Cheese and processed cheese: determination of the total solids content. International Dairy Federation, Brussels, Belgium
19. Bouraoui M, Richard P, Fichtali J (1993) A review of moisture content determination in foods using microwave oven drying. Food Res Inst Int. 26:49–57
20. Mitchell J Jr, Smith DM (1948) Aquametry. Wiley, New York
21. Giese J (1993) In-line sensors for food processing. Food Technol 47(5):87–95
22. Wilson RH, Kemsley EK. On-line process monitoring using infrared techniques. . Food processing automation II proceedings of the American society of agricultural engineers, ASAE Publication 02–92. St. Joseph, MI: American Society of Agricultural Engineers; 1992.
23. Pande A. Handbook of moisture determination and control. New York: Marcel Dekker; 1974.
24. Anonymous. Kett Microwave Moisture Meters Santiago, CA: Kett; 2009 [cited 2016]. Available from: www.kett.com.
25. Reid DS. Water activity: Fundamentals and relationships. In: Barbosa-Canovas BV, Fontana AJ, Schmidt SJ, Labuza TP (eds). Water activity in foods. Ames, IA: Blackwell Publishing; 2007. p. 15–28.
26. Anonymous. Code of Federal Regulations. Title 21 Parts 108, 113, 114. Washington DC: US Government Printing Office; 2015.
27. Scott WJ. Water relations of food spoilage microorganisms. Advances in Food Research, 1957; 7:83–127.
28. International N. Non-potentially hazardous foods. . Ann Arbor, MI: 2000 Nov. 10. Report No.: ANSI/NSF 75–2000.
29. Anonymous. International Standard ISO 21807: Microbiology of food and animal feeding stuffs – Determination of water activity. Switzerland: International Organization for Standardization; 2004.
30. Labuza TP, Tannenbaum, SR, Karel, M. Water content and stability of low moisture and intermediate moisture foods. Journal of Food Technology. 1970; 24:543–50.
31. Scott VN, Clabero RS, Troller JA. Chapter 64. Measurement of water activity (aw), acidity, and brix. In: Downes FP, Ito K, (eds). Compendium of Methods for the Microbiological Examination of Foods, 4th Edn: American Public Health Association; 2001. P. 649–657.
32. Decagon Devices. AquaLab water activity meter operator's manual, Version 4. Pullman, WA: Decagon Devices; 2009.
33. Rodel W. Water activity and its measurement in food (Chapter 16). In: Kress-Rogers E, Brimelow CJB (eds). Instrumentation and Sensors for the Food Industry. Cambridge: Woodhead; 2001. p. 453–83.
34. Airaksinen S, Karjalainen M, Shevchenko A, Westermarck S, Leppanen E, Rantanen J, et al. Role of water in the physical stability of solid dosage formulations. J Pharm Sci. 2005:2147–65.
35. Sing KSW, Everett DH, Haul RAW, Moscou L, Pierotti RA, Rouquerol J, et al. Reporting physisorption data for gas/solid systems with special reference to the determination of surface area and porosity. Pure and Applied Chemistry. 1985; 57(4):603–19.
36. Trussel F, Diehl H. Efficiency of chemical desiccants. Analytical Chemistry. 1963; 35:674–7.
37. Bell LN, Labuza TP. Moisture Sorption: Practical Aspects of Isotherm Measurement and Use. 2nd Edn. St. Paul, MN: American Association of Cereal Chemists; 2000.
38. CRC Handbook of Chemistry and Physics, 96th ed. CRC Press: Boca Raton, FL, 2015.

Ash Analysis

G.Keith Harris (✉)

Department of Food, Bioprocessing & Nutrition Sciences,
North Carolina State University,
Raleigh, NC 27695, USA
e-mail: keith_harris@ncsu.edu

Maurice R. Marshall

Department of Food Science and Human Nutrition,
University of Florida,
Gainesville, FL 32611-0370, USA
e-mail: martym@ufl.edu

S. Nielsen (ed.), *Food Analysis*, Food Science Text Series,
DOI 10.1007/978-3-319-45776-5_16, © Springer International Publishing 2017

16.1   INTRODUCTION

Ashing is an important first step in proximate or specific mineral analysis. **Ash** refers to the inorganic (mineral) residue remaining after the combustion or complete acid-facilitated oxidation of organic matter in food. A basic knowledge of the characteristics of various ashing procedures and types of equipment is essential to ensure reliable results. Two major types of ashing are used: dry ashing and wet ashing. Dry ashing is primarily used for proximate composition and for some types of specific mineral analyses. Dry ashing requires very high (500–600 °C) temperatures, which may be achieved by conventional or microwave heating. Wet ashing (acid-facilitated oxidation) is used as a preparation for the analysis of minerals that might be volatilized and lost during dry ashing. Wet ashing uses lower temperatures than dry ashing and relies on strong acids and chemical oxidizers to rid samples of organic material. Microwave systems can also be used for both dry and wet ashing. In either case, microwave systems tend to speed the ashing process, although sample throughput may be a limiting factor. Most dry samples (i.e., whole grain, cereals, dried vegetables) need no preparation, while fresh vegetables and other high-moisture foods are generally dried prior to ashing. High-fat products such as meats may need to be dried and fat extracted before ashing to prevent smoke generation during heating steps. Ashing is a gravimetric analysis, in which the final (ashed) weight is compared to the original weight of the sample. The ash content of foods can then be expressed on either a wet weight (as is) or dry weight basis. For general and food-specific information on measuring ash content, see references [1–14].

16.1.1   Definitions

Dry ashing refers to the use of a muffle furnace capable of maintaining temperatures of 500–600 °C. Water and volatiles are vaporized, and organic substances are burned in the presence of oxygen to form CO_2 and oxides of N_2. In this way, organic materials are removed from the sample. The remaining minerals are converted to oxides, sulfates, phosphates, chlorides, and silicates. Elements such as Fe, Se, Pb, and Hg may partially volatilize with this procedure, so other methods must be used if ashing is a preliminary step for the analysis of these minerals.

Wet ashing is a procedure used to chemically oxidize and remove organic substances using strong acids, oxidizing agents, or combinations thereof. The acids and oxidizing agents used must also be capable of solubilizing the remaining minerals. Hydrochloric, sulfuric, nitric, and perchloric acids are commonly used. Perchloric acid requires a specially designed hood, since it may leave explosive peroxides behind.

Since wet ashing is conducted either at lower temperatures than dry ashing or in sealed vessels, minerals are not lost due to volatility. For this reason, wet ashing is preferable to dry ashing as a preparation for specific elemental analysis.

16.1.2   Importance of Ash in Food Analysis

Ash content represents the total mineral content in foods. Determining the ash content may be important for several reasons. It is a part of proximate analysis for nutritional evaluation. Ashing is also the first step in preparing a food sample for specific elemental analysis, whether for essential nutrients or for highly toxic heavy metals. Because certain foods are high in particular minerals, ash content can be important from nutritional, toxicological, and food quality standpoints. For example, dairy and beef are known to be rich sources of calcium and iron. Ash content is commonly part of the ingredient specification for flours and whole grains. Rice grown in arsenic-containing soil effectively concentrates it. High levels of transition metals in lipid-rich foods may speed rancidity and limit shelf life. The mineral content of plant foods tends to be more variable than that of animal products.

16.1.3   Ash Contents in Foods

The ash content of food can range from 0 to 12 %, but rarely exceeds 5 % for fresh foods.

The average ash content for various food groups is given in Table 16.1.

16.2   METHODS

Principles, materials, instrumentation, general procedures, and applications are described below for various ash determination methods. Refer to methods cited for detailed instructions of the procedures.

16.2.1   Sample Preparation

Ashing does not require large sample sizes. A 2–10 g sample is generally sufficient for ash determination. Some newer instruments allow for sample sizes as low as 250 mg. For this reason, it is necessary to ensure that a homogenous, representative sample is obtained. Milling, grinding, and other methods used to homogenize samples will not significantly alter ash values for proximate analysis. In contrast, if ashing is used as a preparatory step for specific mineral analyses, mineral contamination from the environment or from grinding equipment is of potential concern and may require the use of sample blanks to account for this. (This is particularly true of metals such as iron, since most grinders and mincers are of steel construction.) Repeated use of

16.1 table   Ash content of selected foods

Food item	Percent ash (wet weight basis)
Cereals, bread, and pasta	
Rice, brown, long grain, raw	1.5
Corn meal, whole grain, yellow	1.1
Hominy, canned, white	0.9
White rice, long grain, regular, raw, enriched	0.6
Wheat flour, whole grain	1.6
Macaroni, dry, enriched	0.9
Rye bread	2.5
Dairy products	
Milk, reduced fat, fluid, 2 %	0.7
Evaporated milk, canned, with added vitamin A	1.6
Butter, with salt	2.1
Cream, fluid, half and half	0.7
Margarine, hard, regular, soybean	2.0
Yogurt, plain, low fat	1.1
Fruits and vegetables	
Apples, raw, with skin	0.2
Bananas, raw	0.8
Cherries, sweet, raw	0.5
Raisins	1.9
Potatoes, raw, skin	1.6
Tomatoes, red, ripe, raw	0.5
Meat, poultry, and fish	
Eggs, whole, raw, fresh	0.9
Fish fillet, battered or breaded and fried	2.5
Pork, fresh, leg (ham), whole, raw	0.9
Hamburger, regular, single patty, plain	1.9
Chicken, broilers or fryers, breast meat only, raw	1.0
Beef, chuck, arm pot roast, raw	1.1

From US Department of Agriculture, Agricultural Research Service (2016) USDA National Nutrient Database for Standard Reference. Release 28. Nutrient Data Laboratory Home Page, http://ndb.nal.usda.gov

glassware can also be a source of contaminants. This can be addressed by soaking crucibles and glassware in an acid bath to solubilize mineral contaminants and rinsing repeatedly with distilled-deionized water. The crucibles themselves can also be "pre-ashed" in a muffle furnace to remove organic contaminants.

16.2.1.1 Plant Materials

Plant materials are generally oven-dried prior to grinding and ashing. The primary goal for this oven-drying step is to remove moisture. Since samples are often used for multiple determinations (e.g., protein, fiber, lipid), it may be necessary to keep oven temperatures

at or below 100 °C to prevent the destruction or alteration of non-mineral analytes. Fresh stem and leaf tissue should be dried in two stages (i.e., first at a lower temperature of 55 °C, then at a higher temperature) to prevent artifact lignin and other undesired products. Plant material with 15 % or less moisture may be ashed without prior drying. If the goal of the analysis is ash or specific mineral analysis alone, low-temperature drying can be accomplished via the use of a muffle furnace equipped with temperature gradient settings. Samples can thus be dried at low temperature and then ashed in the same crucible.

16.2.1.2 Fat and Sugar Products

Animal products, syrups, and spices require special consideration prior to ashing because of high fat, moisture, or high sugar content that may result in loss of sample due to spattering, swelling, or foaming. Depending on the application, crucible covers or the use of sealed crucibles may be used to contain samples. Meats, sugars, and syrups should be evaporated to dryness on a steam bath or with an infrared (IR) lamp prior to ashing. For products that tend to form a crust on heating, one or two drops of olive oil (which contains no ash) may be added to allow steam to escape as the crust forms.

Smoking and burning may occur upon ashing for some products (e.g., cheese, seafood, spices). Allow this process to finish slowly by keeping the muffle door slightly open until combustion stops, then close the door fully, and proceed with the normal ashing protocol. A sample may be ashed after drying and fat extraction. In most cases, mineral loss is minimal during drying and fat extraction. Fat-extracted samples should not be heated until flammable extraction solvents (hexane, ether, etc.) have been evaporated completely to avoid the potential risk of solvent ignition or explosion.

16.2.2 Dry Ashing

16.2.2.1 Principles and Instrumentation

Dry ashing is incineration at high temperature (525 °C or higher). Incineration is accomplished with a muffle furnace. Several models of muffle furnaces are available, ranging from large capacity units requiring either 208- or 240-V electric supplies to small benchtop units utilizing 110-V outlets.

The muffle furnace may have to be placed in a heat-proof room along with drying ovens. It is important to make sure large furnaces of that type are equipped with a double-pole, single-throw switch. Heating coils are generally exposed, and care must be taken when taking samples in and out with metal tongs.

Crucible selection is critical in ashing because the type depends upon the specific use. Primary considerations relate to chemical stability and resistance to high temperatures. **Quartz crucibles** are resistant to acids and halogens, but not alkali, at high temperatures. **Vycor® brand crucibles** are stable to 900 °C, but **Pyrex® Gooch crucibles** are limited to 500 °C. Ashing at a lower temperature of 500–525 °C may result in slightly higher ash values because of less decomposition of carbonates and loss of volatile salts. **Porcelain crucibles** resemble quartz crucibles in their properties but will crack with rapid temperature changes. Since they are relatively inexpensive, porcelain crucibles are often the crucible of choice. **Steel crucibles** are resistant to both acids and alkalis and are inexpensive, but they are composed of iron, chromium, and nickel, which are possible sources of contamination. **Platinum crucibles** are very inert and are probably the best overall, but they are currently far too expensive for routine use for large numbers of samples. **Quartz fiber crucibles** are disposable, unbreakable, and can withstand temperatures up to 1,000 °C. They are porous, allowing air to circulate around the sample and speed combustion. This reduces ashing times significantly and makes them ideal for solids and viscous liquids. Quartz fiber also cools in seconds, virtually eliminating the risk of burns.

All crucibles should be marked for identification. Marks on crucibles with a felt-tip marking pen will disappear during ashing in a muffle furnace. Laboratory inks scribed with a steel pin are available commercially. Crucibles also may be etched with a diamond point and marked with a 0.5 M solution of $FeCl_3$, in 20 % HCl. An iron nail dissolved in concentrated HCl forms a thick brown pigment that is a satisfactory marker. The crucibles should be fired and cleaned prior to use.

The *advantages* of conventional dry ashing are that it is relatively a safe and inexpensive method, that less sample is required than for other methods, that it requires no acids or other added reagents or blank subtraction, that, depending on the size of the oven, many samples may be processed at once, that sample recovery is high, and that little attention is needed once ignition begins [2]. Usually a large number of crucibles can be handled at once, and the resultant ash can be used additionally in other analyses for most individual elements, acid-insoluble ash, and water-soluble and insoluble ash. The *disadvantages* are the length of time required (12–18 h or overnight) and expensive equipment relative to wet ashing. There will be some loss of the volatile elements and interactions between mineral components and crucibles. Volatile elements at risk of being lost include As, B, Cd, Cr, Cu, Fe, Pb, Hg, Ni, P, V, and Zn.

16.2.2.2 *Procedures*

AOAC International has several dry ashing procedures (e.g., AOAC Methods 900.02 A or B, 920.117, 923.03) for certain individual foodstuffs. The general procedure includes the following steps:

1. Weigh a 5- to 10-g sample into a tared crucible (making sure to account for the weight of crucible covers if they are used). Pre-dry if the sample is very moist.
2. Place crucibles in a cool muffle furnace. Use tongs, gloves, and protective eyewear if the muffle furnace is warm.
3. Heat samples for 12–18 h (or overnight) at about 550 °C.
4. Turn off muffle furnace and wait to open it until the temperature has dropped to at least 250 °C, preferably lower. Open the furnace door slowly to avoid losing the powdery ash that may be disturbed by air movement.
5. Using safety tongs, quickly transfer crucibles to a desiccator with a porcelain plate and desiccant. Cover crucibles, close desiccator, and allow crucibles to cool prior to weighing.

Note: Warm crucibles will heat air within the desiccator. With hot samples, a cover may bump to allow air to escape. A vacuum may form on cooling. At the end of the cooling period, the desiccator cover should be removed gradually by sliding to one side to prevent a sudden inrush of air. Covers with a ground-glass sleeve or fitted for a rubber stopper allow for slow release of a vacuum.

The ash content is calculated as follows:

$$\% \, \text{Ash (dry basis)} = \frac{(WAA - TWOC)}{(OSW \times DMC)} \times 100 \quad (16.1)$$

where:

> WAA = weight after ashing
> TWOC = tare weight of crucible
> OSW = original sample weight
> DMC = % solids/100

Using the dry matter coefficient allows for the direct conversion of wet weight percent ash values to dry weight ash values. For example, if corn meal is 87 % dry matter, the dry matter coefficient would be 0.87. If it is necessary to calculate percent ash on an as-received or wet weight basis (includes moisture), delete the dry matter coefficient from the denominator. If moisture content was determined in the same crucible prior to ashing, the denominator becomes (dry sample wt – tared crucible wt).

16.2.2.3 *Special Applications*

Some of the AOAC procedures recommend steps in addition to those listed previously. If carbon is still present following the initial incineration, several drops of water or nitric acid should be added; then the sample should be re-ashed. If the carbon persists, such as with high-sugar samples, follow this procedure:

1. Suspend the ash in water.
2. Filter through ashless filter paper because this residue tends to form a glaze.
3. Dry the filtrate.
4. Place paper and dried filtrate in muffle furnace and re-ash.

Other suggestions that may be helpful and accelerate incineration:

1. High-fat samples should be extracted either by using the crude fat determination procedure or by burning off prior to closing the muffle furnace.
2. Glycerin, alcohol, and hydrogen will accelerate ashing.
3. Samples such as jellies will spatter and can be mixed with cotton wool to avoid this.
4. Salt-rich foods may require a separate ashing of water-insoluble components and salt-rich water extract. Use a crucible cover to prevent spattering.
5. An alcoholic solution of magnesium acetate can be added to accelerate ashing of cereals. An appropriate blank determination is necessary.

16.2.3 Wet Ashing

16.2.3.1 *Principle, Materials, and Applications*

Wet ashing is sometimes called **wet oxidation** or **wet digestion**. Its primary use is preparation for specific mineral analysis. Often, analytical testing laboratories use only wet ashing to prepare samples for certain mineral analyses (e.g., Fe, Cu, Zn, P), because losses would occur by volatilization during dry ashing. There are several *advantages* to using the wet ashing procedure. Minerals will usually stay in solution, and there is little or no loss from volatilization because of the lower temperature. The oxidation time is short and requires a hood, hot plate, and long tongs, plus safety equipment. The *disadvantages* of wet ashing are that it takes virtually constant operator attention, corrosive reagents are necessary, and only small numbers of samples can be handled at any one time. If the wet digestion utilizes perchloric acid, all work needs to be carried out in an expensive special fume hood called a **perchloric acid hood**, since working under a normal fume hood could lead to the deposit of explosive peroxides in the ventilation system.

Unfortunately, using a single acid for wet ashing does not give complete and rapid oxidation of organic material, so a mixture of acids often is used. Combinations of the following acid solutions are used most often: (1) **nitric acid**, (2) **sulfuric acid-hydrogen peroxide**, and (3) **perchloric acid**. Different combinations are recommended for different types of samples. The nitric-perchloric combination is generally faster than the sulfuric-nitric procedure.

While wet digestion with perchloric acid is an AOAC procedure (e.g., AOAC Method 975.03), many analytical laboratories avoid the use of perchloric acid in wet ashing, and instead use a combination of nitric acid with either sulfuric acid, hydrogen peroxide, or hydrochloric acid. This is because perchloric acid can form explosive peroxide by-products; it is *extremely* dangerous to work with. The reason that perchloric acid is still used is that it allows for better extraction of certain minerals from resistant matrices, such as bone. When working with perchloric acid, special fume hoods free of plastic or glycerol-based caulking compounds and with wash-down capabilities must be used. Precautions for use of perchloric acid are found in the AOAC methods under "Safe Handling of Special Chemical Hazards." Cautions must be taken when fatty foods are wet ashed using perchloric acid. While perchloric acid does not interfere with atomic absorption spectroscopy, it does interfere with the traditional colorimetric assay for iron.

16.2.3.2 *Procedures*

The following is a wet ash procedure using concentrated nitric and sulfuric acids (*to be performed in a fume hood*) (John Budin, Silliker Laboratories, Chicago, IL, personal communication):

1. Accurately weigh a dried, ground 1-g sample in a 125-mL Erlenmeyer flask (previously acid washed and dried).
2. Prepare a blank of 3 ml of H_2SO_4 and 5 ml of HNO_3, to be treated like the samples. (Blank is to be run with every set of samples.)
3. Add 3 ml of H_2SO_4 followed by 5 ml of HNO_3 to the sample in the flask.
4. Heat the sample on a hot plate at ca. 200 °C (boiling). Brown-yellow fumes will be observed.
5. Once the brown-yellow fumes cease and white fumes from decomposing H_2SO_4 are observed, the sample will become darker. Remove the flask from the hot plate. Do not allow the flask to cool to room temperature.
6. *Slowly* add 3–5 ml of HNO_3.
7. Put the flask back on the hot plate and allow the HNO_3 to boil off. Proceed to the next step when all the HNO_3 is removed and the color is clear to straw yellow. If the solution is still dark in color, add another 3–5 ml of HNO_3 and boil. Repeat the process until the solution is clear to straw yellow.
8. While on the hot plate, reduce the volume appropriately to allow for ease of final transfer.

Allow the sample to cool to room temperature, and then quantitatively transfer the sample to an appropriately sized volumetric flask.

9. Dilute the sample to volume with ultrapure water, and mix well. Dilute further, as appropriate, for the specific type of mineral being analyzed.

A combination of dry and wet ash procedure is given in AOAC Method 985.35 "Minerals in Infant Formula, Enteral Products, and Pet Foods."

16.2.4 Microwave Ashing

Both **wet ashing** and **dry ashing** can be done using microwave instrumentation, rather than the conventional dry ashing in a muffle furnace and wet ashing in a flask or beaker on a hot plate. The CEM Corporation (Matthews, NC) has developed a series of instruments for dry and wet ashing, as well as other laboratory systems for microwave-assisted chemistry. While the ashing procedures by conventional means can take many hours, the use of microwave instrumentation can reduce sample preparation time to minutes, allowing laboratories to increase their sample throughput significantly (current CEM models have a throughput of 5–24 samples per hour). This advantage has led to widespread use of microwave ashing, especially for wet ashing, within both analytical and quality control laboratories in food companies.

16.2.4.1 *Microwave Wet Ashing*

Microwave wet ashing (acid digestion) may be performed safely in either an open- or closed-vessel microwave system. Choice of the system depends on the amount of sample and the temperatures required for digesting. Due to the ability of the closed vessels to withstand higher pressures (some vessels can handle up to 1,500 psi), acids may be heated past their boiling points. This ensures a more complete dissolution of hard-to-digest substances. It also allows the chemist to use nitric acid with samples that might normally require a harsher acid, such as sulfuric or perchloric. In closed vessels specifically designed for high-temperature/high-pressure reactions, nitric acid can reach a temperature of 240 °C. Thus, **nitric acid** is often the acid of choice, though hydrochloric, hydrofluoric, and sulfuric acids also are used, depending on the sample and the subsequent analysis being performed. **Closed-vessel microwave digestion systems** (Fig. 16.1) can process up to 40 samples at a time, with vessel liners available in Teflon®, TFM™ fluoropolymer, and quartz. These systems allow the input of time, temperature, and pressure parameters in a step-by-step format (ramping). In addition, some instruments enable the user to adjust the power and offer "change-on-the-fly" software, which allows the method to be changed while the reaction is running.

Typically, in a closed-vessel microwave system, samples are placed in vessels with the appropriate amount of acid. The vessels are sealed and set on a carousel where the temperature and pressure sensors are connected to a control vessel. The carousel then is placed in the microwave cavity and the sensors connected to the instrument. Time, temperature, pressure, and power parameters are chosen and the unit is started. Digestions normally take less than 30 min. Because of the pressure generated by raising the temperature of a reaction, vessels must be allowed to cool before being opened. The ability to process multiple samples simultaneously provides the chemist with greater throughput than traditional methods. (Note that some closed-vessel microwave digestion systems may also be used for acid concentration, solvent extraction, protein hydrolysis, and chemical synthesis with the proper accessories.)

Open-vessel digestion systems (Fig. 16.2) are used often for larger sample sizes (up to 10 g) and for samples that generate substantial amounts of gas as they are digested. Open-vessel systems can process up to six samples, each according to its own parameters in

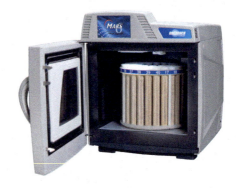

16.1 **figure** Microwave closed-vessel digestion system (Courtesy of CEM Corporation, Matthews, NC)

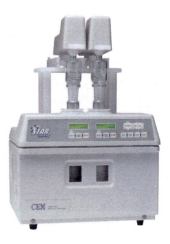

16.2 **figure** Microwave open-vessel system (Courtesy of CEM Corporation, Matthews, NC)

a sequential or simultaneous format. Teflon®, quartz, or Pyrex® vessels are used and condensers are added for refluxing. Acid (reagent) is automatically added according to the programmed parameters. Sulfuric, nitric, hydrochloric, and hydrofluoric acids, as well as hydrogen peroxide, can all be used in open-vessel systems. These instruments do not require the use of a fume hood, because a vapor containment system contains and neutralizes harmful fumes.

Generally, in an open-vessel microwave system, sample is placed in a vessel and the vessel set in a slot in the microwave system. Time, temperature, and reagent addition parameters are then chosen. The unit is started, the acid is added, and the vapor containment system neutralizes the fumes from the reaction. Samples are typically processed much faster and more reproducibly than on a conventional hot plate. (Note that some open-vessel systems may be used for evaporation and acid concentration as well.)

16.2.4.2 *Microwave Dry Ashing*

Compared to conventional dry ashing in a muffle furnace that often takes many hours, **microwave muffle furnaces** (Fig. 16.3) can ash samples in minutes, decreasing analysis time by as much as 97 %. Microwave muffle furnaces can reach temperatures of up to 1200 °C. These systems may be programmed with various methods to automatically warm up and cool down. In addition, they are equipped with exhaust systems that circulate the air in the cavity to help decrease ashing times. Some also have scrubber systems to neutralize any fumes. Any crucible that may be used in a conventional muffle furnace may be used in a microwave furnace, including those made of porcelain, platinum, quartz, and quartz fiber. Quartz fiber crucibles cool in seconds and are not breakable. Some systems can process up to 15 (25-ml) crucibles at a time.

Typically, in microwave dry ashing, a desiccated crucible is weighed, sample is added, and the crucible is weighed again. The crucible then is placed in the microwave furnace, and the time and temperature parameters are set. A step-by-step (ramping) format may be used when programming the method. The system is started

and the program is run to completion. The crucible then is carefully removed with tongs and reweighed. The sample then may be further analyzed, if necessary. Some tests call for acid to be added to a dry-ashed sample, which is then digested for further analysis.

A comparative study [14] showed that dry ashing various plants for 40 min using a microwave system (CEM Corporation, Matthews NC) was similar to the 4-h time in a conventional muffle furnace. Twenty minutes was shown to be adequate for the plant material used except for Cu determinations, which needed 40 min to obtain similar results. Other comparative examples include dried egg yolks, which can be ashed in 20 min in a microwave system, but require 4 h in a conventional muffle furnace. It takes 16 h to ash lactose in a conventional muffle furnace, but only 35 min in a microwave furnace. Though microwave furnaces may not hold as many samples as a conventional furnace, their speed actually allows significantly more samples to be processed in the same amount of time. Also, microwave furnaces do not require fume hood space.

16.2.5 Other Ash Measurements

The following are several special ash measurements and their applications:

1. **Soluble and insoluble ash** (e.g., AOAC Method 900.02). After dry ashing a sample, the ash is solubilized in boiling water and filtered, and the soluble and insoluble fractions are re-ashed. These measurements of soluble and insoluble ash are an index of the fruit content of preserves and jellies. A lower ash in the water-soluble fraction is an indication that extra fruit is added to fruit and sugar products.

2. **Ash insoluble in acid**. After dry ashing a sample, the ash is solubilized in 10 % HCl, boiled, and filtered, and the insoluble material is re-ashed. This is sometimes used as a measure of the surface contamination of fruits and vegetables and wheat and rice coatings. Those contaminants are generally silicates and remain insoluble in acid, except HBr.

3. **Alkalinity of ash** (e.g., AOAC Method 900.02, 940.26). To a dry-ashed sample, 10 mL 0.1*N* HCl and then boiling water are added. This sample is titrated with 0.1*N* NaOH using a methyl orange indicator. The volume of required NaOH in the titration determines the alkalinity of the ash. Ash of fruits and vegetable is alkaline, while ash of meats and some cereals is acid.

4. **Sulfated ash** (AOAC Method 900.02, 950.77). This method, applied mostly to sugars, syrups, and color additives, measures the amount of residual substance not volatilized when the sample is ignited in the presence of sulfuric acid. The sample is moistened with sulfuric acid and heated on a hot plate and then in a

 Microwave muffle furnace (Courtesy of CEM Corporation, Matthews, NC)

muffle furnace. This entire procedure is repeated, with the final weight expressed as % sulfated ash.

16.3   COMPARISON OF METHODS

The four major ashing methods described in this chapter are summarized and compared in Table 16.2. Ash determination by dry ashing requires a muffle furnace, which is relatively more expensive than doing wet ashing on a hot plate. Wet ashing requires a hood (a special hood if perchloric acid is used), corrosive reagents, and constant operator attention. While wet oxidation causes little volatilization, dry ashing will result in the loss of volatile elements. If the intent is further elemental analyses, the specific elements being analyzed will dictate whether wet or dry ashing is used. Some micro- and most volatile elements will require special equipment and procedures. Refer to Chaps. 9 and 21 for specific preparation procedures for elemental analyses.

Both dry and wet ashing can be done using microwave systems that require relatively expensive instrumentation, but they greatly reduce the time for ashing and do not require use of a fume hood. New atomic absorption and mass spectrometry-based methods that allow for direct analysis of fresh sample may eventually replace both dry and wet ashing for proximate and mineral analysis. Given the novelty and expense of these new methods relative to traditional dry and wet ashing methods, it is unlikely that these newer methods will replace them in the short term.

16.4   SUMMARY

The two major types of ashing analysis, dry ashing and wet ashing (chemical oxidation), can be accomplished by conventional means or by the use of microwave systems. The procedure of choice depends upon the use of ash following its determination and its limitations based on cost, time, and sample numbers. Conventional dry ashing is based upon sample incineration at high temperatures in a muffle furnace. Except for certain volatile elements, which may be lost during dry ashing, the residue may be used for both proximate analysis and further specific mineral analyses. Wet ashing is often used as a preparation for specific elemental analysis by atomic absorption, inductively coupled plasma and/or mass spectrometry, since it simultaneously dissolves minerals and removes organic material via oxidation. Wet ashing conserves volatile elements but requires more operator time than dry ashing, is limited to a smaller number of samples, and requires the use of highly caustic solvents. Microwave ashing (dry or wet) is faster than conventional methods and

requires little additional equipment or space, other than a dedicated fume hood. Ashing may be replaced at some point in the future by methods allowing for the direct determination of mineral identity and content from fresh samples. For the moment, however, ashing remains an essential component of proximate analysis and is a key preparation step for specific mineral analyses.

16.5   STUDY QUESTIONS

1. Identify four potential sources of error in the preparation of samples for ash analysis, and describe a way to overcome each.
2. You are determining the total ash content of a product using the conventional dry ashing method. Your boss asks you to switch to a conventional wet ashing method because he/she has heard it takes less time than dry ashing.

   (a) Do you agree or disagree with your boss concerning the time issue and why?
   (b) Not considering the time issues, why might you want to continue using dry ashing, *and* why might you change to wet ashing?

3. Your lab technician was to determine the ash content of buttermilk by conventional dry ashing. The technician weighed 5 g of liquid buttermilk into one weighed platinum crucible, immediately put the crucible into the muffle furnace using a pair of all stainless steel tongs, and ashed the sample for 48 h at 800 °C. The crucible was removed from the muffle furnace and set on a rack in the open until it was cool enough to reweigh. What errors were made in the preceding method (there were several)? What were the likely results of these errors? What instructions should you have given your technician before starting the procedure in order to avoid the errors you noted?
4. How would you recommend to your technician to overcome the following problems that could arise in conventional dry ashing of various foods?

   (a) You seem to be getting volatilization of phosphorus, when you want to later determine the phosphorus content.
   (b) You are getting incomplete combustion of a product high in sugar after a typical dry ashing procedure (i.e., the ash is dark colored, not white or pale gray).
   (c) The typical procedure takes too long for your purpose. You need to speed up the procedure, but you do not want to use the standard wet ashing procedure.

table 16.2 Summary of common ashing methods

Ashing method	Principle	Sources of error	Advantages	Disadvantages	Applications
Dry ashing	Sample heated to very high (500–600 °C) temperature. All organic matter incinerated. Remaining inorganic material quantitated gravimetrically	Microelement contamination (from grinder or water used to clean crucibles). Sample loss during pre-ash drying step. Volatilization of some elements. Incomplete combustion	Can analyze many samples at once. Requires little technician time. Safe. No blanks needed	Slow (takes 12–18 h); some minerals volatilized. Minerals difficult to resolubilize	Total ash content for proximate analysis. May be used as preparation for specific mineral analysis
Wet ashing	Organic matter oxidized using acids and oxidizing agents, leaving inorganic matter	Microelement contamination. Must run blanks to correct for organic matter in acid and oxidizing agent. Sample loss may occur due to spattering	Shorter time (~2 h) than dry ashing. Minerals stay in solution. Little or no volatilization	Requires constant attention. Lower sample throughput than dry ashing. Can be dangerous. Use of strong acids and oxidizers. Sample loss due to spattering	Ashing of samples prior to mineral analysis for official analyses
Microwave ashing (dry)	Microwave energy heats sample to very high temperatures. Incinerates organic matter. Leaves inorganic matter to be quantitated gravimetrically	Microelement contamination (from grinder or water used to clean crucibles). Sample loss during pre-ash drying step. Volatilization of elements. Incomplete combustion	More rapid (~30 min) than regular dry ashing	Costly. Lower sample throughput than regular dry ashing. Some minerals volatilized. May be hard to resolubilize	Determine total ash content for determination of proximate composition or for quality control purposes
Microwave ashing (wet)	Microwave energy and acid (and sometimes oxidizing agent) are used to oxidize and incinerate organic matter, leaving inorganic matter	Microelement contamination. Must runs blanks to correct for any organic matter in acid and oxidizing agent	Takes less time (~30 min) than regular wet ashing. Minerals stay in solution. Little or no volatilization	Expensive. Can handle fewer samples per run than standard wet or dry ashing procedures	Rapid ashing of samples prior to mineral analysis by rapid or official methods

(d) You have reason to believe the compound you want to measure after dry ashing may be reacting with the porcelain crucibles being used.

(e) You want to determine the iron content of some foods but cannot seem to get the iron solubilized after the dry ashing procedure.

5. Identify advantages and disadvantages of using wet and dry microwave ashing compared with conventional wet and dry ashing.

16.6   PRACTICE PROBLEMS

1. A grain was found to contain 11.5% moisture. A 5.2146-g sample was placed into a crucible (28.5053 g of tare). The ashed crucible weighed 28.5939 g. Calculate the percentage ash on (a) an as-received (wet weight) basis and (b) a dry matter basis. Make sure to adjust for significant figures in this and other practice problems.

2. A vegetable (23.5000 g) was found to have 0.0940 g acid-insoluble ash. What is the percentage acid-insoluble ash?

3. You wish to have at least 100 mg of ash from a cereal grain. Assuming 2.5% of ash on average and no sample loss, what is the *minimum* weight of grain required for ashing?

4. You wish to have a coefficient of variation (CV) below 5% with your ash analyses. The following ash data are obtained: 2.15%, 2.12%, and 2.07%. Are these data acceptable, and what is the CV?

5. A sample of hamburger was analyzed. Here are the results: sample weight, 2.034 g; weight after drying, 1.0781 g; weight after ether extraction of the dried sample, 0.4679 g; and weight of ash, 0.0233 g. What is the percentage ash on (a) a wet weight basis, (b) a fat-free basis (including water), and (c) a fat-free basis (without water)?

Answers

1. (a) 1.70%, (b) 1.92%

   Calculate ash from sample:

Crucible + ash	28.5939 g
Tared crucible	28.5053 g
Ash	0.0886 g

   (a) Calculate for ash on a wet weight basis (a):

   $$\frac{0.0886 \text{ g ash}}{5.2146 \text{ g sample}} \times 100\% = 1.70\% \text{ or } 1.7\%$$

(b) Calculate for ash on a dry weight basis (b):

$$\frac{0.0886 \text{ g ash}}{5.2146 \text{ g sample} \times \left(\dfrac{100\% - 11.5\%}{100\%} \text{ dry matter coeff.}\right)} \times 100\%$$
$$= 1.92\%.$$

or

$$5.214 \text{ g sample} \times \frac{11.5 \text{ g water}}{100 \text{ g sample}} = 0.5997 \text{ g water}$$

$$5.214 \text{ g sample} - 0.5997 \text{ g water} = 4.6149 \text{ g sample dry wt}$$

$$\frac{0.0886 \text{ g ash}}{4.6149 \text{ g dry wt sample}} \times 100\% = 1.92\%$$

2. 0.4%

   Calculate % insoluble ash:

   $$\frac{0.0940 \text{ g acid insoluble ash}}{23.5 \text{ g sample}} \times 100\% = 0.4\%$$

3. 4 g

   100 mg = 0.1 g ash.
   2.5% = 2.5 g ash/100 g sample.

   $$\frac{2.5 \text{ g ash}}{100 \text{ g sample}} = \frac{0.1 \text{ g ash}}{x}$$

   $$2.5x = 10$$

   $$x = 4 \text{ g sample}$$

4. Yes, 1.9%

   Calculate the mean:

   $$\frac{2.15\% + 2.12\% + 2.07\%}{3} = 2.11\%$$

   Calculation of mean and standard deviation was done using Excel:

1	2.15%
2	2.12%
3	2.07%
	Average = 2.11%
	Std deviation = 0.0404

   Coefficient of variation (CV)

   $$= \frac{SD}{\overline{X}} \times 100\%$$

   $$CV = \frac{0.0404}{2.11} \times 100\% = 1.91\%$$

   Is it within the 5% level for CV? Yes.

5. (a) 1.1%, (b) 1.64%

   Sample wet wt, 2.034 g
   Sample dry wt, 1.0781 g
   Wt after extraction, 0.4679 g
   Wt of ash, 0.0233 g

(a) Calculate on a wet weight basis:

$$\frac{0.0233 \text{ g ash}}{2.034 \text{ g sample}} \times 100\% = 1.15\%$$

(b) Calculate on a fat-free, wet weight basis:

$$2.034 \text{ g wet sample} - 1.0781 \text{ g solids}$$
$$= 0.9559 \text{ g water} \left(\text{this is 47\% moisture} \right)$$

$$1.0781 \text{ g solids dry wt} - 0.4679 \text{ g solids}$$
$$\text{after extraction} = 0.6102 \text{ g fat}$$

$$2.034 \text{ g wet sample} - 0.6102 \text{ g fat}$$
$$= 1.4238 \text{ g wet sample wt without fat}$$

$$\frac{0.0233 \text{ g ash}}{\left(1.4238 \text{ g wet sample wt without fat} \right)} \times 100\%$$
$$= 1.64\% \text{ ash, fat-free basis}$$

(c) Calculate on fat-free, dry weight basis:

$$\frac{0.0233 \text{ g ash}}{\left(0.4679 \text{ g dry sample wt without fat} \right)} \times 100\%$$
$$= 4.98\% \text{ ash, fat-free dry basis}$$

Acknowledgment The author of this chapter wishes to acknowledge the contributions of Dr. Leniel H. Harbers (Emeritus Professor, Kansas State University) for previous editions of this chapter. Also acknowledged in the preparation of this chapter is the assistance of Dr. John Budin (Silliker Laboratories, Chicago Heights, IL) as well as Michelle Horn, Ruth Watkins, and Anthony Danisi (CEM Corporation, Matthews, NC).

REFERENCES

1. Analytical Methods Committee (1960) Methods for the destruction of organic matter. Analyst 85: 643–656. This report gives a number of methods for wet and dry combustion and their applications, advantages, disadvantages, and hazards.

2. Akinyele, IO, Shokunbi, OS (2015) Comparative analysis of dry and wet digestion methods for the determination of trace and heavy metals in food. Food Chem 173: 682–684.

3. AOAC International (2016) Official methods of analysis, 20th edn. (On-line). AOAC International, Rockville, MD. This reference contains the official methods for many specific food ingredients. It may be difficult for the beginning student to follow.

4. Aurand LW, Woods AE, Wells MR (1987) Food composition and analysis. Van Nostrand Reinhold, New York. The chapters that deal with ash are divided by foodstuffs. General dry ashing procedures are discussed under each major heading.

5. Bakkali K, Martos NR, Souhail B, Ballesteros E (2009) Characterization of trace metals in vegetables by graphite furnace atomic absorption spectrometry after closed vessel microwave digestion. Food Chem 116(2): 590–594.

6. Kuboyama, K, Sasaki, N, Nakagome, Y, Kataoka, M (2005) Wet Digestion. Anal Chem 360: 184–191.

7. Mesko MF, De Moraes DP, Barin JS, Dressler VL, Knappet G (2006) Digestion of biological materials using the microwave-assisted sample combustion technique, Microchem J 82: 183–188.

8. Neggers YH, Lane RH (1995) Minerals, ch. 8. In: Jeon IJ, Ikins WG (eds) Analyzing food for nutrition labeling and hazardous contaminants. Marcel Dekker, New York. This chapter compares wet and dry ashing, and summarizes the following in tables: losses of specific elements during dry ashing; acids used in wet oxidation related to applications; AOAC methods for specific elements related to food applications.

9. Palma M N N, Rocha G C, Valaderes Filho SC, Detman E (2015) Evaluation of Acid Digestion Procedures to Estimate Mineral Contents in Materials from Animal Trials. Asian-Australas J Anim Sci 11: 1624–1628.

10. Pomeranz Y, Meloan C (1994) Food analysis: theory and practice, 3rd edn. Chapman & Hall, New York. Chapter 35 on ash and minerals gives an excellent narrative on ashing methods and is easy reading for a student in food chemistry. A good reference list of specific mineral losses is given at the end of the chapter. No stepwise procedures are given, however.

11. Smith GF (1953) The wet ashing of organic matter employing hot concentrated perchloric acid. The liquid fire reaction. Anal Chim Acta 8: 397–421. This reference gives an in-depth review of wet ashing with perchloric acid. Tables on reaction times with foodstuffs and color reactions are informative and easy for the food scientist to understand.

12. Wehr HM, Frank JF (eds) (2004) Standard methods for the examination of dairy products, 17th edn. American Public Health Association, Washington, DC. This text gives detailed analytical procedures for ashing dairy products

13. Wooster HA (1956) Nutritional data, 3rd edn. H.J. Heinz, Pittsburgh, PA.

14. Zhang H, Dotson P (1994) Use of microwave muffle furnace for dry ashing plant tissue samples. Commun Soil Sci Plant Anal 25 (9/10): 1321–1327.

Fat Analysis

Wayne C. Ellefson

Nutritional Chemistry and Food Safety,
Covance Laboratories,
3301 Kinsman Boulevard, Madison, WI 53714, USA
e-mail: Wayne.Ellefson@covance.com

S. Nielsen (ed.), *Food Analysis*, Food Science Text Series,
DOI 10.1007/978-3-319-45776-5_17, © Springer International Publishing 2017

17.1 INTRODUCTION

17.1.1 Definitions

Lipids, proteins, and carbohydrates constitute the principal structural components of foods. Lipids are a group of substances that, in general, are soluble in ether, chloroform, or other organic solvents but are sparingly soluble in water. However, there exists no clear scientific definition of a lipid, primarily due to the water solubility of certain molecules that fall within one of the variable categories of food lipids [1]. Some lipids, such as triacylglycerols, are very hydrophobic. Other lipids, such as di- and monoacylglycerols, have both hydrophobic and hydrophilic moieties in their molecules and are soluble in relatively polar solvents [2]. Short-chain fatty acids such as C1–C4 are completely miscible in water and insoluble in nonpolar solvents [1].

As implied above, the most widely accepted definition is based on solubility as previously stated. While most macromolecules are characterized by common structural features, the designation of "lipid" being defined by solubility characteristics is unique to lipids [2]. Lipids comprise a broad group of substances that have some common properties and compositional similarities [3]. Triacylglycerols are fats and oils that represent the most prevalent category of the group of compounds known as lipids. The terms lipids, fats, and oils are often used interchangeably.

The term "lipid" commonly refers to the broad, total collection of food molecules that meet the definition previously stated. Fats generally refer to those lipids that are solid at room temperature, and oils generally refer to those lipids that are liquid at room temperature. While there may not be an exact scientific definition, the US Food and Drug Administration (FDA) has established a regulatory definition for nutrition labeling purposes. The FDA has defined total fat as the sum of fatty acids from C4 to C24, calculated as triglycerides. This definition provides a clear path for resolution of any nutrition labeling disputes.

17.1.2 General Classification

The general classification of lipids that follows is useful to differentiate lipids in foods [3]:

1. **Simple lipids:** ester of fatty acids with alcohol (i.e., fats, waxes)
2. **Compound lipids:** compounds containing groups in addition to an ester of a fatty acid with an alcohol (i.e., phospholipids, cerebrosides, and sphingolipids)
3. **Derived lipids:** substances derived from neutral lipids or compound lipids (e.g., fatty acids, long-chain alcohols, sterols, fat-soluble vitamins, and hydrocarbons)

17.1.3 Content of Lipids in Foods

Foods may contain any or all types of the lipid compounds previously mentioned. The lipid content in bovine milk (Table 17.1) illustrates the complexity and variability of lipids in a food system, having lipids that differ in polarity and concentrations.

Foods contain many types of lipids, but those that tend to be of greatest importance are the triacylglycerols and the phospholipids. **Liquid triacylglycerols** at room temperature are referred to as **oils**, such as soybean oil and olive oil, and are generally of plant origin. **Solid triacylglycerols** at room temperature are termed **fats**. Lard and tallow are examples of fats, which are generally from animals. The term *fat* is applicable to all triacylglycerols whether they are normally solid or liquid at ambient temperatures. Table 17.2 shows the wide range of lipid content in different foods.

table   Lipids of bovine milk

Kinds of lipids	Percent of total lipids
Triacylglycerols	97–99
Diacylglycerols	0.28–0.59
Monoacylglycerols	0.016–0.038
Phospholipids	0.2–1.0
Sterols	0.25–0.40
Squalene	Trace
Free fatty acids	0.10–0.44
Waxes	Trace
Vitamin A	(7–8.5 µg/g)
Carotenoids	(8–10 µg/g)
Vitamin D	Trace
Vitamin E	(2–5 µg/g)
Vitamin K	Trace

Adapted from Patton and Jensen [4] with permission of Jenness and Patton [5] *Principles of Dairy Chemistry*. Jenness, R., and Patton, S. Copyright ©1959, John Wiley & Sons, Inc, with permission

17.2 table   Fat content of selected foods

Food item	Percent fat (wet weight basis)
Cereals, bread, and pasta	
Rice, white, long grain, regular, raw, enriched	0.7
Sorghum	3.3
Wheat, soft white	2.0
Rye	2.5
Wheat germ, crude	9.7
Rye bread	3.3
Cracked wheat bread	3.9
Macaroni, dry, enriched	1.5
Dairy products	
Milk, reduced fat, fluid, 2%	2.0
Skim milk, fluid	0.2
Cheddar cheese	33.1
Yogurt, plain, whole milk	3.2
Fats and oils	
Lard, shortening, oils	100.0
Butter, with salt	81.1
Margarine, regular, hard, soybean	80.5
Salad dressing	
Italian, commercial, regular	28.3
Thousand island, commercial, regular	35.1
French, commercial, regular	44.8
Mayonnaise, soybean oil, with salt	79.4
Fruits and vegetables	
Apples, raw, with skin	0.2
Oranges, raw, all commercial varieties	0.1
Blackberries, raw	0.5
Avocados, raw, all commercial varieties	14.7
Asparagus, raw	0.1
Lima beans, immature seeds, raw	0.9
Sweet corn, yellow, raw	1.2
Legumes	
Soybeans, mature seeds, raw	19.9
Black beans, mature seed, raw	1.4
Meat, poultry, and fish	
Beef, flank, separable lean and fat	5.0
Chicken, broilers or fryers, breast meat only	1.2
Bacon, pork, cured, raw	45.0
Pork, fresh, loin, whole, raw	12.6
Finfish, halibut, Atlantic and Pacific, raw	2.3
Finfish, cod, Atlantic, raw	0.7
Nuts	
Coconut meat, raw	33.5
Almonds, dried, unblanched, dry roasted	52.8
Walnuts, black, dried	56.6
Egg, whole, raw, fresh	10.0

From US Department of Agriculture, Agricultural Research Service (2015) USDA National Nutrient Database for Standard Reference. Release 28 Nutrient Data Laboratory Home Page, http://ndb.nal.usda.gov

17.1.4 Importance of Analysis

An accurate and precise quantitative and qualitative analysis of lipids in foods is important for accurate nutritional labeling, to determine whether the food meets the standard of identity, and to ensure that the product meets manufacturing specifications. Inaccuracies in analysis may prove costly for manufacturers and could result in a product of undesirable quality and functionality.

17.1.5 General Considerations

By definition, lipids are soluble in organic solvents and insoluble in water. Therefore, water insolubility is the essential analytical property used as the basis for the separation of lipids from proteins, water, and carbohydrates in foods. Glycolipids are soluble in alcohols and have a low solubility in hexane. In contrast, triacylglycerols are soluble in hexane and petroleum ether, which are nonpolar solvents. The wide range of relative hydrophobicity of different lipids makes the selection of a single universal solvent impossible for lipid extraction of foods. Some lipids in foods are components of complex lipoproteins and lipopolysaccharides; therefore, successful extraction requires that bonds between lipids and proteins or carbohydrates be broken so that the lipids can be freed and solubilized in the extracting organic solvents.

17.2 SOLVENT EXTRACTION METHODS

17.2.1 Introduction

For routine quality control purposes, the total lipid content of a food is commonly determined by simple organic solvent extraction methods or by alkaline (Sect. 17.2.6.1) or acid (Sect. 17.2.6.2) hydrolysis followed by solvent extraction in a Mojonnier flask. For multicomponent food products, acid hydrolysis is often the method of choice. Solvent extraction methods may be used as a first step in the gas chromatographic determination of fatty acid content for nutrition labeling as required by FDA regulation in the United States for the determination of total fat (see Chap. 3).

The accuracy of direct solvent extraction methods (i.e., without prior acid or alkaline hydrolysis) greatly depends on the solubility of the lipids in the solvent used and the ability to separate the lipids from complexes with other macromolecules. The lipid content of a food determined by extraction with one solvent may be quite different from the content determined with another solvent of different polarity. In addition to solvent extraction methods, there are nonsolvent wet extraction methods and several instrumental methods that utilize the physical and chemical properties of lipids in foods for fat content determination.

Many of the methods cited in this chapter are official methods of AOAC International. Refer to these methods and other original references cited for detailed instructions of procedures. There are many methods available for the determination of lipid content. This chapter will focus on some of the primary methods in common use.

17.2.2 Sample Preparation

The validity of the fat analysis of a food depends on proper sampling and preservation of the sample before the analysis (see also Chap. 5). An ideal sample should be as close as possible in all of its intrinsic properties to the material from which it is taken. However, a sample is considered satisfactory if the properties under investigation correspond to those of the bulk material within the limits of the test [7].

The sample preparation for lipid analysis depends on the type of food and type and nature of lipids in the food [8]. The extraction method for lipids in liquid milk is generally different from that for lipids in solid soybeans. To analyze the lipids in foods effectively, knowledge of the structure, the chemistry, and the occurrence of the principal lipid classes and their constituents is necessary. Therefore, there is no single standard method for the extraction of all kinds of lipids in different foods. For the best results, sample preparation should be carried out under an inert atmosphere of nitrogen at low temperature to minimize chemical reactions such as lipid oxidation.

One or more preparatory steps are common in lipid analysis to aid in extraction: (1) removal of water, (2) reduction of particle size, (3) or separation of the lipid from bound proteins and/or carbohydrates through the use of techniques such as alkaline hydrolysis (Sect. 17.2.6.1) or acid hydrolysis (Sect. 17.2.6.2). The first two of these steps are described in sections immediately below.

17.2.2.1 *Predrying Sample*

Lipids cannot be effectively extracted with ethyl ether from moist food because the solvent cannot easily penetrate the moist food tissues due to the hydrophobicity of the solvents used or the hydroscopic nature of the solvents. The ether, which is hygroscopic, becomes saturated with water and inefficient for lipid extraction. Drying the sample at elevated temperatures is undesirable because some lipids become bound to proteins and carbohydrates, and bound lipids are not easily extracted with organic solvents. Vacuum oven drying at low temperature or lyophilization increases the surface area of the sample for better lipid extraction. Predrying makes the sample easier to grind for better extraction, breaks fat-water emulsions to make fat dissolve easily in the organic solvent, and helps to free fat from the tissues of foods [7].

17.2.2.2 *Particle Size Reduction*

The extraction efficiency of lipids from dried foods depends on particle size; therefore, adequate grinding is very important. The classical method of determining fat in oilseeds involves the extraction of the ground seeds with selected solvent after repeated grinding at low temperature to minimize lipid oxidation. For better extraction, the sample and solvent are mixed in a high-speed comminuting device such as a blender. It can be difficult to extract lipids from whole soybeans because of the limited porosity of the soybean hull and its sensitivity to dehydrating agents. The lipid extraction from soybeans is easily accomplished if the beans are broken mechanically by grinding. Extraction of fat from finished products can be a challenge, based on the ingredients (e.g., energy bars with nuts, caramel, protein, granola, soybean oil). Such products may best be ground after freezing with liquid nitrogen.

17.2.3 Solvent Selection

Ideal solvents for fat extraction should have a high solvent power for lipids and low or no solvent power for proteins, amino acids, and carbohydrates. They should evaporate readily and leave no residue, have a relatively low boiling point, and be nonflammable and nontoxic in both liquid and vapor states. The ideal solvent should penetrate sample particles readily, be in single component form to avoid fractionation, and be inexpensive and nonhygroscopic [6, 7]. It is difficult to find an ideal fat solvent to meet all of these requirements. Ethyl ether and petroleum ether are the most commonly used solvents, but pentane and hexane are used to extract oil from soybeans.

Ethyl ether has a boiling point of 34.6 °C and is a better solvent for fat than petroleum ether. It is generally expensive compared to other solvents, has a greater danger of explosion and fire hazards, is hygroscopic, and forms peroxides [6]. **Petroleum ether** is the low boiling point fraction of petroleum and is composed mainly of pentane and hexane. It has a boiling point of 35–38 °C and is more hydrophobic than ethyl ether. It is selective for more hydrophobic lipids, cheaper, less hygroscopic, and less flammable than ethyl ether. The detailed properties of petroleum ether for fat extraction are described in AOAC Method 945.16 [8].

A combination of two or three solvents is frequently used. The solvents should be purified and peroxide-free, and the proper solvent-solute ratio must be used to obtain the best extraction of lipids from foods [7].

17.2.4 Continuous Solvent Extraction Method: Goldfish Method

17.2.4.1 *Principle and Characteristics*

For continuous solvent extraction, solvent from a boiling flask continuously flows over the sample held in a

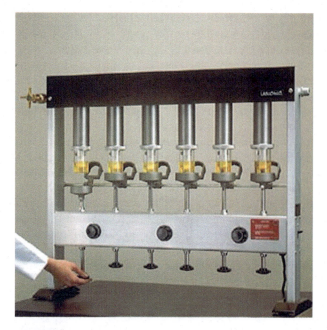

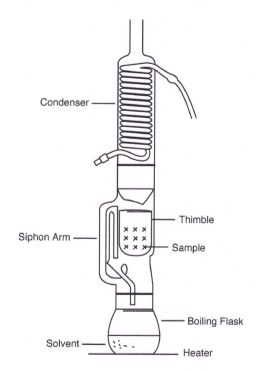

 Goldfish fat extractor (Courtesy of Labconco Corp., Kansas City, MO; www.labconco. com/_scripts/EditItem.asp?ItemID=487)

17.2 **figure** Soxhlet extraction apparatus

ceramic thimble. Fat content is measured by weight loss of the sample or by weight of fat removed.

The continuous methods give faster, more efficient extraction than semicontinuous extraction methods. However, they may cause channeling which results in incomplete extraction. The Goldfish test is an example of a continuous lipid extraction method [6, 7]. This method involves significant fire risk and has been discontinued in many laboratories, but is briefly described here for comparison to semicontinuous and discontinuous extraction methods (Sects. 17.2.5 and 17.2.6).

17.2.4.2 General Procedure
The sample is weighed into a thimble. Fat is extracted from the sample with boiling ethyl ether in the Goldfish apparatus (Fig. 17.1). Fat content is calculated as given in Eq. 17.1:

% Fat on dry weight basis

$$= (\text{g of fat in sample} / \text{g of dried sample}) \times 100 \quad (17.1)$$

17.2.5 Semicontinuous Solvent Extraction Method: Soxhlet Method

The Soxhlet method (AOAC Method 920.39C for Cereal Fat; AOAC Method 960.39 for Meat Fat) [8] is an example of the semicontinuous extraction method and is described below.

17.2.5.1 Principle and Characteristics
For semicontinuous solvent extraction, the solvent builds up in the extraction chamber for 5–10 min and

completely surrounds the sample and then siphons back to the boiling flask. Fat content is measured by weight loss of the sample or by weight of fat removed.

This method provides a soaking effect of the sample and does not cause channeling. However, this method requires more time than the continuous method. Instrumentation for more rapid and automated versions of the Soxhlet method is available (e.g., Ankom XT15 Extractor, Ankom Technology, Macedon, NY; Soxtec™, FOSS in North America, Eden Prairie, MN).

17.2.5.2 General Procedure (See Fig. 17.2)
For the Soxhlet or Goldfish methods, if the sample contains more than 10 % H₂O, dry the sample to constant weight at 95–100 °C under pressure ≤100 mmHg for about 5 h (AOAC Method 934.01). If analyzing a variety of food products, it may be more efficient to dry all samples, regardless of moisture content.

Samples are weighed into thimbles, placed in the Soxhlet apparatus (Fig. 17.2), and extracted with an appropriate solvent. As this procedure uses heat, it is more dangerous to use ethyl ether than other solvents. Many laboratories now use petroleum ether or hexane. Extraction time is 16 h in most cases. Certain products may lend themselves to shorter extraction time. The extract is then evaporated, and fat is determined gravimetrically. Fat content is calculated as in Eq. 17.2:

% Fat on dry weight basis

$$= (\text{g of fat in sample} / \text{g of dried sample}) \times 100 \quad (17.2)$$

17.2.6 Discontinuous Solvent Extraction Methods

17.2.6.1 Alkaline Hydrolysis Method (Mojonnier Method)

17.2.6.1.1 Principle and Characteristics

The terms Mojonnier fat, base hydrolysis, and alkaline hydrolysis are often used interchangeably to describe the following method. The term alkaline hydrolysis will be used in this chapter. After the use of ammonia to precipitate protein and free any bound fat, extraction is conducted with a mixture of ethyl ether and petroleum ether in a Mojonnier flask (Fig. 17.3), and the extracted fat is dried to a constant weight and expressed as percent fat by weight.

The alkaline hydrolysis test does not require removal of moisture from the sample, so it can be applied to both liquid and solid samples. The alkaline hydrolysis method was developed for and is applied primarily to dairy foods. If petroleum ether is used to purify the extracted fat, this method is very similar to the **Roese-Gottlieb method** (AOAC Method 905.02) in both principle and practice. The Mojonnier flasks (Fig. 17.3) are used not only for alkaline hydrolysis methods, but may also be used to conduct hydrolysis (acid, alkaline, or combination) prior to fat extraction and GC analysis for the determination of fat content

via fatty acid profile analysis (Sect. 17.2.6), according to the US nutrition labeling regulations.

17.2.6.1.2 General Procedure

For AOAC Method 989.05 (Fat in Milk), the sample is prepared with appropriate warming and handling to ensure a homogenous sample, and then it is weighed into a Mojonnier flask. Ammonium hydroxide is added to precipitate the milk protein, and then ethanol is added to prevent gel formation. Following these treatments, fat is extracted first with ethyl ether to extract most lipids present, followed by petroleum ether to assist with removal of water from the ethyl ether and to assist with complete extraction of nonpolar lipids. Extractions are commonly repeated twice more to ensure complete extraction of lipids. The ether solution is decanted from the Mojonnier flask into the previously weighed Mojonnier fat dish. The solvent is evaporated, and fat content is determined gravimetrically, as in Eq. 17.3:

$$\% \text{ Fat} = 100 \times \left\{ \left[\left(\text{wt dish} + \text{fat} \right) - \left(\text{wt dish} \right) \right] \right.$$
$$\left. - \left(\text{avg wt blank residue} \right) \right\} / \text{wt sample} \qquad (17.3)$$

A pair of reagent blanks must be prepared every day. For reagent blank determination, use 10 ml of distilled water instead of milk sample. The reagent blank should be <0.002 g.

17.2.6.2 Acid Hydrolysis Procedure

17.2.6.2.1 Principle and Characteristics

A significant portion of the lipids in foods such as dairy, bread, flour, and animal products is bound to proteins and carbohydrates, and direct extraction with nonpolar solvents is inefficient. Such foods must be prepared for lipid extraction by acid hydrolysis. This includes a significant percentage of finished food products. Table 17.3 shows the inaccuracy that can occur if samples are not prepared by acid hydrolysis. Acid hydrolysis can break both covalently and ionically bound lipids into easily extractable lipid forms. Specifically, there are a variety of AOAC methods for

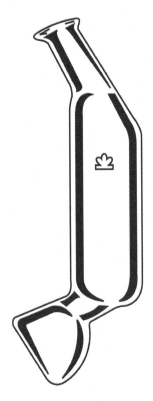

17.3 figure   Mojonnier fat extraction flask (Courtesy of Kontes Glass Co., Vineland, NJ)

17.3 table   Effects of acid digestion on fat extraction from foods

| | Percent fat | |
	Acid hydrolysis	No acid hydrolysis
Dried egg	42.39	36.74
Yeast	6.35	3.74
Flour	1.73	1.20
Noodles	3.77–4.84	2.1–3.91
Semolina	1.86–1.93	1.1–1.37

Adapted from Joslyn [6], p. 154, with permission

fat that involve an acid hydrolysis with HCl, followed by extraction with a combination of ethyl ether and petroleum ether [8]. Ethanol and solid hexametaphosphate may be added to facilitate separation of lipids from other components before food lipids are extracted with solvents [6, 7]. For example, the acid hydrolysis of two eggs requires 10 ml of HCl and heating in a water bath at 65 °C for 15–25 min or until the solution is clear [6].

17.2.6.2.2 General Procedure

For AOAC Method 922.06 (Fat in Flour), the sample is weighed into a Mojonnier flask. Hydrochloric acid digestion is carried out. Fat is extracted first with ethyl ether to extract most lipids present, followed by petroleum ether to assist with removal of water from the ethyl ether and to assist with complete extraction of nonpolar lipids. Extractions are commonly repeated twice more to ensure complete extraction of lipids. The ether solution is decanted from the Mojonnier flask into the previously weighed Mojonnier fat dish. The solvent is evaporated, and then fat is determined gravimetrically using Eq. 17.4:

$$\% \text{ Fat} = 100 \times \left\{ \left[\left(\text{wt dish} + \text{fat} \right) - \left(\text{wt dish} \right) \right] \right.$$
$$\left. - \left(\text{avg wt blank residue} \right) \right\} / \text{wt sample} \qquad (17.4)$$

17.2.6.3 Chloroform-Methanol Procedure

17.2.6.3.1 Principle and Characteristics

The combination of chloroform and methanol has been used commonly to extract lipids. The "Folch extraction" [9] applied to small samples, and the "Bligh and Dyer extraction" [10] applied to large samples of high moisture content both utilize this combination of solvents to recover lipids from foods. These methods have been reviewed, and procedures were modified by Christie [11] and others. The Bligh and Dyer procedure [10] is a modification of the Folch extraction [9], designed for more efficient solvent usage for low-fat samples. The Christie modification [11] of these former methods replaced water with 0.88 % potassium chloride aqueous solution to create two phases.

In both the modified Folch extraction and Bligh and Dyer procedure, food samples are mixed/homogenized in a chloroform-methanol solution, and the homogenized mixture is filtered into a collection tube. A 0.88 % potassium chloride aqueous solution is added to the chloroform-methanol mixture containing the extracted fats. This causes the solution to break into two phases: the aqueous phase (top) and the chloroform phase containing the lipid (bottom). The phases are further separated in a separatory funnel or by centrifugation. After the evaporation of the chloroform, the fat can be quantitated by weight.

The various methanol-chloroform extraction procedures are rapid, well suited to low-fat samples, and can be used to generate lipid samples for subsequent fatty acid compositional analysis. The procedure has been more applied to basic commodities, rather than to finished product samples. For consistent results, the procedures must be followed carefully, including the ratio of chloroform and methanol. A cautionary note is that chloroform and methanol are highly toxic, so the extraction procedure must be done in well-ventilated areas.

17.2.6.3.2 General Procedure

For AOAC Method 983.23 (Fat in Foods), samples are extracted first with methanol, and then chloroform is added on top, and extraction is continued. Up to two additional such extractions may be performed. Potassium chloride is added to aid in separation of the layers. Solvent is evaporated, and fat is determined gravimetrically as in Eq. 17.5:

$$\% \text{ Fat} = 100 \times \left\{ \left[\left(\text{wt dish} + \text{fat} \right) - \left(\text{wt dish} \right) \right] \right.$$
$$\left. - \left(\text{avg wt blank residue} \right) \right\} / \text{wt sample} \qquad (17.5)$$

17.2.7 Total Fat by Gas Chromatography for Nutrition Labeling

17.2.7.1 Principle

After adding an internal standard and an antioxidant, the sample is treated by acid and/or alkaline hydrolysis, and then fat is extracted with ether. Fatty acids are converted to fatty acid methyl esters (FAMEs), then separated by gas chromatography (GC), and quantitated against the internal standard, with the sum equal to total fat (AOAC Method 996.06).

The saturated and monounsaturated fats are calculated as the sum of the respective fatty acids. Monounsaturated fat includes only *cis* form. Trans fat can be quantified utilizing this method in conjunction with identification criteria established by the American Association of Oil Chemists (AOCS Method Ce 1h-05) [12] and Golay et al. [13].

17.2.7.2 General Procedure

The GC method for total fat (AOAC 996.06) is summarized as follows:

1. Add pyrogallic acid to sample.
2. Add internal standard (triundecanoin, $C_{11:0}$) to sample.
3. Subject sample to acid and/or alkaline hydrolysis.
4. Extract fat from sample with ethyl ether and/or petroleum ether.
5. Methylate fatty acids with boron trifluoride (in methanol) to form fatty acid methyl esters (FAMEs).

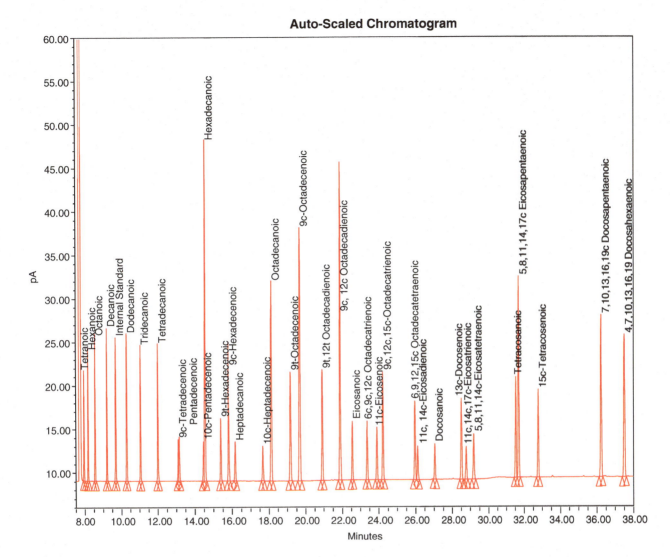

17.4 Fatty acid chromatogram

figure

6. Separate FAMEs with GC capillary column and use detector to quantify peaks, compared to internal standard (Fig. 17.4).

7. Total fat is calculated as the sum of individual fatty acids expressed as triglyceride equivalents.

17.3 NONSOLVENT WET EXTRACTION METHODS

17.3.1 Babcock Method for Milk Fat

17.3.1.1 *Principle*

In the Babcock method, H_2SO_4 is added to a known amount of milk in the Babcock bottle (Fig. 17.5). The sulfuric acid digests protein, generates heat, and releases the fat. Centrifugation and hot water addition isolate fat for quantification in the graduated portion of the test bottle. The fat is measured volumetrically, but the result is expressed as percent fat by weight.

17.3.1.2 *Applications*

The Babcock method, which is a common official method for the determination of fat in milk (AOAC Method 989.04 and 989.10), takes about 45 min, and duplicate tests should agree within 0.1%. The Babcock method does not determine the phospholipids in the milk products. It is not applicable to products containing chocolate or added sugar without modification because of charring of chocolate and sugars by sulfuric acid. A modified Babcock method is used to determine essential oil in flavor extracts (AOAC Method 932.11) and fat in seafood (AOAC Method 964.12).

17.6 figure Gerber fat butyrometer

17.5 figure Babcock milk test bottles for milk (**a**), cream (**b**), and cheese (Paley bottle) (**c**) testing (Courtesy of Kimble Glass Co., Vineland, NJ)

17.3.2 Gerber Method for Milk Fat

17.3.2.1 *Principle*

The principle of the Gerber method is similar to that of the Babcock method, but it uses sulfuric acid and amyl alcohol and a Gerber bottle (Fig. 17.6) (AOAC Method 2000.18). The sulfuric acid digests proteins and carbohydrates, releases fat, and maintains the fat in a liquid state by generating heat.

17.3.2.2 *Applications*

The Gerber method is comparable to the Babcock method but is simpler and faster and has wider application to a variety of dairy products [14]. The isoamyl alcohol generally prevents the charring of sugar found with the regular Babcock method. This test is more popular in Europe than in America.

17.4 INSTRUMENTAL METHODS

Instrumental methods offer numerous attractive features compared to the previously described extraction methods. In general, they are rapid, some are nondestructive, and require minimal sample preparation and chemical consumption. These methods can provide significant labor savings when one must perform analysis of many samples on a daily basis.

However, the equipment can be expensive, and some methods require the establishment of calibration curves specific to various compositions. Despite these drawbacks, several of the following instrumental methods are very widely used in quality control as well as research and product development applications. The following section describes several of these instrumental methods.

17.4.1 Infrared Method

The infrared (IR) method is based on absorption of IR energy by fat at a wavelength of 5.73 μm. The more the energy absorption at 5.73 μm, the higher the fat content of the sample [15]. (See Chap. 8 for a discussion of IR spectroscopy). Both near-infrared (NIR) or mid-infrared spectrophotometers are used in this method. For example, AOAC Method 2007.04 (Fat, Moisture, and Protein in Meat and Meat Products; specifies use of a FOSS Food Scan™ NIR spectrophotometer) is based on correlating NIR data with the values obtained from analysis by conventional methods, to then predict the concentration of fat, moisture, and protein in a sample being tested. Mid-IR spectroscopy is used in infrared milk analyzers to determine milk fat content (AOAC Method 972.16). NIR spectroscopy has been used to measure the fat content of commodities such as meats, cereals, and oilseeds in the laboratory and can be adapted for online measurement. NIR has been applied to near-line batch monitoring for food processing plants. NIR is an analytical technique that provides a prediction of chemical measurement. The NIR models are developed and validated to provide equivalent results to the wet chemical measurement. This method requires a comprehensive database of constituent values to create the model along with ongoing verification.

17.4.2 X-Ray Absorption Method

X-ray absorption is a rapid analysis method that has received interest in application to lipid analysis in meats [16]. This method can be adapted to in-line analysis of meats. Determination of the amount of fat in meat and meat products is based on the fact that the X-ray absorption of lean meat is higher than that of fat. This method has been used for the rapid determination of fat in meat and meat products using the standard curve of the relationship between X-ray absorption and fat content determined by a standard solvent extraction method [7]. For example, the MeatMaster™ II fat analysis instrument (Foss, Eden Prairie, MN), commonly used to rapidly determine the percent fat of meat products, is based on X-ray absorption.

17.4.3 Nuclear Magnetic Resonance

Nuclear magnetic resonance (NMR) can be used to measure lipids in food materials in a nondestructive way. It is one of the most popular methods for use in determining lipid melting curves to measure solid fat content (see Chap. 23), and more affordable instruments are becoming more popular for measuring total fat content. Total fat content can be measured using low-resolution pulsed NMR (AOAC Method 2008.06). (The principles and applications of NMR are described in Chap. 10). NMR analysis is a very rapid and accurate method, and while the principals of NMR are relatively complex, the use of NMR can be quite simple, especially due to the high degree of automation and computer control. Systems such as CEM Smart Trac II rapidly measure total fat and moisture.

17.4.4 Accelerated Solvent Extraction

Accelerated solvent extraction (ASE) was developed to replace Soxhlet and other extraction techniques for many samples. The automation and rapid extraction time of accelerated solvent extraction are its key advantages. Accelerated solvent extraction (Fig. 17.7) uses solvents at increased temperatures and pressures to accelerate the extraction process. Analytes are more soluble and thus dissolve more quickly in hot solvents. Pressurized solvent is heated well above its boiling point and introduced under high pressure to extract the fat from the sample. By using hot, pressurized liquid solvents, accelerated solvent extraction decreases the amount of solvent and the time needed to complete an extraction. Most extractions can be completed in less than 20 min, using less than 20 mL of solvent.

17.4.5 Supercritical Fluid Extraction

Fat extraction can be carried out using **supercritical fluid extraction** (SFE) instruments that use supercriti-

17.7
figure

Accelerated solvent extractor (Courtesy of Thermo Fisher Scientific, Waltham, MA)

cal carbon dioxide as the solvent. Samples are weighed into extraction cells, the extraction chamber is heated and pressurized to a set value, and the supercritical fluid is then pumped through the extraction cell facilitating extraction of the target analytes from the sample matrix (Fig. 17.8). This technique is finding greater use because of the cost and environmental problems associated with the use and disposal of organic solvents. When pressurized CO_2 is heated above a certain critical temperature, it becomes a supercritical fluid which has some of the properties of a gas and some of a liquid. The fact that it behaves like a gas allows it to easily penetrate into a sample and extract the lipids, while the fact that it behaves like a fluid helps it dissolve lipids (especially at higher pressures). The pressure and temperature of the solvent are later reduced, which causes the CO_2 to turn to a gas, leaving the lipid fraction remaining. The lipid content of a food is determined by weighing the percentage of lipid extracted from the original sample [17].

17.5 COMPARISON OF METHODS

Various fat analysis methods are summarized and compared in Tables 17.4 and 17.5. Dried (or very low moisture content) samples are required for fat determination by the Soxhlet extraction method (traditional version or automated), which is applicable to many food commodities. If the samples are moist or liquid foods, the Mojonnier method is generally applicable to determination of the fat content. Acid hydro-

17.8 **figure** Supercritical fluid extractor (Courtesy of Parr Instrument Co., Moline, IL)

lysis or alkaline hydrolysis is widely used on many finished food products. The instrumental methods such as IR and NMR are very simple, reproducible, and fast and are gaining in popularity. The application of instrumental methods for fat determination generally requires a standard curve between the signal of the instrument analysis and the fat content obtained by a standard solvent extraction method. However, a rapid instrumental method can be used as a quality control method for fat determination of a specific food.

Major uses of the Soxhlet and Mojonnier methods include the following: (1) extract fat prior to GC analysis, (2) quality control of formulated products, (3) determine fat content during product development, (4) verify when fat content is <0.5 g/serving (so nutrient content claim can be made), and (5) defat samples prior to fiber analysis. Compared to GC analysis of fat content by AOAC Method 996.06, these three methods are faster and cheaper, but may give a higher fat content (which must be recognized when using these methods for product development).

17.6 SUMMARY

Lipids have often been defined by their solubility characteristics rather than by some common structural feature. Lipids in foods can be classified as simple, compound, or derived lipids. The lipid content of foods varies widely, but quantitation is important because of regulatory requirements, nutritive value, and functional properties. To analyze food for the fat content accurately and precisely, it is essential to have a comprehensive knowledge of the general compositions of the lipids in the foods, the physical and chemical properties of the lipids as well as the foods, and the principles of fat determination. There is no single standard method for the determination of fats in different foods. The validity of any fat analysis depends on proper sampling and preservation of the sample prior to analysis. Predrying of the sample, particle size reduction, and acid hydrolysis prior to analysis also may be necessary. The total lipid content of foods is commonly determined by organic solvent extraction methods, such as semicontinuous (e.g., Soxhlet) and discontinuous (e.g., Mojonnier), or by GC analysis for nutrition labeling. Nonsolvent wet extraction methods, such as the Babcock or Gerber, are commonly used for certain types of food products. Instrumental methods, such as NMR, infrared, ASE, SFE, and x-ray absorption, are also available for fat determination. These methods are rapid and may be useful for quality control but may require additional refinement in order to develop consistent correlation to a standard solvent extraction method.

17.7 STUDY QUESTIONS

1. What are some important considerations when selecting solvents to be used in continuous and noncontinuous solvent extraction methods?
2. To extract the fat from a food sample, you have the choice of using ethyl ether or petroleum ether as the solvent, and you can use either a Soxhlet or a Goldfish apparatus. What combination of solvent and extraction would you choose? Give all the reasons for your choice.
3. Itemize the procedures that may be required to prepare a food sample for accurate fat determination by a solvent extraction method (e.g., Soxhlet method). Explain why each of these procedures may be necessary.
4. You performed fat analysis on a new super energy shake (high carbohydrate and protein) using standard Soxhlet extraction. The value obtained for fat content was much lower than that expected. What could have caused the measured fat content to be low, and how would you modify the standard procedure to correct the problem?

17.4
table

Summary of fat analysis methods

Method	Principle	Advantages	Disadvantages	Applications
Soxhlet	Fat is extracted, semicontinuously, with organic solvent. Fat content is measured by weight loss of sample or weight of fat removed	Provides for soaking effect. Does not cause channeling. Sample remains cool	Time to result is long. Errors low on many products with bound fat. Manual test	Grains, raw meat
Soxhlet (automated)	These methods take the traditional Soxhlet methodology described above and automate the process, removing some of the manual labor necessary to obtain results	Requires less labor. Faster time to result	Higher equipment cost	Same as Soxhlet
Mojonnier	Fat is extracted (discontinuous) with mixture of organic solvents. Extracted fat is dried to constant weight	Faster than Soxhlet	Errors low on many products with bound fat. Manual test	Same as above and on some nonprotein beverages
Mojonnier-base hydrolysis	Protein is precipitated with ammonia. Fat is extracted with mixture of organic solvents. Extracted fat is dried to constant weight	Best traditional method for dairy products	Manual test. Narrow range of scope	Intended for use on milk and milk-derived products
Mojonnier-acid hydrolysis	Organic matter (nonfat) is digested with acid. Fat is extracted with mixture of organic solvents. Extracted fat is dried to constant weight	Wide range of applicability for many food item	Manual test. More labor required than some methods	Applicable to most foods. Not applicable to raw soybeans
Chloroform-methanol	Fat is extracted with a combination of chloroform and methanol. Addition of potassium chloride causes the solution to break into two phases, with the fat in the chloroform phase. Evaporation of the chloroform allows for fat to be quantitated by weight	Wide range of applicability for many food item	Chloroform and methanol are both highly toxic; measures all fat-soluble materials. Manual test, significant labor expenditure. Need to deal with emulsions formed during extraction	Most foods
Gas chromatography (GC)	After adding an internal standard and a reagent to prevent oxidation, sample is treated by acid and/or alkaline hydrolysis, then fat is extracted with ether. Fatty acids are converted to fatty acid methyl esters (FAMEs). FAMEs are separated by GC and quantitated. Sum of fatty acids equals fat content	Measures only true fat, as defined by FDA. (Not other fat-soluble compounds)	Time consuming. Expensive. Many compounds to separate	Official method for nutrition labeling
Babcock	Sulfuric acid added digests protein, generates heat, and releases fat. After centrifugation of sample, fat content is measured volumetrically	Takes only ~45 min	Manual test. Get charring with high sugar products, so difficult to read volume of fat	Commonly applied to milk and other dairy products

(continued)

table 17.4 (continued)

Method	Principle	Advantages	Disadvantages	Applications
Gerber	Similar to Babcock method but uses sulfuric acid and amyl alcohol	Compared to Babcock, it is simpler and faster, and the isoamyl alcohol prevents charring of sugar	Manual test	Commonly applied to milk and other dairy products, especially high sugar products
Infrared spectroscopy	NIR provides a prediction of fat measurement. Based on absorption of IR energy by fat at 5.73 μm. NIR models are developed and validated to provide equivalent results to classical methods	Rapid	Provides only an estimate of fat content; instrument must be calibrated against official method	Mid-IR spectroscopy is commonly used in infrared milk analyzers. NIR commonly used to measure fat content of meats, cereals, and oilseeds
Specific gravity	Fat is extracted from sample with the solvent perchloroethylene. Specific gravity of sample solvent extract is related to fat content	Rapid	Values are estimates, based on correlation chart created using an official method	Meat
NMR	Certain nuclei will absorb and re-emit RF energy over a narrow band of frequencies when placed in a static magnetic field. The frequency at which the NMR effect occurs for a given nuclear isotope is dependent on the magnetic field strength of the magnet, and the phenomenon is caused by the interaction between the nuclear magnetic dipole of a nucleus and the magnetic field it experiences. Fat is determined by pulsed radio frequency (RF) energy while within a static 0.47 T magnetic field. The resulting NMR signal is recorded and analyzed for the total proton activity of fat present in the sample [8]	Rapid. Very low labor cost. Tremendous potential in high-throughput fat testing laboratories	Expensive instrument. Range of applicability has not yet been totally demonstrated	Potential application to many foods
Accelerated solvent extraction (ASE)	Pressurized solvent is heated to well above its boiling point and introduced under high pressure to extract the fat from the sample	Very rapid	Instrument cost. May require some work to match classical results	Use in place of Soxhlet. Use for other fat-related tests
Supercritical fluid extraction (SFE)	The sample is heated and pressurized to a set value, and the supercritical fluid is then pumped through the extraction cell facilitating extraction of the fat from the sample matrix	Very rapid	Instrument cost. May require some work to match classical results	Use in place of Soxhlet. Use for other fat-related tests

17.5 table Comparison of classical methods for fat analysis

Methods	Similarities	Differences
Soxhlet vs. Goldfish	Use for same applications (see list in Table 17.4) Use extraction with organic solvents Measure fat by wt. loss or wt. of fat removed (gravimetric) Both take hours to do; sample must be dry or low in moisture Use same type of cellulose "thimble" to hold sample for extraction	Soxhlet Semicontinuous Soaking effect Sample stays cool No channeling May be slower (takes 4–16 h, depending on drip rate) Goldfish Continuous method No soaking effect Sample gets heated Channeling likely Extract for 4 h (so may be faster)
Babcock vs. Mojonnier	Can use on liquid or solid sample Both developed originally for dairy	Babcock Use no solvents (uses sulfuric acid) Dairy products only Measure fat volume Mojonnier Use multiple solvents Any food product Measure fat weight
Mojonnier vs. Soxhlet	Use for same applications Gravimetric methods	Mojonnier Use multiple solvents Solid or liquid sample Discontinuous method Soxhlet Use single solvent Solid sample only (dry or low in moisture) Semicontinuous method
Babcock vs. Gerber	Same applications (dairy) Volumetric measurement	Babcock Use only sulfuric acid Use sulfuric acid to release fat Would get charring with high sugar dairy product Slower; more complex Gerber Use sulfuric acid plus isoamyl alcohol Can use on high sugar dairy product simpler and faster
Mojonnier vs. GC method	GC method commonly uses Mojonnier for initial fat extraction	Mojonnier Common component of GC method Measure by wt. of fat Used for applications in summary table GC Requires acid and/or alkaline hydrolysis, and extraction is commonly done with Mojonnier-type flasks but has post steps to measure fat as sum of FAs Used for nutrition labeling

5. What is the purpose of the following chemicals used in the Mojonnier method?

 (a) Ammonium hydroxide
 (b) Ethanol
 (c) Ethyl ether
 (d) Petroleum ether

6. What is a key application of the GC method and what does it specifically quantify?

7. What is the purpose of the following procedures used in Babcock method?

 (a) Sulfuric acid addition
 (b) Centrifugation and addition of hot water

8. Which of the following methods are volumetric and which are gravimetric determinations of lipid content: Babcock, Soxhlet, Mojonnier, Gerber?

17.8 PRACTICE PROBLEMS

1. To determine the fat content of a semimoist food by the Soxhlet method, the food was first vacuum oven dried. The moisture content of the product was 25%. The fat in the dried food was determined by the Soxhlet method. The fat content of the dried food was 13.5%. Calculate the fat content of the original semimoist product.

2. The fat content of 10 g of commercial ice cream was determined by the Mojonnier method. The weights of extracted fat after the second extraction and the third extraction were 1.21 g and 1.24 g, respectively. How much of fat, as a percentage of the total, was extracted during the third extraction?

Answers

1. If the sample weight of a semimoist food is 10 g and the moisture content is 25%, the dried weight of the original food is 7.5 g ($10\,g \times 75\% = 7.5\,g$). If the fat content of the dried food is 13.5%, the 7.5 g of dried sample has 1.0125 g fat (7.5 g dried food $\times 13.5\%$ fat $= 1.0125$ g fat). The 10 g of semimoist food contains the same amount of fat, i.e., 1.0125 g. Therefore, the fat content of the semimoist food is 10.125% (1.0125 g fat/10 g semimoist food).

2. $[(1.24-1.21\ g)/10\ g] \times 100 = 0.3\%$

Acknowledgment The author of this chapter wishes to acknowledge Dr. David Min (deceased) who was the key author on this Fat Analysis chapter for the first to fourth editions of this textbook.

REFERENCES

1. O'Keefe SF (1998) Nomenclature and classification of lipids, chap 1. In: Akoh CC, Min DB (eds) Food lipids: chemistry, nutrition and biotechnology. Marcel Dekker, New York, pp 1–36
2. Belitz HD, Grosch W (1987) Food chemistry. Springer, Berlin
3. Nawar WW (1996) Lipids chap. 5. In: Fennema OR (ed) Food chemistry, 3rd edn. Marcel Dekker, New York, pp 225–319
4. Patton S, Jensen RG (1976) Biomedical aspects of lactation. Pergamon, Oxford, p 78
5. Jenness R, Patton S (1959) Principles of dairy chemistry. Wiley, New York
6. Joslyn MA (1970) Methods in food analysis, 2nd edn. Academic, New York
7. Pomeranz Y, Meloan CF (1994) Food analysis: theory and practice, 3rd edn. Van Nostrand Reinhold, New York
8. Official Methods of Analysis, current on line edition. AOAC International, Gaithersburg, MD (2016)
9. Folch J, Lees M, Stanley GHS (1957) A simple method for the isolation and purification of total lipids from animal tissues. J Biol Chem 226: 297–509
10. Bligh EG, Dyer WJ (1959) A rapid method of total lipid extraction and purification. Can J Physiol 37: 911–917
11. Christie WW (1982) Lipid analysis. Isolation, separation, identification, and structural analysis of lipids. 2nd edn. Pergamon, Oxford
12. AOCS (2009) Official methods and recommended practices, 6th edn. American Oil Chemists' Society, Champaign, IL
13. Golay P-A, Dionisi F, Hug B, Giuffrida F, Destaillats F (2006) Direct quantification of fatty acids in dairy powders, with special emphasis on *trans* fatty acid content. Food Chem 106: 115–1120
14. Wehr HM, Frank JF (eds) (2004) Standard methods for the examination of dairy products, 17th edn. American Public Health Association, Washington, DC
15. Cronin DA, McKenzie K (1990) A rapid method for the determination of fat in foodstuffs by infrared spectrometry. Food Chem 35: 39–49
16. Brienne JP, Denoyelle C., Baussart H, Daudin JD (2001) Assessment of meat fat content using dual energy X-ray absorption. Meat Sci 57:235–244
17. LaCroix DE, Wold, WR, Myer LJD, Calabraro R (2003) Determination of total fat in milk- and soy-based infant formula powder by supercritical fluid extraction: PVM 2:2002. J AOAC Int 86:86–95

Protein Analysis

Sam K.C. Chang (✉) • Yan Zhang

Department of Food Science, Nutrition, and Health Promotion,
Mississippi State University,
Mississippi State, MS 39762, USA
e-mail: sc1690@msstate.edu; yzhang@fsnhp.msstate.edu

S. Nielsen (ed.), *Food Analysis*, Food Science Text Series,
DOI 10.1007/978-3-319-45776-5_18, © Springer International Publishing 2017

18.1 INTRODUCTION

18.1.1 Classification and General Considerations

Proteins are an abundant component in all cells, and almost all except storage proteins are important for biological functions and cell structure. Food proteins are very complex. Many have been purified and characterized. Proteins vary in molecular mass, ranging from approximately 5,000 to more than a million Daltons. They are composed of elements including hydrogen, carbon, nitrogen, oxygen, and sulfur. Twenty α-amino acids are the building blocks of proteins; the amino acid residues in a protein are linked by peptide bonds. Nitrogen is the most distinguishing element present in proteins. However, nitrogen content in various food proteins ranges from 13.4 to 19.1 % [1] due to the variation in the specific amino acid composition of proteins. Generally, proteins rich in basic amino acids contain more nitrogen.

Proteins can be classified by their composition, structure, biological function, or solubility properties. For example, simple proteins contain only amino acids upon hydrolysis, but conjugated proteins also contain non-amino acid components. Proteins have unique conformations that could be altered by denaturants such as heat, acid, alkali, 8 M urea, 6 M guanidine-HCl, organic solvents, and detergents. The solubility as well as functional properties of proteins could be altered by denaturants. The analysis of proteins is complicated by the fact that some food components possess similar physicochemical properties. Nonprotein nitrogen could come from free amino acids, small peptides, nucleic acids, phospholipids, amino sugars, porphyrin, and some vitamins, alkaloids, uric acid, urea, and ammonium ions. Therefore, the total organic nitrogen in foods would represent nitrogen primarily from proteins and to a lesser extent from all organic nitrogen-containing nonprotein substances. Depending upon methodology, other major food components, including lipids and carbohydrates, may interfere physically with analysis of food proteins.

Numerous methods have been developed to measure protein content. The basic principles of these methods include the determinations of nitrogen, peptide bonds, aromatic amino acids, dye-binding capacity, ultraviolet absorptivity of proteins, and light scattering properties. In addition to factors such as sensitivity, accuracy, precision, speed, and cost of analysis, what is actually being measured must be considered in the selection of an appropriate method for a particular application.

18.1.2 Importance of Analysis

Protein analysis is important for:

1. **Nutrition labeling**
2. **Pricing:** The cost of certain commodities is based on the protein content as measured by nitrogen content (e.g., cereal grains, milk for making certain dairy products, e.g., cheese).
3. **Functional property investigation**: Proteins in various types of food have unique food functional properties: for example, gliadin and glutenins in wheat flour for bread making, casein in milk for coagulation into cheese products, and egg albumen for foaming. (See Chap. 24, Sect. 24.3.3.)
4. **Biological activity determination**: Some proteins, including enzymes or enzyme inhibitors, are relevant to food science and nutrition: for instance, the proteolytic enzymes in the tenderization of meats, pectinases in the ripening of fruits, and trypsin inhibitors in legume seeds are proteins. To compare between samples, enzyme activity often is expressed in terms of specific activity, meaning units of enzyme activity per mg of protein.

Protein analysis is required when you want to know:

1. Total protein content
2. Content of a particular protein in a mixture
3. Protein content during isolation and purification of a protein
4. Nonprotein nitrogen
5. Amino acid composition (see Chap. 24, Sect. 24.3.1)
6. Nutritive value of a protein (see Chap. 24, Sect. 24.3.2)

18.1.3 Content in Foods

Protein content in food varies widely. Foods of animal origin and legumes are excellent sources of proteins. The protein contents of selected food items are listed in Table 18.1 [2].

18.1.4 Introduction to Methods

Principles, general procedures, and applications are described below for various protein determination methods. See Table 18.2 for a summary of methods described, including more details about their applications and AOAC numbers [3]. Advantages and disadvantages of methods are included in the summary table, rather than in the text. Please refer to the references cited within the text for detailed instructions of the procedures. Many of the methods covered in this chapter are described in somewhat more detail in recent books on food proteins [4–6]. The Kjeldahl, Dumas (N combustion), infrared spectroscopy, and anionic dye-binding methods described are from the *Official Methods of Analysis* of AOAC International [3] and are used commonly in nutrition labeling and/or quality control. The other methods described are used

Protein content of selected foods

Food item	Percent protein (wet weight basis)
Cereals and pasta	
Rice, brown, long-grain, raw	7.9
Rice, white, long-grain, regular, raw, enriched	7.1
Wheat flour, whole-grain	13.7
Corn flour, whole-grain, yellow	6.9
Spaghetti, dry, enriched	13.0
Cornstarch	0.3
Dairy products	
Milk, reduced fat, fluid, 2%	3.2
Milk, nonfat, dry, regular, with added vit. A	36.2
Cheese, cheddar	24.9
Yogurt, plain, low fat	5.3
Fruits and vegetables	
Apple, raw, with skin	0.3
Asparagus, raw	2.2
Strawberries, raw	0.7
Lettuce, iceberg, raw	0.9
Potato, whole, flesh, and skin	2.0
Legumes	
Soybeans, mature seeds, raw	36.5
Beans, kidney, all types, mature seeds, raw	23.6
Tofu, raw, firm	15.8
Tofu, raw, regular	8.1
Meats, poultry, fish	
Beef, chuck, arm pot roast	21.4
Beef, cured, dried beef	31.1
Chicken, broilers or fryers, breast meat only, raw	23.1
Ham, sliced, regular	16.6
Egg, raw, whole, fresh	12.6
Finfish, cod, Pacific, raw	17.9
Finfish, tuna, white, canned in oil, drained solids	26.5

From the US Department of Agriculture, Agricultural Research Service [2]

commonly in research laboratories working on proteins.

18.2 NITROGEN-BASED METHODS

18.2.1 Kjeldahl Method

18.2.1.1 Principle

In the Kjeldahl procedure, proteins and other organic food components in a sample are digested with sulfuric acid in the presence of catalysts. The **total organic nitrogen** is converted to ammonium sulfate. The digest is neutralized with alkali and distilled into a boric acid solution. The borate anions formed are titrated with standardized acid, which is converted to nitrogen in the sample. The result of the analysis represents the crude protein content of the food since nitrogen also comes from nonprotein components (note that the Kjeldahl method also measures nitrogen in any ammonia and ammonium sulfate).

18.2.1.2 Historical Background

In 1883, Johann Kjeldahl developed the basic process of today's Kjeldahl method to analyze organic nitrogen. An excellent book to review the Kjeldahl method for total organic nitrogen was written by Bradstreet [7]. Several important modifications have improved the original Kjeldahl process, but the original method and the current procedure (as described in detail below) both include the same basic steps: (1) digestion, (2) neutralization and distillation, and (3) titration.

18.2.1.3 General Procedures and Reactions

18.2.1.3.1 Sample Preparation

Solid foods are ground to pass a 20-mesh screen. Samples for analysis should be homogeneous. No other special preparations are required.

18.2.1.3.2 Digestion

Place sample (accurately weighed) in a Kjeldahl flask. Add acid and catalyst; digest until clear to get complete breakdown of all organic matter. Nonvolatile ammonium sulfate is formed from the reaction of nitrogen and sulfuric acid.

$$\text{Protein} \xrightarrow[\text{Heat, catalyst}]{\text{Sulfuric acid}} \left(\text{NH}_4\right)_2\text{SO}_4 \qquad (18.1)$$

During digestion, protein nitrogen is liberated to form ammonium ions; sulfuric acid oxidizes organic matter and combines with ammonium formed; carbon and hydrogen elements are converted to carbon dioxide and water.

18.2.1.3.3 Neutralization and Distillation

The digest is diluted with water. Alkali-containing sodium thiosulfate is added to neutralize the sulfuric acid. The ammonia formed is distilled into a boric acid solution containing the indicators methylene blue and methyl red (AOAC Method 991.20).

$$\left(\text{NH}_4\right)_2\text{SO}_4 + 2\text{NaOH} \rightarrow$$
$$2\text{NH}_3 + \text{Na}_2\text{SO}_4 + 2\text{H}_2\text{O} \qquad (18.2)$$

$$\text{NH}_3 + \text{H}_3\text{BO}_3\left(\text{boric acid}\right) \rightarrow$$
$$\text{NH}_4 + \text{H}_2\text{BO}_3^-\left(\text{borate ion}\right) \qquad (18.3)$$

18.2 table Protein analysis method comparison

Method	Chemical basis	Principle	Advantages	Disadvantages	Applications
Kjeldahl	Nitrogen (total organic)	Determine N by method that involves digestion, neutralization, distillation, and titration. Use N content to calculate protein content	Inexpensive (if not automated system). Widely used and accepted method for over a century	Measures total organic N, and not just protein N. Time consuming. Uses corrosive reagents. Lower precision than some other methods	Applicable to all foods. Little used now, due to availability of automated Dumas systems
Dumas	Nitrogen (total organic and inorganic)	N is released upon combustion of sample at very high temperature. N gas is quantitated by gas chromatography using a thermal conductivity detector. Use N content to calculate protein content	Requires no hazardous chemicals. Rapid (few minutes). Automated instruments allow for analyzing many samples without attention	Expensive equipment. Measures total organic and inorganic N, and not just protein N	Applicable to all foods. Widely used now, compared to Kjeldahl method, for both official and quality control purposes
Infrared spectroscopy	Peptide bond	Presence of peptide bond in protein molecules causes absorption of radiation at specific wavelength in mid- or near infrared region	Rapid way to estimate protein content. Requires minimal training	Expensive equipment. Only provides an estimate of protein content. Instrument must be calibrated against results from official methods	Applicable to wide range of food products (grains, cereal, meat, dairy). Used as rapid, quality control method
Anionic dye-binding	Basic amino acid residues (of histidine, arginine, and lysine) and N-terminus of protein molecule	Residues identified react with anionic sulfonic acid dye to form an insoluble complex. Unbound soluble dye is measured by absorbance and related to protein concentration	Rapid (15 min. or less for non-automated method; much less for automated method). Relatively accurate. No corrosive reagents. Does not measure nonprotein N. More precise than Kjeldahl method. Can be used to estimate changes in available lysine content, since the dye does not bind altered, unavailable lysine	Not as sensitive as some other colorimetric methods. Requires a calibration curve for a given food commodity, since proteins differ in basic amino acid content so differ in dye-binding capacity. Not suitable for hydrolyzed proteins due to dye binding to N-terminal amino acids. Some nonprotein components bind dye or protein, to cause error	Automated version used for quality control purposes, especially as a method to compare results against a nitrogen-based method (to check for economic adulteration)

(continued)

18.2 table (Continued)

Method	Chemical basis	Principle	Advantages	Disadvantages	Applications
Bicinchoninic acid	Peptide bond and specific amino acids (cysteine, cystine, tryptophan, and tyrosine)	Peptide bond is complexed with cupric ions under alkaline conditions. Cuprous ions are chelated by BCA reagent to give color measured by spectroscopy	Good sensitivity, and micro-BCA method is even better (0.5–10 ug). Nonionic detergents and buffer salts do not interfere with the reaction, nor do medium concentrations of denaturing reagents	Color is not stable with time. Any compound capable of reducing Cu^{+2} to Cu^{+} will lead to color formation. Reducing sugars and high concentrations of ammonium sulfate interfere. Get color variation among proteins	Widely used method for protein isolation and purification. Has largely replaced other quantitative research colorimetric methods
Absorbance at 280 nm	Tyrosine and tryptophan	Aromatic amino acids, tryptophan and tyrosine, cause proteins to absorb at 280 nm. Absorbance can be used to estimate protein content	Rapid. Relatively sensitive (100 ug protein required). No interference from ammonium sulfate and other buffer salts. Nondestructive (so samples can be used after protein determination)	Nucleic acids can absorb at 280 nm. Aromatic amino acid contents in proteins vary between food sources, so results are qualitative. Requires relatively pure, clear, and colorless samples	Best used in purified protein systems (e.g., postcolumn detection of intact proteins)
Absorbance at 220 nm	Peptide bond	Peptide bonds cause proteins to absorb at 220 nm. Absorbance can be used to estimate protein content	Rapid. Nondestructive (so samples can be used after protein determination)	Many things other than peptide bonds absorb at 220 nm. Requires relatively pure, clear, and colorless samples	Best used with purified, hydrolyzed protein systems (i.e., postcolumn detection of hydrolyzed proteins)
Biuret	Peptide bond	Peptide bond is complexed with cupric ions under alkaline conditions to give color that is quantitated by spectroscopy	Less expensive, faster, and simpler than Kjeldahl method. Does not detect nonpeptide or nonprotein sources. Few interferences		

18.2.1.3.4 Titration

Borate anion (proportional to the amount of nitrogen) is titrated with standardized HCl.

$$H_2BO_3^- + H^+ \rightarrow H_3BO_3 \qquad (18.4)$$

18.2.1.3.5 Calculations

$$\text{Moles of HCl} = \text{moles NH}_3$$
$$= \text{moles N in the sample} \qquad (18.5)$$

A reagent blank should be run to subtract reagent nitrogen from the sample nitrogen.

$$\%N = N\text{HCl} \times \frac{\text{Corrected acid volume}}{\text{g of sample} \times 1000}$$
$$\times \frac{14g\,N}{\text{mol}} \times \frac{100}{1000} \qquad (18.6)$$

where:

N HCl = normality of HCl in moles/1,000 mL
Corrected acid vol. = (mL std. acid for sample) −
    (mL std. acid for blank)
14 = atomic weight of nitrogen

A factor is used to convert percent N to percent crude protein. Most proteins contain 16 % N, so the conversion factor is 6.25 (100/16 = 6.25).

$$\%N / 0.16 = \% \text{ protein}$$
$$\text{OR}$$
$$\%N \times 6.25 = \% \text{ protein} \qquad (18.7)$$

Conversion factors for various foods are given in Table 18.3 [1, 8].

18.2.1.4 Applications

The Kjeldahl method is an AOAC official method for crude protein content and has been the basis for evaluation of many other protein methods. The Kjeldahl method is still used for some applications, but now has limited use in many countries due to the availability and advantages of automated nitrogen combustion (Dumas) systems (Sect. 18.2.2) (see also Table 18.2 for advantages and disadvantages of the methods).

18.2.2 Dumas (Nitrogen Combustion) Method

18.2.2.1 Principle

The combustion method was introduced in 1831 by Jean-Baptiste Dumas. It has been modified and automated to improve accuracy since that time. Samples are combusted at high temperatures (700–1,000 °C) with a flow of pure oxygen. All carbon in the sample is converted to carbon dioxide during the flash combustion. Nitrogen-containing components produced include N_2 and nitrogen oxides. The nitrogen oxides are reduced to nitrogen in a copper reduction column at a high temperature (600 °C). The total nitrogen (including inorganic fraction, i.e., including nitrate and nitrite) released is carried by pure helium and quantitated by **gas chromatography** using a **thermal conductivity detector** (TCD) [9]. Ultrahigh purity acetanilide and EDTA (ethylenediaminetetraacetate) may be used as the standards for the calibration of the nitrogen analyzer. The nitrogen determined is converted to protein content in the sample using a protein conversion factor.

18.2.2.2 Procedure

Samples (approximately 100–500 mg) are weighed into a tin capsule and introduced to a combustion reactor in automated equipment. The nitrogen released is measured by a built-in gas chromatograph. Figure 18.1 shows the flow diagram of the components of a Dumas nitrogen analyzer.

18.3 table Nitrogen to protein conversion factors for various foods

Food	Percent N in protein	Factor
Egg or meat	16.0	6.25
Milk	15.7	6.38
Wheat	18.76	5.33
Corn	17.70	5.65
Oat	18.66	5.36
Soybean	18.12	5.52
Rice	19.34	5.17

Data from [1, 8]

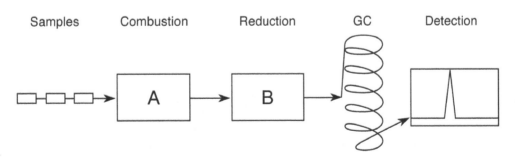

Samples   Combustion   Reduction   GC   Detection

18.1 figure General components of a Dumas nitrogen analyzer. (**A**) the incinerator. (**B**) copper reduction unit for converting nitrogen oxides to nitrogen, gas chromatography (*GC*) column, and detector

18.2.2.3 *Applications*

The combustion method is a faster and safer alternative to the Kjeldahl method [10] and is suitable for all types of foods. As an AOAC method, the Dumas method is widely used for official purposes, but its speed also allows for quality control applications. The industry uses different units/systems, depending on sample size and protein content. Freeze drying can be used to concentrate diluted liquid samples, e.g., waste steam samples.

18.3   INFRARED SPECTROSCOPY

18.3.1 Principle

Infrared spectroscopy measures the **absorption of radiation** (near- or mid-infrared regions) by molecules in food or other substances. Different functional groups in a food absorb different frequencies of radiation. For proteins and peptides, various **mid-infrared** bands (6.47 μm) and **near-infrared** (NIR) bands (e.g., 3,300–3,500 nm, 2,080–2,220 nm, 1,560–1,670 nm) characteristic of the **peptide bond** can be used to estimate the protein content of a food. By irradiating a sample with a wavelength of infrared light specific for the constituent to be measured, it is possible to predict the concentration of that constituent by measuring the energy that is reflected or transmitted by the sample (which is inversely proportional to the energy absorbed) [11].

18.3.2 Procedure

See Chap. 8 for a detailed description of instrumentation, sample handling, and calibration and quantitation methodology.

18.3.3 Applications

Mid-infrared spectroscopy is used in infrared milk analyzers to determine milk protein content, while near-infrared spectroscopy is applicable to a wide range of food products (e.g., grains, cereal, meat, and dairy products) [3, 12, 13], especially as a rapid method to test nonstandard milk.

18.4   COLORIMETRIC METHODS

When protein reacts with specific reagents under certain conditions, colorful compounds are generated, and the absorbance is measured by spectrophotometer. The protein content is expressed on the basis of standard protein such as bovine serum albumin (BSA), and thus this method is not an absolute method. Due to the differences in the composition of proteins, these methods have limited use. However, because of high

sensitivity, these methods have the advantage of requiring a small sample size.

18.4.1 Dye-Binding Methods

18.4.1.1 *Anionic Dye-Binding Method*

18.4.1.1.1   *Principle*

The protein-containing sample is mixed with a known excess amount of **anionic dye** in a buffered solution. Proteins bind the dye to form an insoluble complex. The unbound soluble dye is measured after equilibration of the reaction and the removal of insoluble complex by centrifugation or filtration.

$$\text{Protein} + \text{excess dye} \rightarrow \text{Protein} - \text{dye insoluble complex} \\ + \text{unbound soluble dye} \quad (18.8)$$

The anionic sulfonic acid dye, including acid orange 12, orange G, and Amido Black 10B, binds cationic groups of the **basic amino acid residues** (imidazole of histidine, guanidine of arginine, and ε-amino group of lysine) and the **free amino terminal group** of the protein [14]. The amount of the unbound dye is inversely related to the protein content of the sample [14].

18.4.1.1.2   *Procedure*
1. The sample is finely ground (60 mesh or smaller sizes) and added to an excess dye solution with known concentration.
2. The content is vigorously shaken to equilibrate the dye-binding reactions and filtered or centrifuged to remove insoluble substances.
3. Absorbance of the unbound dye solution in the filtrate or supernatant is measured and dye concentration estimated from a dye standard curve.
4. A straight calibration curve can be obtained by plotting the unbound dye concentration against total nitrogen (as determined by Kjeldahl method) of a given food covering a wide range of protein content.
5. Protein content of the unknown sample of the same food type can be estimated from the calibration curve or from a regression equation calculated by the least squares method.

18.4.1.1.3   *Applications*
Anionic dye-binding has been used to estimate proteins in milk [15, 16], wheat flour [17, 18], soy products [18, 19], and meats [20]. The anionic dye-binding method may be used to estimate the changes in available lysine content of cereal products during processing since the dye does not bind altered, unavailable lysine. Since lysine is the limiting amino acid in cereal products, the available lysine content represents pro-

tein nutritive value of the cereal products [21]. An automated Sprint™ Rapid Protein Analyzer has been developed by the CEM Company (Matthews, NC) based on the anionic dye-binding method. This rapid protein analyzer combines and automates all steps (sample weighing, homogenization, predetermined dye addition, filtration, absorption measurement, and protein calculation) in a compact bench top unit. This dye-binding method has been approved in AOAC Method 2011.04 for raw and processed meats. This analyzer has shown great agreement with Kjeldahl method in protein analysis of milk, cream, soybean, and soymilk [22, 23]. This method is safe (the proprietary dye iTag is nonhazardous), easy to use, and efficient. Because different types of sample matrixes and proteins exhibit different responses on the Sprint Rapid Protein Analyzer, this automated method requires calibration for each type of food protein determined using other official methods.

18.4.1.2 Bradford Dye-Binding Method

18.4.1.2.1 Principle
When **Coomassie Brilliant Blue G-250** binds to protein, the **dye changes color** from reddish to bluish, and the absorption maximum of the dye is shifted from 465 to 595 nm. The change in the absorbance at 595 nm is proportional to the protein concentration of the sample [24]. Like other dye-binding methods, the Bradford relies on the **amphoteric nature of proteins**. When the protein-containing solution is acidified to a pH less than the isoelectric point of the protein(s) of interest, the dye added binds electrostatically. Binding efficiency is enhanced by hydrophobic interaction of the dye molecule with the polypeptide backbone adjoining positively charged residues in the protein [4]. In the case of the Bradford method, the dye bound to protein has a change in absorbance spectrum relative to the unbound dye.

18.4.1.2.2 Procedure
1. Coomassie Brilliant Blue G-250 is dissolved in 95 % ethanol and acidified with 85 % phosphoric acid.
2. Samples containing proteins (1–100 μg/mL) and standard BSA solutions are mixed with the Bradford reagent.
3. Absorbance at 595 nm is read against a reagent blank.
4. Protein concentration in the sample is estimated from the BSA standard curve.

18.4.1.2.3 Applications
The Bradford method has been used successfully to determine protein content in worts and beer products [25] and in potato tubers [26]. This procedure has been improved to measure microgram quantities of proteins [27]. Due to its rapidity, sensitivity, and fewer interferences than the Lowry method, the Bradford method has been used widely for the analysis of low concentrations of proteins and enzymes in their purification and characterizations. Improved Coomassie dye-based protein assays based on the Bradford method have been developed (e.g., CB™ and CB-X™, G-Biosciences, Geno Technology, Inc.) for increased compatibility with buffers and conditions commonly used in protein isolation. Other dye-based protein assays (e.g., RED 660™ Protein Assay, Thermo Scientific Pierce) are being developed to improve the linearity, color stability, and interferences of the Bradford and improved Coomassie/Bradford assays.

18.4.2 Copper Ion-Based Methods

Since the biuret method was first developed to measure protein content based on the reaction with copper ions, several improved methods have been developed. Following the description of the basic biuret method below are the modified Lowry and the bicinchoninic acid (BCA) methods, which are both based in part on the biuret method.

18.4.2.1 Biuret Method

18.4.2.1.1 Principle
A violet-purplish color is produced when **cupric ions** are complexed with **peptide bonds** (substances containing at least two peptide bonds, i.e., oligopeptides, large peptides, and all proteins) under **alkaline conditions** (Fig. 18.2). The absorbance of the color produced is read at 540 nm. The color intensity (absorbance) is proportional to the protein content of the sample [28].

18.4.2.1.2 Procedure
1. A 5-mL biuret reagent is mixed with a 1-mL portion of protein solution (1–10 mg protein/mL).

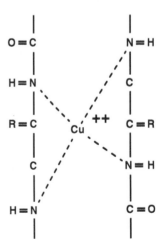

18.2 Reaction of peptide bonds with cupric ions

figure

The reagent includes copper sulfate, NaOH, and potassium sodium tartrate, which is used to stabilize the cupric ion in the alkaline solution.

2. After the reaction mix is allowed to stand at room temperature for 15 or 30 min, the absorbance is read at 540 nm against a reagent blank.

3. Filtration or centrifugation before reading absorbance is required if the reaction mixture is not clear.

4. A standard curve of concentration versus absorbance is constructed using **bovine serum albumin** (BSA).

18.4.2.1.3 Applications

The biuret method has been used to determine proteins in cereal [29, 30], meat [20], soybean proteins [19], and as a qualitative test for animal feed [31]. The biuret method also can be used to measure the protein content of isolated proteins, but it has been largely replaced for this use by more sensitive methods such as the modified Lowry method and especially the BCA methods for the reasons described in Sect. 18.4.2.

18.4.2.2 Lowry Method

18.4.2.2.1 Principle

The Lowry method [32, 33] combines the **biuret reaction** with the reduction of the **Folin-Ciocalteu phenol reagent** (phosphomolybdic-phosphotungstic acid) by **tyrosine** and **tryptophan** residues in the proteins (Fig. 18.3). The bluish color developed is read at 750 nm (high sensitivity for low protein concentration) or 500 nm (low sensitivity for high protein concentration). The original procedure has been modified by Miller [34] and Hartree [35] to improve the linearity of the color response to protein concentration and replace the use of two unstable reagents with one stable reagent.

18.4.2.2.2 Procedure

The following procedure is based on the modified procedure of Hartree [35]:

1. Proteins to be analyzed are diluted to an appropriate range (20–100 μg).

2. K Na Tartrate-Na$_2$CO$_3$ solution is added after cooling then incubated at room temperature for 10 min.

3. CuSO$_4$-K Na Tartrate-NaOH solution is added after cooling then incubated at room temperature for 10 min.

4. Freshly prepared Folin's reagent is added and then the reaction mixture is mixed and incubated at 50 °C for 10 min.

5. Absorbance is read at 650 nm.

6. A standard curve of BSA is carefully constructed for estimating protein concentration of the unknown.

18.4.2.2.3 Applications

Because of its simplicity and sensitivity, the Lowry method has been widely used in protein biochemistry. However, it has not been widely used to determine proteins in food systems without first extracting the proteins from the food mixture.

18.4.2.3 Bicinchoninic Acid Method

18.4.2.3.1 Principle

Proteins and peptides (as short as di-peptides) reduce **cupric ions** to **cuprous ions** under **alkaline conditions** [36], which is similar in principle to that of the biuret reaction. The cuprous ion then reacts with the apple-greenish **bicinchoninic acid reagent** (BCA) to form a purplish complex (one cuprous ion is chelated by two BCA molecules) (Fig. 18.4). The color measured at 562 nm is near linearly proportional to protein concentration over a wide range of concentration from micrograms up to 2 mg/mL. Peptide bonds and four amino acids (cysteine/cystine, tryptophan, and tyrosine) contribute to the color formation with BCA.

18.4.2.3.2 Procedure

1. Mix (one step) the protein solution with the BCA reagent, which contains BCA sodium salt,

a

b

18.3 **figure** Side chains of amino acids tyrosine (**a**) and tryptophan (**b**)

18.4 **figure** Protein reaction with cupric ions under alkaline conditions to form cuprous ions, which react with bicinchoninic acid (*BCA*) to form purple color, measured at 562 nm (Figure courtesy of Pierce Biotechnology Technical Library, Thermo Fisher Scientific, Inc., Rockford, IL)

sodium carbonate, NaOH, and copper sulfate, pH 11.25.

2. Incubate at 37 °C for 30 min, or room temperature for 2 h, or 60 °C for 30 min. The selection of the temperature depends upon the sensitivity desired. A higher temperature gives a greater color response.
3. Read the solution at 562 nm against a reagent blank.
4. Construct a standard curve using BSA.

18.4.2.3.3 Applications

The BCA method is widely used in protein isolation and purification due to its advantage (over the modified Lowry method and any Coomassie dye-based assay) of being compatible with samples containing up to 5 % detergents. While most dye-binding methods are faster, the BCA method is less affected by protein compositional differences, so there is better protein-to-protein uniformity.

18.5 ULTRAVIOLET ABSORPTION METHODS FOR PROTEINS AND PEPTIDES

18.5.1 Ultraviolet 280 nm Absorption for Protein

18.5.1.1 Principle

Proteins show strong absorption in the **ultraviolet** (UV) region at **UV 280 nm**, primarily due to **tryptophan** and **tyrosine** residues in the proteins. Because the content of tryptophan and tyrosine in proteins from each food source is fairly constant, the absorbance at 280 nm could be used to estimate the concentration of proteins, using **Beer's law**. Since each protein has a unique aromatic amino acid composition, the extinction coefficient (E_{280}) or molar absorptivity (E_m) must be determined for individual proteins for protein content estimation.

18.5.1.2 Procedure

1. Proteins are solubilized in buffer or alkali.
2. Absorbance of protein solution is read at 280 nm against a reagent blank.
3. Protein concentration is calculated according to the following equation:

$$A = abc \qquad (18.9)$$

where:

A = absorbance
a = absorptivity
b = cell or cuvette path length
c = concentration

18.5.1.3 Applications

The UV 280 nm method has been used to determine the protein contents of milk [37] and meat products [38]. It has not been used widely in food systems. This technique is better applied in a purified protein system or to proteins that have been extracted in alkali or denaturing agents such as $8 M$ urea. Although peptide bonds in proteins absorb more strongly at 190–220 nm than at 280 nm, the low UV region is more difficult to measure.

18.5.2 Peptide Measurement at 190–220 nm

Peptides without or with low level of tyrosine or tryptophan residues can be quantified at 190–220 nm at which peptide bonds have maximum absorption. The extinction coefficients in the far UV range can be calculated with the consideration of contribution of tyrosine and tryptophan to the absorption [39]. Protein can also be measured in this UV range. For example, NanoDrop™ 2,000/2,000 c from Thermo Scientific can be used for protein and peptide quantification at 205 nm with extinction coefficient being 31.

18.6 NONPROTEIN NITROGEN DETERMINATION

18.6.1 Principle

Protein is precipitated by trichloroacetic acid (TCA) and separated from nonprotein nitrogen (NPN)-containing compounds.

18.6.2 Procedure

1. Add appropriate amount of TCA into a sample solution to make the final TCA concentration around 10 %. Alternatively, prepare a 10 % TCA solution and add dry sample powder into the solution.
2. Mix the reaction mixture thoroughly and let precipitate settle for 5 min.
3. Filter the mixture through Whatman No. 1 filter paper. Alternatively, centrifugation at 30,000 g can be used to separate precipitated protein from the supernatant that contains NPN.
4. Determine nitrogen content of the filtrate by the Kjeldahl method.
5. Convert nonprotein nitrogen to protein equivalent with conversion factor.

18.6.3 Applications

The NPN determination is used to check economic adulteration with nitrogen-rich compounds such as urea, ammonia, and melamine. NPN for milk can be tested by AOAC methods (991.21 and 991.22). The results obtained are not absolute, but are relative to the standardized conditions used in the procedure [40].

18.7　COMPARISON OF METHODS

1. **Sample preparation:** The Kjeldahl, Dumas, and infrared spectroscopy methods require little preparation. Sample particle size of 20 mesh or smaller generally is satisfactory for these methods. Some of the newer NIR instruments can make measurements directly on whole grains and other coarsely granulated products without grinding or other sample preparation. Other methods described in this chapter require fine particles for extraction of proteins from the complex food systems.

2. **Principle:** The Dumas and Kjeldahl methods measure directly the nitrogen content of foods. However, the Kjeldahl method measures only organic nitrogen plus ammonia, while Dumas measures total nitrogen, including the inorganic fraction. (Therefore, Dumas gives a higher value for products that contain nitrates/nitrites.) Other methods of analysis measure the various properties of proteins. For instance, the biuret method measures peptide bonds, and the Lowry and BCA methods measure a combination of peptide bonds and specific amino acids. Infrared spectroscopy is an indirect method to estimate protein content, based on the energy absorbed when a sample is subjected to a wavelength of infrared radiation specific for the peptide bond.

3. **Sensitivity:** Kjeldahl, Dumas, and biuret methods are less sensitive than Bradford, Lowry, BCA, or UV methods.

4. **Speed:** After the instrument has been properly calibrated, infrared spectroscopy is likely the most rapid of the methods discussed. In most other methods involving spectrophotometric (colorimetric) measurements, one must separate proteins from the interfering insoluble materials before mixing with the color reagents or must remove the insoluble materials from the colored protein-reagent complex after mixing. However, the speed of determination in the colorimetric methods and in the Dumas method is faster than with the Kjeldahl method.

5. **Applications:** Although both Kjeldahl and Dumas methods can be used to measure N content in all types of foods, in recent years the Dumas method has largely replaced the Kjeldahl method for nutrition labeling (since Dumas method is faster, has a lower detection limit, and is safer). However, the Kjeldahl method is the preferred method for high fat samples/products since fat may cause an instrument fire during the incineration procedure in the Dumas method. The Kjeldahl method or the nitrogen combustion method is specified to correct for protein content in the most recent AOAC official method to measure the fiber content of foods (see Chap. 19, Sect. 19.5). Melamine, a toxic nitrogen adulterant, is included in the total nitrogen content if measured by the Kjeldahl or Dumas methods. NIR, CEM-Sprint dye binding, and Dumas methods are good for quality control in food processing plants.

18.8　SPECIAL CONSIDERATIONS

1. To select a particular method for a specific application, sensitivity, accuracy, and reproducibility as well as physicochemical properties of food materials must be considered. The data should be interpreted carefully to reflect what actually is being measured.

2. Food processing methods, such as heating, may reduce the extractability of proteins for analysis and cause an underestimation of the protein content measured by methods involving an extraction step.

3. Except for the Dumas and Kjeldahl methods, and the UV method for purified proteins, all methods require the use of a standard or reference protein or a calibration with the Kjeldahl method. In the methods using a standard protein, proteins in the samples are assumed to have similar composition and behavior compared to the standard protein. The selection of an appropriate standard for a specific type of food is important.

4. **Nonprotein nitrogen** is present in practically all foods. To determine **protein nitrogen**, the samples usually are extracted under alkaline conditions then precipitated with trichloroacetic acid or sulfosalicylic acid. The concentration of the acid used affects the precipitation yield. Therefore, nonprotein nitrogen content may vary with the type and concentration of the reagent used. Heating could be used to aid protein precipitation by acid, alcohol, or other organic solvents. In addition to acid precipitation methods used for nonprotein nitrogen determination, less empirical methods such as dialysis, and

ultrafiltration and column chromatography could be used to separate proteins from small nonprotein substances.

5. In the determination of the nutritive value of food proteins, including **protein digestibility** and **protein efficiency ratio** (PER), the Kjeldahl or Dumas method with a 6.25 conversion factor is used to determine crude protein content. The PER could be underestimated if a substantial amount of nonprotein nitrogen is present in foods. A food sample with a higher nonprotein nitrogen content (particularly if the nonprotein nitrogen does not have many amino acids or small peptides) may have a lower PER than a food sample containing similar protein structure/composition and yet with a lower amount of nonprotein nitrogen.

18.9   SUMMARY

Methods based on the unique characteristics of proteins and amino acids have been described to determine the protein content of foods. The Kjeldahl and Dumas methods measure nitrogen. Infrared spectroscopy is based on absorption of a wavelength of infrared radiation specific for the peptide bond. Copper-peptide bond interactions contribute to the analysis by the biuret, Lowry, and BCA methods. Amino acids are involved in the Lowry, BCA, dye-binding, and UV 280-nm methods. The BCA method also utilizes the reducing power of proteins in an alkaline solution. The various methods differ in their speed and sensitivity.

In addition to the commonly used methods discussed, there are other methods available for protein quantification. Because of the complex nature of various food systems, problems may be encountered to different degrees in protein analysis by available methods. Rapid methods may be suitable for quality control purposes, while a sensitive method is required for work with a minute amount of protein. Indirect colorimetric methods usually require the use of a carefully selected protein standard or a calibration with an official method.

18.10   STUDY QUESTIONS

1. What factors should one consider when choosing a method for protein determination?
2. The Kjeldahl method of protein analysis consists of three major steps. List these steps in the order they are done and describe in words what occurs in each step. Make it clear why milliliters of HCl can be used as an indirect measure of the protein content of a sample.
3. Why is the conversion factor from Kjeldahl nitrogen to protein different for various foods, and how is the factor of 6.25 obtained?
4. Differentiate the principles of protein determination by dye binding with an anionic dye such as Amido Black vs. the Bradford method, which uses the dye Coomassie Blue G-250.
5. With the anionic dye-binding method, would a sample with a higher protein content have a higher or a lower absorbance reading than a sample with a low protein content? Explain your answer.
6. For each of the situations described below, identify a protein assay method most appropriate for use and indicate the chemical basis of the method (i.e., what does it really measure?)

   (a) Nutrition labeling
   (b) Intact protein eluting from a chromatography column; qualitative or semiquantitative method
   (c) Intact protein eluting from a chromatography column; colorimetric, quantitative method
   (d) Rapid, quality control method for protein content of cereal grains

7. The FDA found melamine (see structure below) in pet food linked to deaths of pets in the United States. The FDA also found evidence of melamine in wheat gluten imported from China used as one of the ingredients in the production of the pet food. Melamine is a nitrogen-rich chemical used to make plastic and sometimes used as a fertilizer.

   (a) Knowing that each ingredient is tested and analyzed when imported, explain how melamine in wheat gluten could have escaped detection.
   (b) How can the adulteration of wheat gluten be detected (not necessarily detecting melamine specifically) using a combination of protein analysis methods? Explain your answer.

18.11   PRACTICE PROBLEMS

1. A dehydrated precooked pinto bean was analyzed for crude protein content in duplicate using the Kjeldahl method. The following data were recorded:

Moisture content = 8.00 %
Wt of Sample 1 = 1.015 g
Wt of Sample 2 = 1.025 g
Normality of HCl used for titration = 0.1142 N
HCl used for Sample 1 = 22.0 mL
HCl used for Sample 2 = 22.5 mL
HCl used for reagent blank = 0.2 mL

Calculate crude protein content on both wet and dry weight basis of the pinto bean, assuming pinto bean protein contains 17.5 % nitrogen.

2. A 20-mL protein fraction recovered from a column chromatography was analyzed for protein using the BCA method. The following data were the means of a duplicate analysis using BSA as a standard:

BSA mg/mL	Mean absorbance at 562 nm
0.2	0.25
0.4	0.53
0.6	0.74
0.8	0.95
1.0	1.15

The average absorbance of a 1-mL sample was 0.44. Calculate protein concentration (mg/mL) and total protein quantity of this column fraction.

3. A turkey frankfurter was analyzed for crude protein content using the Kjeldahl method. The following data were recorded:

Wt of sample = 0.5172 g
Normality of HCl = 0.1027 N
Vol of HCl used for sample = 8.8 mL
Vol of HCl used for reagent blank = 0.2 mL

Calculate % crude protein content on wet weight basis.

For calculating the conversion factor, assume that amino acids in meat protein contain 16 % nitrogen (or look up the conversion factor).

$$\% \text{ Nitrogen} = \frac{\left(\text{mL HCl sample} - \text{mL HCl blank}\right) \times N \text{ HCl} \times 1.4007}{\text{Sample weight (g)}}$$

$$\% \text{ N} \times \text{Conversion Factor} = \% \text{ Protein}$$

$$100 / \% \text{ N} = \text{Conversion Factor}$$

4. Why is it difficult to accurately determine protein content of a composite food such as a sausage pizza when using the Kjeldahl and combustion methods for protein?

5. A standard procedure for the Dumas combustion method for protein states that a sample should contain between 10 and 50 mg N. Using the USDA Nutrient Database, calculate the weight of sample to use for analysis. Sample: turkey thigh meat.

6. Estimate the weight of sample (Wheat Thins) you should use for Kjeldahl protein determination if you want to titrate the sample with at least 7 mL 0.1 N acid. Show calculation. Use the nutritional label as a guide (serving size: 29 g; protein: 2 g).

7. A stock solution of bovine serum albumin (BSA) was prepared previously by a coworker. Before using the BSA to prepare a standard curve for biuret determination, you need to confirm that the stock solution contains 20 mg protein/mL using UV spectroscopy. The $E^{1\%}$ for BSA is 6.3 at 280 nm. If the stock solution was correctly prepared, calculate the absorbance you would expect to find when the stock solution is diluted tenfold (0.3 mL stock with 2.7 mL water) and absorbance of this solution determined at 280 nm in a 3-mL cuvette with a 0.5-cm path length.

8. Colorimetric protein assay. Calculate the concentration of protein in an unknown solution X using the biuret test for protein. The stock solution contains 20-mg protein/mL. The standard curve was prepared using the stock solution as follows (used 1-cm path length cuvette):

Standard curve	1	2	3	4	5	6
Vol. of H₂O (mL)	1	0.8	0.6	0.4	0.2	0.0
Vol. of protein standard (mL)	0.0	0.2	0.4	0.6	0.8	1.0
Vol. of biuret reagent (mL)	4.0	4.0	4.0	4.0	4.0	4.0
Absorbance at 540 nm	0.0	0.174	0.343	0.519	0.691	0.823

Solution X was prepared for analysis as follows:

Sample	Tube 1	Tube 2
Vol. of H₂O (mL)	0.8	0.2
Vol. of solution X (mL)	0.2	0.8
Vol. of biuret reagent (mL)	4.0	4.0
Absorbance at 540 nm	0.451	0.857

(a) Explain why the absorbance of the Biuret assay is measured at 540 nm.
(b) Calculate the concentration of protein (mg/tube) in each tube used to prepare the standard curve.
(c) Prepare a standard curve (in mg/tube). Label the graph properly.

(d) What sample dilution (tube 1 or tube 2) should you use to determine the concentration of the unknown sample? Why?

(e) Calculate the concentration of protein in solution X. List the concentration as mg protein/cuvette and then express results as mg protein/mL of solution X.

Answers

1. Protein content = 19.75 % on a wet weight basis; 21.47 % on a dry weight basis.

   Calculations:

   $$\%N = N\,HCl \times \frac{\text{Corrected acid volume}}{\text{g of sample}}$$
   $$\times \frac{14\,\text{g N}}{\text{mol}} \times 100 \qquad (18.6)$$

   where:

   $N\,HCl$ = normality of HCl in mol/1,000 mL
   Corrected acid vol. = (mL std. acid for sample) – (mL std. acid for blank)
   14 = atomic weight of nitrogen

   Corrected acid volume for Sample 1 = 22.0 mL – 0.2 mL = 21.8 mL

   Corrected acid volume for Sample 2 = 22.5 mL – 0.2 mL = 22.3 mL

   % N for Sample 1
   $$= \frac{0.1142\,\text{mole}}{1000\,\text{ml}} \times \frac{21.8\,\text{ml}}{1.015\,\text{g}} \times \frac{14\text{g N}}{\text{mol}} \times 100\,\% = 3.433\,\%$$

   % N for Sample 2
   $$= \frac{0.1142\,\text{mole}}{1000\,\text{ml}} \times \frac{22.3\,\text{ml}}{1.025\,\text{g}} \times \frac{14\text{g N}}{\text{mol}} \times 100\,\% = 3.478\,\%$$

   Protein conversion factor
   $= 100\,\%/17.5\,\%\,N = 5.71$

   Crude protein content for Sample 1
   $= 3.433\,\% \times 5.71 = 19.6\,\%$

   Crude protein content for Sample 2
   $= 3.478\,\% \times 5.71 = 19.9\,\%$

   The average for the duplicate data
   $= (19.6\,\% + 19.9\,\%)/2 = 19.75\,\%$
   $= {\sim}19.8\,\%$ wet weight basis

   To calculate protein content on a dry weight basis: Sample contains 8 % moisture, and therefore, the sample contains 92 % dry solids, or 0.92 g out of 1 g sample. Therefore, protein on a dry weight basis can be calculated as follows: 19.75 %/0.92 g dry solids = 21.47 % = ~21.5 % dry weight basis.

2. Protein content = 0.68 mg/mL. Total protein quantity = 6.96 mg
   Calculations:
   Plot absorbance (y-axis, absorbance at 562 nm) versus BSA protein concentration (x-axis, mg/mL) using the data above. Determine the equation of the line ($y = 1.11\,x + 0.058$) and then use this equation and the given absorbance ($y = 0.44$) to calculate the concentration ($x = 0.344$ mg/mL). Since 1 mL of sample gives a concentration of 0.344 mg/mL and we have a total of 20 mL collected from column chromatography, we will have a total of (0.344 mg/mL × 20 mL) = 6.88 mg protein in this collected column fraction.

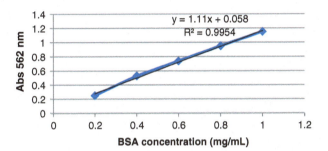

3. $\dfrac{100}{16\%} = 6.25$

   $$\%\,N = \frac{(8.8 - 0.2\,\text{mL})(0.1027\,\text{N})(1.4007)}{0.5172}$$
   $$= 2.39\,\%\,N \times 6.25 = 14.95\,\%\ \text{Protein}$$

4. The composite food contains many different proteins with different proportions of amino acids of varying nitrogen contents. The following is the conversion factors of some ingredients:

   Meat: 6.25
   Dairy: 6.38
   Wheat flour: 5.7

5. Database: 19.27 g protein/100 g
   Estimate:

   $$\%\,N = \frac{19.27\,\%\ \text{protein}}{6.25} = 3.05\,\%\,N$$
   $$\frac{3.08\,\text{g}}{100\,\text{g}} = \frac{0.010\text{g}}{x}$$
   $$x = 0.3247\,\text{g}$$
   $$5x = 5 \times 0.3247\,\text{g} = 1.6234\,\text{g}$$
   Weight of sample = 0.32 – 1.62 g

6. $\%\,\text{Protein} = \dfrac{2\text{g}}{29\text{g}} \times 100 = 6.9\%$

Conversion factor for wheat flour is 5.7,

$$so \; N\% = \frac{6.9\%}{5.7} = 1.2\%$$

$$1.2\% = \frac{(7\,mL)(0.1N)(1.4007)}{sample\;weigh}$$

$$Sample\;weight = 0.82\,g$$

7. Concentration of BSA after tenfold dilution $= 2\,mg/mL = 0.002\,g/g = 0.2\%$

$$A = a \times b \times c = 6.3 \times 0.5 \times 0.2 = 0.63$$

8. (a) The complex has maximum absorbance at this wavelength.
   (b) Tube 1: 0 mg/tube
   Tube 2: 4 mg/tube
   Tube 3: 8 mg/tube
   Tube 4: 12 mg/tube
   Tube 5: 16 mg/tube
   Tube 6: 20 mg/tube
   (c) Standard curve prepared as in Problem 2.
   (d) Tube 1 is used because its absorbance is within the range of the standard curve.
   (e) Based on the standard curve and absorbance of X solution:
   Protein concentration calculated is about 10 mg/tube
   10 mg/0.2 mL = 50 mg/mL

Acknowledgments The authors thank Dr. Denise Smith of the Washington State University for her contribution of several valuable practice problems and answers to this chapter. We also thank Dr. S. Suzanne Nielsen for her valuable suggestions for improvement of this chapter.

REFERENCES

1. Jones DB (1931) Factors for converting percentages of nitrogen in foods and feeds into percentages of proteins. US Dept. Agric. Circular No. 183, August. USDA, Washington, DC
2. US Department of Agriculture, Agricultural Research Service (2016). USDA National Nutrient Database for Standard References, Release 28. Nutrient Data Laboratory Home Page. http://ndb.nal.usda.gov
3. AOAC International (2016) Official methods of analysis, 20th edn. (On-line). AOAC International, Rockville, MD
4. Yada RY, Jackman RL, Smith JL, Marangoni AG (1996) Analysis: quantitation and physical characterization, (Chapter 7). In: Nakai S, Modler HW (eds) Food proteins. Properties and characterization. VCH, New York, pp 333–403
5. Kolakowski E (2001) Protein determination and analysis in food system, (Chapter 4). In: Sikorski ZE (ed) Chemical and functional properties of food protein. Technomic Publishing, Lancaster, PA pp 57–112
6. Owusu-Apenten RK (2002) Food protein analysis. Quantitative effects on processing. Marcel Dekker, New York
7. Bradstreet RB (1965) The Kjeldahl method for organic nitrogen. Academic, New York
8. Mossé J (1990) Nitrogen to protein conversion factor for ten cereals and six legumes or oilseeds. A reappraisal of its definition and determination. Variation according to species and to seed protein content. J Agric Food Chem 38: 18–24
9. Wilson PR (1990) A new instrument concept for nitrogen/protein analysis. A challenge to the Kjeldahl method. Aspects Appl Biol 25: 443–446
10. Wiles PG, Gray I, Kissling RC (1998) Routine analysis of proteins by Kjeldahl and Dumas methods: review and interlaboratory study using dairy products. J AOAC Int 81: 620–632
11. O'Sullivan A, O'Connor B, Kelly A, McGrath MJ (1999) The use of chemical and infrared methods for analysis of milk and dairy products. Int J Dairy Technol 52: 139–148
12. Luinge HJ, Hop E, Lutz ETG, van Hemert JA, de Jong EAM (1993) Determination of the fat, protein and lactose content of milk using Fourier transform infrared spectrometry. Anal Chim Acta 284: 419–433
13. Krishnan PG, Park WJ, Kephart KD, Reeves DL, Yarrow GL (1994) Measurement of protein and oil content of oat cultivars using near-infrared reflectance. Cereal Foods World 39(2): 105–108
14. Fraenkel-Conrat H, Cooper M (1944) The use of dye for the determination of acid and basic groups in proteins. J Biol Chem 154: 239–246
15. Udy DC (1956) A rapid method for estimating total protein in milk. Nature 178: 314–315
16. Tarassuk NP, Abe N, Moats WA (1966) The dye binding of milk proteins. Technical bulletin no. 1369. USDA Agricultural Research Service in cooperation with California Agricultural Experiment Station. Washington, DC
17. Udy DC (1954) Dye-binding capacities of wheat flour protein fractions. Cereal Chem 31: 389–395
18. AACC International (2010) Approved Methods of American Association of Cereal Chemists International. 11th edn. (On-line), American Association of Cereal Chemists, St. Paul, MN
19. Pomeranz Y (1965) Evaluation of factors affecting the determination of nitrogen in soya products by the biuret and orange-G dye-binding methods. J Food Sci 30: 307–311
20. Torten J, Whitaker JR (1964) Evaluation of the biuret and dye-binding methods for protein determination in meats. J Food Sci 29: 168–174
21. Hurrel RF, Lerman P, Carpenter KJ (1979) Reactive lysine in foodstuffs as measured by a rapid dye-binding procedure. J Food Sci 44: 1221–1227
22. Amamcharla JK, Metzger LE (2010) Evaluation of a rapid protein analyzer for determination of protein in milk and cream. J Dairy Sci 93: 3846–3857
23. Ou YQ, Chang SKC (2011) Comparison of a rapid dye-binding method with the Kjeldahl and NIR methods for determining protein content in soybean and soymilk. Annual Meeting of the Institute of Food Technologists. June 11–14, 2011. New Orleans, LA
24. Bradford M (1976) A rapid and sensitive method for the quantitation of microgram quantities of protein utilizing the principle of protein-dye binding. Anal Biochem 72: 248–254

25. Lewis MJ, Krumland SC, Muhleman DJ (1980) Dye-binding method for measurement of protein in wort and beer. J Am Soc Brew Chem 38: 37–41

26. Snyder J, Desborou S (1978) Rapid estimation of potato tuber total protein content with Coomassie Brilliant Blue G-250. Theor Appl Genet 52: 135–139

27. Bearden Jr JC (1978) Quantitation of submicrogram quantities of protein by an improved protein-dye binding assay. Biochim Biophys Acta 533: 525–529

28. Robinson HW, Hodgen CG (1940) The biuret reaction in the determination of serum protein. 1. A study of the conditions necessary for the production of the stable color which bears a quantitative relationship to the protein concentration. J Biol Chem 135: 707–725

29. Jennings AC (1961) Determination of the nitrogen content of cereal grain by colorimetric methods. Cereal Chem 38: 467–479

30. Pinckney AJ (1961) The biuret test as applied to the estimation of wheat protein. Cereal Chem 38: 501–506

31. AOAC (1965) Official methods of analysis, 10th edn. Association of Official Analytical Chemists, Washington, DC

32. Lowry OH, Rosebrough NJ, Farr AL, Randall RJ (1951) Protein measurement with the Folin phenol reagent. J Biol Chem 193: 265–275

33. Peterson GL (1979) Review of the Folin phenol protein quantitation method of Lowry, Rosebrough, Farr, and Randall. Anal Biochem 100: 201–220

34. Miller GL (1959) Protein determination for large numbers of samples. Anal Chem 31: 964

35. Hartree EF (1972) Determination of protein: a modification of the Lowry method that gives a linear photometric response. Anal Biochem 48: 422–427

36. Smith PK, Krohn Rl, Hermanson GT, Mallia AK, Gartner FH, Provensano MD, Fujimoto EK, Goeke NM, Olson BJ, Klenk DC (1985) Measurement of protein using bicinchoninic acid. Anal Biochem 150: 76–85

37. Nakai S, Wilson HK, Herreid EO (1964) Spectrophotometric determination of protein in milk. J Dairy Sci 47: 356–358

38. Gabor E (1979) Determination of the protein content of certain meat products by ultraviolet absorption spectrophotometry. Acta Alimentaria 8(2): 157–167

39. Scopes RK (1974) Measurement of protein by spectrophotometry at 205 nm. Anal Biochem 59: 277–282

40. Regenstein JM, Regenstein CE (1984) Protein functionality for food scientists. In "Food Protein Chemistry." Chapter 27. Academic Press Inc. p. 274–334

Carbohydrate Analysis

James N. BeMiller

Department of Food Science,
Purdue University,
West Lafayette, IN 47907-2009, USA
e-mail: bemiller@purdue.edu

S. Nielsen (ed.), *Food Analysis*, Food Science Text Series,
DOI 10.1007/978-3-319-45776-5_19, © Springer International Publishing 2017

19.1   INTRODUCTION

Carbohydrates are important in foods as a major source of energy, to impart crucial textural properties, and as dietary fiber which contributes to overall health. There is interest in analysis of food products and ingredients for the various types of carbohydrates (not only the different structural types but also types differing in physiological effects, e.g., digestible vs. nondigestible, metabolizable vs. non-metabolizable, caloric vs. reduced caloric vs. noncaloric, prebiotic vs. non-prebiotic). However, definitions of the types are not always agreed upon, and analytical methods do not always measure exactly what is included in the definition, which results in controversies about what should be measured and how. This chapter covers analysis of carbohydrates primarily by structural type.

Digestible carbohydrates are converted into monosaccharides, which are absorbed and provide metabolic energy and satiety. Nondigestible polysaccharides (all those other than starch) comprise the major portion of dietary fiber (Sect. 19.6). Carbohydrates also provide other attributes, including bulk, body, viscosity, stability to emulsions and foams, water-holding capacity, stability to freezing and thawing, browning (including generation of flavors and aromas), and a range of desirable textures (from crispness to smooth, soft gels), and they may lower water activity and thereby inhibit microbial growth. Basic carbohydrate structures, chemistry, and terminology can be found in references [1, 2].

Major occurrences of major carbohydrates in foods are presented by structural classes in Table 19.1. Ingested carbohydrates are almost exclusively of plant origin, with milk lactose being the major exception. Of the **monosaccharides** (sometimes called **simple sugars**), only D-glucose and D-fructose are found in other than minor amounts. Monosaccharides are the only carbohydrates that can be absorbed from the small intestine. Higher saccharides (**oligo-** and **polysaccharides**) must first be digested (i.e., hydrolyzed to monosaccharides) before absorption and utilization can occur. (Note: The Food and Agriculture Organization (FAO) and the World Health Organization (WHO) [3] recommend that carbohydrates be classified by molecular size into sugars [degree of polymerization (DP) 1–2], oligosaccharides (DP 3–9), and polysaccharides (DP >9), but carbohydrate chemists (according to international nomenclature rules) consider an oligosaccharide to be a carbohydrate composed of 2 to 10 (or 2–20) sugar (saccharide) units). Polysaccharides usually contain

from about 30 to 60,000 or more monosaccharide units. Humans can digest only sucrose, lactose, maltodextrins (maltooligosaccharides), and starch. All are digested with enzymes found in the small intestine.

At least 90 % of the carbohydrate in nature is in the form of polysaccharides. As stated above, the starch polymers are the only polysaccharides that humans can digest and use as a source of calories and carbon. All other polysaccharides are nondigestible. **Nondigestible polysaccharides** can be divided into **soluble** and **insoluble** classes. Along with lignin and other nondigestible, nonabsorbed substances, they make up **dietary fiber** (Sect. 19.6.1). As dietary fiber, they regulate normal bowel function, reduce the postprandial hyperglycemic response, and may lower serum cholesterol. However, nondigestible polysaccharides most often are added to processed foods because of the functional properties they impart. Nondigestible oligosaccharides serve as prebiotics and are, therefore, increasingly used as ingredients in functional foods and nutraceuticals. The foods in which dietary fiber components can be used, and particularly the amounts that can be incorporated, are limited because addition above a certain level usually changes the characteristics of the food product. Indeed, as already stated, they are often used as ingredients because of their ability to impart important functional properties at a low level of usage, rather than for a physiological effect.

Carbohydrate analysis is important from several perspectives. Qualitative and quantitative analyses are used to determine compositions of foods, beverages, and their ingredients. **Qualitative analysis** ensures that ingredient labels present accurate compositional information. **Quantitative analysis** ensures that added components are listed in the proper order on ingredient labels. Quantitative analysis also ensures that stated amounts of specific components of consumer interest are proper and that the caloric content can be calculated. Table 19.2 summarizes some of the methods described in this chapter and commonly used for nutrition labeling, quality assurance, or research for food ingredients and/or products. Of increasing importance are analyses to determine authenticity and origin of foods, beverages, and ingredients. Both qualitative and quantitative analysis can be used to authenticate (i.e., to detect adulteration of) food ingredients and products and for quality assurance.

The most commonly used methods of carbohydrate determination are presented here. However, methods often must be made specific to a particular food product because of the nature of the product and the presence of other constituents. Approved methods are referenced, but method approval has not kept pace with methods development; so in some cases, other methods are presented. Methods that have been in longtime use, although not giving as much or as precise information as newer methods, nevertheless may be useful for product standardization in some cases.

In general, evolution of analytical methods for low-molecular-weight carbohydrates has followed the succession: qualitative color tests, adaptation of the color test for reducing sugars based on reduction of Cu(II) to Cu(I) (Fehling test) to quantitation of reducing sugars, qualitative paper chromatography, quantitative paper chromatography, gas chromatography (GC) of derivatized sugars, qualitative and quantitative thin-layer chromatography, enzymic methods, and high-performance liquid chromatography (HPLC). Some older methods are still in use, and multiple official methods for the analysis of mono- and disaccharides in foods are currently approved by AOAC International [4]. Methods employing nuclear magnetic resonance (NMR), Fourier transform infrared (FTIR) spectroscopy (Sect. 19.7.3 and Chap. 8), near-infrared (NIR) spectroscopy (Sect. 19.7.4 and Chap. 8), immunoassays (Chap. 27), fluorescence spectroscopy (Chap. 7), capillary electrophoresis (Sect. 19.4.2.4), and mass spectrometry (MS) (Sect. 19.7.2 and Chap. 11) have been published but are not yet in general use for carbohydrate analysis. Reference [5] also may be consulted for food carbohydrate analysis.

According to the nutrition labeling regulations of the US Food and Drug Administration (FDA) [6] [21 CFR 101.9 (c)(6)(i)–(iv)], the following are details regarding carbohydrates (all declared in relation to a serving, as defined by the FDA):

1. **Total carbohydrate** content of a food must be calculated by subtraction of the sums of the crude protein, total fat, moisture, and ash in a serving from the total weight of the food (i.e., total carbohydrate is determined by difference). (Note that this calculation is not an actual measurement of carbohydrate content. Its accuracy depends on the accuracies of determinations of the other components, but this method is required by US regulations for nutrition labeling. As described in Chap. 3, Sect. 3.2.1.6, caloric content for the label can be calculated with or without taking into account the insoluble dietary content of the food.)

2. **Dietary fiber** (FDA definition given in Table 19.5) content in a serving must also be stated on the label. Declaration of contents of the subcategories **soluble fiber** and **insoluble fiber** is voluntary.

3. **Total sugars** are defined for labeling purposes as the sum of all free monosaccharides and disaccharides (such as glucose, fructose, lactose, and sucrose).

4. **Added sugars**, a required listing on the nutrition label in the 2016 updated regulations, are defined by the FDA as follows: "Added sugars are either added during the processing of foods, or are packaged as such, and include sugars (free, mono- and disaccharides), sugars from syrups and honey, and sugars from concentrated fruit or vegetable juices that are in excess of what would be expected from the same volume of 100 percent fruit or vegetable juice of the same type, except …." The statement continues by stating the exceptions of what is not included in "added sugar."

5. **Sugar alcohols**' declaration on the nutrition label is voluntary, except it is required if a claim is made on the label or in labeling about sugar alcohol or total sugars, or added sugars when sugar alcohols are present in the food. "Sugar alcohols are defined as the sum of saccharide derivative in which a hydroxyl group replaces a ketone or aldehyde group and whose use in the food is listed by FDA (e.g., mannitol or xylitol) or in generally recognized as safe (e.g., sorbitol)." If only one sugar alcohol is present in the food (e.g., xylitol), the specific name of the sugar alcohol may be used in place of "sugar alcohol."

See Table 19.3 [7] for the total carbohydrate, sugar, and total dietary fiber (TDF) content of selected foods.

19.2 SAMPLE PREPARATION

19.2.1 General Information

Sample preparation is related to the specific carbohydrate being determined (because carbohydrates have such a wide range of solubilities) and the specific raw material, ingredient, or food product being analyzed. However, some generalities can be presented (Fig. 19.1).

19.1 table   Occurrences of some major carbohydrates in foods

Carbohydrate	Source	Constituent(s)
Monosaccharides[a]		
D-glucose (dextrose)	Naturally occurring in honey, fruits, and fruit juices. Added as a component of glucose syrups and high-fructose syrups. Produced during processing by hydrolysis (inversion) of sucrose	
D-fructose	Naturally occurring in honey, fruits, and fruit juices. Added as a component of high-fructose syrups. Produced during processing by hydrolysis (inversion) of sucrose	
Sugar alcohol[a]		
Sorbitol (D-glucitol)	Added to food products, primarily as a humectant	
Disaccharides[a]		
Sucrose	Widely distributed in fruit and vegetable tissues and juices in varying amounts. Added to food and beverage products	D-fructose D-glucose
Lactose	In milk and products derived from milk	D-galactose D-glucose
Maltose	In malt. In varying amounts in various glucose syrups and maltodextrins	D-glucose
Higher oligosaccharides[a]		
Maltooligosaccharides	Maltodextrins. In varying amounts in various glucose syrups	D-glucose
Raffinose	Small amounts in beans	D-glucose D-fructose D-galactose
Stachyose	Small amounts in beans	D-glucose D-fructose D-galactose
Polysaccharides		
Starch[b]	Widespread in cereal grains and tubers. Added to processed foods	D-glucose
Food gums/hydrocolloids[c] Algins	Added as ingredients	[d]
Carboxymethylcelluloses		
Carrageenans		
Gellan		
Guar gum		
Gum arabic		
Hydroxypropylmethyl celluloses		
Inulin		
Konjac glucomannan		
Locust bean gum		
Methylcelluloses		
Pectins		
Xanthan		
Cell-wall polysaccharides[c] Pectin (native)	Naturally occurring	
Cellulose		
Hemicelluloses		
Beta-glucan		

[a]For analysis, see Sect. 19.4.2
[b]For analysis, see Sect. 19.5.1
[c]For analysis, see Sect. 19.5.2
[d]For compositions, characteristics, and applications, see [1, 2]

19.2 table   Summary of carbohydrate analysis methods

To determine:	Description of method	Method measures	Advantages/disadvantages
Total carbohydrate for nutrition label	Grams of carbohydrate per serving is calculated as total grams of serving minus (g of moisture + g of protein + g of lipid + g of ash)	Total carbohydrate by difference	Not an actual measurement of carbohydrate. Depends on accuracy of determinations of other components, but this method is required by US regulations. Should not be used to calculate caloric content because carbohydrate components of dietary fiber, such as cellulose, provide essentially no calories
Total carbohydrate[a]	Spectrophotometric, phenol-sulfuric acid	Measures all carbohydrates except sugar alcohols	Solution must be clear, i.e., carbohydrates must be soluble, so the method may not measure all carbohydrates. Method requires a standard curve made with the same exact mixture of carbohydrates in the same ratio that occurs in the sample
Total reducing sugars[a]	Spectrophotometric. Somogyi-Nelson and related methods	Primarily used to measure glucose/dextrose, maltose, and other low-molecular-weight oligosaccharides in glucose syrups	If carbohydrates are not already in solution, requires extraction. Solution must be clear. Fructose gives some response
Glucose/dextrose[a]	(1) Enzymic assay using GOPOD reagent (spectrophotometric)[b] (2) HPLC	Both methods specifically determine the amount of glucose in a mixture of sugars	Extraction required. Enzymic method can be automated
Fructose[a]	HPLC	Specific determination of fructose in a mixture of sugars	Extraction required
Sucrose[a]	(1) Enzymic assay using GOPOD reagent (spectrophotometric)[b] (2) HPLC	Both methods specifically determine the amount of sucrose in a mixture of sugars	Extraction required. Enzymic method can be automated. For enzymic method, solution must be clear
Lactose[a]	(1) Enzymic assay (spectrophotometric)[b] (2) HPLC. Enzymic assay employs galactose oxidase[b]	Both methods specifically determine the amount of lactose in a mixture of sugars, except that the enzymic method will also measure free galactose (uncommon) or other galactose-containing substances	Extraction required. Enzymic method can be automated
Concentrations of syrups[a]	(1) Measurement of specific gravity using a hydrometer (2) Refractive index using a refractometer	Concentration of solids in the solution	Solutions must be pure and of a single substance
Starch[a]	(1) Hydrolysis of starch to glucose using a mixture of amylases and determination of glucose using the (GOPOD) reagent (2) Hydrolysis of starch with glucoamylase and determination of glucose with glucose oxidase[b]	Specific for starch, including modified starches	Does not measure resistant starch[c]. Sample must be free of glucose or a correction made for it. Amylases must be purified to remove any interfering activities

19.2
table   (continued)

To determine:	Description of method	Method measures	Advantages/disadvantages
Pectin[a]	Spectrophotometric. m-Hydroxydiphenyl-sulfuric acid method	Uronic acids	Extraction may be required. Standard curve is required. Other hydrocolloids containing uronic acids will interfere
Dietary fiber	Gravimetric (residue after removal of lipids, digestible starch, and protein and subtraction of ash content)	"Total dietary fiber"	Does not include low-molecular-weight soluble dietary fiber. Is not a measure of the physiological efficacy of the particular dietary fiber. Soluble and insoluble dietary fiber can be determined by specific methods for them

[a]For research and quality assurance. Not required for nutrition labeling
[b]The YSI Life Sciences instrument method uses an electrode that detects hydrogen peroxide
[c]A Megazyme kit includes resistant starch in the measurement

For most foods, the first step is drying, which also can be used to determine moisture content. For other than beverages, drying is done by placing a weighed amount of material in a vacuum oven and drying to constant weight at 55 °C and 1 mm Hg pressure. Then, the material is ground to a fine powder, and lipids are extracted using 19:1 vol/vol chloroform-methanol in a Soxhlet extractor (Chap. 17). (Note: Chloroform-methanol forms an azeotrope boiling at 54 °C with a mole ratio of 0.642:0.358 or a vol/vol ratio of 3.5:1 in the vapor.) Without prior extraction of lipids and other lipid-soluble substances, extraction of water-soluble carbohydrates will likely be incomplete.

Other sample preparation schemes may be required. For example, the AOAC International [4] method for presweetened, ready-to-eat breakfast cereals calls for removal of fats by extraction with petroleum ether (hexane) rather than the method described above and extraction of sugars with 50 % ethanol (AOAC Method 982.14), rather than the method described below.

19.2.2 Extraction and Cleanup for Determination of Mono- and Oligosaccharides

Food raw materials and products and some ingredients are complex, heterogeneous, biological materials. Thus, it is quite likely that they may contain substances that interfere with measurement of the mono- and oligosaccharides present, especially if a spectrophotometric method is used. Interference may arise either from compounds that absorb light of the same wavelength used for the carbohydrate analysis or from insoluble, colloidal material that scatters light, since light scattering will be measured as absorbance. Also, the aldehydic or keto group of the sugar can react with other components, especially amino groups of proteins, a reaction (the **nonenzymatic browning (Maillard) reaction**) that simultaneously produces color and destroys the sugar. Even if chromatographic methods, such as HPLC (Sect. 19.4.2.1), are used for analysis, the mono- and oligosaccharides must be isolated from the other components of the food before chromatography.

For determination of any mono- (glucose, fructose), di- (sucrose, lactose, maltose), tri- (raffinose), tetra- (stachyose), or other oligosaccharides (e.g., **maltodextrins**) present, the dried, lipid-free sample (Sect. 19.2.1) is extracted with **hot 80 % ethanol** in the presence of precipitated calcium carbonate to neutralize any acidity (AOAC Method 922.02, 925.05) (Fig. 19.1). Some of the higher oligosaccharides from added maltodextrins or **fructooligosaccharides** (FOS) may also be extracted. Most carbohydrates (especially those of low molecular weight) are soluble in hot 80 % ethanol. However, much of the composition of a food (other than water) is in the form of polymers, and almost all polysaccharides and proteins are insoluble in hot 80 % ethanol. Thus, this extraction is rather specific. Extraction is done by a batch process. Refluxing 1 h, cooling, and filtering is standard practice. (A Soxhlet apparatus cannot be used because aqueous ethanol undergoes azeotropic distillation as 95 % ethanol.) Extraction should be done at least twice to check for and ensure completeness of extraction. If the foodstuff or food product is particularly acidic, for example, a low-pH fruit, neutralization may be necessary to prevent hydrolysis of sucrose, which is particularly acid labile; thus, precipitated calcium carbonate is routinely added.

The 80 % ethanol extract will contain components other than carbohydrates, in particular ash, pigments, organic acids, and perhaps free amino acids and low-molecular-weight peptides. Because the mono- and oligosaccharides are neutral and the contaminants are

19.3 table

Total carbohydrate, sugars, and total dietary fiber contents of selected foods

Food	Approximate percent total carbohydrate (wet weight basis)	Sugars, %	TDF[a], %
Cereals, bread, and pasta			
Bagels, plain	53	NR[b]	2.3
Bread, white	49	5.7	2.7
Macaroni, dry, enriched	75	NR	4.3
Macaroni, cooked	27	1.1	4.3
Ready-to-eat cereals			
Cheerios	73	4.4	9.4
Corn flakes	84	9.5	3.3
Dairy products			
Ice cream, soft serve, chocolate	22	21	0.7
Ice cream, light chocolate	23	20	0
Milk, reduced fat (2%)	4.8	5.1	0
Milk, chocolate, commercial	1.0	9.5	0.8
Yogurt, plain, low fat (12 g protein/8 oz)	7.0	7.0	0
Fruits and vegetables			
Apples, raw, with skin	14	10	2.4
Apples, raw, without skin	13	10	1.3
Applesauce, canned, sweetened	20	NR	12
Broccoli, raw	6.6	1.7	2.6
Broccoli, cooked	8.8	3.5	3.0
Carrots, raw	9.6	4.7	2.8
Carrots, cooked	8.2	3.5	3.0
Grapes, raw	18	16	0.9
Potatoes, raw, with skin	18	1.0	2.1
Tomato, juice	4.1	2.9	0.8
Meat, poultry, and fish			
Bologna, beef	4.3	2.1	0
Chicken, broilers or fryers, skinless, boneless breast	0	0	0
Chicken, breast, tenders, cooked	18	0	NR
Fish sticks, frozen, prepared	22	1.7	1.5
Other			
Beer, regular	3.6	0	0
Beer, light	1.6	0.1	0
Carbonated beverage, cola, regular	10	9.9	0
Carbonated beverage, cola, diet	0.1	0	0
Cream of mushroom soup	6.8	0.4	0.7
Honey	82	81	0.2
Salad dressing, ranch	5.9	4.7	0
Salad dressing, reduced fat	21	3.8	1.1
Salad dressing, fat free	27	5.6	0.1

From US Department of Agriculture, Agricultural Research Service (2016) USDA National Nutrient Database for Standard Reference. Release 28. Nutrient Data Laboratory Home Page, http://ndb.nal.usda.gov
[a]Total dietary fiber
[b]Not reported

charged, the contaminants can be removed by **ion-exchange** techniques. Because reducing sugars can be adsorbed onto and be isomerized by strong anion-exchange resins in the hydroxide (OH^-) form, a weak anion-exchange resin in the carbonate (CO_3^{2-}) or hydrogencarbonate (HCO_3^-) form is used. [**Reducing sugars** are those mono- and oligosaccharides that contain a free carbonyl (aldehydic or keto) group and, therefore, can act as reducing agents; see Sect. 19.4.1]. Because sucrose and sucrose-related oligosaccharides

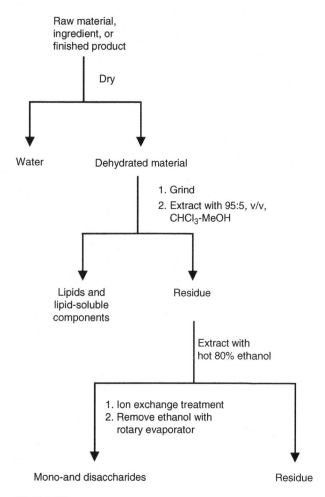

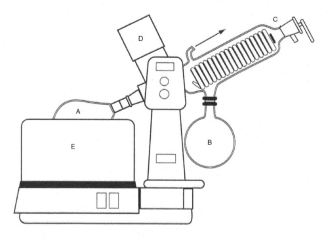

19.2 figure   Diagram of a rotary evaporator. The solution to be concentrated is placed in the round-bottom flask (*A*) in a water bath (*E*) at a controlled temperature. The system is evacuated by means of a water aspirator or vacuum pump; connecting tubing is attached at the arrow. Flask A turns (generally slowly). Because of the reduced pressure, evaporation is relatively rapid from a thin film on the inside walls of flask A produced by its rotation, the large surface area, and the elevated temperature. *C* is a condenser. *D* is the motor. Condensate collects in flask B. The stopcock at the top of the condenser is for releasing the vacuum

 19.1 figure   Flow diagram for sample preparation and extraction of mono- and disaccharides

are very susceptible to acid-catalyzed hydrolysis, the anion-exchange resin should be used before the cation-exchange resin. However, because the anion-exchange resin is in a carbonate or hydrogencarbonate form, the cation-exchange resin (in the H⁺ form) cannot be used in a column because of CO_2 generation. Mixed-bed columns are not recommended for the same reason. AOAC Method 931.02 reads basically as follows for cleanup of ethanol extracts: Place a 50-mL aliquot of the ethanol extract in a 250-mL Erlenmeyer flask. Add 3 g of anion-exchange resin (OH⁻ form) and 2 g of cation-exchange resin (H⁺ form). Let stand 2 h with occasional swirling.

The aqueous ethanol is removed from the extract under reduced pressure using a **rotary evaporator** (Fig. 19.2) and a temperature of 45–50 °C. The residue is dissolved in a known, measured amount of water. Filtration should not be required, but should be used if necessary. Some methods employ a final passage

through a hydrophobic column such as a Sep-Pak C18 cartridge (Waters Associates, Milford, MA) as a final cleanup step to remove any residual lipids, proteins, and/or pigments. However, this should not be necessary if the lipids and lipid-soluble components were properly removed prior to extraction.

19.3 TOTAL CARBOHYDRATE: PHENOL-SULFURIC ACID METHOD

19.3.1 Principle and Characteristics

Carbohydrates are destroyed by strong acids and/or high temperatures. Under these conditions, a series of complex reactions takes place, beginning with a simple dehydration reaction as shown in Eq. 19.1.

$$\underset{\substack{|\\OH}}{-\overset{\overset{H}{|}}{C}}-\underset{\substack{|\\OH}}{\overset{\overset{H}{|}}{C}}- \xrightarrow[H_2O]{} -\underset{\substack{|\\OH}}{\overset{\overset{H}{|}}{C}}=C- \rightleftharpoons -CH_2-\underset{\substack{\|\\O}}{C}- \quad (19.1)$$

Continued heating in the presence of acid produces various furan derivatives that react with various phenolic compounds, such as phenol, resorcinol,

orcinol, α-naphthol, and napthoresorcinol, and with various aromatic amines, such as aniline and *o*-toluidine, to produce colored compounds [1]. The most often used condensation is with phenol itself [8]. This widely used method is simple, rapid, sensitive, accurate, and specific for carbohydrates. The reagents are inexpensive, readily available, and stable. Virtually all mono-, oligo-, and polysaccharides can be determined using the phenol-sulfuric acid method. (Oligo- and polysaccharides react because they undergo hydrolysis in the hot, strong acid, releasing monosaccharides.) However, neither sorbitol nor any other **sugar alcohol (alditol, polyol, polyhydroxyalcohol)** gives a positive test. A stable color is produced, and results are reproducible. Under proper conditions, the phenol-sulfuric method is accurate to ±2 %.

The reaction is not stoichiometric, and the extent of color formation is, in part, a function of the structure of the sugar. Therefore, a standard (calibration) curve (Chaps. 4 and 6) must be used. Ideally, the standard curve will be prepared using mixtures of the same sugars present in the same ratio as they are found in the sample being analyzed. If this is not possible (e.g., if a pure preparation of the sugar being measured is not available or if more than one sugar is present either as free sugars in unknown proportions or as constituent units of oligo- or polysaccharides or mixtures of them), D-glucose is used to prepare the standard curve. In these cases, accuracy is a function of conformity of the standard curve made with D-glucose to the curve that would be produced from the exact mixture of carbohydrates being determined. In any analysis requiring a standard curve, the concentrations used to construct the standard curve must cover a range that begins below the lowest carbohydrate concentration of the samples and extends above the highest concentration of the samples and must be within the limits reported for sensitivity of the method.

19.3.2 Outline of Procedure

1. A clear, aqueous solution of carbohydrate(s) is transferred using a pipette into a small tube.
2. An aqueous solution of phenol is added, and the contents are mixed.
3. Concentrated sulfuric acid is added rapidly to the tube so that the stream produces good mixing. The tube is agitated. (Adding the sulfuric acid to the water produces considerable heat.) A yellow-orange color results.
4. Absorbance is measured at 490 nm.
5. The average absorbance of blanks (sample alone and reagents alone) is subtracted, and the amount of sugar is determined by reference to a standard curve.

19.4 MONO- AND OLIGOSACCHARIDES

19.4.1 Total Reducing Sugar

19.4.1.1 Somogyi-Nelson Method

19.4.1.1.1 Principle
Oxidation is a loss of electrons; reduction is a gain of electrons. Reducing sugars are those sugars that have an aldehydic group (aldoses) which acts as a reducing agent by giving up electrons to an oxidizing agent, which is reduced by receiving the electrons. Oxidation of the aldehydic group produces a carboxylic acid group (Eq. 19.2).

$$R-\overset{\overset{O}{\parallel}}{C}-H + 2Cu(OH)_2 + NaOH \longrightarrow R-\overset{\overset{O}{\parallel}}{C}-O^-Na^+ + Cu_2O + 3H_2O \quad (19.2)$$

The most often used method to determine amounts of reducing sugars is the Somogyi-Nelson method [9], also at times referred to as the Nelson-Somogyi method. This and other reducing sugar methods (Sect. 19.4.1.2) can be used in combination with enzymic methods (Sect.19.4.2.3) for determination of oligo- and polysaccharides. In enzymic methods, specific hydrolases are used to convert the oligo- or polysaccharide into its constituent monosaccharide or repeating oligosaccharide units, whose total amounts are measured using a reducing sugar method. The Somogyi-Nelson method is based on **reduction of Cu(II) ions to Cu(I) ions by reducing sugars**. The reaction is conducted in an alkaline solution containing tartrate or citrate ions, which function to keep the copper ions in solution. The Cu(I) ions then reduce an arsenomolybdate complex prepared by reacting ammonium molybdate $[(NH_4)_6Mo_7O_{24}]$ and sodium arsenate (Na_2HAsO_7) in sulfuric acid. Reduction of the arsenomolybdate complex produces an intense, stable blue color that is measured spectrophotometrically. The extent of color formation is, in part, a function of the sugars present, so the method must be used with a standard curve (Chaps. 4 and 6) of the sugars in the same ratio as they are found in the sample being analyzed or D-glucose (if a constituent sugar is available or the constituents are unknown).

19.4.1.1.2 Outline of Procedure
1. A solution of copper(II) sulfate and an alkaline buffer solution are added by pipettes to a solution of reducing sugars(s) (prepared as per the sample preparation procedure described in Sect. 19.2.2) and a water blank.
2. The resulting solution is heated in a boiling water bath.
3. A reagent prepared by mixing solutions of acidic ammonium molybdate and sodium arsenate is added.

4. After mixing, dilution, and remixing, absorbance is measured at 520 nm.

5. After subtracting the absorbance of the reagent blank, the A_{520} is converted into glucose equivalents using a standard plot of µg of glucose vs. absorbance.

19.4.1.2 Other Methods

An alternative method to the Somogyi-Nelson method that is also based on **reduction of Cu(II) ions in alkaline solution to Cu(I) ions** is the **Lane-Eynon method** (AOAC Method 945.66). To perform the Lane-Eynon method, the solution to be analyzed is added (using a burette) to a flask containing a boiling, alkaline solution of cupric sulfate of known concentration containing potassium sodium tartrate and methylene blue. Any reducing sugars in the solution being analyzed reduce Cu(II) ions to Cu(I) ions. When all the Cu(II) ions have been reduced, further addition of reducing sugars results in the indicator losing its blue color. The volume of the solution required to reach the end point is used to calculate the amount of reducing sugar present in the sample. Again, because this reaction is not stoichiometric and because each reducing sugar reacts differently, this method must be used with a standard curve (Chap. 4).

A keto group cannot be oxidized to a carboxylic acid group, and thus ketoses are not reducing sugars. However, under the alkaline conditions employed, ketoses are isomerized to aldoses [1] and, therefore, are measured as reducing sugars. Because the conversion is not 100%, the response is less with ketoses, so a standard curve (Chap. 4) made with D-fructose as one of the sugars in the mixture of sugars should be used if it is present.

The **dinitrosalicylic acid method** [10] will measure reducing sugars naturally occurring in foods or released by enzymes, but is not much used. In this reaction, 3,5-dinitrosalicylate is reduced to the reddish monoamine derivative.

19.4.2 Specific Analysis of Mono- and Oligosaccharides

Determination of contents of specific mono- and oligosaccharides is often done chromatographically. The most commonly used method is high-performance liquid chromatography (HPLC) (Sect. 19.4.2.1). The method is simple and determines sugar alcohols in addition to reducing sugars. Gas chromatography (GC) (Sect. 19.4.2.2) is more time consuming in that it requires derivatization of the sugars. In GC, sugars are determined as their reduced forms (sugar alcohols).

19.4.2.1 High-Performance Liquid Chromatography

19.4.2.1.1 Overview

HPLC (Chap. 13) is the method of choice for analysis of mono- and oligosaccharides in foods and can be used for analysis of polysaccharides after hydrolysis (Sect. 19.5.2.2) to their constituent monosaccharides. HPLC gives both qualitative analysis (identification of the carbohydrate) and, with peak integration, quantitative analysis. HPLC analyses are rapid, can tolerate a wide range of sample concentrations, and provide a high degree of precision and accuracy. HPLC requires micron-filter filtration prior to injection. Complex mixtures of mono- and oligosaccharides can be analyzed. The basic principles and important parameters of HPLC (the stationary phase, the mobile phase, and the detector) are presented and discussed in Chap. 13. Some details related to carbohydrate analysis are discussed here. The use of HPLC to determine food and other carbohydrates has been reviewed many times; some recent reviews can be found in references [11–20]. Specific details of methods of analysis of specific food ingredients or products should be obtained from the literature. Sample preparation for HPLC analysis is discussed in references [17, 19]. Various column packing materials and detectors have been used. Only the most often employed column packing material and detector are presented here. The reviews should be consulted for other columns and detectors.

19.4.2.1.2 Anion-Exchange HPLC

Separation of carbohydrates by HPLC is most often done using anion-exchange (AE-HPLC) columns. Carbohydrates have pKa values in the pH range 12–14 and are, therefore, very weak acids. In a solution of high pH, some carbohydrate hydroxyl groups are ionized, allowing sugars to be separated on columns of anion-exchange resins. Special resins have been developed for this purpose. The general elution sequence is sugar alcohols (alditols), monosaccharides, disaccharides, and higher oligosaccharides.

19.4.2.1.3 Pulsed Electrochemical Detection

The **pulsed electrochemical detector** (ECD) [formerly called a **pulsed amperometric detector** (PAD)], which relies on oxidation of carbohydrate hydroxyl and aldehydic groups, is universally used with anion-exchange chromatography [11–19, 21–25]. ECD requires a high pH. Both gradient and graded elutions can be used. The solvents employed are simple and inexpensive (sodium hydroxide solutions, with or without sodium acetate). (Water may be used, but when it is, post-column addition of a sodium hydroxide solution is required.) The detector is suitable for both reducing and nonreducing monosaccharides. Lower detection limits are approximately 1.5 ng for monosaccharides and 5 ng for di-, tri-, and tetrasaccharides. ECD responses vary from sugar to sugar and change continuously, so standards must be run and response factors calculated at least daily.

AE-HPLC coupled to an ECD has been used to examine the complex oligosaccharide patterns of

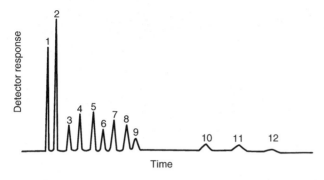

19.3 figure High-performance liquid chromatogram of some common monosaccharides, disaccharides, alditols, and the trisaccharide raffinose at equal wt/vol concentrations separated by anion-exchange chromatography and detected by pulsed electrochemical detection. Peak *1* glycerol, *2* erythritol, *3* L-rhamnose, *4* D-glucitol (sorbitol), *5* mannitol, *6* L-arabinose, *7* D-glucose, *8* D-galactose, *9* lactose, *10* sucrose, *11* raffinose, *12* maltose

19.4 figure Modification of D-galactose in preparation for gas chromatography

many food components and products. The method has the advantage of being applicable to baseline separation within each class of carbohydrates (Fig. 19.3) and of providing separation of homologous series of oligosaccharides into their components [21, 26]. A newer detector with increasing use is the evaporative light-scattering detector [27].

19.4.2.1.4 Other HPLC Methods

There are other HPLC methods for carbohydrate analysis. Among them is what is called normal-phase chromatography, which is also rather widely used. In normal-phase chromatography, the stationary phase is polar, and elution is effected by employing a mobile phase of increasing polarity. Silica gel that has been derivatized with one or more of several reagents to incorporate amino groups is used. These so-called amine-bonded stationary phases are generally used with acetonitrile-water as the eluent. The elution order is monosaccharides and sugar alcohols, disaccharides, and higher oligosaccharides [28]. Amine-bonded silica gel columns have been used successfully to analyze the low-molecular-weight carbohydrate contents of foods [29].

19.4.2.2 Gas Chromatography

19.4.2.2.1 Overview

GC (i.e., **gas-liquid chromatography,** GLC) (Chap. 14), like HPLC, provides both qualitative and quantitative analysis of carbohydrates [17, 30]. For GC, sugars must be converted into volatile derivatives. The most commonly used derivatives are the alditol peracetates [31, 32]. These derivatives are prepared as illustrated in Fig. 19.4 for D-galactose. A **flame ionization**

detector (FID) is the detector most often used for peracetylated carbohydrate derivatives, but mass spectrometers are increasingly used as detectors. A mass spectrometric (MS) detector lowers detection limits, and an MS/MS detector lowers them even more [33]. A technique known as gas chromatography-combustion-isotope ratio mass spectrometry has been applied to determine the origins and adulterations of food and food ingredients [34].

The most serious problem with GC for carbohydrate analysis is that two preparation steps are involved: reduction of aldehydic groups to primary hydroxyl groups and conversion of the reduced sugars (alditol) into volatile peracetate esters; and of course, for the analysis to be successful, each of these steps must be 100% complete. The basic principles and important parameters of GC (the stationary phase, temperature programming, and detection) are presented and discussed in Chap. 14.

19.4.2.2.2 Neutral Sugars: Outline of Procedure [31]

1. **Reduction to Alditols.** Neutral sugars from the 80% ethanol extract (Sect. 19.2.2) or from hydrolysis of a polysaccharide (Sect. 19.5.2.2) are reduced at 40 °C with an excess of sodium or potassium borohydride dissolved in dilute ammonium hydroxide solution. After reaction, glacial acetic acid is added to destroy excess borohydride. The acidified solution is evaporated to dryness. A potential problem exists: If fructose is present, either as a naturally occurring sugar, from the hydrolysis of inulin, or as an additive [from a high-fructose syrup (HFS), invert sugar, or honey], it will be reduced to a mixture of D-glucitol (sorbitol) and D-mannitol (Fig. 19.5).

19.5 figure Reduction of D-fructose to a mixture of alditols

2. **Acetylation of Alditols.** Acetic anhydride and a catalyst are added to a dry mixture of alditols. After 10 min at room temperature, water and dichloromethane are added. After mixing, the dichloromethane layer is washed with water and evaporated to dryness. The residue of alditol peracetates is dissolved in a polar organic solvent (usually acetone) for chromatography.

3. **GC of Alditol Peracetates.** Alditol peracetates may be chromatographed isothermally and identified by their retention times relative to that of inositol hexaacetate, inositol being added as an internal standard prior to acetylation. It is essential to run standards of the alditol peracetates of the sugars being determined with inositol hexaacetate as an internal standard to determine elution times and relative responses.

19.4.2.3 Enzymic Methods

19.4.2.3.1 Overview

Enzymic methods (Chap. 26) generally have great specificity for the carbohydrate being determined, do not require high purity of the sample being analyzed, have very low detection limits, do not require expensive equipment, and are easily automated [35, 36]. However, the methods are spectrophotometric and thereby require clear solutions, so extraction and cleanup is required (Sect. 19.4.2.3.2).

The method of choice for the determination of starch employs a combination of enzymes in sequential **enzyme-catalyzed reactions** and is specific for starch, as long as purified enzyme preparations are used (Sect. 19.5.1.1). Other enzymic methods for the determination of carbohydrates have been developed (Table 19.4). They are often, but not always, specific for the substance being measured. Kits for several enzymic methods have been developed and marketed. The kits contain specific enzymes, other required reagents,

19.4 table Selected enzymic methods of carbohydrate analysis

Carbohydrate	Reference	Kit form[a]
Monosaccharides		
Pentoses		
L-arabinose	[35, 36]	
D-xylose	[35, 36]	
Hexoses		
D-fructose	[35, 36]	x
D-galactose	[35, 36]	x
D-galacturonic acid	[35]	
D-glucose		
Using glucose oxidase	[36], Sect. 19.4.2.3.3	x
Using glucose dehydrogenase	[35, 36]	
Using glucokinase (hexokinase)	[35, 36]	x
D-mannose	[35, 36]	
Monosaccharide derivatives		
D-gluconate/D-glucono-δ-lactone	[35, 36]	x
D-glucitol/sorbitol	[35, 36]	x
D-mannitol	[35, 36]	
Xylitol	[35, 36]	x
Oligosaccharides		
Lactose	[35, 36]	x
Maltose	[35, 36]	x
Sucrose	[35, 36]	x
Raffinose, stachyose, verbascose	[35, 36]	x
Polysaccharides		
Amylose, amylopectin (contents and ratio)		x
Cellulose	[35, 36]	
Galactomannans (guar and locust bean gums)	[35]	
β-Glucan (mixed-linkage)	[35]	x
Glycogen	[35, 36]	
Hemicellulose	[35, 36]	
Inulin	[35, 36]	x
Pectin/poly (D-galacturonic acid)	[35, 36]	
Starch	Sect. 19.5.1.1" [35, 36]	x

[a]Available in kit form from companies such as R-Biopharm, Megazyme, and Sigma-Aldrich

buffer salts, and detailed instructions that must be followed because enzyme concentration, substrate concentration, concentration of other required reagents, pH, and temperature all affect reaction rates and results. A good description of a method will point out any interferences from other substances and other limitations.

19.4.2.3.2 Sample Preparation

It is sometimes recommended that the **Carrez treatment** [37], which breaks emulsions, precipitates proteins, and absorbs some colors, be applied to food products prior to determination of carbohydrates by enzymic and other methods. The Carrez treatment involves addition of a solution of potassium ferrocyanide ($K_4[Fe(CN)_6]$, potassium hexacyanoferrate), followed by addition of a solution of zinc sulfate ($ZnSO_4$), followed by addition of a solution of sodium hydroxide. The suspension is filtered, and the clear filtrate is used directly in enzyme-catalyzed assays. Carrez solutions are commercially available.

19.4.2.3.3 Enzymic Determination of D-Glucose (**Dextrose**)

Glucose oxidase oxidizes D-glucose quantitatively to D-glucono-1,5-lactone (glucono-delta-lactone), the other product being hydrogen peroxide (Fig. 19.6). To measure the amount of D-glucose present, **peroxidase** and a colorless compound (a leuco dye) that can be oxidized to a colored compound are added. In the reaction catalyzed by peroxidase, the leuco dye is oxidized to a colored compound, which is measured spectrophotometrically. Various dyes are used in commercial kits. The method using this combination of two enzymes and an oxidizable colorless compound is known as the **glucose oxidase/peroxidase/dye (GOPOD) method**.

YSI Life Sciences makes a commercial instrument that utilizes the glucose oxidase enzyme immobilized between two membranes and an electrode that measures the released hydrogen peroxide. Results are obtained in less than 60 s. Using other immobilized enzymes, the instrument will determine amounts of D-galactose (using galactose oxidase), sucrose (using invertase and glucose oxidase), lactose (using galactose oxidase), and starch (using glucoamylase/amyloglucosidase and glucose oxidase).

Another **coupled-enzyme** enzymic method, also available in kit form, but less often used, involves reaction of D-glucose with ATP in the presence of hexokinase to form glucose 6-phosphate (G6P) + ADP. The reaction mixture also contains glucose 6-phosphate dehydrogenase (G6PDH) and $NADP^+$. G6PDH catalyzes the oxidation of G6P to D-gluconate 6-phosphate and reduction of $NADP^+$ to NADPH, so the amount of NADPH formed is equivalent to the amount of D-glucose that was present. The amount of NADPH formed is determined by measuring its absorbance at 340 nm (a wavelength which $NADP^+$ does not absorb).

19.4.2.4 Capillary Electrophoresis

Capillary zone electrophoresis (see also Chap. 24, Sect. 24.2.5.3) has also been used to separate and measure carbohydrates, but because carbohydrates lack chromophores, pre-column derivatization and detection with a UV or fluorescence detector are required [11, 16, 38–43].

19.5 POLYSACCHARIDES

19.5.1 Starch

Starch is second only to water as the most abundant component of food. Starch is found in all parts of plants (leaves, stems, roots, tubers, seeds). A variety of commercial starches are available worldwide as food additives. These include corn (maize), waxy maize, high-amylose corn (amylomaize), wheat, rice, potato, tapioca (cassava), yellow pea, sago, and arrowroot starches. In addition, starch is the main component of wheat, rye, barley, oat, rice, corn, mung bean, and pea flours and certain roots and tubers such as potatoes, sweet potatoes, and yams.

19.5.1.1 Total Starch

19.5.1.1.1 Principle and Procedure

The only reliable method for determination of total starch is based on complete conversion of the starch into D-glucose by purified enzymes specific for starch and determination of the D-glucose released by an enzyme specific for it (Sect. 19.4.2.3.3). In the proce-

19.6

figure Coupled enzyme-catalyzed reactions for the determination of D-glucose

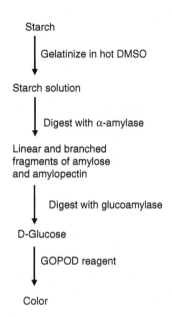

Starch

↓ Gelatinize in hot DMSO

Starch solution

↓ Digest with α-amylase

Linear and branched fragments of amylose and amylopectin

↓ Digest with glucoamylase

D-Glucose

↓ GOPOD reagent

Color

 19.7 figure Flow diagram for determination of total starch

dure outlined in Fig. 19.7, **α-amylase** catalyzes hydrolysis of unbranched segments of 1,4-linked α-D-glucopyranosyl units, forming primarily maltooligosaccharides composed of three to six units. **Debranching enzymes** (both **pullulanase** and **isoamylase** are used) catalyze hydrolysis of the 1,6 linkages that constitute the branch points of starch polysaccharide molecules and molecules of maltooligosaccharides derived from starch polysaccharide molecules and thereby produce short linear molecules. **Glucoamylase** (**amyloglucosidase**) acts at the nonreducing ends of starch oligo- and polysaccharide chains and releases D-glucose, one unit at a time; it will catalyze hydrolysis of both 1,4 and 1,6 α-D-glucopyranosyl linkages. In the assay, glucose and a starch low in protein and lipid content (such as potato starch) are used as standards after determining their moisture contents.

19.5.1.1.2 Potential Problems
Starch-hydrolyzing enzymes (amylases) must be purified to eliminate any other enzymic activity that would release D-glucose (e.g., cellulases, invertase, sucrase, β-glucanase) and catalase, which would destroy the hydrogen peroxide on which the enzymic determination of D-glucose depends. The former contamination would give false high values and the latter, false low values. Even with purified enzymes, problems can be encountered. The method may not be quantitative for a high-amylose or another starch at least partially resistant to enzyme-catalyzed hydrolysis after cooking. **Resistant starch** (RS), by definition, is composed of starch and starch-degradation products that escape

digestion in the small intestine [44]. Four types of starch are generally considered to be resistant to digestion in the small intestine or so slowly digested that they pass essentially intact through the small intestine, i.e., without conversion into D-glucose:

1. Starch that is physically inaccessible to amylases because it is trapped within a food matrix, even though it is gelatinized (**RS1**)
2. Starch that resists enzyme-catalyzed hydrolysis because it is uncooked, i.e., not gelatinized (**RS2**)
3. Retrograded starch (i.e., starch polymers that have recrystallized after gelatinization of the granule (**RS3**); cooled cooked potatoes and other starchy foods, such as pasta, contain resistant starch)
4. Starch that has been modified structurally in such a way as to make it less susceptible to digestion (**RS4**)

RS is at best only partially converted into D-glucose by this method; rather, most of it is included in the analysis for dietary fiber (Sect. 19.6). Methods for the specific determination of RS have been reviewed [45].

One method of starch analysis (AOAC Method 969.39, AACCI Method 76-13.01) overcomes these problems. In it, the starch is dispersed in dimethyl sulfoxide (DMSO) and then is converted quantitatively to D-glucose by first treating the solution with a **thermostable α-amylase** to effect depolymerization and solubilization of the starch (Fig. 19.7). Addition of **glucoamylase** (**amyloglucosidase**) then effects complete conversion of the fragments produced by the action of α-amylase into D-glucose. D-Glucose is determined with a GOPOD reagent (Sect. 19.4.2.3.3). The method determines total starch. It does not reveal the botanical source of the starch or whether it is a native starch or a modified food starch. The botanical source of the starch may be determined by microscopic examination (Chap. 32, Sect. 32.2.2.4) of the material being analyzed before it is cooked.

19.5.2 Non-starch Polysaccharides (Hydrocolloids/Food Gums)

19.5.2.1 Overview
A starch or starches may occur naturally in a fruit or vegetable tissue, in addition to being used as an ingredient in a food product, either as isolated starch or as a component of a flour. Other polysaccharides are almost always added as ingredients, although there are exceptions. These added polysaccharides, along with the protein gelatin, comprise a group of ingredients known as **hydrocolloids** or **food gums**. The non-starch polysaccharides used as additives in food products are

obtained from land plants, seaweeds (marine algae), and microorganisms and by chemical derivatization of cellulose. Their use is widespread and extensive.

Analytical methods are required for these polysaccharides to enable both suppliers and food processors to determine the purity of a hydrocolloid product, to ensure that label declarations of processors are correct, and to confirm that hydrocolloids have not been added to standardized products in which they are not allowed. In addition, it may be desirable to determine such things as the **β-glucan** content of an oat or barley flour or a breakfast cereal for a label claim of a specific dietary fiber or the **arabinoxylan** content of a wheat flour to set processing parameters for bakery products.

Determination of polysaccharides classified as hydrocolloids is problematic because polysaccharides have a variety of chemical structures, solubilities, and molecular weights. Unlike proteins and nucleic acids, the structures of molecules of a single polysaccharide preparation from a plant or microorganism, with very few exceptions, vary from molecule to molecule. In addition, the average structure can vary with the source and the environmental conditions under which the plant or microorganism was grown. Some polysaccharides are neutral; some are anionic. Some contain ether, ester, and/or cyclic acetal groups in addition to sugar units, either naturally or as a result of chemical modification. Some are soluble only in hot water; some are soluble only in room-temperature or colder water; some are soluble in both hot and cold water, and some require aqueous solutions of acids, bases, or metal ion-chelating compounds to release them from plant tissues. And all polysaccharide preparations are composed of a mixture of molecules with a range of molecular weights; so while all molecules of food and beverage components such as D-glucose, D-fructose, maltose, and sucrose have identical structures and molecular weights, each molecule of a polysaccharide preparation probably differs from all other molecules in that sample in structure and/or molecular weight. This structural diversity complicates determination of both the types and amounts of polysaccharides in a food product [46]. As a result, no single approach that will determine all hydrocolloids, either qualitatively or quantitatively, is available. Other potential problems are that hydrocolloids are usually added to foods in very small amounts (0.01–1%), and blends of hydrocolloids are often used to extend functionalities.

Current methods depend on extraction of the hydrocolloid(s), followed by deproteinization of the extract and precipitation of the hydrocolloids by addition of ethanol, acetone, or 2-propanol (isopropanol), but low-molecular-weight (low-viscosity-grade) hydrocolloids may not be precipitated. Because blends of hydrocolloids are often used in food products, fractionation may be required. Fractionation, like extraction and precipitation,

invariably results in some loss of material. Most often, an isolated polysaccharide is identified by identifying and quantitating its constituent sugars after acid-catalyzed hydrolysis. However, sugars are released from polysaccharides by hydrolysis at different rates and are destroyed by hot acids at different rates, so even the exact monosaccharide composition of a polysaccharide preparation may be difficult to determine and may not be achieved. A hydrolytic enzyme specific for the polysaccharide being determined (if available) is useful if the specific hydrocolloid present is known. Analytical strategies for and problems associated with the determination of hydrocolloids in foods have been reviewed [46, 47].

19.5.2.2 Hydrocolloid Content Determination

Most schemes for analysis of food products for food gums are targeted to a specific group of food products, as it is difficult, perhaps impossible, to develop a universal scheme. A general scheme for isolation and purification of non-starch, water-soluble polysaccharides is presented in Fig. 19.8 [48]. Letters in the parentheses below refer to the same letters in Fig. 19.8:

(a) It is usually difficult to extract polysaccharides quantitatively when fats, oils, waxes, and proteins are present. Therefore, lipid-soluble substances are removed first. Before this can be effected, the sample must be dried. Freeze-drying is recommended. If the dried material contains lumps, it must be ground to a fine powder. A known weight of dry sample is placed in a Soxhlet apparatus, and the lipid-soluble substances are removed with 19:1 vol/vol chloroform-methanol. (See note in Sect. 19.2.1) (n-Hexane has also been used.) Solvent is removed from the sample by air-drying in a hood and then by placing the sample in a desiccator, which is then evacuated.

(b) Although not in the published scheme, soluble sugars, other low-molecular-weight compounds, and ash can be removed at this point using hot 80% ethanol as described in Sect. 19.2.2.

(c) Protein is removed by enzyme-catalyzed hydrolysis. The cited procedure [48] uses papain as the protease. However, bacterial alkaline proteases are recommended to prevent the action of contaminating carbohydrases – all of which have acidic pH optima. (Essentially all commercial enzyme preparations, especially those from bacteria or fungi, have carbohydrase activities in addition to proteolytic activity.) In this procedure, proteins are denatured for easier digestion by dispersion of the sample in sodium acetate buffer containing sodium chloride and heating the mixture.

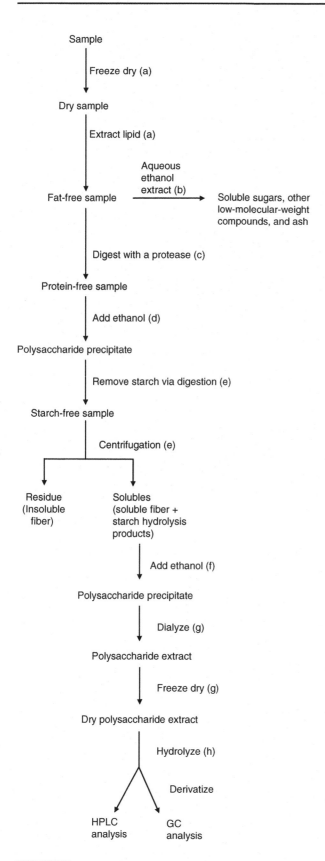

(d) Any solubilized polysaccharides are precipitated by addition of sodium chloride to the cooled dispersion, followed by addition of four volumes of absolute ethanol. The mixture is centrifuged.

(e) The pellet is suspended in acetate buffer. To this suspension is added a freshly prepared solution of glucoamylase in the same buffer. This suspension is then incubated. Just as in the analysis of starch, highly purified enzyme must be used to minimize hydrolytic breakdown of other polysaccharides. [This step may be omitted in future analyses of the same product if no glucose is found in the centrifugate (supernatant) from step f, indicating that no starch is present.] Centrifugation after removal of starch polysaccharides isolates insoluble dietary fiber (IDF) (Sect. 19.6).

(f) Solubilized polysaccharides are reprecipitated by addition of sodium chloride and four volumes of absolute ethanol to the cooled dispersion. The mixture is centrifuged. The precipitate (pellet) of water-soluble polysaccharides (often added hydrocolloids) is soluble dietary fiber (SDF) (Sect. 19.6).

(g) The pellet is suspended in deionized water, transferred to dialysis tubing, and dialyzed against frequent changes of sodium azide solution (sodium azide used to prevent microbial growth). Finally, dialysis against deionized water is done to remove the sodium azide. The retentate is recovered from the dialysis tubing and freeze-dried.

(h) Polysaccharide identification relies on hydrolysis to constituent monosaccharides and identification of these sugars. For hydrolysis, polysaccharide material is added to a Teflon-lined, screw-capped vial. Trifluoroacetic acid solution is added (usually 2 M), and the vial is tightly capped and heated (usually for 1–2 h at 121 °C) (49). After cooling, the contents are evaporated to dryness in a hood using a stream of air or nitrogen. Sugars are determined by HPLC (Sect. 19.4.2.1) or GC (Sect. 19.4.2.2). If GC is used, inositol is added as an internal standard. Qualitative and quantitative analysis of the polysaccharides present can be determined by sugar analysis. For example, guaran, the polysaccharide component of guar gum, yields D-mannose and D-galactose in an approximate molar ratio of 1.00:0.56.

The described acid-catalyzed hydrolysis procedure does not release uronic acids quantitatively. The presence of **uronic acids** can be indicated by the **_m_-hydroxydiphenyl (3-phenylphenol) assay** [50–52]. This and similar methods are based on the same principle as the phenol-sulfuric acid assay (Sect. 19.3), i.e., condensation of dehydration products with a phenolic

figure   Flow diagram for isolation and analysis of polysaccharides

compound to produce colored compounds that can be measured quantitatively by means of spectrophotometry. If present, specific uronic acids can be identified by a specific GC procedure for them.

19.5.2.3 *Pectin*

19.5.2.3.1 *Nature of Pectin*

Even though **pectin** is a very important food hydrocolloid, no official method for its determination has been established. What few methods have been published basically involve its precipitation (by addition of ethanol) from jams, jellies, etc. in which it is the only polysaccharide present.

Even the definition of pectin is somewhat ambiguous. What may be called "pectin" in a native fruit or vegetable is a complex mixture of polysaccharides whose structures depend on the source, including the stage of development (i.e., the degree of ripeness) of the particular fruit or vegetable. Generally, much of this native material can be described as a main chain of α-D-galactopyranosyluronic acid units (some (usually many) of which are in the methyl ester form) interrupted by L-rhamnopyranosyl units (1, 2). Many of the rhamnosyl units have polysaccharide (arabinan, galactan, or arabinogalactan) chains attached to them. Other sugars, such as D-apiose, also are present. In the manufacture of commercial pectin, much of the neutral sugar part is removed. Commercial pectin is, therefore, primarily poly(α-D-galacturonic acid methyl ester) with various degrees of esterification, and sometimes amidation.

Enzyme action during development/ripening or during processing can partially de-esterify and/or depolymerize native pectin. These enzyme-catalyzed reactions are important determinants of the stability of fruit juices, tomato sauce, tomato paste, apple butter, etc. in which some of the texture is supplied by pectin.

19.5.2.3.2 *Pectin Content Determination*

Conditions for extraction of pectin from various plant tissues, followed by its precipitation, have been studied for many years and continue to be investigated – not for analytical interests, but because of pectin's commercial value. Several different extractants, extraction conditions, precipitants, and precipitation conditions (with variations of each) have been investigated and optimized. Product characteristics vary with the source material, isolation conditions, and in the case of fruit sources, the degree of ripeness. The constant, but not the sole, constituent of pectins is **D-galacturonic acid** as the principal component (often at least 80%). However, glycosidic linkages of uronic acids are difficult to hydrolyze, so methods involving acid-catalyzed hydrolysis are generally not applicable. Therefore, pectins are often determined using the

m-**hydroxydiphenyl method** [50–52], following isolation of crude pectin. For reviews of methods for the determination of pectin, see references [53, 54]. A procedure involving methanolysis followed by reverse-phase HPLC has been published [55]. Interference by other hydrocolloids in the determination of pectin has been reviewed [56].

19.6 DIETARY FIBER

19.6.1 Definition

Because labeling of food products for dietary fiber content is required, an official analytical method(s) for its determination is required. The first step in adopting a method must be agreement on what constitutes dietary fiber. Then, there must be a method that measures what is included in the definition. However, no single definition of **dietary fiber** has been agreed upon by all domestic and international organizations that need, or would like, a definition of it [57]. One hurdle is that dietary fiber not only needs a chemical definition for development of an assay procedure, but a measurement of it is important because of its positive physiological effects, which differ in effectiveness from source to source.

Four official definitions are given in Table 19.5. AACC International was the first organization to develop an official definition for dietary fiber [58, 59]. According to the AACC International definition, and that of others, dietary fiber is essentially the sum of the nondigestible components of a food ingredient or product. No polysaccharide other than starch is digested in the human small intestine; so all polysaccharides other than nonresistant starch are included in this definition. Of the oligosaccharides, only sucrose, lactose, and those derived from starch (maltodextrins) are digested. **Analogous carbohydrates** are defined as those carbohydrate-based food ingredients that are nondigestible and nonabsorbable but are not natural plant components. Wax (suberin and cutin) is included within associated substances. The definition also includes some of the health benefits known to be associated with ingestion of dietary fiber. Since adoption of the AACC International definition of dietary fiber, modified versions have been adopted by both governmental and nongovernmental organizations around the world, such as the US Institute of Medicine [60], the FDA, and the Codex Alimentarius Commission [61] (Table 19.5). Definitions adopted by some other regulatory bodies, commissions, and organizations can be found in reference [62].

Formulating a definition acceptable to all is difficult because dietary fiber from different sources is often composed of different mixtures of nondigest-

19.5 table Definitions of dietary fiber

AACC International	"Dietary fiber is the edible parts of plants or analogous carbohydrates that are resistant to digestion and absorption in the human small intestine with complete or partial fermentation in the large intestine. Dietary fiber includes polysaccharides, oligosaccharides, lignin and associated plant substances. Dietary fiber promotes beneficial physiological effects, such as, laxation, and/or blood cholesterol attenuation, and/or blood glucose attenuation" [58, 59]
US Institute of Medicine	"Dietary fiber consists of nondigestible carbohydrates and lignin that are intrinsic and intact in plants. Functional fiber consists of isolated, nondigestible carbohydrates that have beneficial physiologic effects in humans. Total fiber is the sum of dietary fiber and functional fiber" [60]
Codex Alimentarius Commission	"Dietary fiber denotes carbohydrate polymers with 10 or more monomeric units, which are not hydrolysed by the endogenous enzymes in the small intestine of humans and belonging to the following categories: edible carbohydrate polymers naturally occurring in the food consumed; carbohydrate polymers obtained from food raw materials by physical, enzymatic or chemical means; synthetic carbohydrate polymers" [61] (Note: Published definition contains footnotes)
Food and Drug Administration	"Dietary fiber is defined as non-digestible soluble and insoluble carbohydates (with 3 or more monomeric units), and lignin that are instrinsic and intact in plants; isolated or synthetic non-digestible carbohydrates (with 3 or more monomeric units) determined by FDA to have physiological effects that are beneficial to human health" [6]

19.6 table Components of dietary fiber

Insoluble dietary fiber
  Cellulose, including microcrystalline
    and powdered cellulose added as ingredients
  Lignin
  Insoluble hemicelluloses and soluble hemicelluloses
    entrapped in the lignocellulosic matrix
  Resistant starch
Soluble dietary fiber
  Soluble hemicelluloses not entrapped
    in the lignocellulosic matrix
  Native pectin (most)
  Hydrocolloids (most)
  Nondigestible oligosaccharides, such as
    those derived from inulin (FOS)

ible and nonabsorbable carbohydrates and other substances with different effects on human physiology. However, there is general agreement that dietary fiber consists of oligo- and polysaccharides, lignin, and other substances not acted on by the digestive enzymes in the human stomach or small intestine and that most, but not all, dietary fiber is plant cell-wall material (cellulose, hemicelluloses, lignin) and thus is composed primarily of polysaccharide molecules. Because only the amylose and amylopectin molecules in cooked starch are digestible, all other polysaccharides are components of dietary fiber.

Some are components of **insoluble fiber**; some make up **soluble fiber**. Major components of soluble and insoluble dietary fiber are listed in Table 19.6. **Total dietary fiber** (TDF) is the sum of insoluble and soluble dietary fiber.

19.6.2 Methods

19.6.2.1 *Overview*

Measurement of insoluble fiber is important not only in its own right, but also for calculating the caloric content of a food. According to nutrition labeling regulations, one method allowed to calculate calories involves subtracting the amount of insoluble dietary fiber from the value for total carbohydrate, before calculating the calories based on protein, fat, and carbohydrate content (approximately 4, 9, and 4 cal per gram, respectively) (Chap. 3). This scheme ignores the fact that soluble fiber, like insoluble fiber, is also essentially noncaloric. [Fiber components can contribute some calories via absorption of products of fermentation (mostly short-chain fatty acids) from the colon.]

The food component that may be most problematic in fiber analysis is **starch**. In any method for determination of dietary fiber, it is essential that all digestible starch be removed, since incomplete removal of digestible starch increases the residue weight and inflates the estimate of fiber. All fiber methods include a heating step (e.g., 95–100 °C for 35 min) to **gelatinize starch granules** and make them susceptible to hydrolysis, which is effected using a thermostable α-amylase and glucoamylase (Sect. 19.5.1.1.1). Resistant starch granules and/or molecules (Sect. 19.5.1.1.2) remain essentially intact and, therefore, are components of dietary fiber, but some nondigestible products made from starch may not be determined as dietary fiber by approved methods.

Nondigestible oligosaccharides such as those derived from inulin (a fructan), certain maltodextrins designed to be nondigestible, and partially hydrolyzed guar gum [63] may be problematic in an analytical sense since they are in the soluble portion that is not precipitated with 78 % ethanol. They should be measured in AOAC Method 2009.01 (AACCI Method 32-45.01) and AOAC Method 2011.25 (AACCI Method 32-50.01). Methods for determination of fructans in certain products have been reviewed [64].

It is essential that either all digestible materials be removed from the sample so that only nondigestible components remain or that a correction be applied for any remaining digestible contaminants. **Lipids** are removed easily from the sample with organic solvents (Sect. 19.2) and generally do not pose analytical problems. **Protein** and **salts/minerals** that are not removed from the sample during the solubilization steps should be corrected for by Kjeldahl nitrogen analysis (Chap. 18) and by ashing (Chap. 16) on other samples of the fiber residue.

The scheme presented in Fig. 19.9 is designed to separate non-starch, water-soluble polysaccharides from other components for quantitative and/or qualitative analysis. The residue from the filtration step (e) is insoluble fiber, and those components precipitated from the supernatant with alcohol [step (f)] constitute soluble fiber.

(Defat sample if >10% lipid)

↓ (a) Treat mixture with an alkaline protease

↓ (b) Treat hot mixture with a thermostable α-amylase

↓ (c) Treat mixture with glucoamylase → **Insoluble dietary fiber (IDF)** (e) or **Soluble dietary fiber (SDF)** (f)

(d) Add 4 volumes of 95% ethanol

Collect residue + precipitate by filtration. Wash with 78% ethanol, 95% ethanol and acetone. Air dry; then oven dry. Weigh

Subtract weights of protein and ash determined on duplicate samples

Total dietary fiber (TDF)

 Flow diagram of AOAC Method 991.43 (AACCI Method 32-07.01)

19.6.2.2 Sample Preparation
Measures of fiber are most consistent when the samples are low in fat (less than 10% lipid), dry, and finely ground. If necessary, the sample is ground to pass through a 0.3–0.5-mm mesh screen. If the sample contains more than 10% lipid, the lipid is removed by extraction with 25 parts (vol/wt) of petroleum ether or hexane in an ultrasonic water bath. The mixture is then centrifuged, and the organic solvent is decanted. This extraction is repeated. The sample is air-dried to remove the organic solvent. It may then be dried overnight in a vacuum oven at 70 °C if a measure of lipid and moisture content is required. Loss of weight due to fat and moisture removal is recorded, and the necessary correction is made in the calculation of the percentage dietary fiber value determined in the analysis.

If samples contain large amounts of soluble sugars (mono-, di-, and trisaccharides), the samples should be extracted three times with 80% aqueous ethanol in an ultrasonic water bath at room temperature for 15 min. The supernatant liquid is discarded, and the residue is dried at 40 °C.

A variety of methods have been developed and used at different times for different products. *AOAC Official Methods of Analysis* [4] and *AACC International Approved Methods* [65] are listed in Table 19.7. It is obvious from the list that methods are generally specific for the type of fiber or the fiber component desired to be measured. Several methods are available in kit form.

19.6.2.3 Enzymic-Gravimetric Method
Dietary fiber is most often determined **gravimetrically** after digestible carbohydrates, lipids, and proteins are selectively solubilized by chemical reagents or removed by enzyme-catalyzed hydrolysis. After such treatments, non-solubilized and/or undigested materials are collected by filtration, and the fiber residue is recovered, dried, and weighed.

19.6.2.3.1 Total, Soluble, and Insoluble Dietary Fiber
AOAC Method 991.43 (AACCI Method 32-07.01) determines total, insoluble, and soluble dietary fiber in cereal products, fruits, vegetables, processed foods, and processed food ingredients. It contains the features of a general analytical method for dietary fiber:

1. **Principle.** Starch and protein are removed from a sample by treating the sample sequentially with a thermostable α-amylase, a protease, and glucoamylase. The insoluble residue is recovered and washed [**insoluble dietary fiber** (IDF)]. Ethanol is added to the soluble portion to precipitate soluble polysaccharides [**soluble dietary fiber** (SDF)]. To obtain **total dietary fiber**

19.7

table

Some official methods of analysis for dietary fiber in food ingredients and products

AOAC Official Method No. (4)	AACCI Approved Method No. (65)	Description of method and measured substance
994.13	32-25.01	TDF determined as neutral sugar and uronic acid monomer units plus Klason lignin by a gas chromatographic-spectrophotometric-gravimetric method
993.21		Nonenzymic-gravimetric method for TDF applicable to determination of >10% TDF in foods and food products with <2% starch
985.29	32-05.01	Enzymic-gravimetric method for TDF in cereal grains and cereal grain-based products
	32-06.01	A rapid gravimetric method for TDF
991.42, 992.16		Enzymic-gravimetric method for insoluble dietary fiber in vegetables, fruits, and cereal grains
993.19		Enzymic-gravimetric method for soluble dietary fiber
991.43	32-07.01	Enzymic-gravimetric method for total, soluble, and insoluble dietary fiber in grain and cereal products, processed foods, fruits, and vegetables
2002.02	32-40.01	Enzymic method for RS2 and RS3 in food products and plant materials
	32-21.01	Enzymic-gravimetric method for insoluble and soluble dietary fiber in oats and oat products
	32-32.01	Enzymic-spectrophotometric method for total fructan (inulin and FOS) in foods applicable to FOS
999.03		Enzymic-spectrophotometric method for fructan (inulin) in foods (not applicable to FOS)
997.08	32-31.01	AE-HPLC method for fructan in foods and food products applicable to the determination of added inulin in processed foods
2000.11	32-28.02	AE-HPLC method for polydextrose in foods
	32-22.01	Enzymic method for β-glucan in oat fractions and unsweetened oat cereals
	32-23.01	Rapid enzymic procedure for β-glucan content of barley and oats
2001.03	32-41.01	Enzymic-gravimetric and HPLC method for dietary fiber containing added resistant maltodextrin
2001.02	32-33.01	HPLC method for trans-galactooligosaccharides (TGOS) applicable to added TGOS in selected food products
2009.01	32-45.01	Determines high-molecular-weight and low-molecular-weight soluble dietary fiber by an enzymic-gravimetric method and HPLC
2011.25	32-50.01	Determines insoluble, soluble, and total dietary fiber according to the Codex Alimentarius definition by an enzymic-gravimetric method and HPLC

(TDF), alcohol is added after digestion with glucoamylase, and the IDF and SDF fractions are collected together, dried, weighed, and ashed.

2. **Outline of Procedure**. A flow diagram outlining the general procedure for the method is given in Fig. 19.9. Letters in the parentheses refer to the same letters in Fig. 19.9:

(a) To samples devoid of significant lipid solvent-soluble substances is added a basic buffer containing an alkaline protease.

(b) After protein digestion, the pH is adjusted to the acid side, a thermostable α-amylase is added, and the mixture is heated at 95–100 °C to gelatinize any starch so that the α-amylase can break it down. After cooling the mixture to 60 °C, an alkaline protease is added, and the mixture is incubated at 60 °C to break down the protein.

(c) Glucoamylase is added, and the mixture is incubated at 60 °C to complete the digestion of any starch.

The next few steps differ depending on whether total, insoluble, or soluble fiber is to be determined.

(d) To determine TDF, four volumes of 95% ethanol are added (to give an ethanol concentration of 78%). The mixture is vacuum filtered through a pre-weighed, fritted crucible containing pre-washed Celite (a siliceous filter aid). The residue in the crucibles is dewatered by washing with 78% ethanol, 95% ethanol, and acetone in that order. Then, the crucibles are air-dried (to remove all acetone), oven-dried at 103 °C, and weighed. Since some protein and salts/minerals are combined with plant cell-wall constituents, protein (Kjeldahl procedure (Chap. 18)) and ash (muffle furnace procedure (Chap. 15)) are determined on separate duplicate samples, and fiber values are corrected for them. If resistant starch in the fiber residue is to be determined separately, it can be determined using AOAC Method 2002.02 (AACC International Method 32-40.01).

(e) To determine IDF, the mixture obtained from step (c) is vacuum filtered through a tared, fritted crucible containing pre-washed Celite. The residue retained by the filter is washed with water, then dewatered by washing in order with 78% ethanol, 95% ethanol, and acetone. The crucibles are air-dried (to remove all acetone), oven-dried at 103 °C, and weighed. SDF is in the filtrate.

(f) To determine soluble dietary fiber, four volumes of 95% ethanol (to give an ethanol concentration of 78%) are added to the filtrate and water washes from step (e) at 60 °C to precipitate soluble fiber. The precipitate is collected by vacuum filtration through tared, fritted crucibles containing pre-washed Celite. The residues are dewatered by washing with

78% ethanol, 95% ethanol, and acetone in that order. Then, the crucibles are air-dried (to remove all acetone), oven-dried at 103 °C, and weighed.

Duplicate reagent blanks must be run through the entire procedure for each type of fiber determination. Table 19.8 is a form used to calculate fiber percentages. Using the equations shown, percent dietary fiber is expressed on a dry weight basis if the sample weights are for dried samples. Representative values obtained using this method are given in Table 19.9.

Note: Neither this method for TDF nor that for SDF determines SDF that does not precipitate in 78% aqueous ethanol, including some or most inulin, polydextrose, digestion-resistant maltodextrins, and partially hydrolyzed guar gum and all fructo-, arabinoxylo-, xylo-, and galactooligosaccharides. AOAC Method 2009.01 (AACCI Method 32-45.01) incorporates the deionization and HPLC procedures of AOAC Method 2002.02 (AACCI Method 32-40.01) to quantitate these lower-molecular-weight, digestion-resistant materials in the filtrate so that all SDF is measured.

19.8
table   Dietary fiber data sheet[a]

	SAMPLE				BLANK			
	Insoluble Fiber		Soluble Fiber		Insoluble Fiber		Soluble Fiber	
Sample wt (mg)	m_1	m_2						
Crucible + Celite wt (mg)								
Crucible + Celite + residue wt (mg)								
Residue wt (mg)	R_1	R_2	R_1	R_2	R_1	R_2	R_1	R_2
Protein (mg) **P**								
Crucible + Celite + ash wt (mg)								
Ash wt (mg) **A**								
Blank wt (mg) **B**[b]								
Fiber (%)[c]								

Adapted with permission from *The Journal of AOAC International*, 1988, 71:1019. Copyright 1988 by AOAC International

$$^{b}\text{Blank}\,(\text{mg}) = \frac{R_1 + R_2}{2} - P - A$$

$$^{c}\text{Fiber}\,(\%) = \frac{\dfrac{R_1 + R_2}{2} - P - A - B}{\dfrac{m_1 + m_2}{2}} \times 100$$

| 19.9 table | Total, soluble, and insoluble dietary fiber in foods as determined by AOAC Method 991.43 |

Food	Soluble[a]	Insoluble[a]	Total[a]
Barley	5.02	7.05	12.25
High-fiber cereal	2.78	30.52	33.73
Oat bran	7.17	9.73	16.92
Soy bran	6.90	60.53	67.14
Apricots	0.53	0.59	1.12
Prunes	5.07	4.17	9.29
Raisins	0.73	2.37	3.13
Carrots	1.10	2.81	3.93
Green beans	1.02	2.01	2.89
Parsley	0.64	2.37	2.66

Adapted from Official Methods of Analysis, 20th ed. Copyright 2016 by AOAC International
[a]Grams of fiber per 100 g of food on a fresh weight basis

19.6.2.3.2 Dietary Fiber Components as Defined by Codex Alimentarius

The most recent method (AOAC Method 2011.25; AACCI Method 32-50.01) combines aspects of previously approved methods to measure individual components of dietary fiber as defined by Codex Alimentarius. It is outlined in Fig. 19.10.

Note: SDFP is dietary fiber that is soluble in water, but insoluble in 78 % ethanol; it includes most hydrocolloids and some of such compounds as polydextrose, nondigestible maltodextrins, inulin, and partially hydrolyzed guar gum. SDFS is dietary fiber that is soluble in both water and 78 % ethanol; it includes various oligosaccharides, such as low-molecular-weight FOS and galactooligosaccharides.

19.7 PHYSICAL METHODS

19.7.1 Measurements of Sugar Concentrations in Solution

The concentration of a carbohydrate in solution can be determined by measuring the solution's specific gravity, refractive index (Chap. 6), or optical rotation. The specific gravity is the ratio of the density of a substance to the density of a reference substance (usually water) both at a specific temperature. By far the most common way to determine specific gravity is the use of a hydrometer calibrated in °Brix, which corresponds to sucrose concentrations by weight, or in Baumé modulus (°Bé) (AOAC Method 932.14). The obtained values are then converted into concentrations using tables constructed for the substance being measured in a pure solution. Measurement of specific gravity as a means of determining sugar concentration is accurate only for pure sucrose or other solutions of a single pure substance, but it can be, and is, used for obtaining approximate values for liquid products for which appropriate specific gravity tables have been constructed (Chap. 6).

When light passes from one medium to another, it changes direction, i.e., it is bent or refracted. The ratio of the sine of the angle of incidence to the sine of the angle of refraction is called the index of refraction, or refractive index (RI). RI varies with the nature and concentration of the dissolved compound, the temperature, and the wavelength of light used. By holding the nature of the compound, the temperature, and the wavelength constant, the concentration of the dissolved compound can be determined by measuring the RI (Chap. 6). To determine RI, the solution must be clear. Like determination of specific gravity, the use of RI to determine concentrations is accurate only for pure sucrose or other solutions of a single pure substance. Also like specific gravity, it is used for obtaining approximate sugar concentrations in liquid products. Refractometers that read directly in sucrose concentration units are available.

Most compounds that contain a chiral carbon atom have optical activity, i.e., they will rotate the plane of polarization of polarized light. A polarimeter measures the extent to which a compound in solution rotates a plane of polarized light. Carbohydrates have chiral carbon atoms and, thus, optical activity. Carbohydrates can rotate the plane of polarized light through an angle that depends on the nature of the compound, the temperature, the wavelength of light, and the concentration of the compound. The concentration of the compound can be determined from a value known as the specific optical rotation if all other factors are held constant and if the solution contains no other optically active compounds. Determination of specific optical rotation can be used to measure sucrose concentration (AOAC Methods 896.02, 925.46, 930.37). Determination of sucrose concentration by polarimetry requires a clear solution. Instruments are available that read in units of the International Sugar Scale. Determination of specific optical rotation before and after hydrolysis of sucrose into its constituent sugars, D-glucose and D-fructose, a process called inversion, can be used to determine sucrose in the presence of other sugars (AOAC Methods 925.47, 925.48, 926.13, 926.14).

19.7.2 Mass Spectrometry

There are many different variations of mass spectrometry (MS) (Chap. 11). Most of the techniques applied to carbohydrates are used for structural analysis. MS has been used for determination of carbohydrates, but not in a routine manner. Particularly useful is the matrix-assisted laser-desorption time-of-flight (MALDI-TOF) technique for analysis of a homologous series of

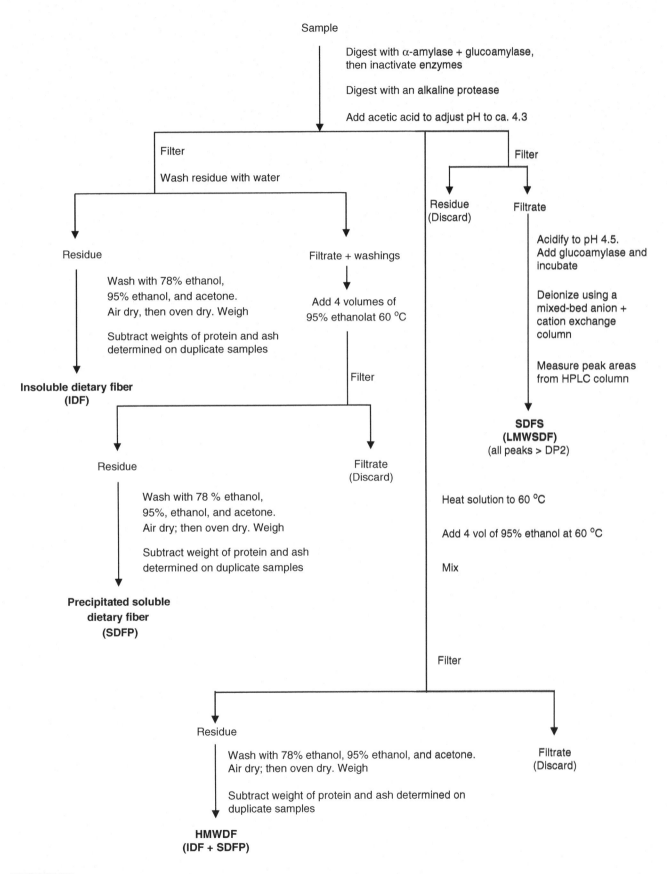

Flow diagram of AOAC Method 2011.25 (AACCI Method 32-50.01)

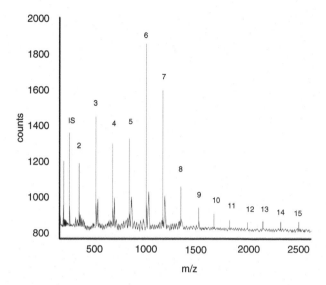

19.11 figure MALDI-TOF mass spectrum of maltooligo-saccharides produced by hydrolysis of starch. *Numbers* indicate DP. *IS* internal standard (From [26], used with permission, Copyright Springer-Verlag, 1998)

oligosaccharides (Fig. 19.11). A comparison between anion-exchange HPLC (Sect. 19.4.2.1) (the most used carbohydrate analysis technique today), capillary electrophoresis (Sect. 19.4.2.4), and MALDI-TOF MS for the analysis of maltooligosaccharides led to the conclusion that that the latter technique gave the best results [27].

19.7.3 Fourier Transform Infrared (FTIR) Spectroscopy

FTIR (Chap. 8, Sect. 8.3.1.2) methods are simple and rapid. Detection limits are greater than those required for most other methods. Spectral libraries have been compiled for several hydrocolloids, including κ-, ι-, and λ-carrageenans [66–70], pectin [66, 68, 71], galactomannans [66, 68], and cellulose derivatives [72].

19.7.4 Near-Infrared (NIR) Spectroscopy

NIR (Chap. 8, Sect. 8.4) spectrometry has been used to determine dietary fiber [73] and sugar [74] contents and to identify cellulose derivatives [72].

19.8 SUMMARY

For determination of low-molecular-weight carbohydrates, older colorimetric methods for total carbohydrate, various reducing sugar methods, and physical measurements have largely been replaced by chromatographic methods. Older chemical methods suffer from the fact that different sugars give different results, which makes them particularly problematic when a mixture of sugars is present. Older physical methods suffer from the fact that they work only with pure substances. However,

some older methods continue to be used for simplicity, quality assurance, and product standardization. Chromatographic methods (HPLC and GC) separate mixtures into their component sugars, identify each component by retention time, and provide a measurement of the quantity of each component. HPLC is widely used for identification and measurement of mono- and oligosaccharides. Enzymic methods are specific and sensitive, but seldom, except in the case of starch, is determination of only a single component desired.

Polysaccharides are important components of many food products. Yet, there is no universal procedure for their analysis. Generally, isolation must precede measurement. Isolation introduces errors because losses of constituents occur with both extraction, recovery, and separation techniques. Identification is done by hydrolysis to constituent monosaccharides and their determination. An exception is starch, which can be digested to glucose using specific enzymes (amylases), followed by measurement of the glucose released and, therefore, can be specifically measured.

Insoluble dietary fiber, soluble dietary fiber, and total dietary fiber are each composed primarily of non-starch polysaccharides. Methods for the determination of total dietary fiber and its components rely on removal of the digestible starch using amylases and often on removal of digestible protein with a protease, leaving a nondigestible residue.

19.9 STUDY QUESTIONS

1. Give three reasons why carbohydrate analysis is important.
2. "Proximate composition" refers to analysis for moisture, ash, fat, protein, and carbohydrate. Identify which of these components of "proximate composition" are actually required on a nutrition label. Also, explain why it is important to measure the non-required components quantitatively if one is developing a nutrition label.
3. Distinguish chemically between monosaccharides, oligosaccharides, and polysaccharides, and explain how solubility characteristics can be used in an extraction procedure to separate monosaccharides and oligosaccharides from polysaccharides.
4. Discuss why 80 % ethanol (final concentration) is used to extract mono- and oligosaccharides, rather than using water. What is the principle involved?
5. What are the principles behind total carbohydrate determination using the phenol-sulfuric acid method? Why is a standard curve employed? Why is a reagent blank used? What are limitations of the method?

6. Define reducing sugar. Classify each of the following as a reducing or nonreducing carbohydrate: D-glucose, D-fructose (Conditions must be described. Why?), sorbitol, sucrose, maltose, raffinose, maltotriose, cellulose, amylopectin, and κ-carrageenan.

7. What is the principle behind determination of total reducing sugars using the Somogyi-Nelson and similar methods?

8. The Somogyi-Nelson and Lane-Eynon methods can be used to measure reducing sugars. Explain the similarities and differences of these methods with regard to the principles involved and the procedures used.

9. Describe the principle behind AE-HPLC separation of carbohydrates.

10. Describe the general procedure for preparation of sugars for GC. What is required for this method to be successful?

11. Why has HPLC largely replaced GC for analysis of carbohydrates?

12. What are the advantages of enzymic methods? What are the limitations (potential problems)?

13. Describe the principles behind the enzymic determination of starch. What are the advantages of this method? What are the potential problems?

14. What is the physiological definition and the chemical nature of resistant starch? What types of foods have relatively high levels of resistant starch?

15. Briefly describe a method that could be used for each of the following:
    (a) To prevent hydrolysis of sucrose when sugars are extracted from fruits via a hot alcohol extraction
    (b) To remove proteins from solution for an enzymic determination of carbohydrates
    (c) To measure total carbohydrate
    (d) To measure total reducing sugars
    (e) To measure glucose enzymically
    (f) To measure simultaneously the concentrations of individual free sugars

16. Describe the principle behind each step in Fig. 19.9. What is the reason for each step?

17. Describe the principles behind separation and analysis of cellulose, water-soluble gums, and starch.

18. Using pectin as an example, explain why the quantitative analysis of hydrocolloids is so difficult.

19. What is a general definition of dietary fiber? Why is the definition of dietary fiber important?

How does the definition of dietary fiber affect development of an analytical procedure for it?

20. List the general component classes of dietary fiber that are usually determined for research purposes.

21. List the constituents of dietary fiber.

22. Explain how measurement of dietary fiber relates to calculating the caloric content of a food product.

23. Explain the purpose(s) of each of the steps in the AOAC Method 991.43 for total dietary fiber listed below as applied to determination of the dietary fiber content of a high-fiber snack food:
    (a) Heating sample and treating with α-amylase
    (b) Treating sample with glucoamylase
    (c) Treating sample with protease
    (d) Adding four volumes of 95 % ethanol to sample after treatment with glucoamylase and protease
    (e) After drying and weighing the filtered and washed residue, heating one duplicate final product to 525 °C in a muffle furnace and analyzing the other duplicate sample for protein

24. What are differences between AOAC Method 994.13 and AOAC Method 2011.26 with regard to what is measured?

25. Describe the principles behind and the limitations of determining sugar (sucrose) concentrations by (a) specific gravity determination, (b) refractive index measurement, and (c) polarimetry.

19.10 PRACTICE PROBLEMS

1. The following data were obtained when an extruded breakfast cereal was analyzed for total fiber by AOAC Method 991.43 (AACCI Method 32-07.01).

Sample wt, mg	1,002.8
Residue wt, mg	151.9
Protein wt, mg	13.1
Ash wt, mg	21.1
Blank wt, mg	6.1
Resistant starch, mg	35.9

What is percent total fiber (a) without and (b) with correction for resistant starch determined to the appropriate number of significant figures?

2. The following tabular data were obtained when a high-fiber cookie was analyzed for fiber con-

tent by AOAC Method 991.43 (AACCI Method 32-07.01).

	Sample			
	Insoluble		Soluble	
Sample wt, mg	1,002.1	1,005.3		
Crucible + Celite wt, mg	31,637.2	32,173.9	32,377.5	33,216.4
Crucible + Celite + residue wt, mg	31,723.5	32,271.2	32,421.6	33,255.3
Protein, mg	6.5		3.9	
Crucible + Celite + ash wt, mg		32,195.2		33,231.0

	Blank			
	Insoluble		Soluble	
Crucible + Celite wt, mg	31,563.6	32,198.7	33,019.6	31,981.2
Crucible + Celite + residue wt, mg	31,578.2	32,213.2	33,033.4	33,995.6
Protein, mg	3.2		3.3	
Crucible + Celite + ash wt, mg		32,206.8		31,989.1

What is the (a) total, (b) insoluble, and (c) soluble fiber content of the cookie determined to the appropriate number of significant figures?

Answers

1. Number of Significant figures = 2 (6.1 mg)

(a) $\dfrac{151.9 - 13.1 - 21.1 - 6.0}{1002.8} \times 100 = 11\%$

(b) $\dfrac{151.9 - 13.1 - 21.1 - 6.1 - 35.9}{1002.8} \times 100 = 7.5\%$

2. (Calculations are done a little differently than those at the bottom of Table 19.8).

Insoluble dietary fiber

Number of significant figures = 2 (6.5 mg, 3.2 mg)
Blank residue = 31,578.2 − 31,563.6 mg = 14.6 mg
$\qquad$ 32,231.2 − 32,198.7 mg = 14.5 mg
$\qquad\qquad$ Ave. = 14.6 mg
Blank ash = 32,206.8 − 32,198.7 = 8.1 mg

First sample residue = 31,723.5 − 31,637.2 mg = 86.3 mg
$\qquad$ ash = 32,195.2 − 32,173.9 mg = 21.3 mg
86.3 mg (residue weight)
−14.6 mg (blank)
− 3.3 mg (protein, 6.5 − 3.2 (blank)
−13.2 *mg* (*ash*, 21.3 − 8.1 (*blank*)
55.2 mg
(55.2 mg ÷ 1,002.1[(sample wt.]) × 100 = 5.5%

Second sample residue = 32,271.2 − 32,173.9 mg = 97.3 mg
97.3 − 14.5 − 3.3 − 13.2 = 66.3 mg
(66.3 ÷ 1005.3 mg (sample wt.)) × 100 = 6.6%
Average of 5.5% and 6.6% = 6.1%

Second sample residue = 32,271.2 − 32,173.9 mg = 97.3 mg
97.3 − 14.5 − 3.3 − 13.2 = 66.3 mg
(66.3 ÷ 1005.3 mg (sample wt.)) × 100 = 6.6%
Average of 5.5% and 6.6% = 6.1%

Soluble dietary fiber

Number of significant figures figures = 2 (3.9 mg, 3.3 mg)
Blank residue = 33,033.4 − 33,019.6 mg = 13.8 mg
$\qquad$ 33,995.6 − 31,981.2 mg = 14.4 mg
$\qquad$ Ave. = 14.1 mg
Blank ash = 31,989.1 − 31,981.2 mg = 7.9 mg

First sample residue = 32,421.6 − 32,377.5 mg = 44.1 mg
$\qquad$ ash = 33,231.0 − 33,216.4 mg = 14.6 mg
44.1 mg (residue weight)
−14.1 mg (blank)
−0.6 mg (protein 3.9 − 3.3 (blank))
−6.7 (ash, 14.6 − 7.9 (blank))
22.7 mg
(22.7 mg ÷ 1,002.1 (sample wt.)) × 100 = 2.3%

Second sample residue = 33,255.3 − 33,216.4 mg = 38.9 mg
38.9 mg (residue weight)
−14.1 mg (blank)
−0.6 mg (protein, 3.9 − 3.3 (blank))
−6.7 mg (ash, 14.6 − 7.9 (blank))
17.5 mg
(17.5 mg ÷ 1,005.3 (sample wt.)) × 100 = 1.7%
Average of 2.3% and 1.7% = 2.0%

Total dietary fiber = 6.1% (insoluble fiber)
$\qquad$ + 2.0% (soluble fiber) = 8.1%

REFERENCES

1. BeMiller JN (2007) Carbohydrate Chemistry for Food Scientists, 2nd edn. AACC International, St. Paul, MN

2. BeMiller JN, Huber K (2017) Carbohydrates (Chap 4). In: Damodaran S, Parkin KL, Fennema OR (eds), Food Chemistry, 5th edn. Marcel Dekker, New York

3. FAO/WHO expert consultation on carbohydrates in human nutrition. 14–18 April 1997, Rome

4. AOAC International (Online) Official Methods of Analysis. AOAC International, Gaithersburg, MD

5. Peris-Tortajada M (2004) Carbohydrates and starch (Chap 13). In: Nollet LML (ed) Handbook of Food Analysis, 2nd edn. Marcel Dekker, New York

6. Anon. (2016) Code of Federal Regulations, Title 21, Part 101.9 – Food Nutrition Labeling of Food. U.S. Government Printing Office, Washington, DC

7. USDA (2016) USDA Nutrient Database for Standard Reference. Release 28, 2016. http://ndb.nal.usda.gov

8. Dubois M, Gilles KA, Hamilton JK, Rebers PA, Smith F (1956) Colorimetric method for determination of sugars and related substances. Analytical Chemistry 28:350

9. Wood TM (1994) Enzymic conversion of cellulose into D-glucose. Methods in Carbohydrate Chemistry 10:219.

10. Miller G, Blum R, Glennon WEG, Burton A. (1960.) Measurement of carboxymethylcellulase activity. Analytical Biochemistry 1:127

11. El Rassi Z (ed) (1995) Carbohydrate analysis. Journal of Chromatography Library vol 58

12. Andersen R, Sørensen A (2000) Separation and determination of alditols and sugars by high-pH anion-exchange chromatography with pulsed amperometric detection. Journal of Chromatography A 897:195

13. Hanko VP, Rohrer JS (2000) Determination of carbohydrates, sugar alcohols, and glycols in cell cultures and fermentation broths using high-performance anion-exchange chromatography with pulsed amperometric detection. Analytical Biochemistry 283:192

14. Cataldi TRI, Campa C, DeBenedetto GE (2000) Carbohydrate analysis by high-performance anion-exchange chromatography with pulsed amperometric detection: The potential is still growing. Fresenius' Journal of Analytical Chemistry 368:7391

15. Zhang Y, Lee YC (2002) High-performance anion-exchange chromatography of carbohydrates on pellicular resin columns. Journal of Chromatography Library 66:207

16. Soga T (2002) Analysis of carbohydrates in food and beverages by HPLC and CE. Journal of Chromatography Library 66:483

17. Montero CM, Dodero MCR, Sánchez DAG, Barroso CG (2004) Analysis of low molecular weight carbohydrates in foods and beverages: A review. Chromatographia 59:15

18. Corradini C, Cavazza A, Bignardi C (2012) High-performance anion-exchange chromatography coupled with pulsed electrochemical detection as a powerful tool to evaluate carbohydrates of food interest: principles and applications. International Journal of Carbohydrate Chemistry 487564

19. Peris-Tortajada M (2012) HPLC determination of carbohydrates in foods (Chap 7) In: Nollet LM, Toldra F (eds) Food analysis by HPLC, 3rd edn. CRC Press, Boca Raton

20. Yan X (2014) High-performance liquid chromatography for carbohydrate analysis (Chap 3) In: Zuo Y (ed) High-performance liquid chromatography (HPLC). Nova Science Publishers, Hauppauge, NY

21. Ammeraal RN, Delgado GA, Tenbarge FL, Friedman RB (1991) High-performance anion-exchange chromatography with pulsed amperometric detection of linear and branched glucose oligosaccharides. Carbohydrate Research 215:179

22. Marioli JM (2001) Electrochemical detection of carbohydrates in HPLC. Current Topics in Electrochemistry 8:43

23. LaCourse WR. (2002) Pulsed electrochemical detection of carbohydrates at noble metal electrodes following liquid chromatographic and electrophoretic separation. Journal of Chromatography Library 66:905

24. Baldwin RP (2002) Electrochemical detection of carbohydrates at constant potential after HPLC and CE separations. Journal of Chromatography Library 66:947

25. LaCourse WR (2009) Advances in pulsed electrochemical detection for carbohydrates. Advances in Chromatography 47:247

26. Kazmaier T, Roth S, Zapp J, Harding M, Kuhn R (1998) Quantitative analysis of malto-oligosaccharides by MALDI-TOF mass spectrometry, capillary electrophoresis and anion exchange chromatography. Fresenius' Journal of Analytical Chemistry 361:473

27. Dvořáčková E, Šnóblová M, Hrdlička P (2014) Carbohydrate analysis: from sample preparation to HPLC on different stationary phases coupled with evaporative light-scattering detection. Journal of Separation Science 37:323

28. Churms SC (1995) High performance hydrophilic interaction chromatography of carbohydrates with polar sorbents (Chap 3). In: reference 11

29. Ball GFM (1990) The application of HPLC to the determination of low molecular weight sugar and polyhydric alcohols in foods: a review. Food Chem 35: 117

30. Hernandez-Hernandez O, Moreno FJ, Sanz ML (2012) Analysis of dietary sugars in beverages by gas chromatography. Food and Nutritional Components in Focus 3:208

31. Ruiz-Matute AI, Hernandez-Hernandez O, Rodriguez-Sanchez S, Sanz ML, Martinez-Castro I (2011) Derivatization of carbohydrates for GC and GC-MS analyses. Journal of Chromatography B: Analytical Technologies in the Biomedical and Life Sciences 879: 1226

32. Fox A, Morgan SL, Gilbart J (1989) Preparation of alditol acetates and their analysis by gas chromatography (GC) and mass spectrometry (MS) (Chap 5). In: Biermann CJ, McGinnis GD (ed) Analysis of carbohydrates by GLC and MS. CRC Press, Boca Raton

33. Fox A (2002) A current perspective on analysis of sugar monomers using GC-MS and GC-MS/MS. Journal of Chromatography Library 66:829

34. van Leeuwen KA, Prenzler PD, Ryan D, Camin F (2014) Gas chromatography-combustion-isotope ratio mass spectrometry for traceability and authenticity in foods and beverages. Comprehensive Reviews in Food Science and Food Safety 13:814

35. BeMiller JN (ed) (1994) Methods in carbohydrate chemistry, vol 10 Enzymic Methods. John Wiley, New York

36. Bergmeyer HU (ed) (1984) Methods of Enzymatic Analysis, vol 6 Metabolites 1: Carbohydrates, 3rd edn. Verlag Chemie, Weinheim, Germany

37. Cabálková, J., Žídková, J., Přibyla, L., and Chmelík, J. (2004) Determination of carbohydrates in juices by capillary electrophoresis, high-performance liquid chromatography, and matrix-assisted laser desorption/ionization-time of flight-mass spectrometry. *Electrophoresis* 25:487

38. Thibault P, Honda S (eds) (2003) Capillary Electrophoresis of carbohydrates, Methods in Molecular Biology, vol 213

39. Cortacero-Ramírez S, Segura-Carretero A, Cruces-Blanco C, Hernáinz-Bermúdez de Castro M, Fernandez-Gutiérrez A (2004) Analysis of carbohydrates in beverages by capillary electrophoresis with precolumn derivatization and UV detection. Food Chemistry 87:471

40. Ramírez SC, Carretero S, Blanco CC, de Castro MHB, Gutiérrez AF (2005) Indirect determination of carbohydrates in wort samples and dietetic products by capillary electrophoresis. Journal of the Science of Food and Agriculture 85:517

41. Ma S, Lau W, Keck RG, Briggs JB, Jones AJS, Moorhouse K, Nashabeh W (2005) Capillary electrophoresis of carbohydrates derivatized with fluorophoric compounds. Methods in Molecular Biology 308:397

42. Momenbeik F, Johns C, Breadmore MC, Hilder EF, Macka M, Haddad PR (2006) Sensitive determination of carbohydrates labelled with p-nitroaniline by capillary electrophoresis with photometric detection using a 406 nm light-emitting diode. Electrophoresis 27:4039

43. Volpi N (ed) (2011) Capillary electrophoresis of carbohydrates. From monosaccharides to complex polysaccharides. Humana Press, New York.

44. Asp N-G, Björck I (1992) Resistant starch. Trends in Food Science and Technology 3:111

45. Perera A, Meda V, Tyler RT (2010) Resistant starch: a review of analytical protocols for determining resistant starch and of factors affecting the resistant starch content of foods. Food Research International 43:1959

46. BeMiller JN (2016) Gums/hydrocolloids: analytical aspects (Chap 6). In: Eliasson A-C (ed) Carbohydrates in Food, 3rd edn. CRC Press, Boca Raton

47. Baird JK (1993) Analysis of gums in foods (Chap 23). In: Whistler RL, BeMiller JN (eds) Industrial Gums, 3rd edn. Academic, San Diego

48. Harris P, Morrison A, Dacombe C (1995) A practical approach to polysaccharide analysis (Chap 18). In: Stephen AM (ed) Food polysaccharides and their applications. Marcel Dekker, New York

49. Biermann CJ (1989) Hydrolysis and other cleavage of glycosidic linkages (Chap 3). In: Biermann CJ, McGinnis GD (ed) Analysis of carbohydrates by GLC and MS. CRC Press, Boca Raton

50. Fillisetti-Cozzi TMCC, Carpita NC (1991) Measurement of uronic acid without interference from neutral sugars. Analytical Biochemistry 197:157

51. Ibarz A, Pagán A, Tribaldo F, Pagán J (2006) Improvement in the measurement of spectrophotometric data in the m-hydroxydiphenyl pectin determination methods. Food Control 17:890

52. Yapo BM (2012) On the colorimetric-sulfuric acid analysis of uronic acids in food materials: potential sources of discrepancies in data and how to circumvent them. Food Analytical Methods 5:195

53. Baker RA (1997) Reassessment of some fruit and vegetable pectin levels. Journal of Food Science 62:225

54. Walter RH (1991). Analytical and graphical methods for pectin (Chap 10). In: Walter RH (ed) Chemistry and technology of pectin. Academic Press, San Diego

55. Quemener C, Marot C, Mouillet L, Da Riz V, Diris J (2000) Ouantitative analysis of hydrocolloids in food systems by methanolysis coupled to reverse HPLC. Part 2. Pectins, alginates, and xanthan. Food Hydrocolloids 14:19

56. Kyriakidis NB, Psoma E (2001) Hydrocolloid interferences in the determination of pectin by the carbazole method. Journal of AOAC International 84:1947

57. Gordon DT (2007) Dietary fiber definitions at risk. Cereal Foods World 52:112

58. McCleary BV, Prosky L (eds) (2001) Advanced dietary fibre technology, Blackwell Science, London

59. Anon. (2001) The definition of dietary fiber. Cereal Foods World 46:112

60. Institute of Medicine, Food and Nutrition Board (2005) Dietary Reference Intakes: energy, carbohydrates, fiber, fat, fatty acids, cholesterol, protein and amino acids. National Academies Press, Washington, DC

61. Codex Alimentarius Commission, Food and Agriculture Organization, World Health Organization (2009) Report of the 30th session of the Codex Committee on Nutrition and Foods for Special Dietary Uses. http://www.codex-alimentarius.net/download/report710/al132_26e.pdf

62. Jones JM (2014) CODEX-aligned dietary fiber definitions help to bridge the "Fiber gap". Nutrition Journal 13:34

63. Juneja LR, Sakanaka S, Chu D-C (2001) Physiological and technological functions of partially hydrolysed guar gum (modified galactomannans). (Chap 30). In: McCleary BV, Prosky L (eds), Advanced dietary fibre technology. Blackwell Science, Oxford, UK

64. Austin S, Bhandari S, Cho F, Christiansen S, Cruijsen H, De GR, Deborde J-L, Ellingson D, Gill B, Haselberger P, et al. (2015) Fructans in infant formula and adult/pediatric nutritional formula. Journal of AOAC International 98:1038

65. AACC International (online) Approved Methods. AACC International, St. Paul, MN

66. Copikova J, Syntsya A, Cerna M, Kaasova J, Novotna M (2001) Application of FT-IR spectroscopy in detection of food hydrocolloids in confectionery jellies and food supplements. Czech Journal of Food Science 19:51

67. Chopin T, Whalen E (1993) A new and rapid method of carrageenan identification by FT IR diffuse reflectance spectroscopy directly on dried, ground algal material. Carbohydrate Research 246:51

68. Cerna M, Barros AS, Nunes A, Rocha SM, Delgadillo I, Copikova J, Coimbra MA (2003) Use of FT-IR spectroscopy as a tool for the analysis of polysaccharide food additives. Carbohydrate Polymers 51:383

69. Tojo E, Prado J (2003) Chemical composition of carrageenan blends determined by IR spectroscopy combined with a PLS multivariate calibration method. Carbohydrate Research 338:1309

70. Prado-Fernandez J, Rodriguez-Vazquez JA, Tojo E, Andrade JM (2003) Quantitation of k-, ι-, and λ-carrageenans by mid-infrared spectroscopy and PLS regression. Analytica Chimica Acta 480:23

71. Monsoor MA, Kalapathy U, Proctor A (2001) Determination of polygalacturonic acid content in pectin extracts by diffuse reflectance Fourier transform infrared spectroscopy. Food Chemistry 74:233

72. Langkilde FW, Svantesson A (1995) Identification of celluloses with Fourier-transform (FT) mid-infrared, FT-Raman and near-infrared spectrometry. Journal of Pharmaceutical and Biomedical Analysis 13:409

73. Kays SE, Barton FE II (2002) Near-infrared analysis of soluble and insoluble dietary fiber fractions of cereal food products. Journal of Agricultural and Food Chemistry 50:3024

74. Mehrübeoglu M, Coté GL (1997) Determination of total reducing sugars in potato samples using near-infrared spectroscopy. Cereal Foods World 42:409

Vitamin Analysis

Ronald B. Pegg (✉) • *Ronald R. Eitenmiller*
Department of Food Science and Technology,
The University of Georgia,
Athens, GA 30602-2610, USA
e-mail: rpegg@uga.edu; eiten@uga.edu

S. Nielsen (ed.), *Food Analysis*, Food Science Text Series,
DOI 10.1007/978-3-319-45776-5_20, © Springer International Publishing 2017

20.1 INTRODUCTION

20.1.1 Definition and Importance

Vitamins are defined as relatively low-molecular-weight compounds which humans, and for that matter, any living organism that depends on organic matter as a source of nutrients, require in small quantities for normal metabolism. With few exceptions, humans cannot synthesize most vitamins and therefore need to obtain them from food and supplements. Insufficient levels of vitamins result in deficiency diseases [e.g., scurvy and pellagra, which are due to the lack of ascorbic acid (vitamin C) and niacin, respectively].

20.1.2 Importance of Analysis

Vitamin analysis of food and other biological samples has played a critical role in determining animal and human nutritional requirements. Furthermore, accurate food composition information is required to determine dietary intakes to assess diet adequacy and improve human nutrition worldwide. From the consumer and industry points of view, reliable assay methods are required to ensure accuracy of food labeling. This chapter provides an overview of techniques for analysis of the vitamin content of food.

20.1.3 Vitamin Units

When vitamins are expressed in units of mg or μg per tablet or food serving, it is very easy to grasp how much is present. Vitamins also can be expressed as **international units** (IU), **United States Pharmacopeia** (USP) **units**, and % **Daily Value** (DV). The IU is a unit of measurement for the amount of a substance, based on measured biological activity or effect. For details about IU and USP units of various vitamins, see the Vitamin Analysis chapter in the fourth edition of this textbook. For details about % DV for vitamins, see Chap. 3 in this textbook. When analysis of a foodstuff or dietary supplement is required for its content of vitamins, as might be the case for labeling and quality control purposes, being able to report the findings on different bases becomes important.

20.1.4 Extraction Methods

With the exception of some biological feeding studies, vitamin assays in most instances involve the extraction of a vitamin from its biological matrix prior to analysis. This generally includes one or several of the following treatments: **heat**, **acid**, **alkali**, **solvents**, and **enzymes**.

In general, extraction procedures are specific for each vitamin and designed to stabilize the vitamin. Some procedures are applicable to the combined extraction of more than one vitamin, for example, for thiamin

and riboflavin as well as some of the fat-soluble vitamins [1, 2, 12]. Typical extraction procedures are as follows:

- *Ascorbic acid*: Cold extraction with *meta*phosphoric acid/acetic acid.
- *Vitamin B1 and B2*: Boiling or autoclaving in acid plus enzyme treatment.
- *Niacin*: Autoclaving in acid (noncereal products) or alkali (cereal products).
- *Folate*: Enzyme extraction with α-amylase, protease, and γ-glutamyl hydrolase (conjugase).
- *Vitamins A, E, or D*: Organic solvent extraction, saponification, and re-extraction with organic solvents. For unstable vitamins such as these, antioxidants are routinely added to inhibit oxidation.

For fat-soluble vitamins, the initial extraction with a hydrophobic organic solvent removes all fat-soluble compounds from the food, including all of the triacylglycerols. The **saponification** step that follows (generally either overnight at room temperature or by refluxing at 70 °C, using an antioxidant that protects the sample from oxidation) renders liberated fatty acids from the triacylglycerols insoluble in an organic solvent (because they now exist as soap, typically as a potassium salt), but the fat-soluble vitamins remain soluble. These vitamins are then re-extracted with a hydrophobic organic solvent and concentrated as needed.

20.1.5 Overview of Methods

Vitamin assays can be classified as follows:

1. **Bioassays** involving humans and animals
2. **Microbiological assays** making use of protozoan organisms, bacteria, and yeast
3. **Chemical assays** that include spectrophotometric, fluorometric, chromatographic, enzymatic, immunological, and radiometric methods

In terms of ease of performance, but not necessarily with regard to accuracy and precision, the three systems follow the reverse order. It is for this reason that bioassays, on a routine basis at least, are very limited in their use to those instances in which no satisfactory alternative method is available.

The selection criteria for a particular assay depend on a number of factors, including accuracy and precision, but also economic factors and the sample load to be handled. Applicability of certain methods for a particular matrix also must be considered. It is important to bear in mind that many official methods presented by regulatory agencies are limited in their applicability to certain matrices, such as vitamin concentrates, milk, or cereals, and thus cannot be applied to other matrices without some procedural modifications, if at all.

Because of the sensitivity of certain vitamins to adverse conditions such as light, oxygen, pH, and heat, proper precautions need to be taken to prevent any deterioration throughout the analytical process, regardless of the type of assay employed. Such precautionary steps need to be followed with the test material in bioassays throughout the feeding period. They are required with microbiological and chemical methods during extraction as well as during the analytical procedure.

Just as with any type of analysis, proper sampling and subsampling as well as the preparation of a homogeneous sample are critical aspects of vitamin analysis. General guidelines regarding this matter are provided in Chap. 5 of this text.

The principles and procedures for select vitamin analysis methods are described in this chapter. Calculations for select vitamins are described with the Practice Problems. Many of the methods cited are official methods of AOAC International [2], the British Standards Institution [3–10], or the US Pharmacopeial Convention [11]. Refer to these methods and other original references cited for detailed instructions on procedures. A summary of commonly used regulatory and other methods is provided in Table 20.1. The sections below on bioassay, microbiological, and chemical methods are not comprehensive, but rather just give examples of each type of analysis.

table Commonly used regulatory methods for vitamin analysis

Vitamin	Method designation	Application	Approach
Fat-soluble vitamins			
Vitamin A (and precursors)			
Retinol	AOAC[a] 992.04	Vitamin A in milk and milk-based infant formula	HPLC[b] 340 nm
Retinol	AOAC 2001.13	Vitamin A in foods	HPLC 328 or 313 nm
All-*trans*-retinol	AOAC 2011.07	Vitamin A in infant formula and adult nutritionals	UHPLC[c] 326 nm
All-*trans*-retinol 13-*cis*-retinol	EN 1283-1 [3]	All foods	HPLC 325 nm or Fluorescence[d] $E_x \lambda = 325$ nm $E_m \lambda = 475$ nm
β-Carotene	AOAC 2005.07	β-Carotene in supplements and raw materials	HPLC 445 or 444 nm
β-Carotene	EN 1283-2 [3]	All foods	HPLC 450 nm
Vitamin D			
Cholecalciferol Ergocalciferol	AOAC 936.14	Vitamin D in foods	Bioassay
Cholecalciferol Ergocalciferol	AOAC 995.05	Vitamin D in infant formula and enteral products	HPLC 265 nm
Cholecalciferol Ergocalciferol	AOAC 2002.05	Vitamin D in selected foods	HPLC 265 nm
Cholecalciferol Ergocalciferol	AOAC 2011.11	Vitamin D in infant formula and adult/pediatric nutritional formula	UHPLC-MS/MS[e]
Cholecalciferol Ergocalciferol	AOAC 2012.11	Simultaneous determination of vitamins D_2 and D_3 in infant formula and adult/pediatric nutritional formula	ESI[f] LC-MS/MS
Cholecalciferol Ergocalciferol	EN 1282172 [5]	Vitamin D in foods	HPLC 265 nm
Vitamin E			
All-racemic α-tocopherol	AOAC 2012.10	Simultaneous determination of vitamins E and A in infant formula and adult nutritionals	NP-HPLC[g] Fluorescence $E_x \lambda = 280$ nm $E_m \lambda = 310$ nm
α-tocopherol	AOAC 2012.09	Vitamins A and E in infant formula and adult/pediatric nutritional formula	HPLC Fluorescence $E_x \lambda = 295$ nm $E_m \lambda = 330$ nm

(continued)

20.1
table   (continued)

Vitamin	Method designation	Application	Approach
R,R,R – tocopherols	EN 12822 [6]	Vitamin E in foods	HPLC Fluorescence $E_x \lambda = 295$ nm $E_m \lambda = 330$ nm
Vitamin K			
Phylloquinone	AOAC 999.15	Vitamin K in milk and infant formulas	HPLC postcolumn reduction, Fluorescence $E_x \lambda = 243$ nm $E_m \lambda = 430$ nm
Phytonadione (K_1)	AOAC 2015.09	Trans-vitamin K1 in infant, pediatric, and adult nutritionals	NP-HPLC postcolumn reduction, Fluorescence $E_x \lambda = 245$ nm $E_m \lambda = 440$ nm
Phylloquinone	EN 14148 [7]	Vitamin K in foods	HPLC postcolumn reduction, Fluorescence $E_x \lambda = 243$ nm $E_m \lambda = 430$ nm
Water-soluble vitamins			
Ascorbic acid (vitamin C)			
Ascorbic acid	AOAC 967.21	Vitamin C in juices and vitamin preparations	2,6-dichloroindophenol titration
Ascorbic acid	AOAC 967.22	Vitamin C in vitamin preparations	Fluorescence $E_x \lambda = 350$ nm $E_m \lambda = 430$ nm
Ascorbic acid	AOAC 2012.21	Vitamin C in infant formula and adult/pediatric nutritional formula	HPLC 254 nm
Ascorbic acid	AOAC 2012.22	Vitamin C in infant formula and adult/pediatric nutritional formula	UHPLC 254 nm
Ascorbic acid	EN 14130 [8]	Vitamin C in foods	HPLC 265 nm
Thiamine (vitamin B_1)			
Thiamine	AOAC 942.23	Thiamine in foods	Thiochrome Fluorescence $E_x \lambda = 365$ nm $E_m \lambda = 435$ nm
Thiamine	AOAC 2015.14	Total vitamins B_1, B_2, and B_6 in infant formula and related nutritionals	Enzymatic digestion and UHPLC-MS/MS
Thiamine	EN 14122 [9]	Thiamine in foods	HPLC Thiochrome Fluorescence $E_x \lambda = 366$ nm $E_m \lambda = 420$ nm
Riboflavin (Vitamin B_2)			
Riboflavin	AOAC 970.65	Riboflavin in foods and vitamin preparations	Fluorescence Ex $\lambda = 440$ nm $E_m \lambda = 565$ nm
Riboflavin	AOAC 2015.14	Total vitamins B_1, B_2, and B_6 in infant formula and related nutritionals	Enzymatic digestion and UHPLC-MS/MS

(continued)

20.1
table (continued)

Vitamin	Method designation	Application	Approach
Riboflavin	EN 14152 [10]	Riboflavin in foods	HPLC Fluorescence $E_x \lambda = 468$ nm $E_m \lambda = 520$ nm

Niacin
  Nicotinic acid
  Nicotinamide
	AOAC 944.13	Niacin and niacinamide in vitamin preparations	Microbiological

Nicotinic acid Nicotinamide	AOAC 985.34	Niacin and niacinamide in ready-to-feed milk-based infant formula	Microbiological

Vitamin B_6
  Pyridoxine
  Pyridoxal
  Pyridoxamine
	AOAC 2004.07	Total vitamin B_6 in infant formula	HPLC
			Fluorescence
			$E_x \lambda = 468$ nm
			$E_m \lambda = 520$ nm

Pyridoxine
  Pyridoxal
  Pyridoxamine
	AOAC 2015.14	Total vitamins B_1, B_2, and B_6 in infant formula and related nutritionals	Enzymatic digestion and UHPLC-MS/MS

Folic acid, folate Total folates	AOAC 2004.05	Total folates in cereals and cereal products – trienzyme procedure	Microbiological
Total folates	AOAC 2011.06	Total folates in infant formula and adult nutritionals	Trienzyme extraction and HPLC-MS/MS
Folic acid 5-methyl tetrahydrofolic acid	AOAC 2013.13	Folate in infant formula and adult/pediatric nutritional formula	Trienzyme extraction and UHPLC-MS/MS
Vitamin B_{12} Cyanocobalamin	AOAC 986.23	Cobalamin (vitamin B_{12}) in milk-based infant formula	Microbiological
Cyanocobalamin	AOAC 2011.10	Vitamin B_{12} in infant and pediatric formulas and adult nutritionals	HPLC 550 nm
Cyanocobalamin	AOAC 2014.02	Vitamin B_{12} in infant and pediatric formulas and adult nutritionals	UHPLC 361 nm
Biotin Biotin	USP29/NF24, dietary supplements official monograph [11]	Biotin in dietary supplements	HPLC 200 nm or microbiological
Pantothenic acid Calcium pantothenate	AOAC 992.07	Pantothenic acid in milk-based infant formula	Microbiological
Calcium pantothenate	AOAC 2012.16	Pantothenic acid (vitamin B_5) in infant formula and adult/pediatric nutritional formula	UHPLC-MS/MS

[a]AOAC method [2]
[b]HPLC, high-performance liquid chromatography (in some methods simply called liquid chromatography)
[c]UHPLC, ultra-HPLC
[d]Fluoremetric test giving excitation (E_x) and emission (E_m) wavelengths
[e]MS/MS, tandem mass spectrometry
[f]ESI, eletrospray ionization
[g]NP, normal phase

20.2 BIOASSAY METHODS

Outside of vitamin bioavailability studies, bioassays at the present are used only for the analysis of **vitamins B_{12}** and **D**, and even for them, the bioassays have very limited use. For vitamin D, the bioassay reference standard method (AOAC Method 936.14) (specified for milk, vitamin preparations, and feed concentrates) is known as the line test, which is based on bone calcification. Rats are initially fed a diet that depletes rats of vitamin D and then groups of the rats are fed a diet with known (for standard curve) or unknown (sample) amounts of vitamin D. The rats are then sacrificed, and the sections of specific bones are stained to show the extent of bone calcification.

20.3   MICROBIOLOGICAL ASSAYS

20.3.1   Principle

The growth of microorganisms is proportional to their requirement for a specific vitamin, if all other nutritional needs of the microorganisms are met. Thus, in microbiological assays the growth of a certain microorganism in an extract of a vitamin-containing sample is compared against the growth of this microorganism in the presence of known quantities of that vitamin. Bacteria, yeast, or protozoans are used as test organisms. **Growth** can be measured in terms of **turbidity**, **acid production, gravimetry**, or by **respiration**. With bacteria and yeast, turbidimetry is the most commonly employed system. If turbidity measurements are involved, clear sample and standard extracts vs. turbid ones are essential. With regard to incubation time, turbidity measurement is also a less time-consuming method. The microorganisms are specified by ATCC™ numbers and are available from the *American Type Culture Collection* (ATCC™) (10801 University Blvd., Manassas, VA 20110).

20.3.2   Applications

Microbiological assays are limited to the analysis of water-soluble vitamins. The methods are very sensitive and specific for each vitamin. The methods are somewhat time consuming, and strict adherence to the analytical protocol is critical for accurate results. All microbiological assays can use microtiter plates (96-well) in place of test tubes. Microplate usage results in significant savings in media and glassware, as well as labor.

20.3.3   Niacin

The microbiological analysis of niacin and nicotinamide, as an example of such an assay, is briefly described here (AOAC Method 944.13, 45.2.04) [2, 13]. *Lactobacillus plantarum* ATCC™ 8014 is the test organism. A stock culture needs to be prepared and maintained by inoculating the freeze-dried culture on Bacto *Lactobacilli* agar followed by incubation at 37 °C for 24 h prior to sample and standard inoculation. A second transfer may be advisable in the case of poor growth of the inoculum culture. The final inoculum is added to tubes of niacin assay medium, that contain added known amounts of a USP niacin reference standard (for standard curve) and unknown amounts of niacin (food sample extract). The tubes are incubated at 37 °C for 16–24 h. The percent transmittance at a specific wavelength is measured to determine microbial growth as indicated by turbidity. Using *Lactobacilli sp.* as the test organism, acidimetric measurements could be used instead of turbidity, but the required incubation time would be 72 h.

20.4   CHEMICAL METHODS

20.4.1   High-Performance Liquid Chromatography (HPLC)

20.4.1.1   *Overview*

Because of their relative simplicity, accuracy, and precision, the chemical methods, in particular the chromatographic methods using HPLC/UHPLC, are preferred (see Chap. 13). Numerous vitamins are now commonly measured by HPLC (e.g., A, D, E, K, C, various B vitamins), many as official methods and some unofficial. Liquid chromatography in combination with mass spectrometry (MS) (see Chap. 11) has added a new dimension to vitamin analysis. In general, LC-MS or electrospray ionization (ESI) LC-MS/MS methods are available for each fat- and water-soluble vitamin. Detection by MS leads to increased sensitivity as well as unequivocal identification and characterization of the vitamin. The LC-MS assays have become a mainstay of accurate, cost-effective vitamin analyses. For example, LC-MS is commonly employed for verification of vitamin D content of products with difficult matrices (i.e., comparing results to those with standard LC analysis, e.g., AOAC Method 2012.11, Simultaneous Determination of Vitamins D_2 and D_3 in Infant Formula and Adult/Pediatric Nutritional Formula) and LC-MS/MS for folate (AOAC Method 2013.13, Folate in Infant Formula and Adult/Pediatric Nutritional Formula by a UHPLC-MS/MS assay vs. the microbiological method).

Standard HPLC is commonly employed as an official method of analysis for vitamins A (e.g., AOAC Method 992.04, 50.1.02), E (e.g., AOAC Method 992.03, 50.1.04), and D (e.g., AOAC Method 2002.05, 45.1.22A) and as a quality control method for vitamin C. While HPLC/UHPLC involves a high capital outlay, it is applicable to most vitamins and lends itself in some instances to simultaneous analysis of several vitamins and/or vitamers (i.e., isomers of vitamins). Implementation of multi-analyte procedures for the analysis of water-soluble vitamins can result in assay efficiency with savings in time and materials. To be useful, a simultaneous assay must not lead to loss of sensitivity, accuracy, and precision when compared to single-analyte methods. In general terms, multi-analyte methods for water-soluble vitamin assay of high-concentration products including pharmaceuticals, supplements, and vitamin premixes are quite easily developed. Though the applicability of HPLC has been demonstrated to a wide variety of biological matrices with no or only minor modifications in some cases, one must always bear in mind that all chromatographic techniques, including HPLC, are separation and not identification methods. Therefore, during adaptation of an existing HPLC method to a new matrix, establishing evidence of peak identity and purity is an essential step of the method adaptation or development.

20.4.1.2 Vitamin A

Vitamin A is sensitive to ultraviolet (UV) light, air (and any prooxidants, for that matter), high temperatures, and moisture. Therefore, steps must be taken to avoid any adverse changes in this vitamin due to such effects. Steps include: (1) using low actinic glassware, nitrogen, and/or vacuum, (2) avoiding excessively high temperatures, (3) working in subdued artificial light, and (4) adding pyrogallol as an antioxidant prior to saponification.

HPLC methods are considered the only acceptable methods to provide accurate food measurements of vitamin A activity. For example, in the HPLC method of vitamin A (i.e., retinol isomers) in milk and milk-based infant formula (AOAC Method 992.04, 50.1.02) [2], the test sample is saponified with ethanolic KOH, vitamin A (retinol) is extracted into organic solvent, and then concentrated. Vitamin A isomers – all-*trans*-retinol and 13-*cis*-retinol – levels are determined by HPLC on a silica column (i.e., normal phase). Vitamin A also can be analyzed using reversed-phase HPLC columns.

20.4.1.3 Vitamin D

Vitamin D is typically analyzed by HPLC with a UV-Vis detector (some version of AOAC Method 2002.05) but by HPLC-MS for verification of analyte presence, as needed. Protection against oxidation is done as described for vitamin A above. For the HPLC-UV-Vis analysis, an internal standard (vitamin D_2) is added to the sample that is subjected to basic hydrolysis then saponified in ethanolic KOH. This sample is extracted with heptane, and the heptane organic phase is evaporated to dryness. The reconstituted sample is subjected to a semi-preparative normal-phase HPLC column, from which the fractions are collected, concentrated, and diluted in acetonitrile-methanol. These samples are subjected to a

reversed-phase HPLC column with UV detection to quantitate the D_3. A separate sample is tested in parallel to confirm the absence of endogenous D_2.

20.4.1.4 Vitamin E (Tocopherols and Tocotrienols)

Vitamin E is present in foods as eight different compounds: all are 6-hydroxychromans. The vitamin E family is comprised of α-, β-, γ-, and δ-tocopherol, characterized by a saturated side chain of three isoprenoid units and the corresponding unsaturated tocotrienols (α-, β-, γ-, and δ-). Like vitamins A and D, vitamin E must be protected from oxidation during sample preparation and is commonly analyzed by HPLC. Typically a normal or reversed-phase HPLC column is connected to a fluorescence detector: E_x $\lambda = 290$ nm and E_m $\lambda = 330$ nm (E_x, excitation; E_m, emission) (for fluorescence spectroscopy, see Chap. 7, Sect. 7.3) [14]. An example chromatogram is depicted in Fig. 20.1. Vitamin E is quantitated by external standards from peak area by linear regression.

20.4.2 Other Chemical Methods

20.4.2.1 Vitamin C

The vitamin (L-**ascorbic acid** and L-**dehydroascorbic acid**) is very susceptible to oxidative deterioration, which is enhanced by high pH and the presence of ferric and cupric ions. For these reasons, the entire analytical procedure needs to be performed at low pH and, if necessary, in the presence of a chelating agent. Mild oxidation of ascorbic acid results in the formation of dehydroascorbic acid, which is also biologically active and is reconvertible to ascorbic acid by treatment with **reducing agents** such as β-mercaptoethanol and dithiothreitol. Two AOAC official methods for vitamin C are described below, but vitamin C also can be analyzed in infant formula and adult/pediatric nutritional formula by HPLC with UV detection (AOAC Method 2012.21) and UHPLC with UV detection (AOAC Method 2012.22).

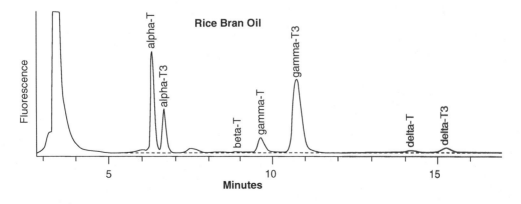

Chromatogram of rice bran oil showing tocopherols and tocotrienols

20.4.2.1.1 2,6-Dichloroindophenol (DCIP) Titrimetric Method

This method is specified as an AOAC official method (AOAC Method 967.21, 45.1.14) [2] for vitamin preparation and juices (i.e., fruits), but it is sometimes used as a secondary method for other foods, because it is a more rapid method than the microfluorometric method (described in Sec. 20.4.2.1.2) applicable to other foods. In the DCIP method, L-ascorbic acid is oxidized to L-dehydroascorbic acid by the oxidation-reduction indicator dye, DCIP. At the endpoint, excess unreduced dye appears rose pink in acid solution (see Figs. 20.2 and 20.3). With colored samples such as red beets or heavily browned products, the rose-pink endpoint is impossible to detect by the human eye. In such cases it, therefore, needs to be determined by observing the change of transmittance using a spectrophotometer with the wavelength set at 545 nm.

20.4.2.1.2 Microfluorometric Method

The vitamin C AOAC microfluorometric (AOAC Method 967.22, 45.1.15) assay is specified for vitamin preparations, but a semiautomated fluorometric AOAC method (AOAC Method 984.26, 45.1.16) is specified as applicable to all food products in the absence of erythorbate [2, 15]. The microfluorometric method mea-

Titrant

2,6-dichloroindophenol
(red)

reduced 2,6-dichloroindophenol
(colorless)

L-ascorbic acid
(colorless)

L-dehydroascrorbic acid
(brown)

20.2 **figure** Chemical reaction between L-ascorbic acid and the indicator dye, 2,6-dichloroindophenol

---

**VITAMIN C ASSAY PROCEDURE
2,6-DICHLOROINDOPHENOL (DCIP) TITRATION**

Sample Preparation

Weigh and extract by homogenizing test sample in *meta*phosphoric acid-acetic acid solution (*i.e.*, 15 g of HPO_3 and 40 ml of HOAc in 500 ml of deionized H_2O). Filter (and/or centrifuge) sample extract, and dilute appropriately to a final concentration of 10-100 mg of ascorbic acid/100 ml.

Standard Preparation

Weigh 50 mg of USP L-ascorbic acid reference standard and dilute to 50 ml with HPO_3-HOAc extracting solution.

Titration

Titrate three replicates each of the standard (*i.e.*, to determine the concentration of the indophenol solution as mg ascorbic acid equivalents to 1.0 ml of reagent), test sample, and blank with the indophenol reagent (*i.e.*, prepared by dissolving 50 mg of DCIP sodium salt and 42 mg of $NaHCO_3$ to 200 ml with deionized H_2O) to a light but distinctive rose pink endpoint lasting ≥ 5 sec.

20.3 **figure** Analysis of vitamin C by the 2,6-dichloroindophenol titration, AOAC Method 967.21, 45.1.14 [2] (Adapted from Pelletier [15])

sures both ascorbic acid and dehydroascorbic acid. All ascorbate forms are oxidized to dehydroascorbic acid (using a boric acid-sodium acetate solution), and then the dehydroascorbic acid is reacted with *o*-phenylene-diamine to produce a **fluorescent quinoxaline compound**. The amount of fluorescence in the sample (compared to a standard and corrected with blanks) is used to quantitate the amount of vitamin C.

20.4.2.2 Thiamine (Vitamin B₁) Thiochrome Fluorometric Method

While thiamine can be quantitated by HPLC, it is still commonly analyzed by the longtime official thiochrome fluorometric procedure (AOAC Method 942.23) [2]. Following sample extraction with dilute acid, enzymatic hydrolysis of thiamine's phosphate esters, and chromatographic cleanup (i.e., purification), thiamine is oxidized to **thiochrome**, which is fluorescent. This method is based on the fluorescence measurement of thiochrome in the test solution compared to that from an oxidized thiamine standard solution.

20.4.2.3 Riboflavin (Vitamin B₂) Fluorometric Method

Like other B vitamins, riboflavin can be analyzed by HPLC, but its natural fluorescence allows for measurement based on this characteristic. Following sample extraction, cleanup, and compensation for the presence of interfering substances, riboflavin is determined fluorometrically, compared to a riboflavin standard (AOAC Method 970.65, 45.1.08) [2].

20.5 COMPARISON OF METHODS

Each type of method has its advantages and disadvantages. In selecting a certain method of analysis for a particular vitamin or vitamins, a number of factors need to be considered, some of which are listed below:

1. Method accuracy and precision
2. The need for bioavailability information
3. Time and instrumentation requirements
4. Personnel requirements
5. The type of biological matrix to be analyzed
6. The number of samples to be analyzed
7. Regulatory requirements – Must official AOAC International methods be used?

At present, the applicability of microbiological assays is limited to water-soluble vitamins (most commonly niacin, B₁₂, and pantothenic acid). Though somewhat time consuming, they generally can be used for the analysis of a relatively wide array of biological matrices without major modifications. Furthermore, less sample preparation is often required compared to chemical assays; yet, with more and more official methods being developed for HPLC and UHPLC, the employment of these microbiological assays is expected to decrease with time.

When selecting a system for analysis, at least initially, it is wise to consider the use of official methods that have been tested through interlaboratory studies and that are published by such organizations as AOAC International [2], the British Standards Institution [3–10], the US Pharmacopeial Convention [11], or the AACC International [16]. Again, one must realize that these methods are limited to certain biological matrices.

20.6 SUMMARY

Three types of methods for the analysis of vitamins – bioassays and microbiological and chemical assays – have been outlined in this chapter, with emphasis on the chemical methods. The methods are, in general, applicable to the analysis of more than one vitamin and several food matrices. However, the analytical procedures must be properly tailored to the analyte in question and the biological matrix to be analyzed; issues concerning sample preparation, extraction, and quantitative measurements are also involved. It is essential to validate any new application appropriately by assessing its accuracy and precision. Method validation is especially important with chromatographic methods such as HPLC, because these methods basically accent separations rather than identification of compounds. For this reason, it is essential to ensure not only identity of these compounds but also, just as important, their purity.

20.7 STUDY QUESTIONS

1. What factors should be considered in selecting the assay for a particular vitamin?
2. To be quantitated by most methods, vitamins must be extracted from foods. What treatments are commonly used to extract the vitamins? For one fat-soluble vitamin and one water-soluble vitamin, give an appropriate extraction procedure.
3. What vitamin must be listed on the US standard nutrition label as of 2018 (see Chap. 3, Sect. 3.2.1.1), and what would be an official method for its analysis?
4. Explain why it is possible to use microorganisms to quantitate a particular vitamin in a food product, and describe such a procedure.
5. There are two commonly used AOAC methods to measure the vitamin C content of foods. Identify these two methods; then compare and

contrast them with regard to the principles involved.

6. Would the vitamin C content as determined by the 2,6-dichloroindophenol method be underestimated or overestimated in the case of heat processed juice samples? Explain your answer.

7. What are the advantages and disadvantages of using HPLC for vitamin analysis?

20.8 PRACTICE PROBLEMS

Please refer to the fourth edition of this Food Analysis textbook for practice problems.

Acknowledgment The author of this chapter wishes to acknowledge W.O. Landen, Jr., who was a coauthor of this chapter for the second to fourth editions of this textbook.

REFERENCES

1. Eitenmiller RR, Ye L, Landen WO Jr (2008) Vitamin analysis for the health and food sciences, 2nd edn., p. 135. Taylor & Francis Group, CRC, Boca Raton, FL

2. AOAC International (2016) Official methods of analysis, 20th edn., AOAC International, Gaithersburg, MD

3. British Standards Institution AW/275 (2014) Foodstuffs. Determination of vitamin A by high performance liquid chromatography.Part 1: Measurement of all-*E*-retinol and 13-*Z*-retinol, BS EN 12823–1:2014

4. British Standards Institution AW/275 (2000) Foodstuffs. Determination of vitamin A by high performance liquid chromatography. Part 2: Measurement of β-carotene, BS EN 12823–2:2000

5. British Standards Institution AW/275 (2009) Foodstuffs. Determination of vitamin D by high performance liquid chromatography. Measurement of cholecalciferol (D_3) and ergocalciferol (D_2), BS EN 12821:2009

6. British Standards Institution AW/275 (2014) Foodstuffs. Determination of vitamin E by high performance liquid chromatography. Measurement of α-, β-, γ- and δ-tocopherol, BS EN 12822:2014

7. British Standards Institution AW/275 (2003) Foodstuffs. Determination of vitamin K1 by HPLC, BS EN 14148:2003

8. British Standards Institution AW/275 (2014) Foodstuffs. Determination of vitamin B_1 by high performance liquid chromatography, BS EN 14122:2014

9. British Standards Institution AW/275 (2014) Foodstuffs. Determination of vitamin B_2 by high performance liquid chromatography, BS EN 14152:2014

10. British Standards Institution AW/275 (2014) Foodstuffs. Determination of vitamin B_6 by high performance chromatography, BS EN 14164:2014

11. United States Pharmacopeial Convention (2016) US Pharmacopoeia National Formulary, USP39–NF34, Nutritional Supplements, Official Monographs, United States Pharmacopoeial Convention, Rockville, MD

12. Blake CJ (2007) Analytical procedures for water-soluble vitamins in foods and dietary supplements: a review. Anal Bioanal Chem 389: 63–76

13. Eitenmiller RR, Ye L Landen WO Jr (2008) Niacin. In: Vitamin analysis for the health and food sciences, Second edition. CRC Press, Boca Raton, Florida pp 361–400

14. Lee J, Ye L, Landen WO Jr, Eitenmiller RR (2000) Optimization of an extraction procedure for the quantification of Vitamin E in tomato and broccoli using response surface methodology. J Food Compos Anal 13: 45–57

15. Pelletier O (1985) Vitamin C (L-ascorbic and dehydro-L-ascorbic acid). In: Augustin J, Klein BP, Becker DA, Venugopal PB (eds) Methods of vitamin assay, 4th edn. Wiley, New York, pp 335–336

16. AACC International (2010) Approved methods of analysis, 11th edn. (On-line), AACC International, St. Paul, MN

Traditional Methods for Mineral Analysis

Robert E. Ward (✉)
Department of Nutrition and Food Sciences,
Utah State University,
Logan, UT 84322-8700, USA
e-mail: robert.ward@usu.edu

Jerrad F. Legako
Department of Animal and Food Sciences,
Texas Tech University,
Lubbock, TX 79409, USA
e-mail: jerrad.legako@ttu.edu

S. Nielsen (ed.), *Food Analysis*, Food Science Text Series,
DOI 10.1007/978-3-319-45776-5_21, © Springer International Publishing 2017

21.1 INTRODUCTION

This chapter describes traditional methods for analysis of minerals involving titrimetric and colorimetric procedures, ion-selective electrodes, and instruments to measure salt content. Other traditional methods of mineral analysis include gravimetric titration (i.e., insoluble forms of minerals are precipitated, rinse, dried, and weighed) and redox reactions (i.e., mineral is part of an oxidation-reduction reaction, and product is quantitated). However, these latter two methods will not be covered because they currently are not used much in the food industry. The traditional methods that will be described have maintained widespread usage in the food industry despite the development of more modern instrumentation such as ion chromatography (Chap. 13), atomic absorption spectroscopy, and inductively coupled plasma-optical emission spectroscopy (Chap. 9). Traditional methods generally require chemicals and equipment that are routinely available in an analytical laboratory and are within the experience of most laboratory technicians. Additionally, traditional methods often form the basis for rapid analysis kits (e.g., AquaChek® for calcium and Quantab® for salt determination) and for automated benchtop analyzers for salt content. In these cases, it is useful to understand the principles employed in the analysis. For additional examples of traditional methods, refer to references [1–6]. For analytical requirements for specific foods, see the *Official Methods of Analysis* of AOAC International [5] and related official methods [6].

21.1.1 Importance of Minerals in the Diet

Dietary **macrominerals** (calcium, phosphorus, sodium, potassium, magnesium, chlorine, and sulfur) are required at more than 100 mg per day by the adult [7–9]. Additional ten minerals (iron, iodine, zinc, copper, chromium, manganese, molybdenum, fluoride, selenium, and silica) are required in milli- or microgram quantities per day and are referred to as **trace minerals. Ultra trace minerals**, including vanadium, tin, nickel, arsenic, and boron, are being investigated for possible biological function, but they currently do not have clearly defined biochemical roles. **Heavy metals** (lead, mercury, cadmium, and arsenic) have been documented to be toxic to the body and should, therefore, be avoided in the diet. Essential minerals such as fluoride and selenium also are known to be harmful if consumed in excessive quantities, even though they do have beneficial biochemical functions at proper dietary levels.

The Nutrition Labeling and Education Act of 1990 (NLEA) mandated labeling of **sodium, iron**, and **calcium** contents largely because of their important roles in controlling hypertension, preventing anemia, and impeding the development of osteoporosis, respectively (see Chap. 3, Fig. 3.1). The newest US food labeling regulations require the addition of **potassium** to the label, due to its numerous health benefits (see Chap. 3). The content of these minerals in several foods are shown in Table 21.1. The content of other minerals may be included on the nutrition label at the producer's option, although this becomes mandatory if the mineral is the subject of a nutrient content claim on the label (Chap. 3, Sect. 3.2.3).

21.1.2 Minerals in Food Processing

Minerals are of **nutritional** and **functional** importance, and for that reason, their levels need to be known and/or controlled. Some minerals are contained at high levels in natural foodstuffs. For example, milk is a good source of calcium, containing about 300 mg of calcium per 8-oz cup. However, direct acid cottage cheese is very low in calcium because of the action of the acid, causing the calcium bound to the casein to be freed and consequently lost in the whey fraction. Similarly, a large portion of the phosphorus, zinc, manganese, chromium, and copper found in a grain kernel is lost when the bran layer is removed in the processing. The enrichment law for flour requires that iron be replaced in white flour to the level at which it occurred naturally in the wheat kernel before removal of the bran.

Fortification of some foods has allowed addition of minerals above levels ever expected naturally. Prepared breakfast cereals often are fortified with minerals such as calcium, iron, and zinc, formerly thought to be limited in the diet. Fortification of salt with iodine has almost eliminated goiter in the USA. In other cases, minerals may be added for functionality. Salt is added for flavor, to modify ionic strength that affects solubilization of protein and other food components, and as a preservative. This significantly increases the sodium content of products such as processed meats, pickles, and processed cheese. Phosphorus may be added as phosphates to increase water-holding capacity in meats and to change the texture of processed cheese. Calcium may be added to promote gelation of proteins and gums. Textural properties of fruits and vegetables can be influenced by the "hardness" or "softness" of the water used during processing.

Water is an integral part of food processing, and **water quality** is a major factor to be considered in the food processing industry. Water is used for washing, rinsing, blanching, cooling, and as an ingredient in formulations. Microbiological safety of water used in food processing is very important. Also important, but generally not appreciated by the consuming public, is the mineral content of water used in food processing. Waters that contain excessive minerals can result in clouding of beverages.

<table>
<thead>
<tr><th rowspan="2">21.1
table</th><th colspan="5">Mineral content of selected foods</th></tr>
</thead>
</table>

Food item	\multicolumn{4}{c}{mg/g (wet weight basis)}			
	Calcium	Iron	Sodium	Potassium
Cereals, bread, and pasta				
Rice, brown, long grain, raw	9	1	5	250
Rice, white, long grain, regular, raw, enriched	28	4	5	115
Wheat flour, whole grain	34	4	2	363
Wheat flour, white, all-purpose, unenriched	15	1	2	107
Pasta, dry, enriched	21	3	6	223
Dairy products				
Milk, whole, 3.25 % milk fat with added vitamin D	113	<1	43	132
Butter, salted	24	<1	643	24
Cheese, cottage, low-fat, 2 % milk fat	111	<1	308	125
Fruits and vegetables				
Apples, raw, with skin	8	<1	1	107
Bananas, raw	5	<1	1	358
Raisins, seedless	50	2	11	749
Potatoes, raw, skin	30	3	10	413
Tomatoes, red, ripe, raw	10	<1	5	237
Meats, poultry, and fish				
Eggs, whole, raw, fresh	56	2	142	138
Fish fillet, battered or breaded, and fried	14	<1	561	251
Pork, fresh, leg (ham), whole, raw	6	1	55	369
Beef, chuck, arm pot roast, raw	16	2	62	290
Bologna, chicken, pork, beef	92	1	1,120	313

From US Department of Agriculture, Agricultural Research Service (2016) USDA National Nutrient Database for Standard Reference. Release 28. Nutrient Data Laboratory Home Page, https://ndb.nal.usda.gov/

Water quality standards in the USA are established by the Environmental Protection Agency as a result of the Safe Drinking Water Act (Chap. 2, Sect. 2.2.5.2), and local water suppliers must ensure regulated substances, such as heavy metals, are below maximum contaminant levels. Tests ensure that levels of total microorganisms, coliforms, inorganic contaminants, disinfection by-products (such as chlorine), and radio-active contaminants are below maximum contaminant levels. In addition, food processors should measure other water quality parameters such as pH, turbidity, hardness, heavy metals, iron, nitrates, and volatiles.

Water hardness is a measure of the dissolved calcium and magnesium salts in water and is commonly expressed in parts per million (ppm). Hardness of water is determined using the following scale: 0–60 ppm is soft, 60–120 ppm is moderately hard, 120–180 ppm is hard, and >180 ppm is very hard. Use of hard water can affect processing in several ways. For example, over time, calcium and magnesium salts may precipitate out of hard water and form scale in pipes and on other surfaces. In addition, hard water reduces the effectiveness of soaps and sanitizers. If the water source at a food production facility contains excess minerals, it can be softened using ion exchange resins. Table 21.2 lists other treatments that may be used to improve water quality for use in food production.

21.2 BASIC CONSIDERATIONS

21.2.1 Nature of Analyses

Mineral analysis is a valuable model for understanding the basic structure of analysis procedures to separate and measure. Separation of minerals from the food matrix is often specific, such as **complexometric titrations** (Sect. 21.3.1) or **precipitation titrations** (Sect. 21.3.2). In these cases of specific separation, nonspecific measurements such as volume of titrant are made and later converted to mass of mineral based on fundamental stoichiometric relationships. In other cases, separation of mineral involves nonspecific procedures such as **ashing** or **acid extraction**. These nonspecific separations require that a specific measurement be made as provided by **colorimetry** (Sect. 21.3.3), **ion-selective electrodes** (ISE) (Sect. 21.3.4), **atomic absorption spectroscopy** (AAS) (Chap. 9), or **inductively coupled plasma-optical emission spectroscopy** (ICP-OES) (Chap. 9).

Because determination mineral mass is the final objective of analysis, measures other than mass are considered to be surrogate, or stand-in, measures. **Surrogate measures** are converted into mineral mass via fundamental stoichiometric and physiochemical relationships or by empirical relationships. Empirical relationships are those associations that need be established by experimentation because they do not follow any well-established physiochemical relationship. An example of a surrogate measurement is the wavelength-specific absorbance of a chromogen-mineral complex (Sect. 21.3.3). It may be possible to convert absorbance into mass of mineral using the fundamental relationships defined by the molar absorptivity and stoichiom-

21.2 table Common water treatments and their rationales for use

| | Treatments | | | | | | |
Quality parameter	Filtration	Membrane filtration	Ion exchange	Chlorination/ ozonization	UV radiation	pH adjustment	Activated carbon
Solids	X	X					
Salts, including hardness		X	X				
pH correction							
Other chemical contaminants, e.g., organic residues	X	X	X				X
Bacteria		X		X	X		
Viruses		X		X	X		
Protozoa		X		X	X		
Algal bloom (toxin)							X

Adapted from [14]

etry of the chromogen-mineral complex. However, it is more commonly required that the absorbance vs. concentration relationship be empirically developed using a series of standards (i.e., a standard curve).

21.2.2 Sample Preparation

Some sample preparation is generally required for traditional methods of mineral analysis to ensure a well-mixed and representative sample and to make the sample ready for the procedure to follow. A major concern in mineral analysis is **contamination** during sample preparation. **Comminution** (e.g., grinding or chopping) and mixing using metallic instruments can add significant mineral to samples and, whenever possible, should be performed using nonmetallic instruments or instruments not composed of the sample mineral. For example, using an aluminum grinder is standard practice for comminution of meat samples undergoing iron analysis. **Glassware** used in sample preparation and analysis should be scrupulously cleaned using acid washes and triple rinsed in the purest water. The latter may necessitate installation of an **ultrapure water system** in the laboratory to further purify the general supply of distilled water.

Solvents, including water, can contain significant quantities of minerals. Therefore, all procedures involving mineral analysis require the use of the purest reagents available. In some cases, the cost of ultrapure reagents may be prohibitive. When this is the case, the alternative is to always work with a reagent blank. A **reagent blank** is a sample of reagents used in the sample analysis, quantitatively the same as used in the sample but without any of the material being analyzed. This reagent blank, representing the sum of the mineral contamination in the reagents, is then subtracted from the sample values to more accurately quantify the mineral.

A method such as near-infrared spectroscopy (Chap. 8, Sect. 8.4) allows for mineral estimation without destruction of the carbon matrix of carbohydrates, fats, protein, and vitamins that make up foods. However, traditional methods generally require that the minerals be freed from this organic matrix in some manner. Chapter 16 describes the various methods used to ash foods in preparation for determination of specific mineral components of the food. In water samples, minerals may be determined without further preparation.

21.2.3 Interferences

Factors such as **pH**, **sample matrix**, **temperature**, and other **analytical conditions** and **reagents** can interfere with the ability of an analytical method to quantify a mineral. Often there are specific interfering substances that must be removed or suppressed for accurate analysis. Two of the more common approaches are to isolate the sample mineral or remove interfering minerals, using selective precipitations or ion exchange resins. Water may need to be boiled to remove carbonates that interfere with several traditional methods of mineral analysis.

If other interferences are suspected, it is a common practice to develop the standard curve using sample mineral dissolved in a background matrix containing interfering elements known to be in the food sample. For example, if a food sample is to be analyzed for calcium content, a **background matrix solution** of the known levels of sodium, potassium, magnesium, and phosphorus should be used to prepare the calcium standards for developing the standard curve. In this manner, the standard curve more closely represents the analysis response to the sample mineral when analyzing a food sample. Alternatively, the standard curve can be developed using a series of sample mineral spikes added to the food sample. A **spike** is a small volume of a concentrated standard that is added to the sample. The volume is small enough so as to not appreciably change the overall composition of the sample, except for the mineral of interest. Thus, measurements of both the standards and the sample are made in the presence of the same background. If the spikes are added before implementation of the analysis protocol, possible effects of incomplete extractions, sample mineral degradation, and other losses are integrated into the standard curve.

21.3 METHODS

21.3.1 EDTA Complexometric Titration

21.3.1.1 *Background Information*
The hexadentate ligand **ethylenediaminetetraacetate** (EDTA) forms stable 1:1 complexes with numerous mineral ions. This gives complexometric titration using EDTA broad application in mineral analysis. Stability of mineral-EDTA complexes generally increases with valence of the ion, although there is significant variation among ions of similar valence due to their coordination chemistry. Endpoints are detected using mineral chelators that have coordination constants lower than EDTA (i.e., indicator has less affinity for mineral ions than does EDTA) and that produce different colors in each of their complexed and free states. **Calmagite** and **Eriochrome Black T** (EBT) are such indicators that are pink when complexed with calcium or magnesium, but blue when no metal ions are complexed. The endpoint of a complexometric EDTA titration using either Calmagite or EBT as the indicator is detected when the color changes from pink to blue.

The pH affects a complexometric EDTA titration in several ways and must be controlled for best performance. The complexation equilibrium is strongly pH dependent. With decreasing pH the chelating sites of EDTA become protonated, thereby decreasing its effective concentration. The pH must be 10 or more for calcium or magnesium to form stable complexes with EDTA. Also, the sharpness of the endpoint increases with increasing pH. However, magnesium and calcium precipitate as their hydroxides at pH 12, so titration pH should probably be no more than 11 to ensure their solubility. Considering all factors, EDTA complexometric titration of calcium and magnesium is specified at pH 10 ± 0.1 using an ammonia buffer [10].

21.3.1.2 *Principle*
Based on the background information above, a summary of the principle of the EDTA complexometric titration method follows (described for a sample containing just calcium): The EDTA in the titrant solution complexes with calcium, 1:1, at pH 10. The EDTA binds to calcium ions stronger than the indicator binds

to the calcium. When all the calcium present in the sample has reacted with the EDTA titrant (i.e., there is no calcium left to react with the indicator), the indicator changes in color from pink to blue. The moles of calcium in the sample are equivalent to the moles of EDTA used to do the titration.

21.3.1.3 *Procedure: Hardness of Water Using EDTA Titration*
Water hardness is determined by EDTA complexometric titration of the total of calcium and magnesium, in the presence of Calmagite, and expressed as the equivalents of calcium carbonate (mg/L) (*Standard Methods for the Examination of Water and Wastewater*, Method 2340, Hardness) [10]. The calcium-Calmagite complex is not stable, and calcium alone cannot be titrated using the Calmagite indicator. However, Calmagite becomes an effective indicator for calcium titration if the buffer solution contains a small amount of neutral magnesium salt and enough EDTA to bind all magnesium. Upon mixing sample into the buffer solution, calcium in the sample replaces the magnesium bound to EDTA. The free magnesium binds to Calmagite, and the pink magnesium-Calmagite complex persists until all calcium in the sample has been titrated with EDTA. The first excess of EDTA removes magnesium from Calmagite and produces a blue endpoint.

21.3.1.4 *Applications*
The major application of EDTA complexometric titration is testing calcium plus magnesium as an indicator of water hardness [10]. However, EDTA complexometric titration is suitable for determining calcium in the ash of fruits and vegetables (AOAC Method 968.31) [5] and other foods that have calcium without appreciable magnesium or phosphorus.

Several of the methods for mineral analysis described in this chapter have been incorporated into portable test strips that can be used to test samples in real time. For example, the water hardness application of the EDTA complexometric titration is made easy using test strips impregnated with Calmagite and EDTA (e.g., **AquaChek**, Environmental Test Systems, Inc., a HACH Company, Elkhart, IN). The strips are dipped into the water to test for total hardness caused by calcium and magnesium. The calcium displaces the magnesium bound to EDTA, and the released magnesium binds to Calmagite, causing the test strip to change color. The color is compared to a reference standard to estimate the calcium concentration.

21.3.2 Precipitation Titration

21.3.2.1 *Principles*
When at least one product of a titration reaction is an insoluble precipitate, it is referred to as **precipitation titrimetry**. Major factors that have limited the usefulness

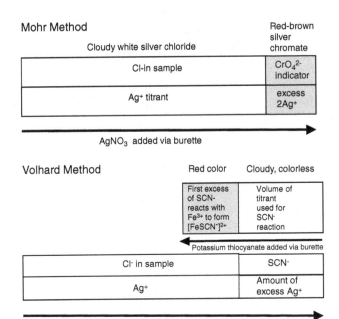

Mohr Method

Cloudy white silver chloride — Red-brown silver chromate

| Cl-in sample | CrO₄²⁻ indicator |
| Ag⁺ titrant | excess 2Ag⁺ |

AgNO₃ added via burette

Volhard Method

Red color — Cloudy, colorless

| First excess of SCN- reacts with Fe³⁺ to form [FeSCN⁻]²⁺ | Volume of titrant used for SCN- reaction |

Potassium thiocyanate added via burette

| Cl- in sample | SCN- |
| Ag⁺ | Amount of excess Ag⁺ |

AgNO₃ added in excess

21.1

figure

Comparison of the Mohr (*forward*) and Volhard (*backward*) chloride titration methods

and accuracy of these methods have been the long times necessary for complete precipitation, failure of the reaction to yield a single product of definite composition, and lack of an endpoint indicator for the reaction. Nonetheless, precipitation titration has resulted in at least two methods that are used widely in the food industry today: Mohr titration and Volhard titration.

The **Mohr method** for chloride determination is a direct or **forward titration** method, in which the chloride (in NaCl) is titrated with silver nitrate in the presence of potassium chromate (Fig. 21.1). The arrow in the figure shows the titrant addition. Initially, the silver reacts with the chloride and the solution is cloudy white. When all the chloride is reacted, the excess silver reacts with chromate to form an orange-colored solid, silver chromate. The volume of silver nitrate titrant and the molarity of that solution are used to calculate the amount of chloride ions, which can be used to calculate the amount of NaCl:

$$Ag^+ + Cl^- \rightarrow AgCl \left(\text{until all } Cl^- \text{ is complexed} \right) \quad (21.1)$$

$$2Ag^+ + CrO_4^{2-} \rightarrow$$
$$Ag_2CrO_4 \text{(orange only after } Cl^- \text{ is}$$
$$\text{all complexed)} \quad (21.2)$$

The **Volhard method** is an indirect or **back-titration** method in which an excess of a standard solution of silver nitrate is added to a chloride-containing sample solution which is shown by the top arrow in Fig. 21.1. The excess silver is then back-titrated (bottom

arrow) using a standardized solution of potassium or ammonium thiocyanate, with ferric ion as an indicator. The volume of thiocyanate solution used in the titration is proportional to the amount of excess silver that did not react with the chloride in the sample. The moles of total silver are equal to the sum of the moles of chloride in the sample and moles of thiocyanate in the titrant:

$$Ag^+ + Cl^- \rightarrow AgCl \text{ (until all } Cl^- \text{ is complexed)} \quad (21.3)$$

$$Ag^+ + SCN^- \rightarrow AgSCN$$
$$\text{(to quantitate silver not}$$
$$\text{complexed with chloride)} \quad (21.4)$$

$$SCN^- + Fe^3 \rightarrow [FeSCN]^{2+}$$
$$\text{(red when there is any } SCN^-$$
$$\text{not complexed to } Ag^+) \quad (21.5)$$

21.3.2.2 *Procedures*

21.3.2.2.1 *Mohr Titration*
Salt in foods may be estimated by titrating the chloride ion with silver. For example, to measure the salt content of butter (AOAC Method 960.29), a butter sample is solubilized in boiling water, potassium chromate is added as the indicator, and then the solution is titrated with a standardized solution of silver nitrate. The orange endpoint in this reaction occurs only when all chloride ion is complexed, resulting in an excess of silver to form the colored silver chromate. The endpoint of this reaction is therefore at the first hint of an orange color. When preparing reagents for this assay, boiled water must be used to avoid interferences from carbonates in the water.

21.3.2.2.2 *Volhard Titration*
To measure the chloride content of cheese by the Volhard method (AOAC 935.43), excess silver nitrate is added to a cheese sample that is boiled in nitric acid solution. After cooling and filtering, the excess silver nitrate is titrated with potassium thiocyanate, using ferric aluminum as the indicator. Once chloride is determined by titration, the chloride weight is multiplied by 1.648 to obtain salt weight (based on molecular weight of Cl vs. NaCl). As in the Mohr titration method, water must be boiled to minimize errors due to interfering carbonates.

21.3.2.3 *Applications*
Precipitation titration methods are well suited for any foods that may be high in chlorides. Because of added salt in processed cheeses and meats, these products should certainly be considered for using this method to detect chloride; then salt content is

estimated by calculation. A second example of a traditional mineral analysis method that has been adapted to a test strip format is the **Quantab® chloride titration** used in AOAC Method 971.19. This method is an adaptation of the principles involved in the Mohr titration method. This test strip adaptation allows for very rapid quantitation of salt in food products and is accurate to ±10 % over a range of 0.3–10 % NaCl in food products.

21.3.3 Colorimetric Methods

21.3.3.1 Principles

Chromogens are chemicals that, upon reaction with the compound of interest, form a colored product. Chromogens are available that selectively react with a wide variety of minerals. Each chromogen reacts with its corresponding mineral to produce a soluble colored product that can be quantified by absorption of light at a specified wavelength. The relationship between concentration and absorbance is given by **Beer's law** as detailed in Chap. 7. Generally, the concentration of a specific mineral in a sample is determined from a standard curve developed during the analysis, although in some cases it is possible to directly calculate concentration based on molar absorptivity of the chromogen-mineral complex.

Samples generally must be ashed or treated in some other manner to isolate and/or release the minerals from organic complexes that would otherwise inhibit their reactivity with the chromogen. The mineral of interest must be solubilized from a dry ash then subsequently handled in a manner that prevents its precipitation. The soluble mineral may need to be treated (e.g., reduced or oxidized) to ensure that all mineral is in a form that reacts with the chromogen [2]. Ideally, the chromogen reacts rapidly to produce a stable product, but if not, absorbance is read at a specific time. As with all mineral analysis of food, special efforts must be put in place to avoid contamination during sampling and analysis.

21.3.3.2 Procedures: Determination of Phosphorus in Milk

The total phosphorus content of foods can be quantified spectrophotometrically using modification of the Murphy-Riley method [11]. In this analysis, food-derived phosphorus reacts with ammonium molybdate at low pH to form phosphomolybdic acid. This product is subsequently reduced by ascorbic acid to a blue color, and antimony potassium tartrate is added to facilitate the reduction. The blue-colored phosphomolybdate complex formed has maximum absorptivity at 880 nm, a wavelength in the infrared range. However, this complex absorbs sufficient radiation at 700 nm, in the visible range, to be measured by most spectrophotometers. A requirement of the assay is that phosphorous must be soluble, and therefore solid food samples must first be ashed. To conduct the measurement, the ashed sample is solubilized in strong acid and then mixed with the Murphy-Riley reagent producing a colored product. A standard curve is prepared along with the samples with known amounts of phosphorus, and a regression equation is used to determine the concentration of phosphorus in the sample.

21.3.3.3 Applications

Colorimetry is used for the detection and quantification of a wide variety of minerals in food, and it is often a viable alternative to atomic absorption spectroscopy and other mineral detection methods. Colorimetric methods generally are very specific and usually can be performed in the presence of other minerals, thereby avoiding extensive separation to isolate the mineral of interest. The assays are particularly robust and often immune to matrix effects that can limit the usefulness of other methods for mineral analysis.

21.3.4 Ion-Selective Electrodes

21.3.4.1 Background Information

Many electrodes have been developed for the selective measurement of various cations and anions, such as bromide, calcium, chloride, fluoride, potassium, sodium, and sulfide [12, 13]. The pH electrode described in Chap. 22 is a specific example of an **ISE**. While the sensor of a pH electrode is specific to hydrogen ions, other sensors available are specific to individual mineral ions. For any ISE, an **ion-selective sensor** is placed such that it acts as a "bridging electrode" between two reference electrodes carefully designed to produce a constant and reproducible **potential**. In this manner, the potential within the sensor remains constant, while potentials develop at the sensor surfaces according to the **Nernst equation** (Chap. 22, Sect. 22.3.2.2), dependent on sample **ion activity** in the solutions contacting each surface. **Ion concentration** is generally substituted for ion activity, which is a reasonable approximation at low concentrations and controlled ionic strength environments. Indeed, this is observed within limitations set by electrode and instrumental capabilities (Fig. 21.2).

21.3.4.2 Principle

The principle for using ISE to measure a specific mineral is the same as for measuring pH (i.e., uses Nernst equation), but the composition of glass in the sensing electrode is made to be specific for the element of interest. When sensing and reference electrodes (often as a combination electrode) are immersed in the sample solution,

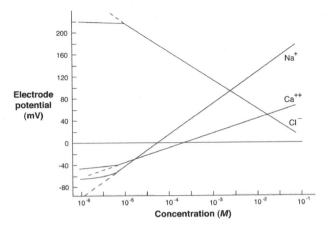

Examples of ion-selective electrode calibration curves for ions important in foods (Courtesy of Van London p Hoenix Company, Houston, TX)

the electrical potential that develops at the surface of the sensing electrode is measured by comparing it to the reference electrode. The voltage that develops between the two electrodes is related to the ion activity for the element of interest, measured as mV. Ion activity is related to ion concentration via the activity coefficient, which is controlled by ionic strength of the sample. The concentration of the element of interest is determined using a standard curve of mV vs. log concentration.

21.3.4.3 General Methodology

For ISE analysis, one simply attaches an ISE for the sample ion to a **pH meter** set on the **mV scale** and follows instructions for determination. Detailed information regarding the performance of specific ISEs is available from vendor catalogs. Typical ISEs likely to be employed for analysis of foods operate in the range of $1–10^{-6}$ M (Fig. 21.2), although the electrode response may be distinctly nonlinear at the lower concentrations. Because the ISE responds to ionic activity, it is important that the **activity coefficient** be kept constant in samples and calibration standards. The activity coefficient (γ) is used to relate ion activity (A) to ion concentration (C) ($A = \gamma C$). Activity coefficient is a function of ionic strength, so **ionic strength adjustment** (ISA) **buffers** are used to adjust the samples and standards to the same ionic strength. These ISA buffers are commercially available. The use of ISA buffers also adjusts the pH, which may be necessary if H^+ or OH^- activities affect the ion-specific sensor or if they interact with the analyte. In the case of metals having insoluble hydroxides, it is necessary to work at a pH that prevents their precipitation. Depending on the selectivity of the ISE, it may be necessary to remove interfering ions from the sample by selective precipitation or complexation.

In view of temperature effects on standard potentials and slopes of electrodes, it is important to keep the electrode and solutions at a constant temperature. This may involve working in a room that is thermostatically controlled to 25 °C (one of the internationally accepted temperatures for electrochemical measurements) and allowing sufficient time for all samples and standards to equilibrate to this temperature. Solutions should be gently stirred during the measurement to attain rapid equilibrium and to minimize concentration gradients at the sensor surface. Finally, it is important to allow sufficient time for the electrode to stabilize before taking a reading. ISEs may not completely stabilize within a practical timeframe, so a decision needs to be made of when to take the reading. The reading may be taken when the rate of change has fallen below some predetermined value or at a fixed time after the electrode was placed in solution. A problem with the latter is that as samples are changed, many ISEs respond more rapidly to an increase in concentration of sample ion as compared to a decrease in concentration of sample ion.

21.3.4.4 Electrode Calibration and Determination of Concentration

In using an ISE, ion concentration can be determined using either a calibration curve or endpoint titration. It is common practice to develop a **calibration curve** when working with an ISE because it allows a large number of samples to be measured rapidly. The electrode potential (volts) is developed in a series of solutions of known concentration and plotted against the standard concentrations on a log scale. Examples of calibration curves for various ions are given in Fig. 21.2. Upon analysis of a test sample, the observed electrode potential is used to determine ion concentration by referring to the calibration curve. Note the nonlinear region of the curve at the lowest concentrations. Total ionic strength and the concentration of interfering ions are especially important factors limiting selective detection of low levels of ions.

ISEs can be used to detect the endpoints of potentiometric titrations by measuring the change in potential as the titration proceeds. One common use is for endpoint detection in the Mohr chloride assay previously described. In potentiometric silver chloride (Mohr) titrations, the ISE used for endpoint detection can either be specific for the titrant species (Ag^+) or the sample ion (Cl^-). If an ISE is used for Ag^+, a T-type titration curve results from the large increase in titrant activity detected at the equivalence point (see Fig. 21.3 for a cation titrant). Sample concentration is calculated from titrant volume to reach the equivalence point, taking into account the stoichiometric relationship between titrant species and sample ion.

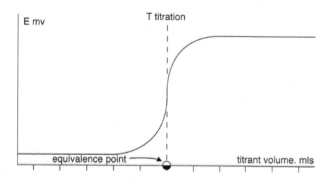

figure

A typical T-type titration (From [2], used with permission)

21.3.4.5 *Applications*

Some examples of applications of ISEs are measuring salt and nitrate in processed meats, salt content of butter and cheese, calcium in milk, sodium in low-sodium ice cream, carbon dioxide in soft drinks, potassium and sodium levels in wine, and nitrate in canned vegetables. An ISE method applicable to foods containing <100 mg sodium/100 g is an official method of AOAC International (Method 976.25). This method employs a sodium combination ISE, pH meter, and magnetic stirrer. Several official methods exist for using ISEs to detect the endpoint in chloride and salt measurement, such as AOAC methods 971.27 (sodium chloride in canned vegetables), 976.18 (salt in seafood), and 986.26 (chloride in milk-based infant formula).

21.4 BENCHTOP RAPID ANALYZERS FOR SALT

There are several rapid methods of salt determination used in food analysis that are not necessarily based on the traditional methods described above in this chapter. Nonetheless, it is useful for food science students to be aware of the methods and to compare the principles to the other methods previously described.

Salt concentration can be measured by either hydrometry or refractometry (Chap. 15, Sects. 15.2.5.2 and 15.2.5.3). A **salometer** is a hydrometer that measures brine strength using a scale that indicates percent of salt saturation. A **salinity refractometer** can be either a manual or automatic device that gives a measurement analogous to that of sugar concentration, or brix, using a typical refractometer. An advantage of the automatic devices, over the less expensive manual models, is that they can correct for changes in the refractive index of salt solutions as a function of temperature.

Several instrument manufacturers sell devices that can be used to determine the salt content of a sample in under a minute using an automatic titration. In these instruments, both the titrant dosing and the endpoint detection are controlled by the instrument which increases the precision over manual titrations where there is subjective error in both determining the titrant dose as well as the endpoint color change. These instruments work on the basic principle of precipitation titrations, as silver ions are still used to precipitate chloride ions, but there are a few differentiating features. For example, some instruments couple automated silver nitrate addition with endpoint determination using an ion-sensitive electrode. Several manufacturers offer models using this technique including Mettler-Toledo (Columbus, OH), Hanna Instruments (Woonsocket, RI), and Metrohm (Herisau, Switzerland). A second type of automatic chloride titrator does not rely on using a silver nitrate titrant but rather generates silver ions in the sample using the principle of coulometry. In instruments manufactured by Nelson-Jameson (Marshfield, WI), Sherwood Scientific (Cambridge, United Kingdom), and DKK-TOA (Toyko, Japan), silver ions are generated from a silver wire in situ via an electric current, and they complex with chloride ions in solution. The potential of the solution is monitored via an ISE, and the reaction is complete when the potential shifts dramatically as in Fig. 21.3. The electric current used to generate silver ions then may be converted to chloride and subsequently salt concentration. One advantage of this method is that no silver nitrate titrant is needed, and thus a standardization step is not necessary.

Another common method of salt determination in commercial laboratories is via conductance (e.g., DiCromat Salt Analyzer, Arrow Scientific, Gladesville, Australia). Salts in solution dissociate into negative and positive ions and thus increase conductance in proportion to their concentration. Conductance-based salt analyzers are rapid and also correct for differences in conductivity as a function of temperature.

21.5 COMPARISON OF METHODS

For labeling, processing, and even practical nutrition, the focus is on only a few minerals that generally can be analyzed by traditional methods. The traditional methods available for mineral analysis are varied, and only a very limited number of examples have been given in this chapter. Choice of methods for mineral analysis must be made considering method performance regarding accuracy, sensitivity, detection limit, specificity, and interferences. Information on method performance is available from the collaborative studies referenced with AOAC official methods. Other factors to be considered include cost per analysis completed, equipment availability, and analytical time compared to analytical volume.

Generally, for a small laboratory with skilled analytical personnel, the traditional methods can be carried out rapidly, with accuracy, and at minimal costs. If a large number of samples of a specific element are to be run, there is certainly a time factor in favor of using atomic absorption spectroscopy or emission spectroscopy, depending on the mineral being analyzed. Atomic absorption spectrophotometers with a graphite furnace or inductively coupled plasma-mass spectrometers are capable of sensitivity in the parts per billion range, which is much lower than the limits of traditional methods. Yet, instrumentation capable of such analysis is expensive and beyond the financial resources of many quality assurance laboratories, and often this degree of sensitivity is not required. This leaves the options of sending samples out to certified laboratories for analysis or utilizing one of the more traditional methods for analysis.

21.6 SUMMARY

The mineral content of water and foodstuffs is important because of their nutritional value, toxicological potential, and interactive effects with processing and texture of some foods. Traditional methods for mineral analysis include titrimetric and colorimetric procedures. The basic principles of these methods are described in this chapter, along with discussion of ISE methodology that has general application for mineral analysis, and some benchtop analyzers for salt content. Table 21.3 gives a summary of many methods covered in this chapter, with comparison to AAS and ICP-OES (Chap. 9).

Procedures are described in this chapter that illustrate use of these traditional methods to quantify minerals of concern in the food industry. These procedures generally require chemicals and equipment routinely available in an analytical laboratory and do not require expensive instrumentation. These methods may be suited to a small laboratory with skilled analytical personnel and a limited number of samples to be analyzed. The traditional procedures will often perform similarly to procedures requiring more instrumentation and may be more robust in actual practice.

Foods are typically ashed prior to traditional analyses because the methods generally require that the minerals be freed from the organic matrix of the foods. Sample preparation and analysis must include steps necessary to prevent contamination or loss of volatile elements, and must deal with a variety of potential interferences. Various approaches are described to account for these possible errors including use of reagent blanks, addition of spikes, and development of standard curves using appropriate mineral matrix background.

Traditional methods for mineral analysis are often automated or adapted to test kits for rapid analysis. Tests for water hardness and the Quantab® for salt determination are examples currently being used. The basic principles involved in traditional methods will continue to be utilized to develop inexpensive rapid methods for screening mineral content of foods and beverages. Familiarity with the traditional principles will allow the food analyst to obtain the best possible performance with the kits and adapt to problems that may be encountered.

21.7 STUDY QUESTIONS

1. What is the major concern in sample preparation for specific mineral analysis? How can this concern be addressed?
2. If the ammonia buffer is pH 11.5 rather than pH 10 in the EDTA complexometric titration to determine the hardness of water, would you expect to overestimate or underestimate the hardness? Explain your answer.
3. In a back-titration procedure, would overshooting the endpoint in the titration cause an over- or underestimation of the compound being quantified? Explain your answer.
4. Describe how and why there is a need to employ standards in background matrix, spikes, and reagent blanks.
5. Explain the principles of using an ISE to measure the concentration of a particular inorganic element in food. List the factors to control, consider, or eliminate for an accurate measure of concentration by the ISE method.
6. You have decided to purchase an ISE to monitor the sodium content of foods produced by your plant. List the advantages this would have over the Mohr/Volhard titration method. List the problems and disadvantages of ISE that you should anticipate.
7. What factors should be considered in selecting a specific method for mineral analysis for a food product?

21.3
table   Summary of mineral analysis methods

Method	Principle	Advantages	Disadvantages	Applications	AOAC method number
EDTA titration method	Description for sample containing just calcium: EDTA of titrant solution complexes with calcium, 1:1, at pH 10; EDTA binds to calcium ions stronger than the indicator binds to calcium. When all the calcium present in the sample has reacted the EDTA titrant, and no calcium is bound to the indicator, the indicator changes from pink to blue. Moles of calcium in the sample are equivalent to moles of EDTA used in titration	Rapid, especially when done with test strip. Inexpensive	Can be interfering compounds. Endpoint is subjectively determined	Testing water hardness. Calcium in ash of fruits and vegetables	AOAC Method 968.31, calcium in canned vegetables
Mohr titration	A forward titration. Chloride (in NaCl) is titrated with silver nitrate in the presence of potassium chromate. Silver reacts with the chloride; when all chloride is reacted, the excess silver reacts with chromate to form an orange-colored solid, silver chromate. Volume and molarity of silver nitrate are used to calculate amount of chloride, which relates to amount of NaCl	Does not require expensive equipment, highly trained personnel, or ashing of sample. Fewer reagents and less time-consuming than Volhard titration. Inexpensive (unless automated equipment)	Subjectivity of determining endpoint of titration (if manual titration)	Salt content of variety of foods. Test strip version and automated instruments available	AOAC Method 960.29, salt in dairy products and butter
Volhard titration	A backward titration. An excess of a standard solution of silver nitrate is added to a chloride-containing solution. Excess silver nitrates is back-titrated with a standardized solution of potassium or ammonium thiocyanate, with ferric ion as an indicator. Volume of thiocyanate solution used is proportional to the excess silver. Moles of total silver is equal to the sum of the moles of chloride in the sample and moles of thiocyanate in the titrant	Does not require expensive equipment, highly trained personnel, or ashing of sample. Rapid. Inexpensive	Subjectivity of determining endpoint of titration. Requires more reagents and time than Mohr titration	Salt content of variety of foods	AOAC Method 935.43 chloride (Total) in cheese; AOAC Method 915.01, chloride in plants
Colorimetric methods	Description for just mineral analysis. Chromogen in reagent reacts with mineral of interest to form a soluble colored compound that can be quantitated by absorption of light at a specific wavelength. Concentration of mineral of interest is determined from standard curve of absorbance vs. concentration, based on Beer's law	Applicable to wide variety of minerals. Very specific. Other minerals usually don't interfere. Less expensive than AAS or ICP-OES, but similar accuracy and precision	Requires significant technician time	Low-cost method for analyzing single element	AOAC Method 944.02, iron in wheat flour

(continued)

Method	Principle	Advantages	Disadvantages	Applications	AOAC Method
Ion-selective electrode	Principle is the same as for measuring pH (i.e., uses Nernst equation), but by varying the composition of the glass in the sensing electrode, the electrode can be sensitive to a specific mineral. Sensing and reference electrodes (often as combination electrode) are immersed in solution with element of interest; electrical potential that develops at surface of sensing electrode is measured by comparing to reference electrode with fixed potential. Voltage between sensing and reference electrode is related to ion activity, measured in mV. Ion activity is related to ion concentration via the activity coefficient, which is controlled by ionic strength. Concentration of element is determined using standard curve of mV vs. log concentration	Can measure many anions and cations directly. Does not require expensive equipment (only pH meter), highly trained personnel, or ashing of sample. Analysis is independent of turbidity, color, or viscosity	Cannot measure at low concentration. Electrode response can be slow. Sensing and reference electrode must be specific to element being measured. High rate of premature failure for some electrodes	Quality control, especially for Na or K	AOAC Method 976.25, sodium in foods for special dietary use
Atomic absorbance spectroscopy (AAS)	Heat energy from a flame converts molecules to atoms and then energy (at specific wavelength, from a hollow cathode lamp (HCL)) raises atoms of a specific element from ground to excited state. Measure absorption of energy, which is linearly related to concentration of element	Can measure many elements (i.e., not just sodium). More sensitive than Mohr or Volhard titration for sodium	Expensive instrument. Requires highly trained personnel. Requires ashing of most types of samples. Requires a different HCL for each element. Use potentially explosive fuel gas. More interferences than ICP-OES. Response is not linear over a broad concentration range	Single-element analysis in a given food or biological sample	AOAC Method 975.03, metals in plants and pet foods; AOAC Method 985.35, minerals in infant formula, enteral products, and pet foods
Inductively coupled plasma-optical emission spectrometry (ICP-OES)	Energy from a plasma converts molecules to atoms and ions and then plasma energy raises atoms from ground to excited state. Measure emission of energy (at specific wavelength) as excited atoms returns to a lower energy state; amount of radiant energy emitted is proportional to concentration of specific element	Can measure many elements (i.e., not just sodium) simultaneously. More sensitive than Mohr or Volhard titration for sodium. Few interferences. Response linear over a large concentration range	Expensive instrument. Requires highly trained personnel. Requires ashing of most types of samples	Multiple elements in a large number of samples. Nutrition labeling	AOAC Method 984.27, calcium, copper, iron, magnesium, manganese, phosphorus, potassium, sodium, and zinc in infant formula; AOAC Method 985.01, metals and other elements in plants and pet foods

21.8   PRACTICE PROBLEMS

1. If a given sample of food yields 0.750 g of silver chloride in a gravimetric analysis, what weight of chloride is present?

2. A 10-g food sample was dried, then ashed, and analyzed for salt (NaCl) content by the Mohr titration method ($AgNO_3 + Cl \rightarrow AgCl$). The weight of the dried sample was 2 g, and the ashed sample weight was 0.5 g. The entire ashed sample was titrated using a standardized $AgNO_3$ solution. It took 6.5 mL of the $AgNO_3$ solution to reach the endpoint, as indicated by the red color of Ag_2CO_4 when K_2CrO_4 was used as an indicator. The $AgNO_3$ solution was standardized using 300 mg of dried KCl. The corrected volume of $AgNO_3$ solution used in the titration was 40.9 mL. Calculate the salt (NaCl) content of the original food sample as percent NaCl (wt/wt).

3. A 25-g food sample was dried, then ashed, and finally analyzed for salt (NaCl) content by the Volhard titration method. The weight of the dried sample was 5 g, and the ashed sample weighed 1 g. Then 30 mL of 0.1 N $AgNO_3$ was added to the ashed sample, the resultant precipitate was filtered out, and a small amount of ferric ammonium sulfate was added to the filtrate. The filtrate was then titrated with 3 mL of 0.1 N KSCN to a red endpoint:

   (a) What was the moisture content of the sample, expressed as percent H_2O (wt/wt)?
   (b) What was the ash content of the sample, expressed as percent ash (wt/wt) on a dry weight basis?
   (c) What was the salt content of the original sample in terms of percent (wt/wt) NaCl? (molecular weight Na=23; molecular weight Cl=35.5).

4. Compound X in a food sample was quantified by a colorimetric assay. Use the following information and Beer's law to calculate the content of Compound X in the food sample, in terms of mg Compound X/100 g sample:

   (a) A 4-g sample was ashed.
   (b) Ashed sample was dissolved with 1 mL of acid and the volume brought to 250 mL.
   (c) A 0.75-mL aliquot was used in a reaction in which the total volume of the sample to be read in the spectrophotometer was 50 mL.
   (d) Absorbance at 595 nm for the sample was 0.543.
   (e) The absorptivity constant for the reaction (i.e., extinction coefficient) was known to be 1574 L M cm.
   (f) Inside diameter of cuvette for spectrophotometer was 1 cm.

5. Colorimetric analysis:

   (a) You are using a colorimetric method to determine the concentration of Compound A in your liquid food sample. This method allows a sample volume of 5 mL. This volume must be held constant but can be comprised of diluted standard solution and water. For this standard curve, you need standards that contain 0, 0.25, 0.50, 0.75, and 1.0 mg of Compound A. Your stock standard solution contains 5 g/L of Compound A. Devise a dilution scheme(s) for preparing the samples for this standard curve that could be followed by a lab technician. Be specific. In preparing the dilution scheme, use no volumes less than 0.5 mL.
   (b) You obtain the following absorbance values for your standard curve:

Sample(mg)	Absorbance (500 nm)
0.00	0.00
0.25	0.20
0.50	0.40
0.75	0.60
1.00	0.80

   Construct a standard curve and determine the equation of the line.
   (c) A 5-mL sample is diluted to 500 mL, and 3 mL of this solution is analyzed as per the standard samples. The absorbance is 0.50 units at 500 nm. Use the equation of the line calculated in part (b) and information about the dilutions to calculate what the concentration is of Compound A in your original sample in terms of g/L.

6. What is the original concentration of copper in a 100-mL sample that shows a potential change of 6 mV after the addition of 1 mL of 0.1 M $Cu(NO_3)_2$?

Answers

1.
$$\frac{x\,g\,Cl}{0.750\,g\,AgCl} = \frac{35.45g/mol}{143.3g/mol}$$
$$x = 0.186\,g\,Cl$$

2.
$$N\,AgNO_3 = \frac{0.300\,g\,KCl}{mL\,AgNO_3 \times 74.555\,g\,KCl/mol}$$

$$0.0984\,N = \frac{0.300\,g}{40.9\,mL \times 74.555}$$

Percent salt=

$$\left(\frac{0.0065\,L\times0.0984\,N\,AgNO_3\times58.5\,g/mol}{10\,g}\right)\times100$$

Percent salt $= 0.37\%$

3. (a) $\dfrac{25\,g\,wet\,sample-5\,g\,dry\,sample}{25\,g\,wet\,sample}\times100=80\%$

   (b) $\dfrac{1\,g\,ash}{5\,g\,dry\,sample}\times100=20\%$

   (c) $mol\,Ag\,added = mol\,Cl^-\,in\,sample +$
   $$mol\,SCN^-\,added$$

   $mol\,Ag = (0.1\,mol/L)\times(0.03\,L)=0.003\,mol$

   $mol\,SCN^- = (0.1\,mol/L)\times(0.003\,L)-0.0003\,mol$

   $0.003\,mol\,Ag = mol\,Cl^- + 0.0003\,mol\,SCN^-$

   $0.0027\,mol = mol\,Cl^-$

   $(0.0027\,mol\,Cl^-)\times\dfrac{58.5\,g\,NaCl}{mol}=0.1580\,g\,NaCl$

   $\dfrac{0.1580\,g\,NaCl}{25\,g\,wet\,sample}=\dfrac{0.00632\,g\,NaCl}{g\,wet\,sample}\times100$

   $$=0.63\%\,NaCl(w/w)$$

4. $A = abc$

   $0.543 = (1574\,L\,g^{-1}cm^{-1})(1\,cm)\,c$

   $c = 3.4498\times10^{-4}\,g/L$

   $c = 3.4498\times10^{-4}\,mg/mL$

   $\dfrac{3.4498\times10^{-4}\,mg}{mL}\times50\,mL=1.725\times10^{-2}\,mg$

   $\dfrac{1.725\times10^{-2}\,mg}{0.75\,mL}\times\dfrac{250\,mL}{4\,g}=1.437\,mg/g$

   $$=143.7\,mg/100\,g$$

5. (a) Lowest dilution volume for 1 mL of stock:

   $\dfrac{0.25\,mg}{0.5\,mL}=\dfrac{1\,mL}{x\,mL}\times\dfrac{5\,mg}{mL}$

   $x = 10\,mL$

Therefore, use a volumetric pipette to add 1 mL of stock to a 10-mL volumetric flask. Bring to volume with ddH$_2$0 to give a diluted stock solution of 0.25 mg/0.5 mL. Use this to make up standards according to the following table.

mg A/5 mL	mL Diluted stock solution	mL H$_2$O
0	0	5.0
0.25	0.5	4.5
0.50	1.0	4.0
0.75	1.5	3.5
1.0	2.0	3.0

(b)
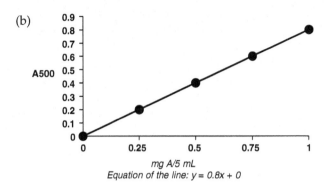
mg A/5 mL
Equation of the line: y = 0.8x + 0

(c) $A_{500} = 0.50 = y$

   $0.50 = 0.8x + 0$

   $x = 0.625$

   $\dfrac{0.625\,mg}{5\,mL}\times\dfrac{5\,mL}{3\,mL}\times\dfrac{500\,mL}{5\,mL}=20.8\,mg/mL$

   $$= 20.8\,g/L$$

6. $C_O = \dfrac{0.001\,L\times\dfrac{0.001\,moles}{L}\times\dfrac{1}{0.100\,L}}{\left(10^{0.006/0.285}-1\right)}=1.6\,mM$

Acknowledgment The author of this chapter wishes to acknowledge Dr. Charles E. Carpenter, who was an author of this chapter for the third and fourth editions of this textbook.

REFERENCES

1. Schwendt G (1997) The essential guide to analytical chemistry, Wiley, New York
2. Kirk RS, Sawyer R (1991) Pearson's composition and analysis of foods, 9th edn. Longman Scientific and Technical, Essex, England
3. Skoog DA, West DM, Holler JF, Crouch SR (2000) Analytical chemistry: an introduction, 7th edn., Brooks/Cole, Pacific Grove, CA
4. Harris DC (1999) Quantitative chemical analysis, 5th edn. W.H. Freeman and Co., New York
5. AOAC International (2016) Official methods of analysis, 20th edn., (On-line). AOAC International, Rockville, MD
6. Sullivan DM, Carpenter DE (eds) (1993) Methods of analysis for nutritional labeling. AOAC International, Arlington, VA
7. Food and Nutrition Board, Institute of Medicine (1997) Dietary reference intakes for calcium, phosphorus, magnesium, vitamin D, and fluoride. National Academy Press, Washington, DC
8. Food and Nutrition Board, Institute of Medicine (2000) Dietary reference intakes for vitamin C, vitamin E, selenium, and carotenoids. National Academy Press, Washington, DC
9. Food and Nutrition Board, Institute of Medicine (2002) Dietary reference intakes for vitamin A, vitamin K, arsenic, boron, chromium, copper, iodine, iron, manganese, molybdenum, nickel, silicon, vanadium, and zinc. National Academy Press, Washington, DC

10. Eaton AD, Clesceri LS, Rice EW, Greenburg AE (eds) (2005) Standard methods for the examination of water and wastewater, 21st edn. Method 2340, hardness. American Public Health Association, American Water Works Association, Water Environment Federation, Washington, DC, pp 2–37 to 2–39

11. Murphy J, Riley JP (1962) A modified single solution method for the determination of phosphate in natural waters. Anal Chim Acta 27:31–36

12. Covington AK (ed) (1980) Ion selective electrode methodology, CRC, Boca Raton, FL

13. Wang J (2006) Analytical electrochemistry, 3rd edn. Wiley, New York

14. European Hygienic Engineering and Design Group (2005). Safe and hygienic water treatment in food factories. Trends in Food Sci Tech 16:568–573

part

Chemical Characterization and Associated Assays

22
chapter

pH and Titratable Acidity

Catrin Tyl (✉)
Department of Food Science and Nutrition,
University of Minnesota,
St. Paul, MN 55108-6099, USA
e-mail: tylxx001@umn.edu

George D. Sadler
PROVE IT, LLC,
Geneva, IL 60134, USA
e-mail: gsadler@proveitllc.com

S. Nielsen (ed.), *Food Analysis*, Food Science Text Series,
DOI 10.1007/978-3-319-45776-5_22, © Springer International Publishing 2017

22.1 INTRODUCTION

There are two interrelated concepts in food analysis that deal with acidity: **pH** and **titratable acidity**. Each of these quantities is analytically determined in separate ways, and each provides its own particular insights on food quality.

Titratable acidity (also called **total acidity**) measures the **total acid concentration** in a food. This quantity is determined by exhaustive titration of intrinsic acids with a standard base. Titratable acidity is a better predictor of acid's impact on flavor than pH. However, total acidity does not give all relevant information on a food. For instance, the ability of a microorganism to grow in a specific food is more dependent on the concentration of free hydronium ions, H_3O^+, than titratable acidity. In aqueous solution, these hydronium ions arise from the combination of water with hydrogen ions (H^+) dissociated from acids. The need to quantify only the free H_3O^+ concentration leads to the second major concept of acidity, that being pH (also called active acidity and discussed in Sect. 22.3). pH is defined as the negative log (base 10) of the hydrogen ion concentration and can span a range of 14 orders of magnitudes. In foods, it is not only a function of the type and concentrations of acids present, but also of the concentration of ionized acid counterparts (their conjugated bases). If present in similar amounts, the system can be described as a buffer (see Sect. 22.4.2.1 and Chap. 2 in Food Analysis laboratory manual).

For general and food-specific information on measuring pH and titratable acidity, see Refs. [1–16]. For the pH and titratable acidity of select foods, see Ref. [10].

22.2 CALCULATION AND CONVERSION FOR NEUTRALIZATION REACTIONS

22.2.1 Concentration Units

This chapter deals with the theory and practical application of titratable acidity calculation and pH determination. To quantitatively measure components of foods, solutions must be prepared to accurate concentrations and diluted into the desired working range (see Food Analysis Laboratory Manual, Chap. 3, Dilutions and Concentrations).

The common terms used for concentration in food analysis should be reviewed (see Food Analysis Laboratory Manual, Chap. 2, Preparation of Reagents and Buffers, Table 2.1). Molarity (M) and normality (N) are the most common SI (International Scientific) terms used in food analysis, but solutions also can be expressed as percentages. It is important that the analyst be able to convert between both systems.

Molarity (M) is a concentration unit representing the number of moles of the solute per liter of solution.

Normality (N) is a concentration unit representing the number of equivalents (Eq) of a solute per liter of solution. In acid and base solutions, the normality represents the concentration or moles of H^+ or OH^- per liter that will be exchanged in a neutralization reaction when taken to completion. For oxidation-reduction reagents, the normality represents the concentration or moles of electrons per liter to be exchanged when the reaction is taken to completion. The following are some examples of molarity vs. normality (equivalents):

Acid-Base Reactions:

$$1M\,H_2SO_4 = 2N\,H_2SO_4$$

Two equivalents of H^+ per mol of acid

$$1M\,NaOH = 1N\,NaOH$$

One equivalent of OH^+ per mol of base

$$1M\,CH_3COOH = 1N\,acetic\ acid$$

One equivalent of H^+ per mol of acid

$$1M\,H_2C_4H_4O_5 = 2N\,malic\ acid$$

Two equivalents of H^+ per mol of acid

Oxidation-Reduction Reactions:

$$HSO_3^- + I_2 + H_2O \rightleftarrows SO_4^{2-} + 2I^- + 3H^+$$

$$1M\,I_2 = 2\,N\,iodine$$

Two equivalents of electrons gained per mol of I_2

$$1M\,HSO_3^- = 2N\,bisulfite$$

Two equivalents of electrons lost per mol of bisulfite

Many analytical determinations in food analysis use the concept of equivalents to measure the normality or equivalency of an unknown solution. Such tests use the stoichiometry or the whole number ratios of atoms that react to produce a particular product. Perhaps the most familiar of these is **acid-base reactions** in which hydrogen ions are exchanged and can be quantified through stoichiometric neutralization with a standard base. Examples for food analyses that use acid-base titration include nitrogen in the Kjeldahl protein determination (see Chap. 18, Sect. 18.2.1.3.4), benzoic acid in sodas, and determining percent titratable acidity. The concept of equivalents also is used in oxidation-reduction problems to quantify unknown analytes that are capable of direct electron transfer.

Equivalent weight can be defined as the molecular weight divided by the number of equivalents in the reactions. For example, the molecular weight of H_2SO_4 is 98.08 g. Since there are two equivalents per

Molecular and equivalent weights of common food acids

Acid	Chemical formula	Molecular weight	Equivalents per mole	Equivalent weight
Citric (anhydrous)	$H_3C_6H_5O_7$	192.12	3	64.04
Citric (hydrous)	$H_3C_6H_5O_7 \cdot H_2O$	210.14	3	70.05
Acetic	$HC_2H_3O_2$	60.06	1	60.05
Lactic	$HC_3H_5O_3$	90.08	1	90.08
Malic	$H_2C_4H_4O_5$	134.09	2	67.05
Oxalic	$H_2C_2O_4$	90.04	2	45.02
Tartaric	$H_2C_4H_4O_6$	150.09	2	75.05
Ascorbic	$H_2C_6H_6O_6$	176.12	2	88.06
Hydrochloric	HCl	36.47	1	36.47
Sulfuric	H_2SO_4	98.08	2	49.04
Phosphoric	H_3PO_4	98.00	3	32.67
Potassium acid phthalate	$KHC_8H_4O_4$	204.22	1	204.22

mole of H_2SO_4, the equivalent weight of H_2SO_4 is 49.04 g. Table 22.1 provides a list of molecular and equivalent weights for acids important in food analysis. In working with normality and milliliters, the term **milliequivalents** (mEq) is usually preferred. Milliequivalent weight is the equivalent weight divided by 1,000.

Percentage concentrations are the mass amount of solute or analyte per 100 mL or 100 g of material. Unlike molarity, normality, and equivalents, percent concentrations neither require nor provide insight into the stoichiometry of a reaction. Percentage can be expressed as volume of solution or for weight of solids and can be on a volume basis (v) or mass basis (w). Therefore the particular basis is normally stated. For example, 10% (w/v), 10% (v/v), and 10% (w/w) express different concentrations in all cases for which the specific gravities of the two components are not the same. When the percentage becomes a number less than 1%, **parts per million** (ppm), **parts per billion** (ppb), and even **parts per trillion** (ppt) usually are preferred. If percentage is defined as the mass of the solute or analyte per mass (or volume) of sample × 100, then ppm is simply the same ratio of mass of solute per mass of sample × 1,000,000 (see Food Analysis Laboratory Manual, Chap. 2, Table 2.2).

22.2.2 Equation for Neutralization and Dilution

There are some general rules in evaluating equilibrium reactions that are helpful in most situations. At full neutralization the equivalents of one reactant in the neutralization equal the equivalents of the other reactant. This can be expressed mathematically as:

$$(\text{vol of } X)(N \text{ of } X) = (\text{vol of } Y)(N \text{ of } Y) \quad (22.1)$$

Equation 22.1 also can be used to solve dilution problems where X represents the stock solution and Y rep-resents the working solution. When Eq. 22.1 is used for dilution problems, any value of concentration (grams, moles, ppm, etc.) can be substituted for N. Units should be recorded with each number since cancelation of units provides a quick check on proper setup of the problem. (See Food Analysis Laboratory Manual, Chap. 2, Practice Problems 1–11.)

22.3   pH

22.3.1 Acid-Base Equilibria

The Brønsted-Lowry theory of neutralization is based upon the following definitions for acid and base:

Acid: A substance capable of donating protons. In food systems the only significant proton donor is the hydronium ion.
Base: A substance capable of accepting protons.

Neutralization is the reaction of an acid with a base to form a salt as shown below:

$$HCl + NaOH \rightleftarrows NaCl + H_2O \quad (22.2)$$

Acids form hydrated protons called **hydronium ions** (H_3O^+), and bases form **hydroxide ions** (OH^-) in aqueous solutions:

$$H_3O^+ + OH^- \rightleftarrows 2H_2O \quad (22.3)$$

At any temperature, the product of the molar concentrations (moles/liter) of H_3O^+ (note: molar concentrations are commonly signified by brackets) and OH^- is a constant referred to as the **ion product constant for water** (K_w):

$$\left[H_3O^+\right]\left[OH^-\right] = K_w \quad (22.4)$$

K_w varies with the temperature. For example, at 25 °C, $K_w = 1.04 \times 10^{-14}$ but at 100 °C, $K_w = 58.2 \times 10^{-14}$.

The above concept of K_w leads to the question of what the $[H_3O^+]$ and $[OH^-]$ are in pure water. Experimentation has revealed that the $[H_3O^+]$ is approximately $1.0 \times 10^{-7} M$, as is that of the $[OH^-]$ at 25 °C. Because the concentrations of these ions are equal, pure water is referred to as being **neutral**.

Suppose that a drop of acid is added to pure water. The $[H_3O^+]$ would increase. However, Kw would remain constant (1.0×10^{-14}), revealing a decrease in the $[OH^-]$. Conversely, if a drop of base is added to pure water, the $[H_3O^+]$ would decrease, while the $[OH^-]$ would increase, maintaining the Kw at 1.0×10^{-14} at 25 °C.

How did the term pH derive from the above considerations? In approaching the answer to this question, one must observe $[H_3O^+]$ and $[OH^-]$ in various foods, as shown in Table 22.2. The numerical values found in Table 22.2 for $[H_3O^+]$ and $[OH^-]$ are bulky and led a Swedish chemist, S.L.P. Sørensen, to develop the pH system in 1909.

pH is defined as the logarithm of the reciprocal of the hydrogen ion concentration, which is equal to the negative logarithm of the molar concentration of hydrogen ions.

$$pH = -\log\left[H^+\right] \qquad (22.5)$$

Thus, a $[H_3O^+]$ of 1×10^{-6} is expressed simply as pH 6. The $[OH^-]$ is expressed as pOH and would be pOH 8 in this case, as shown in Table 22.3.

Calculation of [H+] of a beer with pH 4.30:

Step 1. Substitute numbers into the pH equation:

$$pH = -\log\left[H^+\right]$$
$$4.30 = -\log\left[H^+\right]$$
$$-4.30 = \log\left[H^+\right]$$
$$-4.30 = 0.70 - 5 = \log\left[H^+\right]$$

Step 2. Calculate the antilog of −4.30: $\left[H^+\right] = 5 \times 10^{-5} M$

22.2 table Concentrations of H_3O^+ and OH^- in various foods at 25 °C

Food	$[H_3O^+]^a$	$[OH^-]^a$	K_w
Cola	2.24×10^{-3}	4.66×10^{-12}	1×10^{-14}
Grape juice	5.62×10^{-4}	1.78×10^{-11}	1×10^{-14}
Seven-Up	3.55×10^{-4}	2.82×10^{-11}	1×10^{-14}
Schlitz beer	7.95×10^{-5}	1.26×10^{-10}	1×10^{-14}
Pure water	1.00×10^{-7}	1.00×10^{-7}	1×10^{-14}
Tap water	4.78×10^{-9}	2.09×10^{-6}	1×10^{-14}
Milk of magnesia	7.94×10^{-11}	1.26×10^{-4}	1×10^{-14}

From Pecsok et al. [14], used with permission. Copyright 1971 American Chemical Society
^{a}Moles per liter

22.3 table Relationship of [H+] vs. pH and [OH−] vs. pOH at 25 °C

$[H^+]^1$	pH	$[OH^-]^a$	pOH
1×10^0	0	1×10^{-14}	14
10^{-1}	1	10^{-13}	13
10^{-2}	2	10^{-12}	12
10^{-3}	3	10^{-11}	11
10^{-4}	4	10^{-10}	10
10^{-5}	5	10^{-9}	9
10^{-6}	6	10^{-8}	8
10^{-7}	7	10^{-7}	7
10^{-8}	8	10^{-6}	6
10^{-9}	9	10^{-5}	5
10^{-10}	10	10^{-4}	4
10^{-11}	11	10^{-3}	3
10^{-12}	12	10^{-2}	2
10^{-13}	13	10^{-1}	1
10^{-14}	14	10^0	0

From Pecsok et al. [14], used with permission. Copyright 1971 American Chemical Society
^{a}Moles per liter. Note that the product of $[H^+][OH^-]$ is always 1×10^{-14}

While the use of pH notation is simpler from the numerical standpoint, it is a confusing concept in the minds of many students. One must remember that it is a logarithmic value and that a change in 1 pH unit is actually a tenfold change in the $[H_3O^+]$.

It is important to understand that pH and titratable acidity are not the same. Strong acids such as hydrochloric, sulfuric, and nitric acids are almost fully dissociated at pH 1. Only a small percentage of food acid molecules (citric, malic, acetic, tartaric, etc.) dissociate in solution. This point may be illustrated by comparing the pH of 0.1N solutions of hydrochloric and acetic acid.

$$HCl \rightleftarrows H^+ + Cl^- \qquad (22.6)$$

$$CH_3COOH \rightleftarrows H^+ + CH_3COO^- \qquad (22.7)$$

The HCl almost fully dissociates in solution to produce a pH of 1.02 at 25 °C. By contrast, only about 1 % of CH_3COOH is ionized at 25 °C, producing a significantly higher pH of 2.89. The calculation and significance of partial dissociation on pH is presented in more detail in Sect. 22.4.2.1.

22.3.2 pH Meter

22.3.2.1 *Activity Versus Concentration*

In using pH electrodes, the concept of activity vs. concentration must be considered. **Activity** is a measure of expressed chemical reactivity, while **concentration** is a measure of all forms (free and bound) of ions in solution. Because of the interactions of ions between themselves and with the solvent, the effective concentration or activity is, in general, lower than the actual concentration, although activity and concentration tend to approach each other at infinite dilution. Activity and concentration are related by the following equation:

$$A = \gamma C \qquad (22.8)$$

where:

> A = activity
> γ = activity coefficient
> C = concentration

The **activity coefficient** is a function of ionic strength. Ionic strength is a function of the concentration of, and the charge on, all ions in solution. Activity issues can become significant for hydronium ions below pH 1 and for hydroxyl ions at pH 13 and above. While such extreme pH values are not commonly encountered in food science, ionic strengths may be high, and thus different calculations would need to be used (refer to relevant literature, e.g., Ref. [1]).

22.3.2.2 General Principles

The pH meter is a good example of a **potentiometer** (a device that measures voltage at infinitesimal current flow). The basic principle of **potentiometry** (an electrochemical method of voltammetry at zero current) involves the use of an electrolytic cell composed of two electrodes immersed into a test solution. A voltage develops, which is related to the ionic concentration of the solution. Since the presence of current could alter the concentration of surrounding ions or produce irreversible reactions, this voltage is measured under conditions such that infinitesimal current (10^{-12} amperes or less) is drawn.

Four major parts of the pH system are needed: (1) reference electrode, (2) indicator electrode (pH sensitive), (3) voltmeter or amplifier that is capable of measuring small voltage differences in a circuit of very high resistance, and (4) the sample being analyzed (Fig. 22.1). For convenience most contemporary pH electrodes combine the reference electrode and indicator electrode in a common housing (see Sect. 22.3.2.5).

Each of the two electrodes involved in the measurement is designed carefully to produce a constant, reproducible potential. Therefore, in the absence of other ions, the potential difference between the two electrodes is fixed and easily calculated. However, H_3O^+ ions in solution contribute a new potential across an ion-selective glass membrane built into the indicating electrode. This alters the potential difference between the two electrodes in a way that is proportional to H_3O^+ concentration. The new potential resulting from the combination of all individual potentials is called the **electrode potential** and is readily convertible into pH readings.

Hydrogen ion concentration (or more accurately, activity) is determined by the voltage that develops

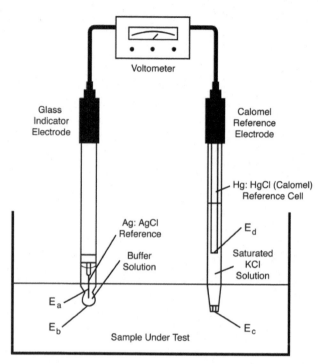

 figure The measuring circuit of the potentiometric system. E_a, contact potential between Ag/AgCl electrode and inner liquid. E_a is independent of pH of the test solution but is temperature dependent. E_b, potential developed at the pH-sensitive glass membrane. E_b varies with the pH of the test solution and also with temperature. In addition to this potential, the glass electrode also develops an asymmetry potential, which depends upon the composition and shape of the glass membrane. It also changes as the electrode ages. E_c, diffusion potential between saturated KCl solution and test sample. E_c is essentially independent of the solution under test. E_d, contact potential between calomel portion of electrode and KCl salt bridge. E_d is independent of the solution under test but is temperature dependent (From Bergveld et al. [4], used with permission)

between the two electrodes. The **Nernst equation** relates the electrode response to the activity where:

$$E = E_o + 2.303 \frac{RT}{zF} \log A \qquad (22.9)$$

where:

> E = measured electrode potential
> E_o = standard electrode potential, a constant representing the sum of the individual potentials in the system at a standard temperature, ion concentration, and electrode composition

R = universal gas constant, 8.313 $\left(\dfrac{J}{mol\ K}\right)$

F = Faraday constant, 96,490 coulombs (C) per mole $\left(\dfrac{C}{mol}\right)$

T = absolute temperature in Kelvin (K)

Z = number of charges on the ion

A = activity of the ion being measured

For monovalent ions (such as the hydronium ion) at 25 °C, the relationship of 2.303 RT/F is calculated to be 0.0591, as follows:

$$\frac{2.303 \times 8.316 \times 298}{96490} = 0.0591 \qquad (22.10)$$

Thus, at 25 °C voltage produced by the electrode system is a linear function of the pH, the electrode potential being essentially + 59 mV (0.059 V) for each change of 1 pH unit. At neutrality (pH 7), the electrode potential is 0 mV. At pH 6, the electrode potential is + 59 mV, while at pH 4, the electrode potential is + 177 mV. Conversely, at pH 8, the electrode potential is – 59 mV (Fig. 22.2).

It must be emphasized that the above relationship between millivolts and pH exists only at 25 °C, and changes in temperature will alter the pH reading in accordance with Eq. 22.8. For example, at 0 °C, the electrode potential is 54 mV, while at 100 °C, it is 70 mV. Modern pH meters have a sensitive attenuator (temperature compensator) built into them in order to account for this effect of temperature.

22.3.2.3 Reference Electrode

The **reference electrode** is needed to complete the circuit in the pH system. This half-cell is one of the most troublesome parts of the pH meter. Problems in obtaining pH measurements are often traced to a faulty reference electrode.

The **saturated calomel electrode** (Fig. 22.1) is the most common reference electrode. It is based upon the following reversible reaction:

$$Hg_2Cl_2 + 2e^- \rightleftarrows 2Hg + 2Cl^- \qquad (22.11)$$

The $E_{0,25\,°C}$ for the saturated KCl salt bridge is +0.2444 V vs. a standard hydrogen electrode; the Nernst equation for the reaction is as follows:

$$E = E_0 - 0.059/2 \log\left[Cl^-\right]^2 \qquad (22.12)$$

Thus, one observes that the potential is dependent upon the chloride ion concentration, which is easily regulated by the use of saturated KCl solution in the electrode.

A calomel reference electrode has three principal parts: (1) a platinum wire covered with a mixture of calomel (Hg_2Cl_2), (2) a filling solution (saturated KCl), and (3) a permeable junction through which the filling solution slowly migrates into the sample being measured. Junctions are made of ceramic or fibrous material. These junctions tend to clog up, causing a slow, unstable response and inaccurate results.

A less widely used reference electrode is the **silver-silver chloride electrode**. Because the calomel electrode is unstable at high temperatures (80 °C) or in strongly basic samples (pH >9), a silver-silver chloride electrode must be used for such application. It is a very reproducible electrode based upon the following reaction:

$$AgCl(s) + e^- \rightleftarrows Ag(s) + Cl^- \qquad (22.13)$$

The internal element is a silver-coated platinum wire, the surface silver being converted to silver chloride by hydrolysis in hydrochloric acid. The filling solution is a mixture of 4M KCl, saturated with AgCl that is used to prevent the AgCl surface of the internal element from dissolving. The permeable junction is usually of the porous ceramic type. Because of the relative insolubility of AgCl, this electrode tends to clog more readily than the calomel reference electrode. More commonly, a double-junction electrode configuration is used in which a separate inner body holds the Ag/AgCl internal element electrolyte and ceramic junction (see Sect. 22.3.2.5). An outer body containing a second electrolyte and junction isolates the inner body from the sample.

22.3.2.4 Indicator Electrode

The **indicator electrode** most commonly used in measuring pH today is referred to as the **glass electrode**. Like the reference electrode, the indicator electrode (Fig. 22.1) also has three principal parts: (1) a silver-silver chloride electrode with a mercury connection that is needed as a lead to the potentiometer; (2) a

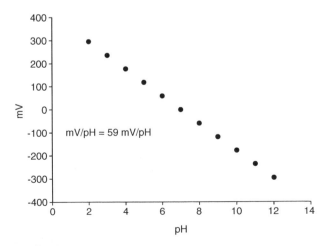

mV/pH = 59 mV/pH

22.2
figure Relationship between mV and pH for a monoprotic acid at 25 °C

buffer solution consisting of $0.01\,N$ HCl, $0.09\,N$ KCl, and acetate buffer used to maintain a constant pH (E_a); and (3) a small pH-sensitive glass membrane for which the potential (E_o) varies with the pH of the test solution. In using the glass electrode as an indicator electrode in pH measurements, the measured potential (measured against the calomel electrode) is directly proportional to the pH as discussed earlier, $E = E_0 - 0.059$ pH.

Conventional glass electrodes are suitable for measuring pH in the range of pH 1–9. However, this electrode is sensitive to higher pH, especially in the presence of sodium ions. Thus, equipment manufacturers have developed modern glass electrodes that are usable over the entire pH range of 0–14 and feature a very low sodium ion error, such as <0.01 pH at 25 °C.

22.3.2.5 *Combination Electrodes*

Today, most food analysis laboratories use **combination electrodes** that combine both the pH and reference electrodes along with the temperature sensing probe in a single unit or probe. These combination electrodes are available in many sizes and shapes from very small microprobes to flat surface probes, from all glass to plastic, and from exposed electrode tip to jacketed electrode tips to prevent glass tip breakage. Microprobes may be used to measure pH of very small systems such as inside a cell or a solution on a microscope slide. Flat surface electrode probes can be used to measure pH of semisolid and high viscosity substances such as meat, cheese, and agar plates and small volumes as low as 10 μl.

22.3.2.6 *Guidelines for Use of pH Meter*

It is very important that the pH meter be operated and maintained properly. One should always follow the specific instructions provided by the manufacturer. For maximum accuracy, the meter should be standardized using at least two buffers (**two-point calibration**). Select two buffers of pH values about 3 pH units apart, bracketing that of the anticipated sample pH. The three standardization buffers used most widely in laboratories are a pH 4.0 buffer, a pH 7.0 buffer, and a pH 9.0 buffer (at 25 °C). These are the typical pink, yellow, and blue solutions found adjacent to pH meters in many laboratories.

When standardizing the pH electrode, follow manufacturer's instructions for **one-point calibration**; rinse thoroughly with distilled water and blot dry. Immerse electrode in the second buffer (e.g., pH 4) and perform a second standardization. This time, the pH meter slope control is used to adjust the reading to the correct value of the second buffer. Repeat these two steps, if necessary, until a value within 0.1 pH unit of the correct value of the second buffer is displayed. If this cannot be achieved, the instrument is not in good working condition. Electrodes should be checked,

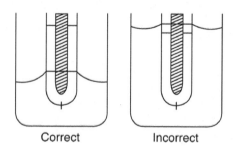

Correct   Incorrect

Correct and incorrect depth of calomel electrodes in solutions (Reprinted with permission from Pecsok et al. [14. Copyright 1971 American Chemical Society)

remembering that the reference electrode is more likely in need of attention. One should always follow the electrode manufacturer's specific directions for storage of a pH electrode. In this way, the pH meter is always ready to be used and the life of the electrodes is prolonged. One precaution that should be followed pertains to a calomel reference electrode. The storage solution level always should be at least 2 cm below the saturated KCl solution level in the electrode to prevent diffusion of storage solution into the electrode (Fig. 22.3).

22.4 TITRATABLE ACIDITY

22.4.1 Overview and Principle

The titratable acidity measures the total acid concentration in a food. Food acids are usually organic acids, with citric, malic, lactic, tartaric, and acetic acids being the most common. However, inorganic acids such as phosphoric and carbonic (arising from carbon dioxide in solution) acids often play an important and even predominant role in food acidulation. The organic acids present in foods influence the flavor (i.e., tartness), color (though their impact on anthocyanin and other pH-influenced pigments), microbial stability (via inherent pH-sensitivity characteristics of organisms), and keeping quality (arising from varying chemical sensitivities of food components to pH). The titratable acidity of fruits is used, along with sugar content, as an indicator of maturity (Sect. 22.4.6). While organic acids may be naturally present in the food, they also may be formed through fermentation, or they may be added as part of a specific food formulation.

Titratable acidity is determined by neutralizing the acid present in a known quantity (weight or volume) of food sample using a standard base. The endpoint for titration is usually either a target pH or the color change of a pH-sensitive dye, typically phenolphthalein. The volume of titrant used, along with the normality of the base and the volume (or weight) of sample, is used to calculate the titratable acidity, expressed in terms of the predominant organic acid.

22.4.2 General Considerations

Many food properties correlate better with pH than with acid concentration. The pH is also used to determine the endpoint of an acid-base titration. The pH determination can be achieved directly with a pH meter, but more commonly using a pH-sensitive dye. In some cases, the way pH changes during titration can lead to subtle problems. Some background in acid theory is necessary to fully understand titration and to appreciate the occasional problems that might arise.

22.4.2.1 *Buffering*

Although pH can hypothetically range from −1 to 14, pH readings below 1 are difficult to obtain due to incomplete dissociation of hydrogen ions at high acid concentrations. At $0.1N$, strong acids are assumed to be fully disassociated. Therefore, for titrations involving strong acids, fully dissociated acid is present at all titrant concentrations; the pH at any point in the titration is equal to the hydrogen ion concentration of the remaining acid (Fig. 22.4).

In contrast to strong acid titration, when weak acids are titrated with strong base, a mixture of this acid and its conjugated base is present in the system. This tends to cushion the solution from abrupt changes in pH. This property of a solution to resist change in pH is termed buffering. **Buffering** occurs in foods whenever a weak acid and its conjugated base are present in the same medium. Because of buffering, a graph of pH vs. titrant concentration is more complex for weak acids than strong acids. However, this

relationship can be estimated by the **Henderson-Hasselbalch equation**:

$$pH = pKa + \log\frac{[A-]}{[HA]} \qquad (22.14)$$

[HA] represents the concentration of undissociated acid. $[A^-]$ represents the concentration of its **conjugated base**. The **pK$_a$** is the pH at which equal quantities of undissociated acid and conjugated base are present. The equation indicates that maximum buffering capacity will exist when the pH equals the pK_a. A graph showing the titration of $0.1N$ acetic acid with $0.1N$ NaOH illustrates this point (Fig. 22.5).

Di- and triprotic acids will have two and three buffering regions, respectively. A pH vs. titrant graph of citric acid is given in Fig. 22.6. If the pK_a steps in polyprotic acids differ by three or more pK_a units, then the Henderson-Hasselbalch equation can predict the plateau corresponding to each step. However, the transition region between steps is complicated by the presence of protons and conjugate bases arising from other disassociation state(s). Consequently, the Henderson-Hasselbalch equation breaks down near the equivalence point between two pK_a steps. However, the pH at the equivalence point is easily calculated. The pH is simply (pK$_a$1 + pK$_a$2)/2. Table 22.4 lists pK_a values of acids important in food analysis.

Precise prediction of pH by the Henderson-Hasselbalch equation requires that all components form ideal solutions. An approximation to ideal solutions occurs at infinite dilution for all active components. However, real solutions may not behave ideally. For such solutions, the Henderson-Hasselbalch equation may only provide a good estimate of pH. For more details about buffers, and for practice problems

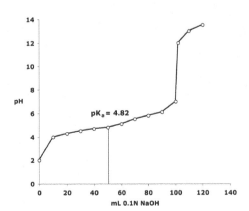

22.4 **figure** Titration of a strong acid with strong base. The pH at any point in the titration is dictated by the hydrogen ion concentration of the acid remaining after partial neutralization with base

22.5 **figure** Titration of a weak monoprotic acid with a strong base. A buffering region is established around the pK_a (4.82). The pH at any point is described by the Henderson-Hasselbalch equation

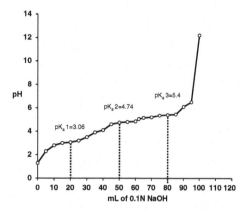

figure Titration of a weak polyprotic acid with a strong base. Buffering regions are established around each pK_a. The Henderson-Hasselbalch equation can predict the pH for each pK_a value if pK_a steps are separated by more than three units. However, complex transition mixtures between pK_a steps make simple calculations of transition pH values impossible

22.4 table pK_a values for some acids important in food analysis[a]

Acid	pK_a1	pK_a2	pK_a3
Oxalic	1.19	4.21	–
Phosphoric	2.12	7.21	12.30
Tartaric	3.02	4.54	–
Malic	3.40	5.05	–
Citric	3.06	4.74	5.40
Lactic	3.86	–	–
Ascorbic	4.10	11.79	–
Acetic	4.76	–	–
Potassium acid phthalate	5.40	–	–
Carbonic	6.10	10.25	–

[a]Note that same values differ from those listed in Table 2.2 of the food Analysis laboratory Manual, Chap. 2. See that chapter for information on factors affecting dissociation constants

using the Henderson-Hasselbalch equation to prepare buffers and calculate pH, see the Food Analysis Laboratory Manual, Chap. 2.

22.4.2.2 Potentiometric Titration

At the **equivalence point** in a titration, the number of acid equivalents exactly equals the number of base equivalents, and total acid neutralization is achieved. As the equivalence point is approached, the denominator [HA] in the Henderson-Hasselbalch equation becomes insignificantly small, and the quotient [A$^-$]/[HA] increases exponentially. As a result, the solution pH rapidly increases and ultimately approaches the pH of the titrant. The exact equivalent point is the halfway mark on this slope of abrupt pH increase. The use of a pH meter to identify the endpoint is called the **potentiometric method** for determining titratable acidity. The advantage of determining the equivalence point potentiometrically is that the precise equivalence point is identified. Since a rapid change in pH (and not some final pH value per se) signals the end of titration, accurate calibration of the pH meter is not even essential. However, in order to identify the equivalence point, a careful record of pH vs. titrant must be kept. This and the physical constraints of pH probes and slow response with some electrodes make the potentiometric approach somewhat cumbersome.

22.4.2.3 Indicators

For simplicity in routine work, an indicator solution is often used to approximate the equivalence point. This approach tends to overshoot the equivalence point by a small amount. When indicators are used, the term **endpoint** or **colorimetric endpoint** is substituted for equivalence point. This emphasizes that the resulting values are approximate and dependent on the specific indicator. Phenolphthalein is the most common indicator for food use. It changes from clear to red in the pH region 8.0–9.6. Significant color change is usually present by pH 8.2. This pH is termed the **phenolphthalein endpoint**.

A review of pK_a values in Table 22.4 indicates that naturally occurring food acids do not buffer in the region of the phenolphthalein endpoint. However, phosphoric acid (used as an acidulant in some soft drinks) and carbonic acid (carbon dioxide in aqueous solution) do buffer at this pH. Consequently, taking the solution from the true equivalence point to the endpoint may require a large amount of titrant when quantifying these acids. Indistinct endpoints and erroneously large titration values may result. When these acids are titrated, potentiometric analysis is usually preferred. Interference by CO_2 can be removed by boiling the sample and titrating the remaining acidity to a phenolphthalein endpoint.

Deeply colored samples also present a problem for endpoint indicators. When colored solutions obscure the endpoint, a potentiometric method is normally used. For routine work, pH vs. titrant data are not plotted as shown in Figs. 22.5 and 22.6. Samples are simply titrated to pH 8.2 (the phenolphthalein endpoint). Even though this is a potentiometric method, the resulting value is an endpoint and not the true equivalence point, since it simply reflects the pH value for the conventional phenolphthalein endpoint which slightly overshoots the equivalence point.

A pH of 7 may seem to be a better target for a potentiometric endpoint than 8.2. This pH, after all, marks the point of true neutrality on the pH scale. However, once all acid has been neutralized, the conjugate base remains. As a result, the pH at the equivalence point is often slightly greater than 7. Confusion also might arise if pH 7 was the target for colored samples and pH 8.2 was the target for noncolored samples.

Dilute acid solutions (e.g., vegetable extracts) require dilute solutions of standard base for optimal accuracy in titration. However, a significant volume of dilute alkali may be required to take a titration from the equivalence point to pH 8.2. Bromothymol blue is sometimes used as an alternative indicator in low-acid situations. It changes from yellow to blue in the pH range 6.0–7.6. The endpoint is usually a distinct green. However, endpoint identification is somewhat more subjective than the phenolphthalein endpoint.

Indicator solutions rarely contain over a few tenths percent dye (w/v). All indicators are either weak acids or weak bases that tend to buffer in the region of their color change. If added too liberally, they can influence the titration by conferring their own buffering to the sample under analysis. Therefore, indicator solutions should be held to the minimum amount necessary to impart effective color. Typically, two to three drops of indicator are added to the solution to be titrated. The lower the indicator concentration, the sharper will be the endpoint.

22.4.3 Preparation of Reagents

22.4.3.1 *Standard Alkali*

Sodium hydroxide (NaOH) is the most commonly used base in titratable acidity determinations. In some ways, it appears to be a poor candidate for a **standard base**. Reagent grade NaOH is very hygroscopic and often contains significant quantities of insoluble sodium carbonate (Na_2CO_3). Consequently, the normality of working solutions is not precise, so each new batch of NaOH must be standardized against an acid of known normality. However, economy, availability, and long tradition of use for NaOH outweigh these shortcomings. Working solutions are normally made from a stock solution containing 50% sodium hydroxide in water (w/v). (See Food Analysis Laboratory Manual, Chap. 2, Preparation of Reagents and Buffers, and Chap. 21, Standard Solutions and Titratable Acidity.) Sodium carbonate is essentially insoluble in concentrated alkali and gradually precipitates out of solution over the first 10 days of storage.

The NaOH can react with dissolved and atmospheric CO_2 to produce new Na_2CO_3. This reduces alkalinity and sets up a carbonate buffer that can obscure the true endpoint of a titration. Water and CO_2 alone can also react to form buffering compounds and generate hydrogen ions, as shown in the following equations:

$$H_2O + CO_2 \leftrightarrow H_2CO_3 \,(\text{carbonic acid}) \quad (22.15)$$

$$H_2CO_3 \leftrightarrow H^+ + HCO_3^- \,(\text{bicarbonate}) \quad (22.16)$$

$$HCO_3^- \leftrightarrow H^+ + CO_3^{-2} \,(\text{carbonate}) \quad (22.17)$$

Therefore, CO_2 should be removed from water prior to making the stock solution. This can be achieved by purging water with CO_2-free gas for 24 h or by boiling distilled water for 20 min and allowing it to cool before use. During cooling and long-term storage, air (with accompanying CO_2) will be drawn back into the container. Carbon dioxide can be stripped from reentering air with a soda lime (20 % NaOH, 65 % CaO, 15 % H_2O) or **ascarite trap** (NaOH-coated silica base). Air passed through these traps also can be used as purge gas to produce CO_2-free water.

Stock alkali solution of 50 % in water is approximately 18 N. A working solution is made by diluting stock solution with **CO_2-free water**. There is no ideal container for strong alkali solutions. Glass and plastic are both used, but each has its drawbacks. If a glass container is used, it should be closed with a rubber or thick plastic closure. Glass closures should be avoided since, over time, strong alkali dissolves glass, resulting in permanent fusion of the contact surfaces. Reaction with glass also lowers the normality of the alkali. These liabilities also are relevant to long-term storage of alkali in burettes. NaOH has a low surface tension. This predisposes to leakage around the stopcock. Stopcock leakage during titration will produce erroneously high acid values. Slow evaporation of titrating solution from the stopcock valve during long periods of nonuse also creates a localized region of high pH with ensuing opportunities for fusion between the stopcock and burette body. After periods of nonuse, burettes should be emptied, cleaned, and refilled with fresh working solution.

Long-term storage of alkali in plastic containers also requires special vigilance because CO_2 permeates freely through most common plastics. Despite this shortcoming, plastic containers are usually preferred for long-term storage of stock alkali solutions. Whether glass or plastic is used for storage, working solutions should be re-standardized weekly to correct for alkalinity losses arising from interactions with glass and CO_2.

22.4.3.2 *Standard Acid*

The impurities and hygroscopic nature of NaOH make it unsuitable as a primary standard. Therefore, NaOH titrating solutions must be standardized against a **standard acid. Potassium acid phthalate** (KHP) is commonly used for this purpose.

KHP's single ionizable hydrogen (pK_a = 5.4) provides very little buffering at pH 8.2. It can be manufactured in very pure form, it is relatively nonhygroscopic, and it can be dried at 120 °C without decomposition or volatilization. Its high molecular weight also favors accurate weighing.

KHP should be dried for 2 h at 120 °C and allowed to cool to room temperature in a desiccator immediately prior to use. An accurately measured quantity of KHP solution is titrated with a base of unknown normality. The base is always the titrant. CO_2 is relatively insoluble in acidic solutions. Consequently, stirring an acid sample to assist in mixing will not significantly alter the accuracy of the titration.

22.4.4 Sample Analysis

A number of official methods exist for determining titratable acidity in various foods [1]. However, determining titratable acidity on most samples is relatively routine, and various procedures share many common steps. An aliquot of sample (often 10 mL) is titrated with a standard alkali solution (often 0.1 N NaOH) to a phenolphthalein endpoint. Potentiometric endpoint determination is used when sample pigment makes use of a color indicator impractical.

Typical titration setups are illustrated in Fig. 22.7 for potentiometric and colorimetric endpoints. Erlenmeyer flasks are usually preferred for samples when endpoint indicators are used. A magnetic stirring bar may be used, but mixing the sample with hand swirling is usually adequate. When hand mixing is used, the sample flask is swirled with the right hand. The stopcock is positioned on the right side. Four fingers on the left hand are placed behind the stopcock

valve and the thumb is placed on the front of the valve. Titrant is dispensed at a slow, uniform rate until the endpoint is approached and then added dropwise until the endpoint does not fade after standing for some predetermined period of time, usually 5–10 s.

The bulkiness of the pH electrode usually demands that beakers be used instead of Erlenmeyer flasks when samples are analyzed potentiometrically. Mixing is almost always achieved through magnetic stirring, and loss of sample through splashing is more likely with beakers than Erlenmeyer flasks. Otherwise, titration practices are identical to those described previously for colorimetric endpoint titrations.

Problems may arise when concentrates, gels, or particulate-containing samples are titrated. These matrices prevent rapid diffusion of acid from densely packed portions of sample material. This slow diffusion process results in a fading endpoint. Concentrates can simply be diluted with CO_2-free water. Titration then is performed and the original acid content calculated from dilution data. Starch and similar weak gels often can be mixed with CO_2-free water, stirred vigorously, and titrated in a manner similar to concentrates. However, some pectin and food gum gels require mixing in a blender to adequately disrupt the gel matrix. Thick foams are occasionally formed in mixing. Antifoam or vacuum can be used to break the foams.

Immediately following processing, the pH values of particulate samples often vary from one particulate piece to another. Acid equilibration throughout the entire mass may require several months. As a result, particulate-containing foods should be liquified in a

figure 22.7 Titratable acidity apparatus

1. Burette

2. Burette Clamp

3. Clamp Support

4. Magnetic Stirring Plate

5. Stopcock

6. pH Meter

7. Combination pH Probe

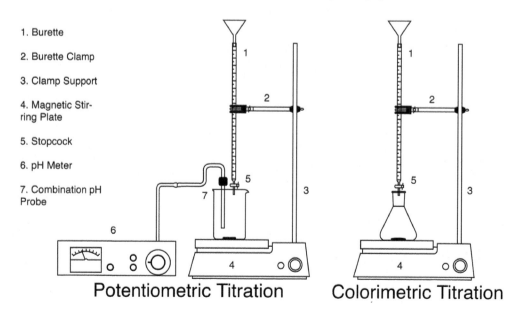

Titratable Acidity Apparatus

Potentiometric Titration   **Colorimetric Titration**

blender before titrating. The comminuting process may incorporate large quantities of air. Air entrapment makes the accuracy of volumetric measurements questionable. Aliquots often are weighed when air incorporation may be a problem.

22.4.5 Calculation of Titratable Acidity

In general chemistry, acid strength is frequently reported in normality (equivalents per liter) and can be calculated using Eq. 22.1. However, food acids are usually reported as percent of total sample weight. Thus, the equation for titratable acidity is as follows:

$$\% \text{ acid} \left(\text{wt} / \text{wt}\right) = \frac{N \times V \times \text{Eq wt}}{W \times 1000} \times 100 \quad (21.18)$$

where:

N = normality of titrant, usually NaOH (mEq/mL)
V = volume of titrant (mL)
Eq. wt. = equivalent weight of predominant acid (mg/mEq)
W = mass of sample (g)
1,000 = factor relating mg to grams (mg/g) $(1/10 = 100/1{,}000)$

Note that the normality of the titrant is expressed in milliequivalents (mEq) per mL, which is a typical way of reporting normality for small volumes. This value is numerically the same as equivalents/liter. Also note that it is easier to report sample mass in grams instead of milligrams, so multiplying sample mass by the factor of 1,000 mg/g allows units to cancel.

For routine titration of fruit juices, milliliters can be substituted for sample weight in grams, as shown in Eqs. 21.18 and 22.19. Depending on the soluble solids content of the juice, the resulting acid values will be high by 1–6%. However, this is common practice:

$$\% \text{ acid} \left(\text{wt} / \text{vol}\right) = \frac{N \times V_1 \times \text{Eq wt}}{V_2 \times 1000} \times 100 \quad (22.19)$$

or

$$\% \text{ acid} \left(\text{wt} / \text{vol}\right) = \frac{N \times V_1 \times \text{Eq wt}}{V_2 \times 10} \quad (22.20)$$

where:

N = normality of titrant, usually NaOH (mEq/mL)
V_1 = volume of titrant (mL)
Eq. wt. = equivalent weight of predominant acid (mg/mEq)
V_2 = volume of sample (mL)
1,000 = factor relating mg to grams (mg/g) $(1/10 = 100/1{,}000)$

For example, if it takes 17.5 mL of 0.085 N NaOH to titrate a 15-mL sample of a juice, the total titratable acidity of that juice, expressed as percent citric acid (molecular weight = 192; equivalent weight = 64), would be 0.635%, wt/vol, citric acid:

$$\% \text{ acid} \left(\text{wt} / \text{vol}\right) = \frac{0.085 \times 17.5 \times 64}{15 \times 10} = 0.635\% \quad (22.21)$$

Notice that the equivalent weight of anhydrous (vs. hydrous) citric acid always is used in calculating and reporting the results of titration.

22.4.6 Acid Content in Food

Most foods are as chemically complex as life itself. As such, they contain the full complement of Krebs cycle acids (and their derivatives), fatty acids, and amino acids. Theoretically, all of these contribute to titratable acidity. Routine titration cannot differentiate between individual acids. Therefore, titratable acidity is usually stated in terms of the **predominant acid**. For most foods this is unambiguous. In some cases, two acids are present in large concentrations, and the predominant acid may change with maturity. In grapes, malic acid often predominates prior to maturity while tartaric acid typically predominates in the ripe fruit. A similar phenomenon is observed with malic and citric acids in pears. Fortunately, the equivalent weights of common food acids are fairly similar. Therefore, percent titratable acidity is not substantially affected by mixed predominance or incorrect selection of the predominant acid.

The range of acid concentrations in foods is very broad. Acids can exist at levels below detection limits or they can be the preeminent substance present in the food. The contribution of acids to food flavor and quality is not told by acid content alone. The tartness of acids is reduced by sugars. Consequently, the **Brix/acid ratio** (often simply called ratio) is usually a better predictor of an acid's flavor impact than Brix (i.e., percent sugar (w/w) in a medium; see Chap. 15, Sect. 15.2.5.3) or acid alone. Acids tend to decrease with the maturity of fruit while sugar content increases. Therefore, the Brix/acid ratio is often used as an index of fruit maturity. For mature fruit, this ratio also can be affected by climate, variety, and horticultural practices. Table 22.5 gives typical acid composition and sugar levels for many commercially important fruits at maturity. Citric and malic acids are the most common acids in fruits and most vegetables; however, leafy vegetables also may contain significant quantities of oxalic acid. Lactic acid is the most important acid in dairy foods for which titratable acidity is commonly used to monitor the progress of lactic acid fermentations in cheese and yogurt production [16].

Organic acids contribute to the refractometer reading of soluble solids. When foods are sold on the

		Typical	
Fruit	Principal acid	percent acid	Typical °Brix
Apples	Malic	0.27–1.02	9.12–13.5
Bananas	Malic/citric (3:1)	0.25	16.5–19.5
Cherries	Malic	0.47–1.86	13.4–18.0
Cranberries	Citric	0.9–1.36	12.9–14.2
	Malic	0.70–0.98	
Grapefruit	Citric	0.64–2.10	7–10
Grapes	Tartaric/malic (3:2)	0.84–1.16	13.3–14.4
Lemons	Citric	4.2–8.33	7.1–11.9
Limes	Citric	4.9–8.3	8.3–14.1
Oranges	Citric	0.68–1.20	9–14
Peaches	Citric	1–2	11.8–12.3
Pears	Malic/citric	0.34–0.45	11–12.3
Pineapples	Citric	0.78–0.84	12.3–16.8
Raspberries	Citric	1.57–2.23	9–11.1
Strawberries	Citric	0.95–1.18	8–10.1
Tomatoes	Citric	0.2–0.6	4

22.5 table Acid composition and °Brix of some commercially important fruits

basis of pound solids, Brix readings are sometimes corrected for acid content. For citric acid, 0.20° Brix is added for each percent titratable acidity.

22.4.7 Volatile Acidity

In acetic acid fermentations, it is sometimes desirable to know how much acidity comes from the acetic acid and how much is contributed naturally by other acids in the product. This can be achieved by first performing an initial titration to measure **titratable acidity** as an indicator of total acidity. The acetic acid is then boiled off, the solution is allowed to cool, and a second titration is performed to determine the **fixed acidity**. The difference between fixed and total acidity is the **volatile acidity**. A similar practice is used sometimes in the brewing industry to separate acidity due to dissolved CO_2 from fixed acids. Fixed acids are titrated after CO_2 is removed by low heat (40 °C) and gentle agitation.

22.4.8 Other Methods

High-performance liquid chromatography (HPLC) (see Chap. 13) and electrochemistry both have been used to measure acids in food samples. Both methods allow identification of specific acids. HPLC uses refractive index, ultraviolet, or for some acids electrochemical detection. Ascorbic acid has a strong electrochemical signature and significant absorbance at 265 nm. Significant absorbance of other prominent acids does not occur until 200 nm or below.

Many acids can be measured with such electrochemical techniques as voltammetry and polarography. In ideal cases, the sensitivity and selectivity of electrochemical methods are exceptional. However, interfering compounds often reduce the practicality of electrochemical approaches.

Unlike titration, chromatographic and electrochemical techniques do not differentiate between an acid and its conjugate base. Both species inevitably exist side by side as part of the inherent food-buffer system. As a result, acids determined by instrumental methods may be 50 % higher than values determined by titration. It follows that Brix/acid ratios can be based only on acid values determined by titration.

New pH probe designs use an ion-sensitive field-effect transistor (ISFET) to measure hydrogen ion concentration. These may use standard calomel or Ag/AgCl reference electrodes. However development of a silicon-based reference electrode has met with mounting success. ISFET electrodes have faster response times than traditional electrodes. Also, compact chip technology makes miniaturized probes possible. ISFET probes do not require a fluid medium, making possible direct pH measurement of gels, meats, cheeses, and other dense foods. The higher cost of a ISFET probe and some tendency for reference electrode drift keep it from replacing traditional pH electrodes for routine analysis.

22.5 SUMMARY

Organic acids have a pronounced impact on food flavor and quality. Unlike strong acids that are fully dissociated, food acids are only partially ionized. Some properties of foods are affected only by this ionized fraction of acid molecules, while other properties are affected by the total acid content. It is impractical to quantify only free hydronium ions in solution by titration. Once the free ions are removed by chemical reaction, others arise from previously undissociated molecules. Indicator dyes, which change color depending on the hydronium ion environment, exist, but they only identify when a certain pH threshold has been achieved and do not stoichiometrically quantify free hydronium ions. The best that can be done is to identify the secondary effect of the hydronium ion environment on some property of the system such as the color of the indicator dyes or the electrochemical potential of the medium. The pH meter measures the change in electrochemical potential established by the hydronium ion across a semipermeable glass membrane on an indicator electrode. The shift in the indicator electrode potential is indexed against the potential of a reference electrode. The difference in millivolt reading between the two electrodes can be converted into pH using the Nernst equation. Indicator dyes are used commonly to identify the endpoint of acidity titrations although pH meters can be used in critical work or when sample color makes indicators impractical.

The hydronium ion concentration can be back-calculated from pH using the definition of pH as the negative log of hydrogen ion concentration. Buffer solutions of any pH can be created using the Henderson-Hasselbalch equation. However, the predictions of all these equations are somewhat approximate unless the activity of acids and conjugate bases is taken into account.

Titratable acidity provides a simple estimate of the total acid content of a food. In most cases, it is only an estimate since foods often contain many acids that cannot be differentiated through titration. Titratable acidity is not a good predictor of pH, since pH is a combined function of titratable acid and conjugate base. Instrumental methods such as HPLC and electrochemical approaches measure acids and their conjugate bases as a single compound and, therefore, tend to produce acid contents that are higher than those determined by titration. Titratable acidity, somewhat curiously, is a better predictor of tartness than the concentration of free hydronium ions as reflected by pH. The perception of tartness is strongly influenced by the presence of sugars.

22.6 STUDY QUESTIONS

1. Explain the theory of potentiometry and the Nernst equation as they relate to being able to use a pH meter to measure H^+ concentration.
2. Explain the difference between a saturated calomel electrode and a silver-silver chloride electrode; describe the construction of a glass electrode and a combination electrode.
3. You return from a 2-week vacation and ask your lab technician about the pH of the apple juice sample you gave him or her before you left. Having forgotten to do it before, the technician calibrates a pH meter with one standard buffer stored next to the meter and then reads the pH of the sample of unpasteurized apple juice immediately after removing it from the refrigerator (40 °F), where it has been stored for 2 weeks. Explain the reasons why this stated procedure could lead to inaccurate or misleading pH values.
4. For each of the food products listed below, what acid should be used to express titratable acidity?

   (a) Orange juice
   (b) Yogurt
   (c) Apple juice
   (d) Grape juice

5. What is a "Brix/acid ratio," and why is it often used as an indicator of flavor quality for certain foods, rather than simply Brix or acid alone?
6. How would you recommend determining the endpoint in the titration of tomato juice to determine the titratable acidity? Why?
7. The titratable acidity was determined by titration to a phenolphthalein endpoint for a boiled and unboiled clear carbonated beverage. Which sample would you expect to have a higher calculated titratable acidity? Why? Would you expect one of the samples to have a fading endpoint? Why?
8. Why and how is an ascarite trap used in the process of determining titratable acidity?
9. Why is volatile acidity useful as a measure of quality for acetic acid fermentation products, and how is it determined?
10. Regarding the use of a standard acid to standardize a NaOH solution to determine titratable acidity, what factors make (a) KHP a good choice and (b) HCl, HNO_3, or H_2SO_4 a poor choice?
11. Could a sample that is determined to contain 1.5% acetic acid also be described as containing 1.5% citric acid? Why or why not?
12. An instructor was grading lab reports of students who had determined the titratable acidity of grape juice. One student had written that the percent titratable acidity was 7.6% citric acid. Give two reasons why the answer was marked wrong. What would have been a more reasonable answer?

22.7 PRACTICE PROBLEMS

1. You are performing a titration on duplicate samples and duplicate blanks that require 4 mL of $1N$ NaOH per titration sample. The lab has 10% NaOH and saturated NaOH. Choose one, and describe how you would prepare the needed amount of NaOH solution.
2. You are performing duplicate titrations on five samples that require 15 mL of $6N$ HCl each. How would you prepare the needed solution from reagent grade HCl?
3. What is the pH of a 0.057 M HCl solution?
4. Vinegar has a $[H^+]$ of 1.77×10^{-3} M. What is the pH? What is the major acid found in vinegar, and what is its structure?

5. Orange juice has a [H⁺] of $2.09 \times 10^{-4}\,M$. What is the pH? What is the major acid found in orange juice and what is its structure?

6. A sample of vanilla yogurt has a pH of 3.59. What is the [H⁺]? What is the major acid found in yogurt and what is its structure?

7. An apple pectin gel has a pH of 3.30. What is the [H⁺]? What is the major acid found in apples and what is its structure?

8. How would you make 1 L of $0.1\,N$ NaOH solution from an $18\,N$ stock solution?

9. A stock base solution assumed to be $18\,N$ was diluted to $0.1\,N$. KHP standardization indicated the normality of the working solution was $0.088\,N$. What was the actual normality of the solution?

10. A 20-mL sample of juice requires 25 mL of $0.1\,N$ NaOH titrant. What would be the percent acid if the juice is (1) apple juice, (2) orange juice, and (3) grape juice?

11. A lab analyzes a large number of orange juice samples. All juice samples will be 10 mL. It is decided that 5 mL of titrant should equal 1% citric acid. What base normality should be used?

12. A lab wishes to analyze apple juice. They would like each milliliter of titrant to equal 0.1% malic acid. Sample aliquots will all be 10 mL. What base normality should be used?

13. Using Table 22.2, calculate the pH of cola.

Answers

1. Four titrations of 4 mL each will be performed requiring a total of about 16 mL of $1\,N$ NaOH. For simplicity, 20 mL of $1\,N$ NaOH can be prepared. Use Eq. 22.1 to solve for the mL of stock solution, e.g., for a 10% NaOH:

$$10g\,\mathrm{NaOH}\,/\,100\,\mathrm{mL} = 100g\,\mathrm{NaOH}\,/\,L$$
$$= 2.5\,N\,\mathrm{NaOH}$$
$$(20\,\mathrm{mL})(1\,N\,\mathrm{NaOH}) = (x\,\mathrm{mL})(2.5\,N)$$
$$x\,\mathrm{mL} = 8\,\mathrm{mL}\,\text{of 10% diluted to 20 mL}$$
$$\text{with distilled water}$$

If saturated NaOH is used, remember from Problem 6 that approximately 8.7 mL of saturated NaOH diluted to 100 mL yields $1.0\,N$. Therefore, 1.87 mL or 2 mL of saturated NaOH diluted to 20 mL with distilled water will yield about $1\,N$ NaOH.

2. A total of (5 samples)(2 duplicates)(15 mL) = 150 mL of $6\,N$ HCl:

$$(150\,\mathrm{mL})(6\,N\,\mathrm{HCl}) = (x\,\mathrm{mL})(12\,N\,\mathrm{HCl})$$
$$x\,\mathrm{mL} = 75\,\mathrm{mL}\,\text{concentrated HCl diluted with}$$
$$\text{distilled water to 150 mL}$$

3. Since HCl is a strong acid, it will be almost completely dissociated. Therefore,

$$\left[\mathrm{HCl}\right] = \left[\mathrm{H}^+\right]$$
$$\left[\mathrm{H}^+\right] = 0.057\,N = 5.7 \times 10^{-2}\,M$$
$$\mathrm{pH} = -\log\left[5.7 \times 10^{-2}\right] = 1.24$$

4. 2.75; acetic acid.

(Use Eq. 22.5, to solve Problems 14–17.)

5. 3.68; citric acid.

6. $1.1 \times 10^{-4}\,M$; lactic acid.

7. $5.0 \times 10^{-4}\,M$; malic acid.

8. Using Eq. 22.1 and solving for volume of concentrate, we get

$$\mathrm{mL\ concentrated\ solution} = \frac{\mathrm{final}\,N \times \mathrm{final\ mL}}{\mathrm{beginning}\,N}$$
$$= \frac{0.1 \times 1000}{18}$$
$$= 5.55\,\mathrm{mL}$$

Consequently, 5.55 mL would be dispensed into a 1-L volumetric flask. The flask would then be filled to volume with distilled CO_2-free water. The normality of this solution will only be approximate since NaOH is not a primary standard. Standardization against a KHP solution or some other primary standards is essential. It is useful sometimes to back-calculate the true normality of the stock solution. Even under the best circumstances, the normality will decrease with time, but back-calculating will permit a closer approximation of the target normality the next time a working standard is prepared.

9. This answer is a simple ratio:

$$\frac{0.088}{0.100} \times 18 = 15.85\,N$$

10. Table 22.5 indicates the principal acids in apple, orange, and grape juice are malic, citric, and tartaric acids, respectively. Table 22.2 indicates the equivalent weight of these acids are malic (67.05), citric (64.04), and tartaric (75.05). The N, v_1 and v_2 are given. Using Eq. 22.20, the percent acid for each of these juices would be:

$$\% \text{ malic acid} = \frac{0.1 \times 25 \times 67.05}{20 \times 10} = 0.84\%$$

$$\% \text{ citric acid} = \frac{0.1 \times 25 \times 64.04}{20 \times 10} = 0.80\%$$

$$\% \text{ malic acid} = \frac{0.1 \times 25 \times 65.05}{20 \times 10} = 0.94\%$$

11. Quality control laboratories often analyze a large number of samples having a specific type of acid. Speed and accuracy are increased if acid concentration can be read directly from the burette. It is possible to adjust the normality of the base to achieve this purpose by rearranging Eq. 22.20:

$$\% \text{ acid} = \frac{N \times V_1 \times \text{Eq wt}}{V_2 \times 10} \qquad (22.20)$$

$$N = \frac{\% \text{ acid} \times V_2 \times 10}{V_1 \times \text{Eq wt}}$$

The desired % acids, v_1, and v_2 are 1%, 5, and 10 mL, respectively. The equivalent weight of citric acid is 64.04 (see Table 22.2). Thus,

$$N = \frac{1 \times 10 \times 10}{5 \times 64.04} = 0.3123$$

In actuality, the standard alkali solution used universally by the Florida citrus industry is 0.3123 N.

12. This can be solved analogously to Problem 11. This time, % acid is 0.1, v_1 is 1 mL, and v_2 is 10 mL. The equivalent weight is 67.05. Therefore,

$$N = \frac{0.1 \times 10 \times 10}{1 \times 67.05} = 0.1491$$

13. Substitute the [H^+] into Eq. 22.5:

$$pH = -\log\left[H^+\right]$$

$$pH = -\log\left[2.24 \times 10^{-3}\right]$$

There are now different ways to obtain the right answer:

(a) Type the concentration into a calculator, take the logarithm, and invert the sign.
(b) Use a computer program such as Excel (the corresponding formula in Excel would be "-LOG10(2.24*10^-3)". While students may be more familiar with a calculator initially, saving results on the computer may be beneficial in the long run, e.g., if similar calculations such as for buffers need to be performed repeatedly. In this case, the same formula could be used, and only certain numbers would need to be changed.
(c) Separate 2.24×10^{-3} into two parts; determine the logarithm of each part:

$$\log 2.24 = 0.350$$
$$\log 10^{-3} = -3$$

Add the two logs together since adding logs is equivalent to multiplying the two numbers:

$$0.350 + (-3) = -2.65$$

Place the value into Eq. 22.1.
All three of these methods should lead to a pH of 2.65.

Acknowledgment The author of this chapter wishes to acknowledge Dr. Patricia A Murphy, who was an author of this chapter for the second to fourth editions of this textbook.

REFERENCES

1. Albert A., Serjeant EP (1984). The determination of ionization constants. A laboratory manual, 3rd edition. Chapman and Hall, New York
2. AOAC International (2012) Official methods of analysis, 19th edn. (Online). AOAC International, Rockville, MD
3. Beckman Instruments (1995) The Beckman handbook of applied electrochemistry. Bulletin No. BR-7739B. Fullerton, CA
4. Bergveld, P (2003) Thirty years of ISFETOLOGY, What happened in the past 30 years and what may happen in the next 30 years. Sensors and Actuators B 88 1–20.
5. Dicker DH (1969) The laboratory pH meter. American Laboratory, February
6. Efiok BJS, Eduok EE (2000) Basic calculations for chemical & biological analysis, 2nd edn. AOAC International, Gaithersburg, MD
7. Thermo Scientific (2009) Thermo Scientific pH electrode handbook, http://iris.fishersci.ca/LitRepo.nsf/0/11CA00B22D99CF698525757600709B39/$file/Thermo%20Scientific%20pH%20Electrode%20Handbook%202009.pdf
8. Gardner WH (1996) Food acidulants. Allied Chemical Co., New York
9. Harris DC, Lucy CA (2016) Quantitative chemical analysis, 9th edn. W.H. Freeman & Co, New York
10. Joslyn MA (1970) pH and buffer capacity, ch. 12, and acidimetry, ch. 13. In: Methods in food analysis: Physical, chemical, and instrumental methods of analysis, Academic, New York
11. Kenkel J (2014) Analytical chemistry for technicians. 4th edn. CRC Press, Boca Raton, FL

12. Mohan C (1997) Buffers. Calbiochem – Novabiochem International, La Jolla, CA
13. Nelson PE, Tressler DK (1980) Fruit and vegetable juice process technology, 3rd edn. AVI, Westport, CT
14. Pecsok RL, Chapman K, Ponder WH (1971) Modern chemical technology, vol 3, revised edn. American Chemical Society, Washington, DC
15. Skogg DA, West DM, Holler JF, Crouch SR (2000) Analytical chemistry: an introduction, 7th edn. Brooks/Cole, Pacific Grove, CA
16. Wehr HM, Frank JF (eds) (2004) Standard method for examination of dairy products, 17th edn. American Public Health Association, Washington, DC

Fat Characterization

Oscar A. Pike (✉)

Department of Nutrition, Dietetics, and Food Science, Brigham Young University, Provo, UT 84602, USA
e-mail: oscar_pike@byu.edu

Sean O'Keefe

Department of Food Science and Technology, Virginia Tech, Blacksburg, VA 24061, USA
e-mail: okeefes@vt.edu

S. Nielsen (ed.), *Food Analysis*, Food Science Text Series,
DOI 10.1007/978-3-319-45776-5_23, © Springer International Publishing 2017

23.1 INTRODUCTION

Methods for characterizing edible **lipids**, fats, and oils can be separated into two categories: (1) those developed to analyze bulk oils and fats and (2) those focusing on analysis of foodstuffs and their lipid extracts. In evaluating foodstuffs, it is usually necessary to extract the lipids prior to analysis. In these cases, if sufficient quantities of lipids are available, methods developed for bulk fats and oils can be utilized.

The methods described in this chapter are divided into four sections. The first is traditional analytical methods for bulk fats and oils, many involving "wet chemistry." Then, two sections discuss methods of measuring lipid oxidation. Some of these methods utilize intact foodstuffs, but most require the lipids to be extracted from foodstuffs. Last addressed are methods for the analysis of lipid fractions, including fatty acids, triacylglycerols, and cholesterol.

Numerous methods exist for the characterization of lipids, fats, and oils [1–13]. This chapter includes methods required for the nutritional labeling of food and others appropriate for an undergraduate food analysis course. Many traditional "wet chemistry" methods have been supplemented or superseded by instrumental methods such as gas chromatography (GC), high-performance liquid chromatography (HPLC), nuclear magnetic resonance (NMR), and Fourier transform infrared (FTIR) spectroscopy. Nonetheless, an understanding of basic concepts derived from traditional methods is valuable in learning more sophisticated instrumental methods.

Many of the methods cited are official methods of the AOAC International [1], American Oil Chemists' Society [2], or International Union of Pure and Applied Chemists [3]. The principles, general procedures, and applications are described for the methods. Refer to the specific methods cited in Table 23.1 for detailed information on procedures.

23.1.1 Definitions and Classifications

As explained in Chap. 17, the term **lipids** refers to a wide range of compounds soluble in organic solvents but only sparingly soluble in water. Chapter 17 also outlines the general classification scheme for lipids. The majority of lipids present in foodstuffs are of the following types: (1) fatty acids and their glycerides, including mono-, di-, and triacylglycerols, (2) phospholipids, (3) sterols (including cholesterol), (4) waxes, and (5) lipid-soluble pigments and vitamins. The commonly used terms monoglyceride, diglyceride, and triglyceride are synonymous with the proper nomenclature terms monoacylglycerol, diacylglycerol, and triacylglycerol, respectively.

In contrast to lipids, the terms fats and oils often refer to bulk products of commerce, crude or refined, that have already been extracted from animal products or oilseeds and other plants grown for their lipid content. The term **fat** signifies extracted lipids that are solid at room temperature, and **oil** refers to those that are liquid. However, the three terms, lipid, fat, and oil, often are used interchangeably.

The FDA has defined **fat content** for nutritional labeling purposes as the total lipid fatty acids expressed as triglyceride, rather than the extraction and gravimetric procedures used in the past (see also Chap. 17, Sect. 17.3.6.1).

"Fat, total" or "Total fat": A statement of the number of grams of total fat in a serving defined as total lipid fatty acids and expressed as triglycerides where fatty acids are aliphatic carboxylic acids consisting of a chain of alkyl groups and characterized by a terminal carboxyl group. Amounts shall be expressed to the nearest 0.5 (1/2) gram increment below 5 grams and to the nearest gram increment above 5 grams. If the serving contains less than 0.5 gram, the content shall be expressed as zero. [21 CFR 101.9(c)(2)]

Fatty acids included in this definition may be derived from triacylglycerols, partial glycerides, phospholipids, glycolipids, sterol esters, or free fatty acids, but the concentration will be expressed as grams of triacylglycerols. This definition for nutritional labeling purpose requires that, rather than determining total fat via extraction and gravimetry (e.g., Soxhlet), total fat be based on GC of **fatty acid methyl esters** (FAMEs). Although this requires more complex and expensive analytical equipment, the analysis provides a better estimation of the fat content of foods.

Saturated fat is defined for nutritional labeling purposes as the sum (in grams) of all fatty acids without double bonds. Saturated fat is expressed to the closest 0.5 g below 5 g/serving and to the closest gram above 5 g. A food with less than 0.5 g of saturated fat per serving has the content expressed as zero. The optional category of **polyunsaturated fat** (PUFA) is defined as *cis, cis*-methylene-interrupted polyunsaturated fatty acids and has the same gram reporting requirements as saturated fats. Another optional category (unless certain label claims are made), **monounsaturated fat,** is defined as *cis*-monounsaturated fatty acids. The requirement that the fatty acids be *cis* prevents including fatty acids that contain *trans* isomers. *Trans* **fatty acids** must now be included on the nutritional label in the United States. The definition for nutritional labeling requires that the *trans* acids are not conjugated. Most *trans* fatty acids found in foods are monounsaturated, and methods for monounsaturated fatty acid and *trans* fatty acid analysis must distinguish between *cis* and *trans* monoenes, as well as the conjugated *cis-trans* fatty acids such as **conjugated linoleic acid** (CLA).

23.1 table   Correlation of selected AOCS [2], AOAC [1], and IUPAC [3] methods

Method	AOCS	AOAC	IUPAC
Bulk fats and oils			
Refractive index	Cc 7-25	921.08	2.102
Melting			
Capillary tube melting point	Cc 1-25	920.157	
Slip melting point	Cc 3-25		
	Cc 3b-92		
DSC melting properties	Cj 1-94		
Dropping point	Cc 18-80		
Wiley melting point	Cc 2-38[a]	920.156	
Smoke, flash, and fire points	Cc 9a-48		
	Cc 9b-55		
Cold test	Cc 11-53	929.08	
Cloud point	Cc 6-25		
Color			
Lovibond	Cc 13e-92		
	Cc 13j-97		
Spectrophotometric	Cc 13c-50		2.103
Iodine value	Cd 1-25[a]	920.159	2.205
	Cd 1c-85		
	Cd 1d-87		
	Cd 1d-92	993.20	
	Cd 1e-01		
Saponification number	Cd 3-25	920.160	2.202
	Cd 3c-91		
	Cd 3a-94		
Free fatty acids (FFAs)	Ca 5a-40	940.28	
	Ca 5d-01		
Acid value	Cd 3d-63	969.17	2.201
Solid fat index (SFI)	Cd 10-57[a]		2.141
Solid fat content (SFC)	Cd 16b-93		2.150
Consistency, penetrometer method	Cc 16-60		
Spreadability	Cj 4-00		
Polar components in frying fats	Cd 20-91	982.27	2.507
Lipid oxidation – present status			
Peroxide value	Cd 8-53[a]	965.33	2.501
	Cd 8b-90		
p-Anisidine value	Cd 18-90		2.504
Hexanal (volatile organic compounds)	Cg 4-94		
Thiobarbituric acid (TBA) test	Cd 19-90		2.531
Conjugated dienes and trienes	Ti 1a-64	957.13	2.206
	Ch 5-91		
Lipid oxidation – oxidative stability			
Oven storage test	Cg 5-97		
Oil stability index (OSI)	Cd 12b-92		
Active oxygen method (AOM)	Cd 12-57[a]		2.506
Oxygen bomb			
Lipid fractions			
Fatty acid composition (including	Ce 1-62	963.22	2.302
saturated/unsaturated, cis/trans)	Ce 1a-13		
	Ce 1h-05	996.06	
	Ce 1i-07		
	Ce 1j-07		
	Ce 1e-91[a]		
	Ce 1f-96		

(continued)

23.1
table (continued)

Method	AOCS	AOAC	IUPAC
Fatty acid methyl esters (FAMEs)	Ce 2-66	969.33	2.301
	Ce 2b-11		
	Ce 2c-11		
trans Isomer fatty acids using IR spectroscopy	Cd 14-95		2.207
	Cd 14d-99		
	Cd 14e-09		
	Cd 14f-14		
Mono- and diacylglycerols	Cd 11-57	966.18	
	Cd 11b-91		
	Cd 11d-96		
Triacylglycerols		986.19	
	Ce 5b-89		
	Ce 5c-93		
Cholesterol (and other sterols)		976.26	2.403

AOCS American Oil Chemists' Society, *AOAC* AOAC International, *IUPAC* International Union of Pure and Applied Chemists
[a]Though no longer current, these methods are included for reference because of their previous common use

There are several CLA isomers identified in ruminant lipids, although the 9 *cis* 11 *trans* isomer appears to be found in greatest abundance [4]. CLA isomers are thought to reduce the risk of cancer and other diseases.

"Trans fat" or "Trans": A statement of the number of grams of trans fat in a serving, defined as the sum of all unsaturated fatty acids that contain one or more isolated (i.e., non-conjugated) double bonds in a trans configuration, except that label declaration of trans fat content information is not required for products that contain less than 0.5 gram of total fat in a serving if no claims are made about fat, fatty acid or cholesterol content. [21 CFR 101.9(c)(2)(ii)].

The presence of *trans* fat in foods has greatly declined due to this label requirement and also because the FDA has tentatively determined that the major source of trans fat in the diet, **partially hydrogenated oils** (PHO), is no longer generally recognized as safe (GRAS). However, some foods will still contain small amounts of *trans* fat because it occurs naturally in various animal products and edible oils.

23.1.2 Importance of Analyses

Such issues as the effect of dietary fat on health and food labeling requirements necessitate that food scientists be able not only to measure the total lipid content of a foodstuff but also to characterize it [4–10]. Health concerns require the measurement of such parameters as cholesterol and phytosterol contents, and amounts of *trans*, n-3/ω3, saturated, and mono- and polyunsaturated fatty acids. Lipid stability impacts not only the shelf life of food products but also their safety, since some oxidation products (e.g., malonaldehyde, cholesterol oxides) have toxic properties. Another area of interest is the

analysis of oils and fats used in deep-fat frying operations [11]. Total polar materials and acid value are used as quality standards in deep-fat frying oil. Finally, the development of food ingredients composed of lipids that are not bioavailable (e.g., sucrose polyesters such as Olestra®) or lipids not contributing the normal 9 Cal/g to the diet (e.g., short- and medium-chain triglycerides such as Salatrim® and Caprenin®) accentuates the need to characterize the lipids present in food.

23.1.3 Lipid Content in Foods and Typical Values

Commodities containing significant amounts of fats and oils include butter; cheese; imitation dairy products such as margarine, spreads, shortening, frying fats, and cooking and salad oils; emulsified dressings such as mayonnaise, peanut butter, and confections; and muscle foods such as meat, poultry, and fish [12, 13]. Information is available summarizing the total fat content of foods (see Chap. 17, Table 17.2) as well as their constituent fatty acids [e.g., 14]. Ongoing studies are refining the quantities of saturated and unsaturated fat, *trans* isomers, cholesterol, cholesterol oxides, phytosterols, and other specific parameters in foods.

Because of their usefulness as food ingredients, it sometimes is important to know the physical and chemical characteristics of bulk fats and oils. Definitions and specifications for bulk fats and oils (e.g., soybean oil, corn oil, coconut oil), including values for many of the tests described in this chapter, can be found in Firestone [14], in the *Merck Index* [15], and in *Fats and Oils* [16]. Table 23.2 gives typical values for several of the tests for some of the common commercial fats and oils. It must be remembered that bulk fats

23.2 table Typical values of selected parameters for fats and oils

Fat/oil source	Refractive index (40 °C)	Melting point (°C)	Iodine value	Saponification value
Butterfat (bovine)	1.453–1.457		25–42	210–254
Cocoa butter	1.456–1.458	30–35	32–42	190–200
Coconut	1.448–1.450	21–26	5–13	242–265
Corn	1.465–1.468	−18 to−10	107–135	156–196
Cottonseed	1.458–1.466	−2	96–121	189–199
Lard			45–168	192–203
Menhaden			150–200	192–199
Olive		−3 to 0	75–94	184–196
Palm (*Elaeis guineensis*)	1.453–1.460	27–42.5	45–56	190–209
Palm kernel (*Elaeis guineensis*)	1.448–1.452	24–30	14–24	230–257
Peanut	1.460–1.465	−5 to −2	73–107	184–196
Rapeseed (canola)	1.465–1.467		110–126	182–193
Safflower seed	1.467–1.470	−5	136–151	186–203
Soybean	1.466–1.470		118–139	188–195
Sunflower seed	1.467–1.469		115–145	186–196
Tallow (beef)	1.450–1.458	45–48	33–50	190–202

Compiled from Firestone [14] with permission from AOCS Press, Copyright 2013

and oils can vary markedly in such parameters due to differences in source, composition, and susceptibility to deterioration. Foods containing even minor amounts of lipids (e.g., <1%) can have a shelf life limited by lipid oxidation and subsequent rancidity.

23.2 GENERAL CONSIDERATIONS

Various fat extraction solvents and methods are discussed in Chap. 17. For lipid characterization, extraction of fat or oil from foodstuffs can be accomplished by homogenizing with a solvent combination such as hexane-isopropanol (3:2, vol/vol) or chloroform-methanol (2:1, vol/vol). The solvent then can be removed using a rotary evaporator or by evaporation under a stream of nitrogen gas. Lipid oxidation during extraction and testing can be minimized by adding antioxidants [e.g., 10–100 mg of t-butylated hydroxytoluene (BHT)/L] to solvents and by taking other precautions such as flushing containers with nitrogen and avoiding exposure to heat and light [17].

Sample preparation is hastened through the use of **solid-phase extraction** (SPE), which consists of passing the lipid extract through a commercially available prepackaged absorbent (e.g., silica gel) that separates contaminants or various fractions based on polarity (see Chap. 14, Sect. 14.2.2.5, and Chap. 33, Sect. 33.2.2.3.2). Constituents present in lipid extractions that may present problems in lipid characterization include phosphatides, gossypol, carotenoids, chlorophyll, sterols, tocopherols, vitamin A, and metals.

Bulk oils such as soybean oil typically undergo the following purification processes: degumming, alkali or physical refining [removal of **free fatty acids** (FFAs)], bleaching, and deodorization after extraction from their parent source. Modifications such as fractionation, winterization, interesterification, and hydrogenation also may be a part of the processing, depending on the commodity being produced. Various methods discussed in this chapter can be used to monitor the refining process.

Changes that lipids undergo during processing and storage include hydrolysis (lipolysis), oxidation, and thermal degradation including polymerization (such as during deep-fat frying operations). These changes are discussed in the following sections on methods.

23.3 METHODS FOR BULK OILS AND FATS

Numerous methods exist to measure the characteristics of fats and oils. Some methods (e.g., titer test) have limited use for edible oils (in contrast to soaps and industrial oils). Other methods may require special apparatus not commonly available or may have been antiquated by modern instrumental procedures [e.g., volatile acid methods (Reichert-Meissl, Polenske, and Kirschner values) have been replaced largely by determination of fatty acid composition using GC]. Methods to determine impurities, including moisture, unsaponifiable material in refined vegetable oil, and insoluble impurities, also are not covered in this chapter. Defined methods exist for the sensory evaluation of fats and oils (see AOCS Methods Cg 1-83 and Cg 2-83) but are outside the scope of this text.

23.3.1 Sample Preparation

Ensure that samples are visually clear and free of sediment. When required (e.g., iodine value), dry the samples prior to testing (AOAC Method 981.11). Because exposure to heat, light, or air promotes lipid oxidation, avoiding these conditions during sample storage will retard rancidity. If samples are solid or semisolid at room temperature, they should be melted and thoroughly mixed before sampling. Sampling procedures are available for bulk oils and fats (AOCS Method C 1-47).

23.3.2 Refractive Index

23.3.2.1 *Principle*

The **refractive index** (RI) of an oil is defined as the ratio of the speed of light in air (technically, a vacuum) to the speed of light in the oil. When a ray of light shines obliquely on an interface separating two materials, such as air and oil, the light ray is refracted in a manner defined by Snell's law, as shown in Eq. 23.1:

$$\theta_1 n_1 = \theta_2 n_2 \qquad (23.1)$$

where:

θ_1 = angle of the incident light
n_1 = refractive index of material 1
θ_2 = angle of the refracted light
n_2 = refractive index of material 2

As can be seen in Fig. 23.1 and from Eq. 23.1, if the angles of incidence and refraction and the refractive index (n) of one of the two materials are known, the refractive index of the other material can be determined. In practice, the θ_1 and n_1 are constant, so n_2 is determined by measuring θ_2.

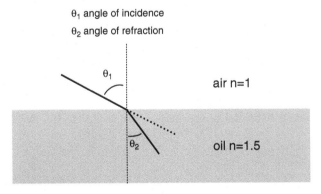

θ_1 angle of incidence

θ_2 angle of refraction

θ_1

air n=1

θ_2

oil n=1.5

23.1 figure

Refraction of light in an air-oil interface

Because the frequency of light affects its refraction (violet light is refracted more than red light), white light can be dispersed or split after refraction through two materials of different refractive indexes (explaining the color separation of diamonds and rainbows). Refractometers often use monochromatic light (or nearly monochromatic light from the sodium doublet D line, that has 589.0 and 589.6 nm wavelengths or light-emitting diodes to provide 589.3 nm) to avoid errors from variable refraction of the different wavelengths of visible light.

23.3.2.2 *Procedure*

Samples are measured with a refractometer at 20 °C for oils and at specified higher temperatures for fats, depending on the temperature at which the fat is completely liquid.

23.3.2.3 *Applications*

RI is related to the amount of saturation in a lipid; the RI decreases linearly as iodine value (a measure of total unsaturation) decreases. RI also is used as a measure of purity and as a means of identification, since each substance has a characteristic RI. However, RI is influenced by such factors as FFA content, oxidation, and heating of the fat or oil. The RI of various lipids is shown in Table 23.2. A relatively saturated lipid such as coconut oil has a different RI ($n = 1.448$–1.450) compared to a relatively unsaturated lipid such as menhaden oil ($n = 1.472$).

23.3.3 Melting Point

23.3.3.1 *Principle*

The melting point may be defined in various ways, each corresponding to a different residual amount of solid fat. The **capillary tube melting point**, also known as the **complete melting point** or **clear point**, is the temperature at which fat heated at a given rate becomes completely clear and liquid in a one-end closed capillary. The **slip melting point** is performed similarly to the capillary tube method and measures the temperature at which a column of fat moves in an open capillary when heated. The **dropping point** (also called the **dropping melting point** or **Mettler dropping point)** is the temperature at which the sample flows through a 0.11-in. hole in a sample cup placed in a specialized furnace. The **Wiley melting point** measures the temperature at which a $1/8 \times 3/8$-in. disk of fat, suspended in an alcohol-water mixture of similar density, changes into a sphere.

23.3.3.2 *Applications*

It appears the predominant method in the United States for measuring melting point is the dropping melting point. The procedure has been automated and therefore is not labor-intensive. The capillary tube method is less useful for oils and fats (in comparison to pure compounds) since they lack a sharp melting point due to their array of various components. The slip melting point often is used in Europe, whereas the Wiley melting point was preferred previously in the United States; the latter is no longer a current AOCS Method. A disadvantage of the Wiley melting point is the subjective determination as to when the disk is spherical. A disadvantage of the slip melting point is its 16-h stabilization time.

23.3.4 Smoke, Flash, and Fire Points

23.3.4.1 *Principle*

The **smoke point** is the temperature at which the sample begins to smoke when tested under specified conditions. The **flash point** is the temperature at which a flash appears at any point on the surface of the sample; volatile gaseous products of combustion are produced rapidly enough to permit ignition. The **fire point** is the temperature at which evolution of volatiles (by decomposition of sample) proceeds with enough speed to support continuous combustion.

23.3.4.2 *Procedure*

A cup is filled with oil or melted fat and heated in a well-lighted container. The smoke point is the temperature at which a thin, continuous stream of bluish smoke is given off. The flash point and fire point are obtained with continued heating, during which a test flame is passed over the sample at 5 °C intervals. For fats and oils that flash at temperatures below 149 °C, a closed cup is used.

23.3.4.3 *Applications*

These tests reflect the volatile organic material in oils and fats, especially FFAs (Fig. 23.2) and residual extraction solvents. Frying oils and refined oils should have smoke points above 200 °C and 300 °C, respectively.

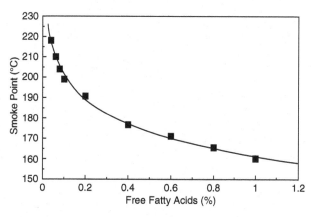

23.2
figure
Effect of free fatty acid content on smoke point of olive oil

23.3.5 Cold Test

23.3.5.1 *Principle*

The **cold test** is a measure of the resistance of an oil to crystallization. The absence of crystals or turbidity indicates proper winterizing.

23.3.5.2 *Procedure*

Oil is stored in an ice bath (0 °C) for 5.5 h and observed for crystallization.

23.3.5.3 *Applications*

The cold test is a measure of success of the winterizing process. It ensures that oils remain clear even when stored at refrigerated temperatures. Winterizing subjects an oil to cold temperatures and then separates the crystallized material from the bulk. This results in an oil that will not cloud at low temperature. This is useful for ensuring that oils remain clear in products, such as salad dressings, that will be stored at refrigeration temperatures after opening.

23.3.6 Cloud Point

23.3.6.1 *Principle*

The **cloud point** is the temperature at which a cloud is formed in a liquid fat due to the beginning of crystallization.

23.3.6.2 *Procedure*

The sample is heated to 130 °C and then cooled with agitation. The temperature of first crystallization is taken to be the point at which a thermometer in the fat is no longer visible.

23.3.7 Color

Two methods for measuring the color of fats and oils are the **Lovibond** method and the **spectrophotometric** method.

23.3.7.1 *Procedure*

In the Lovibond method, oil is placed in a standard sized glass cell and visually compared with red, yellow, blue, and neutral color standards. Results are expressed in terms of the numbers associated with the color standards. Automated colorimeters are available.

For the spectrophotometric method, the sample is heated to 25–30 °C, placed in a cuvette, and absorbance read at the following wavelengths: 460, 550, 620, and 670 nm. The photometric color index is calculated as shown in Eq. 23.2 AOCS Method Cc 13c-50:

$$\text{Photometric color index} = 1.29(A_{460}) + 69.7(A_{550}) + 41.2(A_{620}) - 56.4(A_{670})$$

$$(23.2)$$

23.3.7.2 *Applications*

The color of fats and oils is most commonly evaluated using the Lovibond method. Oils and fats from different sources vary in color. But if refined oil is darker than expected, it is probably indicative of improper refinement or abuse [16]. Though specifically developed for testing the color of cottonseed, soybean, and peanut oils, the spectrophotometric method is probably applicable to other fats and oils as well.

23.3.8 Iodine Value

23.3.8.1 *Principle*

The iodine value (or iodine number) is a measure of degree of unsaturation, which is the number of carbon-carbon double bonds in relation to the amount of fat or oil. **Iodine value** is defined as the grams of iodine absorbed per 100 g of sample. The higher the amount of unsaturation, the more iodine is absorbed and the higher the iodine value.

A common practice is to determine **calculated iodine value** from the fatty acid composition (see Sect. 23.6.1) using AOCS Recommended Practice Cd 1c-85. The calculated iodine value is not meant to be a rapid method but instead gives two results (iodine value of triacylglycerols and FFAs) from one analysis (fatty acid composition).

23.3.8.2 *Procedure*

A quantity of fat or oil dissolved in solvent is reacted, avoiding light, with a measured amount of iodine or some other halogen such as ICl or IBr. Halogen addition to double bonds takes place (Eq. 23.3). A solution of potassium iodide is added to reduce excess ICl to free iodine (Eq. 23.4). The liberated iodine is then titrated with a sodium thiosulfate standard using a starch indicator (Eq. 23.5), and the iodine value is calculated (Eq. 23.6):

$$\underset{\text{(excess)}}{\text{ICl}} + \text{R}-\text{CH}=\text{CH}-\text{R} \rightarrow$$

$$\text{R}-\text{CHI}-\text{CHCl}-\text{R} + \underset{\text{(remaining)}}{\text{ICl}} \quad (23.3)$$

$$\text{ICl} + 2\text{KI} \rightarrow \text{KCl} + \text{KI} + \text{I}_2 \quad (23.4)$$

$$\text{I}_2 + \underset{\text{(blue)}}{\text{starch}} + 2\text{Na}_2\text{S}_2\text{O}_3$$
$$\rightarrow 2\text{NaI} + \underset{\text{(colorless)}}{\text{starch}} + \text{Na}_2\text{S}_4\text{O}_6 \quad (23.5)$$

$$\text{Iodine value} = \frac{(B-S) \times N \times 126.9}{W \times 1000} \times 100 \quad (23.6)$$

where:

  Iodine value = g iodine absorbed per 100 g of sample
  B = volume of titrant (mL) for blank
  S = volume of titrant (mL) for sample
  N = normality of $\text{Na}_2\text{S}_2\text{O}_3$ (mol/L)

126.9 = MW of iodine (g/mol)
W = sample mass (g)
1,000 = conversion of units (mL/L)

Calculated iodine value is obtained from fatty acid composition using Eq. 23.7 for triacylglycerols. A similar equation allows calculation of the iodine value of FFAs:

Iodine value
(triglycerides) = (% hexadecenoic acid × 0.950)
  + (% octadecenoic acid × 0.860)
  + (% octadecadienoic acid × 1.732)
  + (% octadecatrienoic acid × 2.616)
  + (% eicosenoic acid × 0.785)
  + (% docosenoic acid × 0.723)   (23.7)

23.3.8.3 *Applications*

Iodine value is used to characterize oils, to follow the hydrogenation process in refining, and as an indication of lipid oxidation, since there is a decline in unsaturation during oxidation. The calculated value tends to be low for materials with a low iodine value and for oils with greater than 0.5 % unsaponifiable material (e.g., fish oils). The Wijs iodine procedure uses ICl and the Hanus procedure uses IBr. The Wijs procedure may be preferable for highly unsaturated oils as it reacts faster with the double bonds.

23.3.9 Saponification Value

23.3.9.1 *Principle*

Saponification is the process of breaking down or degrading a neutral fat into glycerol and fatty acids by treatment of the fat with alkali (Eq. 23.8):

triacylglycerol   glycerol   potassium salt of fatty acids

(23.8)

The **saponification value** (or saponification number) is defined as the amount of alkali necessary to saponify a given quantity of fat or oil. It is expressed as the milligrams of KOH required to saponify 1 g of the sample. The saponification value is an index of the mean molecular weight of the triacylglycerols in the sample. The mean molecular weight of the triacylglycerols may be divided by 3 to give an approximate mean molecular weight for the fatty acids present; the smaller the saponification value, the longer the average fatty acid chain length.

In common practice, the **calculated saponification value** is determined from the fatty acid composition (see Sect. 23.6.2) using AOCS Recommended Practice Cd 3a-94.

23.3.9.2 *Procedure*

Excess alcoholic potassium hydroxide is added to the sample, and the solution is heated to saponify the fat (Eq. 23.8). The unreacted potassium hydroxide is back-titrated with standardized HCl using phenolphthalein as the indicator and the saponification value is calculated (Eq. 23.9):

$$\text{Saponification value} = \frac{(B-S) \times N \times 56.1}{W} \quad (23.9)$$

where:

> Saponification value = mg KOH per g of sample
> B = volume of titrant (mL) for blank
> S = volume of titrant (mL) for sample
> N = normality of HCl (mmol/mL)
> 56.1 = MW of KOH (mg/mmol)
> W = sample mass (g)

The **calculated saponification value** is obtained from fatty acid composition using Eq. 23.10. The **fractional molecular weight** of each fatty acid in the sample must be determined first by multiplying the fatty acid percentage (divided by 100) by its molecular weight. The **mean molecular weight** is the sum of the fractional weights of all the fatty acids in the sample:

$$\text{Calculated saponification value} =$$
$$\frac{3 \times 56.1 \times 1000}{(\text{mean molecular weight} \times 3) + 92.09 - (3 \times 18)} \quad (23.10)$$

where:

Calculated saponification value = mg KOH per g of sample
> 3 = number of fatty acids per triacylglycerol
> 56.1 = MW of KOH (g/mol)
> 1,000 = conversion of units (mg/g)
> 92.09 = MW of glycerol (g/mol)
> 18 = MW of water (g/mol)

23.3.9.3 *Applications*

Saponification value is useful for determining the average fatty acid chain length of an oil of fat. The calculated saponification value is not applicable to fats and oils containing high amounts of unsaponifiable material, FFAs (>0.1 %), or mono- and diacylglycerols (>0.1 %).

23.3.10 Free Fatty Acids (FFAs) and Acid Value

23.3.10.1 *Principle*

Measures of fat acidity normally reflect the amount of fatty acids hydrolyzed from triacylglycerols (Eq. 23.11):

$$(23.11)$$

FFA is the percentage by weight of a specified fatty acid (e.g., percent oleic acid). **Acid value** (AV) is defined as the mg of KOH necessary to neutralize the free acids present in 1 g of fat or oil. The AV is often used as a quality indicator in frying oils, where a limit is 2 mg KOH/g oil is sometimes used. In addition to FFAs, acid phosphates and amino acids also can contribute to acidity. In samples containing no acids other than fatty acids, FFA and acid value may be converted from one to the other using a conversion factor (Eq. 23.12). Acid value conversion factors for lauric and palmitic are 2.81 and 2.19, respectively:

$$\% \text{FFA (as oleic)} \times 1.99 = \text{acid value} \quad (23.12)$$

Sometimes the acidity of edible oils and fats is expressed as milliliters of NaOH (of specified normality) required to neutralize the fatty acids in 100 g of fat or oil [8].

23.3.10.2 *Procedure*

To a liquid fat sample, neutralized 95 % ethanol and phenolphthalein indicator are added. The sample then is titrated with NaOH and the percent FFA calculated (Eq. 23.13):

$$\% \text{FFA (as oleic)} = \frac{V \times N \times 282}{W \times 1000} \times 100 \quad (23.13)$$

where:

> % FFA = percent free fatty acid (g/100 g), expressed as oleic acid
> V = volume of NaOH titrant (mL)
> N = normality of NaOH titrant (moL/L)
> 282 = MW of oleic acid (g/mol)
> W = sample mass (g)
> 1,000 = conversion of units (mL/L)

23.3.10.3 *Applications*

In crude fat, FFA or acid value estimates the amount of oil that will be lost during refining steps designed to remove fatty acids. In refined fats, a high acidity level means a poorly refined fat or fat breakdown after storage or use. However, if a fat seems to have a high amount of FFAs, it may be attributable to acidic additives (e.g., citric acid added as a metal chelator) since any acid will participate in the reaction [16]. If the fatty acids liberated are volatile, FFA or acid value may be a measure of hydrolytic rancidity.

23.3.11 Solid Fat Content

23.3.11.1 *Principle*

The amount of solids in a fat, termed the **solid fat content** (SFC), is determined using either continuous wave or pulsed NMR. Chapter 10 explains NMR as it relates to measuring the solid content of fats and other foods. Comparison between samples must be made using SFC values taken at the same temperature. Note that the standard temperature used varies from country to country.

Originally, the amount of solids in a fat was estimated using the **solid fat index** (SFI). SFI is measured using dilatometry, which determines the change in volume with change in temperature. As solid fat melts, it increases in volume. Plotting volume against temperature gives a line at which the fat is solid, a line at which it is liquid, and a melting curve in between. The SFI is the volume of solid fat divided by the volume between the upper and lower lines, expressed as a percentage [16]. Though the equipment is expensive, SFC is preferred over SFI because it measures the actual fat content (it is not an estimate), is less subject to error, and takes less time.

23.3.11.2 *Applications*

The amount of solid fat phase present in a plastic fat (e.g., margarine, shortening) depends on the type of fat, its history, and the temperature of measurement. The proportion of solids to liquids in the fat and how quickly the solids melt have an impact on functional properties, such as the mouthfeel of a food. An example of SFI use is shown in Fig. 23.3; the butter with high oleic and low saturated fatty acid composition has lower solid fat and is softer as it melts over the temperature range 10–35 °C and is more easily spreadable at refrigeration temperatures.

23.3.12 Consistency and Spreadability

The textural properties of plasticized fats (e.g., shortenings, margarine, butter) can be measured using such tests as consistency and spreadability. The consistency method described has been used for several decades,

Solid Fat Index

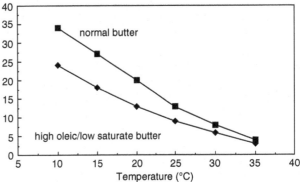

 SFI curves of butter with normal and high oleic/low saturated fatty acid compositions

whereas the spreadability method is a more recently approved method that utilizes texture analysis instruments. The **penetrometer method** of determining consistency measures the distance a cone-shaped weight will penetrate a fat in a given time period. The **spreadability** test delineates the parameters for using a Texture Technologies TA-XT2 Texture Analyzer® (or similar instrument) to determine the force needed to compress a sample. See Chap. 29 for general approaches to characterizing the rheological properties of foods; many aspects can be applied to fats and oils.

The penetrometer method is useful for measuring the consistency of plastic fats and solid fat emulsions. Like SFC, consistency is dependent on the type of fat, its history, and the temperature during measurement. The spreadability method is applicable to lipid-containing solid suspensions, emulsion, and pastes that can maintain their shape at the temperature used for the analysis, including products such as peanut butter and mayonnaise.

23.3.13 Polar Components in Frying Fats

Methods used to monitor the quality of the oil or fat used in deep-fat frying operations are based on the physical and chemical changes that occur, which include an increase in each of the following parameters: viscosity, foaming, FFAs, degree of saturation, hydroxyl and carbonyl group formation, and saponification value. Standard tests used in the evaluation of frying fats include quantitating polar components, conjugated dienoic acids, polymers, and free fatty acids. In addition, there are several rapid tests useful in day-to-day quality assurance of deep-fat frying operations [11].

23.3.13.1 *Principle*

Deterioration of used frying oils and fats can be monitored by measuring the polar components, which include monoacylglycerols, diacylglycerols, FFAs, and oxidation products formed during heating of foodstuffs. Nonpolar compounds are primarily unaltered triacylglycerols. The polar compounds in a sample can be separated from nonpolar compounds using chromatographic techniques.

23.3.13.2 Procedure

Polar components are measured by dissolving the fat sample in light petroleum ether-diethyl ether (87:13), then applying the solution to a silica gel column. Polar compounds are adsorbed onto the column. Nonpolar compounds are eluted, the solvent evaporated, the residue weighed, and the total polar components estimated by difference. Quality of the determination can be verified by eluting polar compounds and separating polar and nonpolar components using thin-layer chromatography (TLC).

23.3.13.3 Applications

A suggested limit of 27 % polar components in frying oil is a guide for when it should be discarded. A limitation of this method is the sample run time of 3.5 h [11]. The acid value is often determined as an alternate indicator of frying oil deterioration; however, rapid procedures based on dielectric constant are becoming more widely used because of their speed.

23.4 LIPID OXIDATION: MEASURING PRESENT STATUS

23.4.1 Overview

The term **rancidity** refers to the off odors and flavors resulting from lipolysis **(hydrolytic rancidity)** or lipid oxidation **(oxidative rancidity)**. **Lipolysis** is the hydrolysis of fatty acids from the glyceride molecule. Because of their volatility, hydrolysis of short-chain fatty acids can result in off odors. Fatty acids shorter than C12 (lauric acid) can produce off odors in foods. Free C12 is often associated with a soapy taste but no aroma. FFAs longer than C12 do not cause significant impairment in taste or odor.

Lipid oxidation (also called autoxidation) as it occurs in bulk fats and oils proceeds via a self-sustaining free radical mechanism. Depending on the fatty acid composition, the prooxidants to which it is exposed, the type and amount of antioxidants present, and other factors, the decomposition of fats and oils can proceed via various mechanisms that produce many different compounds. The traditional explanation is that the initial or primary products are hydroperoxides that undergo scission to form various secondary products including aldehydes, ketones, organic acids, and hydrocarbons [18] (see Fig. 23.4). However, recent reviews [9, 10] suggest secondary products, including epoxides, dimers,

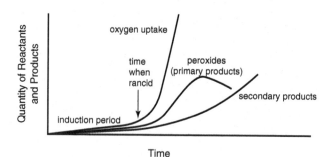

23.4 figure Traditional portrayal of changes in quantities of lipid oxidation reactants and products over time. Recent reviews suggest greater complexity in possible reactions than shown here, where secondary products can be produced simultaneously with hydroperoxides (Adapted from Labuza [18], with permission, Copyright CRC Press, Boca Raton, FL, ©1971)

and polymers, can be produced simultaneously with hydroperoxides.

Many methods have been developed to measure the different compounds as they form or degrade during lipid oxidation. In the early stages of deterioration, effective measures of lipid oxidation include oxygen consumption, conjugated dienes, and hydroperoxide formation using peroxide value. Both conjugated dienes and hydroperoxides decompose, however, and are not valid measures in later stages or isolated samples. Other measures of lipid oxidation include the *p*-anisidine value, measurement of volatile organic compounds (e.g., aldehydes), and thiobarbituric acid reactive substances (TBARS). Some of these procedures have been modified (especially with respect to sample size) for use in biological tissue assays [19]. Other methods that monitor lipid oxidation (and that vary in usefulness) include the iodine value, acid value, Kreis test, and oxirane test, as well as the measurement of conjugated dienes and trienes, total and volatile carbonyl compounds, polar compounds, and hydrocarbon gases [6, 13]. Since the system is dynamic, it is recommended that two or more methods be used to obtain a more complete understanding of lipid oxidation. In addition, free radicals not only oxidize lipids but may co-oxidize other molecules such as proteins, nucleic acids, polysaccharides, vitamins, and pigments. Schaich et al. [10] suggest other measurements be made in addition to monitoring lipid oxidation, methods such as protein oxidation and pigment bleaching.

While quantitating lipid oxidation using methods such as those listed above is usually adequate, in some cases, it may be necessary to *visualize the location* of lipid molecules and lipid oxidation within a food or raw ingredient. **Fluorescence microscopy** with stains specific to lipids can be applied to such a problem (see Chap. 32, Sect. 32.2.2.3). For example, the dye Nile blue (with the active ingredient Nile red) can be combined with a lipid-containing sample and the preparation viewed under a fluorescence microscope [20–22]. Lipids will appear an intense yellow fluorescence, with the intensity of the fluorescence changed by the nature of the lipids and by lipid oxidation. Example applications include localizing oxidized lipids in a cereal product, visualizing interactions between lipids and emulsifiers, and localizing lipids in cheeses, frosting, and chocolates.

23.4.2 Sample Preparation

Most methods require lipid extraction prior to analysis (see Sect. 23.2). However, variations of some methods (e.g., some TBARS tests) may begin with the original foodstuff.

23.4.3 Peroxide Value

23.4.3.1 *Principle*
Peroxide value is defined as the milliequivalents (mEq) of peroxide per kilogram of sample. It is a redox titrimetric determination. The assumption is made that the compounds reacting under the conditions of the test are peroxides or similar products of lipid oxidation.

23.4.3.2 *Procedure*
The fat or oil sample is dissolved in glacial acetic acid-isooctane (3:2). Upon addition of excess potassium iodide, which reacts with the peroxides, iodine is produced (Eq. 23.14). The solution then is titrated with standardized sodium thiosulfate using a starch indicator (Eq. 23.14). Peroxide value is calculated as shown in Eq. 23.15:

$$ROOH + K^+I^- \xrightarrow{H^+, heat} ROH + K^+OH^- + \underset{(excess)}{I_2} \quad (23.14)$$

$$I_2 + \underset{(blue)}{starch} + 2Na_2S_2O_3 \rightarrow 2NaI + \underset{(colorless)}{starch} + Na_2S_4O_6 \quad (23.15)$$

$$Peroxide\ value = \frac{(S - B) \times N}{W} \times 1000 \quad (23.16)$$

where:

> Peroxide value = mEq peroxide per kg of sample
> S = volume of titrant (mL) for sample
> B = volume of titrant (mL) for blank
> N = normality of $Na_2S_2O_3$ solution (mEq/mL)
> 1,000 = conversion of units (g/kg)
> W = sample mass (g)

23.4.3.3 *Applications*
Peroxide value measures a transient product of oxidation (i.e., after forming, peroxides and hydroperoxides break down to form other products). Though peroxide values measured on a given sample stored over time increase and then decrease (see Fig. 23.4), products are

usually determined (using sensory analysis) to be rancid during the period of increasing values. For determination in foodstuffs, a disadvantage of this method is the 5 g fat or oil sample size required; it is difficult to obtain sufficient quantities from foods low in fat. This method is empirical and any modifications may change results. Despite its drawbacks, peroxide value is one of the most common tests of lipid oxidation.

High-quality, freshly deodorized fats and oils will have a peroxide value of zero. Peroxide values >20 correspond to very poor-quality fats and oils, which normally would have significant off flavors. For soybean oil, peroxide values of 1–5, 5–10, and >10 correspond to low, medium, and high levels of oxidation, respectively (AOCS Method Cg 3-91).

23.4.4 *p*-Anisidine Value and Totox Value

23.4.4.1 *Principle*
The *p*-**anisidine value** estimates the amount of α- and β-unsaturated aldehydes (mainly 2-alkenals and 2,4-dienals), which are secondary oxidation products in fats and oils. The aldehydes react with *p*-anisidine to form a chromogen that is measured spectrophotometrically. The **totox value** tends to indicate the total oxidation of a sample using both the peroxide and *p*-anisidine values (Eq. 23.16):

$$\begin{aligned} Totox\ value = p\text{-anisidine value} \\ + (2 \times peroxide\ value) \end{aligned} \quad (23.17)$$

23.4.4.2 *Procedure*
The *p*-**anisidine value** by convention is defined as 100 times the absorbance at 350 nm of a solution containing precisely 1 g of test oil diluted to 100 ml using a mixture of solvent (isooctane) and *p*-anisidine.

23.4.4.3 *Applications*
Since peroxide value measures hydroperoxides (which increase then decrease) and *p*-anisidine value measures aldehydes (decay products of hydroperoxides which continually increase), the totox value usually rises continually during the course of lipid oxidation. Fresh soybean oil should have a *p*-anisidine value <2.0 and a totox value <4.0 (AOCS Method Cg 3-91). Though not common in the United States, *p*-anisidine and totox values are extensively used in Europe [16].

23.4.5 Volatile Organic Compounds

23.4.5.1 *Principle*
Volatile organic compounds (VOCs) present in fats, oils, and foodstuffs are related to flavor, quality, and oxidative stability. These compounds include secondary products of lipid oxidation that can be responsible for the off flavors and odors of oxidized fats and oils (AOCS Method Cg 3-91). The compounds formed will vary depending on the fatty acid composition of the

sample and environmental conditions. Commonly measured compounds include propanal, pentanal, hexanal, and 2,4-decadienal. It is commonly performed using **static (equilibrium) headspace analysis**, which entails the chromatographic analysis of a set volume of vapor obtained from the headspace above a sample held in a closed container (see Chap. 14, Sect. 14.2.2.2). The introduction of **solid-phase microextraction** (SPME) improved the sensitivity of static headspace analysis of VOC because the volatiles are concentrated on the SPME fiber (Chap. 14, Sect. 14.2.2.5).

23.4.5.2 Procedure

General guidelines regarding GC parameters are given in AOCS Method Cg 4-94, under the heading of static headspace; a review of research literature will indicate current practices used with various commodities. Typically, a small sample of the commodity is placed in a container having a septum cap. An internal standard may be added, for example, 4-heptanone [23]. The container is sealed and then heated for a given time. Heating increases the concentration of headspace volatiles [24]. Using a gastight syringe, an aliquot of the vapors in the container headspace then is removed and injected into a GC equipped with a flame ionization detector or mass selective detector. Automated headspace samplers are available that both heat the sample and ensure a constant volume is being analyzed. The quantity of aldehydes, including hexanal and pentanal, and other volatiles of interest then is calculated from peak areas (see Chaps. 12 and 14).

23.4.5.3 Applications

The formation of aldehydes, such as hexanal and pentanal, may correlate well with sensory determination of lipid oxidation since they are major contributors to off flavors in some food commodities. The quantity of other volatile compounds resulting from lipid oxidation can be obtained simultaneously with aldehyde measurement, and may enhance the characterization of lipid oxidation in various food commodities. An advantage of this method is that since volatiles are collected from the headspace of the sample vial, lipid extraction is not required (i.e., intact foodstuffs can be analyzed). However, various SPME fibers are available, and the specific fiber used may not absorb all the volatiles produced.

23.4.6 Thiobarbituric Acid Reactive Substances Test

23.4.6.1 Principle

The **thiobarbituric acid reactive substances** (TBARS) test, also known as the **thiobarbituric acid** (TBA) test, measures secondary products of lipid oxidation, primarily **malonaldehyde**. It involves reaction of malonaldehyde (or malonaldehyde-type products

including unsaturated carbonyls) with TBA to yield a colored compound that is measured spectrophotometrically. The food sample may be reacted directly with TBA but is often distilled to eliminate interfering substances, and then the distillate is reacted with TBA. Many modifications of the test have been developed over the years.

23.4.6.2 Procedure

In contrast to the direct method for fats and oils (see Table 23.1), a commonly used procedure [25, 26] is outlined here that requires distillation of the food commodity prior to determining TBARS. A weighed sample is combined with distilled water and mixed. The pH is adjusted to 1.2 and the sample is transferred to a distillation flask. After addition of BHT (optional), antifoam reagent, and boiling beads, the sample is distilled rapidly and the first 50 ml collected. An aliquot of the distillate is combined with TBA reagent and heated in a boiling water bath for 35 min. Absorbance of the solution is determined at 530 nm, and, using a standard curve, absorbance readings are typically converted to milligrams of malonaldehyde (or TBARS) per kilograms of sample.

23.4.6.3 Applications

The TBA test correlates better with sensory evaluation of rancidity than does peroxide value but, like peroxide value, it is a measure of a transient product of oxidation (i.e., malonaldehyde and other carbonyls readily react with other compounds). An alternative to the spectrophotometric method described is to determine the actual content of malonaldehyde using HPLC analysis of the distillate.

Despite common use, tests for measuring malonaldehyde are only applicable to foods containing fatty acids with at least three double bonds and therefore are not useful in many food systems [10].

23.4.7 Conjugated Dienes and Trienes

23.4.7.1 Principle

Double bonds in lipids are changed from nonconjugated to conjugated bonds upon oxidation. Fatty acid **conjugated diene hydroperoxides** formed on lipid oxidation absorb UV light at about 232 nm and **conjugated trienes** at about 270 nm. Conjugated products of lipid oxidation should not be confused with conjugated linoleic acid isomers, which are produced by incomplete hydrogenation in ruminants.

23.4.7.2 Procedure

A homogeneous lipid sample is weighed into a graduated flask and brought to volume with a suitable solvent (e.g., isooctane, cyclohexane). A 1 % solution (e.g., 0.25 g sample in 25 ml) is made, then diluted or concentrated if necessary to obtain an absorbance between 0.1 and 0.8. The diluted solution must be completely clear. Absorbance is measured with an ultraviolet (UV) spectrophotometer, using the pure solvent as a blank.

A common practice is to calculate the amount of conjugated dienes from chromatographic analysis of fatty acids (see Chap. 14).

23.4.7.3 Applications

The conjugated diene and triene methods are useful for monitoring the early stages of oxidation. The magnitude of the changes in absorption is not easily related to the extent of oxidation in advanced stages. Olive oil grades are based on ultraviolet absorbance.

23.5 LIPID OXIDATION: EVALUATING OXIDATIVE STABILITY

23.5.1 Overview

Because of their inherent properties (e.g., the amount of unsaturation and the presence of natural antioxidants) as well as external factors (e.g., added antioxidants, processing, and storage conditions), lipids and lipid-containing foodstuffs vary in their susceptibility to rancidity. The resistance of lipids to oxidation is known as **oxidative stability**. Inasmuch as determining oxidative stability using actual shelf life determinations at ambient conditions of storage (usually room temperature) requires months or even years, accelerated tests have been developed to evaluate the oxidative stability of bulk oils and fats and various foodstuffs. **Accelerated tests** artificially hasten lipid oxidation by exposing samples to heat, oxygen, metal catalysts, light, or enzymes. A major problem with accelerated tests is assuming reactions that occur at elevated temperatures or under other artificial conditions are the same as normal reactions occurring at the actual storage temperature of the product. An additional difficulty is ensuring that the apparatus is clean and completely free of metal contaminants and oxidation products from previous runs. Therefore, assuming lipid oxidation is the factor that limits shelf life, shelf life determinations at ambient conditions should accompany and hopefully validate the results of accelerated tests of oxidative stability.

The **induction period** is defined as the length of time before detectable rancidity or time before rapid acceleration of lipid oxidation. Induction period can be determined by such methods as calculating the maximum of the second derivative with respect to time or manually drawing tangents to the lines (Fig. 23.5). Measurement of the induction period allows a comparison of the oxidative stability of samples that contain differing ingredients or of samples held at varying

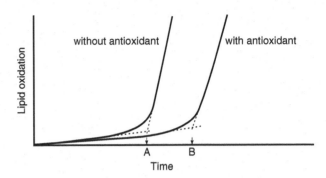

figure

A plot of lipid oxidation over time, showing the effect of an antioxidant on induction period. Time A is induction period of sample without antioxidant and Time B is induction period of sample with antioxidant

storage conditions, and it provides an indication of the **effectiveness of** various **antioxidants** in delaying lipid oxidation and thus prolonging the shelf life of fats, oils, and foodstuffs. Guidelines for conducting such tests have been established (see AOCS Method Cg 7-05). In contrast to measuring the effectiveness of antioxidants in prolonging food shelf life, **antioxidant capacity** refers to the ability of foods and ingredients to counter oxidative reactions in the human body and is discussed in Chap. 25.

23.5.2 Oven Storage Test

As a means of accelerating the determination of oxidative stability, the **oven storage test** is often used. Previously this test was an ill-defined procedure known as the Schaal oven test. It is now a recommended practice of the AOCS (AOCS Method Cg 5-97). This protocol consists of placing a fat or oil of known volume in a forced-draft oven at a temperature above ambient but less than 80 °C, with 60 °C being recommended. Such temperatures are desirable accelerated storage temperatures since the mechanism of oxidation in this range is the same as oxidation at room temperature. Tests should be conducted in the dark, the initial quality of the sample oil should be high, and the surface to volume ratio of the oil must be kept constant for all samples.

To determine an induction period and thus oxidative stability, the oven storage test must be combined with other methods of detecting rancidity, for example, sensory evaluation and peroxide value. More than one method should be used to determine the extent of oxidation; it is recommended that one method measure primary products of lipid oxidation (e.g., peroxide value, conjugated dienes) and the other method measure secondary products (e.g., VOC, p-anisidine value, TBARS, or sensory evaluation). Results of oxidative stability determinations obtained at approximately 60 °C correlate well with actual shelf life determinations [8].

23.5.3 Oil Stability Index

23.5.3.1 *Principle*

The **oil stability index** (OSI) determines induction period by bubbling purified air through an oil or fat sample held at an elevated temperature (often 110 °C or 130 °C), then passing the acidic volatiles (primarily formic acid) into a deionized water trap. The conductivity of the water is measured continuously, resulting in data similar to those shown in Fig. 23.5. Results should specify the air flow rate and temperature used, as well as induction period time. Two instruments that automate this method are the Rancimat® (Metrohm USA, Inc.) and the Oxidative Stability Instrument® (Ultra Scientific, Inc.). The outdated and labor-intensive **active oxygen method** (AOM) is similar to the OSI except induction period is determined by discontinuous measurements of either peroxide value or sensory evaluation of rancid odor.

23.5.3.2 *Applications*

These methods were designed originally to measure the effectiveness of antioxidants. The OSI is determined much faster than tests performed at oven storage test temperatures, but results from the latter may correlate better with actual shelf life. Extrapolating OSI results to actual shelf life is difficult because the high temperatures used result in the formation of compounds not present in samples held at ambient or slightly elevated temperature conditions [8, 27, 28].

Specification sheets for fats and oils often report AOM values to accommodate individuals working in this area who are familiar with AOM values. OSI values can be converted to AOM values.

Applicable to all fats and oils, the OSI has also been researched for applicability to certain low-moisture snack foods (e.g., potato chips and corn chips). Because of the continuous exposure to circulating air, samples that contain more than negligible amounts of water tend to dehydrate during the determination and are not likely to give reliable results because water activity can affect oxidation rates.

23.5.4 Oxygen Bomb

23.5.4.1 *Principle*

Inasmuch as lipid oxidation results in the uptake of oxygen from the surrounding environment (see Fig. 23.4), measuring the time required for the onset of rapid disappearance of oxygen in a closed system provides a means of determining oxidative stability.

23.5.4.2 *Procedure*

The oxygen bomb consists of a heavy-walled container that has a pressure recorder attached. The sample is placed in the container, and oxygen is used to pres-

surize the container to 100 psi. The container then is placed in a boiling water bath. Induction period is determined by measuring the time until a sharp drop in pressure occurs, which corresponds with the rapid absorption of oxygen by the sample.

23.5.4.3 *Applications*

Oxygen bomb results may have a better correlation with rancidity shelf life tests than OSI values. Another advantage, compared to the OSI, is that the oxygen bomb method may be used with intact foodstuffs instead of extracted lipids [16].

23.6 METHODS FOR LIPID COMPONENTS

23.6.1 Overview

The lipid present in food commodities or bulk fats and oils can be characterized by measuring the amount of its various fractions, which include fatty acids, mono-, di-, and triacylglycerols, phospholipids, sterols (including cholesterol and phytosterols), and lipid-soluble pigments and vitamins. Another means of categorizing lipid fractions is inherent in nutrition labeling which involves the measurement of not only total fat but may also require quantification of saturated fat, monounsaturated fat, polyunsaturated fat, and *trans* isomer fatty acids. In addition, foods may contain lipids that do not contribute the same caloric content as normal lipids, for example, sucrose polyesters (e.g., Olestra®), medium-chain triglycerides, and triglycerides that contain short-chain fatty acids (e.g., Salatrim®, Caprenin®). Many of these fractions are determined readily by evaluating the component fatty acids. From the fatty acid composition, calculations are made to determine such parameters as total fat, saturated fat, calculated iodine value, and calculated saponification value.

In contrast, the measurement of total and saturated fat in foods containing Olestra® requires special consideration. AOAC peer-verified Method PVM 4:1995 outlines the use of lipase on the lipid extract, which yields fatty acids and unreacted Olestra®. The fatty acids are converted to calcium soaps and Olestra® is extracted and discarded. The precipitated soaps are converted back to fatty acids, which are subsequently analyzed via capillary GC [29]. Other procedures that have been studied include chromatographic separation using HPLC with an evaporative light-scattering detector (ELSD) [30]. Olestra®, like most lipids, will not absorb UV or VIS light appreciably, preventing analysis by using the common UV-VIS HPLC detectors. For this reason, detectors such as the refractive index (RID), transport flame ionization (TFID), or ELSD are required for analysis of most lipids.

GC (see Chap. 14) is ideal for the analysis of many lipid components. GC can be used for determinations such as total fatty acid composition, distribution and position of fatty acids in lipid, sterols, studies of fat stability and oxidation, assaying heat or irradiation damage to lipids, and detection of adulterants and antioxidants [10]. Methods exist that detail the analysis of various lipid fractions using GC [5]. GC combined with **mass spectrometry** (MS) (see Chap. 11) is a powerful tool used in identification of compounds. **HPLC** (see Chap. 13) also is useful in lipid analyses, especially for components that are not readily volatilized, such as hydroperoxides and triacylglycerols [28]. **Thin-layer chromatography** (TLC) (see Chap. 12) has been used extensively in the past by lipid chemists. Partly due to low cost and ease, TLC is still useful, although many assays may be more quantitative or have better resolution using GC or HPLC.

23.6.2 Fatty Acid Composition and Fatty Acid Methyl Esters (FAMEs)

The **fatty acid composition,** or **fatty acid profile,** of a food product is determined by quantifying the kind and amount of fatty acids that are present, usually by extracting the lipids and analyzing them using **capillary GC** (also described in Chap. 17, Sect. 17.2.7).

23.6.2.1 *Principle*

To increase volatility before GC analysis, triacylglycerols are typically transesterified to form **fatty acid methyl esters** (FAMEs) (Eq. 23.18). Acyl lipids are readily transesterified using base such as sodium hydroxide and methanol. Sodium methoxide produced by this combination will create FAMEs from acyl lipids rapidly but will not react with free fatty acids. Acidic reagents such as methanolic HCL or boron trifluoride (BF$_3$) react rapidly with FFAs but more slowly with acyl lipids. Procedures such as the AOCS Method Ce 1b-89 (this is a joint method with AOAC 991.39) use a two-step methylation, first reacting the lipid with 0.5 N NaOH then with excess BF$_3$/methanol. This allows a rapid methylation of FFAs, acyl lipids, and phospholipids. The sodium hydroxide step is not a saponification procedure (i.e., hydrolysis of acyl groups); it is a direct transmethylation:

$$
\text{triacylglycerol} + CH_3OH \xrightarrow{\text{NAOH then BF}_3} \text{FAMEs}
$$

(23.18)

23.6.2.2 *Procedure*

The lipid is extracted from the food, for example, by homogenizing with a suitable solvent such as hexane-isopropanol (3:2, vol/vol) and then evaporating the solvent. The FAMEs are prepared by combining the extracted lipid with sodium hydroxide methanol and internal standard in isooctane and then heating at 100 °C for 5 min. The sample is cooled and then excess BF$_3$-methanol is added with further heating (100 °C for 30 min). After addition of saturated aqueous sodium chloride, additional isooctane, and mixing, the upper isooctane solution containing the FAMEs is removed and dried with anhydrous Na$_2$SO$_4$, and then diluted to a concentration of 5–10 % for injection onto the GC.

Several methods (see Table 23.1) describe procedures and conditions for using GC to determine fatty acid composition. AOCS Method Ce 1b-89 is specific for marine oils, and AOCS Method Ce 1f-96 is specifically suited for determining *trans* isomer fatty acids.

23.6.2.3 *Applications*

Determination of the fatty acid composition of a product permits the **calculation** of the following categories of fats that pertain to health issues and food labeling: percent saturated fatty acids, percent unsaturated fatty acids, percent monounsaturated fatty acids, percent polyunsaturated fatty acids, CLAs, and percent *trans* isomer fatty acids. Calculation of fatty acids as a percentage is referred to as **normalization**, i.e., the areas of all of the FAMEs are summed and the percent area of each fatty acid is calculated relative to the total area. This is a reasonable procedure because with flame ionization detectors (FID), the weight of fatty acids in a mixture closely parallels the area on the chromatogram. However, this is not absolutely correct. Theoretical correction factors are needed to correct for the FID response which is different depending on the level of unsaturation in FAMEs [31]. A chromatogram showing separation of FAMEs of varying length and unsaturation is shown in Chap. 17, Fig. 17.4. The separation of FAMEs on this SP2560 column is typical of what is seen when using a highly polar (biscyanopropyl polysiloxane) column [32].

The separation of FAMEs on GC columns depends on the polarity of the liquid phase. On nonpolar liquid phases [such as 100 % dimethyl polysiloxane (DB-1, HP-1, CPSil5CB) or 95 % dimethyl, 5 % diphenyl polysiloxane (DB5, HP5, CPSil8CB)], FAMEs are separated largely based on their boiling points. This results in the elution order 18:3n-3>18:3n-6>18:1n-9>18:0>20:0. On phases of medium polarity [such as 50 % cyanopropylphenyl polysiloxane (DB225, HP225, CPSil43CB)], the order of elution is changed because of the interaction of the pi electrons of the double bonds with the liquid phase. The order of elution on these columns would be 18:0>18:1n-9>18:2n-6>18:3n-3>20:0 (first eluted to last). When the polarity of the liquid is increased further with 100 % biscyanopropyl polysiloxane columns (SP2560, CPSil88), the greater interaction of the double bonds with the very polar liquid phase results in an elution pattern 18:0>18:1n-9>18:2n-6>20:0>18:3n-3. As the liquid phase polarity increases, the effect of double bonds on retention time increases. Additionally, *trans* fatty acids interact less effectively with the liquid phase than *cis* acids for steric reasons, so *trans* acids will elute before the corresponding *cis* acid; see Fig. 8.5 where 18:1Δ9 *trans* (elaidate) elutes before 18:1Δ9 (oleic acid), and 18:2Δ9 *trans* Δ12 *trans* (linoelaidate) elutes before linoleic acid (18:2n-6, 18:2 Δ9 *cis* Δ12 *cis*) on this highly polar 100 % biscyanopropyl polysiloxane column. It also can be seen that gamma linolenic (18:3n-6) elutes before linolenic acid (18:3n-3). Because the double bonds are closer to the methyl side of the FAMEs in 18:3n-3, the double bonds can more effectively interact with the liquid phase, resulting in greater retention by the column.

The complexity of the fatty acids found in various foods will affect the details of the GC analysis that can be used. Analysis of FAMEs of a vegetable oil is quite simple and can easily be accomplished in less than 20 min using a column with a medium polarity liquid phase. The fatty acids present in most vegetable oils range from C14 to C24. Coconut and palm kernel oils also contain shorter-chain fatty acids such as C8–C12. Dairy fats contain butyric acid (C4) and other short-chain fatty acids, whereas peanut oil contains C26 at around 0.4–0.5 % of the total FAMEs. Marine lipids contain a much wider range of fatty acids and require care in the separation and identification of FAMEs, many of which have no commercially available standards.

Trans fatty acids in foods originate from three main sources: biohydrogenation in ruminants, incomplete hydrogenation in the conversion of liquid oils to plastic fats, and high-temperature exposure during deodorization. The *trans* fatty acids formed from these three processes are quite different and require careful attention to achieve accurate analysis.

Separation of *trans* FAMEs is facilitated by selection of the most polar column phases available. Currently, Supelco SP2560 and Chrompak CPSil88 are most often used for analysis of *trans* fatty acids [32]. These columns have liquid phases based on 100 % biscyanopropyl polysiloxane. Even with optimized temperature programming and column selection, resolution of *trans* isomers from partially hydrogenated vegetable oil mixtures is incomplete and facilitated by the use of a Fourier transform infrared (FTIR) detector (Chap. 8, Sect. 8.3.1.2) or mass spectrometry (Chap. 11).

23.6.3 *trans* Isomer Fatty Acids Using Infrared Spectroscopy

Most natural fats and oils extracted from plant sources contain only isolated (i.e., methylene interrupted, nonconjugated) *cis* double bonds. Fats and oils extracted from animal sources may contain small amounts of *trans* double bonds. Inasmuch as the *trans* isomer is more thermodynamically stable, additional amounts of *trans* double bonds can be formed in fats and oils that undergo oxidation, or during processing treatments such as extraction, heating, and hydrogenation. Ongoing studies are evaluating the health effects of dietary lipids that contain *trans* fatty acids.

Measurement of *trans* isomer fatty acids is commonly done using GC techniques, such as AOCS Method Ce 1f-96 (see Sect. 23.6.2). However, this section describes the use of infrared (IR) spectroscopy for determining *trans* isomer fatty acids.

23.6.3.1 *Principle*

The concentration of *trans* fatty acids is measurable in lipids from an absorption peak at 966 cm^{-1} in the IR spectrum.

23.6.3.2 *Procedure*

AOCS Method Cd 14-95 requires liquid samples be converted to methyl esters and dissolved in an appropriate solvent that does not absorb in the IR region strongly, due to planar (carbon disulfide) or tetrahedral (carbon tetrachloride) symmetry. The absorbance spectra between 1,050 and 900 cm^{-1} are obtained using an infrared spectrometer (see Chap. 8). Methyl elaidate is used as an external standard in calculating the content of *trans* double bonds. Alternately, AOCS Method Cd 14d-96 determines total *trans* fatty acids using attenuated total reflection-Fourier transform infrared (ATR-FTIR) spectroscopy (see Chap. 8, Sects. 8.3.1.2 and 8.3.2).

23.6.3.3 *Applications*

The methods described will only detect isolated (i.e., nonconjugated) *trans* isomers. This is especially important when oxidized samples are of interest since oxidation results in a conversion from nonconjugated to conjugated double bonds. Also, AOCS Method Cd 14-95 is restricted to samples containing at least 5 % *trans* isomers, and AOCS Method Cd 14d-96 is limited to samples containing at least 0.8 % *trans* isomer fatty acids. For samples containing less than 0.8 % *trans* double bonds, a capillary GC method (AOCS Method Ce 1f-96) is recommended.

23.6.4 Mono-, Di-, and Triacylglycerols

Mono-, di-, and triacylglycerols may be determined using various techniques (see Table 23.1). Older methods use titrimetric approaches, whereas newer methods utilize chromatographic techniques, including HPLC and GC. Short nonpolar columns and very high temperatures are needed for analysis of intact triacylglycerols by GC. Section 23.6.6 describes the use of TLC to separate lipid classes, including mono-, di-, and triacylglycerols.

23.6.5 Cholesterol and Phytosterols

Many methods exist for the quantification of cholesterol and phytosterols in various matrices. Consulting research literature will give an indication of current practice and methods that may be less laborious or adapted for use with specific foodstuffs.

23.6.5.1 *Principle*

The lipid extracted from the food is saponified. The saponification process is a hydrolysis, with acyl lipids being converted to water-soluble FFA salts. Other components (called the unsaponifiable or nonsaponifiable matter) do not change in solubility after hydrolysis and thus remain soluble in organic solvents. Cholesterol (in the unsaponifiable fraction) is extracted and derivatized to form trimethylsilyl (TMS) ethers or acetate esters. This increases their volatility and reduces problems of peak tailing during chromatography. Quantitation is achieved using capillary GC.

23.6.5.2 *Procedure*

AOAC Method 976.26 outlined here is representative of the various procedures available for cholesterol determination. Lipids are extracted from the food and saponified, and the unsaponifiable fraction is extracted. This is accomplished by filtering an aliquot of the chloroform layer through anhydrous sodium sulfate and evaporating to dryness in a water bath using a stream of nitrogen gas. Concentrated potassium hydroxide and ethanol are added and the solution is refluxed. Aliquots of benzene and 1 *N* potassium hydroxide are added and then shaken. The aqueous layer is removed and the process is repeated with 0.5 *N* potassium hydroxide. After several washes with water, the benzene layer is dried with anhydrous sodium sulfate and an aliquot is evaporated to dryness on a rotary evaporator. The residue is taken up in dimethylformamide. An aliquot of this sample is derivatized by adding hexamethyldisilazane (HMDS) and trimethylchlorosilane (TMCS). Water (to react with and inactivate excess reagent) and an internal standard in heptane are added, and then the solution is centrifuged. A portion of the heptane layer is injected into a GC equipped with a nonpolar column. The HMDS and TMCS reagents are rapidly inactivated by water, and thus the reaction conditions must remain anhydrous.

23.6.5.3 *Applications*

GC quantitation of cholesterol is recommended since many spectrophotometric methods are not specific for cholesterol. In the past, samples such as eggs and shrimp have had their cholesterol contents overestimated by relying on less specific colorimetric procedures. Other GC, HPLC, and enzymatic methods are available. For example, cholesterol methods developed for frozen foods [33] and meat products [34] eliminate the fat extraction step, directly saponifying the sample; compared to the AOAC method outlined previously, they are more rapid and avoid exposure to toxic solvents.

Cholesterol oxidation products as well as phytosterols can be quantified using the GC procedure outlined for cholesterol. A wide range of methods for analysis of sterols exist in the literature; most use TMS ether formation to increase volatility of the hydroxide-containing sterols and to improve chromatographic resolution (reduce peak tailing).

23.6.6 Separation of Lipid Fractions by Thin-Layer Chromatography

23.6.6.1 *Procedure*

Thin-layer chromatography (TLC) is performed using silica gel G as the adsorbent and hexane-diethyl ether-formic acid (80:20:2 vol/vol/vol) as the eluting solvent system (Fig. 23.6). Plates are sprayed with 2′,7′-dichlorofluorescein in methanol and placed under ultraviolet light to view yellow bands against a dark background [5].

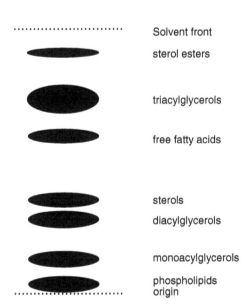

23.6 Schematic thin-layer chromatography (TLC) separation of lipid fractions on silica gel G (Adapted with permission from Christie [5])

23.6.6.2 *Applications*

This procedure permits rapid analysis of the presence of the various lipid fractions in a food lipid extract. For small-scale preparative purposes, TLC plates can be scraped to remove various bands for further analysis using GC or other means. Many variations in TLC parameters are available that will separate various lipids. Thin-layer plates can be impregnated with silver nitrate to allow separation of FAMEs based on their number of double bonds. FAMEs with six double bonds are highly retained by the silver ions on the plate; FAMEs with no double bonds are only slightly retained. This allows a separation of FAMEs based on number of double bonds, which can be useful when identifying FAMEs in complex mixtures (bands are scraped off, eluted with solvent, then analyzed by GC).

23.7 SUMMARY

The importance of fat characterization is evident in many aspects of the food industry, including ingredient technology, product development, quality assur-ance, product shelf life, and regulatory aspects. Lipids are closely associated with health; the cholesterol or phytosterol compositions, and amounts of *trans*, satu-rated, and n-3/ω3 fatty acids are of great concern to consumers.

The methods described in this chapter help to characterize bulk oils and fats and the lipids in food-stuffs. A summary of some of the more common tests is given in Table 23.3. Methods described for bulk oils and fats can be used to determine characteristics such as melting point; smoke, flash, and fire points; color; degree of unsaturation; average fatty acid chain length; and amount of polar components. The per-oxide value, TBA, and hexanal tests can be used to measure the present status of a lipid with regard to oxidation, while the OSI can be used to predict the sus-ceptibility of a lipid to oxidation and the effectiveness of antioxidants. Lipid fractions, including fatty acids, triacylglycerols, phospholipids, and cholesterol, are commonly analyzed by chromatographic techniques such as GC and TLC.

The methods discussed in this chapter represent only a few of the many tests that have been developed

23.3 table Summary of some of the common tests used to characterize fats and oils in bulk products or foodstuffs

Test	Indicated	Actually measured
Iodine value	Degree of unsaturation	Iodine required for absorption by double bonds
Saponification value	Mean molecular weight of triacylglycerols	Alkali required to saponify fat/oil
Free fatty acids	Fatty acid hydrolysis from triglycerides (hydrolytic rancidity)	Potassium hydroxide necessary to neutralize free acids
Solid fat content	Proportion of solid vs. liquid fat	Percent solid fat, by continuous wave or pulsed NMR
Conjugated dienes and trienes	Oxidative rancidity (current status)	Change from nonconjugated to conjugated bonds, due to early stage of lipid oxidation
Peroxide value (PV)	Oxidative rancidity (current status)	Peroxides, due to early stages of lipid oxidation
Totox value	Oxidative rancidity (current status)	Peroxide and p-anisidine values, due to early and later stages of lipid oxidation
Volatile organic compounds	Oxidative rancidity (current status)	Volatile organic compounds (e.g., hexanal, pentanal, pentane), secondary products of lipid oxidation
Thiobarbituric acid reactive Substances (TBARS)	Oxidative rancidity (current status)	Malonaldehyde and similar compounds, as secondary products of lipid oxidation
Oven storage test	Oxidative stability (predicted)	Induction period after exposure to elevated temperature, measured using PV or sensory evaluation
Oil stability index	Oxidative stability (predicted)	Induction period after exposure to air and high temperature, based on acidic volatiles, measured by conductivity
Fatty acid methyl esters (FAMES)	Fatty acid composition of fats and oils; calculated total fat (nutritional labeling)	Chromatographic separation and quantification of individual fatty acids after fat extraction and hydrolysis of triacylglycerols
Trans fatty acids	Trans fat (nutritional labeling)	Chromatographic separation and quantification of fatty acids containing trans bonds after fat extraction and hydrolysis of triacylglycerols
Cholesterol	Cholesterol (nutritional labeling)	Chromatographic determination of cholesterol after fat extraction and saponification

to characterize lipid material. Consult the references cited for additional methods or more detailed explanations. Time, funding, availability of equipment and instruments, required accuracy, and purpose all will dictate the choice of method to characterize oils, fats, and foodstuffs containing lipids.

23.8 STUDY QUESTIONS

1. You want to compare several fat/oil samples for the chemical characteristics listed below. For each characteristic, name one test (give full name, not abbreviation) that could be used to obtain the information desired:

   (a) Degree of unsaturation
   (b) Predicted susceptibility to oxidative rancidity
   (c) Present status with regard to oxidative rancidity
   (d) Average fatty acid molecular weight
   (e) Amount of solid fat at various temperatures
   (f) Hydrolytic rancidity

2. Your analysis of an oil sample gives the following results. What does each of these results tell you about the characteristics of the sample? Briefly describe the principle for each method used:

   (a) Large saponification value
   (b) Low iodine value
   (c) High TBA number
   (d) High FFA content
   (e) High OSI

3. Define solid fat content and explain the usefulness of this measurement.

4. Peroxide value, TBA number, and hexanal content all can be used to help characterize a fat sample:

   (a) What do the results of these tests tell you about a fat sample?
   (b) Differentiate these three tests as to what chemical is being measured.

5. What methods would be useful in determining the effectiveness of various antioxidants added to an oil?

6. You are responsible for writing the specifications for vegetable oil purchased from your supplier for use in deep-fat frying several foods processed by your company. Itemize the tests you should require in your list of specifications (specific values for the tests are not needed). For each test, briefly state what useful information is obtained.

7. The Nutrition Education and Labeling Act of 1990, as amended in 2016 (see Chap. 3), requires that the nutrition label on food products contains information related to lipid constituents. In addition to the amount of total fat (see Chap. 17), the label must state saturated fat, *trans* fat, and cholesterol contents:

   (a) For a product such as traditional potato chips, explain an appropriate method for the analysis of each of these lipid constituents.
   (b) Compared with assays on traditional chips, how would the assays for total fat and saturated fat differ for potato chips made with Olestra®?

8. You have developed a new butter containing added fish oil, which is high in PUFA. Before placing the product on the market, you need to determine its shelf life. What method or methods would you use and why?

9. You work in quality control for a company that makes peanut butter. You received information that between July and August several lots of peanut butter may have been improperly stored and you are concerned about potential lipid oxidation in the product. A total of 50 lots of peanut butter are involved:

   (a) What test(s) would you use to measure lipid oxidation? Include your rationale for selecting the method.
   (b) How would you decide what to do with the peanut butter?

23.9 PRACTICE PROBLEMS

1. A 5.00-g sample of oil was saponified with excess KOH. The unreacted KOH was then titrated with 0.500 *N* HCl (standardized). The difference between the blank and the sample was 25.8 ml of titrant. Calculate the saponification value.

2. A sample (5.0 g) of food grade oil was reacted with excess KI to determine peroxide value. The free iodine was titrated with a standardized solution of 0.10 *N* $Na_2S_2O_3$. The amount of titrant required was 0.60 ml (blank corrected). Calculate the peroxide value of the oil.

3. You analyze the saponification value of an unknown. You use 4.0 g oil and 0.5 *N* HCl to titrate. The difference between the blank and the sample was 43 mL of titrant. What is the average fatty acid molecular weight of the oil?

4. You analyze an oil FAMEs by GC and find the following areas for your identified peaks:

16:0	2,853,369
18:0	1,182,738
18:1n-9	38,999,438
18:2n-6	14,344,172
18:3 n-3	2,148,207

Report the fatty acid composition as % and tentatively identify the oil.

Answers:

1. Saponification value =
$$\frac{25.8 \text{mL} \times 0.500 \text{ mmoL} / \text{mL} \times 56.1 \text{ mg} / \text{mmol}}{5.00\text{g}} = 145$$

2. Peroxide value =
$$\frac{0.60 \text{ mL} \times 0.10 \text{ meq} / \text{mL} \times 1000}{5.0\text{g}} = 12 \text{ meq} / \text{kg}$$

3. Saponification value =
$$\frac{43 \text{mL} \times 0.500 \text{ mol} / \text{mL} \times 56.1 \text{mg} / \text{mmol}}{4.0\text{g}} = 302$$

$$302 = [3 \times 56.1 \times 1000] / \left\{ \begin{array}{c} \left[(\text{Mean mol. wt} \times 3) + 92.09 \right] \\ -(3 \times 18) \end{array} \right\}$$

Mean mol. wt = 172

4. Sum of all areas = 58,416,924

   Area % 16 : 0 = 100 × 2, 853, 369 / 58, 416, 924 = 4.9%

so,

16:0	4.9%
18:0	2.0%
18:1n-9	66.8%
18:2 n-6	22.7%
18:3n-3	3.7%

Based on the fatty acid composition, the oil is probably Canola oil.

REFERENCES

1. AOAC International (2016) Official methods of analysis of AOAC International, 20th edn. (On-line). AOAC International, Rockville, MD
2. AOCS (2013) Official methods and recommended practices of the AOCS, 6th edn. American Oil Chemists' Society, Champaign, IL
3. IUPAC (1987) Standard methods for analysis of oils, fats, and derivatives, and supplements 7th edn. International Union of Pure and Applied Chemistry, Commission on Oils, Fats and Derivatives, Paquot C and Hautfenne A (eds). Blackwell Scientific, Oxford
4. Christie WW (1982) Lipid analysis. Isolation, separation, identification, and structural analysis of lipids, 2nd edn. Pergamon, Oxford
5. Christie WW (1989) Gas chromatography and lipids. A practical guide. The Oily Press, Ayr, Scotland
6. Gray JI (1978) Measurement of lipid oxidation: a review. J Am Oil Chem Soc 55: 539–546
7. Melton SL (1983) Methodology for following lipid oxidation in muscle foods. Food Technol 37(7): 105–111,116
8. Pomeranz Y, Meloan CE (1994) Food analysis: theory and practice, 3rd edn. Chapman & Hall, New York
9. Schaick, K (2013) "Challenges in Analyzing Lipid Oxidation," Ch. 2. In: Lipid oxidation: Challenges in food systems. Logan A, Nienaber U, Pan X (eds.) AOCS Press, Urbana, IL, pp 55–128.
10. Schaich KM, Shahidi F, Zhong Y, and Eskin NAM. (2013) "Lipid Oxidation," Ch 11. In: Biochemistry of Food, 3rd ed., Eskin NAM, ed., Elsevier: London, pp. 419–478.
11. White PJ (1991) Methods for measuring changes in deep-fat frying oils. Food Technol 45(2): 75–80
12. Hui YH (ed) (1996) Bailey's industrial oil and fat products, 5th edn. Wiley, New York
13. McClements DJ, Decker EA (2008) Lipids ch. 4. In: Damodaran S, Parkin KL, Fennema OR (eds) Fennema's Food Chemistry, 4th edn. CRC Press, Boca Raton, FL
14. Firestone, D (2013) Physical and chemical characteristics of oils, fats, and waxes, 3rd ed., AOCS Press, Urbana, IL
15. Anon. (2016) The Merck Index Online. Royal Society of Chemistry, London. https://www.rsc.org/merck-index
16. Stauffer CE (1996) Fats and oils. Eagan press handbook series. American Association of Cereal Chemists, St. Paul, MN
17. Khanal RC, Dhiman TR (2004) Biosynthesis of conjugated linoleic acid (CLA): a review. Pak J Nutr 3: 72–81
18. Labuza TP (1971) Kinetics of lipid oxidation in foods. CRC Crit Rev Food Technol 2: 355–405
19. Buege JA, Aust SD (1978) Microsomal lipid peroxidation. Methods Enzymol 52: 302–310
20. Fulcher RG, Irving DW, de Franciso A (1989) Fluorescence microscopy: applications in food analysis. ch.3. In: Munck L (ed.) Fluorescence analysis in foods. Longman Scientific & Technical, copublished in the U.S. with Wiley, New York, pp 59–109
21. Green FJ (1990) The Sigma-Aldrich handbook of stains, dyes and indicators. Aldrich Chemical, Milwaukee, WI
22. Smart MG, Fulcher RG, Pechak DG (1995) Recent developments in the microstructural characterization of foods, ch. 11. In: Gaonkar AD (ed) Characterization of food: emerging methods. Elsevier Science, New York, pp 233–275
23. Fritsch CW, Gale JA (1977) Hexanal as a measure of rancidity in low fat foods. J Am Oil Chem Soc 54: 225
24. Dupey HP, Fore SP (1970) Determination of residual solvent in oilseed meals and flours: volatilization procedure. J Am Oil Chem Soc 47: 231–233
25. Tarladgis BG, Watts BM, Younathan, MT, Dugan LR (1960) A distillation method for the quantitative determination of malonaldehyde in rancid foods. J Am Oil Chem Soc 37: 1
26. Rhee KS, Watts BM (1966) Evaluation of lipid oxidation in plant tissues. J Food Sci 31: 664–668
27. Frankel EN (1993) In search of better methods to evaluate natural antioxidants and oxidative stability in food lipids. Trends Food Sci Technol 4: 220–225

28. Perkins EG (1991) Analyses of fats, oils, and lipoproteins. American Oil Chemists' Society, Champaign, IL
29. Schul D, Tallmadge D, Burress D, Ewald D, Berger B, Henry D (1998) Determination of fat in olestra-containing savory snack products by capillary gas chromatography. J AOAC Int 81: 848–868
30. Tallmadge DH, Lin PY (1993) Liquid chromatographic method for determining the percent of olestra in lipid samples. J AOAC Int 76: 1396–1400
31. Ackman RG, Sipos JC (1964) Application of specific response factors in the gas chromatographic analysis of methyl esters of fatty acids with flame ionization detectors. J Am Oil Chem Soc 41: 377–378

32. Ratnayake WMN, Hansen SL, Kennedy MP (2006) Evaluation of the CP-Sil88 and SP-2560 GC columns used in the recently approved AOCS Official Method Ce 1h-05: determination of cis-, trans-, saturated, monounsaturated, and polyunsaturated fatty acids in vegetable or non-ruminant animal oils and fats by capillary GLC method. J Am Oil Chem Soc 83: 475–488
33. Al-Hasani SM, Shabany H, Hlavac J (1990) Rapid determination of cholesterol in selected frozen foods. J Assoc Off Anal Chem 73: 817–820
34. Adams ML, Sullivan DM, Smith RL, Richter EF (1986) Evaluation of direct saponification method for determination of cholesterol in meats. J Assoc Off Anal Chem 69: 844–846

Protein Separation and Characterization Procedures

Denise M. Smith

*School of Food Science,
Washington State University,
Pullman, WA 99164-6376, USA
e-mail: denise.smith@wsu.edu*

S. Nielsen (ed.), *Food Analysis*, Food Science Text Series,
DOI 10.1007/978-3-319-45776-5_24, © Springer International Publishing 2017

24.1 INTRODUCTION

Many protein separation techniques are available to food scientists. Several of the separation techniques described in this chapter are used commercially for the production of food or food ingredients, whereas others are used to purify a protein from a food for further study in the research laboratory. Although not the primary focus of this chapter, many methods have been developed to rapidly purify recombinant proteins. In general, separation techniques exploit the biochemical differences in protein solubility, size, charge, adsorption characteristics, and biological affinities for other molecules. These physical characteristics then are used to purify individual proteins from complex mixtures.

The biochemical, nutritional, and functional properties of food proteins can be characterized in a variety of ways. This chapter describes methods of amino acid analysis and several methods for protein nutritional quality analysis. Finally, protein solubility, emulsification, and foaming tests are described, along with gelation and dough formation, for the characterization of protein functional properties. A list of abbreviations is included to summarize terms used throughout this chapter.

24.2 METHODS OF PROTEIN SEPARATION

24.2.1 Initial Considerations

Usually, several separation techniques are used in sequence to purify a protein from a food. In general, the purity of a protein preparation can be improved as more separation steps are added, although this will usually result in a lower recovery or yield. Food ingredients such as protein concentrates may be prepared using only one separation step when high purity is not necessary. Three or more separation steps are often used in sequence to prepare a pure protein for laboratory study. For example, a very common purification procedure for protein includes precipitation by ammonium sulfate, followed by hydrophobic interaction chromatography, ion-exchange chromatography, and lastly gel filtration.

Before starting a separation sequence, it is necessary to learn as much as possible about the biochemical properties of a protein to determine any distinctive characteristics that will make separation easier, such as molecular mass, isoelectric point (pI), solubility properties, denaturation temperature, metal ion binding, and specific ligand recognition. The first separation step should be one that can easily be used with large quantities of material. This is often a technique that utilizes the differential solubility properties of a protein. Each succeeding step in a purification sequence will use a different mode of separation. Some of the most common methods of purification are described in this section and include precipitation, ion-exchange chromatography, hydrophobic interaction chromatography, affinity chromatography, and size-exclusion chromatography. A summary of separation methods described in this chapter can be found in Table 24.1. More detailed information about the various purification techniques can be found in several published sources [1–3].

24.2.2 Separation by Fractional Precipitation

24.2.2.1 Principle

Separation by fractional precipitation exploits the differential solubility properties of proteins in solution. Fractional precipitation is one of the simplest methods for separating a protein of interest from other proteins and contaminants in a mixture. Proteins are polyelectrolytes; thus, solubility characteristics are determined by the type and charge of amino acids in the molecule. Proteins can be selectively precipitated or solubilized by changing **buffer pH, ionic strength, dielectric constant,** or **temperature.** These separation techniques are advantageous when working with large quantities of material, are relatively quick, and are not usually influenced by other food components. Precipitation techniques are used most commonly during early stages of a purification sequence.

24.2.2.2 Procedures

24.2.2.2.1 Salting Out

Proteins have unique solubility profiles in neutral salt solutions. Low concentrations of neutral salts usually increase the solubility of proteins; however, proteins are precipitated from solution as ionic strength is increased. This property can be used to precipitate a protein from a complex mixture. **Ammonium sulfate** $[(NH_4)_2SO_4]$ is commonly used because it is highly soluble, although other neutral salts such as NaCl or KCl may be used to salt out proteins. Generally a two-step procedure is used to maximize separation efficiency. In the first step, $(NH_4)_2SO_4$ is added at a concentration just below that necessary to precipitate the protein of interest. When the solution is centrifuged, less soluble proteins are precipitated while the protein of interest remains in solution. The second step is performed at an $(NH_4)_2SO_4$ concentration just above that necessary to precipitate the protein of interest. When the solution is centrifuged, the protein is precipitated, while more soluble proteins remain in the supernatant. One disadvantage of this method is that large quantities of salt contaminate the precipitated protein and often must be removed before the protein is resolubilized in buffer. Tables and formulas are available in many biochemistry books and online (type "ammonium sulfate calculator" into your web browser) to determine the proper amount of $(NH_4)_2SO_4$ to achieve a specific concentration.

Summary of protein separation methods

Method	Basis of separation	Principle
Addition of ammonium sulfate	Precipitation	Proteins are precipitated from solution as ionic strength increases, using a neutral salt, such as ammonium sulfate
Isoelectric precipitation	Precipitation	Protein has no net charge at pI, so it aggregates and precipitates from solution
Solvent fractionation	Precipitation	Water miscible organic solvents decrease the dielectric constant of an aqueous solution and decrease the solubility of most proteins, so proteins precipitate from solution
Protein denaturation	Precipitation	Proteins heated to high temperature or adjusted to extremes of pH precipitate from solution
Ion-exchange chromatography	Adsorption (based on charge)	Charged protein molecules in solution are reversibly adsorbed to a charged solid support matrix via electrostatic interactions. Bound proteins are eluted from a column by gradually changing the ionic strength or pH of the eluting solution
Affinity chromatography	Adsorption (based on specific biochemical characteristics)	Protein is adsorbed to chromatographic matrix that contains a ligand covalently bound to a solid support; the ligand used has reversible, specific, and unique binding affinity for the protein of interest. Protein that binds to ligand can be unbound by changing the pH, temperature, or concentration of salt or ligand in the eluting buffer
Dialysis	Size	Semipermeable membrane permits passage of small molecules but not larger molecules. Pore sizes of membranes are specified as molecular weight cutoff
Membrane processes (e.g., microfiltration, ultrafiltration, nanofiltration, reverse osmosis)	Size	Pressure is applied to solution sitting on semipermeable membrane with certain molecular weight cutoff. Small molecules pass through, and large ones are retained
Size-exclusion chromatography	Size	Proteins in solution flow through a column packed with beads that have different average pore sizes. Molecules larger than the pores are excluded and pass through the column quickly. Smaller molecules enter the pores of the beads, so elute more slowly from the column, at a rate dependent on their size
SDS-PAGE (sodium dodecyl sulfate-polyacrylamide gel electrophoresis)	Size	Proteins bind SDS to become negatively charged; under a constant voltage they move through an acrylamide gel matrix at a rate based on size alone as all molecules are highly negatively charged
IEF (isoelectric focusing)	Charge	Proteins are separated by charge in an electric field on a gel matrix in which a pH gradient has been generated using ampholytes. Under constant voltage, proteins migrate to the location on the gradient at which pH equals the pI of the protein. Size is not a factor

24.2.2.2.2 Isoelectric Precipitation

The **isoelectric point** (pI) is defined as the pH at which a protein has no net charge in solution. This is determined by the ionizable acidic and basic amino acids making up the protein. Proteins usually aggregate and precipitate at their pI because there is no electrostatic repulsion between molecules. Proteins have different pIs; thus, they can be separated from each other by adjusting solution pH. When the pH of a solution is adjusted to the pI of a protein, the protein precipitates, while proteins with different pIs remain in solution. The precipitated protein can be resolubilized in another solution of different pH.

24.2.2.2.3 Solvent Fractionation

Protein solubility at a fixed pH and ionic strength is a function of the **dielectric constant** of a solution. Thus, proteins can be separated based on solubility differences in **organic solvent-water** mixtures. The addition of water-miscible organic solvents, such as **ethanol** or **acetone**, decreases the dielectric constant of an aqueous solution and decreases the solubility of most proteins. Organic solvents decrease ionization of charged amino acids, resulting in protein aggregation and precipitation. The optimum quantity of organic solvent to precipitate a protein varies from 5 % to 60 %. Solvent fractionation is usually performed at 0 °C or below to prevent protein denaturation caused by temperature increases that occur when organic solvents are mixed with water.

24.2.2.2.4 Denaturation of Contaminating Proteins

Many proteins are **denatured** and precipitated from solution when heated above a certain temperature or by adjusting a solution to highly acid or basic pHs. Proteins that are stable at high temperatures or at extremes of pH may be separated by this technique because many contaminating proteins will precipitate while the protein of interest remains in solution.

24.2.2.3 Applications

All of the above techniques are commonly used to fractionate proteins. The differential solubility of selected muscle proteins in $(NH_4)_2SO_4$ and acetone and temperature stability at 55 °C are illustrated in Table 24.2. These three techniques can be combined in sequence to prepare muscle proteins of high purity.

24.2
table

Conditions for fractionating water-soluble muscle proteins using differential solubility techniques

| | Precipitation range | | |
| | $(NH_4)_2SO_4$, pH 5.5, 10 °C | Acetone, pH 6.5, −5 °C | Stability[a], pH 5.5, 55 °C |
Enzyme	(Percent saturation)	(Percent vol/vol)	
Phosphorylase	30–40	18–30	U
Pyruvate kinase	55–65	25–40	S
Aldolase	45–55	30–40	S
Lactate dehydrogenase	50–60	25–35	S
Enolase	60–75	35–45	U
Creatine kinase	60–80	35–45	U
Phosphoglycerate kinase	60–75	45–60	S
Myoglobin	70–90	45–60	U

Adapted from Scopes [4] with permission of the University of Wisconsin Press. From Briskey, E.J., R.G. Cassens, and B.B. Marsh. The Physiology and Biochemistry of Muscle as Food. Copyright 1970
[a]*U* unstable, *S* stable at heating temperature

One of the best examples of the commercial use of differential solubility to separate proteins is in the production of protein concentrates. Soy protein concentrate can be prepared from defatted soybean flakes or flour using several methods. Soy proteins can be precipitated from other soluble constituents in the flakes or flour using a 60–80 % aqueous alcohol solution, by isoelectric precipitation at pH 4.5 (which is the pI of many soy proteins), or by denaturation with moist heat. These methods have been used to produce concentrates containing greater than 65 % protein. Two or three separation techniques can be combined in sequence to produce soy protein isolates with protein concentrations above 90 %.

24.2.3 Separation by Liquid Chromatography

24.2.3.1 Principle

In addition to fractional precipitation methods, purification of a protein usually includes one or more types of liquid chromatography. Chromatographic separations are based on the differential affinity of a mixture of proteins in solution (mobile phase) for the stationary phase. These chromatographic methods can be performed in columns at atmospheric pressure (by gravity flow) or under applied pressures using centrifugation or **high-performance liquid chromatography** (HPLC) (Chap. 13). Ultra-performance liquid chromatography (UPLC) (Chap. 13, Sect. 13.2.3.3), fast protein liquid chromatography (FPLC), and fast-performance liquid chromatography are variations of HPLC, the primary difference being the pressures applied within the system and the properties of the stationary phase. Ion-exchange chromatography, hydrophobic interaction chromatography, and affinity chromatography are commonly used for protein purification and will be described briefly.

24.2.3.2 Procedures

24.2.3.2.1 Ion-Exchange Chromatography

Ion-exchange chromatography is defined as the reversible adsorption between charged molecules and ions in solution and a charged solid support matrix (Chap. 12, Sect. 12.4.4). Ion-exchange chromatography is the most commonly used protein separation technique and results in an average eightfold purification. A positively charged matrix is called an **anion exchanger** because it binds negatively charged ions or molecules in solution. A negatively charged matrix is called a **cation exchanger** because it binds positively charged ions or molecules. The most commonly used exchangers for protein purification are anionic diethylaminoethyl-derivatized supports, followed by carboxymethyl and phospho-cation exchangers.

The protein of interest is first adsorbed to the ion exchanger under buffer conditions (ionic strength and pH) that maximize the affinity of the protein for the matrix. Contaminating proteins of different charges pass through the exchanger unabsorbed. Proteins bound to the exchangers are selectively eluted from the column by gradually changing the ionic strength or pH of the eluting solution. As the composition of the eluting buffer changes, the charges of the proteins change and their affinity for the ion-exchange matrix is decreased. Protein purification by ion-exchange chromatography can be performed in a column at atmospheric pressure or under applied high pressures. In addition, proteins can be purified using ion-exchange resins in a batch extraction.

24.2.3.2.2 Affinity Chromatography

Affinity chromatography is a type of adsorption chromatography in which a protein is separated in a chromatographic matrix containing a **ligand** covalently bound to a solid support (Chap. 12, Sect. 12.4.5). A ligand can be defined as a substance (molecule or metal) that has a reversible, specific, and unique binding affinity for a receptor site on a protein (Table 24.3). Ligands may be biospecific, such as enzyme inhibitors, enzyme substrates, coenzymes, or antibodies. Other types of ligands include certain dyes and metal ions. Thus, the protein is separated from a complex mixture due to its affinity or specific binding interaction with a ligand immobilized on a solid support.

The protein is passed through a column containing the ligand bound to a solid support, under buffer conditions (pH, ionic strength, temperature, and protein concentration) that maximize binding of the protein to the ligand. Contaminating proteins and molecules that do not bind the ligand are eluted. The bound protein is then desorbed or eluted from the column under conditions that decrease the affinity of the protein for the bound ligand, by changing the pH, temperature, or concentration of salt or ligand in the eluting buffer.

Recombinant proteins are often purified by immobilized metal ion affinity chromatography followed by size-exclusion chromatography. Recombinant proteins can be engineered to contain multiple histidine residues (six to ten molecules of histidine). The proteins containing this polyhistidine tag will bind to an affinity column containing immobilized divalent metal ions, such as nickel. Once contaminants are washed from the column, the tagged protein is eluted using a gradient of imidazole counter ligands (the amino acid histidine contains an imidazole functional group).

Affinity chromatography is a very powerful technique and is a commonly used protein purification procedure. The average purification achieved by affinity chromatography is approximately 100-fold, although 1,000-fold increases in purification have been reported. This technique is more powerful than size exclusion and ion exchange which usually achieve less than a 12-fold purification. Many covalently bound ligands and associated buffers are commercially available as kits. Preactivated solid supports used for covalent bonding of various ligands are also commercially available.

24.2.3.2.3 Hydrophobic Interaction Chromatography

Hydrophobic interaction chromatography is another type of chromatography often used in protein purification schemes (Chap. 12, Sect. 12.4.3; Chap. 13, Sect. 13.3.3). In this technique, proteins are separated due to their reversible interaction with a hydrophobic stationary phase. Hydrophobic interactions are increased at high ionic strength, so this type of chromatography is often used after ammonium sulfate precipitation or after using a salt gradient to elute a protein from an ion-exchange column. The bound protein is usually eluted by decreasing the ionic strength of the mobile phase using a stepwise or gradient elution.

24.2.3.3 Applications

Ion-exchange chromatography is commonly used to separate proteins in the laboratory and can be used for separation and quantification of amino acids as described in Sect. 24.3.1. Ion-exchange chromatography is used to isolate proteins while removing lactose, minerals, and fat from sweet dairy whey. Whey protein isolates (containing greater than 90 % protein) and several protein fractions, such as alpha-lactalbumin, lactoperoxidase, and lactoferrin, are purified from sweet dairy whey using cation-exchange chromatography [5]. Whey protein isolates are used as supplements in nutrition bars and beverages, since the high-quality protein is soluble and digestible.

Affinity chromatography is commonly used for protein purification in the research laboratory and may be used for commercial preparation of proteins by chemical suppliers. It is not generally used for commercial production of food protein ingredients due to the high costs involved, although this technique is used to purify some high-value bioactive peptides and proteins for nutritional applications. Glycoproteins, commonly purified by affinity chromatography, can be separated from other proteins in a complex mixture by utilization of the high carbohydrate-binding affinity of lectins. Lectins, such as concanavalin A, are carbohydrate-binding proteins that can be bound to a

24.3
table

Examples of common biological interactions used to separate proteins via affinity chromatography

Protein target	Ligand
Enzyme	Substrate, inhibitor, cofactor
Antibody	Antigen
Glycoprotein	Lectin, polysaccharide
Hormone	Hormone receptor
Proteins with histidine, cysteine, or tryptophan residues on the surface	Metal ions

solid support and used to bind the carbohydrate moiety of glycoproteins that are applied to the column. Once the glycoproteins are bound to the column, they can be desorbed using an eluting buffer containing an excess of lectin. The glycoproteins bind preferentially to the free lectins and elute from the column.

24.2.4 Separation by Size

24.2.4.1 *Principle*
Protein molecular masses range from about 10,000 to over 1,000,000 Da; thus, size is a logical parameter to exploit for separations. Actual separation occurs based on the **Stokes radius** of the protein, not on the molecular mass. Stokes radius is the average radius of the protein in solution and is determined by protein conformation. For example, a globular protein may have an actual radius very similar to its Stokes radius, whereas a fibrous or rod-shaped protein of the same molecular mass may have a Stokes radius that is much larger than that of the globular protein. Thus, one limitation of these methods is that two proteins of the same molecular mass may occasionally separate differently.

24.2.4.2 *Procedures*

24.2.4.2.1 *Dialysis*
Dialysis is used to separate molecules in solution by the use of semipermeable membranes that permit passage of small molecules but not larger molecules. To perform dialysis, a protein solution is placed into dialysis tubing that has been tied or clamped at one end. The other end of the tubing is sealed, and the bag is placed in a large volume of water or buffer (usually 500–1,000 times greater than the sample volume inside the dialysis tubing) which is slowly stirred. Solutes of low molecular mass diffuse from the bag, while buffer diffuses into the bag. Dialysis is simple; however, it is a relatively slow method, usually requiring at least 12 h and one or more changes of buffer. The protein solution inside the bag is often diluted during dialysis, due to osmotic strength differences between the solution and dialysis buffer. Dialysis can be used to change buffer composition or pH and to remove salt and other impurities of low molecular mass between purification steps or to adjust the buffer composition of a final protein preparation.

24.2.4.2.2 *Membrane Processes*
Microfiltration, ultrafiltration, nanofiltration, and reverse osmosis all are processes that use a semipermeable membrane for the separation of solutes on the basis of size under an applied pressure. These methods are similar to dialysis but are much faster and are applicable to both small- and large-scale separations. Molecules larger than the membrane cutoff are retained and become part of the retentate, while smaller molecules pass through the membrane and become part of the filtrate.

These membrane processes differ mainly in the porosity of the membranes and in the operating pressure used. The porosity of the membrane (membrane pore diameter) sequentially decreases and the pressure used sequentially increases for microfiltration, ultrafiltration, nanofiltration, and reverse osmosis. The approximate pore size of each membrane process relative to the different components present in milk is shown in Fig. 24.1.

Ultrafiltration is commonly used in protein research laboratories, with various commercial units available. A stirred cell ultrafiltration unit is illustrated

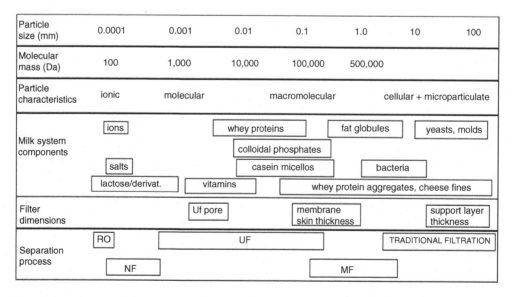

24.1 figure Range of particle sizes used for various membrane filtration techniques compared to particle sizes of milk components and microorganisms. *RO* reverse osmosis, *NF* nanofiltration, *UF* ultrafiltration, *MF* microfiltration (Adapted from Jelen [6], with permission of the International Dairy Federation)

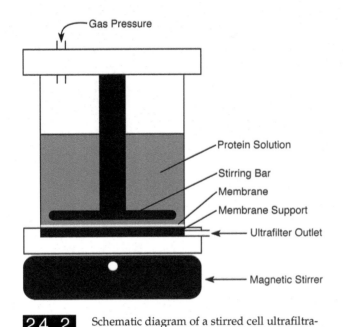

Gas Pressure

Protein Solution

Stirring Bar

Membrane

Membrane Support

Ultrafilter Outlet

Magnetic Stirrer

24.2 figure Schematic diagram of a stirred cell ultrafiltration unit

in Fig. 24.2. The protein solution in the stirred cell is filtered through the semipermeable membrane by gas pressure, leaving a concentrated solution of proteins larger than the membrane cutoff point inside the cell. Disposable centrifugal filtering units are available for small sample volumes with membrane cutoff values ranging from 3,000 to 100,000 Da. Solvents and molecules smaller than the membrane pore size are forced through the membrane by centrifugation resulting in concentration and purification of the protein in the retentate.

24.2.4.2.3 Size-Exclusion Chromatography

Size-exclusion chromatography, also known as gel filtration or gel permeation chromatography, is a column technique that can be used to separate proteins on the basis of size (Chap. 12, Sect. 12.4.6). This technique can also be used for buffer exchange, desalting, and removal of low molecular weight impurities. A protein solution flows through a column packed with a solid support of porous beads made of a cross-linked polymeric material such as agarose or dextran. Beads of different average pore sizes that allow for efficient fractionation of proteins of different molecular masses are commercially available. Molecules larger than the pores in the beads are excluded, moving quickly through the column and eluting from the column in the shortest times. Small molecules enter the pores of the beads and are retarded, thus moving very slowly through the column. Molecules of intermediate sizes partially interact with the porous beads and elute at intermediate times. Consequently, molecules are eluted from the column in order of decreasing size.

Molecular mass can be calculated by chromatographing the unknown protein and several proteins of known molecular mass. Standards of known molecular mass are commercially available and can be used to prepare a standard curve. A plot of the **elution volume** (V_e) of each protein versus **log of the molecular mass** yields a straight line. Size-exclusion techniques generally can be used to estimate molecular mass within ±10%; however, errors can occur if the Stokes radii of the unknown protein and standards are quite different.

24.2.4.3 Applications

Microfiltration can be used to remove particles and microorganisms and has been applied to wastewater treatment and to remove the bacteria from milk and beer. Ultrafiltration and nanofiltration are used to concentrate a protein solution, remove salts, exchange buffer, or fractionate proteins on the basis of size. Ultrafiltration is used to concentrate milk for cheesemaking and to manufacture whey protein products, whereas nanofiltration has been used to remove monovalent ions from salt whey. Ultrafiltration is used to concentrate whole liquid egg and liquid egg white prior to spray drying. Reverse osmosis is often used to purify water and to remove aqueous salts, metal ions, simple sugars, and other small impurities with molecular mass below 2,000. The various membrane systems can be used in combination, for example, ultrafiltration and reverse osmosis in sequence are used to concentrate and fractionate whey proteins, then remove salts and lactose.

Dialysis and size-exclusion chromatography are primarily used in the analytical laboratory in a protein separation sequence. Dialysis may be used to change the buffer to one of the appropriate pH and ionic strength during purification or prior to electrophoresis of a protein sample. Dialysis is used after $(NH_4)_2SO_4$ precipitation of a protein to remove excess salt and other small molecules and to solubilize protein in a new buffer. Size-exclusion chromatography is used to remove salts, change buffers, fractionate proteins, and estimate protein molecular mass.

24.2.5 Separation by Electrophoresis

24.2.5.1 Polyacrylamide Gel Electrophoresis

24.2.5.1.1 Principle

Electrophoresis is defined as the migration of charged molecules in a solution through an electrical field. The most common type of electrophoresis performed with proteins is zonal electrophoresis in which proteins are separated from a complex mixture into bands by migration in aqueous buffers through a solid polymer matrix called a gel. **Polyacrylamide gels** are the most common matrix for zonal electrophoresis of proteins, although other matrices such as starch and agarose may be used. Gel matrices can be formed in glass tubes or as slabs between two glass plates.

Separation depends on the friction of the protein within the matrix and the charge of the protein molecule as described by the following equation:

$$\text{Mobility} = \frac{(\text{Applied voltage})(\text{Net charge on molecule})}{\text{Friction of the molecule}}$$

$$(24.1)$$

Proteins are positively or negatively charged, depending on solution pH and their isoelectric point (pI). A protein is negatively charged if solution pH is above its pI, whereas a protein is positively charged if solution pH is below its pI. The magnitude of the charge and applied voltage will determine how far a protein will migrate in an electrical field. The higher the voltage and stronger the charge on the protein, the greater the migration within the electrical field. Molecular size and shape, which determine the Stokes radius of a protein, also determine migration distance within the gel matrix. Mobility of proteins decreases as molecular friction increases due to an increase in Stokes radius; thus, smaller proteins tend to migrate faster through the gel matrix. Similarly, a decrease in pore size of the gel matrix will decrease mobility.

In nondenaturing or **native electrophoresis**, proteins are separated in their native form based on charge, size, and shape of the molecule. Another form of electrophoresis commonly used for separating proteins is denaturing electrophoresis. **Polyacrylamide gel electrophoresis** (PAGE) with an anionic detergent, **sodium dodecyl sulfate** (SDS), is used to separate protein subunits by size. Proteins are solubilized and dissociated into subunits in a buffer containing SDS and a reducing agent. **Reducing agents**, such as mercaptoethanol or dithiothreitol, are used to reduce disulfide bonds within a protein subunit or between subunits. Proteins bind SDS, become negatively charged, and are separated based primarily on size alone.

24.2.5.1.2 Procedures
A power supply and electrophoresis apparatus containing the polyacrylamide gel matrix and two buffer reservoirs are necessary to perform a separation. A representative slab gel and electrophoresis unit is shown in Fig. 24.3. The power supply is used to generate the electric field by providing a source of constant current, voltage, or power. The electrode buffer controls the pH to maintain the proper charge on the protein and conducts the current through the polyacrylamide gel. Commonly used buffer systems include an anionic tris-(hydroxymethyl)aminomethane buffer with a resolving gel at pH 8.8 and a cationic acetate buffer at pH 4.3.

The polyacrylamide gel matrix is formed by polymerizing **acrylamide** and a small quantity (usu-

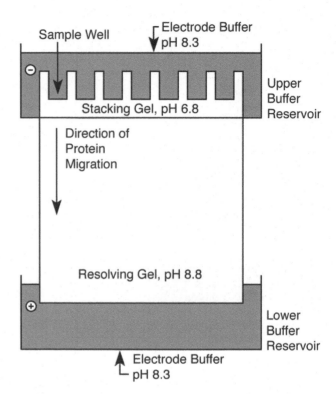

24.3
figure
Schematic diagram of a slab gel electrophoresis unit indicating the pHs of the stacking and resolving gels and the electrode buffer in an anionic discontinuous buffer system

ally 5 % or less) of the **cross-linking reagent, N,N′-methylenebisacrylamide**, in the presence of a **catalyst, tetramethylethylenediamine** (TEMED), and **source of free radicals, ammonium persulfate**. Gels can be made in the laboratory or purchased precast.

A discontinuous gel matrix is usually used to improve resolution of proteins within a complex mixture. The discontinuous matrix consists of a **stacking gel** with a large pore size (usually 3–4 % acrylamide) and a **resolving gel** of a smaller pore size. The stacking gel, as its name implies, is used to stack or concentrate the proteins into very narrow bands prior to their entry into the resolving gel. At pH 6.8, a voltage gradient is formed between the chloride (high negative charge) and glycine ions (low negative charge) in the electrode buffer, which serves to stack the proteins into narrow bands between the ions. Migration into the resolving gel of a different pH disrupts this voltage gradient and allows separation of the proteins into discrete bands.

The pore size of the resolving gel is selected based on the molecular mass of the proteins of interest and is varied by altering the concentration of acrylamide in solution. Proteins are usually separated on resolving gels that contain 4–15 % acrylamide. Acrylamide concentrations of 15 % may be used to separate proteins with molecular mass below 50,000 Da. Proteins greater than 500,000 Da are often separated on gels with acrylamide concentrations below 7%. A **gradient gel** in which the acrylamide concentration increases from the top to the bottom of the gel is often used to separate a mixture of proteins with a large molecular mass range.

To perform a separation, proteins in a buffer of the appropriate pH are loaded on top of the stacking gel. **Bromophenol blue tracking dye** is added to the protein solution. This dye is a small molecule that migrates ahead of the proteins and is used to monitor the progress of a separation. After an electrophoresis run, the separated protein bands in the gels can be visualized using a nonspecific **protein stain** such as **Coomassie Brilliant Blue stain**, **silver stain**, or a **fluorescent gel stain**. Specific enzyme stains or antibodies can be used to detect a particular protein or enzyme.

The electrophoretic or **relative mobility** (R_m) of each protein band is calculated as

$$R_m = \frac{\text{Distance protein migrated from start of resolving gel}}{\text{Distance between start of running gel and tracking dye}} \quad (24.2)$$

Additional procedural details can be found in several sources [1–3].

24.2.5.1.3 Applications

Electrophoretic techniques can be used as one step in a purification process or to aid in the biochemical characterization of a protein. Electrophoresis can be used to help determine the purity of a protein extract. Commercially available preparative electrophoresis units are used to purify large quantities of protein. Small quantities of protein can be eluted and collected from electrophoresis gels using electroelution techniques. Alternatively, proteins can be transferred from electrophoresis gels to a membrane and then stained with antibodies specific to a target protein in a process called electroblotting or western blotting. Western blotting is discussed in more detail in Chap. 27.

Electrophoresis is often used to determine the protein composition of a food product. For example, differences in the protein composition of soy protein concentrates and whey protein concentrates produced by different separation techniques can be detected. The lanes on the left side of Fig. 24.4 illustrate the protein patterns observed when extracts of unheated and heated whey proteins are separated by native PAGE.

SDS-PAGE is used in characterization protocols to determine subunit composition of a protein and to estimate subunit molecular mass. Molecular mass can usually be estimated within an error of ±5%, although highly charged proteins or glycoproteins may be subject to a larger error. Molecular mass is determined by comparing R_m of the protein subunit with R_m of protein standards of known molecular mass (Fig. 24.5). Commercially

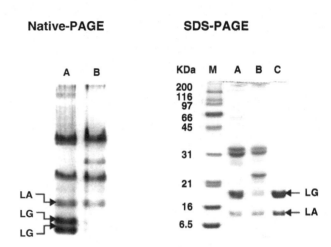

Native-PAGE   **SDS-PAGE**

24.4 figure Electrophoresis of whey proteins fractionated from raw and heated milk. *Left panel*: native-polyacrylamide gel electrophoresis (PAGE). Lane A = raw milk; lane B = heated milk. *Right panel*: sodium dodecyl sulfate-polyacrylamide gel electrophoresis (SDS-PAGE). Lane M = molecular weight markers; lane A = raw milk; lane B = heated milk; lane C = β-lactoglobulin (LG) standard and α-lactalbumin (LA) standard. Ten micrograms of protein were loaded onto each lane. Note a decrease in two isoforms of LG after heating (Reprinted from Chen et al. [7] with permission)

prepared protein standards are available in several molecular mass ranges. To prepare a standard curve, logarithms of protein standard molecular mass are plotted against their corresponding R_m values. The molecular mass of the unknown protein is determined from its R_m value using the standard curve. Electropherograms of unheated and heated whey proteins separated by SDS-PAGE are also illustrated in Fig. 24.4, right side.

24.2.5.2 Isoelectric Focusing

24.2.5.2.1 Principle

Isoelectric focusing, also termed electrofocusing, is a modification of electrophoresis, in which proteins are separated by charge in an electric field on a gel matrix in which a pH gradient has been generated using ampholytes. Proteins are focused or migrate to the location in the gradient at which pH equals the pI of the protein. At this point, the protein has no net charge. Resolution is among the highest of any protein separation technique and can be used to separate proteins with pIs that vary less than 0.02 of a pH unit.

24.2.5.2.2 Procedure

A pH gradient is formed using **ampholytes**, which are small polymers (molecular masses of less than 1,000 Da) containing both positively and negatively charged groups. An ampholyte mixture is composed

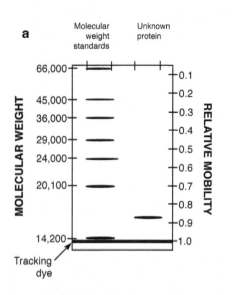

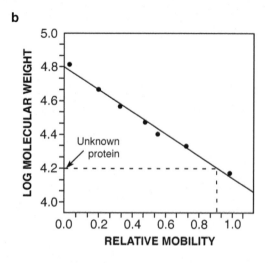

24.5 figure Use of SDS-PAGE to determine the molecular mass of a protein. (**a**) Separation of molecular mass standards and the unknown protein. (**b**) Standard curve for estimating protein molecular mass

of thousands of polymers that exhibit a range of pH values. Ampholyte mixtures are available that cover a narrow pH range (e.g., 1–2 pH units) or a broad range (e.g., 4–6 pH units) and should be selected for use based on properties of the proteins to be separated.

Ampholytes are added to the gel solution prior to polymerization. Once the gel is formed and a current applied, the ampholytes migrate to produce a linear pH gradient; negatively charged ampholytes migrate toward the anode, while positively charged ampholytes migrate toward the cathode. Proteins migrate within this pH gradient depending on their charge until they reach the pH at which they have no net charge (at their pI).

24.2.5.2.3 Applications

Isoelectric focusing is the method of choice for determining the isoelectric point of a protein by comparison to standard marker proteins of known pI. This technique is an excellent method for determining the purity of a protein preparation. Isoelectric focusing can also be used to detect changes in protein composition due to posttranslational modifications such as glycosylation or phosphorylation. Genetic variants of many plant and animal proteins are visualized using isoelectric focusing. Isoelectric focusing is used to differentiate closely related animal and fish species based on protein patterns. The US Food and Drug Administration publishes the Regulatory Fish Encyclopedia [8] which contains isoelectric focusing patterns of about 1,700 finfish and shellfish species. This guide helps state and federal officials to identify species substitution and thus detect product adulteration and economic fraud.

Isoelectric focusing and SDS-PAGE can be combined to produce a two-dimensional electrophoretogram that is extremely useful for separating very

complex mixtures of proteins. This technique is called **two-dimensional electrophoresis**. Proteins are first separated in tube gels by isoelectric focusing. The tube gel containing the separated proteins is then placed on top of an SDS-PAGE slab gel, and proteins are separated. Thus, proteins are separated first on the basis of charge and then according to size and shape. Over 1,000 proteins in a complex mixture have been resolved using this technique. This method is used to verify the genetic purity of hybrid seeds as well as evaluating the up- or downregulation of proteins in various biological processes and disease states in humans and animals. Two-dimensional electrophoresis gels have been used to determine the differences in muscle proteins extracted from raw and cooked pork and goose meat [9].

24.2.5.3 Capillary Electrophoresis

24.2.5.3.1 Principle

Capillary electrophoresis has been described as a hybrid technique incorporating aspects of traditional slab gel electrophoresis and liquid chromatography. Similar principles apply for the separation of proteins by both capillary and conventional electrophoretic techniques; proteins can be separated on the basis of charge or size in an electric field. The primary difference between capillary electrophoresis and conventional electrophoresis (described previously) is that **capillary tubing** is used in place of acrylamide gels cast in tubes or slabs. As the separated proteins migrate along the capillary tubing, they are detected using detectors originally developed for chromatography.

24.2.5.3.2 Procedure

A schematic diagram of a capillary electrophoresis system is shown in Fig. 24.6. A capillary electrophoresis

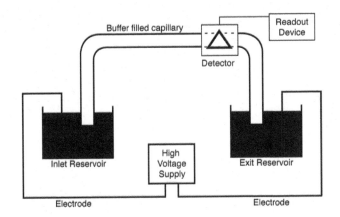

24.6
figure
Schematic diagram of a capillary electrophoresis system

system is comprised of a capillary column, power supply, detector, and two buffer reservoirs. The sample is introduced into the inlet side of the capillary tube by simply replacing the inlet buffer reservoir with the sample solution and applying low pressure or voltage across the capillary until the desired volume of sample has been loaded onto the column. Capillaries are composed of fused silica with internal diameters that commonly range from 10 to 100 µm. Column length varies from a few centimeters to 100 cm. High electric fields ($5–30\,kV/cm$) can be used as the narrow columns dissipate heat very effectively, allowing for short run times of 10–30 min.

Protein bands are not visualized by staining as in conventional electrophoresis. Instead, protein peaks are detected on the column as they migrate past a detector, similar to those used in liquid chromatography. Ultraviolet (UV)-visible detectors are most common, although fluorescence and conductivity detectors are available. The data obtained from a capillary electrophoresis run look like a typical chromatogram from a high-performance liquid chromatograph or gas chromatograph (see Chaps. 13 and 14). Proteins can be labeled with a fluorescent derivative to increase sensitivity when a fluorescent detector is used.

There are three variations of capillary electrophoresis commonly used for protein separations. **Capillary zone electrophoresis** or **free-solution electrophoresis** is the most commonly used form of capillary electrophoresis. It is much like native PAGE, except proteins are separated in free solution inside capillary tubes filled with buffer of the desired pH. Diffusion is prevented within the narrow diameter of the capillaries eliminating the need for a gel matrix. In capillary zone electrophoresis, electroosmotic flow also influences the separation of proteins within capillary tubes. The negatively charged fused silica capillary wall [containing silanol groups

(SiO⁻)] attracts positively charged ions (cations) from the buffer to form a double-ion layer at the interface between the capillary column wall and the buffer. When the electric field is applied, the cations forming the double layer are attracted toward the cathode and "pull" other molecules (independent of charge) in the same direction. Thus, in free-solution capillary electrophoresis, cations, anions, and uncharged molecules can be separated in a single run. Electroosmotic flow can be controlled by changing the pH or ionic strength of the buffer to alter the charge on the capillary wall and change the rate of protein migration.

SDS capillary gel electrophoresis techniques can be used to separate proteins by size and to determine molecular mass. In this technique proteins are denatured and dissociated in the presence of SDS and a reducing agent, then fractionation occurs in polyacrylamide gel-filled capillaries of specific pore sizes. Alternatively, linear polymers, such as methyl cellulose, dextrans, or polyethylene glycol, are added to the buffer within the capillary in a technique called dynamic sieving capillary electrophoresis. These entangled polymers act like the pores of the polyacrylamide gel to slow migration of the larger proteins and allow separation by size.

Proteins also can be separated on the basis of their isoelectric points, in a technique called **capillary isoelectric focusing**. Ampholytes (described in Sect. 24.2.5.2.2) are used to form a pH gradient within the capillary tube. A gel matrix is not needed. In this technique, electroosmotic flow is minimized by coating the capillary walls with buffer additives to prevent undesirable effects caused by surface charge.

24.2.5.3.3 *Applications*
Capillary electrophoresis is used primarily in analytical labs, although the use for routine quality control purposes is increasing. Capillary zone electrophoresis is used for a variety of applications, including the fractionation of milk, cereal, soybean, and muscle proteins [10]. Figure 24.7 illustrates the results obtained by capillary zone electrophoresis when used to fractionate the casein protein fraction of three mammalian milks. Microchip electrophoresis, a miniaturized version of capillary electrophoresis using small volume microchannels, has been developed recently [12]. Electrophoretic separations on microchips can be completed in only a few minutes. This technique has been used to identify 15 different wheat varieties based on variations in the subunit composition of a major wheat protein, glutenin [13]. Results were comparable to those obtained using SDS-PAGE, but with an assay time of less than one minute per analysis.

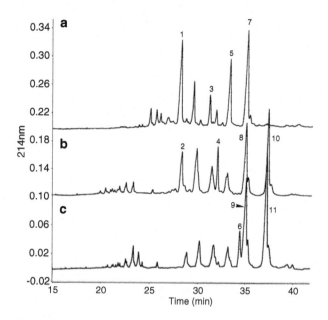

24.7
figure
Electropherograms illustrating differences in casein proteins isolated from (**a**) cow, (**b**) ewe, and (**c**) goat milks when separated by capillary zone electrophoresis. The peaks were identified as (*1*) bovine αS1-casein; (*2*) ovine αS1-casein; (*3*) bovine κ-casein; (*4*) ovine κ-casein; (*5*) bovine β-casein A1; (*6*) caprine κ-casein; (*7*) bovine β-casein A2; (*8*) ovine β2-casein; (*9*) caprine β2-casein; (*10*) ovine β1-casein; and (*11*) caprine β1-casein. Peaks eluting from the capillary column were detected by UV absorption (Reprinted from Molina et al. [11] with permission)

24.3   PROTEIN CHARACTERIZATION PROCEDURES

24.3.1 Amino Acid Analysis

24.3.1.1 *Principle*

Amino acid analysis is used to quantitatively determine the amino acid composition of a relatively pure protein. Amino acid analysis is divided into three steps. First, the protein sample is hydrolyzed to release the amino acids. Amino acids are then separated using chromatographic techniques. Finally, the separated amino acids are detected and quantified. To date, there are no published official standard methods for amino acid analysis [14] in the United States, although some AOAC methods [15] are available describing protein hydrolysis procedures in specific foods. Many new methods are currently under development. Ion-exchange chromatography and reversed-phase liquid chromatography with pre- or post-column derivatization for detection are in widespread use and will be described in this section.

24.3.1.2 *Procedures*

In the most commonly used procedure, a protein sample is **hydrolyzed** in constant boiling 6 *N* HCl at 110 °C for 24 h to release amino acids prior to chromatography. Accurate quantification of some amino acids is difficult because they may be destroyed or converted into other reaction products during hydrolysis. Consequently, special hydrolysis procedures must be used to prevent errors. Research is ongoing to shorten the hydrolysis time, automate the process, and optimize the recovery of all amino acids after hydrolysis.

Tryptophan is completely destroyed by acid hydrolysis. Methionine, cysteine, threonine, and serine are progressively destroyed during hydrolysis; thus, the duration of hydrolysis will influence results. Asparagine and glutamine are quantitatively converted to aspartic and glutamic acid, respectively, and cannot be measured. Isoleucine and valine are hydrolyzed more slowly in 6 *N* HCl than other amino acids, while tyrosine may be oxidized.

In general, losses of threonine and serine can be estimated by hydrolysis of samples for three periods of time (i.e., 24, 48, and 72 h) followed by amino acid analysis. Compensation for amino acid destruction may be made by calculation to zero time assuming first-order kinetics. Valine and isoleucine are often estimated from a 72 h hydrolysate. Cysteine and cystine can be converted to the more stable compound, cysteic acid, by hydrolysis in performic acid and then hydrolyzed in 6 *M* HCl and chromatographed. Tryptophan can be separated chromatographically after a basic hydrolysis or analyzed using a method other than amino acid analysis.

In the original methods developed in the 1950s, amino acids were separated by **cation-exchange chromatography** using a stepwise elution with three buffers of increasing **pH** and **ionic strength**. Variations of this method are still commonly used today and may include the use of gradient elution protocols. In a procedure called **post-column derivatization**, amino acids eluting from the column are derivatized and quantified by reaction with **ninhydrin** (reacts with primary amino group of amino acids) to produce a colored product that was measured spectrophotometrically. The method was automated in the late 1970s and adapted for use with high-performance liquid chromatographs in the 1980s as new ion-exchange resins were developed that could withstand high pressures. Amino acids eluting from the column also may also be derivatized with *o*-phthalaldehyde (OPA) (reacts with the primary amino group of amino acids), then measured with a fluorescence detector. (Note: The ninhydrin and OPA methods can be used not only for amino acid analysis but also to monitor **hydrolysis of proteins** and to assay for **protease activity**, since these result in an increase of free primary amino groups.)

Other methods were developed in the 1980s using **pre-column derivatization** of the amino acids followed by reversed-phase HPLC. The hydrolyzed amino acids are derivatized prior to chromatography with **phenyl isothiocyanate** (reacts with primary amino group of amino acids), OPA, 6-aminoquinolyl-N-hydroxysuccinimidyl carbamate (AQC), or other compounds, separated by reversed-phase HPLC, and quantified by UV or fluorescence spectroscopy. Methods using pre-column derivatizations may be more sensitive and can detect about 0.5–1.5 uM quantities of amino acids. Chromatographic runs usually take 30 min or less.

The quantity of each amino acid in a peak is determined by spiking the sample with a known quantity of internal standard. The **internal standard** is often an amino acid, such as norleucine, not commonly found in a food product. Results are usually expressed as mole percent. This quantity is calculated by dividing the mass of each amino acid (determined from the chromatogram) by its molecular mass, summing the values for all amino acids, dividing each by the total moles, and multiplying the result by 100.

Many new methods of amino acid analysis are currently under development, including procedures using liquid chromatography coupled with mass spectrometry [14], but these are not yet in widespread use.

24.3.1.3 *Applications*

Amino acid analysis is used to determine the amino acid composition of a protein, determine quantities of essential amino acids to evaluate protein quality, identify proteins based on the amino acid profile, detect uncommon amino acids, and corroborate synthetic or recombinant protein structures. Amino acid analysis

provides information for estimating the molecular mass of a protein. Amino acid analysis is also used to meet FDA nutritional labeling regulations for protein. In addition, the Food and Agriculture Organization (FAO) of the United Nations [16] recommends that amino acid analysis be used in place of the Kjeldahl method (N determination) for total protein, although they note that amino acid analysis requires the use of more sophisticated instrumentation and better trained technicians that are not always readily available in some countries. Proteins used in animal diets, infant formulas, sports nutrition products, and therapeutic human diets are often analyzed for protein quality to ensure adequate quantities of essential amino acids. A chromatogram illustrating the separation of amino acids from a casein hydrolysate on a reversed-phase column in under 10 min is shown in Fig. 24.8. The amino acids were derivatized prior to separation (pre-column) using AQC and fractionated by UPLC using an internal standard such as alpha-aminobutyric acid [17].

24.3.2 Protein Nutritional Quality

24.3.2.1 *Introduction*

The nutritional quality of a protein is determined by the **amino acid composition** and the **digestibility** of that protein. **Antinutritional factors** can affect the nutritional quality of a protein. However, foods that contain heat-labile antinutritional factors (e.g., trypsin inhibitors) are usually cooked prior to consumption, thereby inactivating the inhibitor that might reduce protein digestibility. Some foods contain heat-stable antinutritional factors (e.g., tannins) that can decrease the nutritive value of a protein.

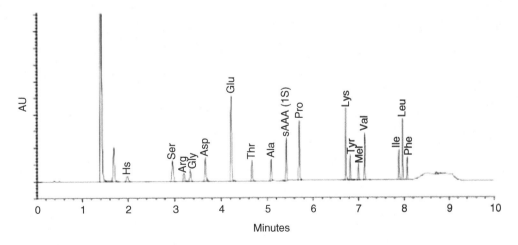

High-performance liquid chromatographic analysis of amino acids from a casein hydrolysate using pre-column derivatization with 6-aminoquinolyl-N-hydroxysuccinimidyl carbamate (AQC) and separated on a reversed-phase column using an Acquity UPLC system (Waters Corp.) with a 2,996 photodiode array detection system at 260 nm. A BEH C18 column (100 × 2.1 mm i.d., 1.7 um) at a flow rate of 0.7 mL/min and column temperature of 55 °C was used. The injection volume was 1 uL. Gradient elution was performed using eluent (**a**) AccQ·Tag$_{ultra}$ eluent A concentrate (5 %, v/v) and water (95 %, v/v) and eluent (**b**) AccQ·Tag$_{ultra}$ (Adapted from Boogers et al. [17] with permission)

Many protein quality assessment methods utilize information about the essential amino acid content of a food. **Essential amino acids** are those that cannot be synthesized in the body and must be present in the diet. Although there are some special cases due to age and medical status of an individual, the amino acids generally categorized as essential (or indispensable) include histidine, isoleucine, leucine, lysine, methionine, phenylalanine, threonine, tryptophan, and valine. Requirements for these amino acids have been determined for various age groups of humans (Table 24.4). The **first-limiting amino acid** of a human food is defined as the essential amino acid present in the lowest amount compared to a reference protein or to human requirements.

A food scientist's concerns regarding protein nutritional quality include meeting the requirements of nutrition labeling, formulating products of high protein quality, and testing the effects of food processing on protein digestibility. The development of valid methods to measure protein quality of foods for humans has been the focus of extensive research efforts over the past 50–60 years [19]. Protein nutritional quality assays may utilize animals in biological tests (*in vivo* assays), chemical or biochemical assays (*in vitro* assays), and/or simply calculations. Because of the time and expense of *in vivo* methods, *in vitro* assays and calculations based on amino acid content often are used to estimate protein quality. This section

24.4 table Amino acid requirements of infants, preschool children, adolescents, and adults (males and females combined)

		His	Ile	Leu	Lys	SAA	AAA	Thr	Trp	Val
Age (years)		Amino acid requirements (mg/kg per day)								
0.5	(Infants)	22	36	73	64	31	59	34	9.5	49
1–2	(Preschool children)	15	27	54	45	22	40	23	6.4	36
11–14	(Adolescent)	12	22	44	35	17	30	18	4.8	29
>18	(Adult)	10	20	39	30	15	25	15	4.0	26
		Scoring pattern (mg/g protein requirement)								
0.5	(Infants)	20	32	66	57	28	52	31	8.5	43
1–2	(Preschool children)	18	31	63	52	26	46	27	7.4	42
11–14	(Adolescent)	16	30	60	48	23	41	25	6.5	40
>18	(Adult)	15	30	59	45	22	38	23	6.0	39

His histidine, *Ile* isoleucine, *Leu* leucine, *Lys* lysine, *SAA* sulfur amino acids, *AAA* aromatic amino acids, *Thr* threonine, *Trp* tryptophan, *Val* valine

Adapted from Food and Agriculture Organization [18]

24.5 table Summary of methods used to measure protein nutritional quality

Method	What is measured	Application
Protein digestibility-corrected amino acid score (PDCAAS)	Amino acid content of first-limiting amino acid compared to requirements of preschool age children and true digestibility based on rat feeding experiment	Nutrition labeling for all but infant foods (to express g protein as % Daily Value)
Protein efficiency ratio (PER)	Weight gain of rats per g protein consumed	Nutrition labeling for infant foods (to express g protein as % Daily Value)
pH shift in vitro digestibility	Change in pH due to enzymatic digestion of protein under standard conditions	Rapid test of protein digestibility
DNFB (1-fluoro-2,3-dinitrobenzene) method of available lysine	Amount of lysine that has not already reacted with other food constituents and thus becomes unavailable as an essential amino acid	Determine effect of heat treatment during food processing on available lysine
Amino acid score (AAS)	Amino acid content of first-limiting amino acid, compared to requirements of preschool age children	Part of PDCAAS assay
Essential amino acid index (EAAI)	Amino acid content of each of nine essential amino acids, compared to their content in a reference protein	Rapid calculation to determine optimum amounts of various proteins in food formulation

of the chapter covers the tests and calculations required for nutrition labeling and mentions briefly several other protein quality methods for specialized applications. Please refer to Table 24.5 which summarizes methods for measuring protein nutritional quality.

24.3.2.2 Protein Digestibility-Corrected Amino Acid Score

24.3.2.2.1 Principle

The **protein digestibility-corrected amino acid score** (PDCAAS method) is used to estimate protein nutritional quality by combining information from (1) a calculation that compares the amount of the first-limiting amino acid in a protein to the amount of that amino acid in a reference protein, and (2) an *in vivo* assay measuring the digestibility of the protein by rats. For nutrition labeling, the PDCAAS must be determined by methods described in 21 CFR 101.9 [20].

24.3.2.2.2 Procedure

1. Determine the amino acid composition of the food.
2. Calculate the **amino acid score** for the first-limiting amino acid, using the requirements of preschool age children as a reference pattern.

Amino acid score

$$= \frac{\text{mg of amino acid in 1 g of test protein}}{\text{mg of amino acid in 1 g of reference protein}}$$

(24.3)

3. Feed male weanling rats standardized diets with 10 % test protein or with no protein, following the procedure for true protein digestibility (AOAC Method 991.29) [15]. **True digestibility** is calculated based on nitrogen ingested and feed intake, corrected for metabolic losses in the feces. If available, published values of true digestibility for the test protein can be used.
4. Calculate PDCAAS:

Amino acid score × % True digestibility   (24.4)

5. For nutritional labeling: (50 g = Daily Value for protein)

% Daily Value

$$= \frac{100 \times (\text{g protein / serving} \times \text{PDCAAS value})}{50\,\text{g protein}}$$

(24.5)

24.3.2.2.3 Applications

The Nutrition Labeling and Education Act (NLEA) requires that the percent Daily Value used on nutrition labels must be determined using the PDCAAS method, except for foods intended for consumption by infants (see also Chap. 3, Sect. 3.2.1.7). Because of the time and cost associated with the PDCAAS method, protein on the nutrition label is often expressed only as amount and not as a percent of the Daily Value. However, if a food label includes any claim regarding the protein, the nutrition label must include protein expressed as a percent of the Daily Value [20] [21CFR 101.9 (c) (7)].

The PDCAAS method generally is thought to better estimate protein quality for humans than the **protein efficiency ratio** (PER) method, which measures rat growth [19]. Rat growth is not comparable to that of adult humans, but it is more comparable to that of human infants. Therefore, the PER method is used to estimate protein quality of only infant foods. The actual digestion of protein by rats is thought to be fairly comparable to that by humans, so the protein digestibility portion of the PDCAAS method utilizes true digestibility as determined with rats. The PDCAAS method includes information on both amino acid composition and protein digestibility, since these are the factors that determine protein nutritional quality. However, perhaps as a limitation to the PDCAAS method, the amino acid score portion of the PDCAAS method includes only information about the first-limiting amino acid and not other essential amino acids. There is no differentiation in amino acid score between two proteins limiting to the same extent in one amino acid, but with the one protein only limiting in that amino acid and another protein limiting in many amino acids.

The continued use of the PDCAAS method is currently under debate by the Food and Agriculture Organization (FAO) of the United Nations [18]. A new protein quality measure, the **digestible indispensable amino acid score** (DIAAS), has been recommended to replace the PDCAAS as the method of choice for dietary protein quality assessment for regulatory purposes. DIAAS is defined as:

DIAAS (%)

$$= 100 \times \frac{(\text{mg of digestible dietary indispensable amino acid in 1g of the dietary protein})}{(\text{mg of the same dietary indispensable amino acid in 1g of the reference protein})}$$

This approach treats each essential amino acid as an individual nutrient, instead of only using the first-limiting amino acid as done in the PDCAAS method.

FAO is also recommending standardization of amino acid analysis methodologies. The DIAAS method is not yet approved by the FDA for nutritional labeling in the United States.

24.3.2.3 Protein Efficiency Ratio

24.3.2.3.1 Principle
The PER method (AOAC Method 960.48) [15] estimates protein nutritional quality in an *in vivo* assay by measuring rat growth as weight gain per gram of protein fed.

24.3.2.3.2 Procedure
1. Determine the nitrogen content of the test protein-containing sample and calculate the protein content.
2. Formulate a standardized test protein diet and a casein control diet to each contain 10% protein.
3. Feed groups of male weanling rats the diet and water ad libitum for 28 days.
4. Record the weight of each animal at the beginning of the assay, at least every 7 days during the assay, and at the end of the 28 days.
5. Record the food intake of each animal during the 28-day feeding trial.
6. Calculate the PER using the average total weight gain and average total protein intake for each diet group at day 28:

$$PER = \text{Total weight gain of test group (g)} / \text{Total protein consumed(g)} \quad (24.6)$$

7. Normalize the PER value for the test protein (i.e., compare the quality of the test protein to that of casein) by assigning casein a PER of 2.5.

$$\text{Adjusted or corrected PER} = \text{PER of test protein} / \text{PER of casein control} \quad (24.7)$$

24.3.2.3.3 Application
The Food and Drug Administration (FDA) requires that the PER method be used when determining protein as a percent of the Daily Value on the nutrition label of foods intended for consumption by infants [20] (21 CFR 101) (see also Chap. 3, Sect. 3.2.1.7). The PER method is limited to this application because the essential amino acid requirements for young rats are similar to those of human infants, but not those of other age groups. The PER method is time consuming, and it does not give any value to a protein for its ability to simply maintain body weight (i.e., a protein that produces no weight gain in the assay has a PER of zero).

24.3.2.4 Other Protein Nutritional Quality Tests

24.3.2.4.1 Essential Amino Acid Index
Essential amino acid index (EAAI) estimates protein nutritional quality based on the content of all essential amino acids compared to a reference protein (or human requirements). The EAAI is a rapid method to evaluate and optimize the amino acid content of food formulations. Unlike the amino acid score component of the PDCAAS method that considers only the first-limiting amino acid, the EAAI method accounts for all essential amino acids. However, EAAI does not include any estimate of protein digestibility, which could be affected by processing method. The essential amino acid content of the test protein (determined by amino acid analysis or from literature values) is compared to that of a reference protein (e.g., casein or human requirements) as follows: (Note: Use values for methionine plus cystine, and phenylalanine plus tyrosine, because they can substitute for one another as essential amino acids.)

$$\text{Essential amino acid index} = \sqrt[9]{\left(\frac{\text{mg of lysine in 1g of test protein}}{\text{mg of lysine in 1g reference protein}} \right) \times (\text{etc. for other 8 essential amino acids})} \quad (24.8)$$

24.3.2.4.2 In Vitro Protein Digestibility
The **pH shift method** is an *in vitro* protein digestibility assay used to estimate the digestibility of a protein by measuring the extent of protein hydrolysis upon reaction under standardized conditions with commercial digestive enzymes. The digestion procedure is designed to simulate human digestion of protein by using the enzymes trypsin, chymotrypsin, peptidase, and a bacterial protease. The pH of the protein solution drops when proteases break peptide bonds, release carboxyl groups, and liberate hydrogen ions. The pH at the end of the digestion period is used to calculate protein digestibility.

The pH shift method is the enzyme digestibility part of the **protein efficiency ratio-calculation method** (C-PER) (AOAC Method 982.30) [15], which combines a calculation of the essential amino acid composition with a calculation based on the *in vitro* assay of digestibility. The C-PER assay is intended for routine quality control screening of foods and food ingredients, to estimate the PER as it would be determined with the rat bioassay method.

The *in vitro* digestibility assay can provide a rapid and inexpensive means to determine and compare the protein digestibility of food products. This assay could be used to determine the effect of processing conditions on the protein digestibility of food products with the same formulation.

24.3.2.5 *Lysine Availability*

If the nutritional quality of a heat-processed product is lower than expected from its amino acid composition, one should assess the availability of lysine in the product. The **free ε-amino group** of the essential amino acid lysine can react with many food constituents during processing and storage to form biologically unavailable complexes with reduced nutritional quality. Lysine can readily complex with reducing sugars in the Maillard browning reaction, oxidized polyphenols, and oxidized lipids. Such reactions are accelerated upon heating and under alkaline conditions. **Lysinoalanine**, often found in alkali-treated proteins, decreases both the digestibility of the protein and the availability of lysine as an essential amino acid. The most commonly used method to measure **available lysine** is a spectrophotometric method that utilizes the reagent 1-fluoro-2,3-dinitrobenzene (DNFB, also referred to as FDNB). The DNFB method, detailed in AOAC Method 975.44 [15], can be used to determine if food processing operations have reduced the availability of lysine in a food.

24.3.3 Assessment of Protein Functional Properties

Protein functionality has been defined as the physical and chemical properties of protein molecules that affect their behavior in food products during processing, storage, and consumption. The functional properties of proteins contribute to the quality attributes, organoleptic properties, and processing yields of food. It is often desirable to characterize the functional properties of food proteins to optimize their use in a food product. Three of the most important protein functional properties in foods include **solubility**, **emulsification**, and **foaming**. This chapter will highlight only a few of the many methods available to measure these three functional properties. Two other functional properties of proteins, **gelation** and **dough formation**, are closely related to viscosity. It should be pointed out that there is no single functional property test that is applicable to all food systems, so careful test selection is imperative.

24.3.3.1 *Solubility*

24.3.3.1.1 *Principle*

One of the most popular tests of protein functionality is solubility. Proteins usually need to be soluble under the conditions of use for optimal functionality in food systems. Many other important functional attributes of proteins are influenced by protein solubility, such as thickening (viscosity effects), foaming, emulsification, water binding, and gelation properties.

Solubility is dependent on the balance of hydrophobic and hydrophilic amino acids that make up the protein, especially those amino acids on the surface of the molecule. Protein solubility also is dependent on the thermodynamic interactions between the protein and the solvent. Protein solubility is influenced by solvent polarity, pH, ionic strength, ion composition, and interactions with other food components, such as lipids or carbohydrates. Common food processing operations, such as heating, freezing, drying, and shearing, may all influence the solubility of proteins in a food system. Proteolysis by endogenous proteases also may alter protein solubility.

24.3.3.1.2 *Procedures*

There are many standardized methods to measure protein solubility including those published by the American Oil Chemists' Society [21] and the AACC International [22]. Other names synonymous with protein solubility include protein dispersibility index and nitrogen solubility index.

In a typical solubility procedure, a protein is dispersed in water or buffer at a specified pH, and the dispersion is centrifuged using defined conditions. Protein insoluble under the test conditions precipitates while soluble protein remains in the supernatant. Buffer and test conditions must be carefully controlled as these parameters have a large influence on results. Total protein and protein in the supernatant are measured, usually by the Kjeldahl method or a colorimetric procedure, such as the Bradford assay or the bicinchoninic acid (BCA) assay (Chap. 18, Sects. 18.4.1.2 and 18.4.2.3). Percentage protein solubility is measured by dividing the protein in the supernatant (soluble protein) by the total protein and multiplying by 100.

24.3.3.2 *Emulsification*

24.3.3.2.1 *Principle*

Food **emulsions** include margarine, butter, milk, cream, infant formulas, mayonnaise, processed cheese, salad dressings, ice cream, and some highly comminuted meat products, such as bologna. Emulsions are mixtures of two or more immiscible liquids, one of which is dispersed as droplets in the other. Oil and water are the two most common immiscible liquids found in food emulsions, although many other food components are usually present. The droplets are collectively called the **discontinuous** or **dispersed phase**, whereas the liquid surrounding the droplets is the **continuous phase**. Energy, in the form of homogenization, blending, or shaking, is used to disperse one immiscible liquid into the other.

Proteins are used as emulsifiers to lower the interfacial energy between phases to facilitate emulsion formation and to improve the stability of emulsions. Proteins migrate to the surface of a droplet during emulsion formation to form a protective layer or membrane on the surface, thus reducing interactions between the two immiscible phases. Emulsions are inherently unstable. The quality of an emulsion is dictated by many factors including droplet size, droplet size distribution, density differences between the two phases, viscosity of the two phases, electrostatic and steric interactions between molecules at the interface, and thickness and viscosity of the adsorbed protein layer. Additional information can be found in textbooks devoted to the topic of emulsification [23].

24.3.3.2.2 Procedures

Food emulsions usually are highly complex systems with multiple ingredients, so industrial scientists often choose to investigate the properties of emulsions in simplified model systems containing only a few of the most important ingredients. For a protein-based emulsion, this model system may contain only water or buffer, oil, and protein. It must be remembered that pH, salt concentration, temperature, type and amount of oil, protein concentration, energy input, and temperature during emulsion formation have a large effect on the properties of the final emulsion. These parameters must be selected prior to establishing a procedure.

The droplet size of the dispersed phase in an emulsion has a large influence on the emulsion quality. Droplet size can influence appearance, stability, and rheological properties of an emulsion, and hence the quality of a food product. Smaller droplets of more uniform size indicate a better emulsion. Droplet size can be determined by turbidimetric techniques, microscopy, laser diffraction, and electrical pulse counting [23].

An efficient emulsifier can prevent the breakdown or phase separation of an emulsion during storage. Emulsions can be stable for a long period of time (months to years), so often test protocols include a destabilization step involving physical or chemical stress. **Emulsion stability** can be tested by centrifugation or agitation of an emulsion at a given speed and time to determine the amount of creaming or oil separation that occurs. This is a fairly rapid test, but may not adequately represent the breakdown of the emulsion during normal storage conditions. Another method involves measuring the change in particle size distribution of the dispersed phase over time [24] and can be performed by laser diffraction or using the LUMiSizer®.

Many other more sophisticated techniques are available to measure the properties of food emulsions and include measures of interfacial properties, measurement of the dispersed phase volume fraction, characterization of emulsion rheology, and investigations into droplet charge [23].

24.3.3.3 Foaming

24.3.3.3.1 Principle

Foams are coarse dispersions of gas bubbles in a liquid or semisolid continuous phase. Like emulsions, foams require energy input during formation and are inherently unstable. Whipping, shaking, and sparging (gas injection) are three common methods of foam formation. Proteins or other large macromolecules in the continuous phase lower the surface tension between the two phases during foam formation and impart stability to films formed around the gas bubbles. Foams are found in cakes, breads, marshmallows, whipped cream, meringues, ice cream, soufflés, mousses, and beer.

24.3.3.3.2 Procedures

Foam volume and foam stability are two important parameters used to evaluate foams. **Foam volume** is dependent on the ability of a protein to lower surface tension between the aqueous phase and gas bubbles during foam formation. The volume of foam generated during a standardized foaming process is recorded and can be compared to other foams made under identical conditions. Often the foam is formed in a blender and then transferred into a graduated cylinder for measurement. Another common approach is to measure foam overrun or foam expansion (i.e., important in ice cream making). This is an indirect measure of the amount of air incorporated into the foam. Percent **overrun** can be calculated as follows:

$$\text{Overrun (\%)} = \frac{(\text{wt. 100 mL liquid} - \text{wt. 100 mL foam})}{(\text{wt. 100 mL foam})} \times 100$$

$$(24.9)$$

Foam stability depends on the properties of the protein film formed around the gas droplets. Free liquid is released as a foam breaks down. A more stable foam usually takes a longer time to collapse. Foam stability is often expressed as a half-life. A sample is whipped for a fixed time under standardized conditions. In one of the simplest methods, the foam is placed in a funnel over a graduated cylinder. The time for half of the original weight or volume of the foam to drain away from the foam is recorded. The greater the half-life, the more stable the foam.

Foam volume and stability are influenced by energy input, pH, temperature, heat treatment, and the type and concentration of ions, sugars, lipids, and proteins in the foam. Hence, all of these variables, except the one under test, must be standardized when designing a procedure to measure foam properties. Other methods can be used to better understand the molecular properties of the foam including: (1) determination of interfacial properties such as surface pressure and film thickness, (2) characterization of the viscoelastic properties of the film, and (3) characterization of the bubble size and distribution.

24.3.3.4 Applications of Testing for Solubility, Emulsification, and Foaming

Proteins perform a variety of functions in food systems. The quality of many foods is dependent on the successful manipulation of protein functional properties during processing. The solubility of a protein under a given set of conditions must be known for optimizing the use of that protein in a food product. For example, proteins must be soluble for optimal function in most beverages. Slight formulation modifications may change product pH and subsequently protein solubility.

Many proteins become insoluble and less functional once they are denatured. Thus, solubility often is used as an index of protein denaturation that may occur during processing. In general, proteins must be soluble to migrate to surfaces during foaming and emulsification. Any change in a process or ingredient that affects solubility also may alter the foaming and emulsification properties of that protein. Freezing, heating, shearing, and other processes can influence the functional properties of proteins in food systems. Thus, it may be necessary to reevaluate solubility, emulsion, and foaming properties of proteins if any processing changes are made.

Food product developers often need to understand the effect of ingredient substitutions on emulsion or foaming properties and subsequent product quality. Variations in raw materials can lead to differences in protein functionality and ultimately product quality. Protein functional tests can be used to compare two ingredients from different manufacturers or to verify the quality of protein ingredients in each batch purchased from a manufacturer. For example, the solubility of different commercial whey and soy protein concentrates may vary, leading to differences in other functional responses and, ultimately, product quality. Also, if a product developer is trying to substitute soy protein for egg protein in a formulation, it might be important to know how the functional attributes of each ingredient are affected by environmental conditions, such as pH or salt concentration. It often is desirable to know how long an emulsion or foam will be stable under a certain set of storage conditions. For example, emulsion stability tests are often used to evaluate the long-term stability of infant formulas and beverages.

24.3.3.5 Gelation and Dough Formation

Two other protein functional properties, gelation and dough formation, are closely related to viscosity and are described here only briefly.

24.3.3.5.1 Gelation

Protein gels are made by treating a protein solution with heat, enzymes, or divalent cations under appropriate conditions. While most food **protein gels** are made by heating protein solutions, some can be made by limited enzymatic proteolysis (e.g., chymosin action on casein micelle to form cheese curd), and some are made by addition of the divalent cations Ca^{2+} or Mg^{2+} (e.g., tofu from soy proteins). Proteins are transformed from the "soluble" state to a cross-linked "gel-like" state in protein gelation. The continuous, cross-linked network structure formed may involve covalent (i.e., disulfide bond) and/or noncovalent interactions (i.e., hydrogen bonds, hydrophobic interactions, electrostatic interactions). Some protein gel networks made by heating a protein solution are thermally reversible (e.g., gelatin gel that is formed by heating then cooling gelatin, but reverts to a liquid when reheated), while some protein gels are thermally irreversible. The stability of a protein gel is affected by a variety of factors, such as the nature and concentration of the protein, temperature, rates of heating and cooling, pH, ionic strength, and the presence of other food constituents. Therefore, these variables must be standardized and controlled to measure and compare gelation properties of proteins. Techniques used to measure rheological properties of foods such as compression, extension, and torsion analysis are applied to determine the properties of protein gels and are described in more detail in Chap. 29. In addition, many empirical tests have been developed to measure gel properties of specific proteins, such as the **Bloom test** designed to measure the strength of gelatin gels and the **folding test** to measure the elasticity of surimi gels.

24.3.3.5.2 Dough Formation

Wheat protein is unique in its ability to form a **viscoelastic dough** suitable for making bread and other bakery products. Gluten, the major storage protein of wheat, is a heterogeneous mixture of the proteins gliadins and glutenins. The unique amino acid composition of these proteins makes possible the formation of a viscoelastic dough, which can entrap carbon dioxide gas during yeast fermentation. Bread-making quality of wheat varieties is commonly tested by measuring dough strength, viscosity, and extensibility. The effect of dough ingredients (e.g., whey or soy proteins, phospholipids, surfactants) can be tested in the same way.

Dough strength is measured under standardized conditions with a variety of commercially available instruments. The **Mixograph**® is used to test the mixing properties of flour to ensure the dough will have the proper consistency, which is essential for automated manufacture of baked products. A **Farinograph**® is a torque meter to determine the water absorption of flour and the mixing properties of dough. The Extensograph®, often used in conjunction with the Farinograph, measures the extensibility of dough and thus can predict baking behavior. The **Rapid Visco Analyser**® (RVA) is a cooking viscometer to rapidly test starch-pasting properties. The RVA incorporates heating, cooling, and variable shear, to test the viscosity of starches, cereals, and other foods. The **Alveograph**® is used to measure gluten strength and extensibility by measuring the force required to blow and break a bubble of dough. Other tests of wheat quality include the **Glutomatic**® which measures the wet gluten content of flour.

24.4   SUMMARY

There are a variety of techniques used to separate and characterize proteins. Separation techniques rely on the differences in the solubility, size, charge, and adsorption characteristics of protein molecules. Ion-exchange chromatography is used to separate proteins on the basis of charge. Affinity chromatography utilizes ligands, such as enzyme inhibitors, coenzymes, or antibodies, to specifically bind proteins to a solid support. Proteins can be separated by size using dialysis, ultrafiltration, and size-exclusion chromatography. Electrophoresis can be used to separate proteins from complex mixtures on the basis of size and charge. SDS-PAGE also can be used to determine the molecular mass and subunit composition of a protein. Isoelectric focusing can be used to determine the isoelectric point of a protein. Capillary electrophoresis is an adaptation of conventional electrophoresis in which proteins are separated in capillary tubes. Chromatographic techniques are used in amino acid analysis to determine the amino acid composition of a protein. The nutritional quality of a protein is determined by the amino acid composition and protein digestibility. The PDCAAS is thought to be a better method than the PER method. The functional properties of proteins are used to characterize a protein for a particular application in a food. Common tests of protein functionality include solubility, emulsification, foaming, and gelation. No single test is applicable to all food systems.

Abbreviations

$(NH_4)_2SO_4$	Ammonium sulfate
AOAC	Association of Official Analytical Chemists International
AQC	6-Aminoquinolyl-N-ydroxysuccinimidyl carbamate
C-PER	Protein efficiency ratio-calculation method
DIAAS	Digestible indispensable amino acid score
EAAI	Essential amino acid index
FAO	Food and Agriculture Organization
FPLC	Fast protein liquid chromatography
HPLC	High-performance liquid chromatography
NFDM	Nonfat dry milk
OPA	o-Phthalaldehyde
PAGE	Polyacrylamide gel electrophoresis
PDCAAS	Protein digestibility-corrected amino acid score
PER	Protein efficiency ratio
pI	Isoelectric point
R_m	Relative mobility
SDS	Sodium dodecyl sulfate
TEMED	Tetramethylethylenediamine
UPLC	Ultra-performance liquid chromatography
UV	Ultraviolet
V_e	Elution volume

24.5   STUDY QUESTIONS

1. You have a protein system with the following characteristics:

Protein	Solubility in $(NH_4)_2SO_4$ (%)	Solubility in ethanol (%)	pI	Denaturation temperature (°C)
1	10–20	5–10	4.6	80
2	70–80	10–20	6.4	40
3	60–75	10–20	4.6	40
4	50–70	5–10	6.4	70

Describe how you would separate protein 4 from the others.
2. Compare and contrast the principles and procedures of SDS-PAGE vs. isoelectric focusing to separate proteins. Include in your explanation how and why it is possible to separate proteins by each method and what you can learn about the protein by running it on each type of system.
3. Explain how capillary electrophoresis differs from SDS-PAGE.
4. Briefly describe what each of the following tells you about the characteristics of the proteins of interest described in the statement (Note: The protein is not the same one in each statement):

(a) When subjected to dialysis using tubing with a molecular mass cutoff of 3,000 Da, a protein of interest is found in the retentate (i.e., not in the filtrate).

(b) When subjected to ultrafiltration using a membrane with a molecular mass cutoff of 10,000 Da, a protein of interest is found in the filtrate (i.e., not in the retentate).

(c) When the protein was subjected to ion-exchange chromatography using an anion-exchange column and a buffer of pH 8.0, a protein of interest is bound to the column.

(d) When a protein of interest was subjected to isoelectric focusing, the protein migrated to a position of approximately pH 7.2 in the pH gradient of the gel.

(e) When a protein of interest was subjected to SDS-PAGE in both the presence and absence of mercaptoethanol, the protein appeared as three bands at molecular mass 42,000, 45,000, and 48,000 Da.

(f) When a solution with various proteins was heated to 60 °C, the protein of interest was found in the precipitate obtained upon centrifugation of the solution.

5. You are submitting a soy protein sample to a testing laboratory with an amino acid analyzer (ion-exchange chromatography) so that you can obtain the amino acid composition. Explain how (a) the sample will be treated initially and (b) the amino acids will be quantified as they elute from the ion-exchange column. Describe the procedures. (Note: You want to quantify all the amino acids.)

(a) How will samples be treated initially?
(b) How will amino acids be quantified?

6. In amino acid analysis, a protein sample hydrolyzed to individual amino acids is applied to a cation-exchange column. The amino acids are eluted by gradually increasing the pH of the mobile phase.

(a) Describe the principles of ion-exchange chromatography.
(b) Differentiate anion vs. cation exchangers.
(c) Explain why changing the pH allows different amino acids to elute from the column at different times.

7. Briefly describe the differences between the following assay procedures:

(a) Amino acid score vs. essential amino acid index
(b) PDCAAS vs. amino acid score

8. You are helping to develop a new process for making a high-protein snack food from cereal grains and soy. You want to determine the protein quality of the snack food under various processing (toasting and drying) conditions. Considering the number of samples to be tested, you cannot afford an expensive *in vivo* assay, and you cannot wait more than a few days to get the results.

(a) What method would you use to compare the protein quality of the snack food made under different processing conditions? Include an explanation of the principles involved.

(b) You suspect that certain time-temperature combinations lead to over-processed products. Your testing from (7a) shows that these samples have a lower nutritional quality. What amino acid(s) in the snack food would you suspect to be the most adversely affected by thermal abuse?

(c) What test(s) could you use to confirm that amino acid(s) have become nutritionally unavailable by the over-processing? How are these tests conducted?

9. Define "protein functionality" and list three important functional properties of foods.

10. Describe a functional test used to measure:

(a) Protein solubility
(b) Emulsion stability
(c) Foam volume

11. You need to reconstitute nonfat dry milk (NFDM) powder for use in a yogurt formula. You have experienced a large amount of variation in the hydration times required for different lots of NFDM. As the quality control manager, you need to develop a test to measure the solubility and hydration characteristics of each lot of NFDM before it is used (in hopes of avoiding future rehydration problems).

(a) How would you measure solubility and what precautions would you take with the method you develop?

(b) Assume that you did the following for the test you developed: (1) mix 20 g of NFDM with 200 g water at 60 °C, (2) blend for 2 min, (3) transfer 50 mL to a centrifuge tube, (4) centrifuge at 10,000 × g for 5 min, and (5) measure the protein content of the supernatant after centrifugation. If the supernatant contains 2.95 % protein and the starting NFDM contained 36.4 % protein, what is the percent protein solubility?

12. You work for a fluid milk manufacturer that sells milk used in steamed coffee products.

With this application, the milk needs to form large amounts of stable foam. Periodically, you have been receiving complaints that the milk does not produce adequate foam when steamed. As a result, you need to design a quality control method to measure foam characteristics of each lot of milk before it is shipped. Describe the test(s) you would perform.

24.6 PRACTICE PROBLEMS

1. Using the data provided in the table below:

   (a) Calculate the EAAI for defatted soy flour.
   (b) Determine the amino acid score for the soy flour.
   (c) Calculate the PDCAAS, using the true digestibility value of 87% for defatted soy flour.

Amino acid	Soy[a] (mg/g protein)	Reference pattern[b] (mg/g protein)
Histidine	26	18
Isoleucine	46	31
Leucine	78	63
Lysine	64	52
Methionine/cystine	26	26
Phenylalanine/tyrosine	88	46
Threonine	39	27
Tryptophan	14	7.4
Valine	46	42

2. You work for a manufacturer of protein supplements sold to body builders. You need to screen several proteins that may be used in a new protein supplement. You have three samples (A, B, and C) to evaluate. The amino acid profiles of these three samples and the reference profile (i.e., amino acid requirements of preschool age children) are shown below.

Amino acid	Reference profile	Sample A	Sample B	Sample C
Histidine	18	26	35	24
Isoleucine	31	50	55	35
Leucine	63	65	46	32
Lysine	52	80	92	80
Methionine/cysteine	26	70	48	50
Phenylalanine/tyrosine	46	70	90	85
Threonine	27	51	40	39
Tryptophan	7.4	16	22	25
Valine	42	60	64	42

(a) Calculate the PDCAAS for each supplement. (Assume that the true digestibilities of Samples A, B, and C are 87%, 93%, and 64%, respectively.)
(b) Which sample would you use if Sample A costs $1.25/lb, Sample B costs $3.25/lb, and Sample C costs $1.15/lb?

Answers

Essential amino acid index

1. (a)
$$= \sqrt[9]{\frac{(26/18)(46/31)(78/63)(64/52)(26/26)}{(88/46)(39/27)(14/7.4)(46/42)}}$$

$$= \sqrt[9]{\frac{(1.44)(1.48)(1.24)(1.23)(1.00)(1.91)(1.44)}{(1.89)(1.10)}}$$

$$= \sqrt[9]{18.5866}$$

$$= 1.38$$

(b) Amino acid score $= 26/26 = 1.00$; lowest ratio represents the limiting amino acid, methionine/cystine.
(c) PDCAAS = amino acid score x true digestibility $= 1.00\,(0.87) = 0.87$

2. (a) The first-limiting amino acid for all three samples is leucine. (Identify by determining the ratio of each amino acid compared to reference profile.)

   PDCAAS = amino acid score for first-limiting amino acid x true digestibility:

   $$\text{Sample A} = (65/63) \times .87 = .90$$
   $$\text{Sample B} = (46/63) \times .93 = .68$$
   $$\text{Sample C} = (32/63) \times .64 = .33$$

   (b) The cost-to-protein quality ratio for each sample is as follows:

   $$\text{Sample A} = (\$1.25/.90) = \$1.39$$
   $$\text{Sample B} = (\$3.25/.68) = \$4.78$$
   $$\text{Sample C} = (\$1.15/.33) = \$3.48$$

   Sample A provides the highest amount of usable protein per dollar, suggesting it would be best to use.

REFERENCES

1. Coligan JE, Dunn BM, Speicher DW, Wingfield PT (eds) (2015) Current Protocols in Protein Science. Wiley, New York
2. Burgess RR, Deutscher MP (eds) (2009) Guide to Protein Purification. Vol. 436. Methods in Enzymology, 2nd edn, Academic Press, San Diego, CA.
3. Janson JC (ed) (2011) Protein Purification. Principles, High Resolution Methods, and Applications. 3rd edn. Methods of Biochemical Analysis. Vol 54 Wiley, Hoboken NJ

4. Scopes RK (1970) Characterization and study of sarcoplasmic proteins. Ch. 22. In: Briskey EJ, Cassens RG, Marsh BB (eds) Physiology and Biochemistry of Muscle as a Food, vol 2., University of Wisconsin Press, Madison, WI, pp 471–492

5. Doultani S, Turhan KN, Etzel MR (2004) Fractionation of proteins from whey using cation exchange chromatography. Process Biochem 39: 1737–1743

6. Jelen P (1991) Pressure-driven membrane processes: principles and definitions. In new applications of membrane processes. Document No. 9201. pp 6–41, International Dairy Federation, Brussels, Belgium

7. Chen WL, Hwang MT, Liau CY, Ho JC, Hong KC, Mao SJ. (2005) β-Lactoglobulin is a thermal marker in processed milk as studied by electrophoresis and circular dichroic spectra. J Dairy Sci 88:1618–1630

8. FDA (2015) Regulatory Fish Encyclopedia (updated 2015 May 12 2015; cited 2015 July 1. Silver Spring, MD. Available from: http://www.fda.gov/Food/FoodScienceResearch/RFE/.

9. Montowska M, Pospiech E. 2013. Species-specific expression of various proteins in meat tissue: Proteomic analysis of raw and cooked meat and meat products made from beef, pork and selected poultry species. Food Chem 136:1461–1469

10. Dolnik V (2008) Capillary electrophoresis of proteins 2005–2007. Electrophoresis 29: 143–156

11. Molina E, Marin-Alvarez PJ, Ramos M. (1999) Analysis of cows', ewes' and goats' milk mixtures by capillary electrophoresis: quantification by multivariate regression analysis. Int Dairy J 9:99–105

12. Breadmore MC (2012) Capillary and microchip electrophoresis: Challenging the common conceptions. J Chromatog A 1221: 42–55

13. Marchett-Deschmann M, Lehner A, Peterseil V, Sovegjarto F, Hochegger R, Allmaier G (2011) Fast wheat variety classification by capillary gel electrophoresis-on-a-chip after single-step one-grain high molecular weight glutenin extraction. Anal Bioanal Chem 400:2403–2414

14. Otter DE (2012) Standardised methods for amino acid analysis of food. British J Nutr 108:S230-S237

15. AOAC International (1990) Official methods of analysis, 15th edn. AOAC International, Gaithersburg, MD (Note: more recent editions available, however this edition is cited in the 21 CFR 101.9 Food Labeling)

16. Food and Agriculture Organization (2003) Food energy – Methods of analysis and conversion factors: report of a technical workshop. FAO Food and Nutrition Paper 77. Rome, Italy

17. Boogers I, Plugge W, Stokkermans YQ, Duchateau ALL (2008) Ultra-performance liquid chromatographic analysis of amino acids in protein hydrolysates using an automated pre-column derivatisation method. J Chromatog A 1189 (1–2):406–409 http://dx.doi.org/10.1016/j.chroma.2007.11.052

18. Food and Agriculture Organization (2013) Dietary protein quality evaluation in human nutrition. FAO Expert Consultation, Food and Nutrition Paper 92. Rome, Italy

19. Boye J, Wijesinha-Bettoni R, Burlingame B (2012) Protein quality evaluation twenty years after the introduction of the protein digestibility corrected amino acid score method. Br J Nutr 108:S183–S211

20. Federal Register (2015) Title 21 code of federal regulations part 101. Food labeling; Superintendent of documents. US Government Printing Office, Washington, DC

21. AOCS (2013) Official methods and recommended practices of the AOCS, 6th edn., 3rd printing. American Oil Chemists Society, Champaign, IL

22. AACC International (2009) Approved methods of analysis, 11th edn. AACC International, St. Paul, MN online

23. McClements DJ (2015) Food emulsions: Principles, Practices, and Techniques. 3rd edn. CRC Press, Boca Raton, Florida

24. Wrolstad RE, Acree TE, Decker EA, Penner MH, Reid DS, Schwartz SJ, Shoemaker CF, Smith D, Sporns P, (eds) (2005) Handbook of food analytical chemistry vol 1. Wiley, Hoboken, NJ

chapter 25

Determination of (Total) Phenolics and Antioxidant Capacity in Food and Ingredients

Mirko Bunzel (✉) • Rachel R. Schendel
Department of Food Chemistry and Phytochemistry,
Karlsruhe Institute of Technology (KIT),
Karlsruhe, Germany
e-mail: mirko.bunzel@kit.edu; rachel.schendel@kit.edu

S. Nielsen (ed.), *Food Analysis*, Food Science Text Series,
DOI 10.1007/978-3-319-45776-5_25, © Springer International Publishing 2017

25.1 INTRODUCTION

The term **phenolics** refers to several thousand aromatic plant metabolites, which possess at least one hydroxyl group attached to the phenyl ring. In the plant, phenolic compounds are important as structural components [1], in particular for the stabilization of the plant cell wall, and as a response to wounding and infestation. In addition, a plethora of nonstructural constituents, which are important as plant defense mechanisms against biotic and abiotic stress, establishing color, etc., has been identified from plants and plant-based foods. In food products, phenolics such as tocopherols and gallates are long known to increase the oxidative stability of lipids. They are also used as food colorants (e.g., anthocyanins) and contribute to the flavor of food products (e.g., vanillin). More recently, interest in phenolics was generated from studies suggesting potential health-promoting effects of phenolics in general or protective effects of specific phenolic compounds against specific diseases such as coronary heart disease and certain forms of cancer [2]. However, a general protective health effect of phenolics is highly unlikely, especially because many toxic phenolic compounds are known, too. Most importantly, many health-promoting effects of specific components were suggested based only on *in vitro* assays.

Currently, phenolics are often classified as either **simple phenolics**, which contain only one phenolic ring, for example, hydroxybenzoic acids and (monomeric) hydroxycinnamic acids, or **polyphenols**, which are made up of at least two phenolic units such as flavonoids, stilbenes, or lignans (Fig. 25.1). Polyphenolic food constituents with more than two phenolic rings are, for example, hydrolyzable and non-hydrolyzable tannins. However, besides these comparably low-molecular-weight phenolic compounds, plant-based foods contain high-molecular-weight phenolic compounds such as lignin or the phenolic domain of suberin. In addition to the naturally occurring phenolics synthesized by the plant, food processing leads to the formation of new phenolic compounds or significant modifications of plant-based phenolics. For example, the Maillard reaction results in the formation of phenolic products, and naturally occurring phenolics are incorporated into melanoidins, with coffee melanoidins (containing modified chlorogenic acids) being one of the most prominent examples [3].

Assays described in this chapter are of four types: (total) phenolics, hydrogen atom transfer-based antioxidant capacity assays, single electron transfer-based antioxidant capacity assays, and accelerated lipid oxidation assays. Many of the methods are given in more detail than in most other chapters in this book because this is a relatively recent field in food science and there are few official methods of analysis. Also, because all of the described methods have limitations, these are discussed in more detail in this chapter.

25.2 ANALYSIS OF (TOTAL) PHENOLICS

Based on the diverse array of phenolic compounds, which are often associated with non-phenolic compounds such as carbohydrates or organic acids as well, it is not possible to determine all phenolic compounds with a single assay as the term "total phenolic assays" may suggest. Also, many tests that are used to measure sum parameters are rather unselective, a fact which is also true for total phenolic assays. Thus, a critical evaluation of the results is necessary instead of claiming specific total phenolic contents of food products and food ingredients. A critical step in the application of total phenolic assays is choosing the extraction procedure, which defines the phenolic compounds to be included in the test results. Also, due to the poor selectivity of the total phenolic assays, specific extraction procedures may be used to exclude matrix compounds otherwise being determined in the assays.

25.2.1 Sample Preparation

Clear beverages such as white wine or clear apple juice usually do not require any further preparation but are either used directly or after dilution depending on the concentration of phenolic compounds. Fresh fruits, vegetables, cereals, and processed food products are directly ground and extracted, or they are lyophilized, ground, and extracted. Direct extraction after grinding requires information about the water content of the food product to reliably adjust the organic solvent/water ratio during the extraction step, a requirement that is often neglected in practice. Although lyophilization is the gentlest way to dry food products, small losses of phenolic compounds due to degradation may still occur during this process. After crushing fresh fruits or vegetables, the mash needs to be processed quickly due to potential losses of phenolics by the oxidizing action of the enzyme polyphenol oxidase found in plant tissues.

Many phenolic compounds are located in vacuoles and can easily be extracted with different solvents after cell wall and membrane rupture. Phenolic compounds that are bound to plant polymers need to be liberated by hydrolytic procedures [4].

25.2.1.1 *Extraction*
Phenolic compounds and their soluble conjugates such as sugar derivatives are extracted with water, polar organic solvents, or mixtures of polar organic solvents and water [5]. Extraction temperatures from room temperature up to 90 °C are used, depending on the analyzed food product and the stability of the phenolic compounds to be extracted. In some cases, less polar organic solvents such as ethyl acetate are used, which extract less polar phenolic compounds but exclude very polar phenolics (especially phenolic glycosides and other polar conjugates). Polar solvents, which are frequently

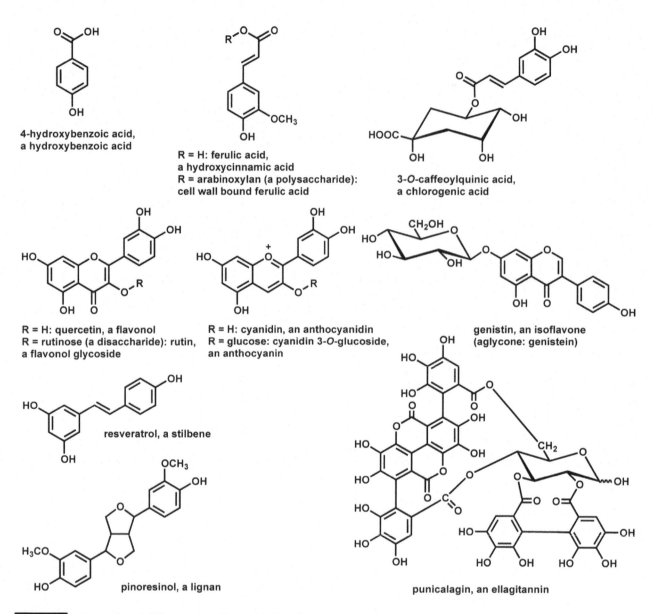

4-hydroxybenzoic acid,
a hydroxybenzoic acid

R = H: ferulic acid,
a hydroxycinnamic acid
R = arabinoxylan (a polysaccharide):
cell wall bound ferulic acid

3-O-caffeoylquinic acid,
a chlorogenic acid

R = H: quercetin, a flavonol
R = rutinose (a disaccharide): rutin,
a flavonol glycoside

R = H: cyanidin, an anthocyanidin
R = glucose: cyanidin 3-O-glucoside,
an anthocyanin

genistin, an isoflavone
(aglycone: genistein)

resveratrol, a stilbene

pinoresinol, a lignan

punicalagin, an ellagitannin

25.1 figure   Examples of different phenolics in food products

used, include methanol, ethanol, and acetone, often as an 80/20 (vol/vol) or 50/50 (vol/vol) mixture with water. Aqueous 80% ethanol or methanol solutions work well for most food products, solubilizing the bulk of phenolic compounds and, at the same time, precipitating out many polymers such as polysaccharides and proteins. Acidifying the extraction solutions may assist in stabilizing certain phenolic compounds such as anthocyanins.

25.2.1.2 Hydrolysis

Depending on the purpose of the analysis, it may be of interest to measure insoluble phenolics in addition to soluble phenolics. Common types of linkages involving phenolic compounds are glycosidic linkages and ester linkages. Especially in cereal products, but also in other plant-based foods, phenolic acids are ester linked. Their liberation requires alkaline hydrolysis (e.g., $2M$ NaOH, room temperature, 16 h) of the solvent-extracted residue (Sect. 25.2.1.1). Alkaline conditions may partially degrade phenolic acids such as hydroxycinnamic acids. To reduce oxidative degradation, NaOH solutions should be purged with nitrogen, and the headspace of the capped hydrolysis tubes should be flushed with nitrogen as well. Liberated phenolic acids are often extracted from the hydrolysate after acidification, which protonates the phenolic acids and makes them suitable for extraction with less polar organic solvents such as ethyl acetate or diethyl ether.

25.2.2 Colorimetric Assays for Determination of "Total" Phenolics

25.2.2.1 *Principles and Characteristics*

Colorimetric assays to determine total phenolic contents are often based on the ability of phenolic compounds to be oxidized. Permanganate and ferric ions were applied as oxidation reagents in the past, with ferric ions still used to measure the antioxidant capacity of plant foods. At present, two other reagents are more commonly used, which were originally developed to measure tyrosine and the aromatic but non-phenolic amino acid tryptophan: the **Folin-Denis reagent** [6, 7] and the **Folin-Ciocalteu reagent** [8]. Both reagents contain complex polymeric ions formed from phosphomolybdic and phosphotungstic heteropoly acids. Phenolates are formed under alkaline conditions and then oxidized, thereby reducing the initially yellow phosphotungstic-phosphomolybdic reagent. Reduction of the phosphotungstic-phosphomolybdic reagent results in the formation of a blue color ("molybdenum-tungsten blue"), which can be measured spectrophotometrically at a wide range of wavelengths, with 750 nm or 760 nm often being used. Because the phenolic compounds are only oxidized under alkaline conditions, but both the oxidizing reagent and the formed molybdenum-tungsten blue were occasionally described as unstable in alkaline solutions, several procedures with different sequences of reagent addition and time periods in between reagent addition and spectrophotometric analysis have been suggested. Also, when using the initially developed Folin-Denis reagent, a precipitate may form. The Folin-Ciocalteu reagent addresses this issue through the addition of lithium sulfate to the reagent. In addition, the Folin-Ciocalteu reagent has been described as being more sensitive than the Folin-Denis reagent [9]. The molar absorptivity (Chap. 7, Sect. 7.2.1) depends on both the reagent used and the phenolic compounds tested. For example, *ortho*-diphenols produce much larger (in many cases nearly twice as high, but varying) absorptivities than monophenolic compounds or *meta*-diphenols, demonstrating the empirical nature of this approach. Color formation by a sample is compared to color formation of a standard compound, preferentially gallic acid, (+)-catechin, or, more traditionally, tannic acid (Fig. 25.2), and the results are reported as gallic acid, (+)-catechin, or tannic acid equivalents.

Poor selectivity is a major shortcoming of both Folin reagent-based total phenolic assays. Several studies demonstrated that other reducing compounds besides phenolics can reduce the Folin reagents and are thus incorrectly determined as phenolic compounds in this assay. Besides ascorbic acid and some other vitamins, many other compounds such as thiols (e.g., cysteine and glutathione), nucleotide bases, and redox-active metal ions were reported to be active in this test [10]. In some reports, reducing sugars were described to be active, too. Due to their reactivity toward other non-phenolic reducing compounds, the Folin-based phenolic assays were also suggested as candidates for measuring the antioxidant or reducing activity of a sample rather than estimating total phenolics.

25.2.2.2 *Outline of Folin-Ciocalteu Procedure*

1. A clear sample solution is added to distilled water in a volumetric flask.
2. The Folin-Ciocalteu reagent (commercially available) is added, and the contents are mixed.
3. After 1 min and before 8 min, 20% sodium carbonate solution is added, and the volume is adjusted (a pH of about 10 should be achieved after mixing with the Folin-Ciocalteu reagent and the sample solution) [11].

25.2 figure   Phenolic compounds used as standard compounds in Folin-based "total" phenolic assays. Tannic acid is often a mixture of different, structurally related compounds

4. After about 2 h (the blue color has been demonstrated to be comparably stable), the generated color is determined spectrophotometrically in a 1-cm cuvette at 760 nm.

5. The average absorbance of phenol-free blanks (a yellow color should fade to colorless) is subtracted.

6. The amount of "total phenolics" is determined by using a standard curve.

7. Depending on the standard compound used (most often gallic acid or (+)-quercetin), the amount of total phenolics is reported, for example, as mg gallic acid equivalents per liter (if liquid samples such as white wine were used).

25.2.3 Chromatographic Methods

Chromatographic methods are widely used to quantify individual phenolic compounds or to monitor phenolic profiles of food products. However, they are not well suited for measuring the total phenolic contents of food products unless they contain only a few phenolic compounds, which are all known and available as standard compounds, a rather unlikely scenario. The choice of the chromatographic method depends on the phenolic compounds to be determined. Whereas all extractable phenolics are theoretically appropriate for liquid chromatographic methods, significantly less phenolic compounds are volatile enough to be analyzed by gas chromatography (GC) without prior derivatization.

25.2.3.1 *High-Performance Liquid Chromatography*

High-performance liquid chromatography (HPLC)/ ultra-performance liquid chromatography (UPLC) (Chap. 13, Sect. 13.2.3.3) is the method of choice for most phenolic compounds. Following extraction of the phenolic compounds as described in Sects. 25.2.1.1 and/or 25.2.1.2, the samples are either directly (after membrane filtration using suitable filter materials or sample centrifugation) injected into the HPLC system, or the extracts are further purified by means of liquid/liquid extraction or solid-phase extraction (SPE) (Chap. 14, Sect. 14.2.2.5). Reversed-phase SPE cartridges such as C18-cartridges are conditioned and loaded with the sample. Unwanted matrix components are eluted with water or mixtures of water with (typically) low amounts of organic modifiers and discarded, and the analytes are eluted with solvents containing large portions of organic modifiers such as methanol. Recoveries depend on the phenolics to be analyzed and the exact outline of the procedure. Capacity of the selected SPE cartridges, the solvent choice for the elution of unwanted matrix components, and eluent volumes are usually critical parameters requiring optimization to achieve high recoveries. If SPE is used to separate phenolic acids from neutral phenolic compounds, the pH of the solvents is another parameter that may be adjusted to optimize SPE conditions [12].

Separation of phenolic compounds is usually achieved on reversed-phase columns (Chap. 13, Sect.

13.3.2), with C18- and phenyl-hexyl-stationary phases being preferentially used in the analysis of phenolics from food products. Phenyl-hexyl-columns may, depending on the gradient system used, provide extra selectivity for aromatic compounds due to π-π interactions (non-covalent attractive interactions that involve π-electron systems, often formed between two aromatic rings, also known as π-π stacking) between the stationary phase and the phenolic analytes. Mobile phases are mostly water and varying amounts of organic modifiers, with methanol and acetonitrile being most often used. Depending on the phenolic compounds to be separated, the addition of small amounts of an acidic modifier, such as trifluoroacetic acid or formic acid, to the eluent may be required to achieve good separation. For example, acid addition to the eluent is necessary to separate phenolic acids. Without acid addition peaks become broad due to partial deprotonation of the carboxyl group, whereas acid addition to the eluent ensures protonated carboxyl groups and results in narrow, symmetric peaks. If mass spectrometric (MS) detection is used (Chap. 11), trifluoroacetic acid should be avoided because it suppresses ionization using electrospray ionization (ESI); formic acid may be used to adjust the pH instead [13].

Due to their phenyl unit, all phenolic compounds are UV active, making UV detection the preferred detection mode [14]. Depending on the conjugation of the π-electron system, the maximum of UV absorption can vary considerably. For example, ferulic acid with its extended π-electron system (propenylic side chain) shows a UV maximum (depending on the solvent and pH) around 325 nm, whereas 3-(3-hydroxy-4-methoxyphenyl) propionic acid has a UV maximum around 275 nm (Fig. 25.3). With the now routine presence of photodiode array detectors (capable of measuring the full UV-visible spectrum simultaneously (Chap. 7, Sect. 7.2.6.3) in analysis laboratories, each compound can be measured at its individual UV absorption maximum simultaneously. When only single-wavelength UV detectors are available, 280 nm or, more traditionally, 254 nm (main UV emission line of mercury vapor lamps) are most often used to monitor all phenolic compounds. Because phenolic compounds can easily be oxidized, their detection with electrochemical detectors is also an option, which, however, has not achieved broad application in the analysis of phenolics. Depending on the analyzed phenolic compounds, fluorescence detection, with its obvious advantages of increased selectivity and sensitivity (Chap. 7, Sect. 7.3), may be used. With state-of–the-art MS detectors becoming more affordable, MS and MS/MS detection (Chap. 11) is not only increasingly used in research but also more routinely in analytical testing laboratories.

25.2.3.2 *Gas Chromatography*

Because many phenolic compounds are not volatile without decomposition, gas chromatographic analysis of phenolics requires prior derivatization (Chap. 14, Sect. 14.2.3). **Silylation** using **derivatization reagents** such as *N,O*-bis(trimethylsilyl)trifluoroacetamide

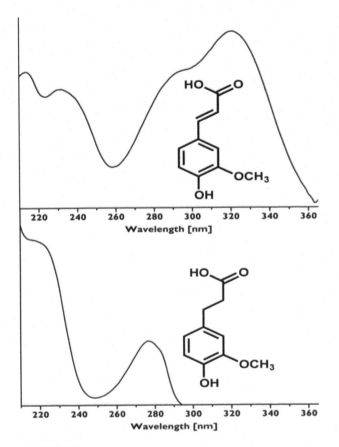

25.3 figure UV spectra and structures of ferulic acid (*top*) and 3-(3-hydroxy-4-methoxyphenyl)propionic acid (*bottom*)

25.4 figure Silylation of ferulic acid using *N,O*-bis(trimethylsilyl)trifluoroacetamide (BSTFA) as derivatization reagent. Carboxyl groups are converted into trimethylsilyl esters; hydroxyl groups are converted into trimethylsilyl ethers

tural information for the compound, respectively, which also helps to ensure peak purity [15].

25.3 ANTIOXIDANT CAPACITY ASSAYS

25.3.1 General Principles and Limitations of Antioxidant Capacity Assays

Traditionally, the antioxidant capacity of individual compounds or plant extracts has been measured to identify suitable antioxidants for the development of food products with enhanced stability against lipid deterioration. Because several mechanisms are involved in oxidative lipid degradation, including photooxidation, lipoxygenase-catalyzed oxidation, and autoxidation, different compounds such as radical scavengers, metal ion chelators, enzyme inhibitors, and singlet oxygen quenchers all have the potential to delay lipid oxidation in food products. However, the vast majority of studies performed in the recent past did not target food products but rather biological processes occurring in the human body [2]. Hundreds of individual, mostly phenolic compounds and, most often undefined, food and plant extracts were analyzed for their *in vitro* antioxidant capacity, with the intention of evaluating antioxidants potentially involved in the prevention of diseases such as atherosclerosis, diabetes, and cancer, i.e., diseases in which oxidative processes are thought to be critically involved. However, the use of simple *in vitro* antioxidant tests (as described below) has been criticized because it is not possible to mimic complex *in vivo* environments with these test systems. Compared to food, the evaluation of antioxidants from the perspective of human health and disease is even more complicated because many phytochemicals are not necessarily effective because of their own antioxi-

(BSTFA) is a commonly used derivatization method for phenolic compounds, replacing an active hydrogen by a trimethylsilyl (TMS) group. Applied in aprotic solvents with pyridine as catalyst, BSTFA reacts with both hydroxyl and carboxyl groups of phenolic compounds, which results in TMS ethers and esters, respectively (Fig. 25.4). Silylation often can be performed directly in the silylation reagent with or without addition of a catalyst [15]. In particular, the silylation of sterically hindered hydroxyl groups may be improved by using aprotic solvents and/or catalysts. Whereas derivatization sufficiently improves volatility of simple phenolic compounds and enables GC separation, larger phenolics, especially if conjugated with sugars, are still not suitable analytes for GC procedures. The TMS derivatives of simple phenolics such as hydroxybenzoic and hydroxycinnamic acids are well separated on nonpolar stationary GC phases (100% methyl-substituted polysiloxane or 5% diphenyl-95% dimethyl polysiloxane). Detection can be carried out nonspecifically using a flame ionization detector (FID) or more specifically using MS detection. Depending on whether MS detection is performed in the selected ion-monitoring mode or in the scan mode (resulting in total ion chromatograms) (Chap. 11), MS detection either shows the additional advantage of higher sensitivity compared to FID detection or supplies additional struc-

dant capacity but, among other factors, due to their stimulation of a phase II response, which includes the induction of classical phase II enzymes (antioxidant and detoxifying enzymes) [16, 17]. Because of these concerns, the USDA's Nutrient Data Laboratory (NDL) removed the "USDA Oxygen Radical Absorbance Capacity (ORAC) Database for Selected Foods" from the NDL website. They concluded that "there is no evidence that the beneficial effects of polyphenol-rich foods can be attributed to the antioxidant properties of these foods. The data for antioxidant capacity of foods generated by in vitro (test-tube) methods cannot be extrapolated to in vivo (human) effects and the clinical trials to test benefits of dietary antioxidants have produced mixed results" [18].

If any of the antioxidant capacity tests described below are used to determine the antioxidant capacity of individual food constituents or extracts, one must clearly define the goal of performing this test and ask whether this assay is suitable to obtain appropriate answers. Critical interpretation of results from these tests must go beyond just producing numbers. It is particularly important to reflect whether the chosen test system simulates a complex food system (or other biological systems) in sufficient detail [19]. A critical factor is whether the food system is an emulsion, with oil in water being most important for food products, or bulk oil. If the food product of interest is an emulsion but the test system uses only water (to test for "hydrophilic antioxidants") or an organic solvent (to test for "lipophilic antioxidants"), the results from these assays are hard to interpret because the distribution of the antioxidants in an emulsion and at the interface is not reflected by the test system. If the test system is able to simulate an emulsion, it should be asked whether the oxidized substrate employed by the assay represents the compounds to be oxidized in the food product. For example, many test systems use free fatty acids, which form micelles and therefore behave very differently from emulsified triglycerides. Other factors that need to be evaluated are radical initiators or radicals used in these test systems.

Several antioxidant capacity assays in numerous variations have been applied in the past [20, 21]. Efforts to standardize these test systems are underway but were not successful for all tests yet. Even where standardization was achieved, many different protocols are still used. To broadly classify assays that are supposed to measure radical scavenging capacities, they were divided into **hydrogen atom transfer** (HAT) reaction-based assays and **single electron transfer** (SET) reaction-based assays [22]. **HAT assays** determine the ability of an antioxidant or a mixture of antioxidants to scavenge free radicals by hydrogen donation. **SET assays** measure the ability of antioxidants to transfer one electron, thereby reducing radicals and also other compounds such as redox-active metal ions. By accepting an electron, the radical is con-

verted into an anion, which can accept a proton to form a stable compound.

25.3.2 Hydrogen Atom Transfer (HAT) Assays

25.3.2.1 Oxygen Radical Absorbance Capacity (ORAC) Assay

25.3.2.1.1 Principle

The ORAC assay utilizes azo compounds (compounds with the functional group R–N=N–R) as radical generators and monitors the progression of oxidation via disappearance of a fluorescent probe. Upon gentle heating (37 °C), the azo compound [2,2'-azobis(2-amidinopropane) dihydrochloride, AAPH, Fig. 25.5] disintegrates, releasing nitrogen gas and two carbon-centered radicals (Fig. 25.6). In an oxygenated environment, these carbon-centered radicals rapidly form peroxyl radicals, which attack the fluorescent probe (usually fluorescein, Fig. 25.5). In the presence of HAT-acting antioxidant compounds that quench the peroxyl radicals, the loss of fluorescein is delayed until the antioxidants are consumed. The loss of the fluorescent probe is monitored over time with a fluorescence spectrophotometer, and fluorescence intensity vs. time is plotted. The antioxidative capacity of tested samples is measured as the total area under the curve for the sample ($TAUC_{sample}$) minus the total area under the curve for the blank ($TAUC_{blank}$). These unitless areas are converted into **Trolox** (a water-soluble vitamin E analog, Fig. 25.5) **equivalents** based on the area generated by Trolox in the same assay. The ORAC assay conditions may be modified to accommodate either lipophilic or hydrophilic antioxidants.

25.3.2.1.2 Measurement Procedure

Early ORAC procedures are based on the methods described by Prior et al. [23] and Wu et al. [24]. A helpful discussion of technical challenges related to the ORAC assay may be found in Schaich et al. [25]. However, the ORAC assay described here largely reflects the AOAC Method 2012.23 for Total Antioxidant Activity [26]:

1. **Sample Extraction.** *Solid samples* should be lyophilized, ground, and extracted with acetone/water (1:1, vol/vol) for hydrophilic ORAC analysis. The acetone/water extract is made up to a defined volume with 0.075 M phosphate buffer (pH 7.4) and analyzed under the hydrophilic ORAC assay conditions. *Liquid samples* are centrifuged, and the supernatant is diluted with 0.075 M phosphate buffer (pH 7.4) and analyzed under the hydrophilic ORAC assay conditions. Sample preparation for lipophilic samples is not clearly outlined in the AOAC method, although the preparation of an lipo-

fluorescein

A-crocin

2,2'-azobis(2-amidinopropane) dihydrochloride, AAPH

Trolox

2,2'-azino-bis(3-ethylbenzothiazoline-6-sulfonic acid), ABTS·+

2,2-diphenyl-1-picryhydrazyl radical, DPPH

25.5
figure

Probes (fluorescein, crocin), radicals (DDPH, ABTS·+), the radical generating azo compound AAPH, and the water-soluble vitamin E analog Trolox used in various antioxidant capacity tests

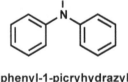

25.6
figure

A simplified mechanism demonstrating the generation of peroxyl radicals from 2,2'-azobis(2-amidinopropane) dihydrochloride (AAPH)

philic ORAC extraction solution made up of hexane and ethyl acetate (3:1, vol/vol) is described.

2. **Preparation of Standard Solutions and Calibration**. A Trolox calibration curve is prepared in 0.075 M phosphate buffer (pH 7.4) for

the hydrophilic ORAC assay. For the lipophilic ORAC assay, a calibration curve is produced by dissolving Trolox in a 1.4 % randomly methylated β-cyclodextrin (RMCD) solution prepared in 0.075 M phosphate buffer (pH 7.4). Fluorescein solution and AAPH solution, which must be freshly made before each sample run, are also prepared in 0.075 *M* phosphate buffer (pH 7.4).

3. **Hydrophilic ORAC Assay.** A fluorescence microplate reader should be set to an excitation wavelength of 485 nm and an emission wavelength of 530 nm, with an incubator temperature of 37 °C. All solutions should be sparged with oxygen immediately before use [25]. Appropriately diluted hydrophilic sample extract, Trolox calibration solutions, or phosphate buffer for blank are pipetted into sample wells, and fluorescein solution is added. The sample is mixed, AAPH solution is added, and the solution is mixed again. Fluorescence is measured at 1-min intervals for 35 min.

4. **Lipophilic ORAC Assay.** The assay is performed using the same conditions as the hydrophilic assay except that samples are diluted with 7 % RMCD solution prepared in acetone/water (50:50, vol/vol).

25.3.2.1.3 Calculation of Results

The area under the curve (AUC) is calculated according to Eq. 25.1:

$$AUC = \begin{pmatrix} 0.5 + f_1 / f_0 + \ldots f_i / f_0 + \\ \ldots f_{34} / f_0 + 0.5 \times (f_{35} / f_0) \end{pmatrix} \quad (25.1)$$

where:

f_0 = initial fluorescence intensity upon addition of all solutions

f_i = fluorescence intensity at time i

Net AUC for a sample or calibration point is calculated by subtracting AUC of the blank from that of the sample or calibration point. Net AUC for the calibration points vs. Trolox concentration is plotted in graphing software, and a linear or quadratic standard curve equation is calculated. ORAC values for samples should be expressed as µmol Trolox equivalents and normalized to dry weight.

25.3.2.1.4 Advantages and Limitations

The ORAC assay utilizes a controlled source of peroxyl radicals, which creates a better model of real-life antioxidant interactions with lipids in food systems than other antioxidant assays that use stable radical compounds. However, the assay is comparatively difficult to perform properly, and several points need to be considered. First, reaction temperature must be exactly controlled at 37 °C to ensure consistent, reproducible peroxyl radical generation. Uniform dissolved oxygen concentrations between analyses must also be ensured. Additionally, the concentrations of fluorescein, AAPH, and sample must be optimized [25]. If the reaction proceeds slowly (>1 h required for loss of fluorescence), the AAPH concentration should

be increased. Fluorescein concentrations that are too high result in fluorescence quenching via π-π-stacking of the aromatic rings and hydrogen bonding between phenol groups; fluorescence quenching can be recognized when a solution's fluorescence intensity remains the same upon dilution or even increases. A range of sample dilutions also should be measured to test for non-radical sample interactions with fluorescein, as well as potential depression of fluorescence by sample components (see Schaich et al. [25]).

25.3.2.2 Crocin Bleaching Assay

25.3.2.2.1 Principle
Similar to the ORAC assay, the crocin bleaching assay also uses a controlled source of peroxyl radicals to assess HAT activity. **Crocin** (Fig. 25.5), a water-soluble carotenoid isolated from saffron, absorbs UV-Vis light strongly at 440–443 nm but loses color ("bleaches") upon radical oxidation and is therefore used as probe.

25.3.2.2.2 Measurement Procedure
The crocin bleaching assay has not been studied as extensively as the ORAC assay, and no standardized method exists in the literature. Although earlier reports utilized photolysis of hydroperoxides to produce alkoxyl radicals [27], most recent studies have used peroxyl radicals generated by heating azo compounds (e.g., AAPH). No specific sample preparation method has been recommended, but the sample preparation and extraction procedures described in Sect. 25.2.1 may be used. For hydrophilic antioxidants, the reaction is typically carried out in buffer (either sodium phosphate buffer or phosphate-buffered saline, pH 7.0 or 7.4) at 37 °C, crocin stock solutions are prepared in methanol, the sample is incubated at 37 °C, and absorption decline following addition of free radical generator (azo compound) is monitored at 440, 443, or 450 nm. For application of the crocin bleaching assay to lipophilic antioxidants, a lipophilic azo compound as free radical initiator in an organic solvent environment [e.g., dimethylformamide (DMF) or a toluene: DMF mixture (1:4, vol/vol)] may be used [28].

25.3.2.2.3 Calculation of Results
The ability of antioxidant compounds to hinder color loss has been varyingly expressed as Trolox or α-tocopherol equivalents [28], "percent inhibition of crocin bleaching value" [29], IC$_{50}$ (dose of substance causing a 50 % inhibition of crocin bleaching) [30], and relative rate constants [27].

25.3.2.2.4 Advantages and Limitations
Advantages of the crocin bleaching assay include its sensitivity and reproducibility [30]. Like the ORAC assay, temperature must be tightly controlled during the crocin bleaching assay to ensure a consistent gen-

eration of peroxyl radicals between samples. Additionally, many food components (e.g., carotenoids) absorb at the same wavelengths used in the crocin bleaching assay. However, the greatest current limitation for this assay is the lack of a standardized, accepted method of analysis and a format for expressing results.

25.3.3 Single Electron Transfer (SET) Assays

25.3.3.1 *Trolox Equivalent Antioxidant Capacity (TEAC) Assay*

25.3.3.1.1 *Principle*
The TEAC assay, also known as the 2,2'-azino-bis(3-ethylbenzothiazoline-6-sulfonic acid) (ABTS$^{•+}$) assay utilizes the stable, nitrogen-based ABTS$^{•+}$ radical. The ABTS$^{•+}$ radical is strongly colored, but ABTS is colorless, which allows for easy monitoring of appearance/disappearance of the radical via UV-Vis absorbance. Early versions of this method used a metmyoglobin-H_2O_2-system to generate hydroxyl radicals, which then oxidized ABTS to its radical form. Antioxidant compounds were added to the system before radical formation, and inhibition of radical formation was related to antioxidant activity. However, this order of reagent addition was confounded by the fact that antioxidants could react with the hydroxyl radicals, metmyoglobin, and ABTS$^{•+}$, which resulted in inaccurate estimation of antioxidant activity [20, 25]. Therefore, current TEAC/ ABTS$^{•+}$ methods first generate high ABTS$^{•+}$ concentrations, followed by antioxidant addition and monitoring ABTS$^{•+}$ decline [31].

25.3.3.1.2 *Measurement Procedure*
1. **Generation of ABTS$^{•+}$.** Fresh stock solutions of ABTS$^{•+}$ should be prepared weekly. Potassium persulfate is added to an ABTS solution in deionized water, and this mixture is allowed to stand for 12–16 h at room temperature until a deep blue-green color develops. Before each analysis session, a working solution is created by diluting aliquots of the ABTS$^{•+}$ stock solution in water until an absorbance of 1.0 at 734 nm is reached.
2. **Sample Measurement.** The ABTS$^{•+}$ working solution is pipetted into a cuvette followed by the addition of the antioxidant solution or extract (see Sect. 25.2.1). The absorbance at 734 nm is monitored and recorded at a fixed time point after addition of the antioxidant solution (6 min is the time point most often used). The absorbance of the ABTS$^{•+}$ working solution without antioxidant solution is measured using the solvent of the sample solution in place of sample. For assessment of more lipophilic antioxidants, the TEAC assay also may be performed in ethanol instead of water.
3. **Measurement of Trolox Standards.** A Trolox stock solution is created in the same solvent as

used for the antioxidant samples and appropriately diluted to create an equidistant (i.e., calibration points are equally spaced), five-point calibration curve. The calibration points are measured using the same method as for samples.

25.3.3.1.3 *Calculation of Results*
The decrease in absorbance for calibration points vs. Trolox concentration is plotted with graphing software, and a linear standard curve equation is calculated. The standard curve equation is used to convert drop in absorbance for antioxidant samples (A_0–A_f) to equivalent Trolox concentrations.

25.3.3.1.4 *Advantages and Limitations*
Although results from the TEAC/ABTS$^{•+}$ method should not be extrapolated into food or biological systems, the assay provides a rapid and easy possibility for an initial screening of samples for antioxidant activity or for monitoring changes in the same samples over time [25, 32]. The TEAC/ABTS$^{•+}$ method is primarily governed by steric considerations of the large nitrogen-based ABTS$^{•+}$ radical molecule. Antioxidants that function primarily via SET or do not have bulky ring systems react rapidly, whereas unwieldy antioxidant structures or antioxidants functioning via HAT react slowly. Therefore, a strong response in the TEAC/ABTS$^{•+}$ method does not necessarily correlate well with radical-quenching capabilities in real-world food or biological systems. Additionally, because effective HAT-acting antioxidants react only slowly in this system, the typical method of measuring absorbance after a given time may underestimate the activity of many extracts.

25.3.3.2 *2,2-Diphenyl-1-Picryhydrazyl Radical (DPPH) Assay*

25.3.3.2.1 *Principle*
The DPPH assay was first proposed by Blois over 50 years ago for "the antioxidant assay of biological materials" [33]. It is now the basis of AOAC Method 2012.04, for antioxidant activity in foods and beverages [26]. DPPH is a stable organic nitrogen radical (Fig. 25.5) and produces deeply blue-violet solutions that strongly absorb at 517 nm. Upon quenching of the radical by antioxidant compounds, which can occur both by electron transfer and hydrogen atom transfer, solutions are decolorized, and the absorbance at 517 nm decreases. Quenching of the DPPH radical was shown to happen rapidly by the SET mechanism, but antioxidant compounds that act primarily by HAT quenching also show activity. However, hydrogen-bonding solvents, such as the methanol or ethanol often used for this assay, impede the HAT mechanism and thus favor SET-acting antioxidant compounds [34, 35].

25.3.3.2.2 Measurement Procedure

The AOAC Method 2012.04 described here adds a DPPH standard solution directly to solid samples without a separate extraction step, which maximizes extraction efficiency, but many methods in the literature first prepare sample extracts, preferably in methanol, from solid samples. A DPPH reagent solution in methanol/water (approx. 50:50, vol/vol) is prepared; the solution should be protected from light, and it is advised to prepare fresh stock solutions daily. DPPH reagent solution is added directly to lyophilized, ground solid samples (which can be appropriately diluted with corn starch) or homogenous liquid samples (which can be appropriately diluted with deionized water). An equidistant, four-point Trolox calibration curve is prepared by mixing aliquots of a Trolox stock solution (also prepared in methanol/water, 50:50, vol/vol) with DPPH solution. All samples are incubated at 35 °C for 4 h on an orbital shaker and filtered (if cloudy). Following incubation, the absorbance at 517 nm is measured against a distilled water blank.

25.3.3.2.3 Calculation of Results

The amount of DPPH radical quenched by the sample extract is calculated as μmol Trolox equivalents/100 g of sample.

25.3.3.2.4 Advantages and Limitations

The DPPH assay is simple to perform and may be useful as an initial qualitative screening tool to identify the presence of SET-acting antioxidant compounds. However, the assay is hampered by major drawbacks. The nitrogen-based, bulky DPPH radical is a poor model for the radicals encountered in food and biological systems. Determining the total amount of DPPH consumed by the sample after several hours (or, as described in other procedures, after reaching a reaction plateau) and its "total antioxidant capacity" ignores the short lifetimes (seconds or less) of free radicals in food systems [36]. However, focusing on reaction rate in the DPPH assay as a means of ranking antioxidants has also proven futile, because reaction rates are almost exclusively governed by the steric accessibility of the antioxidant molecule to the bulky DPPH's radical site [34]. Additionally, both the "total antioxidant capacity" and reaction rate of individual antioxidants are affected by antioxidant concentration, with stoichiometry (moles of DPPH consumed/mole of antioxidant) decreasing with increased antioxidant concentration [34]. Mixtures of antioxidants display more activity depression than can be explained by steric interferences [25], meaning that the assay is completely irrelevant for ranking samples containing multiple components in varying concentrations, such as plant extracts. Finally, the assay does not

capture antioxidants which function primarily via the HAT mechanism. Based on these limitations, the assay should not be used for quantitative ranking of the antioxidant capacity of individual compounds or mixtures.

25.3.3.3 Ferric Reducing Antioxidant Power (FRAP) Assay

25.3.3.3.1 Principle

In contrast to the TEAC/ABTS$^{\bullet+}$ and DPPH assays, which operate under mixed SET and HAT mechanisms, the FRAP assay measures exclusively SET-active compounds. The assay monitors the reduction of a ferric salt (ferric 2,4,6,tripiridyl-s-triazine, TPTZ), producing a colored product which absorbs at 593 nm. The redox potential of Fe(III)-TPTZ (0.7 V) is similar to that of ABTS$^{\bullet+}$ (0.68 V), so compounds that show activity in the TEAC/ABTS$^{\bullet+}$ assay will also react in the FRAP assay. However, because the FRAP assay must be performed at an acidic pH value of 3.6 to protect iron solubility, which increases the redox potential, FRAP values are usually lower than TEAC/ABTS$^{\bullet+}$ assay values [20].

25.3.3.3.2 Measurement Procedure and Calculation of Results

Fresh working solutions of Fe(III)-TPTZ are prepared by mixing acetate buffer (pH 3.6) with TPTZ and FeCl$_3 \bullet$6H$_2$O stock solutions before sample analysis. Samples (see extraction procedure described in Sect. 25.2.1), typically dissolved in aqueous solutions, are measured by mixing FRAP working solution with sample and measuring the increase in absorbance at 593 nm against a FRAP reagent blank after a set time (typically 4 or 8 min). Calibration is calculated via Fe(II) equivalents by preparing equidistant, five-point calibration curves with FeSO$_4 \bullet$ 7H$_2$O solution [37].

25.3.3.3.3 Advantages and Limitations

Because the assay's mechanism is purely SET based, it can be used to screen antioxidants for SET functionality. Additionally, the assay is rapid and robust. However, the active mechanism of the FRAP assay [the ability to reduce Fe(III)] has little correlation to the radical-quenching mechanisms (HAT) displayed by many antioxidants.

25.3.4 Assays Based on Oxidation of Lipids

Besides the assays described in Sects. 25.3.2 and 25.3.3, which were often designed for a high sample throughput and all possess the limitation of being oversimplified test systems, the effects of individual compounds, mixtures of antioxidants, or plant extracts on oxidative stability in more complex systems or the actual food product to be protected from oxidative degradation can be tested in **accelerated lipid oxidation tests**. These test systems are described in detail in Chap. 23, Sect. 23.5. Using, for example, the Rancimat® does not necessarily require a lipid extraction, but direct measurement of food products is possible (e.g., cookies, nuts, and microwave popcorn). However, foods containing high water or protein contents such as salad dressings or sausages require a prior lipid extraction. Using the whole food product is closest to what happens during storage with the limitations arising from using accelerated oxidation conditions as described in Chap. 23, Sect. 23.5.1. If the lipid phase needs to be extracted from an emulsion-based food product such as salad dressings, the same limitations as described earlier occur (i.e., partition of the antioxidants between the phases and at the interface cannot be modeled).

25.4 SUMMARY

Phenolic compounds and other antioxidants are of great interest as they have the potential to stabilize food products against lipid deterioration. Because many diseases such as diabetes and certain forms of cancer are linked to oxidative stress, phenolic compounds from food products and other plant sources are discussed in terms of disease prevention, too. Thus, there is a demand to measure phenolic compounds in food products and to evaluate the antioxidant capacity of individual food constituents, extracts, and ingredients. Methods to determine the total phenolic contents of food products such as the Folin-Ciocalteu assay are unspecific and do not, depending on the overall composition of the food product, necessarily reflect the phenolic contents. Wherever possible, it is therefore advised to study individual phenolic compounds by using HPLC/UPLC or GC approaches. A plethora of antioxidant capacity assays has been described in the past that differ in the applied principle and, more often, in details of the procedure such as probes, radical initiators, solvents, etc. Because several of these tests were designed as high-throughput methods, they can be used to screen many compounds/extracts, but suffer from oversimplification and do not reflect the complexity of oxidative degradation of lipids and other organic compounds in biological systems. Therefore, it is not advised to use these assays to screen for compounds with potential health benefits. If these assays are used to screen compounds to improve the shelf life of food, the assays should be critically evaluated in terms of whether they are suitable models for the food system to be protected from oxidative degradation. The efficacy of antioxidants identified in these test systems to delay lipid oxidation and to improve shelf life needs to be confirmed in real food products under real storage conditions.

25.5 STUDY QUESTIONS

1. Why do compounds such as ascorbic acid and glutathione show a positive response in the Folin-Ciocalteu assay, a test often selected by researchers to measure the total phenolic contents of food products?
2. Name different ways to extract food products with or without prior hydrolysis for the analysis of phenolic compounds and give examples of phenolic compounds that are either included or excluded in the analysis depending on the extraction procedure.
3. Why does RP-HPLC analysis of phenolic acids usually require a pH adjustment of the mobile phase?
4. Why does the GC analysis of phenolic compounds often require a prior derivatization? What is a suitable derivatization procedure for many phenolic compounds?
5. Why did the USDA's Nutrient Data Laboratory remove the USDA ORAC database for Selected Foods from the NDL website?
6. Why do many antioxidant capacity assays fail to model food products that are emulsions?
7. What is the difference between HAT and SET antioxidant capacity assays?
8. How does the ORAC assay aim to measure hydrophilic versus lipophilic antioxidants?
9. What are some major limitations of the DPPH assay due to the structure of the stable radical DPPH used in this assay?
10. What is the difference between preparing the ABTS$^{•+}$ radical before and after addition of the sample solution to be tested?

Acknowledgments The authors wish to thank Baraem Ismail, Andrew Neilson, and Jairam Vanamala for reviewing the chapter and providing helpful comments.

REFERENCES

1. Ralph J, Bunzel M, Marita JM, Hatfield RD, Lu F, Kim H, Schatz PF, Grabber JH, Steinhart H (2004) Peroxidase-dependent cross-linking reactions of *p*-hydroxycinnamates in plant cell walls. Phytochem Rev 3(1):79–96
2. Halliwell B (2007) Dietary polyphenols: Good, bad or indifferent for your health. Cardiovasc Res 73(2):341–347
3. Gniechwitz D, Reichardt N, Ralph J, Blaut M, Steinhart H, Bunzel M (2008) Isolation and characterisation of a coffee melanoidin fraction. J Sci Food Agric 88(12):2153–2160
4. Barberousse H, Roiseux O, Robert C, Paquot M, Deroanne C, Blecker C (2008) Analytical methodologies for quantification of ferulic acid and its oligomers. J Sci Food Agric 88(9):1494–1511

5. Naczk M, Shahidi F (2004) Extraction and analysis of phenolics in food. J Chromatogr A 1054(1–2):95–111

6. Folin O, Denis W (1912) On phosphotungstic-phosphomolybdic compounds as color reagents. J Biol Chem 12:239–243

7. Folin O, Denis W (1912) Tyrosine in proteins as determined by a new colorimetric method. J Biol Chem 12:245–251

8. Folin O, Ciocalteu V (1927) On tyrosine and tryptophane determinations in proteins. J Biol Chem 73:627–650

9. Singleton VL, Rossi Jr. JA (1965) Colorimetry of total phenolics with phosphomolybdic-phosphotungstic acid reagents. Am J Enol Vitic 65(3):144–158

10. Everette JD, Bryant QM, Green AM, Abbey YA, Wangila GW, Walker RB (2010) Thorough study of reactivity of various compound classes toward the Folin-Ciocalteu reagent. J Agric Food Chem 58(14):8139–8144

11. Singleton VL, Orthofer R, Lamuela-Raventós RM (1999) Analysis of total phenols and other oxidation substrates and antioxidants by means of Folin-Ciocalteu reagent. Methods Enzymol 299:152–178

12. Robbins JR (2003) Phenolic acids in foods: An overview of analytical methodology. J Agric Food Chem 51(10):2866–2887

13. Jilek ML, Bunzel M (2013) Dehydrotriferulic and dehydrodiferulic acid profiles of cereal and pseudocereal flours. Cereal Chem 90(5):507–514

14. Dobberstein D, Bunzel M (2010) Separation and detection of cell wall-bound ferulic acid dehydrodimers and dehydrotrimers in cereals and other plant materials by reversed phase high-performance liquid chromatography with ultraviolet detection. J Agric Food Chem 58(16):8927–8935

15. Bunzel M, Ralph J, Marita JM, Hatfield RD, Steinhart H (2001) Diferulates as structural components in soluble and insoluble cereal dietary fibre. J Sci Food Agric 81(7):653–660

16. Stevenson DE, Hurst RD (2007) Polyphenolic phytochemicals – just antioxidants or much more? Cell Mol Life Sci 64(22):2900–2916

17. Dinkova-Kostova AT (2002) Protection against cancer by plant phenylpropenoids: Induction of mammalian anticarcinogenic enzymes. Mini Rev Med Chem 2(6):595–610

18. USDA (2010) Oxygen radical absorbance capacity (ORAC) of selected foods, Release 2 [updated 01/2/2016]. Agricultural Research Service, US Department of Agriculuture. Available from: http://www.ars.usda.gov/Services/docs.htm?docid=15866.

19. Frankel EN, Meyer AS (2000) The problems of using one-dimensional methods to evaluate multifunctional food and biological antioxidants. J Sci Food Agric 80(13):1925–1941

20. Prior RL, Wu X, Schaich K (2005) Standardized methods for the determination of antioxidant capacity and phenolics in foods and dietary supplements. J Agric Food Chem 53(10):4290–4302

21. Moon J-K, Shibamoto T (2009) Antioxidant assays for plant and food components. J Agric Food Chem 57(5):1655–1666

22. Huang D, Ou B, Prior RL (2005) The chemistry behind antioxidant capacity assays. J Agric Food Chem 53(6):1841–1856

23. Prior RL, Hoang H, Gu L, Wu X, Bacchiocca M, Howard L, Hampsch-Woodill M, Huang D, Ou B, Jacob R (2003) Assays for hydrophilic and lipophilic antioxidant capacity (oxygen radical absorbance capacity (ORAC$_{FL}$)) of plasma and other biological and food samples. J Agric Food Chem 51(11):3273–3279

24. Wu X, Beecher GR, Holden JM, Haytowitz DB, Gebhardt SE, Prior RL (2004) Lipophilic and hydrophilic antioxidant capacities of common foods in the United States. J Agric Food Chem 52(12):4026–4037

25. Schaich KM, Tian X, Xie J (2015) Hurdles and pitfalls in measuring antioxidant efficacy: A critical evaluation of ABTS, DPPH, and ORAC assays. J Funct Foods 14:111–125

26. AOAC International (2016) Official methods of analysis, 20th edn., 2016 (online). AOAC International, Rockville, MD

27. Bors W, Michel C, Saran M (1984) Inhibition of the bleaching of the carotenoid crocin a rapid test for quantifying antioxidant activity. Biochim Biophys Acta, Lipids Lipid Metab 796(3):312–319

28. Tubaro F, Micossi E, Ursini F The antioxidant capacity of complex mixtures by kinetic analysis of crocin bleaching inhibition. J Am Oil Chem Soc 73(2):173–179

29. Ordoudi SA, Tsimidou MZ (2006) Crocin bleaching assay step by step: observations and suggestions for an alternative validated protocol. J Agric Food Chem 54(5):1663–1671

30. Lussignoli S, Fraccaroli M, Andrioli G, Brocco G, Bellavite P (1999) A microplate-based colorimetric assay of the total peroxyl radical trapping capability of human plasma. Anal Biochem 269(1):38–44

31. Re R, Pellegrini N, Proteggente A, Pannala A, Yang M, Rice-Evans C (1999) Antioxidant activity applying an improved ABTS radical cation decolorization assay. Free Radical Biol Med 26:1231–1237

32. Tian X, Schaich KM (2013) Effects of molecular structure on kinetics and dynamics of the Trolox equivalent antioxidant capacity assay with ABTS$^{+•}$. J Agric Food Chem 61(23):5511–5519

33. Blois MS (1958) Antioxidant determinations by the use of a stable free radical. Nature 181:1199–1200

34. Xie J, Schaich KM (2014) Re-evaluation of the 2,2-diphenyl-1-picrylhydrazyl free radical (DPPH) assay for antioxidant activity. J Agric Food Chem 62(19):4251–4260

35. Foti MC, Daquino C, Geraci C (2004) Electron-transfer reaction of cinnamic acids and their methyl esters with the DPPH• radical in alcoholic solutions. J Org Chem 69(7):2309–2314

36. Pryor WA (1986) Oxy-radicals and related species: their formation, lifetimes, and reactions. Annu Rev Physiol 48(1):657–667

37. Benzie IFF, Strain JJ (1996) The ferric reducing ability of plasma (FRAP) as a measure of "antioxidant power": the FRAP assay. Anal Biochem 239(1):70–76

26
chapter

Application of Enzymes in Food Analysis

Jose I. Reyes-De-Corcuera (✉)
Department of Food Science and Technology,
The University of Georgia,
Athens, GA 30602-7610, USA
e-mail: jireyes@uga.edu

Joseph R. Powers
School of Food Science,
Washington State University,
Pullman, WA 99164-6376, USA
e-mail: powersjr@wsu.edu

S. Nielsen (ed.), *Food Analysis*, Food Science Text Series,
DOI 10.1007/978-3-319-45776-5_26, © Springer International Publishing 2017

26.1 INTRODUCTION

Enzymes are protein catalysts capable of great specificity and reactivity under physiological conditions. Enzymatic analysis is the measurement of compounds with the aid of added enzymes or the measurement of endogenous enzyme activity to give an indication of the state of a biological system including foods. The fact that enzyme catalysis can take place under relatively mild conditions allows for measurement of relatively unstable compounds not amenable to some other techniques. In addition, the specificity of enzyme reactions can allow for measurement of components of complex mixtures without the time and expense of complicated chromatographic separation techniques.

There are several uses of enzyme analyses in food science and technology. In several instances, enzyme activity is a useful measure for adequate processing of a food product. The thermal stability of enzymes has been used extensively as a measure of heat treatment; for example, peroxidase activity is used as a measure of adequacy of blanching of vegetable products. Similarly, lactoperoxidase has been used to verify the adequacy of milk pasteurization. Enzyme activity assays are also used by the food technologist to assess potency of enzyme preparations used as processing aids such as pectinases and cellulases blends used for the clarification of fruit juices. In other instances, residual enzyme activity affects the flavor or color of products during storage. For example, if not properly blanched, lipoxygenase is responsible for the development of off-flavors in stored frozen vegetables. Polyphenol oxidase catalyzes the discoloration of wheat flour and noodles as well as of several fruit juices and pastes.

The food scientist can also use commercially available enzyme preparations to measure constituents of foods that are enzyme substrates. For example, glucose content can be determined in a complex food matrix containing other monosaccharides by using readily available enzymes. A corollary use of commercially available enzymes is to measure enzyme activity as a function of enzyme inhibitor content in a food. Organophosphate insecticides are potent inhibitors of the enzyme acetylcholinesterase, and hence the activity of this enzyme in the presence of a food extract is a measure of organophosphate insecticide concentration in the food. Also of interest is the measurement of enzyme activity associated with food quality. For example, catalase activity is markedly increased in milk from mastitic udders. Catalase activity also parallels the bacterial count in milk. Another use of enzyme assays to determine food quality is estimation of protein nutritive value by monitoring the activity of added proteases on food protein samples (see Chap. 24). Enzymes can be used to measure the appearance of degradation products such as trimethylamine in fish during storage. Enzymes are also used as preparative tools in food analysis. Examples include the use of amylases and proteases in fiber analysis (Chap. 19) and the enzymatic hydrolysis of thiamine phosphate esters in vitamin analysis.

Food scientists must realize that enzyme activity is greatly affected by its environment and that enzymes can degrade and lose their activity at high temperatures or under other conditions that deviate not too far from optimal. Therefore, processing enzymes must be stored and handled carefully, typically refrigerated or frozen. To successfully carry out enzyme analyses in foods, an understanding of certain basic principles of enzymology is necessary. After a brief overview of these principles, examples of the use of enzymatic analyses in food systems are examined.

26.2 PRINCIPLES

26.2.1 Enzyme Kinetics

26.2.1.1 Overview

Enzymes are biological catalysts that are proteins. A catalyst increases the rate (velocity) of a thermodynamically possible reaction. The enzyme does not modify the equilibrium constant of the reaction, and the enzyme catalyst is not consumed in the reaction. Because enzymes affect rates (velocities) of reactions, some knowledge of **enzyme kinetics** (study of rates) is needed for the food scientist to effectively use enzymes in analysis. To measure the rate of an enzyme-catalyzed reaction, typically one mixes the enzyme with the substrate under specified conditions (pH, temperature, ionic strength, etc.) and follows the reaction by measuring the amount of product that appears or by measuring the disappearance of substrate. Some of the fundamentals of chemical kinetics are summarized on Fig. 26.1A. However, because students of food analysis have typically already studied enzyme kinetics in biochemistry and food chemistry courses, many details of enzyme kinetics such as the derivation of reaction rate equations are not covered here, but can be found in the 4th ed. of this textbook or elsewhere [1].

The formation of the enzyme-substrate complex (Fig. 26.1, Eq. 26.4), called the **pre-steady-state period**, is very rapid (on the millisecond scale) and is not normally seen in the laboratory. The slope of the linear portion of the curve following the pre-steady-state period gives us the **initial velocity** (v_0). After the pre-steady-state period, a **steady-state period** exists in which the concentration of the enzyme-substrate complex is constant. A time course needs to be established experimentally by using a series of points or a continuous assay to establish the appropriate time frame for the measurement of the initial velocity.

FOR A SIMPLE ENZYME-CATALYZED REACTION FOR ONE SUBSTRATE "S" CONVERTED TO PRODUCT "P"

A) PRINCIPLES OF CHEMICAL KINETICS:

For a simple chemical reaction: $A \xrightarrow{k} B$, the rate of reaction is given by:

$$v = -\frac{d[A]}{dt} = \frac{d[B]}{dt} \quad [26.1]$$

By the law of mass action the rate equation is written as:

$$-\frac{d[A]}{dt} = k[A]^n \quad [26.2]$$

Where k is the rate constant of the reaction and n is the order of the reaction. Commonly encountered orders of reaction relevant to this chapter are order zeroth and first that integrated can be written respectively as:

$$[A_0] - [A] = kt \quad and \quad ln\frac{[A]}{[A_0]} = -kt \quad [26.3]$$

$$E + S \underset{k_{-1}}{\overset{k_1}{\rightleftharpoons}} ES \quad [26.4]$$

$$ES \xrightarrow{k_2} E + P \quad [26.5]$$

$$v = \frac{d[P]}{dt} = -\frac{d[S]}{dt} \quad [26.6]$$

B) MICHAELIS-MENTEN PLOT

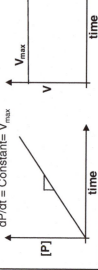

$V_{max} = k_2[E]$

$K_M = \dfrac{k_{-1} + k_2}{k1}$

$$V_0 = \frac{V_{max}[S]}{K_M + [S]} \quad [26.7]$$

MICHAELIS-MENTEN EQUATION

C) DETERMINATION OF MICHAELIS MENTEN PARAMETERS:

- Determine the initial rate of reaction at different [S]
- Alternatively monitor [S] or [P] vs. time (progress curve)
- Calculate the rate of reaction by taking the first derivative of [S] or [P] with respect to time
- Linearize using Lineweaver-Burk approach:

$$\frac{1}{v_0} = \frac{K_M}{V_{max}}\frac{1}{[S]} + \frac{1}{V_{max}} \quad [26.8]$$

Slope = K_M/V_{max}

Intercept = $1/V_{max}$

D) ENZYME ACTIVITY WHEN [S] IS VERY SMALL:

- [S] << K_M (Practically, K_M > 20[S])

$$v = \frac{V_{maz}}{K_M}[S] \quad [26.9]$$

which is **first order** with respect to [S]
- Use of initial rates which are difficult to obtain experimentally because the rate of reaction is not constant

dP/dt = $V_{max}/K_M[S]$

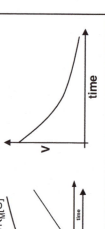

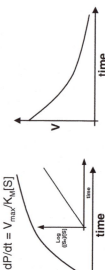

E) ENZYME ACTIVITY IN EXCESS OF SUBSTRATE:

- [S] >> K_M (Practically, [S] > 20 K_M)
- Practically all of the enzyme is bound to the substrate forming the complex ES.

$$v = V_{max} = k_2[E_0] \quad [26.10]$$

which if the enzyme is not losing its activity is a constant and is a the best measurement of **enzyme activity** since it is a **zeroth order** with respect to [S] and **first order** with respect to [E_0]

dP/dt = Constant= V_{max}

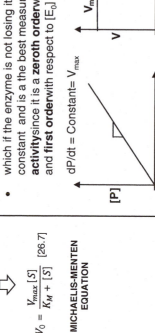

26.1 figure

Principles of chemical and enzyme kinetics

The rate of an enzyme-catalyzed reaction depends on the concentration of the enzyme and on the substrate concentration. With a fixed enzyme concentration, increasing substrate concentration results in an increased velocity (see Fig. 26.1B). As substrate concentration increases further, the increase in velocity slows until, with a very large concentration of substrate, no further increase in velocity is noted. The velocity of the reaction at this very large substrate concentration is the **maximum velocity** (V_{max}) of the reaction under the conditions of that particular assay. The substrate concentration at which one half V_{max} is observed is equal to the **Michaelis constant** (K_M). This constant is an indication of the relative binding affinity of the enzyme for a particular substrate. The lower the K_M, the greater the affinity of the enzyme for the substrate. Both K_M and V_{max} are affected by environmental conditions such as pH, temperature, ionic strength, and the nature of the solvent, i.e., organic or aqueous (although most enzyme reactions occur in aqueous environments, some reaction occur in organic media such as in lipids) (Table 26.1).

26.2.1.2 *Michaelis-Menten Equation*
Michaelis-Menten equation (Fig. 26.1, Eq. 26.7) is that of a hyperbola that relates the reaction velocity to substrate concentration for an enzyme-mediated reaction (Fig. 26.1B).

26.2.1.3 *Apparent Order of Simple Enzyme Reactions*
The velocity of an enzyme-catalyzed reaction increases as substrate concentration increases (see Fig. 26.1B). An apparent **first-order reaction** with respect to substrate concentration (Fig. 26.1, Eq. 26.9) is obeyed in the region of the curve where substrate concentration is small ($[S] \ll K_m$, see Fig. 26.1D), assuming that the active enzyme concentration $[E]$ is constant. This means that the velocity of the reaction is directly proportional to the substrate concentration in this region. When the substrate concentration is further increased, the velocity of the reaction no longer increases linearly, and the reaction is **mixed order**. This is seen in the curvilinear portion of Fig. 26.1B. If substrate concentration is increased further, the velocity asymptotically approaches the maximum velocity (V_{max}). In this linear, nearly zero slope portion of the plot, the velocity is independent of substrate concentration. However, note that at large substrate concentrations ($[S] \gg K_M$, see Fig. 26.1E), the velocity is directly proportional to enzyme concentration (Fig. 26.1, Eq. 26.10). Thus, in this portion of the curve, where $[S] \gg K_M$, the rate of the reaction is **zero order** with respect to substrate concentration (is independent of substrate concentration) but first order with respect to enzyme concentration.

For practical purposes, if we are interested in **measuring the amount of active enzyme** in a reaction mix-

Factors that affect enzyme reaction rate

Factors	Michaelis constant (K_M)	Maximum velocity (V_{max})
Source of enzyme	Yes	Yes
Substrate	Yes	Yes
Enzyme concentration	No	Yes
Substrate concentration	No	No
Temperature	Yes	Yes
pH	Yes	Yes
Inhibitors	Yes	Yes (some)

ture, we should, if possible, work at substrate concentrations so that the observed velocity approximates V_{max}. At these substrate concentrations, enzyme is directly rate limiting to the observed velocity. In practice, and unless not possible, V_{max} is used most of the time as a measure of **enzyme activity** and is expressed as a rate of change in concentration with respect to time. Enzyme activity is often reported in arbitrary units such as change in absorbance per weight of sample per unit time.

Conversely, if we are interested in **measuring substrate concentration** by measuring initial velocity, we must be at substrate concentrations less than K_M in order to have a rate directly proportional to substrate concentration. However, knowledge of K_M and V_{max} is necessary, which in practice would require at least the experimental determination of V_{max} immediately prior to using the enzyme to account for any decrease of enzyme activity over time.

26.2.1.4 *Determination of Michaelis Constant (K_M) and V_{max}*
To properly design an experiment in which velocity is zero order with respect to substrate and first order with respect to enzyme concentration, or conversely an experiment in which we would like to measure rates that are directly proportional to substrate concentration, we must know the K_M. The most popular method for determining K_M is the use of a **Lineweaver-Burk plot** (Fig. 26.1C) drawn by using the reciprocal of the Michaelis-Menten equation (Fig. 26.1, Eq. 26.8).

This equation is that of a straight line where $1/v0$ is the ordinate (vertical axis), $1/[S]$ is the abscissa (horizontal axis), K_M/V_{max} is the slope, and $1/V_{max}$ is the ordinate at the origin. Linear regression is used to determine K_M and V_{max}. Enzyme mechanisms are often more complex than what is presented here. For example, oxidases often require two substrates that bind to the enzyme simultaneously or in sequence. Detailed description of such mechanisms and rate equations is beyond the scope of this chapter. However, in practice, simple Michaelis-Menten kinetics is frequently assumed.

26.2.2 Factors That Affect Enzyme Reaction Rate

The velocity of an enzyme-catalyzed reaction is affected by a number of factors, including enzyme and substrate concentrations, temperature, pH, ionic strength, and the presence of inhibitors or activators.

26.2.2.1 *Effect of Enzyme Concentration*

In excess of substrate, the velocity of an enzyme-catalyzed reaction is directly proportional to (increases linearly with) the enzyme concentration in the reaction mixture (Fig. 26.1, Eq. 26.10). Therefore, if possible, determination of enzyme activity should be done at concentrations of substrate much greater than K_M. Under these conditions, a zero-order dependence of the rate with respect to substrate concentration and a first-order relationship between rate and enzyme concentration exist. It is critical that the substrate concentration is saturating during the entire period the reaction mixture is sampled and the amount measured of product formed or substrate disappearing is linear over the period during which the reaction is sampled. The activity of the enzyme is obtained as the slope of the linear part of the line of a plot of product or substrate concentration vs. time as depicted by the solid lines in Fig. 26.2. However, when the concentration of enzyme is very large, during the course of the reaction, the substrate concentration decreases and is no longer in excess, resulting in deviations from linearity as shown in the dashed lined in Fig. 26.2. If a large number of samples are to be assayed, a single aliquot is often taken and assayed at a single reaction time. This can be risky and will give good results only if the time at which the sample is taken falls on the linear portion of a plot of substrate or product concentration versus time of reaction (see Fig. 26.2). Normally, one designs an experiment in which enzyme concentration is estimated such that no more than 5–10 % of the substrate has been converted to product within the time used for measuring the initial rate.

Sometimes it is not possible to carry out enzyme assays at $[S] \gg K_m$. The substrate may be very expensive or relatively insoluble or K_M may be large (i.e., $K_M > 100$ mM). In these cases, the reaction rate is determined by measuring initial rates of the reactions, in which the change in substrate or product concentration is determined at times as close as possible to time zero. This is shown in Fig. 26.2 by the solid lines drawn tangent to the slopes of the initial parts of the curves. The slope of the tangent line gives an approximation of the initial rate. Enzyme concentration can also be estimated at substrate concentrations much less than K_M. When substrate concentration is much less than K_M, the substrate term in the denominator of the Michaelis-Menten equation can be ignored and $v = (V_{max}[S])/K_m$ which is the equation for a first-order

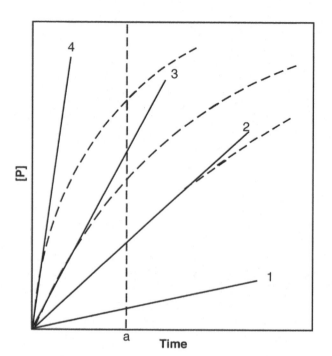

 figure Effect of enzyme concentration on time course of an enzyme-catalyzed reaction. The *dashed lines* are experimentally determined data with enzyme concentration increasing from 1 to 4. The *solid lines* are tangents drawn from the initial slopes of the experimental data. If a single time point, *a*, is used for data collection, a large difference between actual data collected and that predicted from initial rates is seen

reaction with respect to substrate concentration (see Fig. 26.1D). Under these conditions, a plot of product concentration vs. time gives a nonlinear plot (Fig. 26.1D, left plot). A plot of log ($[S_o]/[S]$) vs. time gives a straight-line relationship (Fig. 26.1D, left plot inset). The slope of the line of the log plot is directly related to the enzyme concentration. When the slopes of a series of these log plots are further plotted as a function of enzyme concentration, a straight-line relationship should result. If possible, the reaction should be followed continuously or aliquots removed at frequent time intervals and the reaction allowed to proceed to greater than 10 % of the total reaction.

26.2.2.2 *Effect of Substrate Concentration*

The substrate concentration velocity relationship for an enzyme-catalyzed reaction in which enzyme concentration is constant is shown in Fig. 26.1B. As noted before, the rate of the reaction is first order with respect to substrate concentration when $[S] \ll K_m$ (Fig. 26.1D). At $[S] \gg K_m$, the reaction is zero order with respect to substrate concentration and first order with respect to $[E]$ (Fig. 26.1E). At substrate concentrations between the first-order and zero-order regions, the enzyme-catalyzed reaction is **mixed order** with respect to substrate concentration. When initial rates are obtained, a linear relationship between v_0 and E_o should be seen. However, in practice in most common laboratories, it is very difficult to accurately determine initial rates because it is practically impossible to instantaneously mix substrates and enzyme and instrumentally determine product or substrate concentrations. Therefore, experimental determinations of initial rates typically underestimate the actual value.

26.2.2.3 Environmental Effects

The subsections below describe how various environmental factors affect K_M and V_{max}. K_M is affected by temperature, pH, and the presence of inhibitors, and V_{max} is affected by enzyme concentration, temperature, pH, and the presence of some inhibitors. Although beyond the scope of this chapter, food analysts must also be aware that like temperature, pressure also affects K_M and V_{max}. Furthermore the stability of several enzymes has increased at high pressures up to about 400 MPa [2].

26.2.2.3.1 Effect of Temperature on Enzyme Activity

Temperature can affect observed enzyme activity in several ways. Most obvious is that temperature can affect the stability of enzyme and also the rate of the enzyme-catalyzed reaction. Other factors in enzyme-catalyzed reactions that may be considered include the effect of temperature on the solubility of gases that are either products or substrates of the observed reaction and the effect of temperature on pH of the system. A good example of the latter is the common buffering species Tris (tris [hydroxymethyl] aminomethane), for which the pK_a changes 0.031 per 1 °C change.

Temperature affects both the stability and the activity of the enzyme, as shown in Fig. 26.3. At relatively low temperatures, the enzyme is stable. However, at higher temperatures, denaturation dominates, and a markedly reduced enzyme activity represented by the negative slope portion of line 2 is observed:

$$E_{active} \xrightarrow{k_{inact}} E_{inactive} \qquad (26.11)$$

Line 1 of Fig. 26.3 shows the effect of temperature on the velocity of the enzyme-catalyzed reaction. The velocity is expected to increase exponentially as the temperature is increased. As shown by line 1, the velocity approximately doubles for every 10 °C rise in temperature. The net effect of increasing temperature on the rate of conversion of substrate to product (line 1) and on the rate of the denaturation of enzyme (line 3) is line 2 of Fig. 26.3. The temperature optimum of the enzyme is at the maximum point of line 2. The tem-

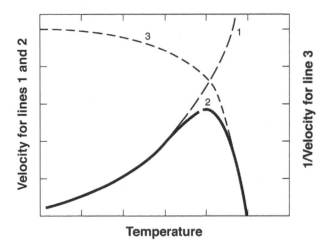

Temperature

26.3 figure Effect of temperature on velocity of an enzyme-catalyzed reaction. Temperature effect on substrate to product conversion is shown by line 1. Line 3 shows effect of temperature on rate of enzyme denaturation (right-hand y-axis is for line 3). The net effect of temperature on the observed velocity is given by line 2 and the temperature optimum is at the maximum of line 2

perature optimum is not unique for any given enzyme. Each enzyme has multiple optima depending on the type of substrate, pH, salt concentration, substrate concentration, and time of reaction. For this reason, investigators should fully describe a system in which the effects of temperature on observed enzyme activity are reported. Each optimum temperature occurs under conditions in which the enzyme is denaturing. Therefore even under excess of substrate, plots of substrate or product concentration vs. time are often not linear since the enzyme concentration is constantly changing. For this reason, for analytical purposes, enzyme activity is assayed at temperatures colder than the optimal for activity.

The data of Fig. 26.3, line 1, can be plotted according to the **Arrhenius equation**:

$$k = Ae^{-E_a/RT} \qquad (26.12)$$

which can be written as:

$$\log k = \log A - \frac{E_a}{2.3RT} \qquad (26.13)$$

where:

k = a specific rate constant at some temperature, T (K)

E_a = activation energy, the minimum amount of energy a reactant molecule must have to be converted to product

R = gas constant

A = a frequency factor (preexponential factor)

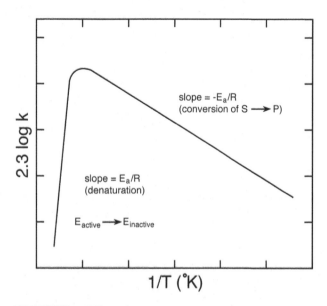

Effect of temperature on rate constant of an enzyme-catalyzed reaction. The data are plotted 2.3 log k versus $1/T$ (°K) according to the Arrhenius equation, $k = Ae-E_aRT$

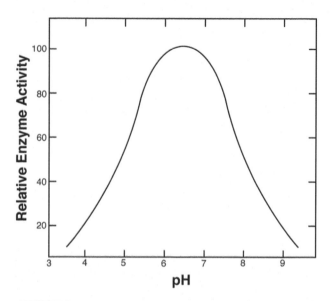

Typical velocity-pH curve for an enzyme-catalyzed reaction. The maximum on the curve is the optimum for the system and can vary with temperature, specific substrate, and enzyme source

The rate constant of inactivation k_{inact} also follows Arrhenius equation. The positive slope on the left side (high temperature) of the **Arrhenius plot** (Fig. 26.4) gives a measure of the **activation energy** (E_a) for the denaturation of the enzyme. Note that a small change in temperature has a very large effect on the rate of denaturation. The slope on the right side of Fig. 26.4 gives a measure of the E_a for the transformation of substrate to product catalyzed by the enzyme. If the experiment is carried out under conditions in which V_{max} is measured ($[S] >> K_M$), then the activation energy observed will be for the catalytic step of the reaction (k_2).

26.2.2.3.2 Effect of pH on Enzyme Activity

The observed rate of an enzyme-catalyzed reaction is greatly affected by the pH of the medium. Enzymes have pH optima and commonly have bell-shaped curves for activity versus pH (Fig. 26.5). This pH effect is a manifestation of the effects of pH on enzyme stability and on rate of substrate to product conversion and may also be due to changes in ionization of substrate.

The rate of substrate to product conversion is affected by pH because pH may affect binding of substrate to enzyme and the ionization of catalytic groups such as carboxyl or amino groups that are part of the enzyme's active site. The stability of the tertiary or quaternary structure of enzymes is also pH dependent and affects the velocity of the enzyme reaction, especially at extreme acidic or alkaline pHs. The pH for maximum stability of an enzyme does not necessarily coincide with the pH for maximum activity of that same enzyme. For example, the proteolytic enzymes

trypsin and chymotrypsin are stable at pH 3, while they have maximum activity at pH 7–8.

To establish the pH optimum for an enzyme reaction, the reaction mixture is buffered at different pHs and the activity of the enzyme is determined. To determine pH enzyme stability relationships, aliquots of the enzyme are buffered at different pH values and held for a specified period of time (e.g., 1 h). The pH of the aliquots is then adjusted to the pH optimum and each aliquot is assayed. The effect of pH on enzyme stability is thus obtained. These studies are helpful in establishing conditions for handling the enzyme and also may be useful in establishing methods for controlling enzyme activity in a food system. Note that pH stability and the pH optimum for the enzyme activity are not true constants. That is to say, these may vary with particular source of enzyme, the specific substrate used, the temperature of the experiment, or even the buffering species used in the experiment. In the use of enzymes for analysis, it is not necessary that the reaction be carried out at the pH optimum for activity, or even at a pH at which the enzyme is most stable (although it should be stable for the duration of the experiment), but it is critical to maintain a fixed pH during the reaction (i.e., use buffer) and to use the same pH in all studies to be compared.

26.2.2.4 Activators and Inhibitors

26.2.2.4.1 Activators

Some enzymes contain, in addition to a protein portion, small molecules that are activators of the enzyme.

Some enzymes show an absolute requirement for a particular inorganic ion for activity, while others show increased activity when small molecules are included in the reaction medium. These small molecules can play a role in maintaining the conformation of the protein, or they may form an essential component of the active site, or they may form part of the substrate of the enzyme.

In some cases, the activator forms a nearly irreversible association with the enzyme. These nonprotein portions of the enzyme are called **prosthetic groups**. The amount of enzyme activator complex formed is equal to the amount of activator present in the mixture. In these cases, activator concentration can be estimated up to concentrations equal to total enzyme concentration by simply measuring enzyme activity.

In most cases, dissociation constants for an enzyme activator complex are within the range of enzyme concentration. Dissociable nonprotein parts of enzymes are categorized as **coenzymes**. When this type of activator is added to an enzyme, a curvilinear relationship similar to a Michaelis-Menten plot results, making difficult the determination of an unknown amount of activator. A reciprocal plot analogous to a Lineweaver-Burk plot can be constructed using standards and unknown activator concentrations estimated from such a plot.

One food-related enzyme reaction involving an activator is measuring pyridoxal phosphate, a form of vitamin B_6. This reaction measures the reactivation of coenzyme-free yeast aminotransferase by coupling the transamination reaction with malate dehydrogenase. This is possible when malate dehydrogenase, NADH, alpha-ketoglutarate, aspartate, and the aminotransferase are in excess, and the pyridoxal 5-phosphate added is rate limiting:

$$\text{alpha-ketoglutarate} + \text{aspartate} \xrightleftharpoons{\text{aminotransferase}} \text{glutamate} + \text{oxalacetate} \quad (26.14)$$

$$\text{oxalacetate} + \text{NADH} + \text{H}^+ \xrightleftharpoons{\text{malate dehydrogenase}} \text{malate} + \text{NAD}^+ \quad (26.15)$$

Another example of an essential activator is the pyridine coenzyme NAD^+. NAD^+ is essential for the oxidation of ethanol to acetaldehyde by alcohol dehydrogenase:

$$\text{ethanol} + \text{NAD}^+ \xrightleftharpoons{\text{alcohol dehydrogenase}} \text{acetaldehyde} + \text{NADH} + \text{H}^+$$

$$(26.16)$$

In the reaction, NAD^+ is reduced to NADH and can be considered a second substrate. Another example of an activator of an enzyme is the chloride ion with α-amylase. In this case, α-amylase has some activity in the absence of chloride. With saturating levels of chloride, the α-amylase activity increases about fourfold. Other anions, including F^-, Br^-, and I^-, also activate α-amylase. These anions must not be in the reaction mixture if α-amylase stimulation is to be used as a method of determining chloride concentration.

26.2.2.4.2 Inhibitors

An enzyme inhibitor is a compound that when present in an enzyme-catalyzed reaction medium decreases the enzyme activity. Enzyme inhibitors can be categorized as **irreversible** or **reversible** inhibitors. Enzyme inhibitors include inorganic ions, such as Pb^{2+} or Hg^{2+}, which can react with sulfhydryl groups on enzymes to inactivate the enzyme, compounds that resemble substrate, and naturally occurring proteins that specifically bind to enzymes (such as protease inhibitors found in legumes):

1. **Irreversible Inhibitors.** When the dissociation constant of the inhibitor enzyme complex is very small, the decrease in enzyme activity observed will be directly proportional to the inhibitor added. The speed at which the irreversible combination of enzyme and inhibitor reacts may be slow, and the effect of time on the reduction of enzyme activity by the addition of inhibitor must be determined to ensure complete enzyme-inhibitor reaction. For example, the amylase inhibitor found in many legumes must be preincubated under specified conditions with amylase prior to measurement of residual activity to accurately estimate inhibitor content [3]. Irreversible inhibitors decrease V_{max} as the amount of total active enzyme is reduced.

2. **Reversible Inhibitors.** Most inhibitors exhibit a dissociation constant such that both enzyme and inhibitor are found free in the reaction mixture. Several types of reversible inhibitors are known: **competitive, noncompetitive**, and **uncompetitive**.

Competitive inhibitors usually resemble the substrate structurally and compete with substrate for binding to the active site of the enzyme, and only one molecule of substrate or inhibitor can be bound to the enzyme at one time. An inhibitor can be characterized as competitive by adding a fixed amount of inhibitor to reactions at various substrate concentrations and by plotting the resulting data by the Lineweaver-Burk method and noting the effect of inhibitor relative to that of control reactions in which no inhibitor is added. If the inhibitor is competitive, the slope and x-intercept of the plot with inhibitor are altered, while the y-intercept ($1/V_{max}$) is unaltered. It can be shown that the ratio of the **uninhibited initial velocity** (v_o) to the **inhibited initial velocity** (v_i) gives:

$$\frac{v_0}{v_i} = \frac{[I]K_M}{K_i(K_M + [S])}$$ (26.17)

where:

K_i = dissociation constant of the enzyme-inhibitor complex

$[I]$ = concentration of competitive inhibitor

Thus, a plot of v_0/v_i vs. inhibitor concentration will give a straight-line relationship. From this plot the concentration of a competitive inhibitor can be found [2]. Soybean trypsin inhibitor, a well-known antinutritional factor, is a polypeptide that acts as reversible competitive inhibitor. Trypsin is a protease enzyme synthesized by humans and other animals needed to digest proteins. Soybean trypsin inhibitor is typically inactivated with heat during processing of soybean products. The inhibitor activity is determined based on the ratio of trypsin activity in the presence of the inhibitor compared to trypsin activity in the absence of inhibitor. Trypsin activity can be determined by following the rate of the following reaction by measuring the change in absorbance at 253 nm:

N-benzoyl-L-arginine ethyl ester +

H$_2$O $\xrightarrow{\text{Trypsin}}$ N-benzoyl-L-arginine + ethanol

(26.18)

A **noncompetitive inhibitor** binds to enzyme independent of substrate and is bound outside the active site of the enzyme. A noncompetitive inhibitor can be identified by its effect on the rate of enzyme-catalyzed reactions at various substrate concentrations and the data plotted by the Lineweaver-Burk method. A noncompetitive inhibitor will affect the slope and the y-intercept as compared to the uninhibited system, while the x-intercept, $1/K_M$, is unaltered. Analogous to competitive inhibitors, a standard curve of v_0/v_i vs. inhibitor concentration may be prepared and used to determine the concentration of a noncompetitive inhibitor [2].

Uncompetitive inhibitors bind only to the enzyme-substrate complex. Uncompetitive inhibition is noted by adding a fixed amount of inhibitor to reactions at several substrate concentrations and plotting the data by the Lineweaver-Burk method. An uncompetitive inhibitor will affect both the x- and y-intercepts of the Lineweaver-Burk plot as compared to the uninhibited system, while maintaining an equal slope to the uninhibited system (i.e., a parallel line will result). A plot of v_0/v_i vs. inhibitor concentration can be prepared to use as a standard curve for the determination of the concentration of an uncompetitive inhibitor [2].

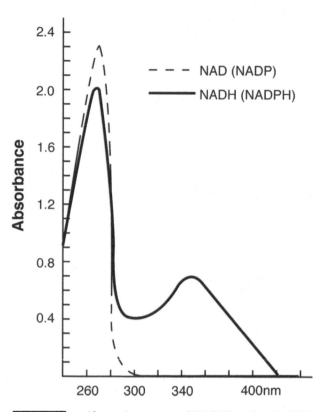

26.6 figure Absorption curves of NAD(P) and NAD(P)H; λ = wavelength. Many enzymatic analysis methods are based on the measurement of an increase or decrease in absorbance at 340 nm due to NAD(H) or NADP(H)

26.2.3 Methods of Measurement

26.2.3.1 Overview

For practical enzyme analysis, it is necessary to be familiar with the methods of measurement of the reaction. Any physical or chemical property of the system that relates to substrate or product concentration can be used to follow an enzyme reaction. A wide variety of methods are available to follow enzyme reactions, including **absorbance spectrometry, fluorimetry, manometric methods, titration, isotope measurement, chromatography, mass spectrometry,** and **viscosity.** A good example of the use of spectrophotometry as a method for following enzyme reactions is the use of the spectra of the pyridine coenzyme NAD(H) and NADP(H), in which there is a marked change in absorbance at 340 nm upon oxidation-reduction (Fig. 26.6). Many methods depend on the increase or decrease in absorbance at 340 nm when these coenzymes are products or substrates in a coupled reaction.

An example of using several methods to measure the activity of an enzyme is in the assay of α-amylase activity [4]. α-Amylase cleaves starch at α-1,4 linkages in starch and is an endoenzyme. An endoenzyme cleaves a polymer substrate at internal linkages. This reaction can be followed by a number of methods, including reduction in viscosity, increase in reducing groups upon hydrolysis, reduction in color of the starch-iodine complex, and polarimetry. However, it is difficult to differentiate the activity of α-amylase from β-amylase using a single assay. β-Amylase cleaves maltose from the nonreducing end of starch. While a marked decrease in viscosity of starch or reduction in iodine color would be expected to occur due to α-amylase activity, β-amylase can also cause changes in viscosity and iodine color if in high concentration. To establish whether α-amylase or β-amylase is being measured, the analyst must determine the change in number of reducing groups as a basis of comparison. Because α-amylase is an endoenzyme, hydrolysis of a few bonds near the center of the polymeric substrate will cause a marked decrease in viscosity, while hydrolysis of an equal number of bonds by the exoenzyme, β-amylase, will have little effect on viscosity.

In developing an enzyme assay, it is wise to first write out a complete, balanced equation for the particular enzyme-catalyzed reaction. Inspection of the products and substrates for chemical and physical properties that are readily measurable with available equipment will often result in an obvious choice of method for following the reaction in the laboratory.

If one has options in methodology, one should select the method that is able to monitor the reaction continuously, is most sensitive, and is specific for the enzyme-catalyzed reaction. For multiple routine analyses, the use of controlled temperature 96-well microplate readers with automatic injection and shaking with absorbance and fluorescence detectors has become widespread. These instruments are particularly useful to characterize enzyme kinetics using multiple enzyme and/or multiple substrate dilutions. These instruments are also commonly used in molecular biology assays such as enzyme-linked immunosorbent assays (Chap. 27, Sect. 27.3.2) and polymerase chain reaction (Chap. 33, Sect. 33.6.2) for the detection of bacteria, allergens, and genetically modified organisms.

26.2.3.2 Coupled Reactions

Enzymes can be used in assays via coupled reactions. **Coupled reactions** involve using two or more enzyme reactions so that a substrate or product concentration can be readily followed. In using a coupled reaction, there is an **indicator reaction** and a **measuring reaction**. For example:

$$S1 \xrightarrow{E1} P1 \qquad (26.19)$$

measuring reaction

$$P1 \xrightarrow{E2} P2 \qquad (26.20)$$

indicating reaction

The role of the indicating enzyme (E2) is to produce P2, which is readily measurable and, hence, is an indication of the amount of P1 produced by E1. Alternatively the same sequence can be used in measuring S1, the substrate for E1. When a coupled reaction is used to measure the activity of an enzyme (e.g., E1 above), it is critical that the indicating reaction be fast, hence not rate limiting: the measuring reaction must always be the slowest, hence rate determining. Consequently, E2 activity should be much greater than E1 activity for an effective assay. Coupled enzyme reactions can have problems with respect to pH of the system if the pH optima of the coupled enzymes are quite different. It may be necessary to allow the first reaction (e.g., the measuring reaction catalyzed by E1 above, Eq. 26.19) to proceed for a time and then arrest the reaction by heating to denature E1. The pH is adjusted, the indicating enzyme (E2, Eq. 26.20) added, and the reaction completed. If an endpoint method is used with a coupled system, the requirements for pH compatibility are not as stringent as for a rate assay because an extended time period can be used to allow the reaction sequence to go to completion.

26.3 APPLICATIONS

As described previously, certain information is needed prior to using enzyme assays analytically. In general, knowledge of K_M, time course of the reaction, the enzyme's specificity for substrate, the pH optimum and pH stability of the enzyme, and effects of temperature on the reaction and stability of the enzyme are desirable. Many times this information is available from the literature. However, a few preliminary experiments may be necessary, especially in the case of experiments in which velocities are measured. A time course to establish linearity of product formation or substrate consumption in the reaction is a necessity. An experiment to show linearity of velocity of the enzyme reaction to enzyme concentration is recommended (see Fig. 26.2).

26.3.1 Substrate Assays

The following is not an extensive compendium of methods for the measurement of food components by enzymatic analysis. Instead, it is meant to be representative of the types of analyses possible. Table 26.2 summarizes substrate assays covered in this section. The reader can consult handbooks published by the manufacturers of enzyme kits (e.g., Megazyme; https://secure.megazyme.com/Dynamic.aspx?control=CSCatalogue&categoryName=AssayKits), the review article by Whitaker [5], the handbook from Boehringer-Manheim [6], and the series by Bergmeyer [7] for a more comprehensive guide to enzyme methods applicable to foods.

26.2
table   Summary of substrate analysis examples

Substrate of interest	Enzyme(s)	Compound measured	Measured property	Uses/comments
Glutamate	Glutamate dehydrogenase (GluDH) and lactate dehydrogenase (LDH)	NADH	Endpoint decrease in A_{340}	Relevant to products containing monosodium glutamate (MSG). Coupled reaction + regeneration system: all glutamate is converted into α-ketoglutarate, and all NADH is consumed with pyruvate + LDH. Enzymes are heat inactivated, and then NADH + GluDH are added
Glutamate	Glutamate dehydrogenase (GluDH) and diaphorase	Formazan	Endpoint increase in A_{492}	Coupled reactions. NADH produced by GluDH reacts with iodonitrotetrazolium chloride
Sulfite	Sulfite oxidase (SO) and NADH peroxidase (POD)	NADH	Endpoint decrease in A_{340}	Used to determine sulfites in wines. Coupled reaction
Glucose	Glucose oxidase (GOx) and POD	Oxidized o-diansidine	Endpoint increase in A_{420}	Used to monitor glucose concentration in yeast fermentations. Coupled reaction
	Hexokinase (HK) and glucose-6-phosphate dehydrogenase (G6PDH)	NADPH	Endpoint increase in A_{340}	Used to determine malic acid in apple juice as alternative to titratable acidity. Coupled reaction
D-Malic acid	D-Malate dehydrogenase	NADH	Endpoint decrease in A340	

26.3.1.1 Sample Preparation

Because of the specificity of enzymes, sample preparation prior to enzyme analysis is often minimal and may involve only extraction and removal of solids by filtration or centrifugation. Regardless, due to the wide variety of foods that might be encountered by the analyst using enzyme assays, a check should be made of the extraction and enzyme reaction steps by standard addition of known amounts of analyte to the food and extract, then measuring recovery of that standard. If the standard additions are fully recovered, this is a positive indication that the extraction is complete, that sample does not contain interfering substances that require removal prior to the enzymatic analysis, and that the reagents are good. In some cases, interfering substances are present but can be readily removed by precipitation or adsorption. For example, polyvinylpolypyrrolidone (PVPP) powder can be used to decolorize juices or red wines. With the advent of small syringe minicolumns (e.g., C18, silica, and ion exchange cartridges), it is also relatively easy and fast to attain group separations to remove interfering substances from a sample extract.

26.3.1.2 Total Change/Endpoint Methods

While substrate concentrations can be determined in rate assays when the reaction is first order with respect to substrate concentration ($[S] \ll K_m$), substrate concentration can also be determined by the total change or endpoint method. In this method, the enzyme-catalyzed reaction is allowed to go to completion so that concentration of product, which is measured, is directly related to substrate. An example of such a system is the measurement of glucose using glucose oxidase and peroxidase, described in Sect. 26.3.1.3.2.

In some cases, an equilibrium is established in an endpoint method in which there is a significant amount of substrate remaining in equilibrium with product. In these cases, the equilibrium can be altered. For example, in cases in which a proton-yielding reaction is used, alkaline conditions (increase in pH) can be used. Trapping agents can also be used, in which product is effectively removed from the reaction, and by mass action the reaction goes to completion. Examples include the trapping of ketones and aldehydes by hydrazine. In this way, the product is continually removed and the reaction is pulled to completion. The equilibrium also can be displaced by increasing cofactor or coenzyme concentration.

Another means of driving a reaction to completion is a regenerating system [7]. For example, in the measurement of glutamate, with the aid of glutamate dehydrogenase, the following can be done:

$$\text{glutamate} + \text{NAD}^+ + \text{H}_2\text{O}$$
$$\xrightleftharpoons{\text{glutamate dehydrogenase}} 2 \cdot \text{oxoglutarate} + \text{NADH} + \text{NH}_4^+$$
$$(26.21)$$

$$\text{pyruvate} + \text{NADH} + \text{H}^+$$
$$\xrightleftharpoons{\text{lactate dehydrogenase}} \text{NAD}^+ + \text{lactate}$$
$$(26.22)$$

In this system, NADH is recycled to NAD^+ via lactate dehydrogenase until all the glutamate to be measured is consumed. The reaction is stopped by heating to denature the enzymes present, a second aliquot of glutamate dehydrogenase and NADH is added, and the α-ketoglutarate (equivalent to the original glutamate) is measured via decrease in absorbance at 340 nm. An example in which the same equilibrium is displaced in the measurement of glutamate is as follows:

$$\text{glutamate} + \text{NAD}^+ + \text{H}_2\text{O}$$
$$\xrightleftharpoons{\text{glutamate dehydrogenase}} \alpha\text{-ketoglutarate} + \text{NADH} + \text{NH}_4^+$$
$$(26.23)$$

$$\text{NADH} + \text{INT} \xrightarrow{\text{diaphorase}} \text{NAD}^+ + \text{formazan}$$
$$(26.24)$$

Iodonitrotetrazolium chloride (INT) is a trapping reagent for the NADH product of the glutamate dehydrogenase catalyzed reaction. The formazan formed is measurable colorimetrically at 492 nm.

26.3.1.3 Specific Applications

26.3.1.3.1 Measurement of Sulfite

Sulfite is a food additive that can be measured by several techniques, including titration, distillation followed by titration, liquid chromatography, and an enzymatic method (Chap. 33, Sect. 33.8.2). In the enzymatic method, sulfite is specifically oxidized to sulfate by the commercially available enzyme sulfite oxidase (SO):

$$\text{SO}_3^{2-} + \text{O}_2 + \text{H}_2\text{O} \xrightarrow{\text{SO}} \text{SO}_4^{2-} + \text{H}_2\text{O}_2 \qquad (26.25)$$

The H_2O_2 product can be measured by several methods including the use of the enzyme NADH peroxidase:

$$\text{H}_2\text{O}_2 + \text{NADH} + \text{H}^+ \xrightarrow{\text{NADH-peroxidase}} 2\text{H}_2\text{O} + \text{NAD}^+$$
$$(26.26)$$

The amount of sulfite in the system is equal to the NADH oxidized, which is determined by decrease in absorbance at 340 nm. Ascorbic acid can interfere with the assay but can be removed by using ascorbic acid oxidase [8].

26.3.1.3.2 Colorimetric Determination of Glucose

The combination of the enzymes glucose oxidase and peroxidase can be used to specifically measure glucose in a food system [9] (see also Chap. 19, Sect. 19.4.2.3.3). Glucose is preferentially oxidized by glucose oxidase to produce gluconolactone and hydrogen peroxide. The hydrogen peroxide plus o-dianisidine in the presence of peroxidase produces a yellow color that absorbs at 420 nm (Eqs. 26.17 and 26.18). This assay is normally carried out as an endpoint assay, and there is stoichiometry between the color formed and the amount of glucose in the extract, which is established with a standard curve. Because glucose oxidase is quite specific for glucose, it is a useful tool in determining the amount of glucose in the presence of other reducing sugars:

$$\beta\text{-D-glucose} + \text{O}_2$$
$$\xrightarrow{\text{glucose oxidase}} \alpha\text{-gluconolactone} + \text{H}_2\text{O}_2$$
$$(26.27)$$

$$\text{H}_2\text{O}_2 + \text{o-diansidine}$$
$$\xrightarrow{\text{peroxidase}} \text{H}_2\text{O} + \text{oxidized dye (colored)}$$
$$(26.28)$$

26.3.1.3.3 Starch/Dextrin Content

Starch and dextrins can be determined by enzymatic hydrolysis using amyloglucosidase, an enzyme that cleaves α-1,4 and α-1,6 bonds of starch, glycogen, and dextrins, liberating glucose (see Chap. 19, Sect. 19.5.1.1). The glucose formed can be subsequently determined enzymatically. Glucose can be determined by the previously described colorimetric method. An alternative method of measuring glucose is by coupling hexokinase (HK) and glucose-6-phosphate dehydrogenase (G6PDH) reactions:

$$\text{glucose} + \text{ATP} \xrightarrow{\text{HK}} \text{glucose-6-phosphate} + \text{ADP}$$
$$(26.29)$$

$$\text{glucose-6-phosphate} + NADP^+ \xrightarrow{\text{G6PDH}}$$
$$\text{6-phosphogluconate} + NADPH + H^+$$

$$(26.30)$$

The amount of NADPH formed is measured by absorbance at 340 nm and is a stoichiometric measure of the glucose originating in the dextrin or starch hydrolyzed by amyloglucosidase. Note that HK catalyzes the phosphorylation of fructose as well as glucose. The determination of glucose is specific because of the specificity of the second reaction, catalyzed by G6PDH, in which glucose-6-phosphate is the substrate. Note as well that hydrolysis methods do not allow determining the degree of polymerization of the starting molecule.

This assay sequence can be used to detect the dextrins of corn syrup used to sweeten a fruit juice product. A second assay would be needed, however, without treatment with amyloglucosidase to account for the glucose in the product. The glucose determined in that assay would be subtracted from the result of the assay in which amyloglucosidase is used.

The same HK-G6PDH sequence used to measure glucose can also be used to measure other carbohydrates in foods. For example, lactose and sucrose can be determined via specific hydrolysis of these disaccharides by β-galactosidase and invertase, respectively, followed by the use of the earlier described HK-G6PDH sequence.

26.3.1.3.4 Determination of D-Malic Acid in Apple Juice

Two stereoisomeric forms of malic acid exist. L-Malic acid occurs naturally, while the D form is normally not found in nature. Synthetically produced malic acid is a mixture of these two isomers. Consequently, synthetic malic acid can be detected by a determination of D-malic acid. One means of detecting the malic acid is through the use of the enzyme decarboxylating D-malate dehydrogenase (DMD) [10]. DMD catalyzes the conversion of D-malic acid as follows:

$$\text{D-malic acid} + NAD^+ \xrightarrow{\text{DMD}} \text{pyruvate}$$
$$+ CO_2 + NADH + H^+$$

$$(26.31)$$

The reaction can be followed by the measurement of NADH photometrically. Because CO_2 is a product of this reaction and escapes, the equilibrium of the reaction lies to the right and the process is irreversible. This assay is of value because the addition of synthetic D/L malic acid can be used to illegally increase the acid content of apple juice and apple juice products.

26.3.2 Enzyme Activity Assays

Like the section above on substrate assays, this section on enzyme assays is not comprehensive, but rather is intended to give select assays important in the food industry. Table 26.3 summarizes the enzyme activity assays covered.

26.3.2.1 Peroxidase Activity

Peroxidase is found in most plant materials and is reasonably stable to heat. A heat treatment that will destroy all peroxidase activity in a plant material is usually considered to be more than adequate to destroy other enzymes and most microbes present. In vegetable processing, therefore, the adequacy of the blanching process can be monitored by following the disappearance of peroxidase activity [11]. Peroxidase catalyzes the oxidation of guaiacol (colorless) in the presence of hydrogen peroxide to form tetraguaiacol (yellow brown) and water (Eq. 47). Tetraguaiacol has an absorbance maximum around 450 nm. Increase in absorbance at 450 nm can be used to determine the activity of peroxidase in the reaction mixture:

$$H_2O_2 + \text{guaiacol}$$
$$\xrightarrow{\text{peroxidase}} \text{tetraguaiacol} + H_2O$$
$$(\text{colored})$$

$$(26.32)$$

26.3.2.2 Lipoxygenase

Lipoxygenase may be a more appropriate enzyme to measure the adequacy of blanching of vegetables than peroxidase [12]. Lipoxygenase refers to a group of enzymes that catalyzes the oxidation by molecular oxygen of fatty acids containing a *cis, cis*, 1,4-pentadiene system producing conjugated hydroperoxide derivatives:

$$(-CH=CH-CH_2-CH=CH-) + O_2$$

$$\xrightarrow{\text{lipoxygenase}} (-COOH-CH=CH-CH=CH-)$$
$$(\text{conjugated})$$

$$(26.33)$$

A variety of methods can be used to measure lipoxygenase activity in plant extracts. The reaction can be followed by measuring loss of fatty acid substrate, oxygen uptake, occurrence of the conjugated diene at 234 nm, or the oxidation of a cosubstrate such as carotene [13]. All these methods have been used, and each has its advantages. The oxygen electrode method is widely used and replaces the more cumbersome manometric method. The electrode method is

Summary of enzyme activity analysis examples

Enzyme of interest	Substrate	Compound measured	Measured property	Uses/comments
Peroxidase	Guaiacol	Tetraguaiacol	Dynamic increase in A_{450}	Used to assess the efficacy of blanching
Lipoxygenase	Linoleic acid	Dissolved O_2	Dynamic decrease in [O2], amperometric measurement	Used to assess the efficacy of blanching. Quantitative and does not require clear solutions
Alkaline phosphatase	Disodium phenyl phosphate	Indophenol	Endpoint measurement of A_{650}	Used to determine the adequacy of dairy products pasteurization. Coupled reaction that requires physical separation of indophenol
α-Amylase	β-Limit dextrin (produced in excess of β-amylase)	Starch-iodine color	Time to match color in a color standard disk after	Used to determine degree of sprouting in wheat flours as well as the activity of enzyme formulations added to flours
	Starch	–	Time for a plunger to fall after heating starch + water (falling number) or measurement of viscosity after ramping the temperature of a starch + water mix	
Rennet	Azocasein (casein to which a dye has been covalently bound)	Fragments of azocasein	Endpoint measurement of A_{345}	To measure the proteolytic activity of rennet by separation of fragments of azocasein that do not precipitate upon addition of trichloroacetic acid
Pectinmethyl-esterase	High methoxyl pectin	H^+	Time to return to initial pH (~7.7) after addition of 0.1 mL of 0.05 N NaOH solution	To determine residual PME activity in citrus juices after pasteurization. Also relevant in other fruit products

rapid and sensitive and gives continuous recording. It is normally the method of choice for crude extracts, but secondary reactions involving oxidation must be corrected for or eliminated. Zhang et al. [14] have reported the adaptation of the O_2 electrode method to the assay of lipoxygenase in green bean homogenates without extraction. The Clark electrode can be used for amperometric measurement of the oxygen. More recently oxygen sensors based on fiber optics and fluorescence quenching have been made commercially available and have the advantage of not needing the maintenance of Clark electrodes. Due to the rapidity of the method (<3 min including the homogenization), online process control using lipoxygenase activity as a control parameter for optimization of blanching of green beans is a real possibility. The formation of conjugated diene fatty acids with a chromophore at 234 nm can also be followed continuously. However, optically clear mixtures are necessary. Bleaching of carotenoids has also been used as a measure of lipoxygenase activity. However, the stoichiometry of this method is uncertain, and all lipoxygenases do not have equal carotenoid bleaching activity. Williams et al. [12] have developed a semiquantitative spot test assay for lipoxygenase in which I- is oxidized to I_2 in the presence of the linoleic acid hydroperoxide product and the I_2 detected as an iodine starch complex.

26.3.2.3 Phosphatase Assay

Alkaline phosphatase is a relatively heat stable enzyme found in raw milk. The thermal stability of alkaline phosphatase in milk is greater than the non-spore-forming microbial pathogens present in milk. The phosphatase assay has been applied to dairy products to determine whether pasteurization has been done properly and to detect the addition of raw milk to pasteurized milk. A common phosphatase test is based on the phosphatase-catalyzed hydrolysis of disodium phenyl phosphate liberating phenol [15]. The phenol product is measured colorimetrically after reaction with CQC (2,6-dichloroquinonechloroimide) to form a blue indophenol. The indophenol is extracted into n-butanol and measured at 650 nm. This is an example of a physical separation of product to allow the ready measurement of an enzyme reaction. More recently, a rapid fluorometric assay was developed and commercialized for measurement of alkaline phosphatase in which the rate of fluorophore production can be monitored directly without butanol extraction used to measure indophenol when phenylphosphate is used as substrate [16]. The fluorometric assay was shown to give greater repeatability compared to the standard assay in which phenylphosphate is used as substrate and was capable of detecting 0.05 % raw milk in a pasteurized milk sample. Similar chemistry has been applied to the measurement of acid phosphatase activity in meats as a means of ensuring adequate cooking via correlation of enzyme activity to endpoint temperature [17].

26.3.2.4 α-Amylase Activity

Amylase activity in malt is a critical quality parameter. The amylase activity in malt is often referred to as diastatic power and refers to the production of reducing substances by the action of α- and β-amylases on starch. The measurement of diastatic power involves digestion of soluble starch with a malt infusion (extract) and following increase in reducing substances by measuring reduction of Fehling's solution or ferricyanide. Specifically measuring α-amylase activity (often referred to as **dextrinizing activity**) in malt is more complicated and is based on using a limit dextrin as substrate. **Limit dextrin** is prepared by action of β-amylase (free of α-amylase activity) on soluble starch. The β-amylase clips maltose units off the nonreducing end of the starch molecule until an α-1,6-branch point is encountered. The resulting product is a β-limit dextrin that serves as the substrate for the endo cleaving α-amylase. A malt infusion is added to the previously prepared limit dextrin substrate and aliquots removed periodically to a solution of dilute iodine. The α-amylase activity is measured by changed color of the starch-iodine complex in the presence of excess β-amylase used to prepare the limit dextrin. The color is compared to a colored disc on a comparator. This is continued until the color is matched to a color on a comparator. The time to reach that color is **dextrinizing time** and is a measure of α-amylase activity, a shorter time representing a more active preparation.

Because α-amylase is an endoenzyme, when it acts on a starch paste, the viscosity of the paste is dramatically reduced, greatly influencing flour quality. Consequently, α-amylase activity is of great importance in whole wheat. Wheat normally has small amounts of α-amylase activity, but when wetted in the field, preharvest sprouting (pregermination) can occur in wheat, with a dramatic increase in α-amylase activity. Preharvest sprouting cannot be easily detected visually, so measurement of α-amylase activity can be used as a sensitive estimate of preharvest sprouting. The **falling number method** is a procedure in which ground wheat is heated with water to form a paste, and the time it takes for a plunger to fall through the paste is recorded [18, 19]. Accordingly, the time in seconds (the falling number) is inversely related to the α-amylase activity and the degree of preharvest sprouting. A rapid visco analyzer (RVA) can be used as refined version of this method. Samples are placed in the RVA where the temperature is ramped and the decrease in viscosity over time is monitored. The viscosity after 20 min is a good indication of α-amylase activity. This method of measuring enzyme activity is a good example of using change in physical property of a substrate as a means of estimation of enzyme activity.

26.3.2.5 Rennet Activity

Rennet, an extract of bovine stomach, is used as a coagulating agent in cheese manufacture. Most rennet activity tests are based on noting the ability of a preparation to coagulate milk. For example, 12 % nonfat dry milk is dispersed in a 10 mM calcium chloride solution and warmed to 35 °C. An aliquot of the rennet preparation is added and the time of milk clotting observed visually. The activity of the preparation is calculated in relationship to a standard rennet. As opposed to coagulation ability, rennet preparations can also be evaluated for proteolytic activity by measuring the release of a dye from azocasein (casein to which a dye has been covalently attached). In this assay, the rennet preparation is incubated with 1 % azocasein. After the reaction period, the reaction is stopped by addition of trichloroacetic acid. The trichloroacetic acid precipitates the protein that is not hydrolyzed. The small fragments of colored azocasein produced by the hydrolysis of the rennet are left in solution and absorbance read at 345 nm [20, 21]. This assay is based on the increase in solubility of a substrate upon cleavage by an enzyme.

26.3.2.6 Pectinmethylesterase Activity

Pectinmethylesterase (PME) is present in citrus and other fruits. It is particularly important to the citrus juice industry because as it cleaves the methoxyl groups of pectin, producing methanol and a proton. It causes the precipitation of the juice cloud which is undesirable. In fact, the extent of thermal treatment of citrus juices is not dictated by microbial kill but by PME inactivation. PME activity is assayed by titration with NaOH at a pH near 7.7 by determining the time to return to initial pH (~7.7) after addition of 0.1 mL of 0.05 N NaOH to a high methoxyl pectin solution to which the enzyme is added. Due to the release of protons, the pH of the pectin solution decreases and, as a result, the activity of the enzyme changes as well. For that reason, the deviation from pH ~7.7 should be small. The activity units are in milliequivalents of NaOH per unit time, per mass of sample [22].

26.3.3 Biosensors/Immobilized Enzymes

The use of immobilized enzymes as analytical tools is currently receiving increased attention. An immobilized enzyme in concert with a sensing device is an example of a biosensor. A biosensor is a device comprised of a biological sensing element (e.g., enzyme, antibody, etc.) coupled to a suitable transducer (e.g., optical, electrochemical, etc.). The most widely used enzyme electrode is the glucose electrode in which glucose oxidase is combined with an oxygen electrode or with electrochemical detection of hydrogen peroxide to determine glucose concentration [23–26]. When the electrode is put into a glucose solution, the glucose diffuses into the membrane where it is converted to gluconolactone by glucose oxidase with the uptake of oxygen and production of hydrogen peroxide. The oxygen uptake and the production of hydrogen peroxide are measures of the glucose concentration. Similar systems have been commercialized in which lactate, ethanol, sucrose, lactose, and glutamate can be measured. In the case of some of these sensors, multiple enzymes are immobilized. For example, for sucrose analysis, invertase, mutarotase, and glucose oxidase are immobilized on the same membrane. A large number of other enzyme electrodes have been reported. For example, a glycerol sensor, in which glycerol dehydrogenase was immobilized, has been developed for the determination of glycerol in wine [27]. NADH produced by the enzyme was monitored with a platinum electrode. Currently, the main limitation of enzyme biosensors is their short operational life due to the lack of stability of most of the oxidases used in their construction [28]. Table 26.3 summarizes substrate assays covered in this section.

26.4   SUMMARY

Enzymes, due to their specificity and sensitivity, are valuable analytical devices for quantitating compounds that are enzyme substrates, activators, or inhibitors. In enzyme-catalyzed reactions, the enzyme and substrate are mixed under specific conditions (pH, temperature, ionic strength, substrate concentration, and enzyme concentrations). Changes in these conditions can affect the reaction rate of the enzyme and thereby the outcome of the assay. The enzymatic reaction is followed by measuring either the amount of product generated or the disappearance of the substrate. Applications for enzyme analyses will increase as a greater number of enzymes are purified and become commercially available. In some cases, gene amplification techniques will make enzymes available that are not naturally found in great enough amounts to be used analytically. The measurement of enzyme activity is useful in assessing food quality

and as an indication of the adequacy of heat processes such as pasteurization and blanching. In the future, as in-line process control (to maximize efficiencies and drive quality developments) in the food industry becomes more important, immobilized enzyme sensors, along with microprocessors, will likely play a prominent role.

26.5   STUDY QUESTIONS

1. The Michaelis-Menten equation mathematically defines the hyperbolic nature of a plot relating reaction velocity to substrate concentration for an enzyme-mediated reaction. The reciprocal of this equation gives the Lineweaver-Burk formula and a straight-line relationship as shown below.

$$\frac{1}{v_0} = \frac{K_m}{V_{max}}\frac{1}{[S]} + \frac{1}{V_{max}}$$

(a) Define what v_0, K_m, V_{max}, and $[S]$ refer to in the Lineweaver-Burk formula.

v_0

K_m

V_{max}

[S]

(b) Based on the components of the Lineweaver-Burk formula, label the y-axis, x-axis, slope, and y-intercept on the plot.

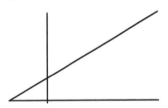

(c) What factors that control or influence the rate of enzyme reactions affect K_m and V_{max}?

K_m

V_{max}

2. Explain, on a chemical basis, why extremes of pH and temperature can reduce the rate of enzyme-catalyzed reactions.
3. Differentiate among competitive, noncompetitive, and uncompetitive enzyme inhibitors.
4. You believe that the food product you are working with contains a specific enzyme inhibitor. Explain how you would quantitate the amount of enzyme inhibitor (I) present in an extract of the food. The inhibitor (I) in question can be purchased commercially in a

purified form from a chemical company. The inhibitor is known to inhibit the specific enzyme E, which reacts with the substrate S to generate product P, which can be quantitated spectrophotometrically.

5. What methods can be used to quantitate enzyme activity in enzyme-catalyzed reactions?

6. What is a coupled reaction, and what are the concerns in using coupled reactions to measure enzyme activity? Give a specific example of a coupled reaction used to measure enzyme activity.

7. Explain how D-malic acid can be quantitated by an enzymatic method to test for adulteration of apple juice.

8. Why is the enzyme peroxidase often quantitated in processing vegetables?

9. Explain the purpose of testing for phosphatase activity in the dairy industry, and explain why it can be used in that way.

10. The falling number value often is one of the quality control checks in processing cereal-based products. What is the falling number test, and what information does it provide? What other tests could be used to assay this quality factor?

11. Explain how glucose can be quantitated using a specific immobilized enzyme.

REFERENCES

1. Cavalieri RP Reyes-De-Corcuera JI (2005) Kinetics of chemical reactions in foods, In Barbosa-Canovas GV (ed) Food Engineering. Encyclopedia of Life Support Systems. UNESCO Publishing, Paris, pp 215–239

2. Powers JR, Whitaker JR (1977) Effect of several experimental parameters on combination of red kidney bean (*Phaseolus vulgaris*) α-amylase inhibitor with porcine pancreatic α-amylase. J Food Biochem 1: 239

3. Whitaker JR (1985) Analytical uses of enzymes. In: Gruenwedel D, Whitaker JR (eds) Food Analysis. Principles and Techniques, vol 3, Biological Techniques, Marcel Dekker, New York, pp 297–377

4. Eisemenger MJ, Reyes-De-Corcuera JI (2009) Enzyme Microb Technol 45: 331

5. Bernfeld P (1955) Amylases, α and β. Methods Enzymol 1: 149

6. Boehringer-Mannheim (1987) Methods of Biochemical Analysis and Food Analysis. Boehringer Mannheim Gmb H Mannheim, W. Germany

7. Bergmeyer HU (1983) Methods of Enzymatic Analysis. Academic Press, New York

8. Beutler H (1984) A new enzymatic method for determination of sulphite in food. Food Chem 15: 157

9. Raabo E, Terkildsen TC (1960) On the enzyme determination of blood glucose. Scand J Clin Lab Inves 12: 402

10. Beutler H, Wurst B (1990) A new method for the enzymatic determination of D-malic acid in foodstuffs. Part I: Principles of the Enzymatic Reaction. Deutsche Lebensmittel-Rundschau 86: 341

11. USDA (1975) Enzyme inactivation tests (frozen vegetables). Technical inspection procedures for the use of USDA inspectors. Agricultural Marketing Service, U.S. Department of Agriculture, Washington, DC

12. Williams DC, Lim MH, Chen AO, Pangborn RM, Whitaker JR (1986) Blanching of vegetables for freezing — Which indicator enzyme to use. Food Technol 40(6): 130

13. Surrey K (1964) Spectrophotometric method for determination of lipoxidase activity. Plant Physiology 39: 65

14. Zhang Q, Cavalieri RP, Powers JR, Wu J (1991) Measurement of lipoxygenase activity in homogenized green bean tissue. J Food Sci 56: 719

15. Murthy GK, Kleyn DH, Richardson T, Rocco RM (1992) Phosphatase methods. In: Richardson GH (ed) Standard methods for the examination of dairy products, 16th edn. American Public Health Association, Washington, DC, p 413

16. Rocco R (1990) Fluorometric determination of alkaline phosphatase in fluid dairy products: Collaborative study. J Assoc Off Anal Chem 73: 842

17. Davis CE (1998) Fluorometric determination of acid phosphatase in cooked, boneless, nonbreaded broiler breast and thigh meat. J AOAC Int 81: 887

18. Delwiche SR, Vinyard BT, Bettge AD (2015) Repeatability precision of the falling number procedure under standard and modified methodologies. Cereal Chem. 92 (2):177

19. AACC International (2010) Approved methods of analysis, 11th edn. (On-line), American Association of Cereal Chemists, St. Paul, MN

20. Christen, GL, and Marshall, R.T. 1984. Selected properties of lipase and protease of pseudomonas fluorescens 27 produced in 4 media. Journal of Dairy Science, 67: 1980

21. Kim, SM, Zayas, JF. 1991. Comparative quality characteristics of chymosin extracts obtained by ultrasound treatment. Journal of Food Science 56:406

22. Shimizu Y, Morita K (1990) Microhole assay electrode as a glucose sensor. Anal Chem 62: 1498

23. Kimball (1996) Citrus Processing, a Complete Guide. 2nd ed. Aspen Publishers, Inc. Gaithersburg, MD, p 259–263

24. Reyes-De-Corcuera JI, Cavalieri RP (2003) Biosensors. In: Encyclopedia of agricultural, food, and biological engineering, p 119–123

25. Guilbault GG, Lubrano GJ (1972) Enzyme electrode for glucose. Anal

26. Borisov SM, Wolbeis OS (2009) Optical biosensors. Chem Rev 108: 423

27. Matsumoto K (1990) Simultaneous determination of multicomponent in food by amperometric FIA with immobilized enzyme reactions in a parallel configuration. In: Schmid RD (ed) Flow injection analysis (FIA) based on enzymes or antibodies, GBF monographs, vol 14, VCH Publishers, New York, pp 193–204

28. Reyes-De-Corcuera JI (2016) Electrochemical Biosensors in In: Encyclopedia of agricultural, food, and biological engineering, 2nd edition. [http.www.Dekker.com] (on-line version only)

Immunoassays

Y.-H. Peggy Hsieh (✉) • Q. Rao

Department of Nutrition, Food and Exercise Sciences,
Florida State University,
Tallahassee, FL 32306-1493, USA
e-mail: yhsieh@fsu.edu; qrao@fsu.edu

S. Nielsen (ed.), *Food Analysis*, Food Science Text Series,
DOI 10.1007/978-3-319-45776-5_27, © Springer International Publishing 2017

27.1 INTRODUCTION

Immunochemistry is a relatively new science that has developed rapidly in the last few decades. One of the most useful analytical developments associated with this new science is immunoassay. Originally immunoassays were developed in medical settings to facilitate the study of immunology, particularly the antibody-antigen interaction. Immunoassays now are finding widespread applications outside the clinical field because they are appropriate for a wide range of analytes ranging from microorganisms to proteins to small organic molecules. In the food analysis area, immunoassays are widely used for chemical contaminants analysis, identification of bacteria and viruses, and detection of proteins in food and agricultural products. Protein detection is important for determination of allergens and meat species content, seafood species identification, and detection of genetically modified plant tissues. While immunoassays of all formats are too numerous to cover completely in this chapter, there are several assay formats that have become standard for food analysis because of their specificity, sensitivity, and simplicity (Fig. 27.1).

27.1.1 Definitions

Immunoassays are analytical techniques based on the specific and high-affinity binding of antibodies with particular target antigens. To fully understand immunoassays, some of these terms need to be defined. The two essential elements of any immunoassay are antigens and antibodies. In an immunoassay, antigens and antibodies are used either as target molecules or capture molecules. In other words, a particular antigen can be used to capture its specific antibody, or a specific antibody can be used to trap the target antigen in a sample. An **antigen** is any molecule that induces the formation of antibodies and can bind to these antibodies. **Antibodies** are **immunoglobulin** (Ig) **proteins** produced by animals in response to an antigen. These antibody proteins are secreted by the activated B cells in immune system and bind the particular antigen responsible for their induction. Generally a molecule must be greater than 5000 Da, abbreviated as Da (unit of molecular mass), to be perceived as an antigen by a mammalian immune system. Almost all proteins are large molecules and have the ability to induce antibody formation in the body of humans and animals. However, many of the molecules analyzed in food are not as large as proteins but are small molecules such as

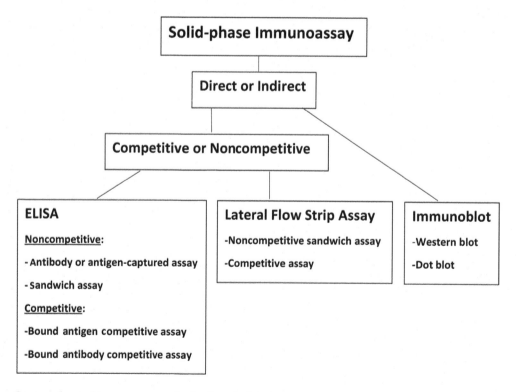

Commonly used immunoassays for food analysis

toxins, or antibiotics and pesticides. When animals are injected with small molecules, they do not develop antibodies against these molecules. To induce specific antibodies to recognize and bind the small target molecule, the solution is to covalently link the small molecule, or some appropriate derivative of the small molecule, to a larger carrier molecule. The small molecule that must be linked to a large carrier protein before it can be used as an immunogen to induce antibodies is called a **hapten**. The carrier protein-linked hapten is called a **conjugate antigen**. Haptens react specifically with the appropriate antibodies but are not immunogenic. The desired molecules used as carriers are proteins that are soluble, unrelated to proteins that may be found in the assay sample, and foreign to the host animal to properly stimulate an immune response. Typical carrier molecules include albumin proteins from a different species, such as bovine serum albumin and hemocyanins that are obtained from crustaceans. Of course, when a conjugate antigen is used for immunization of an animal, its immune system is stimulated to produce antibodies that bind not only the externally attached hapten but also the exposed exterior of the covalently linked carrier protein.

There are five major classes of antibodies, IgA, IgE, IgG, IgM, IgD, according to their heavy chain structure. Animal blood contains trace amounts of IgA and IgD. IgA is mainly found in mucous secretions and plays an important role in mucosal immunity. The unique function of IgD is still unclear. IgM is a very large molecule and can be regarded as a precursor of IgG. IgE is only associated with allergic response in humans and animals. Among these five classes of antibodies, IgG has the highest concentration in blood and is the most important class used in food immunoassay. Since the antibody and antigen are central to any immunoassay, it is useful to better understand the basic structure of the antibody and how it binds the antigen. Figure 27.2 is an idealized diagram of an antibody IgG. The IgG is a Y-shaped molecule made up of four polypeptide chains that are linked by inter- and intra-disulfide bonds. Two of the polypeptide chains are identical and roughly twice as large as the other two identical polypeptide chains. Because of their relative sizes, the former pair is known as **heavy chains** and the latter pair as **light chains**. Overall, an IgG antibody is a very large protein of approximately 150000 Da. Antigen is bound by two identical binding sites made up of the end portions (N terminals) of a heavy and light chain at the top of the Y. These two fragments capable of binding with antigen are called **Fab (fragment antigen binding)**. The third fragment with no antigen-binding capability is called **Fc (fragment crystallizable)**, because it can be crystallized. Different antibodies produced by different B cells can have many variations in amino acid sequences near the binding sites for both the heavy and light chains.

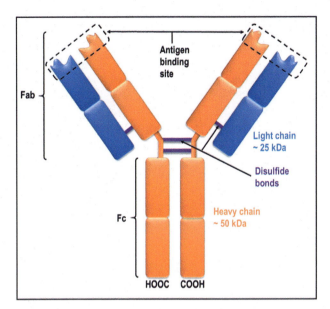

27.2 figure Antibody (IgG) structure

This leads to a tremendous diversity of binding sites for different antibodies. For example, a mouse has 10^7–10^8 different antibodies (and at least this number of different B cells), each with a unique binding site. The rest of the antibody (away from the binding site) is quite consistent, and small variations in this region on the heavy chains result in different antibody classes.

27.1.2 Binding Between Antigen and Antibody

Antibodies can develop remarkably strong binding affinities for their antigens. These affinities are among the strongest noncovalent interactions known between molecules. The binding strength (affinity) between antibody and antigen is one of the most important factors that determines the sensitivity of an immunoassay. The antibody binds to the outside of the antigen molecule in a specific region. This specific region bound by a single antibody binding site is known as an **epitope**. Two types of epitopes on an antigen can be formed. A **linear epitope** is formed by a continuous sequence of amino acid residues, and a **conformational epitope** is formed by noncontiguous amino acid sequences that are folded into close proximity from neighboring or overlapping peptide chain on the surface of the antigen (Fig. 27.3). If this 3-dimensional conformation of the antigen is altered by some kind of the environmental conditions, such as heating or pH changes, the conformational epitope will be destroyed, which means that the antigen cannot bind to the antibody. Moreover, the binding of the antibody to the antigen does not involve covalent bonding, but the same interactions that are responsible for the tertiary

27.3
figure

Linear epitope (**a**) and conformation-dependent epitope (**b**)

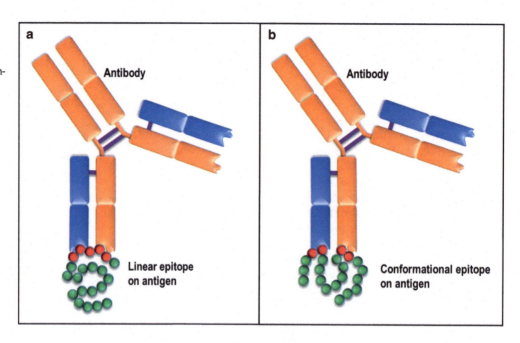

structure of proteins. These interactions include hydrogen bonds, electrostatic and hydrophobic interactions, and van der Waals forces. While the latter interactions, van der Waals, are the weakest, they often can be the most important because every atom can contribute to the antibody-antigen binding as long as the atoms are very close to each other (generally about 0.3–0.4 nm). This requirement for very close proximity is why antibody to antigen binding is considered something like a lock and key interaction, where the surfaces of the antibody binding site and the antigen epitope are mirror complements of each other.

27.1.3 Types of Antibodies

A major variable in an immunoassay is the type of antibody used. When serum antibody is used from any immunized animal, there are many different antibodies that bind different epitopes on the antigen. This collection of different antibodies is known as **polyclonal antibodies**. Scientists knew that individual B cells produced antibodies with only one binding site but were unable to culture B cells outside of the animal. However, in 1975, Köhler and Milstein [1] successfully fused cancer, or myeloma, cells with B cells. The new fused cells, or **hybridomas**, retained the properties of both of the parent cells. That is, they could be cultured, like cancer cells, and produced antibodies like the B cells. Hybridomas thus can be cloned and cultured individually to produce different antibodies with different epitopes. Antibodies produced with this procedure became known as **monoclonal antibodies**. Monoclonal antibodies produced from a single hybridoma are identical in every way and bind antigen with only one type of binding site, that is, a single epitope is bound; therefore, they can be used as standard reagents in immunoassays. Moreover, the

hybridomas were "immortalized" by the procedure and with proper care could produce as much identical antibody as required. It did not take the scientific community long to appreciate the tremendous advantages of these monoclonal antibodies, so Köhler and Milstein were awarded the Nobel Prize for their work on this in 1984. While monoclonal antibodies are initially much more expensive to produce, there is the possibility for limitless identical antibody production, often from non-animal sources such as large-scale production of the hybridomas in cell growth chambers. These advantages outweigh the initial development costs for many immunoassay manufacturers. Detail procedures regarding the antibody production and characterization can be found in the book by Howard and Bethell [2].

27.2 THEORY OF IMMUNOASSAYS

Based on the specific antibody-antigen affinity, various types of immunoassays have been developed to either: (1) use antibody as the capture molecule to search the target antigen or (2) use the antigen as the capture molecule to trap the antibody in a complex sample. The basis of every immunoassay is the detection and measurement of the primary antibody-antigen reaction. In its simplest form, antibody capture of antigen can involve a simple precipitation and be detected visually. Since all antibodies have at least two identical binding sites, they can cross-link epitopes from two identical antigens. If other antigen epitopes are further cross-linked by different antibodies, a large, insoluble network can result which is seen as a precipitate. The immunoprecipitation techniques, including immunodiffusion and agglutination,

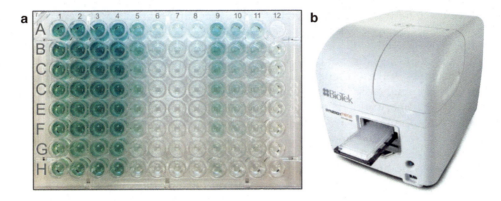

27.4 figure Image of a 96-well plastic microtiter plate (**a**) and plate reader (**b**) used for ELISA

formed the basis of early development of immunoassay techniques and have been used widely for protein and cell identification using antisera. However, these methods only work for antigens with multiple epitopes.

To measure the quantity of soluble antibody-bound antigen molecules in a solution, all immunoassays require two things. The first is that there must be some method to separate or differentiate free antigen from bound antigen. Secondly, these antibody-bound antigens must be quantifiable at low concentrations for maximum sensitivity. Detection at very low concentrations has required very active labels. One of the first successful immunoassay procedures was developed by Yalow and Berson [3] in 1960. This procedure used radioactive iodine, I^{131}, a "hot" radioisotope with a half-life of only 8 days, as a label to reveal the primary antibody-antigen complex. This radioactive label allowed for the second requirement of immunoassays: quantification at low concentrations. Yalow and Berson used paper chromato-electrophoresis to separate their antibody-bound antigen from free antigen, fulfilling the first requirement of an immunoassay. With all the variations in separation and detection techniques in the early stage of immunoassay development, however, the radioactive iodine labeling remained, and these assays became known as **radioimmunoassays** (RIA).

One of the techniques for the separation of unbound from bound molecules in immunoassays involves immobilizing protein on a hydrophobic solid surface. Proteins have large regions that contain hydrophobic amino acid groups that prefer not to be exposed to water. These nonpolar hydrophobic groups include hydrocarbons and aromatic groups that prefer to interact with similar groups, rather than a polar solvent such as water. In aqueous conditions these regions will bind to other hydrophobic surfaces excluding water. Surfaces commonly introduced in immunoassays to take advantage of this type of binding include

charcoal, nitrocellulose, and plastic. Plastic surface in many forms is used commonly for immunoassays. Among the most popular are **microtiter plates** made of plastics such as polystyrene or polyvinyl. These microtiter plates typically are formatted to contain 96 individual wells, each with a maximal capacity of about 300 µl of liquid (Fig. 27.4a). To differentiate the wells, the vertical rows are labeled A to H and the columns numbered 1–12. It is important to realize that proteins bind to the bottom and sides of the wells in these plates randomly through hydrophobic interactions. Other forms of solid surface commonly used in immunoassays include plastic vials, magnetic beads, and nitrocellulose or polyvinylidene difluoride (PVDF) membranes or strips.

27.3 SOLID-PHASE IMMUNOASSAYS

27.3.1 Overview

Every immunoassay technique developed is based on the selection of an **amplification method** that will improve the sensitivity of assays. Labeling of the detecting molecule, either antigen or antibody, is necessary to locate or to quantify the target molecule in a solid-phase immunoassay that uses a solid surface for the separation of unbound from bound molecules. For example, having an easily detected label attached to the antibody allows the target antigen in a complex food sample matrix to be detected. A number of labeling compounds and materials have also been used, such as radioisotopes, fluorochromes, enzymes, biotin, and gold nanoparticles. These different label options become very useful for immunoassay development, or in the use of antibodies for detection in different systems such as examination of tissue under a microscope or proteins separated using electrophoretic techniques followed by an immunoassay.

While RIAs worked well in the clinical field, they were confined to specially equipped laboratories because of the dangers associated with the use of radioactive material. Immunoassays did not develop for more general use, including field use, until enzyme labels were developed. Any immunoassay that uses an enzyme label to reveal the primary antibody-antigen binding is called an **enzyme immunoassay**. Pioneers in this development were Engvall and Perlmann [4] who in 1971 developed a type of enzyme immunoassay that they called an **enzyme-linked immunosorbent assay**, or ELISA. ELISA involves the binding of a soluble antigen or antibody to a solid support (immunosorbent), typically in the form of a 96-well plastic microtiter plate. The bound and unbound molecules can be easily separated by a washing step of the plate. Microtiter plates of hydrophobic plastics are commonly used to immobilize proteins and to separate unbound molecules. Similar immunoassays using a solid support other than plastic microtiter plates also have been developed and are commonly used. Examples are the **dot blot** assay, **Western blot** assay, and more recently developed lateral flow immunochromatographic assay, which use nitrocellulose or polyvinylidene difluoride (PVDF) membranes. These developments have expanded the use of immunoassays to a wider range of applications.

27.3.2 ELISA

27.3.2.1 *Introduction*

Because enzyme immunoassay, especially ELISA, has become the most popular immunoassay for food analysis applications, the general principle and protocols of various formats of ELISA are illustrated in subsequent sections using ELISA as examples.

The enzyme label used in an ELISA converts a colorless substrate to a colored soluble product in the solution, thus generating a detectable signal for the assay. The amount of target protein antigen present in the sample extract is determined from the intensity of color developed in the immunoassay. The ideal enzyme for an enzyme immunoassay is one that is stable, easily linked to antibodies or antigens, and rapidly catalyzes a noticeable change with a simple substrate. With the many enzymes available, two enzymes, **horseradish peroxidase** and **alkaline phosphatase**, are by far the most commonly used in immunoassays. Other enzymes used include β-galactosidase, glucose oxidase, and glucose-6-phosphate dehydrogenase. The use of an enzyme to catalyze a chemical reaction that generates color signal contributes to the sensitivity of the assay because a single enzyme molecule present at the end of the test converts many substrate molecules to detectable colored product, thus amplifying the signal generated by the assay. However, it is more difficult to

achieve a quantitative measure with enzyme immunoassays because the rate of the enzyme reaction involved is difficult to measure and the enzyme-labeled reagents are not homogenous. Minimal laboratory equipment is required to perform ELISA. The color generated from an assay can be visualized to determine the result in a qualitative assay or semi-quantitated spectrophotometrically. The type of spectrophotometer used to quantitatively monitor color development caused by the enzyme action is called a **microplate reader** (Fig. 27.4b). Automated **microplate washer** also is available although most assays can be washed manually.

All ELISA protocols include the following five general steps:

1. *Coating* of antibody or antigen onto the wells of a microfiter plate (solid phase)
2. *Blocking* the remaining uncoated surface on the solid phase with a blocking buffer containing a nonspecific protein such as bovine serum albumin (this is to minimize the nonspecific reactions and also protect the adsorbed antigen or antibody from surface denaturation)
3. *Incubating* with different immunoassay reagents at a specified temperature and time
4. *Washing* the coated surface to separate free, unbound molecules from bound molecules
5. *Detecting* the color developed from the assay visually or spectrophotometrically

Specific procedures vary with the different variations of ELISA. It is important to include both positive and negative controls in an assay along with any analyzed food sample because materials in the food extract can vary widely and these other components can have an effect on the competition for the antibody binding site. This is to ensure that the immunoassay works well (positive control shows positive signal) and that there is no contamination or nonspecific reactions in the assay system (negative control shows negative signal).

27.3.2.2 *Direct Versus Indirect Detection*

All immunoassay signals can be detected directly or indirectly. In the **direct detection method**, the detecting molecules are purified and linked to the label to directly measure the amount of the antibody-antigen complex (Fig. 27.5a). Therefore, more purified immunoreagents need to be used for the label conjugation procedure. In contrast, the **indirect detection method** uses a commercially available intermediate reagent to link the capture molecule. Most often a labeled **anti-species antibody** is used for an antibody-captured assay to indirectly measure the amount of antibody-antigen complex formed (Fig. 27.5b). Although an additional step is involved, indirect assays require less

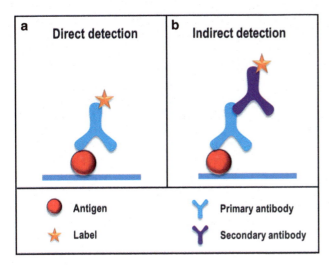

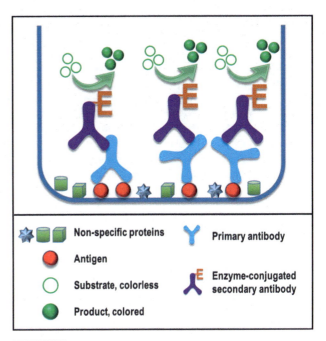

27.5
figure

Direct detection (**a**) and indirect (**b**) detection methods

27.6
figure

Indirect noncompetitive ELISA

immunoreagents and in many cases could be more sensitive because more labeling molecules can be linked to the detection antigen or antibody for enhanced signal production. While the direct detection method is essential when an accurate quantification is required for the assay, the indirect detection method is used in most solid-phase immunoassays.

A simple form of indirect antibody-captured ELISA is demonstrated in Fig. 27.6. The method is often used in the early stage of immunoassay development to detect primary antibodies in antisera or screen hybridoma supernatants for searching desired antibodies. **Primary antibody** refers to the antibody that binds the antigen. Theoretically, the antibody-captured ELISA can be made either as a direct ELISA or an indirect ELISA. However, because the target molecule is the primary antibody that appears in the biological fluid in a very low quantity, it usually is impossible to obtain enough quantity of the primary antibody to prepare the enzyme-conjugated labeling reagent for subsequent detection. Therefore, antibody-captured ELISA is almost always configured in an indirect assay format. The soluble antigen is adsorbed (coating) onto the surface of the microplate wells and incubated. After blocking, diluted samples of antisera or hybridoma supernatants are then added to the wells and incubated to allow the immobilized antigen to bind primary antibodies in the sample. After washing away any unbound molecules, those bound antibodies can be detected by adding an enzyme-linked secondary antispecies antibody, which can easily bind to the constant region of the primary antibody. After another incubation and washing steps, a solution containing substrate is added to generate color in the solution. The color is positively related to the amount of target antibody present in the sample.

The **secondary antibody** used in the indirect ELISA does not have the specificity to bind the antigen but only recognizes the primary antibody; thus it makes a link of the enzyme label to the bound immunomolecules without interfering with the primary antigen-antibody binding. Since antibodies are proteins, they can act as antigens in another animal species. For example, rabbit antibodies injected into a goat can stimulate the goat's immune system to produce goat antibodies that bind to the rabbit antibodies. In this way, goat anti-rabbit antibodies can be produced to bind any antibody produced in a rabbit. There are many advantages to these secondary anti-species antibodies. For example, when antispecies antibodies are used in the above antibody-captured ELISA format (Fig. 27.6), there is no need to label the primary antibody with an enzyme. The primary antibody does not need to be chemically modified; thus the loss of its activity is avoided. After excess material is washed away, goat anti-rabbit antibody labeled with an enzyme can be added to detect the presence of any primary antibody which is produced in rabbit antiserum and binds to the antigen coated on the microtiter plate. Although this procedure adds an additional step, there are many advantages: (1) antispecies antibodies of all types are commercially available from many manufacturers; (2) antispecies antibodies come with a variety of labels such as different enzymes, radioisotopes, or fluorescent compounds for different immunoassay systems; and (3) since the antibody is a very large protein, it has many sites for attachment of a labeled antispecies antibody. This multiplies the labels per antibody, increasing the ability to detect the antibody and resulting in stronger signals with increased sensitivity in an immunoassay, do less primary antibody reagent is needed. Therefore, indirect methods are used in most detection applications.

27.3.2.3 Noncompetitive Versus Competitive Immunoassay Variations

Noncompetitive immunoassay is commonly employed to analyze large molecules such as proteins in a food sample, while **competitive immunoassay** is competitive in nature and is mainly used for small molecule analysis. Both competitive and noncompetitive immunoassays can be detected directly or indirectly. For simplicity, these variations are illustrated in the sections that follow using ELISA as examples. In general, the amount of color development for the noncompetitive ELISA is directly related to the amount of antigen present in the sample Fig. 27.7a. With any competitive ELISA format, there is an inverse relationship

between the amount of color developed and the amount of antigen present in the sample (Fig. 27.7b).

27.3.2.3.1 Noncompetitive ELISA

Noncompetitive ELISA variations involve the revealing of the amount of primary antibody-antigen complex immobilized on the solid phase by the amount of enzyme linked to the detection antigen or antibody molecules to produce a colored product in the assay solution. Therefore, at the end of the assay, the color intensity is positively related to the amount of the target molecules. The absence of the target molecules produces no color, and the presence of high concentration of the target molecules produces strong color. This type of ELISA is used often to detect proteins in a food sample because protein molecules are large enough to link one or more antibodies or to an additional enzyme label on the surface of the protein.

One of the most popular formats for a noncompetitive enzyme immunoassay is the antibody **sandwich immunoassay**. A sandwich ELISA model using both direct and indirect detection method is demonstrated in Fig. 27.8a, b, respectively. The "meat" in the antibody sandwich is the target antigen. In food analysis this can involve identifying a protein adulterant such as undeclared pork meat in a beef product, a protein allergen such as peanut protein, or wheat protein in a product that would be a problem for people suffering from celiac disease.

Generally a primary antibody that binds to the antigen is first immobilized onto a hydrophobic solid phase such as plastic. Excess antibody is removed by washing with a washing solution or simply water

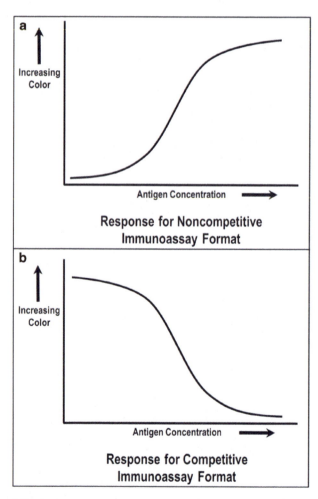

27.7
figure
Relationship between color development and antigen concentration for noncompetitive (**a**) and competitive (**b**) immunoassay formats

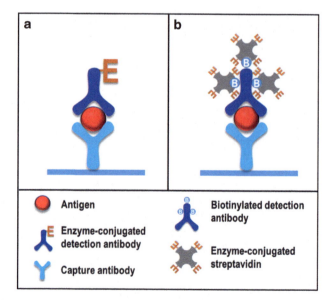

27.8
figure
Direct (**a**) and indirect (**b**) noncompetitive sandwich ELISA

followed by a blocking step, and then the test is ready for analysis of a food extract. The immobilized antibody is called a **capture antibody**. The food extract being tested contains many compounds with or without the target antigen. However, the antibody was prepared by immunization of an animal with a specific, purified protein antigen, and only this protein antigen in the food extract solution will bind to the capture antibody. Now the antigen and the capture antibody are immobilized, and the remaining unbound molecules can be washed away. After the washing step, another primary antibody labeled with an enzyme (Fig. 27.8a) is introduced. This antibody, called the **detection antibody**, also recognizes the antigen, thus forming an antibody-antigen-antibody complex. Again excess detection antibody is washed away and then colorless enzyme substrate is added to develop a color if bound enzyme is present. Enzyme will only be present if the detection antibody has been immobilized by binding to the antigen. The greater the color development, the greater the amount of antigen present. That is, there is a direct proportionality between the amount of color seen in the final step and the amount of antigen present in the extracted food sample. To increase the sensitivity of a sandwich immunoassay, one can use more antibodies for capture of the antigen or link more enzyme molecules to the detection antibody through the use of intermediate reagents, such as biotin and labeled streptavidin. In the latter case, the detection antibody in the indirect sandwich assay (Fig. 27.7b) is purified and linked to biotin. Biotinylation of a primary antibody is relatively simple and rarely affects the antibody activity. Then an additional step of adding enzyme-labeled avidin or streptavidin to the reaction system is needed. Avidin/streptavidin easily and tightly binds to the biotin groups so that more

enzymes are linked to the immobilized molecules (Fig. 27.8b). This indirect sandwich immunoassay format can be made remarkably specific and very sensitive since two antibodies must detect the antigen and more labels are involved in amplifying the reaction signal.

When polyclonal antibodies are used in the sandwich immunoassay, the polyclonal antibody solution is divided into two parts. One part is bound to plastic to become the capture antibody. The second portion of the polyclonal antibody solution is purified and conjugated to an enzyme or biotinylated and becomes the detection antibody. Monoclonal antibodies can also be used, but care must be exercised since a single type of monoclonal antibody cannot be used for both the capture and detection antibodies since only one unique epitope is recognized by any monoclonal antibody. In other words, the antigen must be large enough to bind two antibodies at the same time and therefore must use at least two distinct epitopes recognized by different monoclonal antibodies that recognize two distinct antigen epitopes.

27.3.2.3.2 Competitive Immunoassays

A problem in developing an immunoassay for detecting a small molecule is that a sandwich immunoassay format will not work since two different epitopes on the antigen are required for both antibodies to bind. A small molecule represents only one epitope or even only part of one epitope. The competitive immunoassay format (Fig. 27.9a, b) was, therefore, developed to solve this problem. The first requirement in a competitive ELISA involves immobilizing the small molecule, often as a hapten, or immobilizing the antibody. Subsequent procedures involve the competition between the free small antigen (from a sample) and

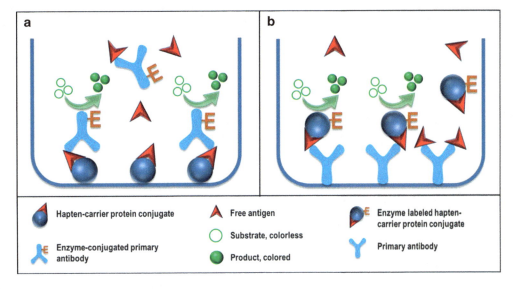

 27.9 Direct competitive ELISA in bound hapten (**a**) and bound antibody (**b**) formats

the hapten (as an added reagent) for the binding of limited amount of the specific antibody. To bind the hapten to a solid surface such as nitrocellulose or plastic, it can again first be linked to a protein that binds to these hydrophobic surfaces. However, the protein used for binding the hapten to the surface is different than the carrier protein binding the hapten used for immunization of the animal, since the animal also has developed antibodies against the carrier protein used for injection, and only the hapten-specific antibodies are desired for the competitive immunoassay. Since all types of competitive immunoassays involve a reduction in absorbance with respect to a control (containing no small molecule or analyte), data often are presented as a ratio of sample absorbance to the absorbance of the control. The concentration of the inhibitor (target antigen) required to reduce the assay absorbance by 50 % (defined as IC_{50}) is a useful value to be determined for a competitive immunoassay because this is the region of greatest change in response compared to concentration changes and therefore the lowest coefficient of variation.

To increase the sensitivity of a competitive immunoassay, the amount of limiting antibody should be reduced. Note that this is the reverse of what one would do to increase the sensitivity of a noncompetitive immunoassay such as a sandwich immunoassay. Theoretically the most sensitive competitive immunoassay would be between one antibody binding site and one hapten, with either of the two labeled with an enzyme. It is for this reason that the ability to detect the presence of the enzyme is so important for a competitive immunoassay. The more sensitive the system is to detect the enzyme, the more sensitive the competitive immunoassay. Two competitive ELISA procedures are described here:

1. **Bound Hapten Format**. In the bound hapten competitive immunoassay format (Fig. 27.9a), the protein-bound hapten is first immobilized to a solid surface by hydrophobic interactions. Excess material is washed away. Next a competition is created between the protein-bound hapten and the free small antigen molecule in a food extract, both competing for binding to the limited binding sites on the antibody labeled with an enzyme. It is important to realize that the free small molecule in the food extract is not completely identical to the immobilized hapten since the latter is covalently linked to a protein. However, if properly designed, the free molecule in the food extract is so chemically similar to the bound hapten that the competition for the limited number of antibody binding sites is nearly equal. The primary antibody bound to immobilized hapten remains after a subsequent washing step. The more small molecules in the food

extract, the more antibody is bound to these free small molecules, and this unbound antibody (and its attached enzyme) will be washed away in the subsequent washing procedure. Finally, the amount of bound antibody is identified by adding the enzyme substrate and observing the amount of color developed. Therefore, there is an inverse relationship between the amount of small molecules or analyte in the food and the amount of color developed in the final step.

2. **Bound Antibody Format**. The other variation for a competitive immunoassay is to bind a limited amount of antibody to the solid phase and create a competition between enzyme-labeled conjugate antigen and free small antigen molecules in the food extract (Fig. 27.9b). It is generally believed that this second format is somewhat superior to the first format for sensitivity although it can require the use of more antibody reagent. Again after a washing step, the final procedure is a color development to determine the amount of antibody-bound antigen-enzyme. This competitive format also results in an inverse relationship between the amount of color and free small molecules in the food extract.

27.3.3 Immunoblots

27.3.3.1 *Western Blot*

As one of the immunoassays that uses the specificity of the antigen-antibody interaction to indicate the presence of particular proteins in a sample, **Western blot** is a laboratory-based method that combines two techniques: **polyacrylamide gel electrophoresis (PAGE)** and **immunoassay**. In the first part of a Western blot, proteins in a complex mixture are separated by PAGE according to their molecular mass. In the second part, the separated proteins are subjected to an immunoassay to detect the presence of antigenic proteins. Using this combination of techniques makes it possible to identify target proteins and confirm their identity by molecular mass. The detection reagent in a typical Western blot is an enzyme-labeled antibody conjugate directly or indirectly labeled to the detection antibody. If the original protein mixture was labeled with a radioactive material, then autoradiography is used to visualize the radioactive signal. The primary antibody used largely determines the specificity and sensitivity of the method. Specific proteins in picogram quantities can be detected in a highly sensitive Western blot.

To prepare protein samples for the initial separation by PAGE, they are typically boiled in a buffer solution containing a reducing agent (usually mercaptoethanol) and detergent (e.g., sodium dodecyl sulfate), to unfold the protein peptide chains. The treated sample is applied to a polyacrylamide gel and separated by electrophoresis based on **molecular mass**. To prepare for the immunoassay portion of the method, the separated protein bands are transferred from the polyacrylamide gel to a **nitrocellulose** or **PVDF membrane**. After blocking the nonspecific binding site, for a direct detection format, this membrane is then incubated with a solution of primary antibody-enzyme conjugate. After washing away the excess conjugate from the membrane, the enzyme substrate is added. A colored band forms at the site on the membrane where the protein that reacted with the antibody was immobilized (Fig. 27.10). The enzyme substrate used in Western blot is different than the ones used in ELISA because the intent is to form an insoluble colored product that stays on the membrane. The color intensity and width of the protein band together indicate the concentration of the target protein in the sample extract. The molecular mass of the protein bound by the antibody can be estimated by its position relative to standard proteins of known molecular weight.

The Western blot procedure is technically complex, so it requires highly trained personnel working in a laboratory setting. However, the Western blot method is especially well suited to analysis of food samples that have been subjected to processing conditions. The Western blot method requires that the antibody used binds to a linear epitope on the protein, or is reactive to **denatured protein** (which would occur under the denaturing conditions in PAGE sample preparation). Because the Western blot method is highly sensitive and specific, and the antibodies used recognize denatured proteins, it is uniquely well suited to detect the presence of target food proteins at low concentrations in processed food.

27.3.3.2 *Dot Blot*

Dot blot represents a simplified version of Western blot. It is usually performed by depositing a drop of sample extract containing the molecule to be detected directly onto a piece of nitrocellulose or PVDF membrane as a dot. Unlike Western blot, protein samples in a dot blot assay do not need to be separated electrophoretically. The target protein molecule is directly detected by specific antibodies. The following general guideline describes a commonly used antigen bound dot blot assay for detecting a particular protein in food samples (Fig. 27.11):

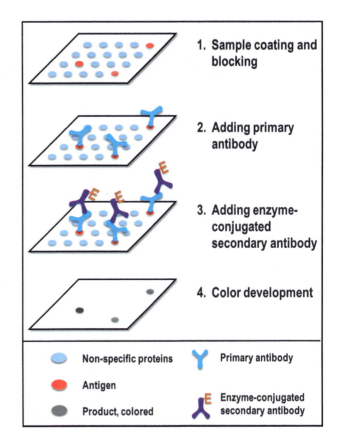

1. **Sample coating and blocking**

2. **Adding primary antibody**

3. **Adding enzyme-conjugated secondary antibody**

4. **Color development**

- Non-specific proteins
- Antigen
- Product, colored
- Primary antibody
- Enzyme-conjugated secondary antibody

27.10 **figure** Western blot technique. Protein profile before (*1*) and after (*2*) transferring and immunodetection of the antigen on the membrane (*3*)

STD: Protein standards; S: sample

1) Polyacrylamide gel after electrophoretic protein separation and staining

2) Protein bands transferred from the gel to a membrane

3) Target antigen detected on the membrane by blotting with the primary antibody

27.11 **figure** Dot blot technique

1. A drop of the food sample extract is applied to the membrane.
2. The remaining binding sites on the membrane are "blocked" using a protein unrelated to the target antigen to minimize the nonspecific binding.
3. Following a rinsing step, the bound molecules in the sample spot are incubated with the primary antibody (for indirect detection) or with the primary antibody-enzyme conjugate (for direct detection). An additional step using an antispecies antibody-enzyme conjugate is required for the indirect assay.
4. After washing and adding the substrate solution, the reaction signal can be revealed either visually by color or measured by chemiluminescence imaging apparatus depending on the enzyme label and the substrates used. The spot signal (i.e., color) intensity positively related to the amount of target protein present in the sample.

This blot dot technique does not offer molecular weight information of the antigen but is useful as a qualitative method for rapid screening of a large number of samples to probe the presence or absence of a protein target in food samples. This technique is also commonly used to evaluate the quality of the antibodies and testing the suitability of experimental design parameters.

27.3.4 Lateral Flow Strip Test

27.3.4.1 *Overview*

The **lateral flow strip** (LFS) test is a simple immunoassay format, commonly used for proteins to determine if their concentration is above or below a specified threshold (cutoff) (Fig. 27.12). The home pregnancy test is the best known LFS method. Results with LFS usually can be visualized in 10–20 min, and no washing steps are required to separate bound and unbound molecules. The characteristics of LFS methods – simplicity, low cost, ease of use, and reliability – make them ideal for use outside a laboratory setting, i.e., **field testing**, where supplies and equipment are limited.

Just as with other immunoassays, LFS methods are configured in a **competitive assay format** for detection of small molecules such as toxins or chemical residues or in a noncompetitive **sandwich immunoassay format**

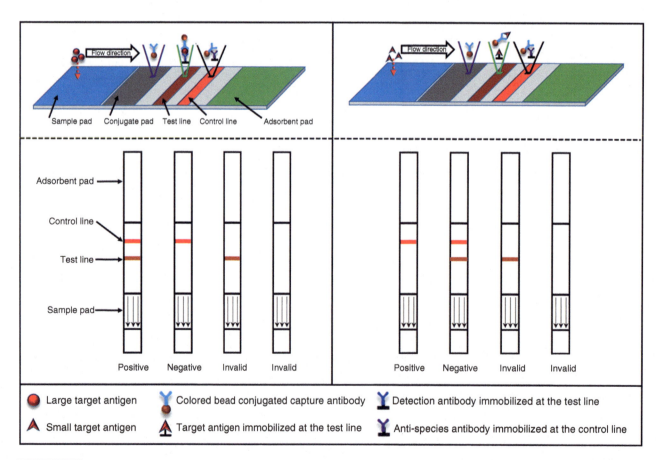

 Schematic view of a sandwich-type (noncompetitive) lateral flow test strip (**a**) and a competitive-type (**b**) lateral flow test strip for detection of large molecules. While the color in an enzyme immunoassay comes from enzyme action on its substrate to create a colored reaction product, the LFS methods instead use very small, spherical, colored particles (colloidal gold or colored latex) attached to antibodies to generate a positive colored signal. The capture antibody is immobilized in a zone on a porous membrane (usually nitrocellulose). By capillary action of the membrane, the test sample travels past the zone of immobilized antibody. The target protein in the test sample binds to the capture antibody. This type of sample movement and separation explains why LFS is called an **immunochromatographic assay**.

27.3.4.2 *Procedure*

Figure 27.12a depicts a typical LFS sandwich immunoassay and shows the various regions of the test strip. A primary antibody capable of binding the target protein is coated onto the surface of very small, colored particles (usually 20–40 nm diameter, colloidal gold or colored latex). These antibody-coated colored particles are dried in a porous pad. When these particles come in contact with liquid samples, they get reconstituted and are able to flow with the sample, moving across the **sample pad** of the strip by capillary action. Any large particulates in the sample liquid can be filtered out by a fiber filter placed at the front of the strip in the sample pad area. A second primary antibody, also capable of binding to the target protein, is immobilized at the **test line** (zone) on the surface of the fibers of the porous nitrocellulose membrane. A **control line** above the test line, containing an antispecies antibody capable of binding antibody-colored particle conjugates, serves to indicate that the test ran appropriately. Sample gets drawn through the test strip by an **absorbent pad** placed at the back of the strip.

Both liquid and solid samples can be analyzed by the LFS assay. A solid sample must be dissolved or dispersed in a liquid solution to extract the target protein (antigen) from the sample. To do the assay, the strip is put in contact with the sample solution at the end of the sample pad. The sample solution is drawn into the test strip by capillary action. The sample first passes through the filter, and the target protein is bound by the colored particle-antibody conjugate. Then the antigen-antibody-colored particle conjugate complex gets drawn into the membrane and is captured by the immobilized antibody in the test zone. A colored line becomes evident as more colored complexes are captured. The color intensity generally correlates with the amount of antigen present in the sample. A dedicated device can be used to measure the color intensity of the test line if a quantitative result is required. Samples containing no target protein show no color at the test line. The control zone binds the excess antibody-colored particle conjugates that pass through the test zone, and then forms a colored line for any complete test. If no colored line is

formed at the control zone, the assay is invalid and needs to be repeated on a new strip.

In the case of competitive LFS immunoassays, the target antigen that is immobilized on the test line and the free antigen from the sample solution will compete for the antibody-colored particle conjugates. Therefore, no colored line is formed for samples containing an antigen concentration above the cutoff value of the LFS test (Fig. 27.11b).

27.3.4.3 *Applications*

Current generation of LFS methods are made into versatile formats and are useful in applications for which their primary attributes (i.e., speed, simplicity, low cost, etc.) are important aspects of the testing (e.g., field testing). LFS methods have been developed for qualitative and to some extent quantitative monitoring of food allergens, foodborne pathogens, food toxins, hormones, and certain food protein ingredients to ensure food safety and quality. They also have been developed to detect the presence of genetically modified organisms in processed foods for consumers' interest, and to detect prohibited ruminant proteins in ruminant feedstuffs for the surveillance of mad cow disease. These easy-to-use rapid tests allow food processors and regulators to comply with regulations governing the labeling of food and feed products with minimal user training and facilities. Various devices of LFS for food allergen detection have been described in a recent handbook [5], and the strengths, weaknesses, opportunities, and threats of LFS methods have been thoroughly discussed in a review article [6].

27.4 IMMUNOAFFINITY PURIFICATION

Besides the use of antibodies in immunoassays as described above, often antibodies are used in food analyses as complements to other analytical methods. This is due to the remarkable specificity of antibodies and their strong binding to antigen. The most common example of this is **immunoaffinity** purification, which is an antigen capture technique. Basically the antibody is immobilized on some support, most often using a covalent linking method so that there is no concern with "bleeding" of the antibody in later steps (Fig. 27.13). The antibody can be bound to a solid phase such as agarose (e.g., Sepharose®) gel. These antibody-bound solid phases can be used later for purification of antigen via a chromatograaphy method or by the use of these phases on the surface of magnetic beads that are separated using a magnet. A simple purification sequence would involve exposing the antibody-bound solid to a food extract to first bind antigen, then washing the solid phase free of all unbound material, and finally releasing the pure antigen. Even though antibodies have such remarkably strong binding constants,

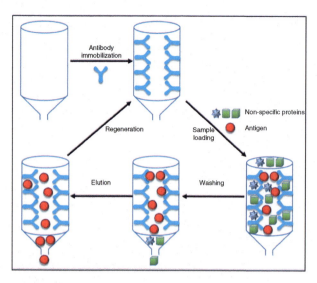

27.13 figure An example of antibody-bound immunoaffinity chromatography

they can be treated to release antigen by simple procedures such as changes in pH or solvent. Since the antibody is a protein, pH changes or solvent changes result in denaturation that changes the conformation of the binding site, causing the release of the antigen. If these changes are carefully selected, denaturation can be reversed by reestablishing moderate conditions so that the valuable antibody-bound solid phase can be reused repeatedly. For sensitive antigens, like enzymes, these elution conditions also can be a concern.

These immunoaffinity purification procedures have been used for small molecules like toxins (e.g., aflatoxins) and even materials as large as cells. Different microorganisms contain unique cell surface antigens that can be selectively bound to aid in purification and differentiation.

27.5   APPLICATIONS

Immunoassays are a well-developed area in food analysis and there is a plethora of literature available. For all sorts of laboratory techniques, Harlow and Lane [7] wrote one of the best books on the use of antibodies. The theory and practice of immunoassays is well handled in several books [8–10]. There are even entire journals, such as *Food and Agricultural Immunology*, devoted to describing methods for preparing food immunoassays.

Because of the speed, simplicity, sensitivity, and selectivity of immunoassays, they are used widely as screening tests for pesticide [11, 12] and drug residues [13, 14] in food (see Chap. 30). Besides chemical analysis, immunoassay techniques are used in microbiology to rapidly detect foodborne pathogens [15, 16] and

bacterial and fungal toxins [17, 18]. Immunoassays also are commonly used for meat and fish species identification [19]. Since immunoassays can easily be developed to detect trace amounts of specific proteins, they are among a number of methods used to detect hidden food allergens [20–22] and genetically modified organisms in foods [23]. In fact, immunoassay can be developed to detect almost any organic substances in a food system. The flexibility is limited only by the availability and quality of the specific antibody used. Immunoassays are being automated for higher analytical throughput and for improved data quality [24]. Food immunoassays can be prepared using very simple and rapid formats, making them ideal for kits used in the field. While every effort is made to control the specificity of these rapid tests, they can suffer from false positives and false negatives. For this reason, immunoassay kits are used most often as rapid screening tests, while food samples that test positive for the target analyte by immunoassay are often confirmed using another, more sophisticated analytical method. Immunoassays are also applicable in situations for which analysis by conventional methods is either not possible or is prohibitively expensive.

The research in the immunoassay area continues to develop the following: (1) specific antibodies, especially monoclonal antibodies for providing the ultimate specificity for the antibody-antigen interaction, (2) different solid-phase configurations to provide simpler means to carry out the assay, (3) new detection systems to further improve the assay sensitivity, and (4) alternative methodologies, such as immunosensors and immunoarray chips for multiple analyte detection and pattern recognition [16, 25].

27.6   SUMMARY

Almost any organic molecule in food can be determined using immunoassays as long as the specific antibodies are available. Both polyclonal antibodies and monoclonal antibodies or a combination of them can be used in an immunoassay. The remarkable selectivity and sensitivity of these assays are the result of the strong binding affinity between antibodies and their antigens. While the precise protocols of immunoassays can vary a great deal, all immunoassays use either a noncompetitive or a competitive format. The competitive format is the only one that can be used for quantification of small (about 1000 Da or less) molecules. ELISA has become the most popular immunoassay that uses an enzyme as the label to reveal the primary antibody-antigen binding through a color reaction catalyzed by the enzyme. In a noncompetitive ELISA with enzyme-derived color development, the more antibody-bound molecules (analyte) in the food sample, the more color develops, while in a competitive ELISA, the reverse is true. The most common labels used for food immunoassays are enzymes, and the most two common enzymes used are alkaline phosphatase and horseradish peroxidase. The reaction signals of both competitive and noncompetitive immunoassays can be detected using direct or indirect labeling methods, although indirect methods are used in most detection applications due to their various advantages over direct methods.

Two other commonly used immunoassays are immunoblot and LFS test. Two types of immunoblots, Western blot and dot blot, are commonly used in food analysis. Western blot is a laboratory-based immunoassay that combines PAGE and immunoassay to reveal the presence and the molecular mass of the antigenic protein on a membrane. Western blot is the most commonly used method for identifying and characterizing an antigen. A positive result also indicates the antigen conferring a linear epitope with the antibody. Dot blot is a simplified version of Western blot without the protein separation procedure, hence does not offer the molecular weight information of the antigen. The LFS, on the other hand, is by far the simplest form of immunoassay. The assay is easy to use and is ideal for outside-the-laboratory applications for which access to equipment and supplies is limited. While the general immunoassay procedure of the immunoblot is similar to ELISA, the LFS assay is a one-step assay based on one-direction movement of the antigen in a sample solution toward immobilized capturing antibodies at different zones. The separation of bound and unbound molecules by washing steps is not required. Because ELISA involves a signal amplification activity of the enzyme, it is generally regarded as inherently more sensitive than LFS methods. However, protein concentrations of less than a part per billion can be detected with a highly sensitive LFS.

Besides being the required constituent in immunoassays, antibodies also can be used to purify specific compounds in food for other analysis methods. These immunoaffinity purification methods allow for rapid purification of analytes from complex food matrices.

27.7 STUDY QUESTIONS

1. What is the relationship between an antigen and an antibody?
2. What is an epitope? What are the two types of epitopes?
3. What is the difference between monoclonal and polyclonal antibodies?
4. All immunoassays have two conditions that they must satisfy; what are they?
5. What is a hapten and what is a conjugate antigen?

6. What are five general steps for an ELISA procedure?
7. What is the rationale for the blocking step in an immunoassay protocols?
8. What is the difference between direct and indirect immunoassays? What are the advantages and disadvantages associated with each type of assay format?
9. Two common immunoassays are the sandwich assay and the competitive assay. Which molecules are best detected by each? Why?
10. What are two types of competitive ELISA procedures? Describe the main differences between these two.
11. Give four common applications of immunoassays in food analysis.
12. What is a Western blot? What is the major difference in reaction signals between Western blot and ELISA?
13. Compare and contrast ELISA and LFS, by identifying the similarities and differences in their characteristics, principles, and applications.
14. Describe, in general terms, how you would use immunoaffinity purification to isolate a protein for which you have developed antibodies.
15. All commercial potatoes contain the toxic glycoalkaloids α-solanine and α-chaconine. Both of these glycoalkaloids have the same large alkaloid portion, known as solanidine. Therefore polyclonal antibodies can be developed in rabbits against solanidine by chemically linking it to a foreign protein (foreign to the rabbit) and injecting the protein-conjugated hapten (solanidine linked using a succinic acid derivative) into rabbits. The antibodies that develop in the rabbit against the hapten bind to the alkaloid portion of both toxic glycoalkaloids. The rabbit antiserum containing primary polyclonal antibodies can be made highly specific to solanidine without cross-reactivity with other similar molecules by removing all cross-reactive antibody components using an immunoaffinity column procedure.

To develop an appropriate competitive ELISA, solanidine is again linked to a protein, but this time a different protein, and this conjugate is used to coat plastic microtiter plates. After excess conjugate is washed away, the plates are ready for the competitive ELISA procedure.

The glycoalkaloids in potatoes are extracted with methanol, and this extract is further diluted with water for use in the ELISA procedure. A standard curve is prepared by diluting standard solutions of α-chaconine at low, medium, and high concentrations with similar aqueous methanol solutions. In addition, a negative control is prepared using methanol and water at similar concentrations to the diluted potato extracts and standards, but without any glycoalkaloid present. Now the various extracts, standards, and negative controls are placed in individual wells with equivalent amounts of diluted rabbit serum containing the specific polyclonal antibodies. After incubation for 30 min at room temperature, all of the wells on the plate are again washed. Next a solution of commercially available goat anti-rabbit antibody conjugated to peroxidase is added to each well. After another 30-min incubation, the wells are again thoroughly washed.

Finally, phenylenediamine substrate solution is added to each well along with peroxide and again the plate is incubated for 30 min. After 30 min, the plate is rapidly read (in under 1 min) using an ELISA plate reader. The wells all contain differing amounts of yellow color:

(a) Tomatidine is a glycoalkaloid found in tomatoes and contains the alkaloid portion tomatine. Would the polyclonal antibodies detect tomatidine?

(b) Why is the protein that the hapten is attached to different for the ELISA procedure than for the injection?

(c) Is the ELISA protocol direct or indirect?

(d) Which wells would you expect to contain the most color, standards, potato extracts, or negative controls?

(e) Would you be concerned if a potato extract gave almost no color at the end of the ELISA procedure?

Acknowledgments The authors thank the following persons from the Institute of Sciences of Food Production (ISPA), National Research Council of Italy, for their helpful comments in revision of this chapter: Michelangelo Pascale, Veronica Lattanzio, and Annalisa De Girolamo.

REFERENCES

1. Köhler G, Milstein C (1975) Continuous cultures of fused cells secreting antibody of predefined specificity. Nature 256: 495–497

2. Howard GC, Bethell DR (2001) Basic methods in antibody production and characterization. CRC, Boca Raton, FL

3. Yalow RS, Berson SA (1960) Immunoassay of endogenous plasma insulin in man. J Clin Inves 39: 1157–1175

4. Engvall E, Perlmann P (1971) Enzyme-linked immunosorbent assay, ELISA III. Quantitation of specific antibodies by enzyme-labeled anti-immunoglobulin in antigen-coated tubes. J Immunol 109: 129–135

5. Baumert JL, Tran DH (2015) Lateral flow devices for detecting allergens in food. In: Flanagan S, editor. Handbook of Food Allergen Detection and Control. Cambridge, UK. pp. 219–228

6. Posthuma-Trumpie GA, Korf J, van Amerongen A (2009) Lateral flow (immuno) assay: its strengths, weaknesses, opportunities and threats. A literature survey. Anal Bioanal Chem 393: 569–82

7. Harlow E, Lane D (1999) Using antibodies: a laboratory manual. Cold Spring Harbor Laboratory Press, Cold Spring Harbor, New York

8. Crowther JR (2010) The ELISA guidebook. 2nd ed. Humana Press. New York

9. Deshpande SS (1996) Enzyme immunoassays: from concept to product development. Chapman and Hall, New York

10. Wild D (2013) The immunoassay handbook: theory and applications of ligand binding, ELISA, and related techniques. 4th ed. Waltham, MA, USA

11. Gabaldón JA, Maquieriera A, Puchades R (1999) Current trends in immunoassay-based kits for pesticide analysis. Crit Rev Food Sci Nutr 39: 519–538

12. Morozova VS, Levashova AI, Eremin SA (2005) Determination of pesticides by enzyme immunoassay. J Anal Chem 60: 202–217

13. Mitchell JM, Griffiths MW, McEwen SA, McNab WB, Yee AJ (1998) Antimicrobial drug residues in milk and meat: causes, concerns, prevalence, regulations, tests and test performance. J Food Prot 61: 742–756

14. Raig M, Toldrá F (2008) Veterinary drug residues in meat: concerns and rapid methods for detection. Meat Sci 78:60–67

15. Swaminathan B, Feng P (1994) Rapid detection of food-borne pathogenic bacteria. Annu Rev Microbiol 48: 401–426

16. Banada PP, Bhunia AK (2008) Antibodies and immunoassays for detection of bacterial pathogens. Ch. 21. In: Zourob M, Elwary S, Turner A (eds) Principles of bacterial detection: biosensors, recognition receptors and microsystems, Springer, New York

17. Pimbley DW, Patel PD (1998) A review of analytical methods for the detection of bacterial toxins. J Appl Microbiol 84: 98S–109S

18. Li W, Powers S, Dai SY (2014) Using commercial immunoassay kits for mycotoxins: 'joys and sorrows'? World Mycotoxin Journal, 7:417–430

19. Hsieh Y-HP (2005) Meat species identification. In: Hui YH (ed) Handbook: food science, technology and engineering. CRC, Boca Raton, FL, pp 30–1–30–19

20. Immer U, Lacorn M (2015) Enzyme-linked immunosorbent assays (ELISAs) for detecting allergens in food. In: Flanagan S, editor. Handbook of Food Allergen Detection and Control. Cambridge, UK. pp. 199–217

21. Owusu-Apenten RK (2002) Determination of trace protein allergens in foods. Ch. 11. In: Food protein analysis. Quantitative effects on processing. Marcel Dekker, New York, pp 297–339

22. Poms RE, Klein CL, Anklam E (2004) Methods for allergen analysis in food: a review. Food Addit Contam 21: 1–31

23. Ahmed FE (2002) Detection of genetically modified organisms in foods. Trends Biotechnol 20: 215–223

24. Bock JL (2000) The new era of automated immunoassay. Am J Clin Pathol 113: 628–646

25. Corgier BP, Marquette CA, Blum LJ (2007) Direct electrochemical addressing of immunoglobulins: Immuno-chip on screen-printed microarray. Biosens Bioelectron 22: 1522–1526

Determination of Oxygen Demand

Yong D. Hang
Department of Food Science and Technology,
Cornell University,
Geneva, NY 14456, USA
e-mail: ydh1@cornell.edu

S. Nielsen (ed.), *Food Analysis*, Food Science Text Series,
DOI 10.1007/978-3-319-45776-5_28, © Springer International Publishing 2017

28.1   INTRODUCTION

Oxygen demand is a commonly used parameter to evaluate the potential effect of organic pollutants on either a wastewater treatment process or a receiving water body. Because microorganisms utilize these organic materials, the concentration of dissolved oxygen is greatly depleted from the water. The oxygen depletion in the environment can have a detrimental effect on fish and plant life.

The two main methods used to measure the oxygen demand of water and wastewater are **biochemical oxygen demand** (BOD) and **chemical oxygen demand** (COD). This chapter briefly describes the principles, procedures, applications, and limitations of each method. Methods described in this chapter are adapted from *Standard Methods for the Examination of Water and Wastewater*, published by the American Public Health Association (APHA) [1]. The book includes step-by-step procedures with equipment for BOD, COD, and other tests for water and wastewater.

28.2   METHODS

28.2.1   Biochemical Oxygen Demand

28.2.1.1 *Principle*

The **biochemical oxygen demand** (BOD) determination is a measure of the amount of oxygen required by microorganisms to oxidize the biodegradable organic constituents present in water and wastewater. The method is based on the direct relationship between the concentration of organic matter and the amount of oxygen used to oxidize the pollutants to water, carbon dioxide, and inorganic nitrogenous compounds. The **oxygen demand** of water and wastewater is proportional to the amount of **organic matter** present. The BOD method measures the biodegradable carbon (carbonaceous demand) and, under certain circumstances, the biodegradable nitrogen (nitrogenous demand).

28.2.1.2 *Procedure*

Place a known amount of a water or wastewater sample that has been seeded with an effluent from a biological waste treatment plant in an airtight BOD bottle, and measure the initial dissolved oxygen immediately. Incubate the sample at 20 °C and, after 5 days, measure the dissolved oxygen content again (APHA Method 4500-0). The dissolved oxygen content can be determined by the membrane electrode method (APHA Method 4500-O G) or the azide modification (APHA Method 4500-0 C), permanganate modification (APHA Method 4500-0 D), alum flocculation modification (4500-0 E), or copper sulfate-sulfamic acid flocculation modification (APHA Method 4500-0 F) of the iodometric method (APHA Method 4500-0 B)

to minimize interference by nitrite or ferrous or ferric iron. The iodometric method is a titrimetric procedure that is based on the oxidizing property of dissolved oxygen, while the membrane electrode method is based on the diffusion rate of dissolved oxygen across a membrane. A dissolved oxygen meter with an oxygen-sensitive membrane electrode made by Fisher, Orion, YSI, or other companies is used to measure the diffusion current, which is linearly proportional to the concentration of dissolved oxygen under steady-state conditions. It is important to change frequently and calibrate the membrane electrode to eliminate the effect of interfering gases such as hydrogen sulfide. The azide-modified iodometric procedure, for example, is used to remove interference of nitrite, which is the most commonly interfering material in water and wastewater. The alum flocculation modification method is commonly used to minimize the interference caused by the presence of suspended solids. The BOD value, which is expressed as mg/L, can be calculated from the difference in the initial dissolved oxygen and the content of dissolved oxygen after the incubation period according to the following equation (APHA Method 5210 B):

$$BOD\,(mg\,/\,L) = 100\,/\,P \times (DOB - DOD) \quad (28.1)$$

where:

DOB = initial oxygen in diluted sample, mg/L
DOD = oxygen in diluted sample after 5-day incubation, mg/L
P = mL sample × 100/capacity of bottle

28.2.1.3 *Applications and Limitations*

The BOD test is used most widely to measure the organic loading of waste treatment processes, to determine the efficiency of treatment systems, and to assess the effect of wastewater on the quality of receiving waters. The 5-day BOD test has some drawbacks because:

1. The procedure requires an incubation time of at least 5 days.
2. The BOD method does not measure all the organic materials that are biodegradable.
3. The test is not accurate without a proper seeding material.
4. Toxic substances such as chlorine present in water and wastewater may inhibit microbial growth.

28.2.2   Chemical Oxygen Demand

28.2.2.1 *Principle*

The **chemical oxygen demand** (COD) determination is a rapid way to estimate the quantity of oxygen used to oxidize the organic matter present in water and wastewater. Most organic compounds are destroyed by refluxing in a strong acid solution with a known quantity of a strong oxidizing agent such as **potassium dichromate**. The excess amount of potassium dichromate left after digestion of the organic matter is measured. The amount of organic matter that is chemically oxidizable is directly proportional to the potassium dichromate consumed.

28.2.2.2 *Procedure*

A known quantity of sample of water or wastewater is refluxed at elevated temperatures for up to 2 h with a known quantity of potassium dichromate and sulfuric acid using an open reflux method (APHA Method 5220 B), a closed reflux titrimetric method (APHA Method 5220 C), or a closed reflux colorimetric method (APHA Method 5220 D). The amount of potassium dichromate left after digestion of the organic matter is titrated with a standard ferrous ammonium sulfate (FAS) solution using orthophenanthroline ferrous complex as an indicator. The amount of oxidizable organic matter, determined as oxygen equivalent, is proportional to the potassium dichromate used in the oxidative reaction. The COD value can be calculated from the following equation (APHA Method 5220 B):

$$COD\,(mg\,/\,L) = (A - B) \times M \times 8000\,/\,D \quad (28.2)$$

where:

A = mL FAS used for blank
B = mL FAS used for sample
M = molarity of FAS
D = mL sample used
8000 = milliequivalent weight of oxygen × 1,000 mL/L

28.2.2.3 *Applications and Limitations*

Potassium dichromate is widely used for the COD method because of its advantages over other oxidizing compounds in oxidizability, applicability to a wide variety of waste samples, and ease of manipulation. The dichromate reflux method can be used to measure the samples with COD values of greater than 50 mg/L.

The COD test measures carbon and hydrogen in organic constituents but not nitrogenous compounds. Furthermore, the method does not differentiate between biologically stable and unstable compounds present in water and wastewater. The COD test is a very important procedure for routinely monitoring industrial wastewater discharges and for the control of waste treatment processes. The test is faster and more reproducible than the BOD method. The obvious disadvantages of the COD method are:

1. Aromatic hydrocarbons, pyridine, and straight-chain aliphatic compounds are not readily oxidized.

2. The method is very susceptible to interference by **chloride**, and thus the COD of certain food processing waste effluents such as pickle and sauerkraut brines cannot be readily determined without modification. This difficulty may be overcome by adding **mercuric sulfate** to the sample prior to refluxing. Chloride concentrations greater than 500–1000 mg/L may not be corrected by the addition of mercuric sulfate. A chloride correction factor can be developed for a particular waste by the use of proper blanks.

28.3 COMPARISON OF BOD AND COD METHODS

The BOD and COD analyses of water and wastewater can result in different values because the two methods measure different materials. As shown in Table 28.1, the COD value of a waste sample is usually higher than its BOD because:

1. Many organic compounds that can be chemically oxidized cannot be biochemically oxidized. For example, cellulose cannot be determined by the BOD method but can be measured by the COD test.
2. Certain inorganic compounds such as ferrous iron, nitrites, sulfides, and thiosulfates are readily oxidized by potassium dichromate. This inorganic COD introduces an error when computing the organic matter of water and wastewater.
3. The BOD test can give low values because of a poor seeding material. The COD test does not require an inoculum.
4. Some aromatics and nitrogenous (ammonium) compounds are not oxidized by the COD method. Other organic constituents such as cellulose or lignin, which are readily oxidized by potassium dichromate, are not biologically degraded by the BOD method.
5. Toxic materials present in water and wastewater that do not interfere with the COD test can affect the BOD results.

28.1 table Oxygen demand of tomato processing wastes

Item	1973	1974	1975
BOD, mg/L	2,400	1,300	1,200
COD, mg/L	5,500	3,000	2,800
TOC, mg/L	2,000	1,100	1,000

From [2]

The COD has value for specific wastes since it is possible to obtain a direct correlation between COD and BOD values. Table 28.2 shows the COD and BOD values of waste effluents from fruit and vegetable processing factories. The BOD/COD ratios of these processing waste effluents varied considerably and ranged from 0.50 to 0.72 [2]. The **BOD/COD ratio** can be a useful tool for rapid determination of the biodegradability of organic matter present in the wastes. A low BOD/COD ratio indicates the presence of a large amount of nonbiodegradable organic matter. Samples of wastewater with high BOD/COD ratios have a small amount of organic matter that is nonbiodegradable.

28.2 table COD and BOD values of selected fruit and vegetable processing wastes

Product	COD (mg/L)	BOD (mg/L)	Mean ratio (BOD/COD)
Apples	395–37,000	240–19,000	0.55
Beets	445–13,240	530–6,400	0.57
Carrots	1,750–2,910	817–1,927	0.52
Cherries	1,200–3,795	600–1,900	0.53
Corn	3,400–10,100	1,587–5,341	0.50
Green beans	78–2,200	43–1,400	0.55
Peas	723–2,284	337–1,350	0.61
Sauerkraut	470–65,000	300–41,000	0.66
Tomatoes	652–2,305	454–1,575	0.72
Wax beans	193–597	55–323	0.58
Wine	495–12,200	363–7,645	0.60

From Splittstoesser and Downing [2]

28.4 SAMPLING AND HANDLING REQUIREMENTS

Samples of water and wastewater collected for oxygen demand determinations must be analyzed as soon as possible or stored under properly controlled conditions until analyses can be made. Samples for the BOD test can be kept at low temperatures (4 °C or below) for up to 48 h. Chemical preservatives should not be added to water and wastewater because they can interfere with BOD analysis. Untreated wastewater samples for the COD test must be collected in glass containers and analyzed promptly. The COD samples can be stored at 4 °C or below for up to 28 days if these are acidified with a concentrated mineral acid (sulfuric acid) to a pH value of 2.0 or below.

28.5 SUMMARY

Oxygen demand is most widely used to determine the effect of organic pollutants present in water and wastewater on receiving streams and rivers. The two important methods used to measure oxygen demand are BOD and COD. Table 28.3 summarizes the principle, advantages, and disadvantages of BOD and COD. The BOD test measures the amount of oxygen required by microorganisms to oxidize the biodegradable organic matter present in water and wastewater. The COD method determines the quantity of oxygen consumed during the oxidation of organic matter in water and wastewater by potassium dichromate.

28.3 table Comparison between biochemical oxygen demand and chemical oxygen demand

	Principal	Advantages	Disadvantages
Biochemical oxygen demand (BOD)	Measure amount of oxygen required by microorganisms to oxidize the biodegradable organic matter present in water and wastewater (i.e., correlation between amount of organic matter and amount of oxygen used to oxidize pollutants)	Measures true compounds of interest (i.e., organic matter). Less expensive. No interference from certain compounds that affect COD results (see text for details)	Slower; less precise. Gives low values if poor seed material. Requires inoculum
Chemical oxygen demand (COD)	Organic compounds are destroyed by refluxing in strong acid solution with a known excess of an oxidizing agent (potassium dichromate) (i.e., correlation between amount of organic matter chemically oxidized and amount of oxidizing agent consumed)	Faster. More precise	Not a direct measurement of organic matter. More expensive. Can overestimate organic matter if sample is high in certain compounds and can underestimate if high in other compounds (see text for details)

Of the two methods used to measure oxygen demand, the BOD test has the widest application in measuring waste loading to treatment systems, in determining the efficiency of treatment processes, and in evaluating the quality of receiving streams and rivers because it most closely approximates the natural conditions of the environment. The COD test can be used to monitor routinely the biodegradability of organic matter in water and wastewater if a relationship between COD and BOD has been established.

28.6   STUDY QUESTIONS

1. In your new job as supervisor of a lab that has previously been using the BOD method to determine oxygen demand of wastewater, you have decided to change to the COD method.

   (a) Differentiate the basic principle and procedure of the BOD and COD methods for your lab technicians.
   (b) In what case would they be instructed to use mercuric sulfate in the COD assay?
   (c) How do you justify making the change from the BOD method to the COD method?

2. In each case described below, indicate if you would expect the COD value to be higher or lower than the results from a BOD test. Explain your answer.

   (a) Poor seed material in BOD test
   (b) Sample containing toxic materials
   (c) Sample high in aromatics and nitrogenous compounds
   (d) Sample high in nitrites and ferrous iron
   (e) Sample high in cellulose and lignin

28.7   PRACTICE PROBLEMS

1. Determine the BOD value of a sample given the following data (see Eq. 28.1):

   DOB = 9.0 mg/L
   DOD = 6.6 mg/L
   P = 15 mL
   Capacity of bottle = 300 mL

2. Determine the COD value of a sample given the following data (see Eq. 28.2):

   mL FAS for blank = 37.8 mL
   mL FAS for sample = 34.4 mL
   Molarity of FAS = 0.025 M
   Sample = 5 mL

Answers

1. BOD = 48 mg/L

   Calculation:

   $$\begin{aligned} BOD(mg/L) &= 100/P \times (9.0mg/L - 6.6mg/L) \\ &= 100/P \times 2.4 \\ &= 240/P \\ &= 240/(15\ mL \times 100/300\ mL) \\ &= 240/5 \\ &= 48 \end{aligned}$$

2. COD = 136 mg/L
   Calculation:

   $$\begin{aligned} COD\ (mg/L) &= (37.8mL - 34.4mL) \times 0.025 \times 8000/D \\ &= 3.4 \times 0.025 \times 8000/D \\ &= 680/D \\ &= 680/5 \\ &= 136 \end{aligned}$$

REFERENCES

1. Rice EW, Baird, RB. Eaton AD, Clesceri LS (eds) (2012) Standard methods for the examination of water and wastewater, 22nd edn. American Public Health Association (APHA), Washington, DC
2. Splittstoesser DF, Downing DL (1969) Analysis of effluents from fruit and vegetable processing factories. NY State Agr Exp Sta Res Circ 17. Geneva, New York

part 6

Analysis of Physical Properties of Foods

chapter 29

Rheological Principles for Food Analysis

Helen S. Joyner (Melito) (✉)
School of Food Science,
University of Idaho,
Moscow, ID 83843-2312, USA
e-mail: hjoyner@uidaho.edu

Christopher R. Daubert
Department of Food, Bioprocessing & Nutrition Sciences,
North Carolina State University,
Raleigh, NC 27695-7624, USA
e-mail: cdaubert@ncsu.edu

S. Nielsen (ed.), *Food Analysis*, Food Science Text Series,
DOI 10.1007/978-3-319-45776-5_29, © Springer International Publishing 2017

29.1 INTRODUCTION

29.1.1 Rheology and Texture

Food scientists are routinely confronted with the need to measure physical properties related to sensory texture and behavior during processing. These properties are determined by **rheological methods**; **rheology** is a science devoted to the deformation and flow of all materials. Rheological properties should be considered a subset of the textural properties of foods, because sensory evaluation of texture encompasses factors beyond rheological properties. Specifically, rheological methods accurately measure force, deformation, and flow, and food scientists and engineers must determine how best to apply this information. For example, the flow of salad dressing from a bottle, the snapping of a candy bar, or the pumping of cream through a homogenizer are each related to the rheological properties of these materials. In this chapter, we describe fundamental concepts pertinent to the understanding of food rheology and discuss typical examples of rheological tests for common foods. A glossary is included to define and summarize the rheological terms used throughout the chapter.

29.1.2 Fundamental and Empirical Methods

Rheological properties are determined by applying and measuring forces and deformations as a function of time. Both fundamental and empirical measurement methods may be used. **Fundamental methods** account for the magnitude and direction of forces and deformations while placing restrictions on acceptable sample shapes and composition. Fundamental tests have the advantage of being based on known concepts and equations of physics. Therefore, fundamental tests performed on different testing equipment with different sample geometries yield comparable results. When sample composition or geometry is too complex to account for all forces and deformations, empirical methods are often used. These **empirical methods** are typically descriptive in nature and ideal for rapid analysis. However, the results of empirical tests are dependent on the equipment and sample geometry, so it may be difficult to accurately compare data among different samples. Empirical tests are of value especially when they correlate with a property of interest, whereas fundamental tests determine true physical properties.

29.1.3 Basic Assumptions for Fundamental Rheological Methods

Two important assumptions for fundamental methodologies are that the material is **homogeneous** and **isotropic**. **Homogeneity** implies a well-mixed and compositionally similar material, an assumption generally valid for fluid foods provided they are not a suspension of large particles, such as vegetable soup. For example, milk, infant formula, and apple juice are each considered homoge-

neous and isotropic. Homogeneity is more problematic in solid foods. For example, frankfurters without skins can be considered homogeneous. However, when particle size is significant, such as fat particles in some processed meats like salami, one must determine if homogeneity is a valid assumption. **Isotropic** materials display a consistent response to a load regardless of the applied direction. In foods such as a steak, muscle fibers make the material anisotropic so the response varies with the direction of the force or deformation.

29.2 FUNDAMENTALS OF RHEOLOGY

Rheology is concerned with how all materials respond to applied forces and deformations, and food rheology is the material science devoted to foods. Basic concepts of **stress** (force per area) and **strain** (relative deformation) are key to all rheological evaluations. Special constants of proportionality, called **moduli**, link stress with strain. Materials that are considered ideal solids (e.g., gelatin gels) obey **Hooke's Law**, where stress is related directly with strain via a modulus. Materials that are considered ideal fluids (e.g., water, honey) obey **Newtonian** principles, and the proportionality constant is commonly referred to as **viscosity**, defined as an internal resistance to flow. These principles for solid and fluid behavior form the foundation for the entire chapter and are described in this section, with additional detail available in Steffe and Daubert [1].

29.2.1 Concepts of Stress

Stress (σ) is always a measurement of **force**. Defined as the force (F, **Newtons**) divided by the area (A, **meters**^{2}) over which the force is applied, stress is generally expressed with units of **Pascals** (Pa). To illustrate the notion of stress, imagine placing a water balloon on a table as opposed to placing it on the tip of a pin (Fig. 29.1). Obviously the tip of the pin has a considerably smaller surface area, causing the stress, or

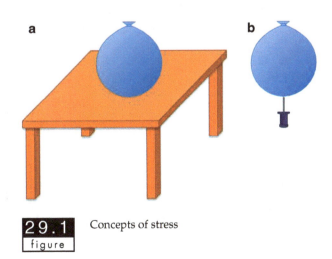

a

b

29.1
figure   Concepts of stress

weight (force due to gravity) per unit area of contact, to be larger when compared with the stress of a balloon resting on a tabletop. Although the force magnitude – the weight of the water balloon – is a constant value for each case, the final outcomes will be very different.

The direction of the force, with respect to the surface area impacted, determines the type of stress. For example, if the force is directly perpendicular to a surface, a **normal stress** results and can be achieved under tension or compression. Should the force act in parallel to the sample surface, a **shear stress** is experienced. Examples of normal stresses include the everyday practices of pressing the edges of a piecrust together and biting through a solid food. Examples of shear forces, on the other hand, occur when spreading butter over a slice of toast, brushing barbecue sauce on chicken, or stirring milk into a cup of coffee.

29.2.2 Concepts of Strain and (Shear) Strain Rate

When a stress is applied to a food, the food deforms or flows. **Strain** is a dimensionless quantity representing the relative deformation of a material, and the direction of the applied stress with respect to the material surface will determine the type of strain. If the stress is normal (perpendicular) to a sample surface, the material will experience **normal strain** (ε). Foods show normal strains when they are compressed (compressive stress) or stretched (tensile stress).

Normal strain (ε) can be calculated as a true strain from an integration over the deformed length of the material (Fig. 29.2):

$$\varepsilon = \int_{L_i}^{L_i + \Delta L} \frac{dL}{L} = \ln\left(1 + \frac{\Delta L}{L}\right) \qquad (29.1)$$

According to Steffe [2], a true strain is more applicable to larger deformations such as may occur during texture profile analysis testing (see Sect. 29.4.1.1.2). Strain calculations result in negative valves for compression and positive values for extension (tensile strains). Rather than expressing a negative strain, many typically record the absolute value of the strain and denote the compressive test mode:

$$\varepsilon = -0.05 = 0.05_{\text{compresion}} \qquad (29.2)$$

On the other hand, when a sample encounters a shear stress, such as the pumping of tomato paste through a pipe, a **shear strain** (γ) is observed. Figure 29.3 shows how a sample deforms when a shear stress is applied. Shear strain is determined from applications of geometry as:

$$\tan(\gamma) = \frac{\Delta L}{h} \qquad (29.3)$$

or:

$$\gamma = \tan^{-1}\left(\frac{\Delta L}{h}\right) \qquad (29.4)$$

where h is the specimen height. For simplification, during exposure to small strains, the angle of shear may be considered equal to the shear strain:

$$\tan(\gamma) \approx \gamma \qquad (29.5)$$

When the material is a liquid, this approach for strain quantification is a bit more challenging. As coffee is

a

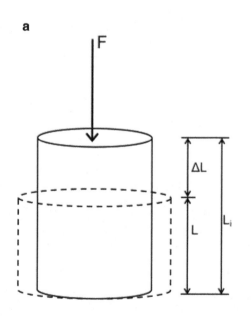

b

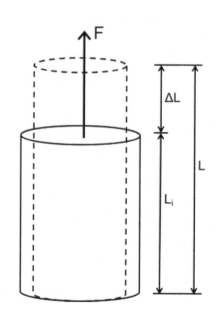

  Normal strain in a cylinder in (**a**) compression and (**b**) tension

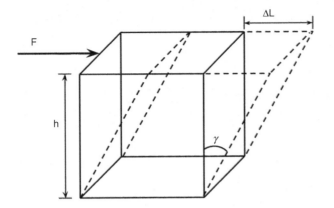

Shear strain in a cube

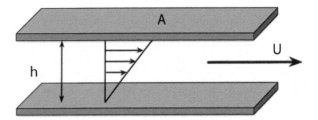

29.4 figure
Shear flow between parallel plates

stirred, water is pumped, or milk is pasteurized, these fluids all are exposed to shear and display irrecoverable deformation. Therefore, a (**shear**) **strain rate** $(\dot\gamma)$, often called shear rate, is typically used to quantify strain during fluid flow. Shear rate is the amount of deformation (strain) per unit time with units of s^{-1}:

$$\gamma = \frac{\Delta L}{h} \qquad (29.6)$$

$$\frac{d\gamma}{dt} = \frac{d\left(\dfrac{\Delta L}{h}\right)}{dt} = \dot\gamma \qquad (29.7)$$

$$\dot\gamma = \frac{1}{h}\frac{d}{dt}(\Delta L) \qquad (29.8)$$

$$\dot\gamma = \frac{U}{h} \qquad (29.9)$$

To picture the shear rate concept, consider a fluid filling the gap between two moveable, parallel plates separated by a known distance, h, as illustrated in Fig. 29.4. Now, set one plate in motion with respect to the other at a constant horizontal velocity, U.

The shear rate for this system can be approximated by dividing the plate velocity by the fluid gap height, producing a value with units of s^{-1}. This shear rate may be more easily understood through the deck of cards analogy. Imagine a stack of playing cards, with each card representing an infinitely thin layer of fluid. When the top card is pushed with some force, the entire deck deforms to some degree proportional with the magnitude of the force. This type of movement is commonly called **simple shear** and may be defined as a laminar deformation along a plane parallel to the applied force.

29.1 table
Elastic and shear moduli for common materials

Material	E, elastic moduli (Pa)	G, shear moduli (Pa)
Apple	1.0×10^7	0.38×10^7
Potato	1.0×10^7	0.33×10^7
Spaghetti, dry	0.27×10^{10}	0.11×10^{10}
Glass	7.0×10^{10}	2.0×10^{10}
Steel	25.0×10^{10}	8.0×10^{10}

29.2.3 Solids: Elastic and Shear Moduli

Hooke's Law states that when a solid material is exposed to a stress, it experiences an amount of deformation or strain proportional to the magnitude of the stress. The constants of proportionality, used to equate stress with strain, are called **moduli**:

$$\text{Stress}(\sigma) \propto \text{Strain}(\varepsilon \text{ or } \gamma) \qquad (29.10)$$

$$\text{Stress} = \text{Modulus} \times \text{Strain} \qquad (29.11)$$

If a normal stress is applied to a sample, the proportionality constant is known as **elastic modulus** (E), often called Young's modulus:

$$\sigma = \frac{F}{A} = E\varepsilon \qquad (29.12)$$

Likewise, if the applied stress is shear stress, the constant is the **shear modulus** (G):

$$\sigma = G\gamma \qquad (29.13)$$

These moduli are inherent properties of the material and have been used as indicators of quality. Moduli of selected foods and materials are provided in Table 29.1.

29.2.4 Fluid Viscosity

For the case of the simplest kind of fluid, the viscosity is constant and independent of shear rate and time. In other words, **Newton's postulate** is obeyed. This postulate states that if the shear stress is doubled, the velocity gradient (shear rate) within the fluid is also

29.2 table	Newtonian viscosities for common materials at 20 °C

Material	Viscosity, μ (Pa s)
Honey	11.0
Rapeseed oil	0.163
Olive oil	8.4×10^{-2}
Cottonseed oil	7.0×10^{-2}
Raw milk	2.0×10^{-3}
Water	1.0×10^{-3}
Air	1.81×10^{-5}

Adapted from [2, 3]

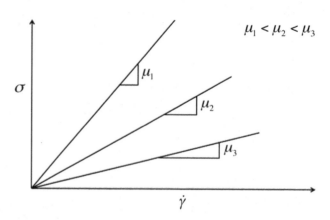

29.5 figure	Rheogram for three different Newtonian fluids

29.3 table	Summary of shear-dependent terminology

	Time independent	Time dependent
Thinning	Pseudoplastic	Thixotropic
Thickening	Dilatent	Rheopectic (anti-thixotropic)

doubled. Typically for fluids, the shear stress is expressed as some function of shear rate and a viscosity term that provides an indication of the internal resistance of a fluid to flow. For **Newtonian fluids**, the viscosity function is constant and called the **coefficient of viscosity** or **Newtonian viscosity** (μ):

$$\sigma = \mu \dot{\gamma} \qquad (29.14)$$

However, for most liquids the viscosity term is not constant, but rather changes as a function of shear rate, and the material is considered **non-Newtonian**. A function called the **apparent viscosity** (η) is defined as the shear-dependent viscosity. Mathematically, the apparent viscosity function is the result of the shear stress divided by the shear rate:

$$\eta = f(\dot{\gamma}) = \frac{\sigma}{\dot{\gamma}} \qquad (29.15)$$

Table 29.2 provides Newtonian viscosities for some common items at 20 °C [2, 3]. Temperature is a very important parameter when describing rheological properties. Viscosity generally decreases as temperature increases.

29.2.5 Fluid Rheograms

The flow behavior of a food can be described with fluid **rheograms** or appropriate rheological models (see Sect. 29.3). A rheogram is a graphical representation of flow behavior, showing the relationship between stress and strain or shear rate. Much can be learned from inspection of rheograms. For example, if a plot of shear stress versus shear rate results in a straight line passing through the origin, the material is a Newtonian fluid, with the slope of the line equaling the Newtonian viscosity (μ), shown in Fig. 29.5. Many common foods exhibit this ideal response, including water, milk, vegetable oils, and honey.

As established earlier, the majority of fluid foods do not show Newtonian flow behavior. The flow changes with shear rate (i.e., mixing speed) or with time at a constant shear rate. Time-independent

deviation from ideal Newtonian behavior will cause the relationship between shear stress and shear rate to be nonlinear. If the viscosity decreases as shear rate increases, the material is referred to as **shear thinning** or **pseudoplastic**. Examples of pseudoplastic food items are applesauce and pie fillings. On the other hand, if the viscosity of the material increases with increased shear rate, the sample is called **shear thickening** or **dilatent**. Cornstarch slurries are well known for dilatent behavior.

Pseudoplasticity and dilatency are **time-independent properties**. Materials that thin and thicken with time are known as **thixotropic** and **rheopectic** liquids, respectively. These fluids are easily detected by monitoring the viscosity at a constant shear rate with respect to time. If pumpkin pie filling is mixed at a constant speed, the material thins (thixotropy) with time due to the destruction of weak bonds linking the molecules. Table 29.3 summarizes the terminology for time-dependent and independent responses.

Many fluids do not flow at low magnitudes of stresses. In fact, a certain catsup brand once staked numerous marketing claims on the "anticipation" of flow from the bottle. Often, additional force was applied to the catsup container to expedite the pouring. The minimum force, or stress, required to initiate flow is known as a **yield stress** (σ_0). Because Newtonian fluids require the stress-shear rate relationship to be a continuous straight line passing through the origin, any material with a yield stress is automatically non-Newtonian. A few common foods possessing yield

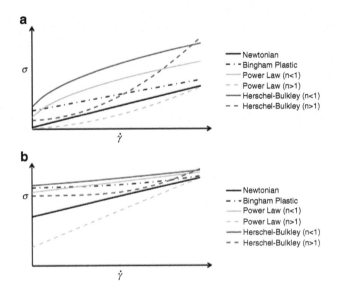

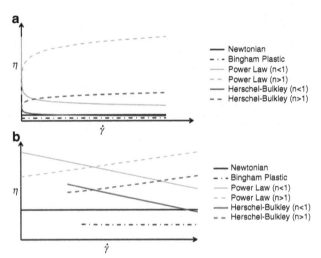

figure 29.6 Typical stress versus shear rate behavior for various flow models in (**a**) normal and (**b**) log scale

Figure 29.7 Typical viscosity versus shear rate behavior flow models in (**a**) normal and (**b**) log scale

stresses are catsup, yogurt, mayonnaise, and salad dressing.

Many foods are explicitly designed to include a certain yield stress. For example, if melted cheese did not have a yield stress, the cheese would flow off a cheeseburger or pizza. If salad dressings flowed at the lowest of applied stresses, the force of gravity would cause the dressing to run off the salad leaves. There are many fascinating rheological features of foods that consumers may never consider!

29.3 RHEOLOGICAL FLUID MODELS

Once data for shear stress and shear rate are collected, **rheological models** can be used to gain a greater understanding of the flow response. Rheological models are mathematical expressions relating shear stress to shear rate, providing a "flow fingerprint" for a particular food. In addition, the models permit prediction of rheological behavior across a wide range of processing conditions. Figures 29.6 and 29.7 show typical rheograms of stress versus shear rate and viscosity versus shear rate for several rheological models.

29.3.1 Herschel-Bulkley Model

For most practical purposes, the **Herschel-Bulkley model** can account for the steady-state rheological performance of many fluid foods:

$$\sigma = \sigma_o + K\dot{\gamma}^n \qquad (29.16)$$

K and n represent material constants called the **consistency coefficient** and **flow behavior index**, respectively. The flow behavior index provides an indication

table 29.4 Manipulation of Herschel-Bulkley model to describe flow behavior

Fluid type	σ_o	n
Newtonian	0	1.0
Non-Newtonian		
Pseudoplastic	0	<1.0
Dilatent	0	>1.0
Yield stress	>0	Any

σ_o yield stress; n flow behavior index

of Newtonian or non-Newtonian flow, provided the material has no yield stress. Table 29.4 illustrates how the Herschel-Bulkley model is used to identify specific flow characteristics.

Table 29.5 displays Herschel-Bulkley model data obtained for a variety of food products. Take, for example, the data for peanut oil. Table 29.5 reports no yield stress and a flow behavior index of 1.00 – the scenario for a Newtonian fluid. Accordingly, the Herschel-Bulkley model does in fact collapse to the special Newtonian case. The following models are considered as simple modifications to the Herschel-Bulkley model [2].

29.3.2 Newtonian Model [$n=1$; $K=\mu$; $\sigma_o=0$]

For Newtonian fluids, Eq. 29.17 is manipulated with the flow behavior index (n) equaling 1.0 and the consistency coefficient (K) equaling the Newtonian viscosity (μ):

$$\sigma = 0 + \mu\dot{\gamma}^1 \qquad (29.17)$$

or

29.5 table Herschel-Bulkley model data for common foods

Product	Temp. (°C)	Shear rate (s⁻¹)	K (Pa sⁿ)	n (–)	σ₀ (Pa s)
Orange juice (13 °Brix)	30	100–600	3.2	0.79	–
			0.06	0.86	
Orange juice (22 °Brix)	30	100–600	3.2	0.79	–
			0.12	0.79	
Orange juice (33 °Brix)	30	100–600	0.14	0.78	–
Apple juice (35 °Brix)	25	3–2,000	0.001	1.00	–
Ketchup	25	50–2,000	6.1	0.40	–
Apple sauce	32	–	200	0.42	240
Mustard	25	–	3.4	0.56	20
Melted chocolate	46	–	0.57	1.16	1.16
Peanut oil	21.1	0.32–64	0.065	1.00	–

Adapted from [2, 4]

$$\sigma = \mu\dot{\gamma} \qquad (29.18)$$

29.3.3 Power Law Model [$\sigma_o = 0$]

Power law fluids show no yield stress (σ_o) and a non-linear relationship between shear stress and shear rate. Pseudoplastic and dilatant fluids may be considered power law fluids, each with different ranges for flow behavior index values; refer to Table 29.4:

$$\sigma = 0 + K\dot{\gamma}^n \qquad (29.19)$$

or

$$\sigma = K\dot{\gamma}^n \qquad (29.20)$$

29.3.4 Bingham Plastic Model [$n = 1$; $K = \mu_{pl}$]

Bingham Plastic materials have a distinguishing feature: a yield stress is present. Once flow is established, the relationship between shear stress and shear rate is linear, explaining why $n = 1.0$ and K is a constant value known as the **plastic viscosity**, μ_{pl}. Caution: The plastic value is *not* the same as the apparent viscosity (η) or the Newtonian viscosity (μ)!:

$$\sigma = \sigma_o + \mu_{pl}\dot{\gamma}^1 \qquad (29.21)$$

or

$$\sigma = \sigma_o + \mu_{pl}\dot{\gamma} \qquad (29.22)$$

29.4 RHEOMETRY

Rheometers are devices used to determine viscosity and other rheological properties of materials. Relationships between shear stress and shear rate are derived from physical values of system configurations, pressures, flow rates, and other applied conditions.

29.4.1 Compression, Extension, and Torsion Analysis

The rheological properties of solid foods are measured by compressing, extending, or twisting the material and can be accomplished by two general approaches called small- or large-strain testing. In **small-strain tests**, the goal is to apply the minimal amount of strain or stress required to measure the rheological behavior while at the same time preventing (or at least minimizing) damage to the sample. The goal in **large-strain** and **fracture tests** is the opposite. Samples are deformed to an extent at which the food matrix is significantly strained, damaged, or possibly fractured. Small-strain tests are used to understand properties of a food network, whereas large-strain tests give an indication of sensory texture or product durability. In general, compressive and extensional tests are generally performed using large strains. Torsional (shear) tests may be performed in using small or large strains.

29.4.1.1 Large-Strain Testing

Compression and tension (i.e., extension) tests are used to determine large-strain and fracture food properties. **Compression tests** are generally selected for solid or viscoelastic solid foods when a tight attachment between sample and the testing fixture is not required, thereby simplifying sample preparation. **Tension** and **torsion tests** are well suited for highly deformable foods when a high level of strain is needed to fracture the sample. The main disadvantage to tension and torsion tests is that the sample must be attached to the test fixture [5]. Hamann et al. [6] provides a detailed comparison and analysis of large deformation rheological testing.

29.4.1.1.1 Determining Stress, Strain, and Elastic Modulus (E) in Compression

There are several assumptions to consider when doing compression testing. Along with the previously mentioned considerations of a homogeneous and isotropic material, the assumption that the food is an **incompressible material** greatly simplifies matters. An incompressible material is one that changes in shape but not volume when compressed. Foods such as frankfurters, cheese, cooked egg white, and other high-moisture, gel-like foods generally are considered incompressible. The calculations for strain are as discussed previously (Eq. 29.1).

During compression, the initial cross-sectional area (A_i) increases as the length decreases. To account for this change, a correction term incorporating a ratio of the cylinder lengths (L/L_i) is applied to the stress calculation:

$$\sigma = \frac{F}{A_i}\left(\frac{L}{L_i}\right) \qquad (29.23)$$

In compression testing one should use a cylindrical shaped sample with a length (L) to diameter ratio of >1.0. The sample should be compressed between two flat plates with diameters exceeding the lateral expansion of the compressed sample (i.e., the entire sample cross-section should be in contact with the plates during testing). The equations are based on the sample maintaining a cylindrical shape when compressed. If this is not the case, the contact surface between the plate and the sample may need lubrication. Water or oil can be used, and one should pick the fluid that provides the desired lubrication without causing any deleterious effects to the sample.

A cylinder of Cheddar cheese 3 cm in length (L_i), with an initial radius (R_i) of 1 cm, was compressed at a constant rate to 1.8 cm (L) and recorded a force of 15 N. Then:

$$\varepsilon = \ln\left[1+(-0.4)\right] = -0.5 = 0.5_{\text{compresion}} \qquad (29.24)$$

$$A_i = \pi R_i^2 = 0.000314\,\text{m}^2 \qquad (29.25)$$

$$\sigma = \frac{F}{A_i}\left(\frac{L}{L_i}\right) = \frac{15\,N}{0.000314\,\text{m}^2} \times \left(\frac{1.8}{3.0}\right) = 28{,}700\,\text{Pa}_{\text{compresion}}$$
$$= 28.7\ \text{kPa}_{\text{compresion}} \qquad (29.26)$$

$$E = \frac{\sigma}{\varepsilon} = \frac{28{,}700\,\text{kPa}}{0.5} = 57.4\,\text{kPa} \qquad (29.27)$$

If the material compressed is a pure elastic solid, the compression rate does not matter. However, if the material is viscoelastic (as is the case with most foods), then the values for stress, strain, and elastic modulus may change with the speed of compression. A complete characterization of a viscoelastic material requires determining these values at a variety of compression rates. Another factor to consider is the level of compression. The sample can be compressed to fracture, or some level below fracture. The goal should be compression to fracture if correlating rheological with sensory properties.

29.4.1.1.2 Texture Profile Analysis

Texture profile analysis (TPA) is an empirical technique using a two-cycle compression test. It is typically performed using a **Universal Testing Machine** (Fig. 29.8a) or a **Texture Analyzer** (Fig. 29.8b). This test

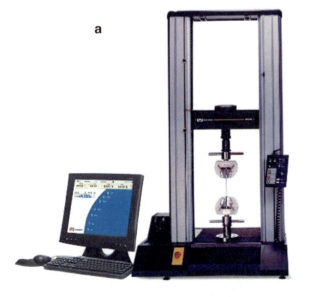

Photos of (**a**) Universal Testing Machine (Courtesy of Instron®, Norwood, MA) and (**b**) Texture Analyzer (Courtesy of Texture Technologies Corp., Hamilton, MA)

was developed by a group of food scientists from the General Foods Corporation and is compiled as force during compression and time. Data analyses correlated numerous sensory parameters, including hardness, cohesiveness, and springiness, with texture terms determined from the TPA test curve. For example, the peak force required to fracture a specimen has been strongly related to sample hardness. Bourne [7] provides a more detailed description of TPA.

29.4.1.2 *Fracture Testing*

Fracture tests are large-strain tests carried out to the point of sample failure. Generally, stress and strain at the failure point are determined by the sudden decrease in stress as the sample fractures. Fracture tests may be performed in compressive, tensile, or shear (torsion) modes. Compressive or shear modes are usually used for food materials, since difficulties in gripping the sample without damaging it make tensile fracture testing difficult.

Several considerations are needed when performing fracture testing. The geometry of the sample must be controlled, especially in fundamental fracture testing. Certain fracture methodologies specify the shape of the sample, for example, a cylinder, capstan, or beam shape. The homogeneity of the sample must also be considered. Anisotropic samples such as meats may have different fracture behaviors depending on their orientation. Testing parameters such as strain rate can affect fracture behavior as well. Hamann et al. [6] provides a more detailed discussion of fracture testing.

29.4.2 Rotational Viscometry

For fluids, the primary mode for rheological measurement in the food industry is rotational viscometry, providing rapid and fundamental information. **Rotational viscometry** involves a known test fixture (geometrical shape) in contact with a sample, and through some mechanical, rotational means, the fluid is sheared by the fixture. Primary assumptions are made for the development of **constitutive equations** (relationships between shear stress and rate) and include the following:

- **Laminar flow**. Laminar flow is synonymous with streamline flow. In other words, if we were to track velocity and position of a fluid particle through a horizontal pipe, the path would only be in the horizontal direction without a shift toward the pipe wall.
- **Steady state**. There are no net changes to the system over time.
- **No-slip boundary condition**. When the test fixture is immersed in the fluid sample, the walls of the fixture and the sample container serve as boundaries for the fluid. This condition assumes that, at

whatever speed either boundary is moving, an infinitely thin layer of fluid immediately adjacent to the boundary is moving at precisely the same velocity.

Rotational rheometers may operate in two modes: **steady shear** or **oscillatory**. The next few sections consider steady shear rotational viscometry. **Steady shear** is a condition in which the sheared fluid velocity, contained between the boundaries, remains constant at any single position. Furthermore, the velocity gradient across the fluid is a constant. Three test fixtures most often used in steady shear rotational viscometry are the **concentric cylinder**, **cone and plate**, and **parallel plates**.

29.4.2.1 *Concentric Cylinders*

This rheological attachment consists of a cylindrical fixture shape, commonly called a **bob** with radius R_b, suspended from a measuring device that is immersed in a sample fluid contained in a slightly larger cylinder, referred to as the **cup** with radius R_c (see Figs. 29.9 and 29.10). **Torque** (M) is an action that generates rotation about an axis and is the product of a force and the perpendicular distance (r), called the **moment arm**, to the **axis of rotation**. The principles involved can be described relative to changing tires on a car. To loosen the lugnuts, a larger tire iron is often required. Essentially, this longer tool increases the moment arm, resulting in a greater torque about the lugnut. Even though you are still applying the same force on the iron, the longer device provides greater torque!

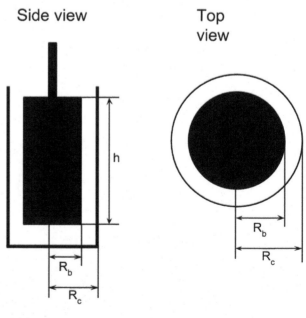

29.9 figure Concentric cylinder geometry

29.10 figure	Photo of cup and bob test fixtures (concentric cylinder geometry) for rheological measurements (Courtesy of TA Instruments, New Castle, DE)

To derive rheological data from experiments, equations for shear stress and shear rate are used. Shear stress at the surface of the bob (σ_b) may be calculated from a force balance as:

$$\sigma_b = \frac{M}{2\pi R_b^2} \qquad (29.28)$$

Therefore, to determine shear stress, all we need to know is the bob geometry (h and Rb) and the torque response (M) of the fluid on the measuring sensor.

A **simple shear approximation** commonly calculates a shear rate at the bob surface and assumes a constant shear rate across the fluid gap. This approximation is valid for small gap widths where $R_c/R_b \leq 1.1$:

$$\dot{\gamma} = \frac{\Omega R_b}{R_c - R_b} \qquad (29.29)$$

This calculation requires the **rotational speed**, or **angular velocity** (Ω), of the bob, typically expressed in radians per second. Converting units of revolutions per minute (rpm) to radians per second is simply achieved by multiplying by $2\pi/60$, and the following example converts 10 rpm to radians per second:

$$\left(\frac{10 \text{ revolutions}}{1 \text{ min}}\right)\left(\frac{1 \text{ min}}{60 \text{ s}}\right)\left(\frac{2\pi \text{ radians}}{1 \text{ revolution}}\right)$$

$$= \frac{1.047 \text{ radians}}{\text{second}} \qquad (29.30)$$

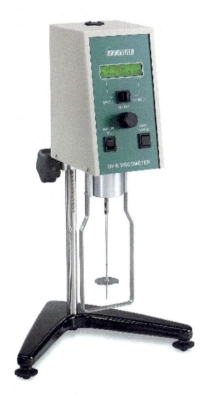

29.11 figure	Photo of Brookfield viscometer (Courtesy of Brookfield AMETEK, Middleboro, MA)

29.12 figure	Photo of Bostwick consistometer (Courtesy of Cole-Parmer, Vernon Hills, IL)

The food industry uses several rheological devices to measure viscosity. These devices include the Brookfield viscometer (Fig. 29.11), the Bostwick consistometer (Fig. 29.12), and Zahn cup (Fig. 29.13). The **Brookfield viscometer** is one of the most common rheological devices found in the food industry. This apparatus uses a spring as a torque sensor. The operator selects a rotational speed (rpm) of the bob, attached to the spring. As the bob moves through the sample fluid, the viscosity impedes free rotation, causing the spring to wind. The degree of spring windup is a direct

29.13
figure

Photo of Zahn cup (Courtesy of Paul
N. Gardner Co., Inc., Pompano Beach, FL)

reflection of the torque magnitude (*M*), used to determine a shear stress at the bob surface. Newer Brookfield
models convert torque to viscosity automatically using
Eq. 29.28, while older models provide conversion factors for calculating viscosity from torque.

The **Bostwick consistometer** uses a given amount
of sample poured into one end of the instrument
behind a gate. A timer is started as the gate is quickly
lifted and the time to reach a certain mark on the ramp
is measured. The distance the fluid travels is called the
consistency. Note that consistency and viscosity are
not interchangeable.

The **Zahn cup** also used a set amount of sample.
The sample is put into the cup and allowed to drain
from a hole in the bottom of the cup. The time from
start of flow to the first break in the stream of liquid is
recorded and can be converted to viscosity using conversion factors and the specific gravity of the fluid.

The Brookfield viscometer, the Bostwick consistometer, and the Zahn cup are suitable for quick quality control measurements, as they provide rapid
measurements and are easy to use and clean. However,
these are empirical instruments and are not as precise
as a rheometer. Furthermore, they are generally used
to measure viscosity at a single shear rate, which can
lead to incorrect assumptions about flow behavior. For
example, what might happen if the Newtonian and
the *n* < 1 Herschel-Bulkley fluid in Fig. 29.7 were tested
only at the shear rate at which their viscosities are
equivalent? It is recommended that these rheological
devices be used with fluids that are Newtonian or
close to Newtonian to remove shear rate effects from
the measurements.

Following progression through a series of rotational speeds, a rheogram can be created showing
shear stress (*σ*) versus shear rate $(\dot{\gamma})$. The importance

29.6
table

Rheological data of tomato catsup collected
using a concentric cylinder test fixture

rpm	Torque (N m)	Shear rate (s⁻¹)	Shear stress (Pa)	Apparent viscosity (Pa s)
1.0	0.00346	2.09	22.94	10.98
2.0	0.00398	4.19	26.39	6.30
4.0	0.00484	8.38	32.10	3.83
8.0	0.00606	16.76	40.18	2.40
16.0	0.00709	33.51	47.02	1.40
32.0	0.00848	67.02	56.23	0.84
64.0	0.01060	134.04	70.29	0.52
128.0	0.01460	268.08	96.82	0.36
256.0	0.01970	536.16	130.63	0.24

of rheograms has been discussed, with a primary significance being apparent viscosity determination
(Eq. 29.15). The following tomato catsup data in
Table 29.6 were collected with a standard cup and bob
system (R_c = 21 mm, R_b = 20 mm, and *h* = 60 mm). Using
Eqs. 29.15, 29.28, and 29.29 you should verify the
results.

29.4.2.2 *Cone and Plate and Parallel Plate*

Another popular system for rotational measurement is
the cone and plate configuration (Figs. 29.14 and
29.15). Its special design permits the shear stress and
shear rate to remain constant for any location of sample in the fluid gap. Test quality is best when the cone
angle (*θ*) is small, and large errors may be encountered
when the gap is improperly set or not well
maintained.

The shear stress may be determined for a cone and
plate configuration as:

$$\sigma = \frac{3M}{2\pi R^3} \qquad (29.31)$$

while the shear rate is calculated as:

$$\dot{\gamma} = \frac{r\Omega}{r\tan\theta} = \frac{\Omega}{\tan\theta} \qquad (29.32)$$

A primary advantage of the cone and plate test
fixture is that shear stress and rate are independent of
position – constant throughout the sample. However,
if there are large particles in the sample, they may
become trapped under the point of the cone, invalidat-
ing the viscosity measurement. In this case, a parallel
plate configuration may be used (Fig. 29.16), as the
standard gap in this configuration is 1.0 mm.

Unlike the cone and plate, shear stress and strain
are not constant between the parallel plates. For consistency, the outer edge of the plate is set as the measurement location. Therefore, care must be taken to
ensure the outer edge of the sample does not change
during testing.

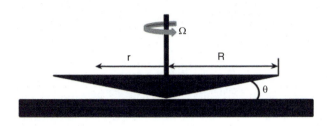

29.14 Cone and plate geometry
figure

29.15 Photos of cone and plate and parallel plate
figure test fixtures for rheological measurements
(Courtesy of TA Instruments, New Castle, DE)

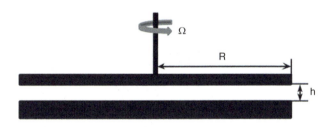

29.16 Parallel plate geometry
figure

The shear stress at the edge of the plate may be determined as:

$$\sigma = \frac{M}{2\pi R^3}\left(3 + \frac{d(\ln M)}{d(\ln \dot{\gamma}_R)}\right) \tag{29.33}$$

The rheometer performs the calculation for the derivative. For a Newtonian fluid, Eq. 29.33 simplifies to:

$$\sigma = \frac{2M}{\pi R^3} \tag{29.34}$$

The shear rate at the edge of the plate is calculated as:

$$\dot{\gamma} = \frac{R\Omega}{h} \tag{29.35}$$

29.4.2.3 Experimental Procedure for Steady Shear Rotational Viscometry

29.4.2.3.1 Test Fixture Selection
Many considerations go into the decision of selecting a fixture for a rheological test. To simplify the process, the information in Table 29.7 should be considered [8].

29.4.2.3.2 Speed (Shear Rate) Selection
When performing a rheological test, it is necessary to have an understanding of the process for which the measurement is being performed. From the earlier example of tomato catsup, the apparent viscosity continuously decreased, exhibiting shear thinning behavior, as the shear rate increased. How would one report a viscosity? To answer that question, the process must be considered. For example, if a viscosity for molten milk chocolate is required for pipeline design and pump specification, a shear rate for this process should be known. All fluid processes administer a certain degree of shear on the fluid, and a good food scientist will consider the processing shear rate for proper rheological property determination. Barnes et al. [9] have prepared a list of common shear rates for typical processes, many of which are shown in Table 29.8.

29.4.2.3.3 Data Collection
Once the test fixture and shear rate ranges have been selected, the experiment can begin. Record values of torque for each viscometer speed.

29.4.2.3.4 Shear Calculations
Values for shear stress and shear rate are solved based on test fixture, fixture geometry, and angular velocity.

29.4.2.3.5 Model Parameter Determination
Shear stress and shear rate can now be inserted into various rheological models previously described in Sect. 29.3. Rheological model parameters such as viscosity (μ, η, μ_{pl}), yield stress (σ_o), consistency coefficient (K), and flow behavior index (n) may be analyzed for an even greater understanding of the flow of the material. For example, one may want to know: Does the material have a yield stress? Is the material shear thinning or shear thickening? What is the viscosity at a specific processing rate? Answering these and similar questions gives the food scientist a greater command of the behavior of the material for process design or quality determination.

29.7 table Advantages and disadvantages of rotational viscometry attachments

Rotational geometry	Advantages	Disadvantages
Concentric cylinder	Good for low-viscosity fluids Good for suspensions Large surface area increases sensitivity at low shear rates	Potential end effects Large sample required
Cone and plate	Constant shear stress and shear rate in gap Good for high shear rates Good for medium- and high-viscosity samples Small sample required Quick-and-easy cleanup	Large particles interfere with sensitivity Potential edge effects Must maintain constant gap height
Parallel plate	Allows measurement of samples with large particles Good for high shear rates Good for medium- and high-viscosity samples Small sample required Quick-and-easy cleanup	Shear stress and strain are not constant in gap Potential edge effects Must maintain constant gap height

29.8 table Predicted shear rates for typical food processes

Process	Shear rate (s^{-1})
Sedimentation of powders in a liquid	10^{-6}–10^{-4}
Draining under gravity	10^{-1}–10^{1}
Extruding	10^{0}–10^{2}
Chewing and swallowing	10^{1}–10^{2}
Coating	10^{1}–10^{2}
Mixing	10^{1}–10^{3}
Pipe flow	10^{0}–10^{3}
Spraying and brushing	10^{3}–10^{4}

Reprinted from [9], with kind permission from Elsevier Sciences – NL, Sara Burgerhartstraat 25, 1055 KV Amsterdam, The Netherlands

29.4.3 Oscillatory Rheometry

The goal in **oscillatory rheometry** is to characterize the **viscoelastic** properties of a material, generally under small stresses and strains. The term viscoelastic implies the material exhibits both fluid-like (viscous) and solid-like (elastic) behaviors simultaneously. Think of a piece of cheese: it springs back (elastic behavior) if you compress it gently, but not quite to its original height (viscous behavior shown as permanent deformation). Viscoelastic behaviors are determined by applying: (1) a stress or strain in **oscillation** and measuring the respective strain or stress and phase angle between stress and strain, (2) a **constant strain** and measuring the decrease (relaxation) in

stress, or (3) a **constant stress** and measuring the rate of deformation (creep). These tests are generally used to evaluate the viscoelastic behaviors of solid and semisolid foods, such as cheese and pudding, although they can also be used to evaluate fluid foods, such as salad dressing. A more detailed description of these techniques may be found in Steffe [2] and Rao [4].

29.5 TRIBOLOGY

Tribology is a subfield of rheology involving the study of friction, lubrication, and wear behaviors of materials. Originally used to study materials such as engine lubricants, tribology has gained interest from the food industry as a way to investigate friction-related aspects of food texture. Sensory attributes such as mouthcoat, chalkiness, or astringency relate poorly to traditional rheological measurements. However, these attributes all involve a sensation of friction. Tribology may be helpful in determining the mechanisms behind these friction-related textural attributes.

Tribological tests are performed by moving one surface against another, fixed surface. A thin layer of the test material is placed between the two surfaces and acts as a lubricant. Different geometries for tribological testing may be seen in Fig. 29.17. The moving surface is slid along the stationary surface at different speeds. Stribeck curves, or plots of friction coefficient versus sliding speed, are gener-

29.17 Tribological testing geometries (Courtesy of
figure   TA Instruments, New Castle, DE)

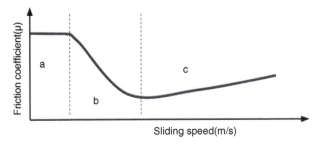

 Stribeck curve
figure

ated from the results (Fig. 29.18). Stribeck curves
consist of three different behavioral regimes: the
boundary regime (a), in which the sliding surfaces
are in contact and friction is relatively constant; the

mixed regime (b), in which the sliding surfaces are
mostly separated by the lubricating material and
friction decreases to a minimum value with
increased sliding speed; and the **hydrodynamic
regime** (c), where the sliding surfaces are fully separated by the lubricating material. Friction behavior in the mouth is generally in the boundary to
mixed regime, with oral sliding speeds estimated at
about 10–30 mm/s.

As a system property rather than a physical property, friction is affected by many aspects of the system. For foods, friction can be impacted by the food
composition and physicochemical properties, food
viscosity, food particle size, oral surface composition
and conditions, and saliva composition and amount.
For example, whole wheat bread has a rougher
mouthfeel (higher friction) than bread made with
refined flour because the large, irregularly shaped
bran particles increase the friction of the whole wheat
bread as it moves against the oral surfaces during
consumption.

29.6   SUMMARY

Rheological testing is simple in that it only requires
the measurement of force, deformation, and time. To
convert these measurements into fundamental phys-
ics-based rheological properties requires an understanding of the material and testing method. Materials
should be homogeneous and isotropic for most fluid
foods and many solid foods. Fundamental rheological
properties are determined based on knowledge of the
stress or strain applied to the sample and the geometry of the testing fixture. Once rheological properties
are determined, they can be described by physical or
mathematical models to gain a more complete understanding of the rheological properties. The advantage
of determining fundamental, rather than empirical,
rheological properties is the use of common units,
independent of the specific instrument, to determine
the rheological property. This approach not only
allows for comparison among values determined on
different instruments but it also permits comparisons
like the flow behavior of honey vs. the flow of paint.
Through rheological methods, food scientists have the
ability to relate theoretical and experimental information from a range of disciplines, including polymer
chemistry and materials sciences, to gain a greater
understanding of the quality and behavior of food
materials.

Glossary

Bostwick consistometer	A rheological device used to measure viscosity in the food industry
Boundary regime	Tribological behavior characterized by high, constant friction due to surface-surface contact
Brookfield viscometer	A rheological device used to measure viscosity in the food industry
Compression	A force acting in a perpendicular (normal) direction toward the body
Concentric cylinder	A test fixture for rotational viscometry frequently called a cup and bob
Cone and plate	A test fixture for rotational viscometry
Constitutive equation	An equation relating stress with strain and sometimes other variables including time, temperature, and concentration
Dilatent	Shear-dependent thickening
Empirical test	Simple tests measuring poorly defined parameters but typically found to correlate with textural or other characteristics
Fundamental test	A measurement of well-defined, physically based rheological properties
Homogeneous	Well mixed and compositionally similar regardless of location
Hydrodynamic regime	Tribological behavior characterized by complete surface-surface separation by the lubricating material
Incompressible	No change in material density
Isotropic	The material response is not a function of location or direction
Kinematic viscosity	The viscosity divided by the density of the material
Laminar flow	Streamline flow
Mixed regime	Tribological behavior characterized by a reduction in friction to a minimum as surface-surface contact decreases
Modulus	A ratio of stress to strain
Newtonian fluid	A fluid with a linear relationship between shear stress and shear rate without a yield stress
Non-Newtonian fluid	Any fluid deviating from Newtonian behavior
No slip	The fluid velocity adjacent to a boundary has the same velocity as the boundary
Oscillatory rheometry	Dynamic test using a controlled sinusoidally varying input function of stress or strain
Parallel plate	A test fixture for rotational viscometry
Pseudoplastic	Shear thinning
Rheogram	A graph showing rheological relationships
Rheology	A science studying how all materials respond to applied stresses or strains
Rheometer	An instrument measuring rheological properties
Rheopectic	Time-dependent thickening of a material
Shear (strain) rate	Change in (shear) strain with respect to time
Simple shear	The relative motion of a surface with respect to another parallel surface creating a shear field within the fluid contained between the surfaces
Simple shear approximation	A prediction technique for shear rate estimation of fluids within a narrow gap
Steady shear	A flow field in which the velocity is constant at each location with time
Steady state	Independent of time
Strain	Relative deformation
Stress	Force per unit area
Tension	A force acting in a perpendicular direction away from the body
Test fixture	A rheological attachment, sometimes called a geometry, which shears the sample material
Thixotropic	Time-dependent thinning of a material
Torque	A force-generating rotation about an axis, which is the product of the force and the perpendicular distance to the rotation axis
Torsion	A twisting force applied to a specimen
Tribology	A branch of rheology involving friction, lubrication, and wear
Viscoelastic	Exhibiting fluid-like (viscous) and solid-like (elastic) behavior simultaneously
Viscometer	An instrument measuring viscosity
Viscosity	An internal resistance to flow
Yield Stress	A minimum stress required for flow to occur
Zahn cup	A rheological device used to measure viscosity in the food industry

Nomenclature

Symbol	Name	Units
A	Area	m^2
A_i	Initial sample area	m^2
E	Modulus of elasticity	Pa
F	Force	N
G	Shear modulus	Pa
h	Height	m
K	Consistency coefficient	Pa s^n
L	Length	m
L_i	Initial length	m
ΔL	Change in length	m
M	Torque	N m
n	Flow behavior index	Unitless
r	Radial distance	m
R	Radius	m
R_i	Initial radius	m
R_b	Bob radius	m
R_c	Cup radius	m
t	Time	s
U	Velocity	m s^{-1}
ε	Normal strain	Unitless
γ	Shear strain	Unitless
γ	Angle of shear	Radians or degrees
$\dot{\gamma}$	Shear (strain) rate	s^{-1}
η	Apparent viscosity	Pa s
θ	Cone angle	Radians or degrees
μ	Newtonian viscosity	Pa s
μ_{pl}	Plastic viscosity	Pa s
σ	Stress	Pa
σ_b	Shear stress at the bob	Pa
σ_o	Yield stress	Pa
Ω	Angular velocity	Radians s^{-1}

29.7 STUDY QUESTIONS

1. How is stress different from force?
2. What is the difference between shear stress and normal stress?
3. What is the definition of apparent viscosity? How does apparent viscosity differ from Newtonian viscosity?
4. Pure maple syrup is a Newtonian fluid and imitation maple syrup is a Power Law fluid. What are the differences in flow behavior of these fluids and how do these differences alter the processing and final texture of these foods?
5. The stress response of applesauce at 26 °C may be described by the following mathematical expression:

$$\sigma = 5.6\dot{\gamma}^{0.45}$$

The stress response of honey at 26 °C obeys a Newtonian model:

$$\sigma = 8.9\dot{\gamma}$$

In both equations, stress is in Pascals and shear rate is in s^{-1}.

(a) Which rheological model is described by the equation for applesauce? What are the consistency coefficient and flow behavior index? Include the units on these quantities.
(b) Calculate the apparent viscosities of applesauce and honey at a shear rate of 0.25, 0.43, 5.10, and 60.0 s^{-1}.
(c) Compare how the viscosities of the two food products change with shear rate. Which food has a higher viscosity?
(d) Explain the importance of multipoint testing when measuring viscosity.

6. You are designing a new chip dip. Describe at least three rheological behaviors that you would like your dip to display.
7. Rheological tests can be empirical or fundamental.

(a) What are the differences between empirical and fundamental rheological tests?
(b) Develop two empirical tests for comparing viscosities of different tomato sauce formulations.
(c) Identify at least one fundamental rheological test that could be used to determine similar properties from your empirical tests.
(d) Explain the advantages of using fundamental rheological tests instead of empirical tests.

8. What sort of tribological differences might you expect between a full-fat product and a low-fat product?

REFERENCES

1. Steffe JF, Daubert CR (2006) Bioprocessing pipelines: rheology and analysis. Freeman, East Lansing, MI
2. Steffe JF (1996) Rheological methods in food process engineering, 2nd edn. Freeman, East Lansing, MI
3. Muller HG (1973) An introduction to food rheology. Crane, Russak, Inc., New York
4. Rao MA (1999) Rheology of fluid and semisolid foods: principles and applications. Aspen, Gaithersburg, MD
5. Diehl KC, Hamann DD, Whitfield JK (1979) Structural failure in selected raw fruits and vegetables. J Texture Stud 10: 371–400
6. Hamann D, Zhang J, Daubert CR, Foegeding EA, Diehl KC (2006) Analysis of compression, tension and torsion for testing food gel fracture properties. J Texture Stud 37: 620–639
7. Bourne MC (1982) Food texture and viscosity: concept and measurement. Academic, New York
8. Macosko CW (1994) Rheology: principles, measurements, and applications. VCH, New York
9. Barnes HA, Hutton JF, Walters K (1989) An introduction to rheology. Elsevier Science, New York

Thermal Analysis

Leonard C. Thomas
DSC Solutions,
27 E. Braeburn Drive, Smyrna, DE 19977, USA
e-mail: LThomas@TAinstruments.com

Shelly J. Schmidt (✉)
Department of Food Science and Human Nutrition,
University of Illinois at Urbana-Champaign,
Urbana, IL 61801, USA
e-mail: sjs@illinois.edu

S. Nielsen (ed.), *Food Analysis*, Food Science Text Series,
DOI 10.1007/978-3-319-45776-5_30, © Springer International Publishing 2017

30.1 INTRODUCTION

Thermal analysis is a term used to describe a broad range of analytical techniques that measure physical and chemical properties as a function of temperature, time, and atmosphere (inert or oxidizing gas, pressure, and relative humidity). Depending on the technique, test temperatures can range from −180 to 1,000 °C or more, allowing investigation into a range of applications, including low-temperature stability and processing (e.g., freezing and freeze-drying) to high-temperature processing and cooking (e.g., extrusion, spray drying, and frying).

Thermal analysis results provide insight into the structure and quality of starting materials, as well as finished products. The physical structure (amorphous, crystalline, semicrystalline) of a material creates a set of physical properties, which in turn define end-use properties, such as texture and storage stability. Areas of application include quality assurance, product development, and research into new materials, formulations, and processing conditions [1–4].

Specific instruments are typically used for characterization of a particular property. Instrumentation includes a transducer, used to measure the property of interest, and a temperature-measuring device, such as a thermocouple, thermopile, or platinum resistance thermometer, used to record the sample's temperature. Experiments are performed while heating, cooling, or at a constant temperature (isothermal), and measured signals are stored for analysis.

Thermal analysis techniques of major interest to the food researcher and the properties they measure are listed in Table 30.1, with the first three widely used techniques discussed in detail in Sect. 30.2.

table 30.1 Thermal analysis techniques of major interest to the food researcher and the properties they measure

Techniques	Abbreviation	Property measured
Thermogravimetric analysis	TGA	Weight change
Differential scanning calorimetry	DSC	Heat flow
Modulated temperature DSC	MDSC®	Heat flow and heat capacity
Thermomechanical analysis	TMA	Dimensional change
Dynamic mechanical analysis	DMA	Stiffness and energy dissipation
Rheology	Rheometer	Viscosity/flow behavior
Moisture sorption analysis	MSA	Moisture sorption

Thermogravimetric analysis (TGA) is typically the first thermal analysis measurement done when characterizing a new material. TGA data can detect and quantify the presence of bulk water and/or associated water and identify the temperature at which molecular decomposition (chemical change) begins. The change in weight measured by TGA is quantitative; however, no information on the chemistry of evolved gases is obtained. If chemical knowledge of evolved gases is desired, TGA can be coupled to a mass spectrometer (MS) (see Chap. 11) and/or Fourier transform infrared (FTIR) spectrometer (see Chap. 8) [5].

Once composition and thermal stability are obtained from TGA results, the physical structure or "form" of the material is typically determined using **differential scanning calorimetry** (DSC) and/or **modulated temperature DSC** (MDSC®). Structure and the temperature(s) at which the structure changes [transition(s)] significantly influence physical and chemical properties of a material. By understanding structure and related physical properties, formulations can be developed that provide desired end-use properties, such as crispness, fast dissolution rate, and extended shelf life.

30.2 MATERIALS SCIENCE

Since the pioneering work of Slade and Levine [e.g., 6, 7], food scientists have been actively applying the principles of materials science to the study of food materials [e.g., 8, 9]. One of the main driving forces underlying the application of materials science principles to foods is that the end-use properties (functionality) of a material at a specific temperature are dependent on the structure of the components at that temperature. Therefore, it is necessary to measure structure as a function of temperature. The primary use of thermal analysis is to determine structure by measuring the physical properties (e.g., heat capacity, flow, expansion, rigidity) associated with that structure.

30.2.1 Amorphous Structure

Many food products are amorphous or have a high amorphous content (semicrystalline). Examples include extruded snacks and breakfast cereals, low-moisture cookies and crackers, sugar-based hard candies, and spray-dried powdered drink mixes. **Amorphous structure** has no regular or systematic molecular order, which means that it has the highest energy and entropy content, the highest molecular mobility, and the fastest rate of dissolution, compared to crystalline structure. A potential problem with amorphous structure is that physical properties can change by orders of magnitude at a specific temperature, termed the **glass transition temperature** (Tg).

At temperatures below Tg, an amorphous material acts like a glass. It is rigid, has low molecular mobility, and has very high viscosity (low ability to flow). Above Tg, these materials act like a rubbery material, viscous liquid, or gel with greater free volume and much higher molecular mobility. Some amorphous materials (e.g., fats, oils, and water) have the ability to crystallize when cooled to lower temperatures. However, these materials may associate (e.g., hydrogen bond) with another material in the formulation and not crystallize, even at temperatures well below their freezing point. In the case of water, it is well known that not all the water in a food material freezes, even at very low storage temperatures [10].

Since properties change so significantly at Tg, it is important to be able to measure and control Tg by selecting appropriate ingredients in the correct weight ratio to other ingredients in a formulation or recipe. In foods, Tg can change by 50 °C or more via changes in moisture content. Crisp snack foods, such as crackers and potato chips, are classic examples of foods whose Tg and resulting physical properties are affected by changes in water content. For example, the Tg of freshly processed potato chips is much higher than room temperature; thus, the chips are crisp and have a pleasant texture. If the chips are exposed to ambient temperature and high humidity conditions for several hours, they become soft and pliable, which are textural characteristics typical of a low-moisture food held above Tg. Once the package of chips is opened, the low-moisture chips begin to absorb moisture from the air, which lowers Tg and creates a different set of physical properties at the consumption temperature.

Tg can be measured using most thermal analysis techniques due to significant changes in physical properties, such as heat capacity (for DSC), coefficient of thermal expansion (for **thermomechanical analysis**, TMA), and stiffness (for **dynamic mechanical analysis**, DMA), that accompany Tg [8]. However, DSC is usually the technique of choice because of easy sample preparation, short test times, straightforward data interpretation, and the ability to use sealed pans (hermetic), which prevent loss of moisture as the sample is heated. Because of the significant increase in molecular mobility and heat capacity that occurs as the sample is heated to a temperature above Tg, there is a corresponding increase in the heat flow rate measured by DSC. Data analysis software measures the temperature and magnitude of the change in heat flow that occurs, which is proportional to the amount of amorphous material in the sample (Fig. 30.1).

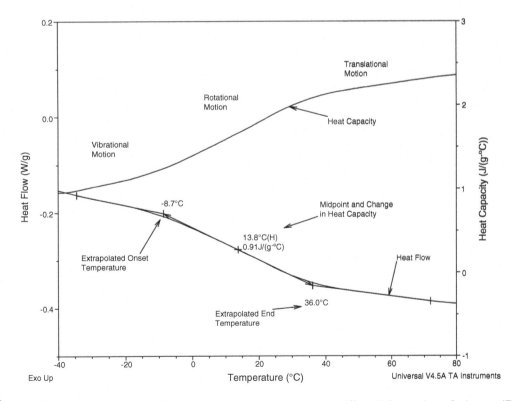

The glass transition of amorphous structure can be measured by differential scanning calorimetry (*DSC*) due to the significant increase in heat capacity that occurs as the material is heated to a temperature above the glass transition temperature (*Tg*). Typical *Tg* analysis includes the extrapolated onset, midpoint (temperature of one-half of the heat capacity change), and end point temperatures, as well as the difference in heat capacity (ΔCp, J/g°C)

30.2.2 Crystalline Structure

Crystalline structure is different from amorphous structure in many ways. Molecules have long-range order, lower energy (heat content) and entropy, higher density, and a different set of physical properties. Molecular mobility is low, which means that heat capacity is low. Melting of the crystalline material creates an amorphous liquid. Upon cooling of the amorphous liquid, some materials (e.g., fats) will crystallize, while others (e.g., sucrose and fructose) transition into an amorphous glass below their T_g.

Because of the increased density and molecular order within the crystal, there is a reduction in the ability of crystalline material to form hydrogen bonds and thus a reduced tendency to "absorb" moisture into their bulk structure from the atmosphere. However, crystalline materials "adsorb" water onto their surface quite quickly, mirroring the % relative humidity of the surrounding environment [11]. The melting of crystalline material occurs at a higher temperature than the glass transition associated with amorphous material, which makes crystalline material more stable and much more rigid (often gritty like table sugar and salt) than amorphous material over a wide temperature range. Because the crystalline structure is more stable, physical properties change less with time.

Since crystalline material has lower heat content than amorphous material, crystalline material must absorb heat (endothermic process) to become amorphous. The absorption of heat during a DSC experiment is seen as an endothermic peak, termed the melting peak. Data analysis software can measure the onset, peak, and end temperatures of the endothermic melting peak and calculate the heat (J/g) required to melt the sample by determining the area under the peak. The area of the melting peak (J/g) increases as the percent crystallinity of the material increases. Figure 30.2 shows the DSC data for melting of the crystalline sugar alcohol (polyol) mannitol, which is commonly used in confectioneries, such as "breath-freshening" mints and gums.

Normally, crystalline structure is converted to amorphous structure by heating the sample to a temperature (an energy level) that is high enough to overcome the energy associated with the crystalline lattice (termed thermodynamic melting). However, crystalline structure can also be lost due to processes such as dissolving of crystals in a solvent (dissolution), dehydration of a hydrated crystalline form, chemical interaction of functional groups between two materials in a mixture, and breaking of chemical bonds (decomposition) at temperatures below the true melting point of the material. Since these are time-dependent (kinetic) processes, the endothermic peak observed in DSC data shifts to a higher temperature as heating rate increases. A material that illustrates this behavior is the mono-

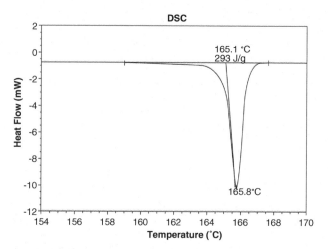

DSC data showing melting of the crystalline sugar alcohol (polyol) mannitol. Analysis of the data shows the extrapolated onset and peak melting temperatures and heat required to melt the crystalline structure (heat of fusion, joules/gram, J/g)

saccharide sugar fructose (Fig. 30.3). Figure 30.3 contains multiple Y-axes, as do other figures in this chapter. Since heat flow rate (W/g) is proportional to heating rate, it is necessary to scale the heat flow curves to different sensitivities to visually compare curves run at different heating rates, resulting in the need for more than one Y-axis.

30.2.3 Semicrystalline Structure

Many foods contain both amorphous and crystalline structures. In some cases, such as a lipid that melts over a temperature range, one ingredient can exist in both phases. A term that is sometimes used to describe a mixture of phases in a lipid is **solid-fat content** (SFC) (see Chap. 23, Sect. 23.3.11). At a particular temperature (usually room temperature, 72 °F or 22 °C), a certain fraction or percentage of the lipid material is solid (crystalline), and the remainder is liquid (melted). One lipid that illustrates this property at room temperature is cocoa butter, a common ingredient in chocolate. Cocoa butter can crystallize into six known forms (termed **polymorphs**), with the lowest stability form (form I) melting at approximately 17 °C and the highest stability form (form IV) melting at approximately 35 °C. Form V is the most desirable crystalline form of cocoa butter in chocolate, yielding a dark brown, glossy appearance, a satisfying snap when broken, and a melting temperature near mouth temperature.

Figure 30.4 shows the broad melt of a sugar-coated chocolate candy. There are several small overlapping melting peaks below 22 °C that are due to melting of the less stable lipid crystal forms in the sample. By measuring the percentage of melting below and

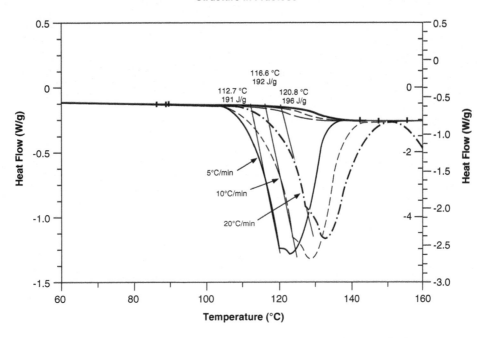

**Effect of Heating Rate on Loss of Crystal
Structure in Fructose**

30.3 figure The endothermic peak associated with loss of crystalline structure in fructose shifts to higher temperatures at higher heating rates, indicating the influence of a time-dependent (kinetic) process

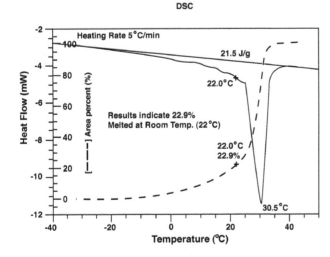

30.4 figure DSC data for sugar-coated chocolate candy, containing cocoa butter, shows several small overlapping melting peaks below 22 °C, due to melting of the less stable crystal lipid forms in the sample. In this particular sample, results show that 22.9 % of the cocoa butter is liquid (*melted*) at 22 °C; the remainder is crystalline (*solid*)

above 22 °C, the ratio of liquid to solid phases can be determined for room temperature. A feature of the DSC data analysis software, termed the "running integral," can plot percent melted versus temperature.

Results show that 22.9 % of the cocoa butter is liquid (melted) at room temperature in this particular sample. The presence of the liquid lipid provides a creamy texture to the chocolate.

30.2.4 Thermodynamic and Kinetic Properties

As discussed in Sect. 30.3.2.1, a natural limitation of DSC, and other thermal analysis techniques used to measure structure, is that heating of the sample is typically required. As temperature increases, molecular mobility increases, which permits the structure to change in ways that are not always obvious in the data. Since the purpose of the experiment is typically to measure the existing structure in the material, users of thermal analysis instrumentation must be able to recognize whether observations are due to changes in structure (e.g., phase transitions) or composition (e.g., solvent evaporation).

Thermodynamic properties (e.g., heat capacity, enthalpy, and density) have absolute values as a function of temperature, while kinetic properties are always a function of time and temperature. Some common examples of kinetic processes observed in foods are freezing of water during cold storage, crystallization of fats, adsorption/desorption of water, staling of bread, and decomposition/oxidation during processing, such as deep-frying.

The easiest way to determine if an observed event in the data is due to thermodynamics or kinetics is to change the heating rate. Since heating rate has the units of °C/min, the reciprocal is min/°C. The higher the heating rate, the less time the sample experiences at each temperature. Therefore, high heating rates decrease the probability of structural change, while low heating rates increase the probability of structural change. If the onset temperature of the observed event remains relatively constant (changes <1 °C) with a tenfold change in heating rate, it is typically a thermodynamic event, while an increase in event temperature with heating rate indicates the influence of a kinetic process. An example of substantial heating rate dependence is shown in Fig. 30.4 for fructose, for which the onset temperature of the endothermic DSC peak shifts from 112.7 °C at a heating rate of 5 °C/min to 120.8 °C at a heating rate of 20 °C/min. As discussed previously, the kinetic event associated with the observed heating rate dependence in fructose is thermal decomposition [12].

30.3 PRINCIPLES AND METHODS

This section describes the working principles of the most frequently used thermal techniques and makes recommendations on optimum experimental conditions for characterizing common materials.

30.3.1 Thermogravimetric Analysis

30.3.1.1 Overview

Thermogravimetric analysis (TGA) should be the first thermal analysis technique used to characterize a new material. TGA provides information about the composition (number of components) of the material and its thermal or oxidative stability (decomposition in inert and oxidizing atmospheres, respectively). TGA instruments use a specially designed and very sensitive analytical balance to measure weight changes, since samples are typically heated from room temperature to 1,000 °C or more. A thermocouple is located close to the sample to continuously record the temperature as weight changes occur. The heated sample chamber is typically purged with an inert gas, such as nitrogen or helium; however, air or oxygen can be used when measuring oxidative stability. Most weight changes are weight losses due to volatilization or decomposition, but weight gain may be observed during early stages of oxidation. A specialized version of TGA is designed with humidity control so that the rate of moisture sorption (both absorption and desorption) can be measured as a function of time, temperature, and relative humidity.

Figure 30.5 shows a schematic of possible designs for both conventional TGA and a humidity-controlled sorption analysis TGA. Both contain sample and reference pans attached to the balance. The reference pan is empty, since its purpose is to offset the weight of the

Thermogravimetric Instrumentation/Technology

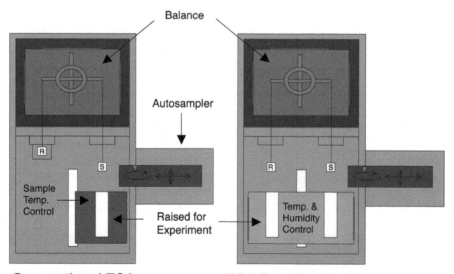

Conventional TGA   **TGA Sorption Analyzer**

30.5 figure Thermogravimetric instrumentation measures weight change using a sensitive analytical balance. The sample container is typically suspended into a temperature- and atmosphere-controlled chamber (Adapted from figure by TA Instruments, New Castle, DE)

sample pan. With conventional TGA, the reference pan is typically not heated. The reference pan for sorption analysis is exposed to the same temperature and relative humidity as the sample. This greatly improves the stability and baseline performance of the measurement. Most modern instruments have autosamplers so that many samples can be run sequentially without need of operator presence.

Most TGA instruments have several natural limitations. Although TGA provides a quantitative measurement of weight change, it is often difficult to quantify the weight of a specific component because weight losses typically overlap in temperature. Improved temperature resolution of these weight losses can usually be obtained by slowing the heating rate and reducing the sample weight. However, the lower heating rate increases test time (reduced productivity), and the smaller sample size reduces accuracy of small weight changes. Another limitation of TGA is that it cannot identify the chemistry of gases evolved from the sample. Knowledge of the gas composition helps to distinguish between water loss and loss of low molecular weight additives, such as flavors and fragrances, and helps to determine the chemical mechanisms involved in cooking and decomposition. Most manufacturers of mass spectrometers and Fourier transform infrared instruments offer interfaces between their products and the TGA instrument so that off-gases can be chemically identified.

30.3.1.2 *Experimental Conditions*

Although TGA experiments can be performed over a wide range of conditions, a good starting point for most materials is as follows:

- Sample weight: 10–20 mg (larger samples improve sensitivity for detection of minor components)
- Pan type: Platinum
- Purge gas: Nitrogen
- Start temperature: Room temperature, typically 20 °C
- Heating rate: 10 °C/min (lower rates improve resolution of overlapping weight losses)
- Final temperature: 300 °C

30.3.1.3 *Common Measurements*

TGA experiments are primarily heating experiments; however, isothermal (constant temperature) conditions can be used to determine drying rates or follow weight changes at processing/cooking temperatures. The most common measurements include the following:

- Temperature of weight change
- Free (or bulk) moisture content

- "Bound" or associated water content (part of the structure)
- Composition (multiple components)
- Decomposition temperature

It should be noted that in reality, there is no such thing as a decomposition temperature. Decomposition is a kinetic process, which means that it is a function of both time and temperature. Therefore, the temperature of weight loss due to decomposition increases if the heating rate is increased.

Figure 30.6 shows TGA data for sugar-frosted cornflakes. Because it contains multiple minor components, a relatively large sample size of 58.6 mg was used to improve sensitivity to detect those components. The Y1-axis is expanded to focus on only the weight loss from 90 % to 100 %. In addition, the Y2- and Y3-axes show the derivative signal, which is the rate of weight loss, at two sensitivities in order to better detect and measure the first two components that contribute to sample weight loss. The first weight loss starts immediately and is typical of unassociated or free water within the sample. The peak in the derivative signal occurs well above 100 °C (boiling point of water) because time is required for diffusion of the water through the sample. Loss of weight due to the second component starts near 178 °C. At this temperature, the sample has lost about 2.7 % of its weight. Since decomposition of some component starts near 178 °C, the chemistry of the sample is changing, making it difficult to measure any meaningful structure by other techniques, such as those described in Sect. 30.3.2.

30.3.2 Differential Scanning Calorimetry

30.3.2.1 *Overview*

Differential scanning calorimetry (DSC) is the most frequently used thermal analysis technique and probably accounts for 70 % of all thermal analysis measurements. Since *every* change in structure (transition) either absorbs or releases heat, DSC is the universal detector for measuring structure. The only major limitation of the technique is the sensitivity of the instrument, which is its ability to detect small transitions or very slow kinetic processes, where the rate of heat flow is similar to or less than the signal noise of the instrument.

The ability of DSC to measure very small rates (microwatts, μJ/s) of heat flow is greatly enhanced because it uses a differential signal. An empty reference pan is subjected to the same thermal environment as the sample pan, and the measured signal from the DSC is the difference in heat flow rate between the sample and reference. This effectively eliminates signal noise and drift caused by heat exchanges with the environment or atmosphere around the pans.

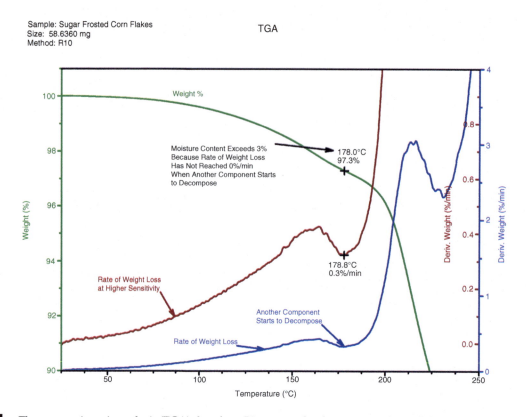

Sample: Sugar Frosted Corn Flakes
Size: 58.6360 mg
Method: R10

TGA

30.6
figure

Thermogravimetric analysis (TGA) data for a 58 mg sample of sugar-coated cornflakes. The derivative signals show rate of weight loss and are plotted at two sensitivities to better illustrate loss of moisture and the first stage of thermal decomposition. At the start of decomposition (178 °C), the weight % curve shows that the sample contains at least 2.7 % water

There are two general approaches to making the differential measurement. One approach uses a common furnace for both the sample and reference (heat flux design), while the other uses individual furnaces (power compensation design). Each design has theoretical advantages and limitations, but performance is mostly based on the manufacturer's ability to build a completely symmetric system so that instrumental effects on the measurement are minimized. This provides the highest sensitivity (ability to detect weak transitions), resolution (ability to separate transitions close in temperature), accuracy, and precision. Heat flux is the most common approach to making DSC measurements and is discussed below.

Figure 30.7 is a cross-sectional view of a heat flux DSC. A common furnace is used for the sample and reference pans, and is typically purged with high purity, ultra-dry nitrogen. The furnace and cooling accessory provide a wide temperature range from −180 % to 725 °C. The sample and reference sensors are temperature sensors, such as thermocouples, thermopiles, or platinum resistance thermometers. They provide the basis of the differential heat flow measurement and directly measure temperature.

Even though DSC is the most useful of the thermal techniques for measuring structure and changes in structure, it has a number of natural limitations, including the following:

- DSC measures the *sum* of all heat flows within the calorimeter. It is sometimes difficult to interpret data because of overlapping events (multiple transitions occurring at the same time and temperature).
- Most measurements involve heating the sample to higher temperatures. As temperature increases, mobility increases, permitting the structure to change in ways that are not always obvious. The measured structure may not be the original structure at the start of the experiment.
- DSC uses a single heating rate. However, higher heating rates provide better sensitivity, while lower heating rates provide better resolution. Therefore, it is not possible to optimize both sensitivity and resolution in a single DSC experiment.
- DSC cannot measure heat capacity under isothermal conditions. Therefore, DSC cannot use heat capacity as a way to follow changes in structure at constant temperature.

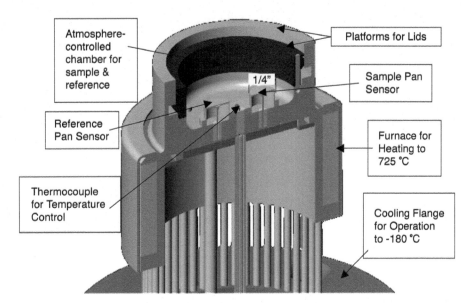

Cross-sectional view of a heat flux DSC. The sample and reference pans are located in a common chamber that is temperature controlled over the range of −180 to 725 °C. The chamber is typically purged with high purity, ultra-dry nitrogen gas (Courtesy of TA Instruments, New Castle, DE)

30.3.2.2 *Experimental Conditions*
Significantly different conditions are used depending on the measurement. However, larger sample weights and higher heating rates always improve sensitivity (ability to detect an event), while smaller sample weights and slower heating rates improve resolution. The conditions listed below provide a good starting point, but require optimization for the best results. The first decision in selecting conditions is the type of pan that will be used. If the sample is dry (less than 0.5% volatile components at 100 °C in TGA), then standard aluminum crimped pans will provide the best results. Crimped pans have lower mass and typically provide better heat transfer between the sample and sensor. However, they are not sealed, which permits evaporation of volatile components as the sample is heated. Hermetic pans are sealed and recommended for samples with volatile components. However, they are heavier than crimped pans and generally provide poorer contact between the sample and sensor, which can lower sensitivity and resolution. Recommended starting conditions are as follows:

- Sample weight: 8–12 mg
- Pan type: Aluminum hermetic
- Start temperature: At least 25 °C below the first transition of interest. This gives the baseline time to stabilize before the temperature of the event is reached, and permits better quantification of the change in heat content (*enthalpy*) resulting from the change in structure.
- Purge gas: Dry nitrogen
- Heating rate: 10 °C/min

- Final temperature: Temperature of 5% weight loss due to decomposition in TGA data should be the maximum temperature for a DSC experiment. In general, it is bad to decompose samples in the DSC cell because decomposition products can condense and affect the quality of future data.

30.3.2.3 *Common Measurements*
DSC can be used to measure the properties and structure of most ingredients used in the food industry. Typically, heating experiments are performed; however, measurements also are made while cooling or under constant temperature (isothermal). Examples of common measurements include the following:

- Glass transition temperature of amorphous structure
- Melting temperature of crystalline structure
- Percent crystallinity of semicrystalline material
- Crystallization of amorphous material
- Denaturation of proteins
- Gelatinization of starch
- Analysis of frozen solutions used for freeze-drying
- Oxidative stability of fats and oils

Changes in structure (transitions) are either **endothermic**, i.e., the sample absorbs additional energy (heat), or **exothermic**, i.e., heat is released. Figure 30.8 shows endothermic transitions associated with denaturation of albumin from chicken egg at concentrations of 1 and 10% (w/w) solution with water. Unfolding of the protein results in an increase in enthalpy, free volume, and molecular mobility, and

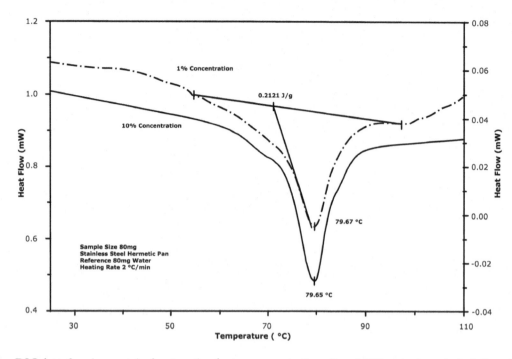

30.8 DSC data showing protein denaturation for two concentrations (1 and 10 % w/w in water) of albumin from
figure chicken egg. Large samples (>50 mg) in high-volume pans are recommended to improve sensitivity of the weak
transitions associated with protein denaturation and starch gelatinization

this requires heat to occur. Denaturation and gelatinization are low-energy processes and therefore require
large sample sizes to provide sufficient sensitivity. In
the case of Fig. 30.8, sample weights of approximately
80 mg were used in high-volume stainless steel pans.

30.3.3 Modulated DSC®

30.3.3.1 Overview
Modulated DSC® (MDSC®) is a special type of DSC
that applies two simultaneous heating rates to the
sample [13]. A linear rate is used to obtain the same
information as provided by conventional DSC, while a
sinusoidal rate superimposed on the linear rate permits measurement of the **heat capacity** component of
the total heat flow signal. As described above, one of
the natural limitations of the DSC technique is that it
can only measure the sum of all heat flows, and this
often makes interpretation of the data difficult. This is
illustrated by a brief review of Eq. 30.1, which is often
used to describe the heat flow signal from DSC:

$$\frac{dH}{dt} = Cp\frac{dT}{dt} + f(T, t) \qquad (30.1)$$

where:

dH/dt = measured heat flow rate (mW = mJ/s)
Cp = heat capacity (J/°C), product of specific
heat (J/g°C) × sample weight (g)
dT/dt = heating rate (°C/min)
$f(T,t)$ = heat flow rate due to time-dependent,
kinetic processes (mW)

As indicated in Eq. 30.1, the heat flow signal measured by conventional DSC has two components: one
associated with heat capacity and the other with
kinetic processes that are a function of both time and
temperature. DSC only measures the sum of the two
components. By applying two simultaneous heating
rates, MDSC® can separate the total signal into its individual components. Figure 30.9 shows temperature
versus time and heating rate versus time for an MDSC®
experiment. The MDSC® average temperature and linear heating rate would be typical of a DSC experiment,
while the modulated temperature and sinusoidal heating rate only occur with MDSC®.

As with every analytical technique, MDSC® also
has limitations, including:

- Slow average heating rates (typically 1–5 °C/min)
  must be used to obtain good separation of overlapping events. This decreases the productivity (number of samples per day) of MDSC® as compared to
  DSC.
- MDSC® is more complex because it requires additional experimental parameters and creates more
  signals than DSC.
- Separation of overlapping events requires the ability to modulate the sample's temperature during
  the events. This is not possible during melting of
  relatively pure materials that melt over just a few
  degrees.
- The probability of structural change is increased in
  the sample while heating due to the slower heating
  rates of MDSC®.

DSC

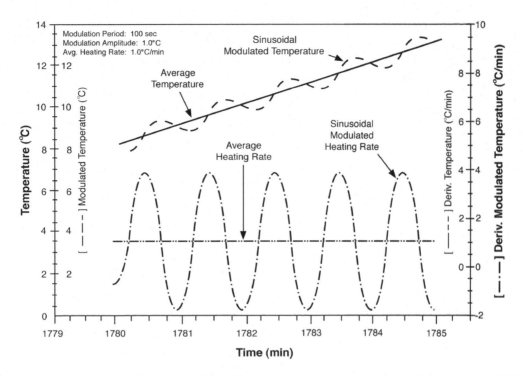

 Modulated temperature DSC (MDSC®) applies two simultaneous heating rates (linear and sinusoidal) to separate the total heat flow (equivalent to conventional DSC) signal into the heat capacity and kinetic components

30.3.3.2 *Experimental Conditions*

As seen in Fig. 30.9, the applied temperature of MDSC® has both linear and sinusoidal components. Therefore, it is necessary to specify conditions for both. Recommended starting conditions that will work for most samples include the following:

- Average linear heating rate = 2 °C/min
- Temperature modulation period = 60 s (use longer periods for larger samples)
- Temperature modulation amplitude = ±1.0 °C
- Other conditions: same as DSC

30.3.3.3 *Common Measurements*

MDSC® is used to make the same measurements as DSC, but has the significant advantage of being able to separate the heat flow signal into the heat capacity and kinetic components. The benefit of this can be seen in Fig. 30.10, which is an analysis of the structure of sugar-frosted cornflakes. Many cereal products exhibit a broad endothermic peak (most likely a relaxation process, also termed physical aging) in DSC data at temperatures above room temperature; however, the expected and very important glass transition is not visible. The **total heat flow** signal of the MDSC® data in Fig. 30.10 (equivalent to conventional DSC) shows such a peak. However, MDSC® also shows the kinetic and heat capacity components of the total signal, allowing for straightforward analysis of the glass transition seen in the **reversing heat flow** signal. In general, the reversing heat flow signal includes heat capacity, changes in heat capacity, and most melting. All kinetic events such as crystallization, decomposition, evaporation, and physical aging appear in the **nonreversing heat flow** signal.

30.4 APPLICATIONS

This section will examine some of the difficulties associated with applying thermal analysis to food products, as well as illustrate some additional uses of the thermal techniques discussed previously to determine food system composition and structure.

30.4.1 Sample Preparation Challenges

- **Poor Sample Contact with Pan.** It is necessary to contain the sample in a DSC pan (typically aluminum hermetic sealed) to perform a DSC experiment. When preparing the sample pan, keep in mind that

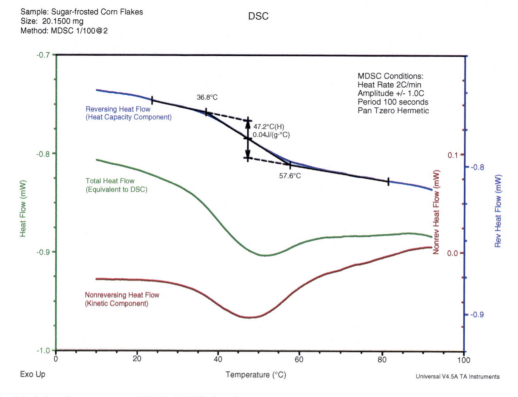

Sample: Sugar-frosted Corn Flakes
Size: 20.1500 mg
Method: MDSC 1/100@2

30.10 **figure** Modulated temperature DSC (MDSC®) data for a 20 mg sample of sugar-coated cornflakes. A large sample was used to improve sensitivity, and a hermetically sealed pan was used to prevent moisture loss during heating. The total signal shows the broad endothermic peak typically observed in DSC data, while the reversing signal (heat capacity component) shows the underlying glass transition. Since the glass transition temperature (T_g) is above room temperature, the cornflakes will be crispy at room temperature

heat transfer between the sample and DSC sensor takes place through the bottom of the pan. Therefore, it is important to provide good contact between the sample and the bottom of the DSC pan to avoid artifacts in the data due to sample movement during the experiment. This can be done by lightly compacting the sample in the pan or using an inner crimping lid to force the solid sample to the bottom of the pan prior to sealing with the hermetic lid.

- **Lack of Homogeneity.** Most food products are heterogeneous, having large variations in composition and structure in their cross sections. Since thermal analysis techniques use relatively small sample sizes, users must look for ways to obtain representative samples. Two simple techniques to address this problem are to use the largest sample possible (that still fits in pan) and do all initial measurements in triplicate to detect compositional inconsistencies. Note that to better detect the weak glass transition in the sugar-frosted cornflakes of Fig. 30.10, a sample weight of 20 mg was used. Although large samples negatively affect resolution, the first objective should be to detect the transition of interest.
- **Controlling Moisture Content.** Food products typically have high surface area and amorphous con-

tent. This can result in moisture being absorbed or lost during sample storage and/or preparation. Moisture exchange with the atmosphere needs to be kept to a minimum, since just a few percent change in water content can change the glass transition temperature by tens of degrees. In addition, hermetic DSC pans should be used to prevent moisture loss during heating, and open samples should not be left in a TGA autosampler tray.

- **Previous Thermal History.** It is often difficult to analyze a glass transition obtained during the first heat scan of a DSC experiment. This is due to numerous kinetic processes caused by previous thermal history (time and temperature) that can occur in the temperature region of the glass transition, including enthalpic relaxation and recovery, stress relaxation, and flow. To eliminate interference from these kinetic processes, it is often useful to heat the sample to a temperature of about 25 °C above the end of the glass transition and then cool it back to the desired starting temperature. This is typically called a heat-cool-heat experiment. An initial experiment will be required to determine this temperature, but it is important to not change the crystalline structure or cause decomposition during the first heat.

30.4.2 Additional Applications

Figure 30.11 is a comparison of the TGA results from the center (crumb) and crust of a loaf of bread. As expected, both samples show a similar decomposition temperature starting just above 200 °C, which can be seen in the derivative signal. Clear differences can be observed in the DSC data in regard to moisture content, the first peak in the derivative signal, and minor components (second minor peak). These differences result in the substantially different physical properties between the center and the crust of the loaf of bread.

Figure 30.12 is a comparison of DSC results from a sugar-coated chocolate candy and the sugar coating alone. Note that the melting of the sugar is different in the presence of the melted chocolate. This is due to the sugar starting to dissolve in the liquid chocolate.

Many materials can exist in either an amorphous or crystalline form depending on how the material was processed. Since the structure or form of the material is a function of its previous thermal history (e.g., time, temperature, relative humidity, pressure), the structure can change as it is heated during the DSC experiment. This can be seen in Fig. 30.13, a DSC experiment on amorphous sucrose that was prepared by freeze-drying. The first observed transition is the glass transition near 53 °C. The Tg of dry (<0.1 % mois-

ture content) sucrose is approximately 68 °C, which decreases as moisture content increases. A Tg of 53 °C indicates that this sample has moisture content of approximately 1.5–2.0 % (dry basis) [14]. The second useful piece of information is the size of the glass transition, which is the magnitude of the step change in heat capacity and is typically expressed in heat capacity units of J/g°C. A value of 0.75 J/g°C for this sample indicates that it is very close to 100 % amorphous, since the value for a 100 % amorphous melt-quenched sample, given in the literature, is 0.78 J/g°C [15]. An endothermic peak is seen at the end of the glass transition. The presence of this peak is common to materials with a glass transition slightly above room temperature and is due to the process of physical aging. The material relaxes over time to a lower energy state and must reabsorb this energy (termed enthalpic recovery) to achieve an equilibrium state above Tg, thus producing the observed endothermic peak [16].

The second transition near 100 °C is an exothermic peak caused by crystallization (commonly termed thermally induced crystallization or cold crystallization) of the amorphous material. The size of the peak (J/g) provides information on the amount of amorphous material that was able to crystallize. By dividing the measured area of the peak (88.57 J/g) by the heat of crystallization for 100 % conversion (approximately 131 J/g obtained from a separate DSC

Comparison of TGA Weight Losses for Samples of
Bread Taken from the Center and Crust of the Loaf

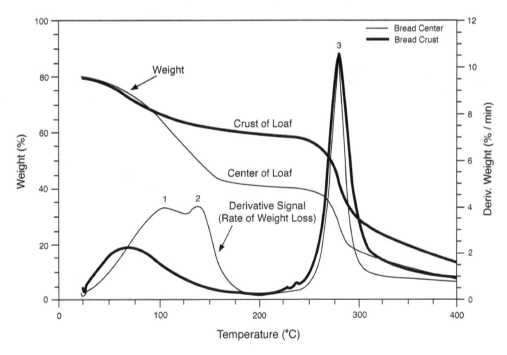

 A comparison of TGA data on the center (crumb) and crust of a loaf of bread baked at 190 °C shows substantial differences at temperatures below 200 °C. This is because the center of the loaf remains much cooler than the crust during baking due to the evaporation of water

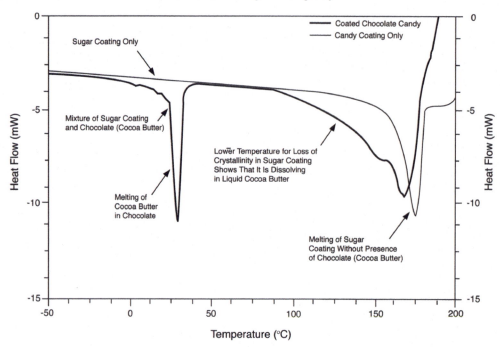

30.12 figure Comparison of DSC data on sugar-coated chocolate candy and the sugar coating alone. Melting of the sugar changes in the presence of the liquid chocolate due to its solubility in the chocolate

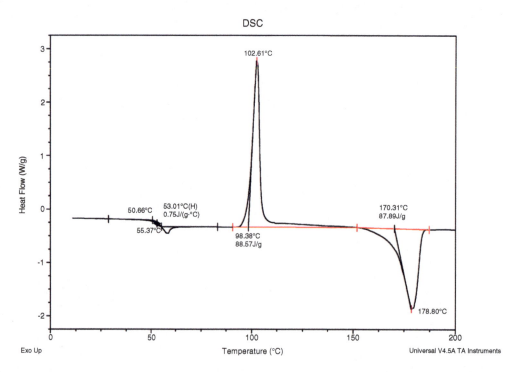

30.13 figure DSC data obtained using a 10 °C/min heating rate shows that a sample of freeze-dried sucrose is 100 % amorphous based on the size of the glass transition and a comparison of the energies (J/g) of crystallization and melting

experiment on 100 % crystalline sucrose at 10 °C/min, reference [15]), it appears that only about 68 % of the amorphous sucrose crystallized during heating.

The last observed transition is an endothermic peak between approximately 150 and 180 °C caused by conversion of the crystalline structure to an amorphous form, typically termed melting. The shape of the peak is not very symmetric, because sucrose begins to decompose (chemical transformation) at the same temperature [12]. Thus, the overlapping of the two processes creates the nonsymmetric peak. As with thermally induced crystallization, the size of the peak (J/g) is a quantitative measure of the amount of structural change. In this case, the size of the melting peak (87.89 J/g) is within experimental error (typically ±2 %) of the size of the cold crystallization peak (88.57 J/g), indicating that all observed melting is the result of the crystallization that occurred during heating and that the original structure was 100 % amorphous.

For the interested reader, additional applications of thermal analysis techniques to food materials, such as gelatinization of starch, oxidation of vegetable oils, and denaturation of meat protein, can be found in the Thermal Analysis Food Application Handbook [17].

30.5 SUMMARY

Thermal analysis is a series of laboratory techniques that measures physical and chemical properties of materials as a function of temperature and time. In a thermal analysis experiment, temperature is either held constant (isothermal) or programmed to increase or decrease at a linear rate. Since temperature and time are controlled in all food preparation processes, thermal analysis instruments can simulate these processes on a very small scale (milligrams) and measure the response of the material. The most frequently used techniques include TGA and DSC.

The utility of thermal analysis to the food scientist is due to the fact that end-use properties (functionality) at a specific temperature are dependent on the structure of the components at that temperature. Therefore, it is necessary to measure structure as a function of temperature. Structure can be defined as amorphous (no molecular order), crystalline, or semicrystalline, with most food products having both amorphous and crystalline components.

Amorphous structure is characterized by analysis of the glass transition, which involves a significant change in the material's heat capacity and molecular mobility. Crystalline structure melts as temperature increases, creating an endothermic peak that provides information on the quantity of crystalline structure and its melting temperature. This chapter has illustrated application of thermal analysis to a variety of food materials.

30.6 STUDY QUESTIONS

1. You have two brands of caramels, one that is hard and one that is chewy.

    (a) What structural transition would you expect to detect in the DSC data for each type of caramel?
    (b) How would the DSC data be useful in understanding the differences in texture of the two caramel types?

2. If amorphous material crystallizes during heating (called cold crystallization), would you expect it to crystallize below or above its glass transition temperature? Explain your choice.
3. Explain in your own words why a cold crystallization peak and a melting peak show opposite DSC heat flow signals. See Fig. 30.13 for an example of this behavior.
4. In Sect. 30.3.3.3, the glass transition is referred to as "very important" – why?
5. Assume that you have been asked to develop a recipe for a crisp cookie based on your knowledge of material properties. To develop the recipe, you need to answer the following questions:

    (a) Should the finished cookie be in the glassy or rubbery state at room temperature? Explain your choice.
    (b) In selecting TGA experimental conditions, would you use a sealed or unsealed sample pan and why?
    (c) In selecting DSC experimental conditions, would you typically use a sealed or unsealed pan and why?
    (d) If you were to run a MDSC® experiment, which MDSC® signal would contain the glass transition and why?
    (e) You find out that high-fructose corn syrup (HFCS) is less expensive than sucrose. Can you substitute the sucrose in your crisp cookie formula for HFCS? Be sure to give reasons for your response.
    (f) While performing a DSC or MDSC® experiment, an amorphous component in the sample begins to crystallize. Would this cause an increase or decrease in the measured heat capacity?
    (g) As stated in Sect. 30.3.2, a limitation of DSC is that it cannot measure heat capacity under isothermal conditions. However, MDSC® can measure heat capacity during an isothermal experiment. Explain this difference between DSC and MDSC® using Eq. 30.1 in Sect. 30.3.3.

REFERENCES

1. Farkas J, Mohácsi-Farkas C (1996) Application of differential scanning calorimetry in food research and food quality assurance. J Therm Anal 47: 1787–1803
2. Raemy A (2003) Behavior of foods studied by thermal analysis. J Therm Anal Calorim 71: 273–278
3. Ievolella J, Wang M, Slade L, Levine H (2003) Application of thermal analysis to cookie, cracker, and pretzel manufacturing, Ch. 2. In: Kaletunc G, Breslauer KJ (eds) Characterization of cereals and flours: properties, analysis, and applications. CRC Press, Boca Raton, FL, pp 37–63
4. Sahin S, Sumnu SG (2006) Physical properties of foods. Springer, New York, p 257
5. Kamruddin M, Ajikumar PK, Dash S, Tyagi K, Baldev RAJ (2003) Thermogravimetry-evolved gas analysis–mass spectrometry system for materials research. Bull Mater Sci 26(4): 449–460
6. Schmidt SJ (2004) Water and solids mobility in foods. Advances in Food and Nutrition Research, vol 48. Academic Press, London, UK, pp 1–101
7. Slade L, Levine H (1988) Non-equilibrium behavior of small carbohydrate-water systems. Pure Appl Chem 60(12):1841–1864
8. Slade L, Levine H (1991) Beyond water activity: recent advances based on an alternative approach to the assessment of food quality and safety. Crit Rev Food Sci Nutr 30(2–3): 115–360
9. Aguilera JM, Lillford PJ (2007) Food materials science: principles and practice. Springer, New York, p 622
10. Sun D-W (2005) Handbook of frozen food processing and packaging. CRC Press, Raton, FL, p 760
11. Schmidt SJ (2012) Exploring the sucrose-water state diagram: Applications to hard candy cooking and confection quality and stability. Manufacturing Confectioner, January: 79–89
12. Lee JW, Thomas LC and Schmidt SJ (2011) Investigation of the heating rate dependency associated with the loss of crystalline structure in sucrose, glucose, and fructose using a thermal analysis approach (Part I). J Ag Food Chem (59): 684–701
13. Thomas L (2006) Modulated DSC technology manual. TA Instruments, New Castle, DE
14. Yu X, Kappes SM, Bello-Perez LA, Schmidt SJ (2008) Investigating the moisture sorption behavior of amorphous sucrose using a dynamic humidity generating instrument. J Food Sci 73(1): E25–E35
15. Magoń A, Wurm A, Schick C, Pangloli P, Zivanovic S, Skotnicki M, Pyda M (2014) Heat capacity and transition behavior of sucrose by standard, fast scanning and temperature-modulated calorimetry. Thermochim Acta 589: 183–196
16. Wungtanagorn R and Schmidt SJ (2001) Thermodynamic properties and kinetics of the physical aging of amorphous glucose, fructose, and their mixture. J Therm Anal Calorim 65: 9–35
17. Widmann G and Oberholzer T (2014) Thermal Analysis Application Handbook, Food Collected Applications. Mettler Toledo, Columbus, OH, p. 66

Color Analysis

R.E. Wrolstad (✉) • *D.E. Smith*
Department of Food Science and Technology,
Oregon State University,
Corvallis, OR 97331-6602, USA
e-mail: ron.wrolstad@oregonstate.edu; dan.smith@oregonstate.edu

31.1 INTRODUCTION

Color, flavor, and texture are the three principal quality attributes that determine food acceptance, and color has a far greater influence on our judgment than most of us appreciate. We use color to determine if a banana is at our preferred ripeness level, and a discolored meat product can warn us that the product may be spoiled. The marketing departments of our food corporations know that, for their customers, the color must be "right." The University of California Davis scorecard for wine quality designates 4 points out of 20, or 20 % of the total score, for color and appearance [1]. Food scientists who establish quality control specifications for their product are very aware of the importance of color and appearance. While subjective visual assessment and use of visual color standards are still used in the food industry, instrumental color measurements are extensively employed. Objective measurement of color is desirable for both research and industrial applications, and the ruggedness, stability, and ease of use of today's color measurement instruments have resulted in their widespread adoption.

Color can be defined as the sensation that is experienced by an individual when radiant energy within the visible spectrum (380–770 nm) falls upon the retina of the eye [2], and a **colorant** is a pigment that is used to color a product. For the phenomenon of color to occur, there must be: (1) a colored object, (2) light in the visible region of the spectrum, and (3) an observer. All three of these factors must be taken into account when assessing and measuring color. When white light strikes an object, it can be absorbed, reflected, and/or scattered. Selective absorption of certain wavelengths of light is the primary basis for the color of an object. Color, as seen by the eye, is an interpretation by the brain of the character of light coming from an object. **Colorimetry** is the science of color measurement [3]. It is possible to define color in mathematical units; however, those numbers do not easily relate to the observed color. A number of color-ordering systems and color spaces have been developed that better agree with visual assessment. In food research and quality control, instruments are needed which provide repeatable data that correspond to how the eye sees color. This chapter will provide a brief description of human physiology of vision and an overview of the different color-ordering and color-measuring systems. The chapter is limited to presenting the basic underlying principles that will hopefully allow for an understanding of how color of food products should be measured. Color measurement is a very complex subject, and for more detailed exploration of the subject, the following references are recommended [2–7].

31.2 PHYSIOLOGICAL BASIS OF COLOR

Humans have excellent color perception and they can detect up to 10,000,000 different colors [8]. They have very poor color memory, however, and cannot accurately recall colors of objects previously observed [5, 9], hence the need for objective measurement of color. While color perception varies somewhat with humans, it is much less variable than that for the senses of taste and smell. Color perception is comparatively uniform for people with normal color vision; however, 8 % of males and 0.5 % of females have physiological defects and perceive colors in a markedly different way [2, 5].

Figure 31.1 is a simplified diagram of the human eye. Light enters the eye through the cornea, passes through the aqueous and vitreous humor, and is focused on the **retina**, which contains the receptor system [10]. The **macula** is a small (approximately 5 mm in diameter) and highly sensitive part of the retina that is responsible for detailed central vision. It is located roughly in the center of the retina. It is yellow-orange colored and contains a high concentration of the carotenoid pigments, lutein and zeaxanthin. It is believed that these dietary antioxidants may protect the retina from photo damage [11]. Age-related macular degeneration results in loss of central vision and is a major health issue in our aging population. The **fovea**, the very center of the macula, is about 2 mm in diameter and contains a high concentration of **cones**, which are responsible for daylight and color vision, known as "**photopic**" vision. The cones contain receptors that

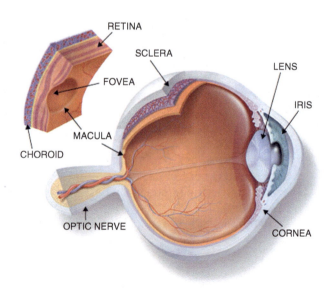

 Diagram of the human eye. http://www.amdcanada.com/template.php?lang=eng&section=4&subSec=2d&content=4_2

are sensitive to red, green, and blue light. Figure 31.2 shows the spectral sensitivity curves for the three respective cones. **Rods** are more widely distributed in the retina and are sensitive to low-intensity light. They have no color discrimination and are responsible for night or "**scotopic**" vision. Figure 31.3 shows the spectral sensitivity curves for scotopic (rod) and photopic (cone) vision, the latter being an integration of the curves shown in Fig. 31.2. Note the sensitivity maximum is at 510 nm for scotopic vision and 580 nm for photopic vision. This accounts for blues appearing to be brighter and reds darker at twilight when both scotopic vision and photopic vision are functioning.

Signals are sent via the optic nerve to the brain, where "vision" occurs. According to the "Color Opponent Theory" [4], the signals from the red, green, and blue receptors are transformed to one brightness signal indicating darkness and lightness and two hue signals, red vs. green and blue vs. yellow. Figure 31.4 shows a diagram of the opponent color model. The

Opponent-Colors Theory

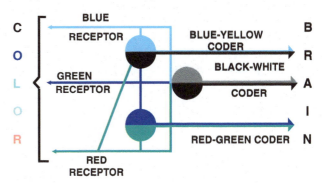

31.4 figure — The color opponent model (Courtesy of HunterLab, Reston, VA)

brain's interpretation of signals is a complex phenomenon and is influenced by a variety of psychological aspects. One such aspect is **color constancy**. The same sheet of white paper will appear white when seen in bright sunlight and also when it is viewed indoors under dim light. The physical stimuli in each case are obviously quite different, but the brain knows that the paper should be white and draws on its experience. A second aspect occurs when a large expanse of color appears brighter than the same color in a small area. One only needs the experience of painting a whole wall of a room and then seeing how different it appears from the small color chip obtained from the paint store.

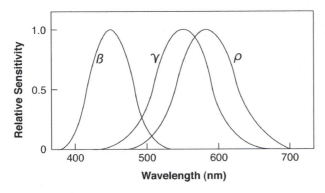

31.2 figure — Spectral sensitivity curves of the three types of cones comprising photopic vision (From Hutchings [5], with kind permission of Springer + Business Media)

31.3 COLOR SPECIFICATION SYSTEMS

There are verbal, visual matching, and instrumental methods for describing and specifying color. Color is three-dimensional, and any color-order system will need to address **hue**, what we instinctively think of as color (e.g., red, blue, green); **value**, which represents lightness and darkness; and **chroma** or **saturation** which indicates intensity. When attempting to verbally describe a color defect or problem, one should attempt to use these three qualities in formulating a color description.

31.3.1 Visual Systems

The **Munsell** system is probably the best known and most widely used visual color-ordering system. It was developed by A.H. Munsell, a Boston art teacher, in 1905. In this system red, yellow, green, blue, and purple plus five adjacent pairs, green yellow, yellow red, red purple, purple blue, and blue green, describe **hue**. **Value** is that quality of color described by lightness and darkness, from white to grey to black. Value is designated from 0 (absolute black) to 10 (absolute white). **Chroma** is that

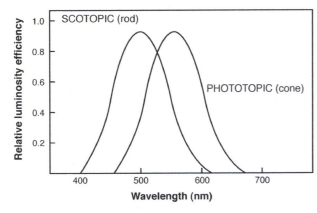

31.3 figure — Spectral sensitivity curves for scotopic (rods) and photopic (cones) vision (From Hutchings [5], with kind permission of Springer + Business Media)

quality that describes the extent a color differs from a gray of the same value. It is designated in increasing numbers starting with 0 (neutral grey) and extending to /16 or even higher. A change from pink to red is an example of an increase in chroma. In Munsell notation, hue is listed first and designated by a number and letter combination. Numbers run from 1 to 100, and the letters are taken from the ten major hue names, e.g., 10 GY. Value follows with a number from 0 to 10 followed by a slash mark, which is followed by a number for chroma (e.g., 5R 5/10).

One of Munsell's objectives was to develop a system based on equal visual perception, with equal steps of perception for each of the coordinates. For example, the difference in value between 2 and 3 is visually equivalent to the difference between 5 and 6. This visual linearity applies to the other coordinates as well. The Munsell systems' visual linearity undoubtedly contributes to its success and wide popularity in many different fields. Figure 31.5 illustrates the Munsell color system, showing a circle of hues at value 5 and chroma 6, the neutral values from 0 to 10, and the chroma of purple-blue (5PB) at value 5. The ten named hues are shown with additional intermediate hues interspersed. The distance from the core to the edge shows increasing chroma, the maximum chroma differing considerably for different hues (e.g., R5 has a maximum of 12 and yellow has a maximum of 6). Interactive kits that demonstrate the relationships between Munsell hue,

value, and chroma are available for purchase [12]. Also available is the Munsell Book of Color with 1,605 colored chips, each with a numerical designation.

Assessing color of foods by visual comparison with color standards is an option for a number of food products. USDA color standards are available for honey, frozen French fried potatoes, peanut butter, and canned ripe olives, for example [12]. This method is simple, convenient, and easy to understand; however, it is subjective.

31.3.2 Instrumental Measurement of Color

31.3.2.1 *Historical Development*

For a more detailed discussion of the historical development, refer to the 4th edition of this text [14]. The **CIE** (**Commission Internationale de l'Eclairage**, or the **International Commission on Illumination**) is the main international organization concerned with color and color measurement [3]. Standard illuminants for color measurement were first established in 1931 by the CIE. Figure 31.6 shows the spectral power distribution curves of three standard CIE illuminants, A, C, and D_{65}. **Illuminant C** was adopted in 1931 and represents overcast daylight, while **illuminant D_{65}**, which was adopted in 1965, also represents average daylight but includes the ultraviolet wavelength region. **Illuminant A**, adopted in 1931, represents an incandescent light bulb. Objects will appear to have different colors when viewed under illuminants A and C. Because of the

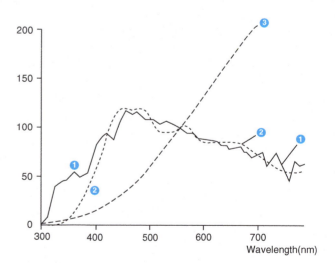

31.5 **figure** Diagram of the Munsell color system, showing a circle of hues at value 5 chroma 6, the neutral values from 0 to 10, and the chromas of purple-blue (5PB) at value 5 (Source, Wikipedia: Jacobolus (http://en.wikipedia.org/wiki/User:Jacobolus))

31.6 **figure** The spectral power distribution curves of three standard CIE illuminants. Standard illuminant D_{65}, average daylight including ultraviolet wavelength [1]; standard illuminant C, average daylight (not including ultraviolet wavelength region) [2]; and standard illuminant A, incandescent light [3] (Courtesy of Konica Minolta Sensing Americas, Inc., Ramsey, NJ)

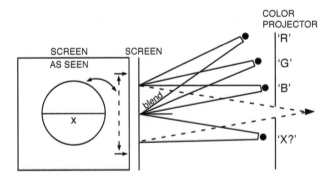

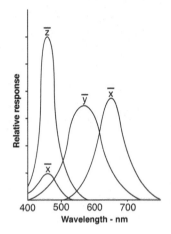

 31.7
figure

Diagram showing three projectors focused on the upper half of a *circle* on the screen. The color to be measured is projected on the lower half and the eye can see both halves simultaneously

predominance of long wavelength light and lesser amounts of shorter wavelength light of illuminant A, one can predict that objects will appear to have a "warmer" color under illuminant A than under other illuminants. **Metamerism** occurs when two objects appear to have the same color under one light source but exhibit different colors under another source.

Scientists knew that a color sensation could be matched by mixing three colored lights [3]. W.D. Wright in 1928 and J. Guild in 1931 conducted independent experiments in which people with normal color vision visually matched spectral (single wavelength) light by mixing different amounts of three primary lights (red, green, and blue) using rheostats (Fig. 31.7). The process was repeated for test colors covering the entire visible spectrum. The **field of view** for these experiments is described as **2°**, which is similar to viewing a dime at an arm's length. The purpose of these viewing conditions was to have primary involvement of the fovea, the retinal area of greatest visual acuity. The red, green, and blue response factors were averaged and mathematically converted to x, y, and z functions that quantify the red, green, and blue cone sensitivity of the average human observer. The observer functions were standardized and adopted by CIE in 1931 as the **CIE 2° Standard Color Observer**. The standard observer curves provide human sensory response factors that are used in color measurement worldwide (Fig. 31.8). Subsequently it was realized that more realistic data could be obtained from a larger field of view. The experiment was repeated using a 10° field of view and adopted by CIE in 1964 as the **10° Standard Observer**. Both sets of data are used today, but the 10° standard observer is preferable because it better correlates to visual assessments.

31.3.2.2 The CIE Tristimulus System

With the adoption of standard observer functions and standard illuminants, it became possible to convert the spectral transmission or reflectance curve of any object to three numerical values. These numbers are known

31.8
figure

Standard observer curves showing the relationship between the *red* (x), *blue* (z), and *green* (y) cone sensitivity and the visible spectrum

as the **CIE tristimulus values**, X, Y, and Z, the amounts of red, green, and blue primaries required to give a color match. The data values for a standard illuminant and the standard observer functions are multiplied by the % reflectance or % transmission values for the object at selected wavelengths. Summation of the products for the wavelengths in the visible spectrum (essentially integrating the areas under the three curves) gives the resulting X, Y, and Z tristimulus values. This can mathematically be represented as follows:

$$X = \int_{380}^{750} RE\bar{x}\,dx \qquad (31.1)$$

$$Y = \int_{380}^{750} RE\bar{y}\,dy \qquad (31.2)$$

$$Z = \int_{380}^{750} RE\bar{z}\,dz \qquad (31.3)$$

where:

R = sample spectrum
E = source light spectrum
$\bar{x}, \bar{y}, \bar{z}$ = standard observer curves.

With the objective of plotting the three coordinates in two dimensions, the CIE converted the X, Y, and Z tristimulus values to x, y, and z coordinates by the following mathematical operation:

$$x = \frac{X}{X+Y+Z} \qquad (31.4)$$

$$y = \frac{Y}{X+Y+Z} \qquad (31.5)$$

$$z = \frac{Z}{X+Y+Z} \qquad (31.6)$$

Since $x + y + z = 1$, only two coordinates are needed to describe color as $z = 1-(x + y)$.

Figure 31.9 shows the 1931 **chromaticity diagram** where x vs. y are plotted to give the horseshoe-shaped locus. Spectral colors lie around the perimeter and white light (illuminant D_{65}) has the coordinates $x = 0.314$, $y = 0.331$. With the aid of a ruler, a line can be drawn from the coordinates for white light through the object coordinates to the edge, which gives the **dominant wavelength**, λd. Dominant wavelength is analogous to hue in the Munsell system. The distance from the white light coordinates to the object coordinates, relative to the distance from the white light coordinates to λ_d, is described as % **purity** and is analogous to chroma in the Munsell system. The standard observer curve for y (green) shown in Fig. 31.8 is very similar to the sensitivity curve for human photopic vision shown in Fig. 31.3. Because of this, tristimulus value Y is known as **luminosity** and is analogous to value in the Munsell system.

Manual calculation of XYZ tristimulus values from reflectance/transmission spectra is a tedious operation. Modern colorimetric spectrophotometers measure the light reflected or transmitted from an object, and the data are sent to a processor where it is multiplied by standard illuminant and standard observer functions to give the XYZ tristimulus values. Since objects with identical XYZ tristimulus values will provide a color match, they find application in the paper, paint, and textile industries. Unfortunately, the XYZ numbers do not easily relate to observed color, and they have the limitation of not having equivalent visual spacing. [Referral to Fig. 31.9 reveals that the wavelength spacing in the green region (500–540 nm) is much larger than that in the red (600–700 nm) or blue (380–480 nm) regions.] The same numerical color differences between colors will not equate to the same visual difference for all colors. This is a severe limitation in measurement of color of food products, as major interest is in how food product color deviates from a standard or changes during processing and storage. Statistical analysis of color data for which numerical units were nonequivalent would be problematic.

31.3.3 Tristimulus Colorimeters and Color Spaces

Richard S. Hunter, Deane B. Judd, and Henry A. Gardner were among the pioneering scientists who in the 1940s were working to develop color-measuring instruments that would overcome the disadvantages of the CIE spectrophotometric tristimulus system [2, 5, 6]. Light sources that were similar to illuminant C were used, along with filter systems that approximated the sensitivity of the cones in the human eye. Empirical approaches were taken to get more equivalent visual spacing. In an effort to get numerical values that better related to observed color, a system that applied the color opponent theory of color perception was developed [3].

The **Hunter color solid** (Fig. 31.10) was first published in 1942 where L indicated **lightness**; a, the **red** (+) or **green** (−) coordinate; and b, the **yellow** (+) or **blue** (−) coordinate. The **Hunter L a b color space** has been widely adopted by the food industry. It is very effective for measuring color differences. The *Lab* system was subsequently improved to give more uniform color spacing. In 1976, the CIE officially adopted the modified system as **CIELAB** with the parameters $L^*a^*b^*$. L^* indicates **lightness** (0–100) with 0 being black and 100 being white. The coordinate a^* is for **red** (+) and **green** (−), and b^* is for **yellow** (+) and **blue** (−). The limits for a^* and b^* are approximately + or − 80. Figure 31.11 shows a portion of the a^*, b^* chromaticity diagram where a^* and b^* are both positive, representing a color range from red to yellow. Point A

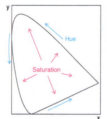

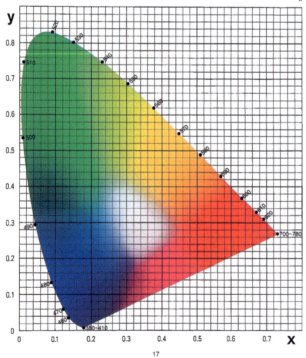

31.9 figure The 1931 x, y chromaticity diagram (Courtesy of Konica Minolta Sensing Americas, Inc., Ramsey, NJ)

Hunter L,a,b Color Space

L = 100

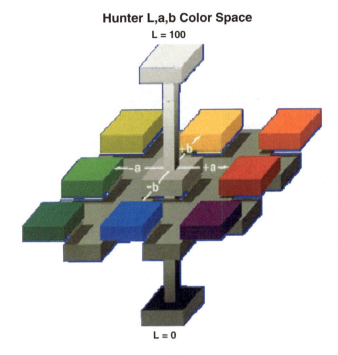

L = 0

31.10 figure The Hunter L, a, b color solid (Courtesy of HunterLab, Reston, VA)

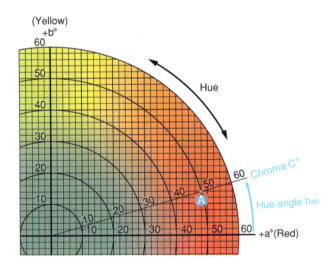

$$\text{Chroma } C^* = \sqrt{(a^*)^2 + (b^*)^2}$$

$$\text{Hue angle } h_{ab} = \tan^{-1}\left\{\frac{b^*}{a^*}\right\}$$

31.11 figure A portion of an a^*, b^* chromaticity diagram showing the position A for a red apple (Courtesy of Konica Minolta Sensing Americas, Inc., Ramsey, NJ)

is the plot of a^* and b^* for a red apple. The angle from the start of the $+a^*$ axis to point A can be calculated as **arctan** b^*/a^* and is known as **hue angle**, h or H^*.

The distance from the center to point A is **chroma**, which is calculated as the hypotenuse of the right triangle formed by the origin and the values of coordinates a and b. $\left(a^{*2} + b^{*2}\right)^{1/2} = C^*$.

The CIE has also recommended adoption of this color scale known as **CIELCH** or $L^*C^*H^*$. This color space (which is illustrated in Fig. 31.12) designates hue (H^*) as one of the three dimensions, the other two being lightness (L^*) and chroma (C^*), which have an obvious parallel to Munsell hue, value, and chroma. This color space is advantageous as hue is most critical to humans with normal color vision for perception and acceptability. In this system, 0° represents red, 90°— yellow, 180°—green, and 270°—blue. Figure 31.13 shows plots of a^* and b^* for three hypothetical objects

Polar CIE L*,C*,h

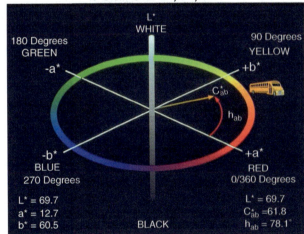

31.12 figure The CIE $L^*C^*H^*$ color space showing the location of "school bus yellow" (Courtesy of HunterLab, Reston, VA)

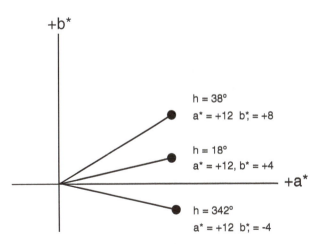

31.13 figure Plots of a^* and b^* for three hypothetical samples

having the following a^*b^* coordinates: $a^* = +12$ and $b^* = +8$; $a^* = +12$ and $b^* = +4$; $a^* = +12$ and $b^* = -4$. While all objects have identical a^* values, their colors range from purplish red ($H^* = 342°$) to red ($H^* = 18°$) to orange ($H^* = 34°$). A common error in interpretation of color measurements is to use only the coordinate a^* as a measure of "redness." Monitoring color change is more understandable if one measures lightness (L^*), hue angle (H^* from 0 to 360°), and chroma. Chroma will increase with increasing pigment concentration and then decrease as the sample becomes darker. Thus, it is possible for one light and one dark sample to have the same hue angle and the same chroma. They will readily be distinguished, however, because of their different L^* values.

The colorimeters that are available in the market today have vastly improved from earlier models with respect to stability, ruggedness, and ease of use. There are handheld instruments that are portable for use in the field, online instruments for process control, and specialized colorimeters for specific commodities. They vary with respect to operating in transmission or reflectance mode and size of sample viewing area. Colorimeters have a high degree of precision, but do not have a high degree of accuracy with respect to identifying or matching colors. Most colorimeters used in research are color spectrophotometers with a diffraction grating for scanning the visible spectrum, with the data being sent to a microprocessor for conversion of reflectance or transmission data to tristimulus numbers. In operating the instrument, choices must be made as to **illuminant, viewing angle** (2° or 10°), and data presentation as **XYZ**, *Lab*, **CIEL*a*b***, or $L^*C^*H^*$. Illuminant D_{65}, 10° viewing angle, and $L^*C^*H^*$ are appropriate for most food applications. It should be obvious that different numbers will be obtained with different illuminants, viewing angles, and color scales. It is critical that the illuminant, viewing angle, and color scale used in color measurement be specified in technical reports and research publications.

31.4 PRACTICAL CONSIDERATIONS IN COLOR MEASUREMENT

Choice of an appropriate instrument, sample preparation, sample presentation, and handling of data are issues that must be dealt with in color measurement.

31.4.1 Interaction of Light with Sample

When a sample is illuminated with light, a number of things occur that are illustrated in Fig. 31.14. Light for which the angle of reflection is equal to the angle of incidence is described as **specular light**. Smooth polished surfaces will appear **glossy** because of the high degree of **specular reflection**. Rough surfaces will

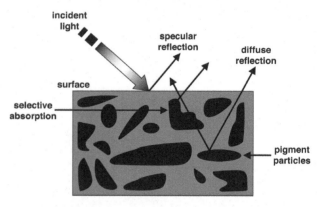

31.14 figure Interaction of light with an object (From Loughry [4], used with permission)

have a great deal of **diffuse reflection** and will have a dull or **matte** appearance. Selective absorption of light will result in the appearance of color. **Opaque** samples will **reflect** light. **Transparent** samples will primarily **transmit** light, and **translucent** samples will both **reflect** and **transmit** light. Ideal samples for color measurement will be flat, smooth, uniform, matte, and either opaque or transparent. A brick of colored Cheddar cheese is one of the few food examples that come close to having those characteristics.

31.4.2 Instrument Choice

Instrument geometry refers to the **arrangement of light source**, **sample placement**, and **detector**. The CIE recognizes the following instrument geometries: 45°/0° where the specimen is illuminated at 45° and measured at 0° and, the inverse, **0°/45°** where the specimen is illuminated at 0° and measured at 45°. Diffuse reflectance is measured since specular light is excluded. These are illustrated in Fig. 31.15. **Diffuse sphere geometry** is the third type where a white-coated sphere is used to illuminate a sample. With some sphere geometry instruments, measurements can either include or exclude specular reflectance. These instruments are versatile in that they can measure in transmission for transparent samples and in reflectance for opaque samples. Some can also measure the amount of light scattering, turbidity or haze in liquid samples, and the amount of gloss in solid samples. Instruments with 45°/0° and 0°/45° geometries can only measure reflectance.

31.4.3 Color Difference Equations and Color Tolerances

When colorimeter measurements are conducted under carefully controlled conditions, data with a high degree of precision can be obtained. In both industrial and research applications, the interest is primarily in how color dimensions deviate from a standard

45°/0° and 0°/45° Specular Excluded Geometry

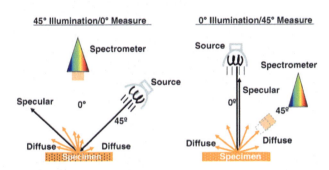

31.15 figure CIE standardized geometries for 45°/0° and 0°/45° instruments (Courtesy of HunterLab, Reston, VA)

or how they change from batch to batch, year to year, or during processing and storage. Color differences are calculated by subtracting $L*a*b*$ and $L*C*H*$ values for the sample from the standard, e.g.,

Delta L* $= L*_{sample} - L*_{standard}$. Positive $\Delta L*$ numbers will be lighter than the standard, and negative $\Delta L*$ numbers will be darker.

Delta a* $= a*_{sample} - a*_{standard}$. Positive $\Delta a*$ numbers will be more "red" (or less "green") than the standard, and negative $\Delta a*$ numbers will be more "green" (or less "red").

Delta b* $= b*_{sample} - b*_{standard}$. Positive $\Delta b*$ numbers will be more "yellow" (or less "blue"), and negative $\Delta b*$ numbers will be more "blue" (or less "yellow").

Delta C* $= C*_{sample} - C*_{standard}$. Positive $\Delta C*$ numbers mean the sample has greater intensity or is more saturated, and negative $\Delta C*$ numbers mean that the sample is less saturated.

Delta H* $= H*_{sample} - H*_{standard}$. Positive $H*$ numbers indicate the hue angle is in the counterclockwise direction from the standard, and negative numbers are in the clockwise direction. If the standard has a hue angle of 90°, a positive $\Delta H*$ is a shift in the green direction, and a negative $\Delta H*$ number is a shift in the red direction.

A single number is often desired in industry for establishing pass/fail acceptability limits. **Total color difference** ($\Delta E*$) is calculated by the following equation:

$$\Delta E* = \left(\Delta L*^2 + \Delta a*^2 + \Delta b*^2 \right)^{1/2} \quad (31.7)$$

A limitation of $\Delta E*$ is that the single number will only indicate the magnitude of color difference, not the direction. Samples with identical $\Delta E*$ numbers will not necessarily have the same visual appearance.

Polar $\Delta L*$, $\Delta C*$, $\Delta H*$ Color Space

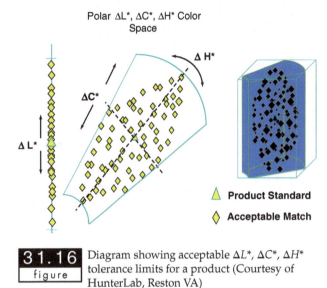

△ **Product Standard**

◇ **Acceptable Match**

31.16 figure Diagram showing acceptable $\Delta L*$, $\Delta C*$, $\Delta H*$ tolerance limits for a product (Courtesy of HunterLab, Reston VA)

In establishing color tolerances, $\Delta L*$, $\Delta C*$, and $\Delta H*$ numbers are preferred since they correlate well with visual appearance. A diagram showing acceptable tolerances based on $\Delta L*$, $\Delta C*$, and $\Delta H*$ numbers is shown in Fig. 31.16. The elliptical shape of the solid arises since tolerances for $\Delta H*$ are considerably narrower than for $\Delta C*$ and $\Delta L*$.

31.4.4 Sample Preparation and Presentation

For color measurement data to be at all useful, the numbers must be consistent and repeatable. Sampling of product must be done so that it is representative of the product and prepared so that it represents the product's color characteristics. Many food samples are far from ideal in that they may be partially transmitting and partially reflecting. Rather than being uniform, they may be mottled or highly variable in color. The number of readings that need to be taken for acceptable repeatability is dependent on the nature of the sample. Another problem is that often the only instrument available is one that is less than ideal for the sample. Gordon Leggett [15] provides some practical tips and a systematic protocol for consistent color measurement of different food categories. Transparent liquids should be measured with a sphere instrument, using a clear glass or plastic cell. A cell filled with distilled water can be used as a blank to negate the effects of cell and solvent. Cell path length is selected based on color intensity. A 20 mm cell is used for most colored liquids, with 10 mm cells for highly absorbing liquids. A very thin 2 mm cell may be appropriate for highly absorbent transparent liquids such as soy sauce. For nearly colorless liquids, a 50 mm cell may be necessary. For clear transparent liquids, a single measurement using a viewing area of 15 mm diameter or greater may be sufficient for good repeatability.

For hazy transparent liquids, two to four readings with replacement of the liquid between readings is necessary to get acceptable repeatability when using a 10 mm path length cell and a sphere instrument.

Liquid samples with high solids are translucent rather than transparent. They can be measured by transmission using a very thin 2 mm path length cell, or measured in reflectance. Here it is necessary to control the thickness of the sample so that it is effectively opaque. Solid foods vary with respect to size, geometry, and uniformity. With some colorimeters, reflectance measurements can be taken directly on the sample. Ideally the surface should be flat. Readings of an apple or orange may be distorted because of the "pillowing" effect, which is a result of the distorted reflectance values from the uneven surface. Pureeing nonuniform materials such as strawberries will give a uniform sample; however, the incorporation of air renders a color extremely different from the sample of interest. For opaque foods, instruments with 45*/0° and 0°/45° geometries are recommended as the measurements correlate better with visual assessment than those obtained with sphere instruments. Instruments with a large area of view, e.g., 25–50 mm, are helpful for area-averaging nonuniform color. For powders, two readings with replacement of the powder between readings may be sufficient, but for flakes, chunks, and large particulates, a large field of view (40 mm or larger) with three to six readings and sample replacement between readings is recommended.

Different commodities present their own peculiarities when it comes to measuring color and appearance. In the proceedings of an American Chemical Society symposium [16], various authors discuss methodology for color measurement of meat, fish, wine, beer, and several fruits and vegetables.

31.5 SUMMARY

Color is three-dimensional, and any color-ordering or color-measuring system needs to address that fact. The Munsell system is a visual system that designates color in terms of hue, value, and chroma. Each of these dimensions has equivalent visual spacing, which is advantageous. The physiology of color vision has been long understood, and it provided the necessary background information for development of the CIE tristimulus system. Standardization of illuminants and experiments using humans with normal color version was necessary to develop color-matching functions that corresponded to the color sensitivity of the human eye. The system permits calculation of numerical XYZ tristimulus values that can accurately represent a color and are useful in color matching. The system does not have equivalent visual spacing, which is a disadvantage when measuring how a sample differs from a standard

or changes during processing and storage. Color-order systems have been developed that are more suitable for measuring color differences. These include the HunterLab system, the CIEL*a*b* system, and the L*C*H* system. The latter two systems are recommended by the CIE, the International Association with responsibility for standardization and measurement of light. They have been widely adopted by the food industry for color measurement. There are many colorimeters available for industrial and research applications that are rugged, easy to standardize, and user friendly. They vary with respect to presentation of sample, size of viewing area, portability, and the ability to measure by transmittance or reflectance. Many food samples are less than ideal for color measurements because they may be partial transmitting and partial reflecting, nonuniform, and of varying size and shape. A number of factors need to be considered with respect to sample preparation and presentation to get measurements that are repeatable and that correspond to visual appearance.

There are a number of excellent illustrative tutorials dealing with color measurement that are available on various websites that have been developed by organizations and commercial companies. The following are recommended: HunterLab [17], Konica Minolta [18], CIE [19], Munsell [20], Color Models Technical Guides [21], A Review of RGB Color Spaces [22], and Beer Color Laboratories [23].

31.6 STUDY QUESTIONS

1. Dominant wavelength (λ_d), % purity, and luminosity (Y) in the CIE XYZ system correspond to what indices in the Munsell system? In the CIE L*C*H* system?
2. Using a calculator, determine hue angle and chroma for the following sets of a*, b* data: a* = +12 and b* = +8, a* = +12 and b* = +4, and a* = 12 and b* = -4.
3. If one wants to use a colorimeter to measure of the amount of browning in maple syrup, what indices would you expect to correspond well with visual assessment?
4. How variable is human color perception when compared with that of taste and smell? What are the human capabilities for color perception and color memory?
5. Why is CIE tristimulus Y used as a measure of luminosity?
6. Give examples where it is appropriate to use a colorimeter with diffuse sphere geometry and, conversely, a colorimeter with 0°/45° reflectance geometry.
7. How can you determine how many readings should be taken for a given sample?

Acknowledgment The authors of this chapter wish to acknowledge Dr. Jack Francis, a legend in the area of color analysis and the person who wrote the chapter on this topic in two previous editions of this book. Ideas for the content or organization, along with some of the text, came from his chapter. Dr. Francis offered the use of his chapter contents.

REFERENCES

1. Amerine MA, Roessler EB (1976) Wines: their sensory evaluation. W. H. Freeman & Co., San Francisco, CA

2. Berns RS (2000) Billmeyer and Saltzman's principles of color technology, 3rd edn. Wiley, NY

3. Loughrey K (2005) Overview of color analysis. Unit F5.1. In: Wrolstad RE, Acree TE, Decker EA, Penner MH, Reid DS, Schwartz SJ, Shoemaker CF, Smith D, Sporns P. (eds) Handbook of food analytical chemistry—pigments, colorants, flavors, texture, and bioactive food components. Wiley, NY

4. Loughry K (2000) The measurement of color, Chap 13. In: Francis FJ, Lauro GJ (eds) Natural food colorants. Marcel Dekker, NY

5. Hutchings JB (1999) Food color and appearance, 2nd edn. Aspen Publishers, Gaithersburg, MD

6. Hunter RS, Harold RW (1987) The measurement of appearance, 2nd edn. Wiley, NY

7. Wright WD (1971) The measurement of color. Van Nostrand Reinhold, NY

8. Francis FJ (1999) Colorants – Eagen Press Handbook Series. Eagan Press, St. Paul, MN

9. Bartleson CJ (1960) Memory colors of familiar objects. J Opt Soc Am 50: 73-77

10. Campbell NA, Reece JB, Mitchell LG (1999) Biology, 5th edn. Benjamin/Cummings, Addison Wesley Longman, Menlo Park, CA

11. Krinsky NI, Landrum JT, Bone RA (2003) Biologic mechanisms of the protective role of lutein and zeaxanthin in the eye. Ann. Rev Nutr. 23:171-201

12. X-rite (2015) Munsell products. Available from https://www.xrite.com/top_munsell.aspx?action=products Accessed 18 Dec. 2015

13. Yurek J. (2012) Color space confusion. Dot Color.com. Available fromhttp://dot-color.com/tag/chromaticity-diagram/. Accessed 18 December 2015

14. Wrolstad, RE, Smith DE (2010) Color Analysis, chap 32. In: Nielsen SS (ed) Food Analysis (Fourth Edition), Springer, NY.

15. Leggett GJ (2008) Color measurement techniques for food products, chap 2. In: Culver CA, Wrolstad RE (eds) Color quality of fresh and processed foods. ACS Symposium Series No. 983, American Chemical Society, Washington DC

16. Culver CA, Wrolstad RE (2008) Color quality of fresh and processed foods. ACS Symposium Series No. 983, American Chemical Society, Washington DC

17. Hunter Lab (2015) Glossary of terms. Available from http://www.hunterlab.com/glossary.html. Accessed 18 Dec 2015

18. Konica Minolta (2015) Precise color communication. Available from http://www.konicaminolta.com/instruments/knowledge/color/ .. Accessed 18 Dec 2015

19. CIE International Commission on Illumination (2015) Available from http://cie.co.at/. Accessed 18 Dec. 2015

20. Munsell (2015) available from http://munsell.com/. Accessed 18 Dec 2015.

21. Color Models Technical Guides (2000) http://dba.med.sc.edu/price/irf/Adobe_tg/models/main.html. Accessed 18 Dec 2015.

22. A Review of RGB Color Spaces (2003) http://www.babelcolor.com/download/A%20review%20of%20RGB%20color%20spaces.pdf. Accessed 18 Dec. 2015

23. Beer Color Laboratories (nd) http://www.beercolor.com/glossary_of_selected_light_and_c.htm. Accessed 18 Dec. 2015

Food Microstructure Techniques

Jinping Dong (✉) • *Var L. St. Jeor*
Research and Development,
Cargill, Inc.,
14800 28th Ave. N, Plymouth, MN 55447, USA
e-mail: Jinping_Dong@cargill.com; Var_StJeor@cargill.com

S. Nielsen (ed.), *Food Analysis*, Food Science Text Series,
DOI 10.1007/978-3-319-45776-5_32, © Springer International Publishing 2017

32.1 INTRODUCTION

While the main function of food is to provide enjoyment (e.g., flavor, aroma, texture) and nutrition (including energy), some other functionalities (e.g., desired shelf life, improved health benefit, friendly labeling) are also necessary for consumer appeal. One of the trends in food research is to be able to design and produce the food with any desired functionality. To achieve this, many approaches are being taken by different researchers and food manufacturers, with one prevailing approach being to understand and control food structures, especially at the microscopic scale. The study of food microstructures typically involves three aspects – visualization, identification, and quantification – with each requiring the aid of different tools.

Food is a complex and often heterogeneous system. Both fresh (typically dominated with cellular structure) and processed (microscopic domains formed by mixed ingredients) foods contain structures that cannot be directly seen with naked eyes. Human eyes are capable of seeing things down to about 1 µm with proper lighting (compared to ~20 to 50 µm being the diameter of a human hair). Many biological cells are a few micrometers in size, but they are not usually visible to our eyes because good illumination is not always available. They appear big and clear under a regular light microscope. Some good salad dressing emulsions have oil or water droplets well below 1 µm in diameter. Elemental plant fibers have a diameter of only a few nanometers. All these require a higher resolving power microscope such as a scanning electron microscope (SEM).

Determining the morphology of the food with the help of microscopes is only one part of the microstructure elucidation. Identifying the distribution of ingredients with different chemical and physical characteristics gives another insight, and this often requires a chemical imaging tool [e.g., Fourier transform infrared (FTIR), Raman, fluorescence, or confocal laser scanning microscope] or a physical structural tool for understanding molecular arrangement or crystallinity (e.g., x-ray diffractometer).

Modern microscopy and chemical imaging tools allow users to not only determine food morphology and ingredient distribution but also to quantify dimension parameters, concentrations, fractions, and kinetic constants. Using x-ray computed tomography, quantification can be done in 3-D with nondestructive imaging. Microscale interactions and forces also can be probed among the food ingredients with the help of instruments such as atomic force microscope. Values obtained from quantitative measurements are the last piece of the puzzle to solve food microstructures.

All of the microstructure techniques listed above will be briefly introduced in this chapter. Many of them were not initially invented for food applications but have been borrowed by food researchers to understand food and to correlate structure to its functionality. This chapter is not intended to be comprehensive in either breadth or depth regarding these techniques but rather to provide an overview, with references for more details. Note that the chapter will not cover measuring particle size and shape (see Chap. 5, Sect. 5.5.2.3) or color (see Chap. 31) and will refer to a number of spectroscopy chapters (Chaps. 7 and 10) as they relate to techniques for characterizing food microstructure. Because there are so many acronyms associated with instrumentation to evaluate food microstructure, a listing of acronyms used in this chapter is given.

32.2 MICROSCOPY

32.2.1 Introduction

One of the most frequently used family of instrumental techniques for analyzing foods and food microstructures is **imaging**. Imaging has traditionally been referred to as microscopy, but microscopy is quickly becoming only one aspect of the much larger field now being identified as imaging. **Microscopy** is the art and science of using microscopes as scientific investigative tools. Since the original development of light microscopes [1, 2], drastic improvements have been achieved in our ability to see in magnification and to differentiate in feature contrast. Example imaging agents include light (photons), x-rays (high-energy photons), electrons, ultrasound, microwaves, radio waves, etc. Most of the imaging agents are based in, and are a part of, the electromagnetic spectrum. Each of these techniques is subdivided into various imaging methods.

Microscopes were invented to visualize objects that cannot be seen by human eyes. How small of an object that can be clearly identified with the microscope determines the resolving power of the instrument. **Resolution** is the ability to distinguish or resolve two small points, which are very close together, as two separate entities. Factors that affect a microscope's resolution include the properties of the imaging agent (e.g., wavelength of the light) and the focusing power of the instrument (e.g., numerical aperture of the objective for light microscope). These are taken into account with a simple equation to calculate the **theoretical resolution limit** for a given microscope's primary or objective lens:

$$R = \lambda / 2NA \qquad (32.1)$$

where:

R = resolution (theoretical resolution limit, minimum distance of the two adjacent objects)
λ = wavelength of the visualizing agent
NA = numerical aperture of the lens [proportional to the refractive index and sin (θ), where θ is the half angle of the incidence of the incoming light to the lens]

Inherent to any lens are imperfections, referred to as **aberrations**, which can cause images to appear distorted, out of focus, with colored fringes, etc. Correction of optical aberrations includes improvement of lens fabrication and grinding techniques, optimization of glass formulations, application of antireflection coatings, control of optical pathways, and combination of multiple lens elements.

Misalignment of the optical illumination is another factor that impacts a lens' optimum resolving power. Specific procedures and the care from the microscopist are needed to achieve perfect alignment and focus of the light beam to give uniform and bright illumination of the specimen. The alignment procedure is called **Köhler alignment**, named after August Köhler, a German physicist and microscopist [3].

32.2.2 Light Microscopy

32.2.2.1 *Introduction*

Light microscopy, as referred to in its name, employs light (or photons) as the imaging agent and magnifying lenses to visualize objects that cannot be seen with naked eyes. Light can be either reflected off of or transmitted through the sample and is then directed through the objective lens to the eye piece(s). Depending on the instrument design, there are two major categories of LMs, namely, stereo and compound microscopes. **Stereo** is a word related to **parallax**, or the difference between the angle of light arriving at your two eyes. These differing angles of view allow the brain to interpret the two differing views [each a two-dimensional (2-D) image] as though one is seeing a single three-dimensional (3-D) image. Stereo microscope typically has low magnification (2–100×) but has long working distance and large viewing depth, which make it easy for visualizing large and odd-shaped specimen. **Compound** references the compound nature of multiple lenses working together to achieve a clear, in focus, magnified image. It usually works in higher magnification range (40–1200×), with the combination of the lenses from objectives and eyepieces.

32.2.2.2 *Contrasting Modes*

In LM there are many ways to manipulate the light both before it reaches the specimen and after it interacts with the specimen to highlight certain features. Many of these are commonly used in the food industry and constitute what are called **contrasting** or **imaging modes** of light microscopy. A partial list of imaging modes includes bright-field, dark-field, phase contrast, birefringence (or cross polarization), differential interference contrast (DIC), and oblique lighting. With the exception of bright-field mode (the simplest lighting mode), each of these alternant modes requires additional special fixtures to be attached to the microscope to accomplish the desired effect. When in use,

they often need to be aligned so they interact with the light correctly. What the various imaging modes produce in image can be referred to as special effects. They do not usually increase resolution per sé, but they can allow one to distinguish structures that may not be easily evident using other imaging modes. Starch is a classic example of a food component that can be examined using any of the imaging modes listed (Fig. 32.1).

32.2.2.3 *Fluorescence Microscopy*

Fluorescence is the light emitted from an atom, a molecule, or a material excited to the electronically active state (see Chap. 7, Sect. 7.3). The wavelength or energy of the excitation light is characteristic to the chemical bonds within a molecular or to the chemical/physical state of a material. By applying proper light sources and optical filters, fluorescence can be used to give unique contrast in light microscopy. Optical filters are added to a bright light source to select target wavelength for excitation, or **excitation spectrum**, and fluorescence filters are added after the sample to capture **emission spectrum**. Fluorescent stains, which contain functional groups that fluoresce upon excitation (called fluorochrome), also can be added to a sample to contrast components that they have strong affinity with. Staining can be done positively (e.g., staining the target structure) or negatively (e.g., staining nontarget structures). Many food materials contain natural fluorochrome. When a proper excitation wavelength is selected, bright color from fluorescence emission can be easily visualized under the light microscope. Figure 32.2 shows the aleurone layer within a wheat kernel that auto-fluoresces bright blue, making that particular layer of cells become obvious to the viewer, over any other cell types within the wheat kernel.

32.2.2.4 *Histology*

Histology, which refers to sample staining combined with light microscopy [4], is often used in biology and medicine to study cells and tissues. Samples are typically sectioned in some ways to create relatively thin slices with even thickness of the materials. This procedure is known as **microtomy** and usually involves a microtome or other mechanical cutting tools [5]. Histology has been applied in food research for a long time owing to its unique contrasting capability. The numerous stains available on the market [6–8] are each designed to stain different things different colors. Stains interact with food ingredients mostly through physical interactions, which make the affiliated ingredients appear with the color of the stain itself. However, the stain-ingredient interaction sometimes can shift the color spectrum. For example, although common cornstarch stains blue with iodine, waxy cornstarch stains red (Fig. 32.3). This is the result of amylopectin dominating the starch in the waxy

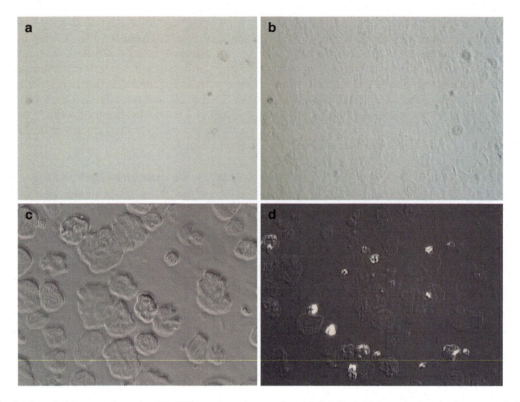

32.1 figure Light microscopy of starch with different imaging modes. (**a**) Bright field image of cooked out starch (all or most of the crystallinity is gone). Low contrast associated with this type of sample makes it nearly impossible to see structure. (**b**) Oblique lighting. (**c**) Phase contrast. (**d**) Cross polarized image of *partially* cooked out starch. Bright granules are not cooked out and are retaining their crystallinity, or at least a portion of it

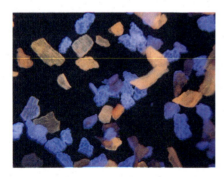

32.2 figure Particles of wheat fiber demonstrating various colors of auto-fluorescence

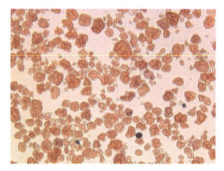

32.3 figure Partially cooked out waxy cornstarch, stained with iodine. *Red* color is due to the amylopectin content of the starch. A few *dark*, common cornstarch granules are also seen in this image, which stain *blue* rather than *red* because they contain amylose rather than amylopectin. Iodine, being a metachromatic stain, can stain these various types of starch different colors, which aids in starch identification

species, as opposed to amylose in the common species. Iodine is referred to as a **metachromatic stain**, meaning one stain can stain different things different colors. Other specialty stains, called **polychromatic stains**, are combinations of stains that do not react with each other but will stain differing food ingredients different colors.

32.2.3 Electron Microscopy

Electron microscopy (EM), which uses electrons as the imaging agent, is of two common types: **scanning electron microscopy** (SEM) and **transmission elec-** tron microscopy (TEM). SEM finds more frequent applications in the food industry, as described further below. Although TEM may provide an order of magnitude of higher resolution, it is not commonly used in food research mainly because it is very time consuming and requires delicate sample preparation (e.g., extremely thin sections of the sample material, typically 60–80 nm).

The five main ways EMs differ from LMs are: (1) the imaging agent (photon vs. electron); (2) their resolving power (electrons have up to 100000 times shorter wavelength than photons, capable of resolving individual atoms); (3) their magnifying power (wide range from 20× to 1000000×); (4) LMs can provide images with visible colors to the eyes, while EMs do not give contrast in color, although false colors can be applied to SEM images to differentiate brightness levels; and (5) EMs work in high vacuum, while LMs tend to work in ambient conditions.

SEMs employ an electron beam to scan across the surface of interest. After interacting with the materials, electrons come out in several different forms. The **secondary electrons** are produced by inelastic collisions of the probe electrons (**primary electrons**) with sample electrons. Sample electrons are ejected at slower speeds than probe electrons, and with lower energy levels. They are collected at the detector for quality imaging with good depth perception. This is the primary imaging mode of SEMs. **Backscattered electrons** (BSE) involve imaging with a high-energy probe electron interacting with the sample elastically. They arrive at the detector with nearly the same speed and energy as they had in the electron probe, resulting in an image that is material dependent. Dense materials, like metals, appear bright in the BSE-SEM image, while less dense materials appear darker (e.g., carbon-based biologicals). BSE imaging is proven to be more valuable than secondary electron imaging because of the details it can provide and because it is less subject to specimen charging (Fig. 32.4).

During SEM imaging, electrons accumulate on the surface (**surface charging**), especially where it is low in conductivity. Applying a thin metal or conductive coating to the specimen surface has been an essential sample preparation step. However, low vacuum and **environmental SEM** (E-SEM) modes use water vapor or other gases within the specimen chamber to help diminish sample charging artifacts. Water vapor can, under the right conditions, also help preserve tender biological samples (including many foods and food ingredients) within the otherwise very high vacuum condition of an SEM.

32.2.4 Energy Dispersive X-Ray Spectroscopy

When scanning electrons of SEM interact with the atoms of the specimen, it may cause electrons to migrate in between electron shells, from which an x-ray emission could be induced. The energies of the x-ray are characteristic to the atomic structures, allowing identification of the elements through an energy-dispersive spectrometer, hence the name of this

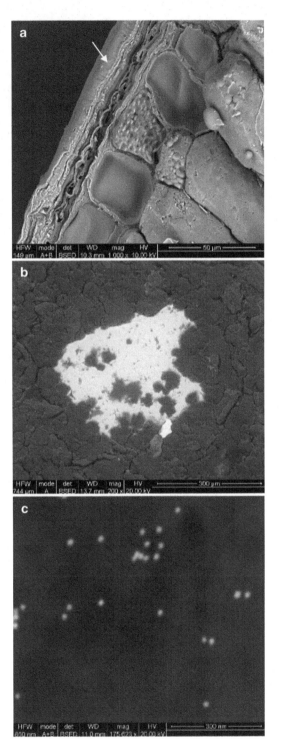

BSE-SEM imaging of various materials. (**a**) A fracture through a seed in cross section. The seed coat is evident at *arrow.* Cellular structure dominates the tissue below the seed coat 1000×. (**b**) A contaminating particle (*white*) within a pharmaceutical product 200×. (**c**) 20 nm gold particles (*white dots*) used to label and identify molecular structure within a biological system 175000×

technique [**energy-dispersive spectroscopy** (EDS)]. The peaks from the EDS spectrum (x-ray energy in KeV in the X-axis and total x-ray counts in the Y-axis) represent specific elements from a periodic table (Fig. 32.5a). Additionally, both qualitative and quantitative analyses are possible for whatever elements exist within the sample.

In contrast to EDS is another technique known as **wavelength dispersive x-ray spectroscopy** (WDS), in which x-rays are diffracted to allow analysis at a certain wavelength. WDS has traditionally been better when one analyzes for a specific element with high frequency and when one needs to know the exact concentration of that one element.

Since the SEM works in a scanning fashion, restoring its electron probe in a predictable and reproducible way, we can choose specific elements from the spectrum and record a 2-D array of element dots, from which an elemental map can be generated to show distribution of compositions (Fig. 32.5b). From EDS maps, we can know not only what elements are present but how much of that element is present and precisely where that element is in the sample.

32.2.5 Atomic Force Microscopy

Atomic force microscopy (AFM), invented in 1986 [9], has found applications in most scientific disciplines, including food science [10]. The basic principle of AFM is depicted in Fig. 32.6. A very sharp tip is attached to the end of a soft **cantilever**. The tip is

brought into contact to the sample, and a piezoelectric **scanner** moves the cantilever to raster scan across the surface. The cantilever deflects (up or down) as the surface topography or tip-surface interaction changes. A **laser beam** bounced off the back of the cantilever moves up and down (or left and right if there is lateral deformation of the cantilever) on a **position-sensitive photodiode detector** (PSPD) with the cantilever deflection. The photoelectric signal is then recorded and transferred into a colored map with contrast corresponding to the cantilever deflection or surface height change. Very often, to avoid large forces from the cantilever/tip damaging the scanned surface, a constant force mode is enabled in which a **feedback** signal is sent to the piezo scanner to compensate for the height change of the sample, thus maintaining a constant tip-to-sample interaction. Depending on the sharpness of the tip (**radius of curvature**) and the **spring constant** of the cantilever (usually in the range of 0.1 to a few nN/nm), the resolution of the acquired image can be in the sub-0.1 nm range, which is sufficient to resolve atoms on a surface.

Although proving to have similar minimum resolution to electron microscopy, AFM finds many advantages over EM and other microscopy techniques. First, AFM measures the true height from the sample surface, which is needed to generate a 3-D surface profile. Second, AFM is not limited by the environment it operates in. Conditions such as vacuum, ambient air, different humidity, elevated or depressed temperature, and even in liquid medium all work well for

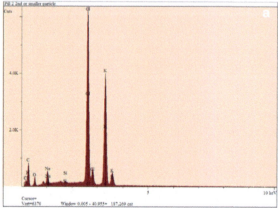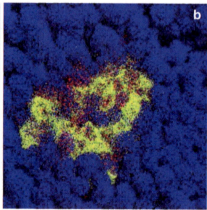

32.5 figure EDS spectrum and EDS mapping. (**a**) An EDS elemental spectrum, with the various peaks labeled for the elements they represent. (**b**) An x-ray dot map (or x-ray-based elemental image) of an organic powder displaying a contaminating material. *Blue* represents where carbon is found in the sample. Powder particles are detected in approximate shape within the *blue* field. The contamination is seen as *yellow* (represents the element sulfur) and *red* (represents the element chlorine). *Black* represents x-ray shadowing (meaning none of those x-ray are making it to the detector)

AFM. Especially in liquid media, many biological macromolecules and living organism studies are possible [11]. Third, because the tip physically touches the sample, a variety of interactions and microscale forces can be measured [12].

Imagine when a tip is brought close to the surface, forces between the tip and surface constantly change as the tip gets closer (repulsive, long range electrostatic, etc.), touches the surface (capillary, van der Waals attraction, etc.), indents into the surface (various nanomechanical, etc.), and breaks away from the surface (adhesive, etc.). As the tip scans laterally, frictional forces also can be detected by quantifying the torsion of the cantilever.

AFM was first introduced to the food science field in 1993 with research work focusing on monitoring food protein changes. Since then, AFM applications have expanded to many areas of food science and technology, such as qualitative imaging of polysaccharides and proteins to understand their conformation and organization in certain environments, quantitative structure analysis on complex food systems (e.g., mechanical strength of food gels)

to correlate with functionality, probing molecular interactions (e.g., interaction between protein and surfactant as co-emulsifiers), and molecular manipulations to observe the reactions among food macromolecules [13].

32.3 CHEMICAL IMAGING

32.3.1 Introduction

Chemical imaging is a set of analytical techniques that can generate a contrast-based image to show the distribution of the chemical or molecular composition of an object (surface or bulk). Chemical images are usually created from a **data cube** as shown in Fig. 32.7. A full spectrum (e.g., from FTIR or Raman) containing chemical signatures of all components, at a single point, is acquired. Then, either the sample is moved through a motorized mechanical or piezoelectric stage or the probing beam [e.g., focused **infrared** (IR) light for **Fourier transform infrared** (FTIR), visible laser for Raman, or electrons for SEM] scans to the next point of interest, at which a second spectrum is collected. Many points are collected on one line, and many lines of data are acquired in a raster scanning fashion. After a 2-D array of spectra is saved, various data processing mechanisms are applied to each individual

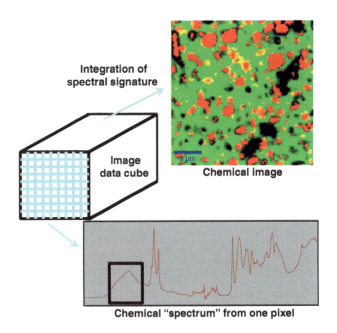

Integration of spectral signature

Image data cube

Chemical image

Chemical "spectrum" from one pixel

32.6 figure   Schematic of a typical AFM instrument

32.7 figure   Schematic of chemical imaging

spectrum. Figure 32.7 shows a typical Raman spectrum, with all peaks corresponding to different functional groups, at certain vibrational modes. The peak inside the blue box is the OH stretching vibration from water in this case, and its area corresponds to the concentration of water at the current measurement location. The colored image is created by integrating this peak, for all collected spectra, with the contrast level proportional to the integrated peak area. If other components are of interest, integration of the corresponding functional peaks from the spectra can be easily applied to generate as many chemical images as possible [14].

32.3.2 Fourier Transform Infrared Microscopy

Chemical imaging with FTIR microscopy employs IR light (mid- or near-infrared) as the incident radiation. Because of the potential absorption of IR by optical components along the beam path, no transmission type of lenses is used in FTIR microscopy (like a regular light microscope, as described in Sect. 32.2.2). Spherical or parabolic mirrors are used to focus the light, and the same optical resolution limits apply as described in Eq. 32.1. Therefore, if imaging in mid-IR (2.5–50 µm), FTIR microscopy can provide chemical maps with theoretical spatial resolution of less than 1.5 µm. However, in practical application, the resolution limit of FTIR micros-

copy is close to 5–10 µm when selecting functional peaks with different wavelength from the spectra.

Most FTIR microscopes adopt a point-to-point scan design, meaning the detector records only one spectrum at a time when the sample is being raster scanned. A full chemical map (e.g., 100×100 spectra) could take 30 min – 2 h (or longer depending on the spectral resolution and pixel settings), including the time of scanning and data recording. In 1995, the introduction of the **focal plane array detector** (FPA) to the FTIR microscope brought IR imaging to a whole new level [15]. A FPA detector contains a 2-D array of **photo-sensitive elements** (e.g., mercury cadmium telluride, MCT) with each capable of capturing a complete but totally separate IR spectrum. The detector array can be anywhere from 16×16 to 128×128 elements, enabling chemical image acquisition of up to 16,384 pixels. The Digilab Stingray FPA system had the best theoretical spatial resolution of 5.5 µm using a 36× objective. Because all spectra are collected simultaneously, the time it takes to complete a whole chemical map equals the acquisition of one single spectrum, which can be from a few seconds to a few minutes, dramatically faster than traditional IR imaging.

FTIR microscopy has been widely applied in food science and technology [14]. For example, Fig. 32.8 shows a study on moisture migration through a model sugar film, mimicking a cereal coating [16].

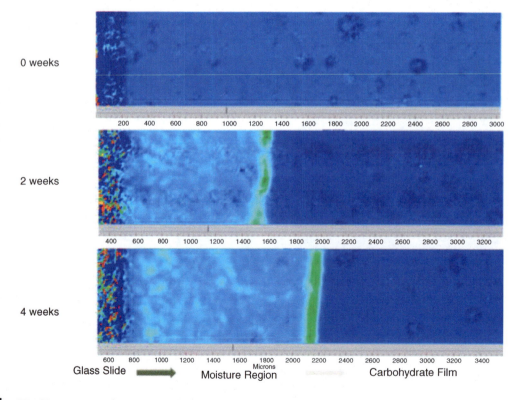

 32.8 **figure** FT-NIR mapping of moisture sorption in a model sugar film

32.3.3 Confocal Raman Microscopy

As introduced in Chap. 8 (Sect. 8.5), Raman scattering is intrinsically weak. For Raman microscopy, because the available radiation is focused into a much smaller area, to avoid thermal damage in the sample from the laser, the effective Raman scattering is even weaker, making Raman microscopy a much less appealing technique. A great leap was seen recently in Raman microscopy technology with the development of ultra-high throughput spectrometers and high quantum efficiency **charge coupled devices** (CCDs) detectors. A high-resolution Raman image (e.g., a 200×200 pixel scan or a 40000 spectral collection) could be acquired within a few minutes or even faster, comparing to a few hours with the old generation Raman microscope [17].

As described in Chap. 8, Raman scattering is independent of excitation wavelength, meaning visible lasers can be employed for Raman experiments, which makes the design of a Raman microscope much easier than FTIR. A regular light microscope can be readily equipped with introducing an incident laser and a mechanism to collect back scattering to enable Raman measurement.

Because visible light (much shorter wavelength than IR) can be used for Raman microscopy, image resolution can be much higher than FTIR imaging. With a green laser (e.g., 532 nm) and an oil immersion objective (e.g., $NA = 1.3$), the resolution limit of a Raman image can be in the 200–300 nm range.

Confocal Raman microscopy has been used extensively in food research [18–20]. However, as described in Chap. 8, fluorescence is often troublesome in image acquisition, especially in food systems with natural ingredients. Using a longer wavelength laser (e.g., 785 nm), defocusing of the incident beam, and applying pre-imaging bleaching are common ways to reduce fluorescence, but image resolution may be deteriorated.

32.3.4 Confocal Laser Scanning Microscopy

Confocal laser scanning microscopy (CLSM) is one of the four types of **confocal microscopy** techniques (the other three are spinning-disk, micro-lens enhanced, and programmable array microscopes), which provide high-resolution chemical images based on fluorescence emission from the sample.

Figure 32.9 shows a typical setup of a confocal microscope. The incident light (a laser or white light source) enters the microscope objective through a dichroic mirror and is focused on the sample through the objective lens. The reflected or backscattered light goes back through the lens and converges at the detector surface, through which the radiation signal is recorded. With a confocal setup, a light blocking plate with a small pinhole is placed right in front of the detector at the converging point to still allow the back-focused light to reach the detector. Imaging light also may be reflected back from locations below (e.g., green

lines) or above the focal plane, but it will converge before or after the pinhole, which will be blocked by the pinhole plate and is not able to reach the detector.

With this configuration, only light coming from the focal plane can be recorded by the detector. No background radiation from below or above the focal plane reaches the detector, which gives the recorded images a much sharper appearance. However, one limitation of the confocal setup is that the field of view is significantly smaller due to the size of the pinhole. Therefore, a scanning mechanism, typically through a set of vibrating mirrors, is often incorporated to provide images with a large field of view. The mirrors vibrate through piezoelectric elements, capable of providing scanning speed of 1800 Hz (lines/s) or faster. The entire focusing system (anything above the sample) is usually capable of moving in the vertical direction, allowing image acquisition at different focal planes. A 3-D image then can be generated by stacking images acquired from many focal planes.

The contrast of CLSM images is based on the fluorescence intensity from the laser's illuminating point. As described in Sect. 32.2.5, there are many ways to label the target object to give fluorescence emission. Samples are often labeled with multiple fluorophores to allow co-localization studies of a complex system.

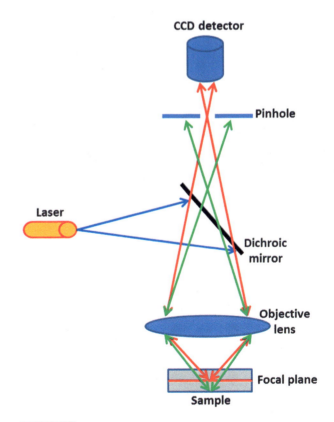

32.9 figure Schematic of a confocal microscope

CLSM has wide applications in biological and medical science disciplines but also finds popular applications in foods (e.g., imaging fat crystal structures, dairy products, food emulsions, food gels, plant materials, etc. [21]).

32.4   X-RAY DIFFRACTION

Molecules, atoms, or particles arrange differently in a solid material. When they pack in an ordered fashion, with symmetrical and repetitive pattern, the solid is called a **crystal**. Crystalline materials are rigid, have fixed melting point, can be cleaved along a definite plane, and have anisotropic physical properties such as electrical conductivity, refractive index, and thermal expansion. In contrast, **amorphous** materials have short range, or no order of molecular alignment. They are less rigid, usually do not have sharp melting points, and have isotropic physical properties.

Within a three-dimensional crystalline structure, a volume element that repeats its geometry and orientation in all directions can usually be identified as a **unit cell**, with the smallest volume often considered as the representative unit (Fig. 32.10a). The unit cell can be defined with six parameters, with three being the edge length (e.g., Fig. 32.10b) and three being the angles in between the edges (denoted as α, β, and γ, not shown). For a given crystal, when these six parameters are measured, the unit cell, or the crystal structure can be elucidated [22].

The structural elements (molecules, atoms, or particles) in a unit cell form parallel planes (dotted lines in Fig. 32.10c) which intersect the unit cell in many different ways. When electromagnetic radiation such as x-rays strikes the crystals, they are reflected by these planes. Assuming the incident x-ray has parallel beams (e.g., Fig. 32.10c, beam 1 and 2), to have a constructive interference between the beams reflecting off of the crystal planes, the additional distance that beam 2 travels ($2 \times d \times \sin\theta$, where d is the spacing of crystal planes and θ is the incident angle of incoming radiation) has to equal $n \times \lambda$ (n is an integer and λ is the wavelength of the radiation):

$$n\lambda = 2d \sin \theta \qquad (32.2)$$

Equation 32.2 is called **Bragg's equation**. Under this condition, the reflected beams add up to give a diffraction pattern. Those with destructive interference are subtracted from each other, which produces much weaker radiation at the detector. With λ known from the x-ray source, and θ measured from x-ray/sample geometry, d can be obtained for a certain set of crystal planes. However, for a single crystal, within the scan range of θ, not all crystal planes can give **constructive diffraction**. A polycrystalline sample (powdered crystals) is usually used as an alternative to capture all diffraction patterns/peaks to resolve the complete crystal structure, and the technique is therefore referred as **powder x-ray diffraction**.

A simple schematic of an **x-ray diffractometer** (XRD) is depicted in Fig. 32.11a. The detector usually

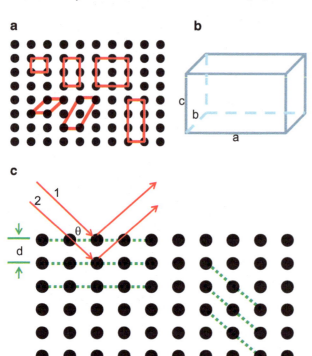

 Crystal unit cell and x-ray diffraction. (**a**) Repeating volume elements, or unit cells in a crystal, (**b**) Unit cell dimension parameters, (**c**) Parallel planes intersecting a unit cell, and reflection of x-rays off the planes

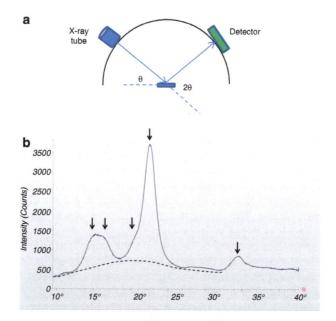

 Schematic of powder diffraction. (**a**) Simple schematic of an x-ray diffractometer. (**b**) XRD spectrum of crystalline cellulose. Dotted line shows the residual amorphous phase.

scans along the arc to vary 2θ, and the XRD spectrum is generated by plotting it with x-ray intensity as a function of 2θ [23].

X-ray diffraction patterns are characteristic of unit cell types and other structural properties of the crystals (e.g., lattice parameters, phase identity, crystallinity, composition, etc.). Figure 32.11b is a typical XRD spectrum for crystalline cellulose which is isolated from plant cell wall materials (from cotton linter in this case).

X-ray diffraction is a powerful tool for food research. Wide applications are found in carbohydrates (e.g., sugar, polysaccharides, etc.) and lipid/fats areas since these materials can easily form crystalline structures under application conditions [24, 25].

32.5 TOMOGRAPHY

32.5.1 Introduction

The word **tomography** is based on the two Greek words, *tomos* (meaning slice or section) and *graphe* (meaning to draw or write). The process of tomography, then, is the reconstructing of the original material, or specimen, from either a series of physical sections (through microtomy or sequential removal from the bulk) of that specimen or from a series of "optical" sections. Though tomography was originally done methodically by hand, the general practice now is to use computer assistance, hence, the term **computer-assisted tomography**. Tomography can be done using any of several possible mechanisms, such as confocal light microscopy, electron microscopy, radio frequency waves (MRI), or x-ray microscopy.

32.5.2 X-Ray Computed Tomography

X-ray computed tomography (CT) [older term is **computerized axial tomography** (CAT)] employs x-rays as the imaging agent to generate a 3-D digital image of the specimen. As the x-ray strikes through the object, a 2-D shadow image is collected on the detector, with contrast proportional to the x-ray attenuation or the physical density of the sample. The sample is rotated in single axis, and multiple images are collected with the orientation. A special computing algorithm (digital geometry processing) is applied to the series of 2-D images to construct a 3-D image that shows the inside of the specimen.

CT is both qualitative and quantitative. CT images can be rendered in both surface and volume modes. Segmentation can be applied to the images to render structures with similar density. The resolution of CT is mostly dependent on the x-ray beam size and the detector parameters. With a microscopic CT system, image resolution can be achieved in the range of a few micrometers.

CT is predominantly used in medical field but has found more and more applications in food research,

e.g., to visualize fillings inside a chocolate cookie, to quantify air cell size and cell wall thickness of a gluten-free bread, to monitor bubble formation of an icing, and to study salt dissolution in a cookie dough.

32.6 CASE STUDIES

32.6.1 Fat Blends

One trend in fat-/oil-related food research is to reduce the amount of saturated fat. Saturated fat forms unique crystal structures at room temperature, which not only acts as the framework of the blends to give desired physical functionality but also provides a network to hold liquid oil. Reducing or replacing the saturated fat, but keeping the same functionality, demands a thorough understanding of the fat crystal structure. Except for ingredients, processing conditions also may be adjusted to allow manipulation of fat crystal formation.

In a fat blend case study, high oleic oil was the liquid oil fraction and palm stearin was the saturated fat fraction. The ratio of the two was varied from 90:10 to 80:20. Structuring agents, which affect crystal formation and growth, were added in small amounts, 2–4% (e.g., monoglyceride, wax, lecithin, etc.). Functional properties such as oil-holding capacity, mechanical strength (e.g., Young's modulus), and plasticity were evaluated through different measurements. Crystallization was characterized by DSC, and crystal polymorphs were measured by XRD. Microstructure was imaged by confocal Raman microscopy and CLSM. Figure 32.12 shows some CLSM images of the fat blends with varying processing conditions and compositions. Drastic differences were observed in crystal structures (Nile red-stained liquid oil, with dark contrast corresponding to fat crystals), which correlated well with other functionality parameters.

32.6.2 Food Emulsions

To make a stable emulsion, emulsifiers with amphiphilic nature are added to prevent liquid droplets from coalescence. Common food emulsifiers are lecithin (from soy or egg yolk), mono- or diglycerides, and proteins. When concentration in a liquid is higher than its critical micelle concentration (CMC), lecithin molecules start to aggregate. The size and packing structure of the aggregates are indications of its packing ability at the oil/water interface when being used as emulsifiers. Figure 32.13 shows the microstructure characterization of emulsions prepared in a study comparing the performance of two different plant-source lecithins. TEM images show that one type of lecithin formed spherical micelles (smaller radius of curvature) and the other formed worm-like micelles. Light microscopy suggested significant differences in droplet size and morphology. CLSM shows that the emul-

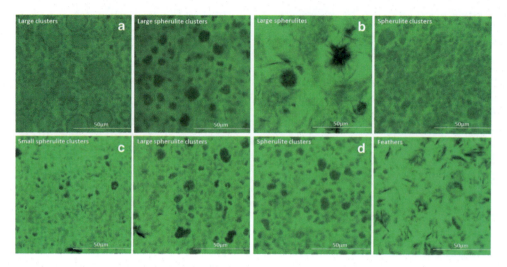

32.12 **figure** CLSM of fat crystal structures. (**a**) Varying cooling rate. (**b**) Different structuring agent. (**c**) Varying liquid/saturated fat ratio. (**d**) Same structuring agent but with varying concentration

sion on the left had lecithin nicely ordered at the interface of the water droplets, while the one on the right did not form the water-in-oil emulsion as desired. Lecithin stayed with the oil in the droplet instead of stabilizing at the interface. Emulsion stability measurement (through centrifuge), particle size measurement [through time-domain nuclear magnetic resonance (TD-NMR) (Chap. 10, Sect. 10.3) and particle size analyzer (Chap. 5, Sect. 5.5.2.3)], and conductivity measurement gave similar microstructure predictions.

32.7 SUMMARY

The future of food research and food production resides in controlling and manipulating food structures to give desired functionality. This chapter briefly introduced some direct and indirect microstructure characterization tools that were mostly invented or developed in and for other science disciplines. They have been borrowed by food scientists and manufacturers for many years to help better understand food systems. Some techniques have become essential tools in food science and technology. More detailed discussions of each technique can be easily found in the references cited in the text. Table 32.1 summarizes some key characteristics of these techniques.

One must be aware that none of the techniques included in Table 32.1 are universal in their applications. Foods can be very complex systems, with a very broad spectrum of types and modifications. Very often, combinations of several scientific techniques are needed to grasp the full picture of the structure of just one food

32.1 **table** Key characteristics of tools to examine food microstructure

Microstructure tool	Imaging agent	Typical resolution	Key information provided
Light microscope	Visible light	~200 nm	Morphology. Composition with histology
Fluorescence microscope	Visible light	~200 nm	Composition
SEM	Electrons	Sub-nm	Surface morphology
EDS	X-rays	Sub-nm	Elemental composition. Elemental distribution
AFM	Sharp tip	Sub-nm	Surface morphology. Interfacial forces
FTIR microscope	IR light	5–10 mm	Chemical distribution
Confocal Raman microscope	Visible laser	~200 nm	Chemical distribution
CLSM	Visible laser	~200 nm	Chemical distribution
XRD	X-rays	20–50 μm	Crystallinity
CT	X-rays	5–10 μm	3-D morphology

type. The purpose of learning food microstructure is to be able to control and design new structures. Progress is continuously being made in this developing field, and huge opportunities are just ahead of us.

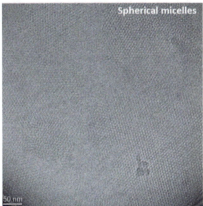

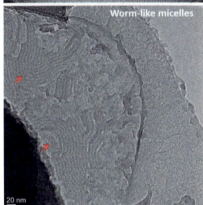

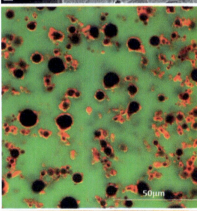

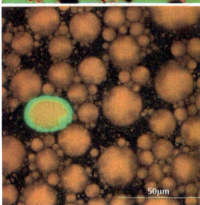

 Imaging characterization of emulsions with different lecithin: *Top two*-TEM; *Bottom two*-CLSM

Acronyms

2-D	Two dimensional
3-D	Three dimensional
AFM	Atomic force microscopy
CAT	Computerized axial tomography
CCD	Charge coupled device
CLSM	Confocal laser scanning microscopy
CMC	Critical micelle concentration
CT	Computed tomography
DIC	Differential interference contrast
EDS	Energy dispersive spectroscopy
EM	Electron microscopy
E-SEM	Environmental scanning electron microscopy
FPA	Focal plane array
FTIR	Fourier transform infrared
IR	Infrared
LM	Light microscopy
SEM	Scanning electron microscopy
TEM	Transmission electron microscopy
XRD	X-ray diffraction

32.8 STUDY QUESTIONS

1. What is the definition of "resolution" with regard to microscopy? How is the resolution of an optical microscope determined?
2. What is the main reason that electron microscopes can work at much higher magnifications (or resolution) than light microscopes?
3. Why do we apply a conductive coating to samples examined by an SEM? Why is a conductive coating usually not needed when examining samples using either a low vacuum SEM or an environmental SEM?
4. When using energy-dispersive x-ray spectroscopy (EDS), what carry the information telling us the elemental content of the sample, and where do they come from?
5. FT-Raman is often an option when sample has strong fluorescence. Why is it not used in Raman microscope?
6. What is "tomography" and what advantages does it have over other imaging techniques? How does computer-assisted x-ray tomography create a 3-D image?
7. You are working on a project to reduce the sugar content in gummy candies. You replaced some sugar with high-intensity sweetener and added some starch to try to maintain the elastic property. However, the candy does not look and taste like the normal gummy candy. You are asked to understand the candy's microstructure. What would be your experimental plan to do the characterization?

REFERENCES

1. Van Helden A, Dupre S, van Gent R, Zuidervaart H (eds) (2010) The origins of the telescope. KNAW Press (now Aksant Academic Publishers), Amsterdam, Netherland

2. James PJ, Thorpe N (1994) Ancient inventions. Random House Publishing, London, UK

3. Delly JG (1988) Photography through the microscope. Eastman Kodak Co., Rochester, NY

4. Green FJ (1991) The Sigma-Aldrich handbook of stains, dyes & indicators. Aldrich Chemical Co., Inc., St. Louis, MO

5. Peter G (1964) Handbook of basic microtechnique, 3rd edn. McGraw Hill, New York

6. Clark G (1981) Staining procedures. Williams & Wilkins, Essex, UK

7. Lillie RD (1977) H. J. Conn's biological stains, 9th edn. Williams & Wilkins Co., Reprinted by Sigma Chemical Co., St. Louis, MO

8. Hergert W, Wriedt T (2012) The Mie theory basics and applications (Chapter 2 authored by Wriedt T), Springer, New York

9. Binnig G, Quate CF, Gerber CH (1986) Atomic force microscope, Physical Review Letters 56:930

10. Haugstad G (2012) Atomic force microscopy: understanding basic modes and advanced applications. Wiley, New York

11. Baró AM, Reifenberger RG (2012) Atomic force microscopy in liquid: biological applications. Wiley, New York

12. Sarid D (1994) Scanning force microscopy: with applications to electric, magnetic, and atomic forces. Oxford University Press, Oxford, UK

13. Yang H., Wang Y, Lai S, An H, Li Y, Chen F (2007) Application of atomic force microscopy as a nanotechnology tool in food science. J Food Sci 72:R65

14. Sasic S, Ozaki Y (2011) Raman, infrared, and near-infrared chemical imaging. Wiley, New York

15. Lewis EN, Treado PJ, Reeder RC, Story GM, Dowrey AE, Marcott C, Levin IW (1995) Fourier transform spectroscopic imaging using an infrared focal-plane array detector, Anal Chem 67:3377

16. Nowakowski CM, Aimutis WR, Helstad S, Elmore DL, Muroski A (2015) Mapping moisture sorption through carbohydrate composite glass with Fourier transform near-infrared (FT-NIR) hyperspectral imaging. Food Biophysics 10:207

17. Dieing T, Hollricher O, Toporski J (2011) Confocal Raman microscopy. Springer, New York

18. Thygesen LG, Lokke MM, Micklander E, Engelsen SB (2003) Vibrational microspectroscopy of food. Raman vs. FT-IR. Trends Food Sci & Tech 14:50

19. Roeffaers MBJ, Zhang X, Freudiger CW, Saar BG, van Ruijver M, van Dalen G, Xiao C, Xie XS (2011) Label-free imaging of biomolecules in food products using stimulated Raman microscopy. J Biomed Optics 16(2):021118

20. Gierlinger N, Schwanninger M (2007) The potential of Raman microscopy and Raman imaging in plant research. Spectroscopy 21:69

21. Lagali N (ed) (2013) Confocal laser microscopy: principles and applications in medicine, biology, and sciences. InTech (open access), Rijeka, Croatia

22. Yoshio W, Eiichiro M, Kozo S (2011) X-ray diffraction crystallography. Springer, New York

23. Jenkins R, Snyder R (1996) Introduction to x-ray powder diffractometry. Wiley-Interscience, New York

24. Putaux JL (2005) Morpholody and structure of crystalline polysaccharides: some recent studies. Macromolecular Symposia 229:66

25. Idziak SHJ (2012) Powder x-ray diffraction of triglycerides in the study of polymorphism, ch. 3. In: Marangoni AG (ed) Structure-function analysis of edible fats, AOCS Press, Urbana, IL

Analysis of Objectionable Matter and Constituents

33 chapter

Analysis of Food Contaminants, Residues, and Chemical Constituents of Concern

Baraem P. Ismail (✉)
Department of Food Science and Nutrition,
University of Minnesota,
St. Paul, MN 55108-6099, USA
e-mail: bismailm@umn.edu

S. Suzanne Nielsen
Department of Food Science,
Purdue University,
West Lafayette, IN 47907-2009, USA
e-mail: nielsens@purdue.edu

S. Nielsen (ed.), *Food Analysis*, Food Science Text Series,
DOI 10.1007/978-3-319-45776-5_33, © Springer International Publishing 2017

33.1 INTRODUCTION: CURRENT AND EMERGING FOOD HAZARDS

The food chain that starts with farmers and ends with consumers can be complex (Fig. 33.1, Ref. [1]). Pesticide treatment, agricultural bioengineering, veterinary drug administration, environmental and storage conditions, processing, transportation, economic gain practices, use of food additives, and/or choice of packaging material may lead to contamination with or introduction (intentionally and non-intentionally) of hazardous substances. Legislation and regulation to ensure food quality and safety are in place and continue to develop to protect the stakeholders, namely, farmers, consumers, and industry. (Refer to Ref. [2] for information on regulations of food contaminants and residues.)

Contaminants/food hazards often have threshold levels (**tolerance levels**) below which no adverse effects are observed. The US Environmental Protection Agency (EPA) establishes these tolerance levels, and the **Food and Drug Administration** (FDA) and **US Department of Agriculture** (USDA) enforce them. However, food safety incidents, microbial, biological, and chemical in nature, continue to occur. The Rapid Alert System for Food and Feed (RASFF) reported a total of 3,049 alerts/notifications in 2015, a value that is 5 % higher than what was reported in 2014 [3]. Notifications were categorized as follows: chemical (36 %), mycotoxins (16 %), microbial (27 %), and other hazards (21 %). Within the chemical category, the most frequently reported hazards include allergens (e.g., histamine and sulfite), heavy metals (e.g., mercury, lead, and cadmium), pesticides (e.g., omethoate and isofenphos-methyl), and veterinary drugs (e.g., β-lactam and chloramphenicol). Microbial contaminants include molds, viruses, and bacteria (discussion and methods of analysis for this category are beyond the scope of this chapter). Examples of

current and emerging chemical hazards include fraud and food adulterants (e.g., melamine), packaging chemicals (e.g., bisphenol A and 4-methylbenzophenone), degradation metabolites (e.g., acrylamide, heterocyclic amines, and furan), naturally occurring substances (e.g., cyanide and solanine), and other chemical contaminants (e.g., 3-monochloropropane-1,2-diol, benzene, and perchlorate). Another category that can be of concern includes the genetically modified organisms (GMO) and their products. Introduction and usages of GMO in food products resulted in the development of legal requirements of safety and labeling.

Given the extent of concerns with contaminants, there is a strong need for adequate and reliable methods of detection and analysis to ensure food quality, safety, and fair trade. There are several well-established and reliable methods of analysis to detect certain food hazards. Development and validation of methods to detect and analyze emerging food hazards is a continuous effort. This chapter will cover some of the screening methods and quantitative methods that are commonly used for the detection and quantification of several food hazards, in addition to some recently developed methods for the detection of newly identified and emerging food hazards. The focus in this chapter on chemical contamination is intended to compliment that of Chap. 35 on Food Forensic Investigations.

33.2 ANALYTICAL APPROACH

As with the analysis of any food constituent, there is an array of methodological approaches and techniques to choose from for the analysis of food hazards. Multiple factors influence the method of choice as listed in Chap. 1, Tables 1.1, 1.2, 1.3, 1.4, and 1.5, including in this case complexity of the food matrix, characteristics of the analyte (e.g., polarity, hydrophobicity, volatility, thermal stability, and chemical reactivity), and suspected level of the contaminant. The objectives for the analysis of contaminants can vary in complexity from mere detection of several suspected contaminants belonging to the same family to the determination of the exact level of a particular contaminant or, in more complex cases, detection of unknown adulterants. The general trend for regulatory institutions and industry is to implement inexpensive and rapid **screening methods**. However, depending on the objective of the analysis, **quantitative methods** that require sophisticated equipment might be needed. In this case, the industry may choose to send their samples to specialized laboratories. Once a method of analysis is chosen, appropriate consideration with regard to sampling and sample preparation needs to be made.

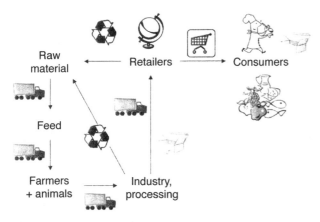

33.1 figure An illustration of the food chain (From Nielen and Marvin [1], with permission)

33.2.1 Choice of Analytical Method

The complexity of the food matrix (see Chap. 1, Sect. 1.4.4) and the characteristics of the analyte significantly influence the choice of extraction, separation, detection, and quantification techniques, as will be discussed in subsequent sections. Accuracy, precision, specificity, and sensitivity of the analytical method (see Chap. 4) are also important considerations. There are some official methods for the analysis of contaminants and residues, others are validated, and some are currently being developed and validated. Methods for the analysis of food hazards can be either qualitative, semiquantitative, or quantitative. A summary of analyses for some of the main contaminants is given in Table 33.1.

33.1 **Summary of analyses for contaminants, residues, and compounds of concern in foods**

Contaminant	Quantitative	Semiquantitative or qualitative (screening methods)
Pesticides	Multiresidue (MRMs) GC (mostly) HPLC Single residue (SRMs) GC (mostly) HPLC Immunoassay	TLC Enzyme inhibition Immunoassay
Mycotoxins	HPLC (mostly) GC Capillary electrophoresis Immunoassays (mostly)	TLC Immunoassay
Antibiotics	HPLC (mostly) GC Immunoassays	Microbial growth inhibition Receptor assays Enzyme-substrate assays Immunoassays
GMOs	PCR (mostly) ELISA	LFS
Allergens	ELISA PCR LC-MS	LFS
Sulfites	Monier-Williams method Ion chromatography Enzyme method HPLC	"Ripper method"
Nitrites	Colorimetric methods Ion chromatography	Ion-selective electrode

HPLC high-performance liquid chromatography, *GC* gas chromatography, *TLC* thin-layer chromatography, *PCR* polymerase chain reaction, *ELISA* enzyme-linked immunosorbent assay, *LFS* lateral flow strip

33.2.1.1 Qualitative or Semiquantitative Methods

Qualitative and **semiquantitative** methods, also known as **screening methods**, are usually used to assay a large number of samples for the presence of one or more contaminants belonging to the same family (e.g., antibiotic residues; see Sect. 33.5.2.1). These methods are fast, of low cost, simple, and robust; they are often less sensitive to small changes in experimental and/or environmental conditions and are not limited to a highly controlled lab environment. While qualitative methods detect the presence of certain contaminants at or above a specified threshold, semiquantitative methods provide an estimate of the concentration of a detected contaminant. These methods include techniques such as **thin-layer chromatography** (TLC), **enzyme inhibition**, and **immunoassays** (see later sections of this chapter).

33.2.1.2 Quantitative Methods

For simultaneous quantitation and structural identification of chemical food contaminants and residues, **gas chromatography** (GC; Chap. 14) and **high-performance liquid chromatography** (HPLC; Chap. 13) are the two main analytical methodologies employed. The combination of GC with mass spectrometry (MS), and the availability of relatively affordable benchtop GC-MS instruments, gave preference to GC analysis for multicomponent contaminant and residue analysis, in spite of having to derivatize polar analytes. However, thermally labile and/or large analytes that cannot be easily volatilized, such as mycotoxins, polar pesticides, and most of the veterinary drug residues, must be analyzed using HPLC. Major advances in HPLC-mass spectrometry (LC-MS) have facilitated direct, selective, and sensitive analysis of the polar analytes. For example, LC-MS has largely replaced microbial and immunochemical methods used for the analysis of veterinary drugs [4, 5]. Additionally, LC-MS is being used for multiclass, multiresidue analysis of pesticides [6, 7], due to the transition from the use of persistent and less polar compounds to the more readily degradable, more polar, thermolabile pesticides.

Immunoassays (Chap. 27) are rapid, simple, and cost-effective means for the detection and quantification of both single and multiple contaminants or residues, such as pesticides, antibiotics, and mycotoxins. Of the various immunoassay techniques, **enzyme-linked immunosorbent assay** (ELISA) (Sect. 27.3.2) and **lateral flow strip** (LFS) tests (Sect. 27.3.4) provide a sensitive detection of toxic analytes, whereas **immunoaffinity chromatography** (Sect. 27.4) is used for concentration and cleanup of the analyte of interest. One drawback of immunoassays is that the antibodies used in the assay may have **cross-reactivity** (affinity) for related chemical structures.

33.2.2 Sample Preparation

33.2.2.1 *Introduction*

Food samples are usually too dilute (e.g., beverages) or too complex (e.g., meat) for direct analysis of trace contaminants and residues. Therefore, sample preparation, including homogenization, extraction, fractionation/cleanup, concentration, and/or derivatization, normally precedes the analysis of food contaminants and residues (Fig. 33.2). Analysts are faced with continuous decrease of legislative limits for food contaminants and residues, resulting in the need for more sensitive, precise, and accurate measurements. Sample preparation techniques for the analysis of food contaminants and residues are continuously improved to guarantee high recovery and reproducibility (see Ref. [8] as an example). Fortunately, there also have been significant advances in analytical methods and equipment. The advanced technology of mass spectrometers (see Chap. 11), specifically, their enhanced "recognition features" when used in tandem, allows them to take over the "selectivity" of classical sample preparation methods. However, for quantification purposes, cleaner extracts are preferred. In both target and multiresidue methods, isotope dilution and addition of internal standard provide the most accurate results, compensating for sample matrix effects and ion suppression in MS. Also, faster and more efficient extraction methods are now available for the analysis of food contaminants and residues, as will be discussed in subsequent sections.

Analytical chemists often focus on perfecting the analytical technique (e.g., chromatographic analysis), overlooking the importance of sampling, sample storage, and sample preparation. Sampling and sample preparation are labor intensive and time consuming but are essential prerequisites for acquiring meaningful analytical data. There is a high probability of error and contamination during sampling and sample preparation, and these cannot be corrected at any point during the analysis. Therefore, an adequate plan for sampling, storing, and preparing samples should be implemented and validated by statistical analysis (see Chap. 5 for more details). Obtaining a sample that is representative of the level of trace substances in a heterogeneous mixture is not an easy task. Sampling for the analysis of a particular contaminant or residue will be briefly discussed in relevant sections of this chapter.

33.2.2.2 *Sample Homogenization*

Food contaminants and residues in many cases are unequally distributed in a food system. For instance, most pesticides are not translocated within plants and are expected to be located on the surface of fresh produce. Thus, removal of inedible parts (outer skins, stems, and cores), when applicable, and homogenization are necessary for reliable and accurate analysis. Homogenization can be achieved by chopping or grinding, followed by blending and mixing. Apart from homogenization, grinding reduces the structural features of the sample and enhances extraction efficiency. Contaminating the sample or exposing it to unnecessary heat, which can cause volatilization or degradation of contaminants or residues, should be avoided. Sometimes, cryogrinding (see Chap. 5, Sect. 5.5.2.1) is preferred for soft samples and to avoid thermal degradation.

33.2.2.3 *Extraction and Cleanup*

33.2.2.3.1 Introduction

Almost all food samples, with the exception of samples directly soluble in an organic solvent (e.g., vegetable oil), require a solvent extraction step (Chap. 12, Sect. 12.2; Chap. 14, Sect. 14.2.2.4) to isolate the target analytes from the matrix. Defatting of lipid-rich food matrices is often required, using hexane or isooctane, prior to the extraction of target contaminants. Extraction of contaminants and residues is traditionally done by solubilizing them in a suitable organic solvent, generally acetonitrile or acetone. **Anhydrous salt** (NaCl or Na_2SO_4) can be added to absorb water. In some cases, water is added so that the crude extract can be purified using a subsequent partitioning step with a second, water-immiscible solvent. When the extraction process is complete, the solvent is separated from insoluble solids by filtration.

Often, the crude sample extract is purified and concentrated before the separation/determination

Sampling ⎰ Collection
         ⎱ Storage

Sample preparation ⎰ Extraction
                    ⎟ Fractionation/ Clean-up
                    ⎟ Concentration
                    ⎱ Derivatisation

Chromatographic analysis   Separation

Data handling ⎰ Identification
              ⎱ Quantification

Reporting

33.2 figure Flow diagram for the analysis of contaminants and residues (From Sandra et al. [10], with permission)

steps. The degree of cleanup required depends on the method of analysis such as the chromatography mode and detector type. The objective of cleanup is to separate the target analytes from coextractives that can interfere with their detection. Often, the preliminary cleanup step is followed by a preparative chromatography step (see Chap. 12 for basic information on chromatography). Cleanup of crude water-acetone or water-acetonitrile extract, for example, can be done by partitioning with a relatively nonpolar organic solvent. The residues then can be further purified by column chromatography (either adsorption or size-exclusion chromatography). Separate fractions of column eluate can be analyzed. The choice of packing material is both analyte and matrix dependent. In some cases, such as determining aflatoxin in milk [9], monoclonal antibody affinity chromatography is used as a one-step column cleanup.

Great efforts have been made to develop extraction techniques that are faster and more efficient and require less solvent. Extraction and purification techniques have been developed based on the characteristics of the target analyte(s), which can be categorized into three classes: (1) **volatile compounds** (VC) that can be analyzed via headspace techniques, after derivatization if needed, (2) **semi-volatile compounds** (SVC) that are GC amendable (thermally stable), and (3) the **nonvolatile** (thermally labile) **compounds** (NVC), which are mostly analyzed by HPLC after extraction. Some of the widely used extraction techniques will be covered briefly in subsequent sections. References [10–14] contain more details on the various extraction techniques.

33.2.2.3.2 Solid-Phase Microextraction

All three classes of compounds listed above can be extracted using **solid-phase microextraction** (SPME), which is a type of **solid-phase extraction** (SPE) (Chap. 14, Sect. 14.2.2.5). VCs are best analyzed by headspace techniques (see Sect. 14.2.2.2). Headspace sampling applications include the analysis of pesticides, furan, and residues from packaging materials. Headspace sampling can be done by immersing the SPME polymer-coated fibers into the headspace. SPME also can be used for SVCs and NVCs, with the polymer-coated fiber directly immersed in the aqueous samples, or placed inside a hollow cellulose membrane.

33.2.2.3.3 The QuEChERS

QuEChERS method [15, 16], which stands for quick, easy, cheap, effective, rugged, and safe, is a dispersive SPE (DSPE) technique that is by far the best method to extract multiple pesticide residues (Sect. 33.3.2) (see Chap. 14, Sect. 14.2.2.5, for SPE). A single sample preparation method and one chromatography method linked to MS (preferably in tandem) to determine as

many pesticides as possible are very much desired. The advantages of the QuEChERS method (outlined in Fig. 33.3) include speed, ease, minimal solvent use, and lower cost as compared to conventional SPE. The types of adsorbents and solvents, as well as the pH and polarity of the solvents, can be adjusted based on the sample matrix and types of analytes. QuEChERS kits are now commercially available (e.g., Sigma Aldrich/ Supelco, Restek, and United Chemical Technologies).

33.2.2.3.4 Microwave-Assisted Solvent Extraction

Microwave-assisted solvent extraction (MASE), which utilizes electromagnetic radiation to desorb organics from their solid matrix, can decrease significantly the extraction time and the quantity of solvent required to efficiently extract target analytes (see Ref. [17] for example application). The efficiency of MASE is attributed to the elevated temperatures that exceed the boiling point of the solvent(s) and the rapid transfer of the analyte(s) to the solvent phase. Commercial systems are available that incorporate the capacity for simultaneously extracting multiple samples within closed, lined (perfluoroalkoxy), pressurized vessels (up to 8–12), using microwave absorbing solvents. Disadvantages of MASE include lack of selectivity and loss of thermolabile analytes.

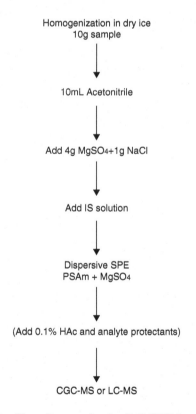

 33.3 figure Flow diagram for the QuEChERS procedure (From Sandra et al. [10], with permission)

33.2.2.3.5 *Accelerated-Solvent Extraction*

Accelerated-solvent extraction (ASE) (Chap. 17, Sect. 17.4.4) utilizes limited quantities of organic solvents at elevated temperature (up to 200 °C) and pressure (1,500–2,000 psi) to statically or dynamically extract solid samples for short periods of time (often <10 min) [18]. See Ref. [19] for example application of ASE for pesticides and antibiotic residues. Disadvantages of ASE include diluting effect and lack of selectivity, requiring further cleanup and concentration.

33.2.2.4 *Derivatization*

The chemical structure of analytes may need to be modified to become suitable for separation and detection by specific chromatographic techniques and detected via available and/or desirable detectors. The process of structural modification through multiple chemical reactions is known as **derivatization** (see Chap. 12, Sect. 12.3.4.2.1; Chap. 13, Sect. 13.2.4.7; Chap. 14, Sect. 14.2.3). Many types of derivatization reactions have been used in the analysis of contaminants and residues, such as selected pesticide residues as discussed in Ref. [20] and the antibiotic nitrofurans as discussed in Ref. [10].

33.3 PESTICIDE RESIDUE ANALYSIS

33.3.1 Introduction

A pesticide is any substance or a mixture of substances formulated to destroy or control "pests" including weeds, microorganisms (e.g., fungi or bacteria), insects, and even mammals. Currently, there are over 1,300 registered pesticides [21], main classes of which include: (1) **herbicides** [for weed control, e.g., triazine (e.g., atrazine)], (2) **insecticides** [e.g., organochloride (OC) (e.g., dichlorodiphenyltrichloroethane (DDT)), organophosphates (OP) (e.g., malathion, dimethoate, omethoate), and methyl carbamates (e.g., aldicarb)], and (3) **fungicides** [e.g., phthalimide (e.g., captan)]. Other types of pesticides may include acaricides, molluscicides, nematicides, pheromones, plant growth regulators, repellents, and rodenticides. Reference [5] lists active substances used in various pesticides products that belong to different chemical groups, and FDA provides a glossary of pesticide chemicals [22].

Pesticide residues may occur in food as a result of the direct application to a crop or a farm animal or as a postharvest treatment of food commodities. Pesticide residues also can occur in meat, milk, and eggs as a result of the farm animal consumption of feed from treated crops. Additionally, pesticide residues can occur in food from environmental contamination and spray drift.

If pesticides were not applied, an estimated one-third of the crop production would be lost. However, pesticides may have adverse effects on human health, including, but not limited to, cancer, acute neurologic toxicity, chronic neurodevelopment impairment, and dysfunction of the immune, reproductive, and endocrine systems. Accordingly, there are strict regulations for pesticide registration and use all over the world.

Risk assessment studies, required by the EPA, are done to determine the nature and extent of toxic effects and to establish the level at which no adverse effects are observed [**no observed adverse effect level** (NOAEL)]. Based on risk assessment studies, **tolerance levels** [or **maximum residue level** (MRL)] of pesticides (the active ingredients as well as toxic metabolites, and transformation products) in food are set and enforced by government agencies, as mentioned in Sect. 33.1, to protect stakeholders and regulate international trade. In fact, tolerance levels must be established prior to registration. In general, the tolerance levels in foods range between 0.01 and 10 mg/kg, depending on the commodity and the pesticide used. Low tolerance levels have necessitated the development of more accurate and sensitive analytical methods to meet the requirements in food.

33.3.2 Types of Analytical Methods

Several factors make the analysis of pesticide residues in food a complex task. These factors include the complexity of the food matrix, the possibility of having pesticide levels as low as pictogram or femtogram amounts, and the considerable differences in the physical and chemical properties of pesticides.

Analytical methods employed for the analysis of pesticides can be categorized into either **single-residue methods** (SRM) or **multiple-residue methods** (MRM). SRMs are designed to measure a single analyte and, often, its toxic metabolites. The majority of SRMs have been developed for purposes of registration and setting tolerance levels, or for investigating the metabolism and environmental fate of a specific pesticide. Sampling, extraction, purification, and determination in SRMs are optimized for the target pesticide. The majority of the SRMs currently in use, which have undergone EPA review or have been published in peer-reviewed scientific journals, are found in Volume II of the *Pesticide Analytical Manual* (*PAM* II) [23]. For purposes of monitoring quality and safety, given the large number of pesticides and the considerable differences in their physical and chemical characteristics (e.g., acidic, basic, or neutral, polar or nonpolar, and volatiles or nonvolatiles), MRMs can determine various pesticides in a single run and are much preferred. Many of the MRMs are found in Volume I of the *Pesticide Analytical Manual* (*PAM* I) [24]. AOAC International [25] also has developed an MRM for pesticide residues, the AOAC Pesticide Screen (AOAC Method 970.52). The identification and quantification in MRMs currently used by the FDA and USDA are based on GC and HPLC analysis.

Prior to chromatographic analysis, sampling, extraction, and fractionation/cleanup in MRMs are optimized to ensure an efficient transfer of most if not all of the pesticide residues present in the sample matrix to the organic phase. To achieve efficient transfer of as much of the pesticide residues as possible from the sample matrix into the organic phase, partitioning using water-miscible solvent is performed. This is followed by partitioning with nonpolar solvent that is miscible with the polar solvent, yet immiscible with water. After the extraction step, a cleanup step is performed using adsorption columns, in particular Florisil, alumina, and silica gels, with solvent mixtures of low polarity for elution. Commercial kits based on solid-phase extraction chromatography are available for pesticide cleanup. Reference [26] includes detailed descriptions of different MRM extraction and cleanup protocols specific for pesticide analysis.

33.3.3 Analytical Techniques Used for the Detection, Identification, and/or Quantification

A wide range of analytical techniques can be used in pesticide analysis. This section highlights some of the current analytical techniques used for the detection and determination of pesticide residues.

33.3.3.1 *Biochemical Techniques*
Biochemical techniques, such as enzyme inhibition assays and immunoassays, are widely used for the detection of pesticides. Commercially available test kits include enzyme inhibition assays. The principle of these assays is based on the inhibition of a specific enzyme, essential for vital functions in insects, by pesticides present in the sample. If no pesticides are present, the enzyme will be active and act on a substrate to cause a change in color. If no change in color occurs, then the test is positive, and further confirmation can be performed following more sophisticated analysis, such as HPLC and GC, to identify and quantify the specific pesticides present.

While enzyme inhibition assays are mostly used as screening methods, with limited sensitivity and selectivity, immunoassays can be tailored to particular purposes ranging from simple screening tests (field portable [27]) to quantitative laboratory tests. Immunoassays, which can be either class or compound specific, are simple and sensitive, have high throughput, and are cost effective as compared with other conventional methods. Additionally, extensive cleanup of extracts is not necessary, unless cross-reactivity exists. ELISA (Chap. 27, Sect. 27.3.2) accounts for almost 90 % of the immunoassays used for pesticide

residue analysis. References [26] and [27] include more information on the application of immunoassays in the analysis of pesticides.

33.3.3.2 *Chromatographic Techniques*

33.3.3.2.1 *Thin-Layer Chromatography*
Thin-layer chromatography (TLC) (Chap. 12, Sect. 12.3.4.2) can be used for screening purposes in the analysis of pesticides. Because of its low resolving capacity, low precision, and limited detection relative to GC and HPLC, it is not used as a quantitative method. However, TLC can be used as a semiquantitative method that precedes more accurate detection and quantification. An example application is the detection and estimation of pesticides that inhibit insect enzymes such as cholinesterases. Once a crude extract is separated by TLC, the plate is sprayed with a solution containing the enzyme(s), followed by a solution containing a specific substrate, which releases a colored product. The lack of color change indicates enzyme inhibition, due to the presence of pesticide residues, and the zone of inhibition is proportional to the quantity of pesticide present.

33.3.3.2.2 *Gas Chromatography*
With the development of fused silica capillary columns (Chap. 14, Sect. 14.3.4.2), a large number of pesticides with similar physical and chemical properties can be separated and detected. In general, GC is the preferred method for the determination of volatile and thermally stable pesticides, such as the OC and OP classes. Choice of columns and detectors is made based on the nature of the pesticides. For example, 5 % diphenyl, 95 % dimethylpolysiloxane stationary phase columns are commonly used in MRMs.

Pesticides often contain heteroatoms, such as O, S, N, Cl, Br, and F, in a single molecule. Therefore, element-selective detectors are often used, such as a flame photometric detector (FPD), which is suited for the detection of P-containing compounds. The FPD (see Chap. 14, Sect. 14.3.5.4) is widely used for the detection of OP pesticides in various crops, without extensive cleanup required. For the determination of OC, the electron capture detector (ECD; see Chap. 14, Sect. 14.3.5.3) is used extensively due to its high sensitivity to organic halogen compounds. With a MRM approach for multiclass detection, and using these selective detectors, several GC injections are required, which is a limitation for conventional GC analysis. Additionally, identification in conventional GC analysis is highly dependent on retention time, which is not an absolute confirmation of identity, due to matrix interferences. Coupling of fused capillary columns to MS detection enhances not only the confirmation process but also the quantitative determination (see Sect. 33.3.3.3).

33.3.3.2.3 High-Performance Liquid Chromatography

Development of HPLC analysis for the separation and detection of pesticides became a necessity as the number of pesticides with poor volatility, relatively high polarity, and thermal instability increased. Classes such as N-methyl carbamate (NMC), urea herbicides, benzoylurea insecticides, and benzimidazole fungicides are typically analyzed by HPLC. These compounds are often analyzed by reversed-phase chromatography (Chap. 13, Sect. 13.3.2) with C18 or C8 columns and aqueous mobile phase, followed by UV absorption, fluorescence, or MS detection (Chaps. 7 and 11). When the sensitivity of UV and fluorescence detection is poor, postcolumn derivatization can be employed [6]. However, interference from other fluorescent compounds is a major disadvantage.

Analysis of pesticides in complex systems following conventional HPLC analysis using fluorescence or UV is often inadequate. Even the use of diode array detection might not be specific enough to resolve spectral differences, which are often too small. Utilization of MS detection widened the scope of HPLC analysis of pesticides. LC-MS is becoming one of the most powerful techniques for the analysis of polar, ionic, and thermally labile pesticides (see Sect. 33.3.3.3.2).

33.3.3.3 Mass Spectrometry Detection

Details on MS instrumentation, modes of ionization, and mass analyzers are provided in Chap. 11. The subsequent sections will provide sample applications of GC-MS and LC-MS used in pesticide residue analysis. References [6, 7, 16, 26, 28–31] describe more application examples and details on method development.

33.3.3.3.1 Gas Chromatography-Mass Spectrometry

The most common GC-MS technique for the analysis of pesticide residue involves single quadrupole instruments with electron impact (EI) ionization. The selected ion monitoring (SIM) option enhances selectivity, increases sensitivity, and minimizes interferences from co-extracted compounds. Ion-trap detectors (ITD) are also used in the analysis of pesticide residues. Using ITD, the analyzing in full-scan mode provides higher sensitivity as compared to single quadrupole analysis and allows for confirmation by library searches (NIST library spectrum search). Additionally, the ITD enables tandem MS analysis (MS/MS) by means of collision-induced dissociation (CID). The use of tandem MS improves selectivity and significantly reduces background without loss of identification capability, thus enabling the analysis of pesticides at trace levels in the presence of many interfering compounds. Figure 33.4 provides an illustration of enhanced compound identity confirmation upon the use of MS/MS vs. only MS. For the determination of multiclass residues, triple quadrupole (QqQ)

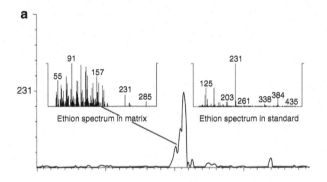

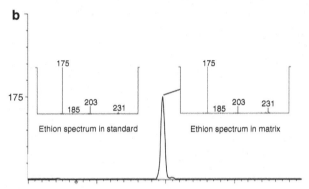

33.4 figure Pesticide residue analysis in vegetable extract using (**a**) EI full scan, showing noisy baseline and no spectral match; vs. (**b**) Electron impact (EI) MS/MS showing clean baseline, symmetrical peaks, and excellent spectral match (Used with permission from Varian Inc., Palo Alto, CA)

MS is becoming a powerful and fast analytical tool, requiring minimum sample preparation. The chromatogram separation becomes less important when the analysis is carried out by GC/QqQ-MS/MS, since the analyzer is able to monitor simultaneously a large number of co-eluting compounds. Reliable quantitation and confirmation can be easily achieved, even at trace concentration levels. Time-of-flight (TOF) MS instruments also are gaining popularity for the simultaneous analysis of multiclass pesticides [16, 30].

33.3.3.3.2 High-Performance Liquid Chromatography-Mass Spectrometry

Similar to GC-MS, the use of LC-MS in the analysis of pesticide has undergone great development. Tandem MS methods, using atmospheric-pressure ionization (API), atmospheric-pressure chemical ionization (APCI), and electrospray ionization (ESI), have been developed for many pesticides such as OP, carbamate, and sulfonylurea pesticides [29]. Specifically, the ESI coupled with MS/MS was found to have high sensitivity and selectivity for a wide range of pesticides in foods. Because of HPLC solvent interferences, single quadrupole instruments are not as widely used in

LC-MS as compared to GC-MS for pesticide analysis. When using LC-MS for pesticide residue analysis, QqQ are the most widely used mass analyzers. LC/TOF-MS has also gained popularity due to its high speed, sensitivity, and selectivity [31].

33.4 MYCOTOXIN ANALYSIS

33.4.1 Introduction

Molds, which are filamentous fungi, can develop on food commodities and produce various types of chemical toxins, collectively known as **mycotoxins**. The main producers of mycotoxins are fungal species belonging to the genera *Aspergillus*, *Fusarium*, and *Penicillium*. Crops can be directly infected with fungal growth and subsequent mycotoxin contamination as a result of environmental factors such as temperature, humidity, weather fluctuations, mechanical damage of crops, and pest attack. In addition, plant stress due to extreme soil dryness or lack of a balanced nutrient absorption can induce fungal growth. Fungal infection and mycotoxin contamination can occur at any stage of the food chain (Sect. 33.1). The United Nation's Food and Agriculture Organization estimated that up to 25% of the world's total crops per year are affected with "unacceptable" levels of mycotoxins [32].

Mycotoxin contamination may occur in food as a result of the direct mold infection of plant-origin commodities, such as cereals, dried fruit, spices, grape, coffee, cocoa, and fruit juices (especially apple based). Mycotoxins also can occur in milk, eggs, and, to a minor extent, meat, as a result of the farm animal consumption of feed from contaminated crops. Additionally, humans and animals can be exposed to mycotoxins through heavily contaminated dust, as in the case of harbors and warehouses.

More than 300 mycotoxins, belonging to various chemical classes, are known. However, the major classes of mycotoxin with a toxicological impact on human health include **aflatoxins** (B1, B2, M1, M2, G1, and G2), **ochratoxins** (e.g., ochratoxin A, OTA), **trichothecenes** [e.g., deoxynivalenol (DON), T2, and HT-2], **fumonisins** (FBs, e.g., FB1, FB2, and FB3), **patulin** (a mycotoxin that occurs mainly in apples and apple products), and **zearalenone** (ZEA). The chemical classification and occurrence of mycotoxins (the major as well as minor ones) in raw commodities and processed foods is described in Ref. [32]. The toxic effects of the aforementioned mycotoxins include, but are not limited to, genotoxicity, carcinogenicity, mutagenicity, and immunotoxicity. Genotoxic compounds have a probability of inducing an effect at any dose; therefore, no threshold dose should be considered and they should not be present in food. For example, there is no postulated safe dose for aflatoxin B1, which is genotoxic and is the most potent naturally occurring carcinogenic

substance known. However, to provide risk managers with necessary basis for making decisions, threshold doses and "safe" **total daily intake** or **total weekly intake** (TDI or TWI) have been set for other mycotoxins (Table 33.2). Based on the threshold doses, tolerance levels for mycotoxin were set in the USA (e.g., 0–35 µg/kg for aflatoxins, 2.5–50 µg/kg for OTA, 300–2000 µg/kg for DON, and 5–50 µg/kg for patulin).

While molds can be destroyed by natural causes or processing of food, mycotoxins can survive. To limit human exposure to mycotoxins, a key critical control point is avoiding the processing of raw materials with unacceptable mycotoxin levels. Therefore, implementing periodic testing is a necessity, including the adoption of reliable sampling procedures and validated methods of analysis. Organizations such as the AOAC International, American Oil Chemists' Society (AOCS), AACC International, and the International Union of Pure and Applied Chemistry (IUPAC) have method validation programs for mycotoxin analysis. References [33–37] provide detailed information on mycotoxin occurrence, health effects, control, sampling and sample preparation, and analysis.

33.4.2 Sampling

Compared to the analysis of other types of residues, the sampling step in the analysis of mycotoxins is by

33.2 table | Total daily intake (TDI)/total weekly intake (TWI) for the major mycotoxins

Mycotoxin[a]	TDI (ng/kg bw per day)	Organization[b]
OTA	4	Health Canada (1989, 1996)
	5	Nordic Council (1991)
	5	EU (1998)
	14	JECFA (1996, 2001)
	120 (TWI)	EFSA (2006 [81])
FBs	2000	EU (2000)
	2000	JECFA (2000)
	400	Health Canada (2001)
DON	3000	Health Canada (1985)
	1000	Health Canada (2001)
	1000	Nordic Council (1998)
	1000	EU (1999)
	1000	JECFA (2000)
ZEA	100	Health Canada (1987)
	100	Nordic Council (1998)
	500	JECFA (2000)
	200	EU (2000)
Patulin	400	JECFA (1996)
	400	EU (2000)
	400	Health Canada (1996)

From Brera et al. [32], used with permission
[a]*OTA* ochratoxin, *FBs* fumonisins, *DON* deoxynivalenol, *ZEA* zearalenone
[b]*EU* European union, *JECFA* Joint FAO/WHO expert committee on food additives

far the largest contributor to the total error. The possible variability is associated with the level and distribution of mycotoxins in the food commodity. An unevenly distributed 0.1 % of the lot is usually highly contaminated, resulting in an overall level above the tolerance limit. Due to this heterogeneous distribution, an appropriate sampling plan is needed to ensure that the concentration in a sample is the same as that in the whole lot. An incorrect sampling protocol can easily lead to false conclusions, often false negative, leading to undesirable health, economic, and trade impacts.

The Commission Regulation 401/2006 [36] provided specifications for the sampling of regulated mycotoxins in various food commodities including cereals, dried fruits, nuts, spices, milk and derived products, fruit juice, and solid apple products. Also, AOAC International, AOCS, and the FDA in collaboration with the USDA have developed detailed sampling plans for separate commodities. In general, an acceptable plan involves obtaining a large number of samples from multiple locations throughout the lot, creating a composite sample, grinding or slurring the composite sample (to reduce particle size and/or increase homogeneity), and subsampling for laboratory analysis. The number and size of collected samples and laboratory subsamples are dependent on the matrix and the size of the lot.

An example of sampling for mycotoxin analysis is the sequential sampling of raw shelled peanuts (with a tolerance level of less than 15 µg/kg for aflatoxins); a bulk sample of approximately 70 kg is randomly accumulated (at a rate of one incremental portion per 225 kg of lot weight). This bulk sample is divided randomly into three 21.8-kg samples using a Dickens mechanical rotating divider and then ground separately. A subsample (1,100 g) is taken from one of the samples, formed into a slurry, and analyzed for aflatoxins, in duplicate. If the average of the two determinations is ≤8 µg/kg, the lot is passed and no further testing is performed. If the average is ≥ 45 µg/kg, the shipment is rejected. For averages >8 µg/kg and <45 µg/kg, a second 21.8-kg sample is analyzed in duplicate, and the average of the four results is used to decide whether to accept (≤12 µg/kg) or reject (≥23 µg/kg) the lot. If the average falls between the second set of determining values, the third 21.8-kg sample is analyzed. If the average of the six determinations is <15 µg/kg, the lot is accepted.

33.4.3 Detection and Determination

Post sampling, sample preparation commonly includes extraction, cleanup, and concentration. Sample preparation steps (discussed in Sect. 33.2.2) that are specific for mycotoxin analysis have been developed. Shaking or blending often is used for mycotoxin extraction, while SPE often is used for cleanup, especially when multi-mycotoxin analysis is needed. For a specific mycotoxin, immunoaffinity LC (IAC) is used for cleanup.

Tremendous effort has been made to develop and optimize qualitative and quantitative methods for mycotoxin analysis. A list of official methods of mycotoxin analysis is included in Ref. [32] and others in Refs. [33–35, 37]. Mycotoxin analysis kits, both for cleanup (for GC, HPLC, or TLC analysis) and for detection (e.g., ELISA, LFS, immunoaffinity with HPLC), are commercially available. The sections below briefly discuss some of the current and newly developed analytical techniques used for the detection and determination of mycotoxins in food.

33.4.3.1 Rapid Methods of Detection

33.4.3.1.1 TLC
A large number of TLC methods for the analysis of mycotoxins have been accepted by AOAC International, including DON in barley and wheat, aflatoxin in peanuts and corn, aflatoxin M1 in milk and cheeses, OTA in barley and green coffee, and ZEA in corn. Conventional TLC techniques are commonly used for screening purposes, with detection limits reaching 2 ng/g. When results are positive, confirmatory and more sensitive quantitative analysis follows. The overall performance of TLC for the analysis of mycotoxin is improved when used in combination with IAC. Additionally, a microcomputer interfaced with a fluorodensitometer and semiconductor-based detection improves the data handling.

33.4.3.1.2 Immunoassays
Three main types of immunoassays can be used for the analysis of mycotoxins: **radioimmunoassay** (RIA), ELISA (see Chap. 27), and **fluorescence polarization immunoassay** (FPIA). The use of RIA, where the mycotoxin (e.g., aflatoxin) is radiolabeled, has gradually been replaced by ELISA. Since all mycotoxins are relatively small molecules (MW <1,000), competitive ELISA is used (see Chap. 27, Sect. 27.3.2). In FPIA, fluorescein-labeled mycotoxin competes with unlabelled mycotoxin analyte in a sample for binding with the antibodies. FPIA has gained popularity since it involves no coating of the plate and less analysis time as compared to ELISA. Results of FPIA were comparable to that of ELISA for the analysis of DON, ZEA, and OTA [32].

Immuno-based testing methods, including membrane-based immunoassays, LFS assays, and biosensors, also have been developed for online control or field testing of mycotoxins. Membrane-based immunoassays are based on the principle of direct competitive ELISA, with the anti-mycotoxin antibody being coated on the membrane surface. For example, test kits based on membrane-based immunoassays have been validated for OTA determination in wheat, rye, maize, and barely [38]. LFS, which is an immunochromatographic test (see Chap. 27, Sect. 27.4), is rapid and capable of detecting many mycotoxins simultaneously. However, any positive test requires confirmation by a reference method such as HPLC (see Sect. 33.4.3.2.1). Biosensors, on the other hand, are compact analytical devices that use biological components such as nucleic acids, enzymes, antibodies, or cells, associated with a transduction system. The transduction system processes the signal produced by the interaction between the target molecule and the biological component. The use of biosensors for the detection of several mycotoxins, such as aflatoxins DON and OTA, is rapidly increasing [32].

33.4.3.2 Quantitative and Confirmative Chemical Methods

33.4.3.2.1 HPLC

For quantitative determination, HPLC is the methodology of choice for most mycotoxins, specifically, aflatoxins, DON, OTA, ZEA, FBs, T-2 and HT-2 toxins, and patulin. Aflatoxins, OTA, and ZEA exhibit a native fluorescence and are detected directly by fluorescence detector. Pre- or postcolumn derivatization (with trifluoroacetic acid, iodine, or bromine) is required to enhance the fluorescence detection of some aflatoxins (i.e., AFB1 and AFG1), whereas precolumn derivatization with the OPA reagent is required for fluorescence detection of FBs. However, direct UV detection is used for DON and patulin. Reversed-phase chromatographic separation is normally employed for multi-mycotoxin analysis. A multifunctional column, which consists of a mixture of reversed-phase, size exclusion, and ion-exchange stationary phases, also can be used (e.g., AOAC Method 49.2.19A for aflatoxin analysis). Methods using IAC-HPLC are reported for the determination of aflatoxins in peanut butter, pistachio paste, fig paste, and paprika powder [39]. Validated HPLC methods for the analysis of various mycotoxins are periodically reported in Methods Committee Reports published in *Journal of AOAC International.*

As is the case with pesticide residue analysis, coupling of HPLC with MS analysis, especially LC-MS/MS, provides greater sensitivity and selectivity, and allows for simultaneous analysis of multiclass mycotoxins. Additionally, the use of LC-MS/MS allows for the detection of conjugated mycotoxins (i.e., the toxin bound to a polar compound such as glucose or another sugar), which are referred to as "masked" or "modified" mycotoxins because they escape routine detection. A method has been developed for the determination of 39 mycotoxins, including conjugated DON, FBs, ZEA, and aflatoxins, in wheat and maize, using LC-ESI-triple quadrupole MS, without cleanup [40].

33.4.3.2.2 GC

GC is not widely used for the detection of mycotoxins, except in the case of trichothecenes. Trichothecenes do not strongly absorb in the UV-Vis range and are non-fluorescent; therefore, GC methods were developed for their determination. Capillary column GC is commonly employed for the simultaneous detection of different trichothecenes, e.g., DON, T2 toxin, and HT-2 toxin, using trifluoroacetyl, heptafluorobutyry, or trimethylsilyl derivatization coupled with electron capture detection. GC is often linked to MS for peak confirmation. GC-MS also can be used for the confirmation of patulin in apple juice [32]. Validated and accepted methods for the determination of trichothecenes using GC are reported as AOAC Official Methods and by the American Society of Brewing Chemists.

33.4.3.2.3 Capillary Electrophoresis

Capillary electrophoresis (CE; see Chap. 24 Sect. 24.2.5.3), generally used as a chromatographic technique, can be employed to separate mycotoxins from matrix components using electrical potential. Methods are available for the determination of patulin in apple juice [41], simultaneous determination of ochratoxins (A and B) and aflatoxins [42], and determination of ZEA using cyclodextrins (for enhancing the native fluorescence) [43].

33.4.3.3 Other Methods of Analysis

Other methods for the detection of mycotoxins have been reported such as **near-infrared** (NIR) and **mid-infrared** (mid-IR) spectroscopy, especially **Fourier transform infrared** (FTIR) spectroscopy (a type of MIR) (see Chap. 8, Sect. 8.3.1.2). FTIR spectroscopy was used for the detection of mycotoxins in infected corn by gathering information from mid-IR absorption spectra [44]. Rapid and accurate identification of mycotoxigenic fungi and their mycotoxins (e.g., FB1 and DON) was achieved using MIR and NIR spectroscopy [45, 46]. Calibration for these methods is based on reference HPLC or GC methods. Testing and validation of IR methods for the detection of mycotoxins are currently being pursued.

33.5 ANTIBIOTIC RESIDUE ANALYSIS

33.5.1 Introduction

Animals intended for human consumption may not only be given drugs (e.g., antibiotics, antifungals, tranquilizers, and anti-inflammatory drugs) at therapeutic levels to combat diseases, but also some countries allow use of drugs (mainly antibiotics) at subtherapeutic levels to reduce the incidence of infectious diseases and for weight gain advantages. The **Center for Veterinary Medicine** (CVM), a branch of the FDA, regulates the manufacture, distribution, administration, and withdrawal period (time between the last drug treatment to the animal and the slaughter or use of milk or eggs by humans) of drugs given to animals, including animals from which human foods are derived. Most drug residues of concern for human food are antibiotic residues and thus are the focus of this section. Some of the major families/types of antibiotics, referred to in later sections, include β-lactams, sulfa drugs, cephalosporins, tetracyclines, and chloramphenicol.

Residual levels of any antibiotics given to animals used for human consumption are of concern for a variety of reasons, including the fact that some consumers are allergic to certain antibiotics, and the possibility that some microbes may develop antibiotic resistance due to constant exposure. Some antibiotics can be carcinogenic as well, such as nitrofuran compounds [47]. Also, on a practical level for dairy products such as cheese made with starter cultures, antibiotic residues would reduce the intended microbial growth and therefore reduce acid production. For all these reasons and more, the FDA has strict regulations on antibiotic residues in human food. Due to these regulations, antibiotic-contaminated meat and milk (including milk products) are considered to be adulterated. The FSIS monitors antibiotic residues in meat products from cattle, swine, lamb, goat, and poultry. Number of violations vary from year to year, for example, 38 antibiotics violations were reported in 2004, while 8 were reported in 2011 [48]. FDA's Center for Food Safety and Nutrition monitors antibiotic residues in milk and its products, including milk from pick-up tanks, pasteurized fluid milk, cheeses, etc. In 2014, out of four million samples tested, 703 were reported to have antibiotic residues including β-lactam and sulfonamide [49].

Samples are tested for antibiotic residues using rapid screening methods. A positive result from a screening test suggests that one or more types of antibiotic are present, so further testing is required to identify and quantify the specific antibiotics present.

33.5.2 Detection and Determination

Often, procedures such as defatting, protein hydrolysis (in case of meat or egg samples), protein precipitation (in case of dairy samples), and aqueous wash (in case of honey to remove excess sugar) precede the extraction of antibiotic residues. For the extraction of many antibiotics, liquid-liquid extraction and SPE are commonly used, followed by a partial purification step often using ion-exchange cleanup systems that take advantage of the acid/base character of the antibiotics. Sample preparation steps, including extraction and isolation (as discussed in Sect. 33.2.2) specific for drug residues, are reviewed in Ref. [50].

Tolerance levels are established for some antibiotics, while others have a zero tolerance (as is the case for nitrofurans and chloramphenicol). Chloramphenicol, for instance, is an antibiotic of considerable current concern in the USA, European Union, and other countries. It is used in some parts of the world in producing shrimp and has been found in imported seafood products (e.g., shrimp, crayfish, and crab). Because of the adverse health effects on humans, the FDA has banned the use of chloramphenicol in animals raised for food production and set a zero tolerance in human food [21 CFR 522.390 (4)]. Therefore, the analytical methods need to be as sensitive and selective as possible.

A wide variety of analytical methods have been developed and optimized for the analysis of antibiotics, categorized as screening or determinative and confirmatory. Reference [4] lists the methods commonly used in the analysis of several antibiotics, and Ref. [51] gives the process used by the AOAC to validate test kits. One excellent resource for antibiotic test methods, specific for milk and dairy products, is Ref. [52]. The sections below briefly discuss some of the analytical techniques used for the detection and determination of antibiotics in susceptible food commodities.

33.5.2.1 *Screening Methods*

Some of the major categories for rapid screening assays, some of which are quantitative, are: (1) microbial growth inhibition, (2) receptor assays, (3) enzyme-substrate assays, and (4) immunoassays. Some screening methods are specific to individual antibiotics, some are specific to a class of antibiotics, and some have no specificity. Screening assays for antibiotic residues in test samples initially relied mostly on inhibition of microbial growth, but now many use other principles for detection.

In **microbial growth inhibition** assays, turbidity, zone of inhibition, or acid production can be measured. In a **turbidity assay**, an indicator organism growing in a clear liquid culture will cause an increased turbidity; growth is inhibited, so turbidity is reduced if antibiotics are present. In a **zone of inhibition assay**, the test material diffuses through an agar-based nutrient medium that has been uniformly inoculated with spores of a susceptible organism. Any antibiotics present in the test material will inhibit the germination and growth of the organism, creating clear zones. In the **acid production assays**, the acid produced when microbes grow causes a color change in the medium. No color change means that the test sample contained an inhibitory substance. While microbial growth inhibition assays are more time consuming than many newer screening tests, they are inexpensive and applicable to testing large numbers of sample and provide some sensitivity to multiple antibiotic categories [52]. AOAC Official Methods include a nonspecific microbiological method for antibiotics and numerous microbiological methods for specific antibiotics [25].

An example of a **receptor assay** is the Charm II® test (Charm Science, Lawrence, MA), which has different versions, designed to detect different groups of antibiotics. The assay involves a competition between labeled antibiotics (labeled using ^{14}C or ^{3}H, depending on the specific type of Charm II® test system) and antibiotic residues in a milk sample for a limited number of specific binding sites on the surface of bacteria added to the test sample. The greater the concentration of antibiotic residue in the sample, the less radiolabeled tracer will become bound to the microorganism. The method can be applied to milk, certain dairy products, honey, and meat.

Enzyme-substrate assays measure the inhibition of an enzyme acting on a substrate, caused by the presence of an antibiotic. An example used for testing raw milk is the Penzyme® III commercial kit (Neogen, Lansing, MI), which is specific for β-lactam antibiotics, and inhibits D,D-carboxypeptidase on an equimolar basis. When this enzyme acts on a specific substrate, it causes release of D-alanine, which can be measured in additional steps of the assay resulting in a color change [52].

Immunoassays, namely, ELISA and LFS (see Chap. 27, Sects. 27.3.3 and 26.3.4, respectively), also can be used for screening of antibiotic residues. The Charm ROSA (Rapid One Step Assay) MRL assay, intended for milk and cream testing, is an example that uses a lateral flow strip. Another example of an immunoassay marketed for testing milk is the SNAP® kit (IDEXX Laboratories, Inc., Westbrook, ME). The assay is based on a competition between residual antibiotics in a milk sample and enzyme-labeled antibiotics in the test kit. The enzyme acts on a substrate to cause a color change; any antibiotics in the milk will result in a decrease in color development. Competitive ELISA can be used for the detection of a specific antibody, for example, the Veratox® assay (Neogen Corporation, Lansing, MI) used to detect chloramphenicol.

33.5.2.2 Determinative and Confirmatory Methods

Quantitative determination of antibiotic residues in food products follows the same general steps as for other trace analytes. After sample preparation steps, the partially purified extract is subjected to chromatographic separation, detection, and quantification. The most commonly used chromatography system is HPLC (mostly using reversed-phase separation mode) coupled with UV, fluorescence, chemiluminescence, or postcolumn reaction detectors. For confirmatory and identification purposes at trace levels, LC-MS and LC-MS/MS are increasingly used in the analysis of antibiotics [5, 19].

The FDA provides regulatory LC-MS/MS methods to detect fluoroquinolones in honey [53]. For

fluoroquinolones in salmon, the FDA has LC method with fluorescence detection [54] and a confirmatory LC/MS method intended for both salmon and shrimp [55]. The confirmatory FDA method for fluoroquinolones in catfish is an electrospray LC/MS method [56]. Additionally, LC-MS/MS has been used for confirmation of β-lactam residues in milk [57]. The detection of 14 different types of sulfonamides, in milk [58] and in condensed milk and soft cheeses [59], at levels below 10 ng/mL, also was achieved using LC-MS/MS. LC-MS/MS has been compared with ultra-high-performance liquid chromatography/quadrupole time-of-flight MS (UHPLC/Q-Tof MS) for analysis of macrolide antibiotic residues in a variety of foods [60]. LC-MS/MS gave a lower limit of detection and better precision, but the UPLC/Q-Tof MS provided better confirmation of positive findings. (See Chap. 13, Sect. 13.2.3.3, for UHPLC.)

33.6 ANALYSIS OF GMOS

33.6.1 Introduction

Agriculturally important plants may be genetically modified by the insertion of DNA from a different organism (**transgene**) into the plant's genome, conferring novel traits that it would not have otherwise, such as herbicide tolerance or insect resistance. The modified plant is termed a **genetically modified organism**, or **GMO**. GMO production relates to a number of food crops, including the following: corn, soybeans, cotton, canola (a cultivar of rapeseed), rice, sugar beets, and papaya. Although protection from pests and herbicide tolerance are the most common GMO traits, other GMO crops include plants that have been modified to improve postharvest quality or enhance the nutritional makeup of the food. Examples include vegetables with an extended shelf life, apples that have reduced browning rate (Granny Smith apples), and "golden rice," which produces the precursor to vitamin A.

Despite the prevalence of GMO crops, there are concerns about limiting crop variation (i.e., monocropping); unintended impacts on other plants, insects, wildlife, and nearby communities; and possible allergies. The debate about the use of GMO crops has resulted in government regulation for application and labeling. There are specific guidelines put forth by various governments regarding the analyses used. Therefore, even producers must be able to accurately test ingredients and other products.

Many companies produce high-quality, easy-to-use test kits, many of which are specific for a certain GMO protein or gene. These kits generally fall into two categories, based on the methodology: **polymerase chain reaction** (PCR) kits that specifically amplify the DNA of the GMO gene (specific or shared by many GMOs) so that it can be detected and **immunoassays** (ELISA and LFS) that are specific for the proteins. Further reading on the GMO and GMO detection topics is listed in the Refs. [61–63].

33.6.2 DNA Methods

Detection of the transgene DNA is an effective method of testing for GMO material in a sample. The analysis involves three distinct steps: (1) **extraction** of the DNA from the sample, (2) **amplification** of the DNA by PCR, and (3) **identification and quantification** of the amplified DNA (note: for **real-time quantitative PCR analyses**, amplification and detection/quantification occur at the same time). While all three steps are important, the PCR amplification is critical to the specificity and success of the analysis.

33.6.2.1 *DNA Extraction*

Application of extreme levels of sheer and heat can damage the DNA, rendering the subsequent PCR and detection ineffective. Extraction of the DNA from the food matrix, therefore, should be performed prior to significant processing of the material, preferably on raw ingredients. DNA extraction protocols differ somewhat, but all include disrupting the matrix to release the DNA, usually by grinding the sample to a fine powder. This is followed by dispersal of the ground material into an extraction solution and removal of unwanted components. For example, lipids may be removed by solvent extraction and protein by the addition of a protease. A final step may involve the precipitation of the DNA with cold alcohol, such as ethanol or isopropanol.

33.6.2.2 *PCR Amplification*

PCR is a cyclic method that exponentially increases the copy number of a specific DNA sequence, by means of enzymatic replication. It uses thermal cycling to alternately replicate the target sequence and then melt the DNA into single strands in order to repeat the process. The method is based on the use of two synthetic DNA fragments that are complementary to opposite ends of the target sequence. These fragments are termed the **primers**, which are commonly 18–35 bases in length, and they can be produced only when the target sequence is known. If a general identification of any GMO material is desired, then the primers used would be complementary to the promoter sequence, which is common to all transgenic crop plants that are normally grown for industrial food production. If a specific GMO product is to be identified, then the primers would consist of a sequence that includes the transgene DNA and plant DNA. This is necessary to avoid the detection of DNA that originates from bacteria that may have been on or in the plants.

In addition to the specific primers, the PCR mixture also includes a heat-stable **DNA polymerase**, such as Taq polymerase (from *Thermus aquaticus*); the **nucleotide bases** that comprise DNA, in the form of **deoxynucleoside triphosphates** (dNTPs); and a buffer solution to maintain the optimal conditions for the reactions. All of these components are present in the commercial kits. The reaction vials are then placed in a **thermal cycler**. Once the PCR system is started, the mixture is first put through a high-temperature cycle to melt the DNA (separate it into single strands), then a lower-temperature cycle to allow the primers to anneal to the single-stranded target DNA, and finally an intermediate-temperature cycle to allow the DNA polymerase to synthesize a new DNA strand complementary to the target strand by adding dNTPs, starting at the primers. The process is then repeated, usually for 30–50 cycles, which is sufficient to produce millions of copies of the DNA.

33.6.2.3 *DNA Analysis*

When sufficient DNA has been generated by PCR, the sample can be analyzed by agarose gel electrophoresis. The sample and standards migrate through the gel, and after the run, the gel is stained and the presence and abundance of DNA can be identified, by comparison to the position and degree of staining of the standards.

Some of the commercial test kits contain reagents to specifically label the double-stranded DNA with tags, such as fluorescein. These kits are meant to be used with special PCR equipment that deposits the finished, labeled mixture into a capillary where the intensity of the fluorescence, which is proportional to the abundance of DNA, is determined by high sensitivity fluorescence spectroscopy, rendering the electrophoresis step unnecessary. In this case the specificity is contingent on the sequence of the primers, which directly determine the nature of the double-stranded DNA present. These kits also include standards and other reagents necessary for accurate quantification of the target DNA abundance in the sample. If real-time PCR is used, the fluorescence is read in the reaction tube after each cycle of amplification, yielding a curve containing multiple data points rather than a single data point for each sample.

33.6.3 Protein Methods

The protein methods for GMOs are immunoassays, primarily ELISA (Chap. 27, Sect. 27.3.2) and LFS (Chap. 27, Sect. 27.3.4), generally used for testing raw agricultural products and not processed products. Both qualitative and quantitative ELISA test kits for GMOs are commercially available, with the latter having typical detection limits of 0.01–0.1%. The LFSs are generally less sensitive (0.1–1.0%), but their speed and ease of use make them ideal for testing in the growing field, at storage areas, and at points of transport.

33.7 ALLERGEN ANALYSIS

33.7.1 Introduction

Food allergens are food proteins that trigger an allergic response. Symptoms of an allergic response include hives, face and tongue swelling, and difficulty breathing and can include the severe, life-threatening allergic reaction called anaphylactic shock. It is important to note that food allergy, which triggers an immune system reaction, is distinct from other adverse responses to food, such as food intolerance (e.g., lactose intolerance), pharmacologic reactions (due mainly to food additives such as sulfites and benzoate), and toxin-mediated reactions (due to residues such as pesticides and mycotoxins).

A significant percentage of the population has food allergies, and the prevalence is rising. More than 160 foods can cause allergic reactions in people with food allergies, but over 90 % of the food allergic reactions in the USA are caused by the eight most common allergenic foods, referred to as '**the big eight**': milk, eggs, fish, crustacean shellfish, tree nuts, peanuts, wheat, and soybeans [64]. In the USA, the Food Allergen Labeling and Consumer Protection Act of 2004 targets these eight food allergens [65]. It applies to all foods regulated by the FDA (both domestic and imported), requiring that labels list all ingredients by their common names and identify the source of all ingredients derived from the eight most common food allergens. Food allergies are an issue around the world; numerous other countries have or are considering labeling regulations. Since there is no cure for food allergies, strict avoidance of food allergens is the only effective action. All of these facts point to the importance of both screening and quantitative methods for analysis of allergenic foods.

Methods available for the detection of food allergens are mainly based on protein or DNA detection, as discussed in subsequent sections. A commercially available rapid screening method that is not based on protein or DNA detection is the swab and adenosine triphosphate (ATP)-sensitive detection method. This method is based on the detection of ATP present on the surface of multiple allergenic foods, e.g., peanut butter, whole egg, soybeans, and milk. It is used mainly to prevent food allergen cross contact during cleaning of processing equipment. Test kits for food allergen analysis (protein or DNA based and others) are commercially available. A review of food allergen analytical methods is given in Ref. [66].

33.7.2 Protein Methods

33.7.2.1 General Considerations

Similar to the analysis of many other food constituents of concern present in small amounts, sampling adequacy and the detection limits are of concern with analysis for allergens. Another concern is the adequate extraction of the different allergens. Unlike the trace analytes of concern mentioned thus far, since the analytes are proteins, the extraction solution is normally a buffer at various pHs and salt concentration. Extraction buffers differ in their ability to extract food allergens, with some solutions extracting the same allergens in different concentrations, while others cannot extract all the allergens. For example, phosphate buffer fails to extract the major peanut allergen (Ara h 3); however, the efficiency of extracting this allergen is greatly enhanced upon the addition of salt (Fig. 33.5). Additionally, it is crucial that the extraction solution is compatible with the assay used (e.g., immunoassay) and does not alter the chemical structure of the analyte. The choice of extraction procedure should also take into consideration the food processing conditions employed. Upon processing, the solubility of the proteins can be reduced due to denaturation and aggregation, resulting in reduced protein recovery [67]. Therefore, to obtain reliable and accurate results, it is crucial to select the right extraction solution for the target analyte(s).

33.7.2.2 Protein-Based Analytical Techniques

Classical protein-based methods used for the analysis of food allergens usually involve antibody-based assays (immunoassays) due to their specificity and sensitivity. Immunoassays target the offending allergen by use of monoclonal or polyclonal antibodies, or a combination of both. Most of the commercially available test kits, which differ in specificity and the number of proteins they target, use polyclonal antibodies. Immunoassay-based methods used for the analysis of food allergens include ELISA (Chap. 27, Sect. 27.3.2), LFS (Chap. 27, Sect. 27.3.4), Western blot (Chap. 27, Sect. 27.3.3.1), biosensor immunoassays (antibodies immobilized on a biosensor chip), and dot immunoblotting (Chap. 27, Sect. 27.3.3.2) [68].

Western blot and dot immunoblotting are mostly used for qualitative and screening purposes. The most commonly used immunoassays for the quantitative analysis of food allergens are ELISA methods, using a competitive or, more commonly, sandwich format (see Chap. 27, Sect. 27.3.2.3). The competitive ELISA is used for the small protein allergens, with a molecular weight less than 5 KDa. Numerous sandwich and competitive ELISA methods have been developed for several food allergens [66]. Lateral flow test strips are useful for screening purposes because they are very rapid and inexpensive and do not require instrumentation.

There are several drawbacks when using immunoassays, including cross-reactivity of antibodies with other proteins, food matrix interferences, structural changes due to food processing, destruction of epitopes upon adsorption onto solid matrices, and variability between manufactures due to use of different extraction buffers and antibodies. Often

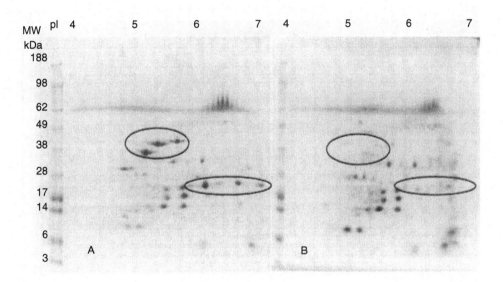

33.5 figure Evaluation, by two-dimensional electrophoresis, of the ability of (**a**) high salt buffer and (**b**) phosphate buffer to extract the peanut allergen Ara h 3 from raw peanuts (From Westphal [67], with permission)

times, 1-D or 2-D electrophoresis is used to isolate target protein to avoid cross-reactivity. In spite of the drawbacks, immunoassays are most commonly used for the detection of allergens. However, advanced proteomic techniques are evolving including mass spectrometry-based proteomic methods, most notably LC coupled with tandem MS, capable of detecting and quantifying multiple allergens simultaneously [66].

33.7.3 DNA Methods

There are several advantages and disadvantages for the use of DNA-based methods in the analysis of food allergens. DNA-based methods do not target the allergen in the sample; therefore, the detection of the allergen-encoding DNA does not always correlate with the presence of the allergen, especially when the food has been fortified with purified protein. Upon processing, for example, production of protein isolates, protein, and DNA could be separated resulting in false conclusions regarding the presence of the allergen in the sample. Regardless of these disadvantages, DNA-based methods are very specific and sensitive techniques with the advantage that targeted DNA is less affected by several processing and extraction conditions as compared to proteins. The choice to use DNA-based methods is dependent on the type of sample being analyzed.

DNA-based methods involve the extraction of the DNA (see Sect. 33.6.2.1) followed by amplification by PCR using a thermostable polymerase (see Sect. 33.6.2.2). The amplified sample is then visualized by fluorescence staining or by Southern blotting following agarose gel electrophoresis. This procedure

normally provides qualitative data, or semiquantitative data if internal standards were used. Quantification can be achieved if real-time PCR (see Sect. 33.6.2.1 and Ref. [66]) or PCR-ELISA was used. PCR-ELISA method involves linking the amplified DNA fragment of an allergenic food to a specific protein-labeled DNA probe, which then is coupled with a specific enzyme-labeled antibody. Quantification of the DNA is based on the enzyme-substrate color producing reaction. Reference [60] includes a list of commercially available real-time PCR and PCR-ELISA test kits; for the analysis of allergens, readers are referred to Ref. [66].

33.8 ANALYSIS OF OTHER CHEMICAL CONTAMINANTS AND UNDESIRABLE CONSTITUENTS

33.8.1 Introduction

Pesticide residues, mycotoxins, antibiotic residues, and allergens in foods have been of concern for many years. However, in any given time period, there are numerous additional chemical hazards that must be addressed. Many of these substances fall into the following categories: (1) banned (e.g., coumarin) or allowed in some countries but not others, (2) legally limited (e.g., sulfites, benzene), (3) intentional contaminants (e.g., melamine), (4) approved for use but of concern (e.g., monosodium glutamate), or (5) natural constituents of concern (e.g., acrylamide, furan). This section would be very lengthy to cover in detail the screening and quantitative methods for many of these chemical hazards. Instead, only a summary (Table 33.3) is given of select current and emerging hazards, with

33.3
table
Select compounds of concern and their methods of analysis

Compound	Nature of compound and reason in food	Reason of concern	Major foods/ ingredients identified	Major methods
Acrylamide	Formed in carbohydrate-rich foods cooked at high temperature	Neurotoxin and carcinogen, and other health risks	Fried and oven-based foods high in carbohydrates	LC-MS/MS
Benzene	Can form at low levels in some beverages containing both ascorbic acid and benzoate salts	Carcinogen	Soft drinks	Headspace GC-MS
Bisphenol A (BPA)	Used to make polycarbonate plastic, but can leach from plastic	Can mimic body's own hormones. Associated with multiple health problems. Banned or amounts are limited by various countries	Foods/beverages in polycarbonate plastic	GC- MS, GC-MS/MS, LC-MS, LC-MS/MS
Cyanide	Naturally found in certain seeds, fruit stones, and cassava roots	Many cyanides are highly toxic to humans	Apple seeds, cherry pits, cassava	UV colorimetry (after reflux distillation)
Furans	Volatile liquid that seems to form during traditional heat treatments	Carcinogen	Variety of foods	Headspace GC-MS
Heterocyclic amines	Formed in meat products heated at high temperatures, with creatine being a necessary precursor with free radicals formed as a result of the Maillard reaction initiating their production	Neurotoxin, mutagen, and carcinogen	Grilled/fried meat and fish to well done stage	Reversed-phase LC coupled with fluorescence detection, LC-MS
Melamine	A trimer of cyanamide (with 66 % N) that has been added illegally to increase apparent protein content of foods and ingredients	Causes kidney failure. Legal limits have been set	Added previously to wheat gluten, milk, and infant formula	HPLC-MS/MS (GC-MS for screening)
4-Methylbenzophenone (4MBP)	Metabolite of a chemical component of ink used for food packaging	Concern for health risk with long exposure	Boxes of cereal	LC-MS/MS, GC-MS
4-Methylimidazole	Formed via Maillard browning during normal cooking of certain foods. Formed in manufacturing of caramel coloring	Some concern about whether it is a possible carcinogen to humans	Roasted meats and coffee, caramel coloring, beverages with caramel coloring	LC-MS/MS (ion chromatographs), GC
3-Monochloropropane 1,2-Diol (3-MCPD)	Formed when proteins (soy) ae hydrolyzed by heat and food-grade acids to create hydrolyzed vegetable protein	Carcinogen	Soups, savory snacks, and gravy mixes flavored with acid-hydrolyzed vegetable protein	GC-MS, LC-MS, SCF-MS/MS

(continued)

33.3
table (continued)

Compound	Nature of compound and reason in food	Reason of concern	Major foods/ ingredients identified	Major methods
Nitrosamines	Produced from nitrites in the presence of proteins upon processing at high temperatures, as in frying	Carcinogen	Cured meats, primarily cooked bacon, beer, some cheeses, nonfat dry milk, fish	GC-MS for volatile nitrosamines, LC-MS for nonvolatile nitrosamines
Perchlorate	Component of rocket fuel and formed naturally	Can interfere with iodide uptake into the thyroid gland, leading to hypothyroidism	Bottled water, milk, lettuce	LC-MS/MS (ion chromatograph)
Solanine	A glycoalkaloid found in potatoes that helps protect the plant, but causes bitter taste when at certain level in potato tuber. Peeling potatoes decreases content	Toxic to humans above certain levels	Potatoes	LC-MS/MS, HPTLC, GC

their major method(s) of analysis [69]. To reach the limit of detection (LOD) desired and to positively identify the compounds, many of the major analyses rely on gas or liquid chromatography, with mass spectrometry. Described in more detail below is the analysis for two compounds of concern, sulfites and nitrates/nitrites, which are not as commonly analyzed by chromatography as by other methods. There are legal limits in foods for both of these compounds, so they are monitored regularly in select foods for quality control purposes.

33.8.2 Sulfites

Although sulfites are classified as allergens, they are covered in this separate section because the nature of the chemical compound, symptoms, and methods of analysis are quite different than for other allergens. Sulfites and sulfiting agents are a group of chemical compounds that include sulfur dioxide (SO_2), sulfurous acid (HSO_3), and the following inorganic sulfite salts that can liberate SO_2: sodium (Na) and potassium (K) sulfite, Na and K bisulfite, and Na and K metabisulfite [70]. In some foods they occur naturally, but in other foods they are added for a variety of reasons, including preventing microbial growth and browning. Sulfites naturally occur to some extent in all wines but are commonly added to stop fermentation at the appropriate time and to prevent spoilage and oxidation. Dried fruits and vegetable products are sometimes treated with sulfites to reduce browning. Shrimp, lobster, and related crustaceans can be treated with

sulfite to prevent "black spot." Some consumers are highly intolerant to sulfite residues in food, most commonly resulting in asthma attacks [71]. Therefore, the FDA forbids sulfites from being added to foods intended for raw consumption (e.g., salad bar foods) and requires the phrase "contains sulfites" on the label of foods that contain greater than 10 ppm sulfites, whether naturally occurring or added during manufacturing [21 CFR 101.100 (a)(4)] [72]. Many countries besides the USA have set strict limits on the residual levels of sulfites in various foods.

Reactions of sulfites with other food components make analysis challenging, often resulting in decreased levels during storage. Most methods of analysis detect free forms of sulfite plus some bound forms. However, none of the available methods, alone, measures all forms of sulfite in foods, which includes free inorganic sulfite plus the many sulfite bound forms. It is not known which forms of sulfite cause the adverse responses in sulfite-sensitive consumers, so the focus has been on measuring as much as possible of the residual sulfite, both free and bound forms [63].

One of the long-time, quantitative methods for sulfite analysis of foods is the **Monier-Williams procedure** (AOAC Method 990.28) [25], which measures "total" SO_2 (actually, free sulfite plus reproducible portion of bound sulfites, such as carbonyl addition products). The FDA refers to this method in regulations for labeling sulfite-containing foods. In this method, the test sample is heated with HCl, converting sulfite to SO_2. Nitrogen gas bubbled through the sample sweeps SO_2 through a condenser and a hydrogen peroxide solution, oxidizing

SO_2 to H_2SO_4. The sulfite content of the sample is directly related to the amount of H_2SO_4 generated, measured by either a gravimetric or turbidimetric procedure.

Other analytical methods for sulfites include the following:

1. The **"Ripper" method**, long used by the wine industry as a rapid screening method, compared to the more time-consuming Monier-Williams method; sulfite is titrated with an iodide-iodate solution, using a starch endpoint indicator; measures "free" SO_2 (APHA Standard Method 4500 – SO_3^{-2} B) [73].
2. Enzymatic method; sulfite is oxidized to sulfate, generating hydrogen peroxide that is further reacted with NADH-peroxidase, to produce NAD which is measured by absorption at 340 nm (Sect. 26.3.1.3.1).
3. Ion chromatography, using amperometric detector (AOAC Method 990.31) [25].
4. HPLC, with ultraviolet [74] or fluorometric [75] detection.

33.8.3 Nitrates/Nitrites

Sodium nitrate and sodium nitrite are both food preservatives commonly used to cure processed meat products, but there are legal limits on the level of sodium nitrite in the finished product. Nitrates (NO_3) and nitrites (NO_2) are very similar chemically, and sodium nitrate is readily converted to sodium nitrite. Sodium nitrate is naturally occurring in some vegetables, and it converts to sodium nitrite when it comes in contact with your saliva. The naturally occurring high levels of nitrates in celery and Swiss chard make powders of them common ingredients in some cured meat products. Both sodium nitrate and sodium nitrite give cured meat an appealing pinkish-reddish color and prevents it from turning brown. More importantly, sodium nitrite works with sodium chloride to prevent the growth of *Clostridium botulinum*. However, when products with high levels of residual nitrites are cooked at high temperatures (e.g., frying of bacon), the nitrites react with amines naturally present in the meat to form nitrosamines, which are suspected carcinogens. Also, high levels of nitrates or nitrites in the diet may induce the disorder methemoglobinemia, which can be fatal due to the reduced ability for oxygen transport in the blood. In addition to the residual nitrates in processed meat products, there is concern about the levels of nitrates and nitrites in drinking water, and in the water and soil used to grown vegetables, especially when consumed by infants and young children [76, 77].

Health concerns have led to the following EPA- and FDA-regulated limits:

1. Drinking water – 10 ppm nitrates, 1 ppm nitrite (EPA; 40 CFR 141) [77]

2. Bottled water – 10 ppm nitrate, 1 ppm nitrite, and 10 ppm total nitrates and nitrites (FDA; 21 CFR 165.110) [78]
3. Finished meat products – 200 ppm sodium nitrite (FDA; 21 CFR 172.175) [79]
   There are also FDA legal limits on sodium nitrate and sodium nitrite for certain cured fish products [79].

During the curing of processed meat products, processors ensure the legal limits of sodium nitrite in the finished product largely by carefully monitoring the amount of sodium nitrate used in the formulation of the product. The levels of nitrates/nitrites may be monitored throughout the production system with relatively rapid tests for nitrates/nitrites. There are both official methods and more rapid rapids of analysis used for testing both processed meat products and water, as described below.

Three of the AOAC International official methods [25] for nitrates/nitrites include the following:

1. Xylenol Method (AOAC Method 935.48, Nitrates and Nitrites in Meat)
   In the xylenol method, the sample, with sulfuric acid added, is treated with 2,4-xylenol to produce 6-nitro-2,4-xylenol, which is distilled into a water-isopropanol alcohol-ammonia hydroxide mixture. The ammonia salt of 6-nitro-2,4-xylenol is a yellow color measured at 450 nm, compared to a nitrate N standard curve.
2. Colorimetric Method (AOAC Method 973.31, Nitrites in Cured Meat)
   In the colorimetric method, sodium nitrite is extracted from the sample then reacted with two reagents, sulfanilamide (sulfa) and naphthylethylenediamine (NEDA). These compounds react with nitrite to produce a purple dye, which is proportional to the concentration of nitrite ions.
3. Ion Chromatographic Method (AOAC Method 993.30, Inorganic Anions in Water)
   This method is intended for measuring nitrate-N, Nitrite-N, and several other inorganic anions in drinking water, but is also used within the meat industry with meat extracts as the sample. Anions in the sample are separated using an ion chromatography system that includes a guard column, separator column, and suppressor device and quantitated using a conductivity detector.

Rapid methods of analysis include ion-selective electrodes (ISEs) and test strips. Regarding ISE, both nitrate and nitrite ISEs are commercially available for testing food products (liquid samples) (see Chap. 21, Sect. 21.3.4, for details of ISEs). Regarding the test strips, various companies make test strips intended for checking the nitrate content of water. For example, AquaChek nitrite/nitrate test strips (Hach, Loveland, CO) are able to test for either nitrates or nitrites, or both, using two pad areas on the test strip, with one area measuring nitrates and the other pad area measuring nitrites. The nitrate test area contains a combination of chemicals to reduce nitrates to nitrites. Nitrites, at an acid pH, react with sulfanilic acid to form a diazonium compound, which couples with an indicator to produce a pink color. The color intensity is proportional to the concentration of nitrate. Color blocks for interpretation of results are provided for both nitrate N (0–50 ppm) and nitrite N (0–3 ppm).

33.9 SUMMARY

Consumer concerns and government regulations focused on the safety of foods dictate the need for analysis of various food contaminants, residues, and chemical constituents of concern. These compounds include pesticide residues, mycotoxins, antibiotic residues, GMOs, allergens, food adulterants, packaging material hazardous chemicals, environmental contaminants, and certain other chemicals. Both rapid screening methods and more time-consuming quantitative methods are required to meet the needs of industry and government, in an effort to ensure a safe and reliable food supply. A positive result from a screening method usually leads to further testing to confirm and quantify the presence of the compound of concern. Sampling and sample preparation can be a significant challenge due to the low levels of the chemicals and the complex food matrices. Sample preparation often includes homogenization, extraction, and cleanup and sometimes requires derivatization. Screening methods increasingly utilize immune-based techniques, such as ELISA, LFS, immunosensors, and immunoaffinity chromatography columns. Some immunoassays can be considered quantitative, rather than just screening methods. Other screening methods commonly used include enzyme inhibition assays, thin-layer chromatography, and inhibition of microbial growth. While GC is a common chromatographic technique used for the quantitative analysis of some pesticides, for many other compounds of concern covered in this chapter, the predominant chromatographic method is HPLC. Both GC and HPLC analysis are now commonly coupled with mass spectrometry detection, often using MS tandem systems. The testing for GMOs and allergens typically involves either protein-based methods (e.g., immunoassays) or DNA methods using PCR. Work continues to improve and develop various methods of analysis for chemical residues and compounds of concern, focusing largely on speed, cost, and reliability of screening methods and detection limits of quantitative methods.

33.10 STUDY QUESTIONS

1. Explain the importance of each of the following steps in sample preparation for the analysis of contaminants and residues of concern:

   (a) Grinding/homogenization
   (b) Extraction
   (c) Cleanup/purification
   (d) Derivatization

2. In the analysis of contaminants and residues of concern, compare and contrast:

   (a) GC vs. LC analysis
   (b) MS vs. MS/MS analysis
   (c) LC with fluorescence or UV detection vs. LC-MS
   (d) TLC vs. automated chromatography (LC or GC)
   (e) Microplate ELISA vs. LFS

3. The "tolerance level" for residues of the pesticide chlorpyrifos on corn grain is 0.05 ppm:

   (a) What is meant by "tolerance level"?
   (b) What federal agency sets that tolerance level?
   (c) What federal agency enforces that tolerance level?
   (d) What is "ppm" equivalent to in terms of (a) weight per volume, and (b) weight per weight units commonly used to express concentration?

   wt/vol: wt/wt:

   (e) In Volumes I and II of the *Pesticide Analytical Manual*, you find described "multiresidue" methods and "single-residue" methods. You also have found numerous screening methods. Which one of these three types of methods (i.e., multiresidue, single residue, screening) would you use to ensure your compliance with the tolerance level for this pesticide? Briefly explain the nature of this type of method, and why you chose this method over the other two types of methods.

4. Mycotoxins are of potential concern in corn, especially in certain growing and storage conditions:

   (a) Sampling is a major contributor to error in the analysis for mycotoxins. Why is sampling for mycotoxin analysis such a challenge?
   (b) Identify the most commonly used quantitative chromatographic method for mycotoxins. Justify the preference of this method.

5. Regarding antibiotic residues, briefly explain the following:

   (a) How might these get into foods?
   (b) What types of foods most likely contain them?
   (c) Why are these antibiotic residues a problem?
   (d) Why have techniques such as immunoassays largely replaced microbial growth inhibition assays for screening antibiotic residues?
   (e) What method is most commonly used for accurate quantitative determination and confirmation?

6. You want to identify GMOs in a soybean field where you do not have access to a sophisticated analytical laboratory. What food constituent would you be analyzing, and which analytical technique would you use? Explain the principle of this technique.

7. You are analyzing unwanted chemical components in the foods/raw materials listed below. Identify one possible unwanted chemical component that would likely be of concern with this specific food, and state an appropriate method for quantitative analysis. Also give one appropriate screening method for each chemical component identified. (Note: For each food, give a different quantitative likely unwanted chemical component and a different method of analysis.)

Likely unwanted chemical component	Quantitative analysis method	Screening method

   (a) Oats
   (b) Peanuts
   (c) Milk
   (d) Wine
   (e) Cured meat

8. Describe the DNA-based method and identify the two protein-based methods for GMO detection and quantitation. Describe the strengths and weaknesses of each.

9. For the analysis of food allergens, protein-based or DNA-based methods are commonly used:

   (a) Give an example of when you would choose a protein-based method over a DNA-based method. Justify your choice and provide the principle of the method of choice.
   (b) Give an example of when you would choose a DNA-based method over a protein-based

method. Justify your choice and provide the principle of the method of choice.

10. Regarding the compounds described in Table 33.3:

    (a) Identify five compounds associated with the heat of cooking/frying of foods.
    (b) Identify two compounds typically analyzed by GC (vs. LC), and explain the preference for GC methods with these compounds.
    (c) Identify two compounds associated with packaging material.
    (d) Identify a compound that is an economic adulterant.
    (e) Identify two compounds naturally present in specific foods, but toxic to humans at certain levels.

11. Regarding sulfites, explain how this "allergen" differs from the food allergens described in Sect. 33.7 in the response of sensitive humans, how they differ in the nature of quantitative methods, and why quantitative determination is relatively difficult for sulfites compared to food allergens.

Acknowledgments The authors of this chapter wish to acknowledge Bradley L. Ruehs, who contributed some content to this chapter in the 4th edition of the textbook. The content and organization of the current chapter also benefited from two chapters in previous editions: "Analysis of Pesticide, Mycotoxin, and Drug Residues in Foods" by William D. Marshall; and "Agricultural Biotechnology (GMO) Methods of Analysis" by Anne Bridges, Kimberly Magin, and James Stave. The authors also would like to thank the following persons from the Institute of Sciences of Food Production (ISPA), National Research Council of Italy, for their helpful comments in revision of this chapter from the 4th ed.: Michelangelo Pascale, Veronica Lattanzio, Vincenzo Lippolis, and Annalisa De Girolamo.

REFERENCES

1. Nielen MWF, Marvin HJP (2008) Challenges in chemical food contaminants and residue analysis, ch. 1. In: Pico Y (ed) Food contaminants and residue analysis. Comprehensive analytical chemistry, volume 51. Elsevier, Oxford, UK
2. Arvanitoyannis IS (2008) International regulations on food contaminants and residues, ch. 2. In: Pico Y (ed) Food contaminants and residue analysis. Comprehensive analytical chemistry, volume 51. Elsevier, Oxford, UK
3. European Union (2015) Rapid alert system for food and feed. Preliminary annual report. http://ec.europa.eu/food/safety/docs/rasff_annual_report_2015_preliminary.pdf
4. Turnipseed SB, Andersen WC (2008) Veterinary drug residues, ch. 10. In: Pico Y (ed) Food contaminants and residue analysis. Comprehensive analytical chemistry, volume 51. Elsevier, Oxford, UK

5. Díaz-Bao M, Barreiro R, Miranda JM, Cepeda A, Regal P (2015) Fast HPLC-MS/MS method for determining penicillin antibiotics in infant formulas using molecularly imprinted solid-phase extraction. J Anal Methods Chem 2015: 1–8

6. Sannino A (2008) Pesticide Residues, ch. 9. In: Pico Y (ed) Food contaminants and residue analysis. Comprehensive analytical chemistry, volume 51. Elsevier, Oxford, UK

7. Stachniuk A, Fornal E (2016) Liquid chromatography-mass spectrometry in the analysis of pesticide residues in food. Food Anal Methods 9: 1654–1665

8. Lehotay SJ, Cook JM (2015) Sampling and sample processing in pesticide and residue analysis. J Agric Food Chem 63: 4395–4404

9. Hansen TJ (1990) Affinity column cleanup and direct fluorescence measurement of aflatoxin M_1 in raw milk. J Food Prot 53: 75–77

10. Sandra P, David F, Vanhoenacker G (2008) Advanced sample preparation techniques for the analysis of food contaminants and residues, ch. 5. In: Pico Y (ed) Food contaminants and residue analysis. Comprehensive analytical chemistry, volume 51. Elsevier, Oxford, UK

11. Handley A (ed) (1999) Extraction methods in organic analysis, Scheffield Academic, Scheffield, England

12. Mitra S (ed) (2003) Sample preparation techniques in analytical chemistry, Wiley, Hoboken, New Jersey

13. Ramos L, Smith RM (eds) (2007) Advances in sample preparation, part I. J Chromatogr A 1152: 1

14. Ramos L, Smith RM (eds) (2007) Advances in sample preparation, part II J Chromatogr A 1153: 1

15. Anastassiades M, Lehotay SJ, Stajnbaher D, Schenck FJ (2003) Fast and easy multiresidue method employing acetonitrile extraction/partitioning and "dispersive solid-phase extraction" for the determination of pesticide residues in produce. J AOAC Int 86: 412–431

16. Niell S, Cesio V, Hepperle J, Doerk D, Kirsch L, Kolberg D, Scherbaum E, Anastassiades M, Heinzen H (2014) QuEChERS-based method for the multiresidue analysis of pesticides in beeswax by LC-MS/MS and GC × GC-TOF. J Agric Food Chem 62: 3675–3683

17. Jiao, Z, Guo Z, Zhang S, Chen H (2015) Microwave-assisted micro-solid-phase extraction for analysis of tetracycline antibiotic in environmental samples. Int J Environ An Ch 95: 82–91

18. Richer BE, Jones BA, Ezzel JL, Porter NL, Avdalovic N, Pohl C (1996) Accelerated solvent extraction: a technique for sample preparation. Anal Chem 68: 1033–1039

19. Tao Y, Yu G, Chen D, Pan Y, Liu Z, Wei H, Peng, D, Huang L, Wang Y, Yuan Z (2012) Determination of 17 macrolide antibiotics and avermectins residues in meat with accelerated solvent extraction by liquid chromatography-tandem mass spectrometry. J Chromatogr B Analyt Technol Biomed Life Sci 897: 64–71

20. Cairns T, Sherma J (1992) Emerging strategies for pesticide analysis. In: Modern methods for pesticide analysis, 9th edn. CRC, Boca Raton, FL

21. EPA (2016) Pesticide Product Information System. https://www.epa.gov/ingredients-used-pesticide-products/pesticide-product-information-system-ppis

22. FDA (2005) FDA Glossary of Pesticide Chemicals. http://www.fda.gov/Food/FoodborneIllnessContaminants/Pesticides/ucm113891.htm

23. FDA (2002) Pesticide analytical manual volume II (updated January, 2002). http://www.fda.gov/food/foodscienceresearch/laboratorymethods/ucm113710.htm

24. FDA (1999) Pesticide analytical manual volume I (PAM), 3rd edn (1994, updated October, 1999). http://www.fda.gov/food/foodscienceresearch/laboratorymethods/ucm111455.htm

25. AOAC International (2016) Official methods of analysis, 18th edn. Official methods of analysis, 20th edn. AOAC International, Gaithersburg, MD

26. Tadeo JL (2008) Analysis of pesticides in food and environmental samples. CRC, Taylor and Francis Group, Boca Ranton, FL

27. Lee NA, Kennedy IR (2001) Environmental monitoring of pesticides by immunoanalytical techniques: Validation, current status, and future perspectives. J AOAC Int 84: 1393–1406

28. Soderberg D (2005) Committee on residues and related topics; pesticides and other chemical contaminants. General referee reports. J AOAC Int 88: 331–345

29. Pico Y, Blasco C, Font G (2004) Environmental and food applications of LC tandem mass spectrometry in pesticide-residues analysis: an overview. Mass Spectrom Rev 23: 45–85

30. Patel K, Fussell RJ, Goodall DM, Keelye BJ (2004) Evaluation of large volume-difficult matrix introduction-gas chromatography-time of flight-mass spectrometry (LV-DMI-GC-TOF-MS) for the determination of pesticides in fruit-based baby foods. Food Addit Contam 21: 658–669

31. Gilbert-López B, García-Reyes JF, Ortega-Barrales P, Molina-Díaz A, Fernández-Alba AR (2007) Analyses of pesticide residues in fruit-based baby food by liquid chromatography/electrospray ionization time-of-flight mass spectrometry. Rapid Commun Mass Spectrom 21(13): 2059–2071

32. Brera C, De Santis B, Debegnach F, Miraglia M (2008) Mycotoxins, ch. 12. In: Pico Y (ed) Food contaminants and residue analysis. Comprehensive analytical chemistry, volume 51. Elsevier, Oxford, UK

33. Pitt JI, Wild CP, Baan RA, Gelderblom WCA, Miller JD, Riley RT, Wu F (2012) Improving public health through mycotoxin control. International Agency for Research on Cancer Publication No 158, Geneva, Switzerland

34. Miller JD, Schaafsma AW, Bhatnagar D, Bondy G, Carbone I, Harris LJ, Harrison G, Munkvold GP, Oswald IP, Pestka JJ, Sharpe L, Sumarah MW, Tittlemier SA, Zhou T (2014) Mycotoxins that affect the North America agri-food sector:state of the art and directions for the future. World Mycotoxin Journal 7: 63–82

35. Rahmani A, Jinap S, Soleimany F (2009) Qualitative and quantitative analysis of mycotoxins. Comprehensive Reviews in Food Science and Food Safety 8: 202–251

36. Commission Regulation (EC) N 401/2006 of 23 February 2006. Laying down the methods of sampling and analysis for the official control of the levels of mycotoxins in food stuff

37. Siantar DP, Trucksess MW, Scott PM, Herman EM (eds) (2008) Food contaminants: mycotoxins and food allergens. ACS symposium series 1001. American Chemical Society, Washington, DC

38. De Saeger S, Sibanda L, Desmet A, Van Peteghem C (2002) A collaborative study to validate novel field immunoassay kits for rapid mycotoxin detection. Int J Food Microbiol 75: 135–142

39. Garner RG, Whattam MM, Taylor PJ, Stow MW (1993) Analysis of United Kingdom purchased spices for aflatoxins using an immunoaffinity column clean-up

procedure followed by high performance liquid chromatography, J Chromatogr A 648: 485–490

40. Sulyok M, Berthiller F, Krska R, Schuhmacher R (2006) Development and validation of a liquid chromatography/tandem mass spectrometric method for the determination of 39 mycotoxins in wheat and maize. Rapid Commun Mass Spectrom 20: 2649–2659

41. Tsao R, Zhou T (2000) Micellar electrokinetic capillary electrophoresis for rapid analysis of patulin in apple cider. J Agric Food Chem 48: 5231–5235

42. Peña R, Alcaraz MC, Arce L, Ríos A, Valcárcel M (2002) Screening of aflatoxins in feed samples using a flow system coupled to capillary electrophoresis. J Chromatogr A 967: 303–314

43. Maragos CM, Appell M (2007) Capillary electrophoresis of the mycotoxin zearalenone using cyclodextrin-enhanced fluorescence. J Chromatogr A 1143: 252–257

44. Greene RV, Gordon SH, Jackson MA, Bennett GA (1992) Detection of fungal contamination in corn: potential of FTIR-PAS and DRS. J Agric Food Chem 40: 1144–1149

45. Berardo N, Pisacane V, Battilani P, Scandolara A, Pietri A, Marocco A (2005) Rapid detection of kernel rots and mycotoxin in maize by near-infrared reflectance spectroscopy. J Agric Food Chem 53: 8128–8134

46. Pettersson H, Åberg L (2003) Near infrared spectroscopy for determination of mycotoxin in cereals. Food Control 14: 229–232

47. De la Calle MB, Anklam E (2005) Semicarbazide: occurrence in food products and state-of-the-art in analytical methods used for its determination. Anal Bioanaly Chem 382: 968–977

48. United States Department of Agriculture Red book archives. USDA Food Safety Inspection Service. http://www.fsis.usda.gov/wps/portal/fsis/topics/data-collection-and-reports/chemistry/red-books/archive

49. FDA (2014) National milk drug residue database. Food and Drug Administration, Center for Food Safety and Nutrition. http://www.fda.gov/downloads/Food/GuidanceRegulation/GuidanceDocumentsRegulatoryInformation/Milk/UCM434757.pdf

50. Fedeniuk RW, Shand P (1998) Theory and methodology of antibiotic extraction from biomatrices. J Chromatogr A 812: 3–15

51. AOAC (2016) The AOAC International rapid methods validation process. http://jornades.uab.cat/workshopmrama/sites/jornades.uab.cat.workshopmrama/files/AOAC_RI.pdf

52. Bulthaus M (2004) Detection of Antibiotic/Drug Residues in Milk and Dairy Products, ch. 12. In: Wehr HM, Frank JF (eds) Standard methods for the examination of dairy products, 17th edn. American Public Health Association, Washington, DC

53. FDA (2006) Preparation and LC/MS/MS analysis of honey for fluoroquinolone residues. 29 Sept, 2006 (last updated 8/11/2015). http://www.fda.gov/Food/FoodScienceResearch/LaboratoryMethods/ucm071495.htm

54. FDA (2003) Concurrent determination of four fluoroquinolones; ciprofloxacin, enrofloxacin, sarafloxacin and difloxacin in Atlantic salmon tissue by LC with fluorescence detection (Oct. 24, 2003). http://www.fda.gov/downloads/Food/FoodScienceResearch/ucm071499.pdf

55. FDA (2003) Confirmation of fluoroquinolone residues in salmon and shrimp tissue by LC/MS: evaluation of single quadrupole and ion trap instruments. Laboratory Information Bulletin 4298. http://www.fda.gov/downloads/Food/FoodScienceResearch/ucm071504.pdf

56. FDA (1997) Confirmation of fluoroquinolones in catfish tissue by electrospray LC/MS: Laboratory Information Bulletin 4108. http://www.fda.gov/downloads/Food/FoodScienceResearch/ucm071507.pdf

57. Holstege DM, Puschner B, Whitehead G, Galey FD (2002) Screening and mass spectral confirmation of B-lactam antibiotic residues in milk using LC-MS/MS. J Agric Food Chem 50(2): 406–411

58. Cavalier C, Curini R, Di Corcia A, Nazzari M, Samperi R (2003) A simple and sensitive liquid chromatography-mass spectrometry confirmatory method for analyzing sulfonamide antibacterials in milk and egg. J Agric Food Chem 51: 558–566

59. Clark SB, Turnipseed SB, Madson MR (2005) Confirmation of sulfamethazine, sulfathiazole, and sulfadimethoxine residues in condensed milk and soft-cheese products by liquid chromatography/tandem mass spectrometry. J AOAC Int 88: 736–743

60. Wang J, Leung D (2007) Analyses of macrolide antibiotic residues I eggs, raw milk, and honey using both ultra-performance liquid chromatography/quadrupole time-of-flight mass spectrometry and high-performance liquid chromatography/tandem mass spectrometry. Rapid Commun Mass Spectrom 21(19): 3213–3122

61. Ahmed FE (2004) Testing of genetically modified organisms in foods. CRC, Boca Raton, FL

62. Heller KJ (2003) Genetically engineered food: methods and detection. Wiley-VCH, Weinheim, Germany

63. Jackson JF, Linskens HF (2009) Testing for genetic manipulation in plants (molecular methods of plant analysis). Springer, New York

64. FDA (2007) Food allergies: What you need to know. February 2007 (last updated 6/30/2009). http://www.fda.gov/Food/ResourcesForYou/Consumers/ucm079311.htm

65. FDA (2004) Food allergen labeling and consumer protection act of 2004 (Public Law 108–282, Title II). 2 Aug, 2004 http://www.fda.gov/Food/GuidanceRegulation/GuidanceDocumentsRegulatoryInformation/Allergens/ucm106187.htm

66. Prado M, Ortea I, Vial S, Rivas J, Calo-Mata P, Velázquez J (2015) Advanced DNA- and protein-based methods for the detection and investigation of food allergens. Crit Rev Food Sci Nutr. doi:10.1080/10408398.2013.873767

67. Westphal CD (2008) Improvement of immunoassays for the detection of food allergens, ch. 29. In: Siantar DP, Trucksess MW, Scott PM, Herman EM (eds) Food contaminants: mycotoxins and food allergens. ACS symposium series 1001, American Chemical Society, Washington, DC

68. Yman IM, Eriksson A, Johansson MA, Hellenas K-E (2006) Food allergen detection with biosensor immunoassays. J AOAC Int 89(3): 856–861

69. FDA (2015) Drug and chemical residue methods. (last undated 06/29/2015) http://www.fda.gov/Food/FoodScienceResearch/LaboratoryMethods/ucm2006950.htm

70. Taylor SL, Bush RK, Nordlee JA (2003) Sulfites, ch. 24. In: Metcalfe DD, Simon RA (eds) Food allergy: adverse reactions to foods and food additives, 3rd edn. Blackwell, Malden, MA, p 324–341

71. Bush RK, Montalbano MM, (2014) Asthma and Food Additives, ch. 27. In: Metcalfe DD, Sampson HA, Simon RA, Lack G (eds) Food allergy: adverse reactions to foods and food additives, 5th edn. Wiley-Blackwell, Hoboken, NJ, p 361–374

72. Anonymous (2016) Code of federal regulations. Food; exemptions from labeling. 21 CFR 101.100 (a) (4). US Government Printing Office, Washington, DC

73. Eaton AD, Clesceri LS, Rice EW, Greenberg AE, (eds) (2005) Standard methods for the examination of water and wastewater, 21st edn., Method 4500 – SO_3^{-2} B. American Public Health Association, Washington, DC

74. McFeeters RF, Barish AO (2003) Sulfite analysis of fruits and vegetables by high-performance liquid chromatography (HPLC) with ultraviolet spectrophotometric detection. J Agric Food Chem 51: 1513–1517

75. Chung SWC, Chan BTP, Chan ACM (2008) Determination of free and reversibly-bound sulfite in selected foods by high-performance liquid chromatography with fluorometric detection. J AOAC Int 91(1): 98–102

76. Sindelar JJ, Miklowski AL (2011) Sodium nitrite in processed meat and poultry meats: a review of curing and examining the risk/benefit of its use. American Meat Science Association 3:1–14

77. Katan MB (2009) Nitrate in foods: harmful or healthy? Amer J Clin Nutr 90(1):11–12

78. Anonymous (2016) Code of federal regulations. Maximum contaminant levels for inorganic chemicals. 40 CRF 141.11. US Government Printing Office, Washington, DC

79. Anonymous (2016) Code of federal regulations. Bottled water. 21 CFR 165.110. US Government Printing Office, Washington, DC

Analysis for Extraneous Matter

Hulya Dogan (✉) • Bhadriraju Subramanyam
Department of Grain Science and Industry,
Kansas State University,
Manhattan, KS 66506, USA
e-mail: dogan@ksu.edu; sbhadrir@ksu.edu

S. Nielsen (ed.), *Food Analysis*, Food Science Text Series,
DOI 10.1007/978-3-319-45776-5_34, © Springer International Publishing 2017

34.1 INTRODUCTION

Analysis for extraneous matter is an important element both in the selection of raw materials for food manufacturing and for monitoring the quality of processed foods. The presence of extraneous material in a food product is unappealing and can pose a serious health hazard to the consumer. It also represents lack of good manufacturing practices and sanitary conditions in production, storage, or distribution. The presence of extraneous materials in the product ingredients may render the final product adulterated and not suitable for human food.

34.1.1 Federal Food, Drug, and Cosmetic Act

The **Federal Food, Drug, and Cosmetic Act** (FD&C Act) of 1938 with amendments administered and enforced by the US Food and Drug Administration (FDA) [1] defines a food as **adulterated** "if it consists in whole or in part of any filthy, putrid, or decomposed substance, or if it is otherwise unfit for food [Section 402 (21 USC 342) (a)(3)]; or if it has been prepared, packed, or held under unsanitary conditions whereby it may have become contaminated with filth, or whereby it may have been rendered injurious to health" [Section 402 (21 USC 342) (a)(4)]. The filthy, putrid, or decomposed substances referred to in the law include the extraneous matter addressed in this chapter. In addition, extraneous matter includes adulterants that may be encountered in processing systems, such as lubricants, metal particles, or other contaminants (animate or inanimate), that may be introduced into a food intentionally or because of a poorly operated food processing system. These aspects are not covered in this chapter.

34.1.2 Good Manufacturing Practices

The **Current Good Manufacturing Practice in Manufacturing, Packing, or Holding Human Food** (cGMPs) was published in 1969 by the Food and Drug Administration (FDA) (21 CFR Part 110) to provide guidance for compliance with the FD&C Act [2] (see also Chap. 2). That regulation provides guidelines for operating a food processing facility in compliance with Section 402 (a)(4), and these guidelines have not been revised since 1986. Currently, the cGMPs are being amended to make the compliance guidelines more risk based. Paramount to complying with the FD&C Act and cGMPs is the thorough inspection of raw materials and routine monitoring of food processing operations to ensure protection of the consuming public from harmful or filthy food products.

34.1.3 Defect Action Levels

Most of our foods are made from or consist in part of ingredients that are obtained from plants or animals and are mechanically stored, handled, and transported in large quantities. It would be virtually impossible to keep those materials completely free of various forms of contaminants. In recognition of that, the FDA [3] has established **defect action levels** (DALs) that reflect current maximum levels for natural or unavoidable defects in food for human use that present no health hazard. They reflect the maximum levels that are considered unavoidable under good manufacturing practices. They apply mainly to contaminants that are unavoidably carried over from raw agricultural commodities into the food processing system. The manner in which foods are manufactured may lead to their contamination with extraneous materials if strict controls in processing are not maintained. This latter type of contamination leads to food safety issues and DALs are not used to determine compliance. Other actionable levels of contaminants may be found in the FDA *Compliance Policy Guide (CPG) Manual* [4].

DALs are published in the FDA handbook and represent limits at which the food product is considered "adulterated" and subject to enforcement action. If there is no published DAL for a product, FDA evaluates and decides on a case-by-case basis. FDA's technical and regulatory experts in filth and extraneous materials use a variety of criteria to determine the significance and regulatory impact of the findings. The most current information of FDA laws and regulations relevant to extraneous matter, including cGMPs, DALs, and CPGs, can be found on the Internet:

Federal Food, Drug, and Cosmetic Act (FD&C Act)	http://www.fda.gov/RegulatoryInformation/Legislation/FederalFoodDrugandCosmeticActFDCAct/default.htm
Current Good Manufacturing Practices (cGMPs)	http://www.fda.gov/food/guidanceregulation/cgmp/
Food Defect Action Levels (DALs)	http://www.fda.gov/food/guidanceregulation/guidancedocumentsregulatoryinformation/sanitationtransportation/ucm056174.htm
Compliance Policy Guidance (CPG)	http://www.fda.gov/ICECI/ComplianceManuals/CompliancePolicyGuidanceManual/

34.1.4 Purposes of Analyses

The major purposes for conducting analyses for extraneous matter in foods are to ensure the protection of the consuming public from harmful or filthy food products, to meet regulatory requirements of the FD&C Act Sections 402 (a)(3) and 402 (a)(4), and to comply with DALs.

34.2 GENERAL CONSIDERATIONS

34.2.1 Definition of Terms

Terms used by **AOAC International** (AOAC Method 970.66) to classify or characterize various types of extraneous materials are defined as follows:

- **Extraneous Materials** – Any foreign matter in a product associated with objectionable conditions or practices in production, storage, or distribution; included are various classes of filth, decomposed material (decayed tissues due to parasitic or non-parasitic causes), and miscellaneous matter such as sand and soil, glass, rust, or other foreign substances. Bacterial counts are not included.
- **Filth** – Any objectionable matter contributed by animal contamination, such as rodent, insect, or bird matter, or any other objectionable matter contributed by unsanitary conditions.
- **Heavy Filth** – Heavier material separated from products by sedimentation based on different densities of filth, food particles, and immersion liquids. Examples of such filth are sand, soil, insect and rodent excreta pellets and pellet fragments, and some animal excreta pellets.
- **Light Filth** – Lighter filth particles that are oleophilic and are separated from product by floating them in an oil-aqueous liquid mixture. Examples are insect fragments, whole insects, rodent hairs and fragments, and feather barbules.
- **Sieved Filth** – Filth particles of specific size ranges separated quantitatively from product by use of selected sieve mesh sizes.

34.2.2 Diagnostic Characteristics of Filth

There are certain qualities characteristic to extraneous materials that serve as proof of presence of foreign or objectionable matter in food. Examples include specific diagnostic characteristics of **molds** (i.e., parallel hyphal walls, septation, granular appearance of cell contents, branching of hyphae, blunt ends of hyphal filaments, non-refracted appearance of hyphae); diagnostic characteristics of **insect fragments** (i.e., recognizable shape, form, or surface sculpture, an articulation or joint, setae or setal pits, sutures), **rodent hairs** (i.e., pigment patterns and structural features), and **feather barbules** (i.e., structural features); diagnostic characteristics of **insect-damaged kernels** (IDK) and packaging materials; and chemical identification of **animal urine** and **excrement**. These diagnostic characteristics are outlined by AOAC International (formerly Association of Official Analytical Chemists) for positive identification of extraneous matter or filth [5].

The **AACC International** (AACCI) (formerly American Association of Cereal Chemists, AACC) pub-lishes a methods book that includes a section on extraneous matter, containing descriptive material helpful in identifying insect and rodent contaminants [6]. Several microscopic and radiographic illustrations are provided by the AACCI as authentic reference materials to help analysts to identify filth. AACCI Method 28-95.01, "Insect, Rodent Hair, and Radiographic Illustrations," provides a series of colored pictures representative of insect fragments commonly found in cereal products and pictorial examples of rodent hair structure.

Kurtz and Harris [7] provide a virtual parts catalog of insect fragments with a series of micrographs. Gentry et al. [8], in an updated version of the Kurtz and Harris publication, include colored micrographs of common insect fragments. Also included in AACCI Method 28-95.01 are radiographic examples of grain kernels that contain internal insect infestation. AACCI Method 28-21.02, "X-ray Examination for Internal Insect Infestation," provides an outline of the apparatus and procedure for X-ray examination of internal insect infestation in grain [9].

34.3 OFFICIAL AND APPROVED METHODS

There are various laboratory methods for separating (isolating) extraneous materials from foods and for identifying and enumerating them. The FDA and the AOAC International have published reference articles, books, and methods on the analysis of extraneous materials. The most authoritative source, and that generally considered official by the FDA, is the *Official Methods of Analysis of AOAC International*, Chap. 16, "Extraneous Materials: Isolation" [5]. This chapter includes methods for extraneous matter isolation in various food categories (Table 34.1). The AOAC International "Extraneous Materials: Isolation" chapter contains a subchapter dealing with **molds**. This includes identification of molds and methods for isolation of molds in fruits and fruit products and vegetables and vegetable products.

The AACCI [6] has established methods for isolating and identifying extraneous matter in cereal grains and their products (AACCI Method 28, listed in Table 34.2). In most instances, the AACCI methods are based on FDA or AOAC methods, but the format is slightly different. The AACCI presents each procedure in an outline form that includes the scope, apparatus, and reagents required and the procedure in itemized steps, while the AOAC methods use a narrative paragraph form (Table 34.2).

A valuable resource on analysis for extraneous matter is *Principles of Food Analysis for Filth, Decomposition and Foreign Matter* (FDA Technical Bulletin No. 1) [10]. The FDA *Training Manual for Analytical Entomology in the Food Industry* [11] is

34.1 table	Official methods of AOAC international for analysis of extraneous materials

Section	Title
16.	Extraneous materials: isolation
16.1	General
16.2	Beverages and beverage materials
16.3	Dairy products
16.4	Nuts and nut products
16.5	Grains and their products
16.6	Baked goods
16.7	Breakfast cereals
16.8	Eggs and egg products
16.9	Poultry, meat, and fish and other marine products
16.10	Fruits and fruit products
16.11	Snack food products
16.12	Sugars and sugar products
16.13	Vegetables and vegetable products
16.14	Spices and other condiments
16.15	Miscellaneous
16.16	Animal excretions
16.17	Mold
16.18	Fruits and fruit products
16.19	Vegetables and vegetable products

34.2 table	Approved methods of the AACC international for analysis of extraneous materials

Number	Title
28	Extraneous matter
28-01.01	Apparatus or materials for extraneous matter methods
28-02.01	Reagents for extraneous matter methods
28-03.02	Special techniques for extraneous matter methods
28-06.01	Cinder and sand particles in farina – counting method
28-07.01	Cinder and sand particles in farina – gravimetric method
28-10.02	Macroscopic examination of external contamination in whole grains
28-19.01	External filth and internal insect infestation in whole corn
28-20.02	Microscopic examination of external contamination in whole grains
28-21.02	X-ray examination for internal insect infestation
28-22.02	Cracking-flotation test for internal insects in whole grains
28-30.02	Macroscopic examination of materials hard to hydrate
28-31.02	Pancreatin sieving method, for insect and rodent filth in materials hard to hydrate
28-32.02	Sieving method, for materials hard to hydrate
28-33.02	Pancreatin nonsieving method for insect and rodent filth in materials easy to hydrate
28-40.01	Acid hydrolysis method for insect fragments and rodent hairs – wheat-soy blend
28-41.03	Acid hydrolysis method for extracting insect fragments and rodent hairs – light filth in white flour
28-43.01	Glass plate method, for insect excreta
28-44.01	Iodine method, for insect eggs in flour
28-50.01	Decantation method, for rodent excreta
28-51.02	Flotation method, for insect and rodent filth
28-60.02	Tween-versene method, for insect fragments and rodent hairs in rye flour
28-70.01	Defatting-digestion method, for insect fragments and rodent hairs
28-75.02	Sieving method, for light filth in starch
28-80.01	Flotation method, for insect and rodent filth in popped popcorn
28-85.01	Ultraviolet light examination, for rodent urine
28-86.01	Xanthydrol test, for urea
28-87.01	Urease-bromothymol blue test paper, for urea
28-93.01	Direction of insect penetration into food packaging
28-95.01	Insect, rodent hair, and radiographic illustrations

prepared to facilitate the orientation of food analysts to the basic techniques they will need for filth analysis. A recent, more advanced resource is *Fundamentals of Microanalytical Entomology: A Practical Guide to Detecting and Identifying Filth in Foods* [12]. Most chapter authors of this resource are, or have been, FDA personnel "involved in the forensic aspect of piecing together the etiological puzzles of how insect filth gets into processed food products" [12]. The authors share their experience gained in gathering and developing evidence used to document violations of the law that the FDA is mandated to enforce.

34.4 BASIC ANALYSIS

Various methods for isolation of extraneous matter are suggested in Sect. 34.3, which defines different types of filth: separation on the basis of differences in **density, affinity for oleophilic solvents**, and **particle size**; **diagnostic characteristics** for identification of filth; and **chemical identification** of contaminants. Since all methods of analysis for extraneous matter for all categories of food cannot be discussed in this chapter, only the underlying principles of the methods are summarized below. Readers may need to refer to the specific AOAC methods cited for detailed instructions of the procedures.

The AOAC and the AACCI methods for analysis of extraneous materials involve the use of one or more of the following basic methods: filtration, sieving, wet sieving, gravimetry, sedimentation/flotation, cracking flotation, heat, acid or enzyme digestion, macroscopic and microscopic methods, and mold counts.

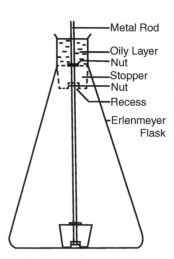

 Fisher US standard test sieve. Range of mesh sizes is available for various particle size separations

 Wildman trap flask. Stopper on shaft is lifted up to neck of flask to trap off floating layer (Adapted from [5], AOAC Method 945.75, Extraneous Materials in Products)

34.4.1 Sieving Method

Separation is based on difference in particle size of product and contaminant using **standard test sieves** (Fig. 34.1). For instance, the insects are (larger) separated from spices (smaller) using a 20-mesh sieve, and wheat grains (larger) are separated from insects (smaller) using a 10- or 12-mesh sieve. Then the contaminant is identified using a **widefield stereomicroscope**.

34.4.2 Sedimentation Method

Separation is based on different densities of product, contaminant, and immersion fluid. Specific gravity of immersion solution (carbon tetrachloride/chloroform) allows heavier shell, sand, glass, metal, or excreta contaminants to settle; less dense product floats. Apparent lower **specific gravity** of internally infested wheat kernels, for instance, allows them to float, while sound wheat kernels settle in 1.27 specific gravity solution. Contaminants are then identified using a microscope.

Analysis of high-fat-containing samples such as nuts requires defatting using petroleum ether prior to filth analysis (AOAC Method 968.33A). The chloroform and chloroform/carbon tetrachloride solvents allow pieces of shell, sand, and soil to settle at the bottom of the beaker on the basis of specific gravity and cause the defatted nut meats to float and be decanted. Essentially, the same procedure is suggested to isolate pieces of rodent excreta from corn grits, rye and wheat meal, whole wheat flour, farina, and semolina in AOAC Method 941.16A. It should be noted that the use of the more toxic solvents such as carbon tetrachloride, chloroform, and petroleum ether is avoided in most contemporary analytical methods.

34.4.3 Flotation Methods

Flotation methods are designed to isolate microscopic filth by floating the filth upward, typically in an **oil/**

water-phased system. Insect fragments, mites, and hairs are lipophilic and like to be in the oil phase, thus they float to the surface with the oils. Plant tissues and most related tissues are hydrophilic, and they tend to stay in the water phase. Therefore, separation is based on the principle of affinity for oleophilic solvents. Gravity further helps this process, and larger particles sink. To accomplish the separation of filth from food, a number of solution systems are used to ensure that the majority of the product sinks, while the oils with trapped filth float. The oil phase is trapped off with a **Wildman trap flask** (Fig. 34.2), filtered, collected on a filter paper, and examined microscopically to determine the amount and kinds of filth present [13].

Flotation is a common method used to determine insect fragments, rodent hairs, and other forms of light filth in wheat flour (AOAC Method 972.32). The acid digestion is used to break down the starch in the flour and allows the other flour constituents to more cleanly separate from the dilute acid solution. Although the AOAC method calls for digestion by autoclaving, AACCI Method 28-41.03 provides for an alternative hotplate digestion, which might be more convenient for some laboratories. The oleophilic property of insect fragments, rodent hairs, and feather barbules allows them to be coated by the mineral oil and trapped in the oil layer for separation and collection on ruled filter paper. The heavier sediments of the digestion are washed and drained from the funnel. Fragments and rodent hairs are reported on the basis of 50 g of flour.

34.4.3.1 Cracking-Flotation Method

Internal infesting insects (such as in grains) can be determined using an oleophilic method. First, any external insects are removed by sieving. Grain sample is coarsely cracked to free insects from kernels. Cracked grain sample is digested in 3–5 % HCl solution and sieved with water to remove hydrolyzed starch and acid. The sample is transferred to a Wildman trap flask and boiled in 40 % alcohol solution to deaerate. Tween 80 (polyoxyethylenesorbitan monooleate) and Na₄EDTA (tetrasodium salt of ethylenediaminetetraacetic acid) solutions are added to cause light bran particles to remain in solution during oil extraction of insect material. Light mineral oil is added to the solution to form a floating layer in which insect material is attracted due to its oleophilic nature. The oil layer is filtered through a ruled paper to collect the contaminating insects. The filter paper is examined microscopically.

34.4.3.2 Light Filth Flotation Method

Oleophilic filth is defined as light filth. Examples of light filth include insects, insect fragments, hairs, and feather barbules which can be detected in a food product by separating them from the food in the oil phase of an oil-aqueous mixture. The analysis of light filth is accomplished through a series of steps, starting with a pretreatment that removes fats, oils, soluble solids, and fine particulate matter to enhance the wettability of the food. The second step requires mixing the food with a water and oil mixture. The food will remain in the aqueous phase and the light filth will rise to the top with the oil phase. In the third step, the extract with filth elements is poured onto a ruled filter paper using a filter flask and funnel (Fig. 34.3) and examined line by line under a stereomicroscope (Fig. 34.4). After identification and enumeration, the results are reported to provide the following information: (a) whole or equivalent insects (adults, pupae, maggots, larvae, cast skins); (b) insect fragments, identified; (c) insect fragments, unidentified; (d) aphids, scale insects, mites, spiders, etc. and their fragments; and (e) rodent hairs (state the length of the hairs).

Insect fragments, rodent hairs, and other light filth can be isolated from flour samples by an acid hydrolysis method. Sediment products of the digestion are allowed to settle out in a separatory funnel and are drained away. The remaining oil layer is filtered through a ruled filter paper and the contaminants identified microscopically. However, certain grain products such as whole wheat flour contain amounts of bran particles that may result in excessive amounts of material being trapped in the oil layer, making it difficult to identify particles of filth. The "Tween-Versene Method for Insect Fragments and Rodent Hairs in Rye Flour" (AACCI Method 28-60.02) utilizes two chemical agents that tend to suppress bran accumulation in the heptane recovery layer. Tween 80 (polyoxyethylene sorbitan monooleate) is a nonionic agent that appears to have certain surface-active properties that make it a useful adjunct to Na₄EDTA (tetrasodium salt of ethylenediaminetetraacetic acid). In the presence of Tween 80, Na₄EDTA appears to be a depressor for food materials

34.3 figure

Filter flask and funnel. Funnel has a collar (partially raised) that holds ruled filter paper in place on the funnel base for trapping filth for examination. Suction is applied with a water aspirator. (www.whatman.com)

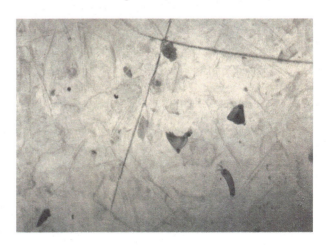

34.4 figure

Microscopic view of insect fragments and rodent hairs on a filter paper

(such as bran and other light plant matter), which otherwise tend to float. It has been suggested that the chelating properties of Na₄EDTA may result in its adsorption onto the surfaces of food particles along with the surfactant Tween 80, thereby preventing an attraction of food particles to oils used to isolate light filth. By preventing plant material from being collected in the heptane layer that is trapped off, contaminants such as oleophilic insect parts (exoskeleton) that are contained in the separating oil are much easier to distinguish and identify. AACCI Method 28-95.01 provides a description of insect fragments and rodent hair characteristics with illustrations [6].

34.4.4 Objectivity/Subjectivity of Methods

Insect parts, rodent hairs, and feather barbules in food products are generally reported as the total number of filth elements counted of each kind encountered per sample unit. They are identified on the basis of objective criteria. However, identifying insect fragments is not a simple task. Training and supervised practice are required to achieve competence and consistency. Some fragments are easily identified on the basis of structural shape and form. Mandibles, for example, are quite distinctive in their shape and configuration; certain species of insects can be determined on the basis of this one structure. In other instances, fragments may be mere chips of insect cuticle that have neither distinctive shape nor form but can be identified as being of insect origin if they have one or more of the characteristics given in Sect. 34.2.2. Experienced analysts should rarely misinterpret fragments.

Isolation of extraneous material from a food product so that it can be identified and enumerated can be a very simple procedure or one that requires a series of several rather involved steps. In the process of isolating fragments from flour by the acid hydrolysis method, for instance, the sample is transferred from the digestion container to the separatory container and then to the filter paper for identification and enumeration. At each of those transfers, there is an opportunity for loss of fragments. Although the analyst may have made every effort to maintain the isolation "quantitative," there are opportunities for error. Both fragment loss and analyst variation are minimized by common use of standard methods and procedures and by proper training and supervised practice.

Another concern involves the significance of **insect fragment counts** (as well as particles of sand, pieces of rodent excreta, rodent hairs, etc.) in relation to **fragment or particle size**. Fragment counts are reported on a numerical basis; they do not reflect the total contaminant biomass that is present. A small fragment is counted the same as a large fragment. The size of the fragment may be a reflection of the process to which a common raw material (e.g., wheat) has been subjected; a more vigorous process produces more and smaller fragments than a less vigorous process. The state of insects may also be a factor. Dead (dried) forms produce greater numbers of fragments than live forms. These factors have been of concern to food processors for some time and have prompted the search for more objective means of determining insect contamination.

34.5   OTHER TECHNIQUES

34.5.1   Overview

Methods described in Sect. 34.3 are directed primarily at routine quality control efforts to determine if the level of natural or unavoidable defects is below the defect action level. To a certain extent, those routine methods can be used to identify the source of contaminants in processed foods. For example, the identification of certain insect fragments can indicate infestation in the raw commodity rather than in the processing system. However, other more sophisticated techniques offer opportunities to pinpoint the nature and source of other contaminants that may exist unavoidably or due to mistakes, accidents, material or equipment failures, or intentional adulteration.

The detection of insects in stored grain and the quantitation of insect parts present in grain products represent a serious and continuing problem for the grain industry. Approved methods of detection primarily involve visual and microscopic inspection and X-ray analysis which require trained personnel and are time consuming, difficult to standardize, and expensive. The assays for insect contamination are preferred to be highly specific, sensitive, rapid, and inexpensive. Moreover, it ideally should be employable by persons having minimal training, particularly in nonlaboratory settings such as at grain elevator and processing sites.

There are several attempts to develop rapid and efficient methods including the use of nuclear magnetic resonance, sound amplification, and infrared spectrometry as alternatives to presently used chemical techniques mentioned in preceding sections. Most of these techniques are expensive and challenging due to difficulty in quantification and identification of specific infestations. Immunological assays, which have found widespread use in clinical diagnostic settings and also in home use, have been explored to detect insect contamination. These methods are described below as they relate to detecting an infestation.

34.5.2   X-Ray Radiography

X-ray radiography is widely used as a test reference method [14]. Grain processors use it as a means of inspecting wheat for internal insect infestation, which is the main the source of insect fragments in processed cereal products (Fig. 34.5). The existing X-ray techniques enable the classification of at least four stages of insect development by measuring the area occupied by the insect, and an accurate classification is also possible based on visible insect morphology [15]. The use of real-time digital imaging instead of X-ray radiographs to discriminate the infested kernels significantly shortened the X-ray procedures. Conventional film observations give, however, better accuracy (3 % error rate) than the digital images (11.7 %) for infestation by third-larval instars, while the error is less than 1 % for both methods with a more advanced stage of larval development [16].

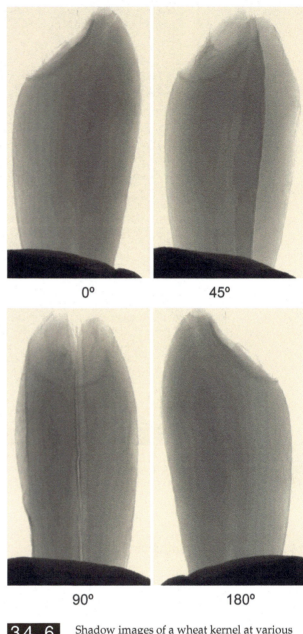

34.6
figure

Shadow images of a wheat kernel at various rotation angles

34.5
figure

X-ray radiograph of infested wheat.
(**a**) Lesser grain borer pupae. (**b**) Rice weevil pupae (Courtesy of Moses Khamis)

34.5.3 X-Ray Microtomography

X-ray microtomography (XMT) is an emerging 3-D imaging technique that operates on the same basic principles as medical computed tomography (CT) scanners but has much higher resolution. It is very effective in characterizing various internal structural features, which are not possible with conventional 2-D imaging methods (Chap. 32, Sect. 32.5.2). Conventional imaging techniques such as light microscopy, scanning electron microscopy (SEM), and digital video imaging have some limitations: they are destructive in nature, as sample preparation involves cutting to expose the cross section to be viewed. High-resolution XMT has a wide range of applications in science and engineering for which accurate 3-D imaging of internal structure of objects is crucial. XMT is able to capture several features of the internal structures of grain kernels, which are not possible with the conventional imaging methods [17]. Figure 34.6 is demonstration of shadow images taken at several step angles during scanning. A typical scan creates 200–400 of those images that are then used to create axial, sagittal, and coronal views shown in Fig. 34.7.

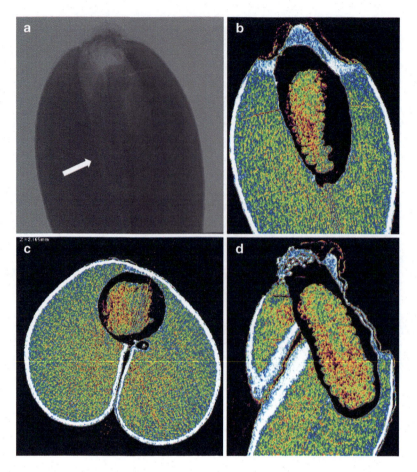

34.7 figure X-ray microtomography (XMT) images. (**a**) Shadow image, (**b**) sagittal view, (**c**) axial view, and (**d**) coronal view of lesser grain borer in a wheat kernel

34.5.4 Electrical Conductance Method

The **electrical conductance** method is based on monitoring the conductance signals for each single kernel during milling in a single-kernel characterization system (SKCS) which is commonly used for wheat hardness determination [15]. This method is highly accurate for detecting older developmental stages of insects: the percentage of properly classified cases for small, medium, and large larvae and pupae are 24.5, 62.2, 87.5, and 88.6, respectively. The accuracy of this method depends also on insect species (rice weevil and lesser grain borer) and wheat type (soft or hard red winter wheat).

34.5.5 Impact-Acoustic Emission

Acoustic detection is a very promising method for early detection of insects in bulk grain. Acoustic identification of insects is based on their ability to generate sound by eating, flying, egg laying, or locomotion [18]. Despite the fact that it has been a known pest detection method in storage facilities for a long time, reliability

and efficacy of acoustic pest detection have been improved only recently as a result of the development of improved acoustic devices and signal processing methods [18].

Impact-acoustic emissions are used as a nondestructive, real-time method for detection of damaged grains and shelled nuts [19]. Kernels are impacted onto a steel plate and the resulting acoustic signal is analyzed to detect damage using different methods: modeling of the signal in the time domain, computing time-domain signal variances and maximums in short-time windows, analysis of the frequency spectrum magnitudes, and analysis of a derivative spectrum. Features were used as inputs to a stepwise discriminant analysis routine, which selected a small subset of features for accurate classification using a neural network. Pearson et al. [19] reported that impact-acoustic emission is a feasible and promising method for detection of IDK, sprout damage, and scab damage. More study is needed to improve accuracy on kernels infested with insects that have not yet emerged from the kernels. The computational cost of classifying a kernel using this technique is very low, allowing inspection of large numbers of wheat kernels very rapidly, ~40 kernels/s. Grain inspectors usually use a 100 g (3,000 kernel) sample to inspect for IDK. This takes an inspector approximately 20 min to analyze manually but can be accomplished in about 75 s with an acoustic system.

According to official standards for wheat grain, a sample as a whole is considered infested when it contains two or more live weevils, or one live weevil and one or more other live insects injurious to stored grain, or two or more other live insects injurious to stored grain [20]. Eliopoulos et al. [18] evaluated the efficacy of bioacoustics in detecting the presence of adult beetles inside the grain mass using a piezoelectric sensor and an acoustic emission amplifier. The accuracy of this system has been reported to be 72–100% in detecting one to two insects per kilogram of hard wheat, which is the standard threshold for classifying a grain mass "clean" or "infested." Also, the detection threshold of this system is very low, 0.1–0.5 insects per kilogram of grain, which is not possible with many other detection methods [18].

34.5.6 Microscopy Techniques

Microscopy techniques including light microscopy, fluorescence microscopy, and scanning electron microscopy (SEM) are used to study the structure/function relationships of food but also can be applied to questions of extraneous matter. For example, SEM with energy dispersive spectroscopy (EDS) can be used to determine the nature of metals in products that may be due to equipment failure or intentional adulteration due to tampering [21]. Light microscopy in a polarized mode can be used to distinguish between plastics, glass, and other fiber or crystalline contaminants [22].

34.5.7 Near-Infrared Spectroscopy

Near-infrared spectroscopy (NIRS) is a relatively fast, accurate, and economical technique available to the grain industry for compositional analysis such as water, oil, fiber, starch, and protein in grains and seeds. It also has relatively recent applications in analysis of extraneous matter. NIRS has been used to identify several coleopteran species [23], and to detect parasitized weevils in wheat kernels [24] and external and internal insect infestation in wheat [25–27]. Berardo et al. [28] reported that NIRS predicts the percentage of *Fusarium verticillioides* infection in maize kernels and the content of ergosterol and fumonisin B1 in meals. In the same way, promising results were obtained when NIRS methodology was applied to detect scab-damaged kernels [29] and estimate deoxynivalenol, ergosterol, and fumonisin in single kernels of wheat [23] and corn [30].

Near-infrared spectroscopy used with a single-kernel characterization system is able to detect later stages of internal insect infestation in wheat with a 95% confidence [27]. In contrast to other procedures, this system is capable of being automated and incorporated into the current grain inspection process. NIRS also has been compared with the current standard insect fragment flotation method for its ability to detect insect fragments in flour [31]. Fragment counts with both techniques were correlated; however, the flotation method was more sensitive below the FDA DAL of 75 insect fragments per 50 g of wheat flour. NIR spectroscopy was able to predict accurately whether flour samples contained less than or more than 130 fragment/50 g.

Hyperspectral imaging is a recent development that has proven to be effective in grain quality inspection. It is an integration of conventional imaging and spectrometry which involves acquiring images at more than 20 wavelengths within the electromagnetic spectrum [32, 33]. Since it is a complex indirect method, hyperspectral data should be confirmed for homogeneity before statistical modeling through various preprocessing techniques, as described in detail by Ravikanth et al. [32]. In the area of cereal grain quality, hyperspectral imaging has been successfully used to identify fungus-damaged kernels [33–35], insect-damaged kernels [36, 37], defective wheat [29, 38], and wheat classes [39, 40]; to detect contaminants from different origins (animals, other cereal grains, botanical impurities, other contaminants) in cereal grains; [41], and to quantify ergot bodies in cereals [42].

34.5.8 Enzyme-Linked Immunosorbent Assays

To develop an optimal immunological assay for an insect contamination of foodstuffs, antibodies are required that are directed against an insect-specific antigen, preferably protein, likely to be present in any life stage of the contaminating insect or in insect remains. Antigens and antibodies are two key parts of any immunoassay (see also Chap. 27).

For an immunoassay with broad specificity, it is required to use an insect-specific protein such as myosin. Myosin is ubiquitous in insects; it is present in large quantities in adult insect tissue and is also present in appreciable quantities in other life stages [43]. An **enzyme-linked immunosorbent assay** (ELISA) method has been developed to measure quantitatively the amount of insect material in a sample [43]. It is also possible to develop an immunoassay specific for a particular species of insect contamination using antibodies having a unique species specificity. Kitto et al. [44] developed such techniques (patented in 1992) for detecting the amount of insect contamination in foodstuffs. The method comprises the following steps:

1. Preparing an aqueous solution or suspension of a homogenized grain sample
2. Substantially affixing at least a portion of solution or suspension to a solid surface
3. Applying to solid surface a specifically binding insect antigen (or antibody) and enzyme to form an antibody-enzyme conjugate, resulting in formation of a colored product when the enzyme reacts with a substrate
4. Washing unbound conjugate from the solid surface
5. Incubating the solid surface with an enzyme substrate under conditions allowing colored product to be formed when enzyme is present
6. Correlating amounts of color formed with an amount of insect contamination

Recent research [45] showed that the myosin in fourth instars of the lesser grain borer developing within kernels of wheat degraded within the first 2 weeks when larvae were killed with phosphine, a fumigant commonly used to manage insect infestations in stored grain. Myosin degradation resulted in underestimating insect fragment estimates by about 58 %.

34.6   COMPARISON OF METHODS

A number of methods that have been developed to detect insects in commodity samples (Table 34.3) are described here in general terms:

1. Density separation based on infested kernels being lighter in weight and floating in a liquid
2. Staining kernels to detect weevil egg plugs
3. Detection of carbon dioxide or uric acid produced by the internally feeding insects
4. Detection of insects hidden inside kernels using near-infrared spectroscopy (NIRS)
5. Detection by use of nuclear magnetic resonance (NMR)
6. Detection by X-ray images and digital image analysis techniques
7. Acoustical sensors to hear sounds from insects feeding inside kernels
8. Enzyme-linked immunosorbent assays (ELISA) to detect myosin in insect muscle

Some of the recent methods have been developed by adapting the single-kernel characterization system (SKCS), computed tomography (CT), acoustic-impact emissions, and use of an electrical conductive roller mill [15, 46, 47].

The choice of method depends on several factors: (1) type of infestation (inside or outside food grains, in the surrounding premises or inside bulk grain), (2) required level of inspection (macroscopic vs. microscopic, qualitative vs. quantitative), (3) availability of equipment and facilities, and (4) required sensitivity [48]. Most of the methods aim to detect the presence of live insects directly or indirectly. External insects are detected by visual inspection, sampling, sieving, and heat-extraction methods, while internal (hidden) insects are detected by radiography, staining techniques to identify egg plugs, and near-infrared and fragment count methods. Determination of uric acid or CO_2 level serves as an indirect way of detecting and estimating internally feeding insects, and these methods may be suitable if infestations are restricted to one insect species. Depending on some storage conditions, grain may contain molds and insects, and in such cases CO_2 produced by molds may interfere in accurately detecting or estimating insects. Fragment count and ELISA methods can be used for the detection of both living and dead insects. In general, problems encountered with these detection methods are that the most accurate methods, such as X-ray and computed tomography (CT), are laborious and expensive, while rapid, automated methods may not be suitable for detecting eggs and young larvae [45].

34.7   ISOLATION PRINCIPLES APPLIED TO FOOD PROCESSING

Examination of stored-product insects often requires extracting them from the commodity. An intensive summary of literature survey on insect extraction and detection methods can be found in reference [49]. Isolation principles, such as particle size and density, discussed in the preceding sections are designed to identify extraneous materials in finished food products, monitoring quality, and compliance with DALs. In addition, some of these principles of isolation are used in a **proactive** way during processing to prevent extraneous matter from being incorporated into finished food products.

Wheat that contains hidden internal insect infestation is the primary source of insect fragments in processed cereal products. The current DAL for internal insect infestation in wheat is 32 **IDK** per 100 g of wheat [3]. IDK are those **visually** determined to have insect tunneling or emergence holes. Most processors rely on much lower levels of IDK ($\leq$6 IDK/100 g) to produce flour that meets customer tolerances and the FDA's DAL for insect fragments in flour. In addition, to prevent adulteration of flour with filth, entoleters and infestation destroyers in the milling process break up insect-damaged kernels, and these broken kernels along with the insect fragments are aspirated out of the milling stream. As previously indicated,

34.3
table Insect detection methods applicable for commodity samples

Test method	Applicability	Comments
Visual inspection	Whole grains, milled products	Qualitative. Only high-level infestation detected
Sampling and sieving	Whole grains, milled products	Commonly practiced. Hidden infestation not detected
Heat extraction	Whole grains	Adults and larvae detected
Acoustics	Whole grains	Feeding sounds: active stages detected Impact-acoustic emissions: Nondestructive. Real time. Detect insect, sprout, and scab damage
Breeding out	Whole grains	Time consuming
Imaging techniques		
X-ray method	Whole grains	Nondestructive. Highly accurate. Able to detect both live and dead insects inside grain kernels. Cannot detect insect eggs. Prohibitive capital cost
Near-infrared spectroscopy	Whole grains, milled products	Rapid. Sensitive. Can be automated. No sample preparation. Cannot detect low levels of infestation. Sensitive to moisture content. Calibration of equipment is complex and frequent
Nuclear magnetic resonance	Whole grains	Less sensitive
Serological techniques	Whole grains, milled products	Highly sensitive, species specific. Shows infestation from unknown past to till date
Uric acid determination	Whole grains, milled products	Shows infestation from unknown past to till date
CO_2 analysis	Whole grains	Simple, time consuming. Indicates current level of infestation. Not suitable for grains having >15 % moisture
Specific gravity methods	Whole grains	Simple and quick. Not suitable for oats and maize
Cracking and flotation method	Whole grains	Variable results noted
Fragment count	Whole grains, milled products	Highly variable results noted. Shows infestation from unknown past to till date
Staining techniques		
Egg plugs	Whole grains	Specific for *Sitophilus* spp.
Ninhydrin method	Whole grains	Eggs and early larvae not indicated

Adapted from Rajendran [48]

X-ray radiography is used by some as a means for selecting grains for processing or for research purposes to age-grade internally developing stages of stored-grain insects. More recently, NIR spectroscopy has provided a new tool for assessing internal insect infestation in wheat. By selectively milling only wheat that has minimal or no evidence of internal insect infestation, grain processors can effectively limit insect fragments in their products. In like manner, bakers and other users of processed grain products can selectively monitor for insect fragments in their raw materials using one of the approved methods for extraction and enumeration of fragments or by sending samples to a private laboratory for fragment analysis.

Most food processing systems that deal with agricultural products generally apply some type of cleaning operations as an initial step. In flour milling, for example, wheat is passed through a system called the "cleaning house" which consists of a series of machines that apply the principles of particle size and density separation. Sieves remove contaminants larger than wheat kernels as well as finer contaminants such as sand. In addition, air (aspiration) is used to remove plant material that is lighter than the grain. Current equipment to remove stones and other dense materials the same size as grain kernels uses air passed upward through an inclined, tilted table. This causes the grain to "float" off the side of the table and the heavier material to continue and "tail" over the end of the table. In earlier systems, grains were passed through washers in which water separated the grain from heavier material (such as stones) much like fluming of potatoes or fruit does. Impact with rotating disks and steel pegs (entoleters and infestation destroyers) or grinding operations are used prior to milling to break open kernels of wheat containing internal insect infestation. As a means of reducing insect fragments in the finished product, this process is followed by aspiration to lift out any light insect contaminants released in the operation.

As a final step in wheat milling, flour is generally passed through sieves fine enough to remove insect eggs and any other contaminants that might be present. This is to assure that when flour leaves the mill, it is free of any viable form of insect contamination [50]. Where flour is used in large quantities, such as commercial bakeries, prior to use, flour is again sieved to ensure that no contamination has occurred in transport and storage of the flour.

Metal contamination has been a major concern of all food processors. Although metal detection methods are not specifically among the isolation techniques represented in AOAC International or AACCI methods manuals, they serve the purpose of isolating contaminants from food products. Magnets of various types have been used on raw materials and processing systems to prevent the passage of metal into handling and processing equipment where both equipment damage and product contamination are concerns. **Metal detectors** are employed in many food processing operations and on finished product packaging lines to detect ferrous and nonferrous metal fragments and to prevent contaminated products from entering consumer food channels.

Recent X-ray technology suggests that X-rays may have an advantage over other methods for detecting metal and that they can also be used to detect glass, wood, plastic, and bone chips in foods. Detection of these extraneous materials also can be automated with rejection systems in packaging lines [51].

34.8 SUMMARY

Extraneous matter in raw ingredients and in processed foods might be unavoidable in the array of foods that are stored, handled, processed, and transported. DALs are established for amounts considered unavoidable and of no health hazard. A variety of methods are available to isolate extraneous matter from foods. Those methods largely prescribed by AOAC International employ a series of physical and chemical means to separate the extraneous material for identification and enumeration. Major concerns in the analysis of food products for extraneous matter are the objectivity of methods and the availability of ade-

quately trained analysts. Some "principles" of isolation are applied in a proactive way in food processing operations.

Currently available methods (both macroscopic and microscopic) show varying degrees of efficiency in analysis of extraneous matter and filth in foods. Some techniques are time consuming, require trained personnel, and are difficult to implement in real time. Some techniques have not been found feasible to be implemented in food inspection systems because of their cost, unreliability, and the varying degrees of success obtained in detecting infestations. Macroscopic and microscopic procedures for characterizing defects in foods tend to supplement each other and together provide a comprehensive evaluation of defects in the product. It is important that the analyst realizes the close association of complementary methods for use as a joint approach in solving analytical problems.

Examining products for extraneous matter is quite different than performing product evaluations for chemical and microbiological contaminants. Typically, the extraneous matter becomes more difficult to detect following its introduction to the process; thus, implementation of comprehensive preventative measures are the ultimate cost-effective safety measures.

34.9 STUDY QUESTIONS

1. Indicate why the FDA has established DALs.
2. Explain why practicing cGMPs has no impact on DALs.
3. List three major reasons for conducting analysis for extraneous matter in foods.
4. What two resources provide methods for separating extraneous matter from cereal grains and their products?
5. There are several basic principles involved in separating (isolating) extraneous matter from foods. List five of these principles and give an example of each principle.
6. Briefly describe the major constraint(s) to currently accepted methods for analyses of extraneous matter in foods.
7. Explain how some of the more recent analytical techniques can assist in identifying sources of extraneous matter in foods.
8. What are some likely sources of error with the various analytical methods?

Acknowledgments The authors of this chapter wishes to acknowledge Dr. John R. Pedersen, who was an author of this chapter for the first to fourth editions of this textbook.

This contribution is paper number 17-101-B of the Kansas Agricultural Experiment Station, Kansas State University, Manhattan, KS 66506.

REFERENCES

1. FDLI (1993) Federal food drug and cosmetic act, as amended. In: Compilation of food and drug laws. The Food and Drug Law Institute, Washington, DC

2. FDA (2009) Current good manufacturing practice in manufacturing, packing, or holding human food. Part 110, Title 21: food and drugs. In: Code of federal regulations. Office of the Federal Register National Archives and Records Administration, Washington, DC

3. FDA (1995) The food defect action levels – current levels for natural or unavoidable defects for human use that present no health hazard (revised 1998). Department of Health and Human Services, Food and Drug Administration. Washington, DC

4. FDA (2000) Compliance policy guide manual. Food and Drug Administration, Office of Regulatory Affairs, Washington, DC

5. AOAC International (2016) Extraneous Materials: isolation. In: Official methods of analysis, 20th edn., 2016 (online). AOAC International, Rockville, MD

6. AACC International (2010) AACCI method 28 extraneous matter. In: Approved methods of the American association of cereal chemists, 11th edn. AACC International, St. Paul, MN

7. Kurtz OL, Harris KL (1962) Micro-Analytical entomology for food sanitation control. Association of Official Analytical Chemists, Washington, DC

8. Gentry JW, Harris KL (1991) Microanalytical entomology for food sanitation control, vols 1 and 2. Association of Official Analytical Chemists, Melbourne, FL

9. AACC International (2010) X-ray examination for internal insect infestation, AACCI method 28–21.02. In: Approved methods of the American association of cereal chemists, 11th edn. AACC International, St. Paul, MN

10. FDA (1981) Principles of food analysis for filth, decomposition, and foreign matter. FDA technical bulletin no. 1, Gorham JR (ed), Association of Official Analytical Chemists, Arlington, VA

11. FDA (1978) Training manual for analytical entomology in the food industry. FDA technical bulletin no. 2, Gorham JR (ed), Association of Official Analytical Chemists, Arlington, VA

12. Olsen AR (ed) (1995) Fundamentals of microanalytical entomology – a practical guide to detecting and identifying filth in foods. CRC, Boca Raton, FL

13. FDA (1998) Introduction and apparatus for macroanalytical methods. In: FDA technical bulletin number 5, macroanalytical procedures manual (MPM), FDA, Washington, DC

14. Pedersen JR (1992) Insects: identification, damage, and detection, ch. 12. In: Sauer DB (ed) Storage of cereal grains and their products. American Association of Cereal Chemists, St. Paul, MN, pp 635–689

15. Pearson TC, Brabec DL, Schwartz CR (2003) Automated detection of internal insect infestations in whole wheat kernels using a Perten SKCS 4100. Appl Eng Agric 19: 727–733

16. Haff RP, Slaughter DC (2004) Real-time X-ray inspection of wheat for infestation by the granary weevil, *Sitophilus granarius* (L.). Trans ASAE 47: 531–537

17. Dogan H (2007) Non-destructive Imaging of agricultural products using X-ray microtomography. Proc Microsc Microanal Conf, 13(2): 512–513

18. Eliopoulos PA, Potamitis I, Kontodimas DC, Givropoulou EG (2015) Detection of adult beetles inside the stored wheat mass based on their acoustic emissions, J Econ Entomol 108(6): 1–7

19. Pearson TC, Cetin AE, Tewfik AH, Haff RP (2007) Feasibility of impact-acoustic emissions for detection of damaged wheat kernels. Digital Signal Process 17: 617–633

20. FGIS (2015) Official United States standards for grain, 7 CFR Part 810. Federal Grain Inspection Service. USDA, Washington, DC.

21. Goldstein JI, Newbury DE, Echlin P, Joy DC, Romig AD Jr, Lyman CE, Fiori C, Lifshin E (1992) Scanning electron microscopy and X-ray microanalysis. A text for biologists, materials scientists, and geologists, 2nd edn. Plenum, New York

22. McCrone WC, Delly JG (1973) The particle atlas, 2nd edn. Ann Arbor Science, Ann Arbor, MI

23. Dowell FE, Ram MS, Seitz LM (1999) Predicting scab, vomitoxin, and ergosterol in single wheat kernels using near-infrared spectroscopy. Cereal Chem 76(4): 573–576

24. Baker JE, Dowell FE, Throne JE (1999) Detection of parasitized rice weevils in wheat kernels with near-infrared spectroscopy. Biol Control 16: 88–90

25. Ridgway C, Chambers J (1996) Detection of external and internal insect infestation in wheat by near-infrared reflectance spectroscopy. J Sci Food and Agric 71: 251–264

26. Ghaedian AR, Wehling RL (1997) Discrimination of sound and granary-weevil-larva-infested wheat kernels by near-infrared diffuse reflectance spectroscopy. J AOAC Int 80: 997–1005

27. Dowell FD, Throne JE, Baker JE (1998) Automated nondestructive detection of internal insect infestation of wheat kernels by using near-infrared reflectance spectroscopy. J Econ Entomol 91: 899–904

28. Berardo N, Pisacane V, Battilani P, Scandolara A, Pietro A, Marocco A (2005) Rapid detection of kernel rots and mycotoxins in maize by near-infrared reflectance spectroscopy. J Agric Food Chem 53: 8128–8134

29. Delwiche SR, Hareland GA (2004) Detection of scabdamaged hard red spring wheat kernels by near-infrared reflectance. Cereal Chem 81(5): 643–649

30. Dowell FE, Pearson TC, Maghirang EB, Xie F, Wicklow DT (2002) Reflectance and transmittance spectroscopy applied to detecting fumonisin in single corn kernels infected with *Fusarium verticillioides*. Cereal Chem 79(2): 222–226

31. Perez-Mendoza P, Throne JE, Dowell FE, Baker JE (2003) Detection of insect fragments in wheat flour by near-infrared spectroscopy. J Stored Prod Res 39: 305–312

32. Ravikanth L, Singh CB, Jayas, DS, White NDG (2015) Classification of contaminants from wheat using near-infrared hyperspectral imaging, Biosystems Engineering 135: 73–86

33. Bauriegel E, Giebel A, Geyer M, Schmidt U, Herppich WB (2011) Early detection of Fusarium infection in wheat using hyperspectral imaging. Computers and Electronics in Agriculture 75(2): 304–312

34. Choudhary R, Mahesh S, Paliwal J, Jayas DS (2009) Identification of wheat classes using wavelet features from near infrared hyperspectral images of bulk samples. Biosystems Engineering 102(2): 115–127

35. Singh CB, Jayas DS, Paliwal J, White NDG (2007) Fungal detection in wheat using near-infrared hyperspectral imaging. Trans ASABE 50(6): 2171–2176

36. Kaliramesh S, Chelladurai V, Jayas DS, Alagusundaram K, White N, Fields P (2013) Detection of infestation by *Callosobruchus maculatus* in mungbean using near-

infrared hyperspectral imaging. Journal of Stored Product Research 52: 107–111

37. Singh CB, Jayas, DS, Paliwal, J, White, NDG (2009) Detection of insect-damaged wheat kernels using near-infrared hyperspectral imaging. Journal of Stored Product Research 45(3): 151–158

38. Singh CB, Jayas DS, Paliwal J, White NDG (2010) Detection of midge-damaged wheat kernels using short-wave near-infrared hyperspectral and digital color imaging. Biosystems Engineering 105(3): 380–387

39. Mahesh S, Manickavasagan A, Jayas DS, Paliwal J, White NDG (2008) Feasibility of near-infrared hyperspectral imaging to differentiate Canadian wheat classes. Biosystems Engineering 101(1): 50–57

40. Mahesh S, Jayas DS, Paliwal J, White NDG (2011) Identification of wheat classes at different moisture levels using near-infrared hyperspectral images of bulk samples. Sensing and Instrumentation for Food Quality and Safety 5(1): 1–9

41. Pierna JAF, Vermeulen P, Amand O, Tossens A, Dardenne P, Baeten V (2012) NIR hyperspectral imaging spectroscopy and chemometrics for the detection of undesirable substances in food and feed. Chemometrics and Intelligent Laboratory Systems 117: 233–239

42. Vermeulen P, Pierna JAF, Van Egmond HP, Dardenne P, Baeten V (2011) Online detection and quantification of ergot bodies in cereals using infrared hyperspectral imaging. Food Additives and Contaminants. Part A, Chemistry, Analysis, Control, Exposure and Risk Assessment 29(2): 232–240

43. Quinn FA, Burkholder WE, Kitto GB (1992) Immunological technique for measuring insect contamination of grain. J Econ Entomol 85: 1463–1470

44. Kitto GB, Quinn FA, Burkholder W (1992) Techniques for detecting insect contamination of foodstuffs, US Patent 5118610

45. Atui MB, Flin PW, Lazzari SMN, Lazzari FA (2007) Detection of *Rhyzopertha dominica* larvae in stored wheat using ELISA: the impact of myosin degradation following fumigation. J Stored Prod Res, 43: 156–159

46. Toews MD, Pearson TC, Campbell JF (2006) Imaging and automated detection of *Sitophilus oryzae* (Coleoptera: Curculionidae) pupae in hard red winter wheat. J Econ Entomol 99(2): 583–592

47. Pearson TC, Brabec DL (2007) Detection of wheat kernels with hidden insect infestations with an electrically conductive roller mill. Appl Eng Agric 23(5): 639–645

48. Rajendran S (2005) Detection of insect infestation in stored foods. In: Taylor SL (ed) Advances in food and nutrition research, volume 49. Elsevier Academic, UK, pp 163–232

49. Hagstrum DW, Subramanyam Bh (2006) Fundamentals of stored product entomology. AACC International, St. Paul, MN

50. Mills R, Pedersen J (1990) A flour mill sanitation manual. Eagan, St. Paul, MN

51. FMC FoodTech (2001) X-ray technology. Solutions 2: 20, 21

Food Forensic Investigation

William R. Aimutis (✉)
Intellectual Asset Management,
Cargill, Inc.,
15407 McGinty Road, Wayzata, MN 55391, USA
e-mail: bill_aimutis@cargill.com

Michael A. Mortenson
Global Food Research, Cargill Research and Development Center,
Cargill, Inc.,
14800 28th Ave. N, Plymouth, MN 55447, USA
e-mail: michael_mortenson@cargill.com

S. Nielsen (ed.), *Food Analysis*, Food Science Text Series,
DOI 10.1007/978-3-319-45776-5_35, © Springer International Publishing 2017

35.1 INTRODUCTION

Modern food manufacturing has evolved to consistently offer food products to consumers that are safe and free from microbial contamination and foreign materials, as well as having a pleasing odor and flavor. However, since most commercial food products are produced on machinery operating at a very high speed, there are incidences of something physically wearing or breaking and occasionally placing unwanted foreign material into a food product. Additionally, extraneous environmental odors may contaminate food raw materials or processed foods and be carried over into the finished product, causing an off-odor and/or taint in the product. There are also occasions when someone has deliberately and fraudulently added something to mixing vessels or to a package to discredit a company or potentially cause physical harm to the consumer. This type of maleficence has been termed **"food tampering."** Investigating the origins of all these mishaps necessitates special handling and testing of the affected products using food forensic investigation. Once contaminants have been identified in either the company's own laboratory or by an accredited outside third-party laboratory, further investigation must be conducted to identify where, when, and how the contamination occurred.

Food forensic investigation is a logical process for the investigation of root causes for a product that is perceived to be physically objectionable to a customer or consumer because it may have an off-odor/taint, have been contaminated by foreign material during processing or at some later time, or have been intentionally tampered with by an outside party. Products under these circumstances must be analyzed from a perspective different from understanding the quality or composition of a product. Food ingredients or finished products with off-odors/taint or containing foreign material contamination could be the subject of litigation as customers sue a supplier for damages as well as facing large fines from the government for not complying with rules and regulations. The food industry has not typically had to comply with such stringent rules for foreign material contamination in the past, and therefore most suppliers are ill-prepared to meet the legislated rules concerning this matter.

Many companies have traditionally relied on bulk testing of their products to meet established quality standards that were often guaranteed by a certificate of analysis (see Chap. 1, Sect. 1.2.4.3). However, bulk testing will miss the what, when, where, and how of foreign material contamination. Therefore, when a complaint is registered that a food is contaminated by foreign material or has an off-odor/taint, a food company must react with appropriate procedures.

Testing to identify the what, when, where, and how can be either handled by the implicated company or by an accredited third party. When a third party is used, it is still imperative for the implicated company to have standardized procedures for how the products will be received, documented, and delivered to the outside party for analysis. This chapter provides an introduction to the craft of food forensic investigation.

35.2 TYPICAL/ATYPICAL ISSUES REQUIRING FORENSIC ANALYSIS

Foreign material or extraneous matter contamination in foods accounts for a large majority of the customer complaints received by food manufacturers, retail outlets, and regulatory authorities [1]. **Foreign material** is defined as anything perceived by the consumer as alien to the food. Consumer perception is important because not all foreign material will be deleterious to those consuming the material and may have originated from the food itself, such as bone fragments or crystals of salt, sugar, or mineral. These are often mistaken by the consumer for glass pieces and will trigger a negative response from consumers, resulting in customer complaints. The majority of foreign material issues can be broadly classified as intrinsic or extrinsic materials to a particular food. **Intrinsic foreign materials** are derived from the raw materials, usually plant or animal origin, used as starting material for an ingredient or processed food product that may have unknowingly contaminated the finished food product. For example, plant stems and leaves, pits and seeds, animal hair/fur, corn cobs, and bone chips are commonly identified from products that were investigated after a consumer complaint and are not foreign to the food being analyzed. These are often referred to as extraneous matter as outlined in Chap. 34. Most intrinsic foreign material complaints will be in solid foods and have entered the food product either at a primary raw material production site (fruit or vegetable processor, slaughter house, or food ingredient manufacturer) or during a consumer's preparation of the food [1]. The types of foods most often involved with a foreign material complaint are vegetables and vegetable products, followed by cereals and cereal products.

Extrinsic foreign material is much more varied and arises in a food usually inadvertently during processing from the farm to the package or often after it has been opened. Common examples of these materials include stones, broken glass, metal fragments and shavings, nuts and bolts, dental materials, human hair, plastics, string, packaging materials, insect parts, grease globules, wood chips and sawdust, gasket materials, or filter fibers. This chapter on food forensic investigation goes beyond typical analysis of extraneous matter but is limited to the causes making a food product physically objectionable to a consumer, i.e., not covering chemical contaminants, which are the subject of Chap. 33.

Food companies may also need to conduct a forensic analysis of products that are suspected to have been tampered with by outside parties with intent to discredit the company or to cause physical harm to consumers. In these cases it is imperative to carefully handle received samples to identify the foreign material and how it may have found its way into the food. If foul play is suspected when a complaint is received, it should be a standard operating procedure (SOP) for the food company to contact local and national law enforcement agencies for assistance in analyzing the materials. Meticulous sample handling and documentation of the sample at all times is imperative due to the possibility or likelihood that the sample becomes part of a lawsuit or criminal investigation. It is not unusual to be advised of a lawsuit months after completing a seemingly routine investigation.

Food fraud is the deliberate and intentional substitution, misrepresentation, addition, or tampering of food, food ingredients, or packaging for economic gain. Food fraud has become much more prevalent during the last decade as prices for some crops and ingredients have risen. Unscrupulous companies will qualify ingredients with food processors by sending pure ingredients for initial analysis and process efficacy but will deliver fraudulent material after they make the qualification and sale. For example, as olive oil prices have risen, some global suppliers began intentionally adding canola oil (which is much less expensive and clean flavored) to consciously be price competitive. Unfortunately, these oils were represented as 100% olive oil so the act of diluting the oil is fraud. Recent estimates indicate greater than 70% of the olive oil sold in the United States (US) is fraudulently contaminated with other oils [2]. Some other common fraudulent foods are the following: milk, honey, saffron, orange juice, coffee, apple juice, tea, fish, and black pepper [2]. Fortunately many of the same analytical methods described elsewhere in this book and in this chapter are useful to identify fraudulent products. US Pharmacopeia (USP) maintains an online database as a repository for food ingredient fraud reports and associated analytical detection methods [3]. USP also includes in the *Food Chemicals Codex* appendix, *Food Fraud Mitigation Guidance* [4].

At the writing of this chapter, there has not been an international effort to standardize food forensic methods for sample handling and testing, despite new regulations being enacted internationally that require food companies to conduct forensic analysis and account for its origin. For example, the US Food and Drug Administration has enacted the *Food Safety Modernization Act* [5], and the Canadian Food Inspection Agency has a similar law, *Safe Food for Canadians Act* [6].

The scientific bases for analytical methods used to date have been well accepted, but there is still a greater need to identify the range of scientific needs to continue this field's development. Food forensic investigation has additional specialized needs because of the demand for "proof" and "evidence" in the context of legal proceedings or law enforcement. These needs may relate more to the quality of the *process* by which material is collected and analyzed as opposed to the science and technology employed.

35.3   ESSENTIAL ELEMENTS OF FOOD FORENSIC TEAMS

35.3.1   Nature of Food Forensic Teams

Food forensic investigation is a multidisciplined approach drawing on knowledge from a number of fields including biology, metallurgy, crystallography, and forensic science. An individual responsible for forensic analysis should have investigative experience especially in identifying foreign material contamination or off-odors/flavors, broad food industry experience (including knowledge about food harvesting, supply chain logistics, agricultural harvesting, and food processing unit operations), and a strong intuition toward detective work. Although many scientific tools and techniques have emerged to conduct food analysis and forensic investigation, the most powerful tools investigators have are their observation skills and the ability to ask the right questions to put the puzzle together.

Many food companies are forming food forensic teams, much like their product recall teams, to handle these instances. As shown above, many intracompany disciplines (including a corporate legal representative) and responsibilities will need to participate in an investigation. Documentation development and training of individuals in following established processes are important to maintain a constant state of readiness to conduct a forensic investigation [7]. Fortunately these investigations are not routine, but it is still an expectation of regulatory agencies that skilled personnel are available to be responsible for identifying and finding the source of contamination. This will require personnel who have the necessary basic education, experience, and skills. Proactive companies will allow their food forensic personnel to routinely practice their art when not under the pressure of an investigation so that when the need arises, they are comfortable with the standardized procedures they will need to follow. Furthermore, continuing education for forensic personnel should be a facet of the overall food forensic program. Proficiency testing of investigators should be part of a company's quality assurance operating procedures [8].

As standard operating procedures (SOPs) are documented by a company for its food forensic team and program, these must include approaches for: (1)

planning and design, (2) documentation, (3) protocols for quality assurance and quality control, and (4) sampling and logistics (Table 35.1). Each of these topics is covered in sections that follow.

35.1
table

Important points to consider in a codified food forensic plan for sampling, collection, and analysis

Develop a mechanism for quickly formulating a "consensus" analytical plan when a new sample (or set of samples) arises

Keep and update a set of standard operating procedures and validation data for analyzing case samples

Maintain a set of documented guidelines, requirements, and procedures for sample preparation for each analytical procedure

Maintain approved procedures for handling and storing samples

Develop standardized methods for data analysis, reporting, and presentation

Maintain reliable logistic plans for sending and receiving samples

Maintain clear and secure lines of communication for data, information, and discussion

Develop a mechanism for formulating an on-the-fly validation plan for a new procedure

35.3.2 Planning and Design

Planning and design means that the company has considered and documented the exact steps to be followed when it has discovered or been alerted to a problem necessitating activation of the food forensic team. Development of a very high-level schematic of the process flow is a good starting point (see an example in Fig. 35.1). Once this is codified, the next steps are to go into more depth on every step of the process and detail the questions on what, where, when, and how. This document may contain a set of questions that need to be answered before proceeding in the standardized process. An example of the initial steps and questions to initiate an investigation is presented in Fig. 35.2.

35.3.3 Documentation

Documentation refers to a collection of documents and records [9]. A food forensic laboratory that has proactively written all of its SOPs, policies, and instructions into a codified form has met the guidelines for quality management [10]. Documents should explain in a written or graphical form what a food forensic laboratory plans to do and how it will be accomplished, as well as instructing scientists how to perform tasks. Codified procedures have been reviewed and accepted by a company's leadership team. A good forensic laboratory will have all of its documents in as complete a

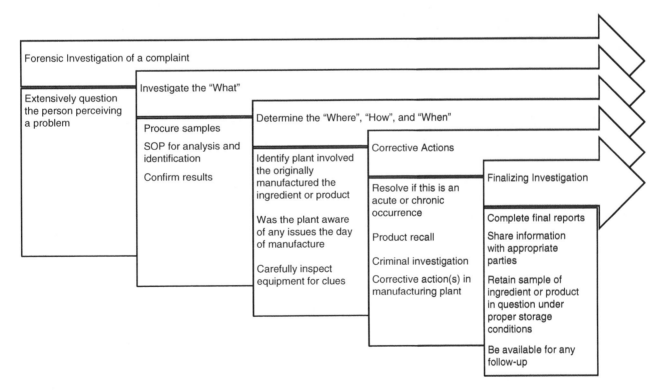

35.1
figure

Generalized process flow of a food forensic investigation

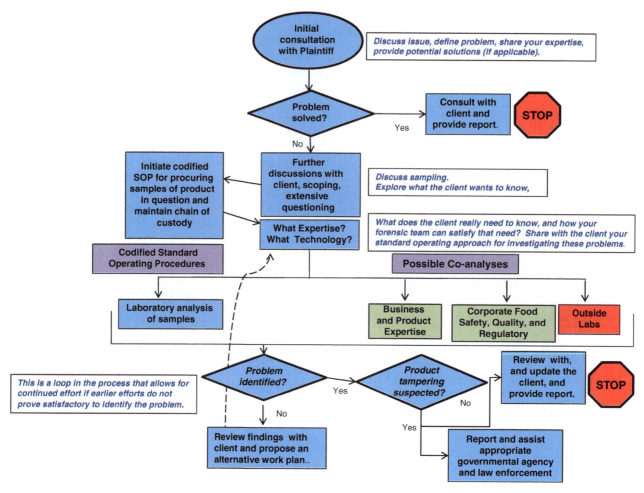

35.2 Decision tree used in a food forensic investigation

figure

form as possible before they are needed in an investigation. Once an investigation is initiated, the documents will provide guidance, explanations, and instructions about how to operate. Records are recorded information about an activity and, thus, do not exist until after the activity has been performed. It is important to maintain records (including who conducted the work) from all stages of a forensic investigation for a successful outcome when a regulatory agency audits.

35.3.4 Quality Assurance and Quality Control

SOPs for quality assurance and quality control are also important for an accused company to be able to defend itself in litigation cases. Yet this is an area that is often either taken for granted or ignored. **Quality assurance** (QA) is different than **quality control** (QC). In a forensic investigation program, QA encompasses all the activities undertaken by the laboratory to assure itself (and its customers) that reliable and accurate testing will be undertaken at all times according to the

codified standard operating procedures [7]. QC in a forensic laboratory are the activities undertaken to confirm that test results are accurate and reliable [7].

QA programs also should be developed for laboratory consumables, such as stock solutions, laboratory standards, and carrier gases, all of which can impact results being generated in a laboratory. Therefore, food forensic teams developing their protocols must think in advance as to what they may need for a number of different circumstances, and accordingly codify where to purchase these materials so they are on an approved list of suppliers that have developed a reputation for quality and are willing to certify the materials purchased. Scientists conducting a forensic investigation must be disciplined to follow protocols the same way every time, including using only preapproved consumables.

Critical components of the QA program are presented in Table 35.1. Food forensic investigators will encounter exceptions in many investigations that require modification of the SOP. Codified standard operating procedures are not meant to cover every possible situation, and it is therefore important to also have in place procedures to help make adjustments as an outlying situation develops (Table 35.1).

35.3.5 Sampling and Logistics

The first step in nearly every forensic investigation involves collecting samples and preserving them in a manner to prevent or at least minimize degradation or contamination from other substances. Sampling and preservation is as important to food forensic analysis as the scientific methods employed. Sample collection and preservation for food forensic investigations and potentially to be used as evidence are critical for efficient and successful investigation and identification. Moreover, a relatively small number of samples may be the basis for an investigation, but yet they could be highly significant in being able to identify the contaminant and its source.

Food is a biological sample that can undergo degradation if not properly handled, including how it is stored and shipped to the forensic laboratory. Any mishandling during this phase could have ramifications in ultimately identifying what the problem is and how it got in the food in the first place. Sampling and logistics refers to the process or procedures used to procure a sample for testing and how a sample is shipped to the forensic laboratory. The fewer number of people who handle a contaminated sample, the less likely a sample will be corrupted or damaged in some manner. This is referred to as the **chain of custody** in criminal forensic cases but is also applicable to food forensic investigation. A "chain of custody" is a system of unbroken control of a sample. It involves isolating a sample and packaging it securely while documenting carefully what has been done. The person carrying out this work formally takes control (custody) of the sam-

ple and is responsible for its safe keeping until it is transferred and signed over to the custody of the next person in the chain and so on.

Sampling and logistics is a fairly small part of the overall process and often is not under the manufacturer's control but can be particularly important for maintaining chain of custody. It would be somewhat unreasonable to expect a company to send its food forensic personnel to the offended customer to prepare the samples and return them to the food forensic laboratory, but in the case of a suspected tampering case that might be required or at least considered. Minimally, a form should be included with any packaging materials sent to return the product in question, and the method for shipping product back to the manufacturer should be indicated. This form should be completed by the person returning the product and include information similar to those asked in a telephone interview as outlined below but should also indicate anyone who has contacted the product in question.

35.4 ASK QUESTIONS BEFORE ANALYSIS BEGINS

The initial stages of any food forensic investigation should start by asking questions – many questions! The types of questions being asked will be different depending where the problem was discovered – at a food manufacturing site or by a consumer (Table 35.2). If the problem is perceived by a consumer, it is very important to get details quickly to record them while they are fresh in the mind. Additionally, speed is important to differentiate if this is a once-off problem or more widespread. If the problem is widespread, it may be necessary for a company to require a product

table Examples of questions to ask a person perceiving a problem product

Questions for food manufacturers	Questions for a consumer
Problem description? Details	What did you perceive as a problem?
What do we know about the problem?	How many members of your family perceived a problem?
Chronic, sporadic, acute?	Where was the product purchased?
Urgency – do you suspect this is dangerous to the public or is a tampered product	Lot code of the product?
Is a customer (consumer) involved?	How was the product prepared with details?
Description of samples to be analyzed? Get more details	Was anyone injured by the problem?
What are the lot numbers of ingredients that may be involved?	Do you still have any of the product remaining?
What is the extent or breadth of the problem?	Would you carefully follow instructions to return it to our investigation team?
Have samples been obtained? Who collected the samples, and how?	How much sample is available?
Where were the samples collected?	Who collected the samples, and how?
How much sample is available?	Where were the samples collected?
Is there a control?	
Can we get access to your manufacturing records for the period of time the product in question was manufactured?	

recall to protect the consuming public. If a consumer perceived the problem, start by communicating directly with the plaintiff by asking about what was observed or perceived. This line of questioning should be conducted by one individual, preferably the scientist leading the investigation. Many cases of foreign material contamination can be traced back to consumers themselves, and they should be questioned about how they have handled the product from the time they purchased it until they perceived a problem (usually when they are consuming the product). An investigator cannot be timid about asking very difficult questions to the plaintiff without seemingly placing blame. Details gathered during early stages of an investigation are often important in identifying where to begin looking for answers, i.e., where the contaminant came from initially and how it entered the ingredient or processed food. Finally, it is imperative that answers to all questions be honestly captured and recorded in written detail since they may be important later if a company needs to defend itself in a lawsuit or if there is criminal prosecution for product tampering.

35.5   ANALYZING "PROBLEM SAMPLES"

When the sample is received in the food forensic laboratory, it is important to document all the details. Photographs and written descriptions should be taken in every step from the preopening of the package to the contents inside. Forensic laboratories will then follow their preestablished decision trees (examples are in Figs. 35.3 and 35.4) to initiate analysis and decide on appropriate methods (Table 35.3). Many food ingredient companies and processors also retain limited quantities of materials from each production lot in the event they need to conduct follow-up analysis. These samples also should be readied for further inspection and/or analysis during a food forensic investigation to compare to suspected products.

35.5.1   Foreign Material Contamination

35.5.1.1   *Introduction*
The majority of issues to be investigated can be handled by personnel routinely conducting extraneous matter analysis (Chap. 34), but inevitably there will be investigations that must be handled by internal or

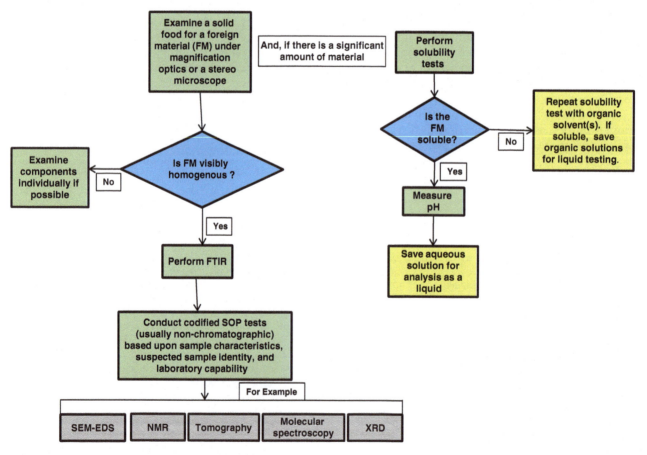

35.3 figure   Flowchart for analysis of an unknown solid sample

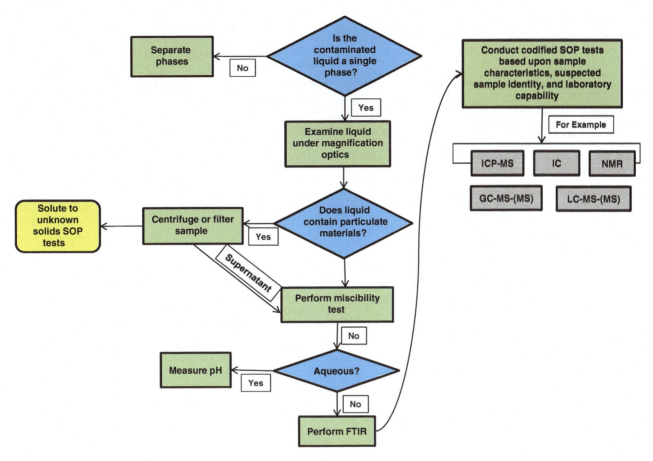

35.4 Flowchart for analysis of an unknown liquid sample
figure

external specialists due to the nature of the contaminant or testing and instrumentation required. A forensic analyst should be familiar with identifying the most commonly encountered foreign materials (Table 35.4).

Visual observation of a contaminated product and comparing to a reference sample will often reveal quite a bit of information to begin identification (Figs. 35.3 and 35.4). Some contaminants are discreet particles or residues visible even by the naked eye. For example, metal products will usually appear as gray, brown, orange, or green specks in a product matrix. Glass is a little more difficult to see since it is usually clear, but indirect lighting can give a glimpse of reflection at times. After visual analysis, the next steps to be taken should consider use of nondestructive techniques to preserve physical integrity of the particulate. This is particularly important if limited quantities of the contaminated material are available. Biological, chemical, and some physical testing should only be considered after nondestructive tests (microscopy, chemical spectroscopy/spectrometry, x-ray microtomography, and x-ray diffraction) have been exhaustively eliminated. The sections below will provide insight on how some methods already covered in other chapters are used for forensic investigations and describe a few additional methods used for this purpose.

35.5.1.2 *Microscopy*

Light microscopy is an easy method to potentially identify minute particles without any sample preparation under magnification using an optical stereo microscope (Chap. 32, Sect. 32.2.2). Crystalline particles can be examined using a polarizing microscope to determine refractive indices, birefringence, optic sign, and interference figures. However, there will be foreign materials in the nanometer size range that require advanced analytical instruments to visualize and assist in identification.

Scanning electron microscopy (SEM) (Chap. 32, Sect 32.2.3) allows specimens to be magnified more than 60,000× and when used with **energy-dispersive x-ray spectrometry** (EDS) (Chap. 32, Sect. 32.2.4) will allow elemental identification qualitatively and quantitatively of unknown inorganic substances especially metal. EDS can distinguish the degree of metal oxidation and differentiate alloy type [11]. A food forensic

35.3 table Instrumentation and techniques typically used in food forensic analysis

Analysis focus

Lipids analysis

Extractions, composition, volatiles, shelf life, antioxidants, unknown ID, mass spectrometry

Spectroscopy

FTIR, Raman, EDS, XRF, μ-XRF, fluorescence, chemical and elemental imaging, ICP

Thermal and mechanical

DSC, moisture isotherms, texture analysis, DMA, RVA, rheometers

Liquid chromatography

HPLC, UHPLC, ion chromatography (IC), mass spectroscopy, mass analysis, fractionation

Gas chromatography (GC)

1D & 2D, flame ionization, olfactory, mass spectroscopy

Microscopy

Light microscopy, SEM, confocal, x-ray microscopy, histological staining, fluorescence, immunostaining, thin sectioning, TEM

X-ray tomography

3-D imaging, surface analysis, nondestructive segmentation, ingredient ID, and localization

35.4 table Foreign materials typically identified in food ingredients, raw foods, and processed foods[a,b]

Glass

Plastics

Animal origin (bone or hair usually)

Metals

Extraneous vegetative matter (including wood)

Minerals (usually as crystals)

Arthropods (insect parts)

Chemicals (usually cleaning materials)

Food material (scorch particles, ingredient clumps that are undissolved, specks, etc.)

Yeast flocculation (especially in carbonated beverages)

Dental materials (teeth, fillings, crowns, etc.)

Fibers (usually from filters)

[a]In relative order of occurrence
[b]Modified from Edwards [1]

laboratory should have an established library of all metal types in its manufacturing plants, and use this as a fingerprint to confirm if metal contamination could have possibly not originated from its plant. EDS is also useful for fingerprinting different types of glass based on their boron, sodium, aluminum, magnesium, lead, potassium, and calcium contents.

New microscopy technologies are also being used to detect foreign bodies in foods that have been typically difficult to visualize. For example, confocal three-dimensional (3-D) micro x-ray scatter imaging technology (Chap. 32, Sect. 32.2.4) is being used with polycapillary x-ray optics to create point-to-point image analysis of low-density soft foreign materials such as plastics [12]. An instrument that uses sub-terahertz and terahertz wave methods can be used to image and identify low-density foreign materials. Low-density organic materials contaminating dry foods are notoriously difficult to image but can be readily visualized using raster-scan imaging devices using Gaussian beam focusing [13]. An additional benefit is that it provides a safe inspection due to using nonionizing radiation.

35.5.1.3 *Chemical Spectroscopy and Mass Spectrometry*

Fourier transform infrared microspectrophotometers (FTIR) (Chap. 32. Sect. 32.3.2) are used to identify the chemical composition of every minute particles of unknown substances. This instrumental method is also used to identify food materials such as wheat grains, to show distribution of protein bodies, starch granules, and cell walls in instances where the foreign body may have originated from the food material itself. Reference spectra in searchable FTIR libraries help to identify most pure organic compounds using this technique. FTIR can be used to differentiate multi-laminate plastic packaging materials. **Raman microscopy/spectroscopy** (Chap. 8, Sect. 8.6; Chap. 32, Sect. 32.3.3) is complimentary to FTIR and assists in identifying unknowns that are infrared inactive. For example, if organic components are suspected contaminants, isolated particles or materials can be pressed onto potassium bromide crystals for Raman analysis to identify the unknown. For example, Raman spectroscopy can quickly identify whether a red smudge on food packaging is blood or rather traces of ink used to print the package label.

Gas chromatography-mass spectroscopy (GC-MS) and **liquid chromatography-mass spectroscopy** (LC-MS) (see Chaps. 11, 13, and 14) are effective at accurately separating and identifying small amounts of an unknown complex/sample. These methods are probably better suited for determining off-odor and taint compounds. Additionally, they are best suited when appreciable amounts of contaminated material are available because sample preparation is destructive since the material must be solvent extracted.

35.5.1.4 *X-Ray Microtomography*

Most foods are porous structures and can be easily examined using **x-ray microtomography** (XMT) (Chap. 32, Sect. 32.5; Chap. 34, Sect. 34.5.3) to reveal a 3-D image [14]. This instrument is also finding use in food forensic investigation to detect electron-dense foreign material contaminants, and it does so in a nondestructive manner. Therefore, if sample size is limited, this can be an ideal option. Unfortunately, this is not a common instrument in most food quality and research laboratories, and it may necessitate using a third party to conduct the analysis. Metal fragments, glass chips, and bone chips are easily differentiated using XMT.

35.5.1.5 X-Ray Diffraction

Crystalline compounds are identifiable using **x-ray diffraction** (XRD) (Chap. 32, Sect. 32.4) even when present in mixtures. A well-trained food forensic investigator will usually be able to identify a crystalline species simply by observing it under a microscope. However, the diffraction patterns obtained using micro XRD must be searched against an XRD powder library to identify specific crystalline phases that may constitute the unknown crystalline particle or confirm the speculation from the analyst. A typical consumer complaint with cheese is perception of broken glass upon mastication. XRD will usually confirm the presence of calcium lactate crystals.

35.5.1.6 Other Emerging Nondestructive Instrumental Methods

Several new nondestructive imaging technologies are emerging from the fields of medicine, pharmaceuticals, and agriculture to be used in food research and forensic investigation laboratories. For example, **hyperspectral imaging** (HIS) integrates conventional imaging with spectroscopy to deliver 3-D data cubes, called hypercubes, of both spatial and spectral information about a sample [15]. In food forensic investigation, HIS is used to detect fecal contamination [16] and pesticide residues [17].

Soft x-ray imaging (SXI) (see Chap. 32, Sect. 32.5.2 for regular x-ray imaging) works similar to traditional x-ray instruments but by using lower photon energy and does not pollute the food like hard x-rays would [15]. This is the same technology being used in airport security screening. As soft x-rays penetrate a food, they lose energy and as the x-ray exits, a sensor produces an image of the interior. Foreign material can be easily detected because it will form a darker shade of gray on the resulting image. SXI is being used to detect fish bone contamination in seafood products [18].

Ultrasound, thermal, and fluorescence imaging are still being investigated as potential tools in food quality and forensic laboratories, but today these instruments find more use in research laboratories. Some of these technologies have potential to be used in processing environments to continually monitor product quality. The current limitation for many of these technologies is budget constraints but that must be weighed against a risk to benefit factor. If these technologies consistently improve finished product quality, there might be a cost benefit.

35.5.1.7 Microchemistry

Earlier chapters in this textbook cover enzyme application in food analysis (Chap. 26), immunoassays (Chap. 27), and food contaminant analysis (Chap. 33). Many of these methods are being miniaturized and require much smaller quantities of test material for analysis. Microchemical analysis is becoming more commonly used to supplement instrumental analysis and clarify ambiguous spectral data. The drawback to using some microchemical tests is that small amounts of the contaminated material must be sacrificed because most of these methods are destructive and involve extractions of some sort. This technology has evolved from the electrical semiconductor industry to enable putting laboratories on microfluidic chips. On the chips are very small quantities of antibodies, enzymes, fluorescent probes, and other chemicals used to confirm identities of unknowns or to confirm their presence. Microchemical tests are some of the best methods to identify substances such as blood and saliva, gluten, and lignins. This technology will continue to become of more importance as "omics" technologies further develop.

Nucleic acids extracted from foods can be analyzed to confirm the authenticity of products. These methods are not routinely used in food forensic laboratories yet, but if an investigator senses they are testing a product for fraudulent activity, it is best to send these products to a laboratory specializing in these types of analysis.

35.5.2 Off-Flavors and Taints

At some point during their careers, all food professionals will experience an ingredient or product exhibiting an off-flavor or a taint. **Off-flavors** result from food degradation or reaction of the ingredient or product, whereas **taints** result from an external contamination [19–21]. The identification and remediation of off-flavors and taints are a blend of science (e.g., sensory, analytical chemistry, food chemistry, food processing, packaging, transportation, etc.) and art (e.g., tactful communication, creative extraction and isolation techniques, application of previous experiences, etc.). Given these numerous facets, the remainder of this section focuses on a practical strategy a food professional might take when presented with an off-flavor or taint. The strategy does not include technical details or specific methodology; for this information, the reader is referred to earlier chapters in this textbook and to the works of others skilled in the field of off-flavors and taints [20, 22–25].

35.5.2.1 Strategy and Supportive Examples

As mentioned earlier in this chapter, forensic teams often follow a high-level schematic to direct process flow. A schematic for elucidation of off-flavors and taints is presented in Fig. 35.5. This flowchart begins with initial consultation and ends with rigorous instrumental analysis; some example techniques are

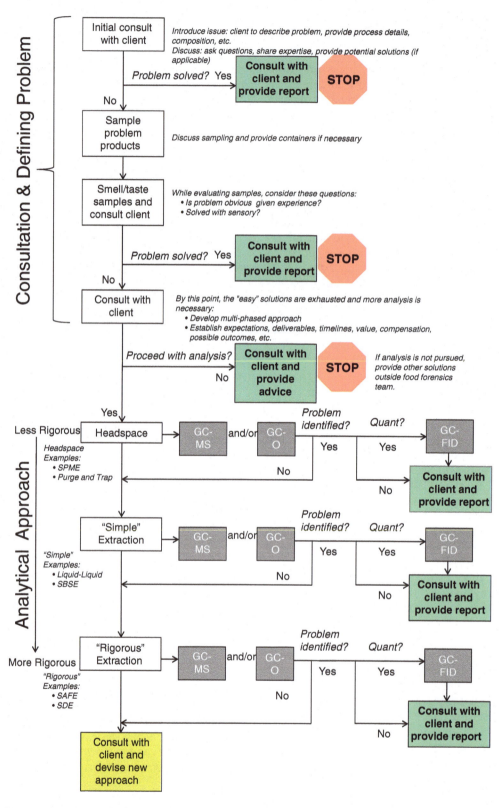

35.5 figure Decision tree for off-flavor/taint forensic investigation

referenced throughout the figure. One underlying theme throughout the schematic is the use of sensory evaluation to guide the process. There are several reasons to use traditional sensory testing procedures for off-flavors and taints (e.g., determining the nature of the problem, threshold, diagnostic, quality control, shelf life, etc.). However, for this strategy, it is assumed the problem is established, and sensory (visual and smelling) evaluation is used to confirm the causative stimuli through the entire forensics process (from consultation, through analysis, and finally linkage to the cause). If the forensic scientists do not experience the off-flavor or taint, there is no chance of identifying the cause of the problem. The following examples illustrate how the forensic scientist's olfaction or taste system might be the most important tool in the strategy.

Consider a "smoky" off-aroma in a food product reported by a client during an initial telephone consultation. To the scientist, the descriptor "smoky" meant "fire" and the logical thought process pointed toward a taint issue. However, the scientist knew the complex nature of "smoky" descriptors and requested a sample for simple sensory evaluation. After smelling the sample, the odor described as "smoky" by the customer was clearly "toasted bread" to the scientist and not "fire" as originally thought. This confusion was cleared by the scientist smelling a sample of the affected product and ultimately guided the investigation toward an off-flavor (volatiles from Maillard browning reaction) instead of a taint (absorbed volatiles from pyrolysis). In this example, the problem was not appropriately defined until the scientist's experience was applied to the sample.

Now, consider the analysis portion of Fig. 35.5. The sensory character of the off-flavor or taint must be considered for any analytical approach chosen. For example, the scientist may have a good idea of a compound that is "musty" in odor character (e.g., in potato products) and wants instrumental confirmation (e.g., GC with retention index or MS). The scientist also should use a **GC-O** technique (i.e., gas chromatography with an olfactory detector outlet) (see Chap. 14, Sect. 14.3.5) on the problem sample and compare data to a compound database containing chromatographic and odor character information (e.g., Flavornet [26]) and, whenever possible, a chemical standard. In other cases of off-flavors or taint, sensory can serve to guide the forensic scientist through challenging odor isolation experiments. The scientist may begin the process with a simple headspace analysis. A GC-MS chromatogram may not show a peak at the same time the offending aroma is detected in a corresponding GC-O aromagram; this is not uncommon since the human nose is more sensitive than MS for certain compounds. From this experience, the scientist learns the compound of interest appears to be captured by the headspace technique but might require larger sample sizes to boost MS signal. Perhaps

the headspace technique might be abandoned for a more rigorous approach (e.g., solvent extraction and/ or distillation); sensory evaluation can be used throughout this process as well. There are several different solvents used for aroma extraction, but these have compound biases [20]. Smelling a few drops of an extract from a sniff strip might give early insight into appropriate solvent choice; if the off-aroma is present, the solvent choice appears to be appropriate. Additionally, if the scientist decided to use a solvent extraction followed by high-vacuum distillation to remove nonvolatile components, it is a good idea to use a sniff strip to make sure the offending aroma is present before spending the time on distillation and subsequent solvent concentration.

35.5.2.2 Off-Odors and Taint Determination

Identifying a problem molecule (or set of molecules) is the first step in determining the origin of an off-flavor or taint. As challenging as identification can be, linkage of the offending molecule(s) to a mechanism of formation or point of contamination could prove more difficult than identification. An off-flavor or taint may begin at any point of the food production process. A number of examples illustrate some possibilities. The raw material may create a problem from the beginning of the supply chain (e.g., feedy, barny, cowy, weedy compounds in milk). Improper handling of materials prior to the production facility may accelerate detrimental reactions (e.g., corn germ sitting in a hot rail car accelerates lipid oxidation). The composition of a food matrix may make it more susceptible to detrimental reactions before, during, and after processing (e.g., lipid oxidation or Maillard browning). The processing environment (e.g., residual cleanser) or unit operations (e.g., microbial outgrowth in pipes) may taint product. Inks from packaging materials may penetrate into the product. Environmental exposure during transportation may taint the product (e.g., diesel fuel/exhaust contamination, mustiness from halogenated anisoles on pallets, or volatile organic compounds (VOCs) from a freshly painted shipping container). The key message is that there are numerous possibilities for off-flavor formation or taint exposure, and the forensic scientist must think broadly to solve a problem. They must tap into the knowledge and experience of colleagues involved in the process (from farm to fork) and from experts in the field [20, 22–25].

Established volatiles isolation, separation, and identification techniques are covered in Chap. 14 and in other references [20, 24, 25, 27]. For efficiency and cost saving, the off-flavor and taint strategy begins by using the simplest approaches then going to more rigorous, time- and labor-intensive techniques only when necessary. Within this strategy, established isolation techniques may require creative modifications to capture the problem before any chromatography occurs. For example, an off-aroma may be detected in a production facility but not yet in a food product; the client wants to be proactive and address the situation before it taints the product. Suddenly the food forensic scientist is now considering the best approach to capture the problem from an entire production plant. The same strategy applies but with a modified approach. Instead of receiving a sample to evaluate, the scientist visits the facility and smells the air. Instead of preparing an ingredient or food product in a 20-mL GC vial for a headspace experiment, the scientist might use a gas sampling pump and a number of adsorbent traps to sample air in the facility. Also, the forensic scientist may need to creatively recreate a situation in which the off-aroma presents itself to customers (e.g., during cooking).

In some cases, the scientist will be asked to quantify an offending compound. The scientist must carefully consider how the quantitative data will be used (e.g., general curiosity, regulatory issues, safety, legal proceedings, etc.). Then the scientist will need to determine if it is appropriate to run tests in a house or submit samples to a third-party laboratory. Either way, quantitative analysis brings an entirely new level of analytical rigor (i.e., method development and validation, procurement or synthesis of chemical standards, strict documentation practices, etc.).

35.6 IDENTIFYING THE WHAT, WHERE, WHEN, AND HOW OF AN ISSUE

Once the unknown foreign material or off-odor is identified, a second phase begins to investigate where, when, and how it contaminated a food ingredient or processed food. This step is important since it will help a company decide if this is an isolated incident; if more widespread, the company may initiate a large-scale recall. If public safety is jeopardized, there is never a question if a recall should be initiated. Identifying what went wrong may require close examination of each step in the manufacturing process, but with experience the investigator will often know where to begin looking for the source of a problem [28] (Table 35.5).

In a large majority of forensic investigations, the source of contamination by a foreign material has originated from the consumer's handling of the food, and the culprit is usually glass particles [1] (Fig. 35.6a). Extraneous matter analysis is covered in Chap. 34, but more details are offered here in relation to a forensic investigation. Glass is particularly bothersome to a consumer because perceptually it can cause great harm to an ingesting consumer. This point emphasizes the need to carefully document conversations with the consumer in the initial investigatory stages, including how they prepared the food. Most glass chips originate from the rim of glasses or casserole dishes used in preparation. Occasionally glass shelves in retailer display cases, such as in a delicatessen, may chip and contaminate a finished or processed food product. Other

35.5 table	Other issues addressed by food forensic analysts: requests to determine what went wrong

Black specks
Broken emulsions
Cause of hysteresis
Crystallization
Degradation products or byproducts
Flock issues
Freeze/thaw damage
Ingredient-ingredient interactions
Missing ingredients
Off color
Poor product functionality
Poor product performance
Processing damage
Processing equipment cleaning issues
Quality of cook
Sediments
Starch analysis (cook, application)
Substituted ingredients
Undercook or overcook
Viscosity issues
Water migration issues
Water activity
Wrong ingredient ID

objects can be mistaken by the consumer as glass fragments. These include calcium lactate crystals from cheeses, magnesium ammonium phosphate (clear and colorless crystalline material commonly known as struvite) from canned fish products like tuna or salmon, and rock salt (sometimes from the consumer's own salt grinder). Another common complaint area is metal from hard or very chewy foods that the consumer finds during chewing. Often times the forensic laboratory will have identified these as metal fragments characteristic of dental appliances including amalgam fillings and root canal posts. Occasionally, a tooth chip also will be identified. These are almost always from the oral cavity of the consumer (usually one with a preexisting dental problem) [1]. However, on occasion, there are instances of contamination by animal teeth, either coming with the meat or from a family pet. It is not unknown for dogs to have restorative dentistry!

Some contaminants will be easily identified as to where they originated since they may be part of the intrinsic materials used to manufacture a food ingredient or processed food. Extrinsic material contaminants will require more investigative work, usually at the plant that manufactured the implicated ingredient or processed food.

Food forensic investigators will need to work closely with operations, sanitation, engineering, and maintenance leaders who oversee unit operations of an

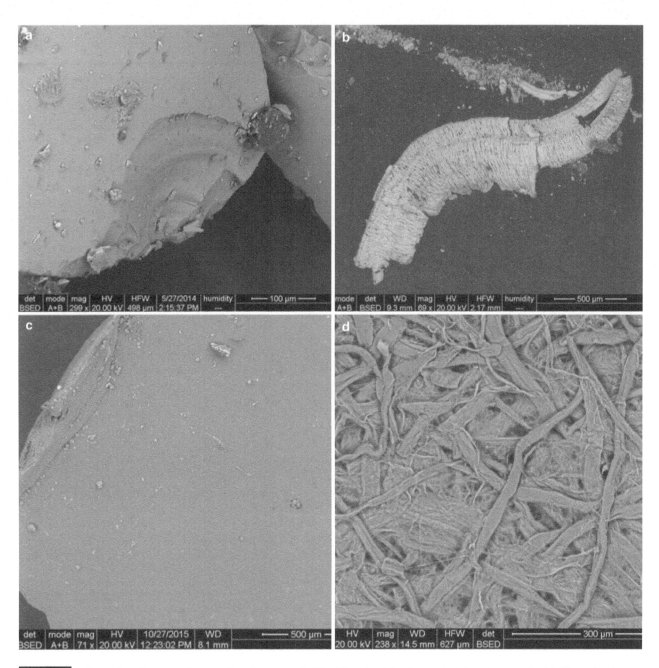

Scanning electron micrographs of commonly identified food foreign material contaminants. (**a**) Glass shard isolated from a processed food product (magnification 299×). (**b**) Stainless steel metal fragment isolated from flour after passing through a sieve (magnification 69×). (**c**) Polymeric gasket material recovered from a processed food product (magnification 71×). (**d**) Cellulosic fibers from in-process filters recovered from corn syrup (magnification 238×)

implicated plant. The initial investigation in identifying the unknown contaminant may provide clues as to where to begin looking for its origin. For example, metal shavings (Fig. 35.6b) that indicate stainless steel contamination (Fig. 35.7) suggest looking initially at the processing equipment itself for scores etched in the metal or an item broken in the processor. Wires contaminating food often can be traced back to milling and sieving equipment. Gasket material is easily traced

back to every gasket in the process (Fig. 35.6c). Contaminants identified as grease globules or oil spots will indicate an investigator should look at every joint and seal that is lubricated. Fine fibers (Fig. 35.6d) are often traced back to filters used in the process. Plastic or string in a contaminated product is often traced back to early staging and premixing steps before processing begins when ingredients are being emptied from sacks into a mixing vessel. Plastic also can contaminate a

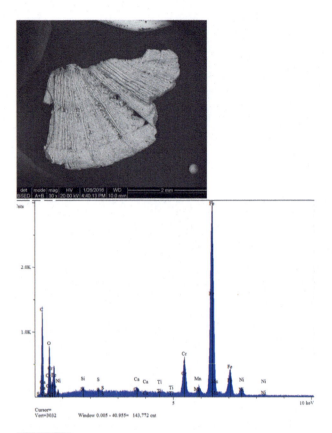

 figure

Scanning electron microscopy (SEM) micrograph and energy-dispersive x-ray spectrometry (EDS) spectrum of a metallic foreign material contamination

product when a package is being manufactured onsite immediately prior to filling, such as in some bottling operations.

A forensic investigator will need to determine, after identifying the contaminant, if it makes sense that it could have gotten into the food at some point from the time it was harvested and processed until it left its manufacturing facility. If it does not make sense, you should suspect product tampering and engage the proper authorities including a regulatory body and law enforcement. For example, if ground glass is identified in a product and there is no glass anywhere in the process, you should become speculative. However, do not overlook the fact that glass can originate from light bulbs, light fixtures, or gauge lenses near the processor. The same concerns should be raised if the identified contaminant is a foreign material with no relationship to food processing. Product tampering is often associated with disgruntled employees or individuals upset with a company for various reasons. Fortunately, product tampering to cause widespread disease or harm to consumers is rare.

35.7   INTERPRETING AND REPORTING DATA

The goals for forensic investigations are to construct explanations as to how a food material was contaminated by foreign materials or developed an off-odor/off-flavor by uncovering origins of the stated problem using accepted scientific protocols and then to understand how it happened. Ultimately, discoveries made in this process will identify steps that can be taken to minimize this happening again in the future. The process of sharing data and results requires careful record keeping throughout the process which when reviewed by others can be replicated and validated.

Reported data must be restricted to exactly what is known, and an investigator should not speculate or guess in reporting results. This is especially important in the fast pace of a potential product recall or product tampering event when there will be intense pressure being exerted to deliver results that may be inaccurate and/or incomplete. Wrongfully reported data could result in falsely reassuring the public or causing unnecessary panic [10]. Food forensic teams should have a single point of contact to the customer, regulatory, or law enforcement to discuss results to avoid confusion or improper release of information.

A forensic investigator's role is not finished at reporting data. Samples will need to be properly stored, and chain of control protocols must be followed. An analyst may need to testify in a lawsuit or criminal proceeding, offer interpretation of results including those that are not conclusive, and recommend any changes to SOPs [10].

35.8   SUMMARY

Much of the food we consume daily comes from crops and livestock that are further processed by food companies using a broad range of processes and production systems. Seemingly unexplainable circumstances do happen in manufacturing environments and can go undetected until a consumer bites into a piece of food and notices a problem. The importance of food forensic investigation is becoming greater given the thorough investigation of problem situations required by new regulatory requirements instituted globally and given the damage to a company's reputation caused by such problems.

Food forensic investigation requires experienced individuals to analyze samples that have something wrong with them either caused by an error in processing or by deliberate acts of individuals trying to discredit a company or cause physical harm to the consumer. When a product is reported with a problem, it is important to follow codified SOPs to analyze the product to identify what the problem is, where and when it occurred in the supply chain, and how it occurred, so corrective action can be taken. Various specialized nondestructive and destructive techniques are critical to investigate such problems of foreign matter contamination. Sensitive instrumental techniques and simple sensory evaluation are critical to identify contaminants causing off-flavors/odors and taints. Food forensic tools and experienced personnel are critical to determine the root cause of both product failures and food tampering.

35.9 STUDY QUESTIONS

1. What is the difference between forensic quality control and forensic quality assurance?
2. Explain what is meant by what, where, when, and how in relation to a food forensic investigation.
3. Explain why it is important for a food forensic investigator to be familiar with unit operations in manufacturing plants.
4. What are potential sources of glass as foreign material contaminants in a food plant that does not use glass packaging?
5. What is meant by chain of custody in a food forensic investigation?
6. The phrase "codified standard operating procedures" was used throughout this chapter. Explain what is meant by that phrase?
7. When you receive a package with samples of a suspect product, what should be your first observations? How will you record these?
8. What is "taint"?
9. A consumer files a complaint that a piece of glass was found in cheese manufactured from your plant. Draw a decision tree outlining the steps you will need to take in a forensic investigation to identify what the customer actually perceived.

Acknowledgments The authors and editor wish to thank the following persons who reviewed this new chapter in the *Food Analysis* textbook and provided very helpful comments: Baraem Ismail (Univ. Minnesota), Patricia Murphy (Iowa State Univ.), Oscar Pike (Brigham Young Univ.), Tom Vennard (Covance), and Jill Webb (Cairngorm Scientific Services), and all students in the Spring 2016 Food Analysis class at Purdue University. We also want to thank Var St. Jeor for the photomicrographs.

REFERENCES

1. Edwards MC, Stringer MF (2007) The breakdowns in food group. Observations on patterns in foreign material investigations. Food Control 18:773–82
2. Hsieh D (2015) Food fraud: a criminal activity implementing preventative measures that increase difficulty in carrying out the crime. Food Qual Safety 2 (2):18–23
3. U.S. Pharmacopeia: USP's Food Fraud Database Available from: http://www.usp.org/food-ingredients/food-fraud-database
4. USP – Food Chemicals Codex, Appendix XVII; Food Fraud Mitigation Guidance. Available from: http://www.usp.org/sites/default/files/usp_pdf/EN/fcc/food-fraud-mitigation-guidance.pdf
5. United States Food and Drug Administration (Internet). Washington D.C.: Food Safety Modernization Act. Available from: http://www.fda.gov/Food/Guidance Regulation/FSMA/
6. Canadian Food Agency (Internet). Quebec: Safe Food for Canadians Act. Available from: http://www.inspection.gc.ca/food/
7. Levy S, Bergman P, Frank A (1999) Quality assurance in forensic evidence. Accredited Quality Assurance 4:253–55
8. Meek T (2015) Engineering defensibility in food labs. Food Quality and Safety 21 (4):35–37
9. Pyzdek T, Keller, P. (2013) The Handbook for Quality Management: A Complete Guide to Operational Excellence, 2nd edn, McGraw-Hill Publishers, London, UK
10. Magnusun ML, Satzger RD, Alcarez A, Brewer J, Fetterolf D, Harper M, Hrynchuk R, McNally MF, Montgomery M., Nottingham E, Peterson J, Rickenbach M, Seidel JL, Wolnik K (2012) Guidelines for the identification of unknown samples for laboratories performing forensic analysis for chemical terrorism. J Forensic Sci 57:636–642
11. Schwandt CS (2016) Forensic analysis: strategy for identifying contaminants while complying with FSMA. Food Quality and Safety 21 (6):14–5
12. Li F, Liu Z, Sun T, Ma Y, Ding X (2015) Confocal three-dimensional micro X-ray scatter imaging for non-destructive detecting foreign bodies with low density and low-Z materials in food products. Food Control 54:120–125
13. Ok G, Kim HJ, Chun HS, Choi SW (2014) Foreign-body detection in dry food using continuous sub-terahertz wave imaging. Food Control 42:284–289
14. Maire E (2012) X-ray tomography applied to the characterization of highly porous materials. Ann Rev Material Sci 42:163–78
15. Chen Q, Zhang, C, Zhao J, Ouyang Q (2013) Recent advances in emerging imaging techniques for non-destructive detection of food quality and safety. Trends Anal Chem 52:261–74
16. Yoon SC, Park B, Lawrence, KC, Windham WR, Heitschmidt GW (2011) Line-scan hyperspectral imaging system for real-time inspection of poultry carcasses with fecal material and ingesta. Comput Electron Agric 79:159–68
17. Hu S, Liu M, Lin H (2006) A study on detecting pesticide residuals on fruit surface using laser imaging. Acta Agri Univ Jiangxi, Manuscript 013
18. Mery D, Lillo I, Loebel H, Riffo V, Soto A, Cipriano A, Aguilera JM (2011) Automated fish bone detection using x-ray imaging. J Food Eng 105:485–92
19. Kilcast D (1996) Sensory evaluation of taints and off-flavours. In: Saxby MJ (Ed) Food Taints and Off-flavors. Chapman and Hall, London, UK, p. 1–40
20. Reineccius G. (2006) Flavor chemistry and technology. Taylor & Francis, Boca Raton, FL, p. 161–200

21. Baigrie B (2003) Introduction. In: Baigrie B (Ed) Taints and off-flavours in food, Woodhead, Boca Raton, FL, p.1–4
22. Saxby MJ (Ed) (1996) Food taints and off-flavors. Chapman and Hall, London, UK
23. Baigrie B (Ed) (2003) Taints and off-flavours in food. Woodhead, Boca Raton, FL
24. Marsili R (Ed) (2007) Sensory-directed flavor analysis. Taylor & Francis, Boca Raton, FL
25. Marsili R (Ed) (1997) Techniques for analyzing food aroma. Marcel Dekker, New York
26. Acree T, Arn H (2004) Flavornet. Available from: http://www.flavornet.org/flavornet.html
27. 27 Maarse H, Grosch W (1996) Analysis of taints and off-flavours. In: Saxby MJ (Ed) Food taints and off-flavors. Chapman and Hall, London, UK, p. 72–106
28. Stringer MF, Hall MN (2007) The breakdowns in food group. A generic model of the integrated food supply chain to aid the investigation of food safety breakdowns. Food Control 18:755–65

Index

Printed by Printforce, the Netherlands